Digital Signal Processing with Kernel Methods

Digital Signal Processing with Kernel Methods

José Luis Rojo-Álvarez
Department of Signal Theory and Communications
University Rey Juan Carlos
Fuenlabrada (Madrid)
and
Center for Computational Simulation
Universidad Politécnica de Madrid, Spain

Manel Martínez-Ramón
Department of Electrical and Computer Engineering
The University of New Mexico
Albuquerque, New Mexico
USA

Jordi Muñoz-Marí
Department of Electronics Engineering
Universitat de València
Paterna (València), Spain

Gustau Camps-Valls
Department of Electronics Engineering
Universitat de València
Paterna (València), Spain

This edition first published 2018

Registered Offices
John Wiley & Sons, Inc., 111 River Street, Hoboken, NJ 07030, USA
John Wiley & Sons Ltd, The Atrium, Southern Gate, Chichester, West Sussex, PO19 8SQ, UK

Editorial Office
The Atrium, Southern Gate, Chichester, West Sussex, PO19 8SQ, UK

For details of our global editorial offices, customer services, and more information about Wiley products visit us at www.wiley.com.

Wiley also publishes its books in a variety of electronic formats and by print-on-demand. Some content that appears in standard print versions of this book may not be available in other formats.

Library of Congress Cataloging-in-Publication data

Names: Rojo-Álvarez, José Luis, 1972– author. | Martínez-Ramón, Manel, 1968– author. | Muñoz-Marí, Jordi, author. | Camps-Valls, Gustau, 1972– author.
Title: Digital signal processing with kernel methods / by Dr. José Luis Rojo-Álvarez, Dr. Manel Martínez-Ramón, Dr. Jordi Muñoz-Marí, Dr. Gustau Camps-Valls.
Description: First edition. | Hoboken, NJ : John Wiley & Sons, 2018. | Includes bibliographical references and index. |
Identifiers: LCCN 2017033418 (print) | LCCN 2017044705 (ebook) | ISBN 9781118705827 (pdf) | ISBN 9781118705834 (epub) | ISBN 9781118611791 (cloth)
Subjects: LCSH: Signal processing–Digital techniques.
Classification: LCC TK5102.9 (ebook) | LCC TK5102.9 .R65 2017 (print) | DDC 621.382/20285–dc23
LC record available at https://lccn.loc.gov/2017033418

Cover design by Wiley
Cover image: ©AF-studio/Gettyimages

Set in 10/12pt Warnock by SPi Global, Pondicherry, India

10 9 8 7 6 5 4 3 2 1

Contents

About the Authors

José Luis Rojo-Álvarez received the Telecommunication Engineering degree in 1996 from University of Vigo, Spain, and a PhD in Telecommunication Engineering in 2000 from the Polytechnic University of Madrid, Spain. Since 2016, he has been a full Professor in the Department of Signal Theory and Communications, University Rey Juan Carlos, Madrid, Spain. He has published more than 90 papers in indexed journals and more than 150 international conference communications. He has participated in more than 60 projects (with public and private fundings), and directed more than 10 of them, including several actions in the National Plan for Research and Fundamental Science. He was a senior researcher at the Prometeo program in Ecuador (Army University, 2013 to 2015) and research advisor at the Telecommunication Ministry. In 2016 he received the Rey Juan Carlos University Prize for Talented Researcher.

His main current research interests include statistical learning theory, digital signal processing, and complex system modeling, with applications to cardiac signals and image processing. Specifically, he is committed to the development of new electrocardiographic imaging systems, long-term cardiac monitoring intelligent systems, and big data for electronic recording and hospital information analysis at large scales.

He joined Persei vivarium, an eHealth company, as Chief Scientific Officer in 2015. Currently, he is running a pioneer degree program on Biomedical Engineering, involving hospitals and companies in the electro-medicine and eHealth fields. In 2017, he also joined the Center for Computational Simulation (Universidad Politécnica de Madrid) for promoting eHealth technology transfer based on multivariate data processing.

Manel Martínez-Ramón received an MsD in Telecommunications Engineering from Universitat Politècnica de Catalunya in 1996, and a PhD in Communications Technologies from Universidad Carlos III de Madrid (Spain) in 1999. In 2004 he spent a 20-month postdoctoral period at the MIND Research Network (New Mexico, USA). He was an Associate Professor at Universidad Carlos III de Madrid until 2013. There, he held various positions from Associate Dean of the School of Engineering to Associate Vice-Chancellor for Infrastructures. He has taught more than 30 different undergraduate and graduate classes in different universities.

Since August 2013 he has been a full professor with the Department of Electrical and Computer Engineering at the University of New Mexico, where he was permanently appointed Prince of Asturias Endowed Chair of the University of New Mexico, later renamed to King Felipe VI Endowed Chair, which is sponsored by the Household of the King of Spain. He is head of the machine learning track of this department and he is the Associate Director of the Center of Emerging Energy Technologies of this university. He is currently a principal investigator of several projects funded by the National Science Foundation and other agencies.

He has co-authored more than 40 journal papers and about 60 conference papers, and several books and book chapters. His research interests are in applications of machine learning to cyberphysical systems, including first-responders systems, smart grids, and cognitive radio.

Jordi Muñoz-Marí was born in València, Spain, in 1970, and received a BSc degree in Physics (1993), a BSc degree in Electronics Engineering (1996), and a PhD degree in Electronics Engineering (2003) from the Universitat de València. He is currently an associate professor in the Electronics Engineering Department at the Universitat de València, where he teaches electronic circuits, digital signal processing, and data science. He is a research member of the Image and Signal Processing (ISP) group. His research activity is tied to the study and development of machine-learning algorithms for signal and image processing.

Gustau Camps-Valls received BSc degrees in Physics (1996) and in Electronics Engineering (1998) and a PhD degree in Physics (2002), all from the Universitat de València. He is currently an Associate Professor (hab. Full Professor) in the Department of Electronics Engineering. He is a research coordinator in the Image and Signal Processing (ISP) group. He is interested in the development of machine-learning algorithms for geoscience and remote-sensing data analysis. He is an author of 130 journal papers, more than 150 conference papers, 20 international book chapters, and editor of the books *Kernel Methods in Bioengineering, Signal and Image Processing* (IGI, 2007), *Kernel Methods for Remote Sensing Data Analysis*" (John Wiley & Sons, 2009), and *Remote Sensing Image Processing* (MC, 2011). He holds a Hirsch's index $h = 47$, entered the ISI list of Highly Cited Researchers in 2011, and Thomson Reuters ScienceWatch identified one of his papers on kernel-based analysis of hyperspectral images as a Fast Moving Front research. In 2015, he obtained the prestigious European Research Council (ERC) consolidator grant on Statistical Learning for Earth Observation Data Analysis. Since 2007 he has been a member of the Data Fusion Technical Committee of the IEEE GRSS, and since 2009 of the Machine Learning for Signal Processing Technical Committee of the IEEE SPS. He is a member of the MTG-IRS Science Team (MIST) of EUMETSAT. He is Associate Editor of the *IEEE Transactions on Signal Processing, IEEE Signal Processing Letters, IEEE Geoscience and Remote Sensing Letters,* and invited guest editor for *IEEE Journal of Selected Topics in Signal Processing* (2012) and *IEEE Geoscience and Remote Sensing Magazine* (2015).

Valero Laparra Pérez-Muelas received a BSc degree in Telecommunications Engineering (2005), a BSc degree in Electronics Engineering (2007), a BSc degree in Mathematics (2010), and a PhD degree in Computer Science and Mathematics (2011). Currently, he has a postdoctoral position in the Image Processing Laboratory (IPL) and an Assistant Professor position in the Department of Electronics Engineering at the Universitat de València.

Luca Martino obtained his PhD in Statistical Signal Processing from Universidad Carlos III de Madrid, Spain, in 2011. He has been an Assistant Professor in the Department of Signal Theory and Communications at Universidad Carlos III de Madrid since then. In August 2013 he joined the Department of Mathematics and Statistics at the University of Helsinki. In March 2015, he joined the Universidade de Sao Paulo (USP). Currently, he is a postdoctoral researcher at the Universitat de València. His research interests include Bayesian inference, Monte Carlo methods, and nonparametric regression techniques.

Sergio Muñoz-Romero earned his PhD in Machine Learning at Universidad Carlos III de Madrid, where he also received the Telecommunication Engineering degree. He has led pioneering projects where machine-learning knowledge was successfully used to solve real big-data problems. Currently, he is a researcher at Universidad Rey Juan Carlos. Since 2015, he has worked at Persei vivarium as Head of Data Science and Big Data. His present research interests are centered around machine-learning algorithms and statistical learning theory, mainly in dimensionality reduction and feature selection methods, and their applications to bioengineering and big data.

Adrián Pérez-Suay obtained his BSc degree in Mathematics (2007), a Master's degree in Advanced Computing and Intelligent Systems (2010), and a PhD degree in Computational Mathematics and Computer Science (2015) about distance metric learning, all from the Universitat de València. He is currently a postdoctoral researcher at the Image Processing Laboratory (IPL) working on feature extraction and classification problems in remote-sensing data analysis, and has worked as assistant professor in the Department of Mathematics at the Universitat de València.

Margarita Sanromán-Junquera received the Technical Telecommunication Engineering degree from Universidad Carlos III de Madrid, Spain, in 2007, the Telecommunication Engineering degree from Universidad Rey Juan Carlos, Spain, in 2009, an MSc in Biomedical Engineering and Telemedicine from the Universidad Politécnica de Madrid, Spain, in 2010, and a PhD in Multimedia and Communication from Universidad Rey Juan Carlos and Universidad Carlos III de Madrid, in 2014. She is currently an Assistant Professor in the Department of Signal Theory and Communications, Telematics, and Computing at Universidad Rey Juan Carlos. Her research interests include statistical learning theory, digital processing of images and signals, and their applications to bioengineering.

Cristina Soguero-Ruiz received the Telecommunication Engineering degree and a BSc degree in Business Administration and Management in 2011, and an MSc degree in Biomedical Engineering from the University Rey Juan Carlos, Madrid, Spain, in 2012. She obtained her PhD degree in Machine Learning with Applications in Healthcare in 2015 in the Joint Doctoral Program in Multimedia and Communications in conjunction with University Rey Juan Carlos and University Carlos III. She was supported by FPU Spanish Research and Teaching Fellowship (granted in 2012, third place in TEC area).

She won the Orange Foundation Best PhD Thesis Award by the Spanish Official College of Telecommunication Engineering.

Steven Van-Vaerenbergh received his MSc degree in Electrical Engineering from Ghent University, Belgium, in 2003, and a PhD degree from the University of Cantabria, Santander, Spain, in 2010. He was a visiting researcher with the Computational Neuroengineering Laboratory, University of Florida, Gainesville, in 2008. Currently, he is a postdoctoral associate with the Department of Telecommunication Engineering, University of Cantabria, Spain, where he is the principal researcher for a project on pattern recognition in time series. His current research interests include machine learning, Bayesian statistics, and signal processing.

Preface

Why Did We Write This Book?

In 2001 we were finishing or had just finished our PhD theses in electronics and signal processing departments in Spain. Each of us worked with complicated and diverse datasets, ranging from the analysis of signals from patients in cooperation with hospitals, to satellite data imagery and antenna signals. All of us had grown in an academic environment where neural networks were at the core of machine learning, and our theses also dealt with them. However, support vector machines (SVMs) had just arrived, and we enthusiastically adopted them. We were probably the Spanish pioneers using them for signal processing. It took a bit to understand the fundamentals, but then everything became crystal clear. It was a clean notation, a neat methodology, often involved straightforward implementations, and admitted many alternatives and modifications. After understanding the SVM classification and regression algorithms (the two first ones that the kernel community delivered), we saw the enormous potential for writing other problems than maximum margin classifiers, and to accommodate the particularities of signal and image features and models.

First, we started to write down some support vector algorithms for problems using standard signal models, the ones that we liked most, such as spectral analysis, deconvolution, system identification, or signal interpolation. Some concepts from both the signal and the kernel worlds seemed to be naturally connected, including the concept of signal autocorrelation, being closely related to the solid theory of reproducing kernel functions. Then, we started to send our brand-new algorithms to good machine-learning journals. Quite often, reviewers criticized that the approaches were trivial, and suggested resubmission to a signal-processing journal. And then, signal-processing reviewers apparently found no novelty whatsoever in redefining old concepts in kernel terms. It seemed that the clarity of the kernel methods methodology was playing against us, and everything was apparently obvious. Nevertheless, we were (and still are) convinced that signal processing is much more than just filtering signals, and that kernel methods are much more than maximum margin classifiers as well. Our vision was that kernel methods should respect signal features and signal models as the only way to ensure model–data integration.

For years we worked in growing and designing kernel algorithms guided by the robustness requirements for the systems in our application fields. We studied other works around these fields, and some of them were really inspiring and useful in our signal-processing problems. We even wrote some tutorials and reviews along these lines, aiming to put together the common elements of the kernel methods design under signal-processing perspectives. However, we were not satisfied with the theoretical tutorials, because our algorithms were designed according to our applications, and the richness of the landscape given by the data was not reflected in these theoretical tutorials, or even not fully conveyed by more application-oriented surveys. This is why we decided to write a book that integrated the theoretical fundamentals, put together representative application examples, and, if possible, to include code snippets and links to relevant, useful toolboxes and packages. We felt that this could be a good way to help the reader work on theoretical fundamentals, while being inspired by real problems. This is, in some sense, the book we would have liked in the 2000s for ourselves. This book is not intended to be a set of tutorials, nor a set of application papers, and not just a bunch of toolboxes. Rather, the book is intended to be a learning tour for those who like and need to write their kernel algorithms, who need these algorithms for their signal-processing applications in real data, and who can be inspired by simple yet illustrative code tutorials.

Needless to say, completing a book like this in the intersection of signal processing and kernel methods has been an enormous challenge. The literature of kernel methods in signal processing is vast, so we could not include all the excellent contributions working in this cross-field during recent years. We tried our best in all chapters, though, by revising the literature of what we feel are the main pillars and recent trends. The book only reflects our personal view and experience, though.

Structure and Contents

This book is divided into three parts: one for fundamentals, one focused on signal models for signal estimation and adaptive filtering, and another for classification, detection, and feature extraction. They are summarized next.

Part One: Fundamentals and Basic Elements

This consists of an introductory set of chapters and presents the necessary set of basic ideas from both digital signal processing (DSP) and reproducing kernel Hilbert spaces (RKHSs). After an introductory road map (*Chapter 1*), a basic overview of the field of DSP is presented (*Chapter 2*). Then, data models for signal processing are presented (*Chapter 3*), on the one hand including well-known signal models (such as sinusoid-based spectral estimation, system identification, or deconvolution), and on the other hand summarizing a set of additional fundamental concepts (such as adaptive filtering, noise, or complex signal models). *Chapter 4* consists of an introduction to kernel functions and Hilbert spaces, including the necessary concepts on RKHS and their properties for being used throughout the rest of the book. This chapter includes the elements of the SVM algorithm for regression, as an instantiation of a kernel algorithm, which in turn will be formally used as a founding optimization algorithm for the algorithms in the next parts.

Part Two: Function Approximation and Adaptive Filtering

This presents a set of different SVM algorithms organized from the point of view of the signal model being used from Chapter 2 and its role in the structured function estimation. The key in this part is that SVM for estimation problems (not to be confounded with the standard SVM for classification) raises a well-structured and founded approach to develop new general-purpose signal processing methods. *Chapter 5* starts with a simple and structured explanation of the three different sets of algorithms to be addressed, which are primal signal models (linear kernel and signal model stated in the primal problem, in the remainder of *Chapter 5*), RKHS signal models (signal model in the RKHS, conventional in kernel literature, to which *Chapter 6* is devoted), and dual signal models (signal model in the dual solution, closely related to function basis expansion in signal processing, to which *Chapter 7* is devoted). These three chapters represent the main axis along which the kernel trick is used to adapt the richness of signal processing model data, with emphasis on the idea that, far from using black-box nonlinear regression/classification model with an ad-hoc signal embedding model, one can actually adapt the SVM equations to the signal model from digital signal processing considerations on the structure of our data. Ending this part, *Chapter 8* provides an overview on the wide variety of kernel methods for signal estimation which can benefit from these proposed concepts for SVM regression, which include such widespread techniques as least-squares SVM, kernel signal-to-noise regression, or Bayesian approaches. A signal processing text for kernel methods in DSP must cover the adaptive processing field, which after some initial basic proposals, seems to be reaching today a state of maturity in terms of theoretical fundamentals; all of them are summarized in *Chapter 9*.

Part Three: Classification, Detection, and Feature Extraction

This presents a compendium of selected SVM algorithms for DSP which are not included in the preceding framework. Starting from the state of the art in SVM algorithms for classification and detection problems in the context of signal processing, the rationale for this set of existing contributions is quite different from Part Two, given that likely the most fundamental concept of SVM classifiers, namely the maximum margin, holds in SVM classification approaches for signal processing. *Chapter 10* revises the conventional SVM classifier and its variants, introduces other kernel classifiers beyond SVMs, and discusses particular advanced formulations to treat with semi-supervised, active, structured-output, and large-scale learning. Then, *Chapter 11* is devoted specifically to clustering, anomaly detection, and one-class kernel classifiers, with emphasis in signal- and image-processing applications. Finally, *Chapter 12* is fully devoted to the rich literature and theoretical developments on kernel feature extraction, revisiting the classical taxonomy (unsupervised, supervised, semi-supervised, and domain adaptation) from a DSP point of view.

Theory, Applications, and Examples

References on theoretical works and additional applications are stated and briefly commented in each chapter. The book can be considered as self-contained, but still it assumes some necessary previous and very basic knowledge on DSP. In the application examples, references are given for the interested reader to be able to update or refresh those concepts which are to be dealt with in each chapter.

Supporting material is also included in two forms. On the one hand, simple examples for fundamental concepts are delivered, so that the reader gains confidence and gets familiar with the basic concepts, some readers may find them trivial for some chapters, on the other hand, real and more advanced application examples are provided in several chapters. Scripts, code, and pointers to toolboxes are mostly in MATLAB™. The source code and examples can be downloaded from GitHub at the following link:

http://github.com/DSPKM/

In this dedicated repository, many links are maintained to other widely used software toolboxes for machine learning and signal processing in kernel methods. This repository is periodically updated with the latest contributions, and can be helpful for the data analysis practitioner. The reader can use the code provided with this book for their own research and analysis. We only ask that, in this case, the book is properly cited:

Digital Signal Processing with Kernel Methods
José Luis Rojo-Álvarez, Manel Martínez-Ramón, Jordi Muñoz-Marí, and Gustau Camps-Valls
John Wiley & Sons, 2017.

When the code for the examples is simple and didactic, it is included in the text, so that it can be examined, copied, and pasted. Those scripts and functions with increased complexity are further delivered in the book repository. Owing to the multidisciplinary nature of the examples, they can be of different difficulty for each reader according to the specific background, so that examples which seem extremely easy for some will be harder to work out for others. The reader is encouraged to spend some time with the code with which they are more unfamiliar and to skip the examples which are already well known.

Acknowledgements

We would like to acknowledge the help of all involved in the collation and review process of the book, without whose support the project could not have been satisfactorily completed. A further special note of thanks goes also to all the staff at John Wiley & Sons, Inc., whose contributions throughout the whole process, from inception of the initial idea to final publication, have been valuable. Special thanks also go to the publishing team at John Wiley & Sons, Ltd., who continuously prodded via e-mail, keeping the project on schedule. It has been really a pleasure to work with such a professional staff.

Our special thanks goes to the coauthors of several of the chapters, who made it possible to cover material that has largely enriched those chapters, and who are listed in Table 1. We would like to express our deepest gratitude to them.

We also wish to thank all of the coauthors of the papers during these years. Without them this work would not have been possible at all: J.C. Antoranz, J. Arenas-García, A. Artés-Rodríguez, T. Bandos, O. Barquero-Pérez, J. Bermejo, K. Borgwardt, L. Bruzzone, A.J. Caamaño-Fernández, S. Canú, C. Christodoulou, J. Cid-Sueiro, P. Conde-Pardo, E. Everss, J.R. Feijoo, M.J. Fernández-Getino, A.R. Figueiras-Vidal, C. Figuera, R. Flamary, A. García-Alberola, A. García-Armada, V. Gil-Jiménez, F.J. Gimeno-Blanes, L. Gómez-Chova, V. Gómez-Verdejo, A.E. Gonnouni, R. Goya-Esteban, A. Guerrero-Curieses, E. Izquierdo-Verdiguier, L.K. Hansen, R. Jenssen, M. Lázaro-Gredilla, N. Longbotham, J. Malo, J.D. Martín-Guerrro, M.P. Martínez-Ruiz, F. Melgani, I. Mora-Jiménez, N.M. Nasrabadi, A. Navia-Vázquez, A.A. Nielsen, F. Pérez-Cruz, K.B. Petersen, M. de Prado-Cumplido, A. Rakotomamonjy, I. Santamaría-Caballero, B. Schölkopf, E. Soria-Olivas, D. Tuia, J. Verrelst, J. Weston, R. Yotti, and D. Zhou.

This book was produced without any dedicated funding, but our research was partially supported by research projects that made it possible. We want to thank all agencies and organizations for supporting our research in general, and this book indirectly. José Luis Rojo-Álvarez acknowledges support from the Spanish Ministry of Economy and Competitiveness, under projects PRINCIPIAS (TEC2013-48439-C4-1-R), FINALE (TEC2016-75161-C2-1-R), and KERMES (TEC2016-81900-REDT), and from Comunidad de Madrid under project PRICAM (S2013/ICE-2933). Manel Martínez-Ramón acknowledges support from the Spanish Ministry of Economy and Competitiveness under project TEC2014-52289-R, from Comunidad de Madrid under project PRICAM (S2013/ICE-2933), and from the National Science Foundation under projects S&CC #1637092 and SBIR Phase II #1632498. Jordi Muñoz-Marí acknowledges support from the Spanish Ministry of Economy and Competitiveness and the European Regional Development Fund, under project TIN2015-64210-R, Gustau Camps-Valls and Luca

Table 1 List of coauthors in specific chapters.

Margarita Sanromán-Junquera	Universidad Rey Juan Carlos	Chapters 3 and 7
Sergio Muñoz Romero	Universidad Rey Juan Carlos	Chapters 3, 5, and 7
Cristina Soguero-Ruiz	Universidad Rey Juan Carlos	Chapters 6 and 7
Luca Martino	Universitat de València	Chapter 8
Steven Van Vaerenbergh	Universidad de Cantabria	Chapter 9
Adrián Pérez-Suay	Universitat de València	Chapter 11
Valero Laparra	Universitat de València	Chapter 12

Martino acknowledge support by the European Research Council (ERC) under the ERC-CoG-2014 project 647423. Steven Van Vaerenbergh is supported by the Spanish Ministry of Economy and Competitiveness, under PRISMA project (TEC2014-57402-JIN).

José Luis Rojo-Álvarez, Manel Martínez-Ramón,
Jordi Muñoz-Marí, and Gustau Camps-Valls
Leganés, Albuquerque, and València, December 2016

List of Abbreviations

ACD	anomalous change detection
AL	active learning
ALD	approximate linear dependency
AP	access point
AR	autoregressive
ARCH	autoregressive conditional heteroscedasticity
ARMA	autoregressive and moving average
ARX	autoregressive exogenous
AUC	area under the (ROC) curve
AVIRIS	Airborne Visible Infrared Imaging Spectrometer
BCI	brain–computer interface
BER	bit error rate
BG	Bernouilli–Gaussian
BRT	bootstrap resampling techniques
BSS	blind source separation
BT	breaking ties
CCA	canonical correlation analysis
CDMM	color Doppler M mode
CESNI	continuous-time equivalent system for nonuniform interpolation
CG	conjugate gradient
CI	confidence interval
CNS	cardiac navigation system
COCO	constrained covariance
CS	compressive sensing
CV	cross-validation
DCT	discrete cosine transform
DFT	discrete Fourier transform
DMGF	double modulated Gaussian function
DOA	direction of arrival
DSM	dual signal model
DSP	digital signal processing
DWT	discrete wavelet transform
EAM	electroanatomical map
EC	elliptically contoured

ECG	electrocardiogram
EEC	error correction code
EEG	electroencephalogram
EM	expectation–maximization
ESD	energy spectral density
FB-KRLS	fixed budget kernel recursive least squares
FFT	fast Fourier transform
FIR	finite impulse response
FT	Fourier transform
GM	Gaussian mixture
GMM	Gaussian mixture model
GP	Gaussian process
GPR	Gaussian process regression
GRNN	generalized regression neural network
HRCN	high reliability communications network
HRV	heart rate variability
HSCA	Hilbert–Schmidt component analysis
HSIC	Hilbert–Schmidt independence criterion
i.i.d.	independent and identically distributed
ICA	independent component analysis
ICF	incomplete Cholesky factorization
IIR	infinite impulse response
IMSE	integrated mean square error
IPM	interior point method
IRWLS	integrated reweighted least squares
KACD	kernel anomaly change detection
KAF	Kalman adaptive filtering
KDE	kernel density estimation
KDR	kernel dimensionality reduction
KECA	kernel entropy component analysis
KEMA	kernel manifold alignment
KF	Kalman filter
KFD	kernel Fisher discriminant
KGV	kernel generalized variance
KICA	kernel independent component analysis
KKT	Karush–Kuhn–Tucker
KL	Kullback–Leibler
KLMS	kernel least mean squares
kMI	kernel mutual information
KMM	kernel mean matching
KNLMS	kernel normalized least mean squares
KOSP	kernel orthogonal subspace projection
KPCA	kernel principal component analysis
KRLS	kernel recursive least squares
KRLS-T	kernel recursive least square tracker
KRR	kernel ridge regression
KSAM	kernel spectral angle mapper

KSNR	kernel signal-to-noise regression/ratio
KTA	kernel–target alignment
LapSVM	Laplacian support vector machine
LASSO	least absolute shrinkage and selection operator
LDA	linear discriminant analysis
LFD	linear Fisher discriminant
LI	linear interpolation
LMF	large margin filtering
LMS	least mean squares
LOO	leave-one-out
LS	least squares
LS-SVM	least-squares support vector machine
LTI	linear time invariant
LUT	look-up table
MA	moving average
MAE	mean absolute error
MAO	most ambiguous and orthogonal
MAP	maximum a posteriori
MCLU	multiclass level uncertainty
MCMC	Markov chain–Monte Carlo
MERIS	medium resolution imaging spectrometer
MIMO	multiple input–multiple output
MKL	multiple kernel learning
ML	maximum likelihood
MLP	multilayer perceptron
MMD	maximum mean discrepancy
MMDE	maximum mean discrepancy embedding
MMSE	minimum mean square error
MNF	minimum noise fraction
MPDR	minimum power distortionless response
MRI	magnetic resonance imaging
MS	margin sampling
MSE	mean square error
MSSF	modulated squared sinc function
MSVR	multioutput support vector regression
MUSIC	multiple signal classification
MVA	multivariate analysis
MVDR	minimum variance distortionless response
NN	neural network
NORMA	naive online regularized risk minimization algorithm
NW	Nadayara–Watson
OA	overall accuracy
OAA	one against all
OAO	one against one
OC-SVM	one class support vector machine
OFDM	orthogonal frequency division multiplexing
OKECA	optimized kernel entropy component analysis

OSP	orthogonal subspace projection
PCA	principal component analysis
PCK	probabilistic cluster kernel
pdf	probability density function
PLS	partial least squares
PSD	power spectral density
PSM	primal signal model
PSVM	parallel support vector machine
QAM	quadrature amplitude modulation
QKLMS	quantified kernel least mean squares
QP	quadratic programming
QPSK	quadrature-phase shift keying
RBF	radial basis function
RHSIC	randomized Hilbert–Schmidt independence criterion
RKHS	reproducing kernel in Hilbert space
RKS	random kitchen sink
RLS	recursive least squares
RMSE	root mean square error
ROC	receiver operating characteristic
RSM	reproducing kernel in Hilbert space signal model
RSS	received signal strength
RV	relevance vector
RVM	relevance vector machine
S/E	signal to error
SAM	spectral angle mapper
SDP	semi-definite program
SE	squared exponential
SMO	sequential minimal optimization
SNR	signal-to-noise ratio
SOGP	sparse online Gaussian process
SOM	self-organizing map
SR	semiparametric regression
SRM	structural risk minimization
SSL	semisupervised learning
SSMA	semisupervised manifold alignment
STFT	short-time Fourier transform
SVC	support vector classification
SVD	singular value decomposition
SVDD	support vector domain description
SVM	support vector machine
SVR	support vector regression
SW-KRLS	sliding window kernel recursive least squares
TCA	transfer component analysis
TFD	time–frequency distribution
TSVM	transductive support vector machine
WGP	warped Gaussian process
WGPR	warped Gaussian process regression

Part I

Fundamentals and Basic Elements

1

From Signal Processing to Machine Learning

Signal processing is a field at the intersection of systems engineering, electrical engineering, and applied mathematics. The field analyzes both analog and digitized signals that represent physical quantities. Signals include sound, electromagnetic radiation, images and videos, electrical signals acquired by a diversity of sensors, or waveforms generated by biological, control, or telecommunication systems, just to name a few. It is, nevertheless, the subject of this book to focus on *digital* signal processing (DSP), which deals with the analysis of digitized and discrete sampled signals. The word "digital" derives from the Latin word *digitus* for "finger," hence indicating everything ultimately related to a representation as integer countable numbers. DSP technologies are today pervasive in many fields of science and engineering, including communications, control, computing and economics, biology, or instrumentation. After all, signals are everywhere and can be processed in many ways: filtering, coding, estimation, detection, recognition, synthesis, or transmission, are some of the main tasks in DSP.

In the following sections we review the main landmarks of signal processing in the 20th century from the perspective of algorithmic developments. We will also pay attention to the cross-fertilization with the field of statistical (machine) learning in the last decades. In the 21st century, model and data assumptions as well as algorithmic constraints are no longer valid, and the field of *machine-learning signal processing* has erupted, with many successful stories to tell.

1.1 A New Science is Born: Signal Processing

1.1.1 Signal Processing Before Being Coined

One might argue that processing signals is as old as human perception of the nature, and you would probably be right. Processing signals is actually a fundamental problem in science. In ancient Egypt, the Greek civilization and the Roman Empire, the "men who knew" (nowadays called scientists and philosophers) measured and quantified river floods, sunny days, and exchange rates digitally. They also tried to predict them and model them empirically with simple "algorithms." One might say that system modeling, causal inference, and world phenomena prediction were matters that already existed at that time, yet were treated at a philosophical level only. Both the mathematical tools and the intense data exploitation came later. The principles of what we actually call

Digital Signal Processing with Kernel Methods, First Edition. José Luis Rojo-Álvarez, Manel Martínez-Ramón, Jordi Muñoz-Marí, and Gustau Camps-Valls.

signal processing date back to the advances in classical numerical analysis techniques of the 17th and 18th centuries. Big names of European scientists, like Newton, Euler, Kirchhoff, Gauss, Cauchy, and Fourier, set up the basis for the latter development of sciences and engineering, and DSP is just the most obvious and particular case out of them. The roots of DSP can be found later in the digital control systems of the 1940s and 1950s, while their noticeable development and adoption by society took place later, in the 1980s and 1990s.

1.1.2 1948: Birth of the Information Age

The year 1948 may be regarded as the birth of modern *signal processing*. Shannon published his famous paper "A mathematical theory of communication" that established bounds for the capacity of a band-limited channel and created the discipline of information theory (Shannon, 1949). Hartley and Wiener fed Shannon's mind with their statistical viewpoint of communication; and others, like Gabor, developed the field enormously. In that year, Shannon also motivated the use of pulse code modulation in another paper. This year was also the year when modern digital methods were introduced: Bartlett and Tukey developed methods for spectrum estimation, while Hamming introduced error correcting codes for efficient signal transmission and recovery. These advances were most of the time motivated by particular applications: audio engineering promoted spectral estimation for signal analysis, and radar/sonar technologies dealt with discrete data during World War II that needed to be analyzed in the spectral domain. Another landmark in 1948 was the invention of the transistor at Bell Labs, which was still limited for commercial applications. The take-off of the signal processing field took place also because Shannon, Bode, and others discussed the possibility of using digital circuits to implement filters, but no appropriate hardware was available at that time.

1.1.3 1950s: Audio Engineering Catalyzes Signal Processing

DSP as we know it nowadays was still not possible at that time. Mathematical tools (e.g., the z-transform) were already available thanks to established disciplines like control theory, but technology was ready only to deal with low-frequency signal processing problems. Surprisingly, the field of audio engineering (boosted by the fever of rock 'n' roll in radio stations!) was the catalyst for new technological developments: automobile phonographs, radio transistors, magnetic recording, high-quality low-distortion loudspeakers and microphone design were important achievements.

The other important industry was telephony and the need for efficient repeaters (amplifiers, transistors): the transatlantic phone cable formed a huge low-pass filter introducing delays and intersymbol interferences in the communications, and time-assignment speech interpolators appeared as efficient techniques to exploit the pauses in speech during a phone conversation. Efficiency in communications took advantage of Shannon's theory of channel capacity, and introduced frequency-division multiplexing. Transmission capacity was alternatively improved by the invention of the coaxial cable and the *vocoder* in the 1930s by Dudley.

This decade is also memorable because of work on the theory of wave filters, mostly developed by Wagner, Campbell, Cauer, and Darlington. A new audio signal representation called *sound spectrograms*, which essentially shows the frequency content of speech as it varies through time, was introduced by Potter, Wigner, Ville, and other researchers. This time–frequency signal representation became widely used in signal processing some time later. Communications during World War II were quite noisy; hence, there was a notable effort in constructing a mathematical theory of signal and noise, notably by Wiener and Rice. The field of seismic data processing witnessed an important development in the early 1950s, when Robinson showed how to derive the desired reflection signals from seismic data carrying out one-dimensional deconvolution.

1.2 From Analog to Digital Signal Processing

1.2.1 1960s: Digital Signal Processing Begins

In the late 1950s, the introduction of the integrated circuit containing transistors revolutionized electrical engineering technology. The 1960s made technology ready for DSP. Silicon integrated circuits were ready, but still quite expensive compared with their analogical counterparts. The most remarkable contributions were the implementation of a digital filter using the bilinear transform by Kaiser, and the work of Cooley and Tukey in 1965 to compute the discrete Fourier transform efficiently, which is nowadays well known as the fast Fourier transform (FFT). DSP also witnessed the introduction of the Viterbi algorithm in 1967 (used especially in speech recognition), the chirp z-transform algorithm in 1968 (which widened the application range for the FFT), the maximum likelihood (ML) principle also in 1968 (for sensor-array signal processing), and adaptive delta modulation in 1970 (for speech encoding).

New and cheaper hardware made digital filters a reality. It was possible to efficiently implement long finite-impulse response (FIR) filters that were able to compete with analog infinite-impulse response (IIR) filters, offering better band-pass properties. But perhaps more crucial was that the 1960s were a time for numerical simulation. For instance, Tukey developed the concept of the *cepstrum* (the Fourier transform of the logarithm of the amplitude spectrum) for pitch extraction in a vocoder. In early 1961, Kaiser and Golden worked intensively to transfer filters from the analog to the digital domain. Digital filters also offered the possibility to synthesize time-varying, adaptive and nonlinear filters, something that was not possible with analog filters. Kalman filters (KFs) took advantage of the statistical properties of the signals for filtering, while Widrow invented the least mean squares (LMS) algorithm for adaptive filtering, which is the basis for neural networks (NNs) training. Bell Labs also developed adaptive equalizers and echo cancellers. Schroeder introduced the adaptive predictive coding algorithm for speech transmission of fair quality, while Atal invented linear predictive coding, which was so useful for speech compression, recognition, and synthesis.

In the 1960s, image processing stepped in the field of DSP through applications in space sciences. The topics of image coding, transmission, and reconstruction were in their infancy. In 1969, Anderson and Huang developed efficient coders and later introduced the famous discrete cosine transform (DCT) for image coding. Other two-dimensional (2D) and three-dimensional (3D) signals entered the arena: computerized tomography scanning, interferometry for high-resolution astronomy and geodesy, and radar imaging contributed with improvements in multidimensional digital filters for image restoration, compression, and analysis. The wide range of applications, usually involving multidimensional and nonstationary data, exploited and adapted previous DSP techniques: Wiener filtering for radar/sonar tracking, Kalman filtering for control and signal-detection systems, and recursive time-variant implementation of closed-form filter solutions, just to name a few.

1.2.2 1970s: Digital Signal Processing Becomes Popular

The 1970s was the time when video games and word processors appeared. DSP started to be everywhere. The speech processing community introduced adaptive differential pulse code modulation to achieve moderate savings in coding. Subband coding divided the signal spectrum into bands and adaptively quantized each independently. The technique was also used for image compression, with important implications in storage and processing.

Filter design continued introducing improvements, such as McClellan's and Parks's design of equiripple FIR filters, the analog-to-digital design procedures introduced by Burrus and Parks, Galand's quadrature mirror filters, and Darlington's multirate filters. Huang pioneered in developing filters for image processing. State-space methods and related mathematical techniques were developed, which were later introduced into fields such as filter design, array processing, image processing, and adaptive filtering. FFT theory was extended to finite fields and used in areas such as coding theory.

1.2.3 1980s: Silicon Meets Digital Signal Processing

The 1980s will be remembered as the decade in which personal computers (PCs) became ubiquitous. IBM introduced its PC in 1981 and standardized the disk operating system. PC clones appeared soon after, together with the first Apple computers and the new IBM PC-XT, the first PC equipped with a hard-disk drive.

The most important fact in the 1980s for the signal processing point of view was the design and production of single-chip DSP. Compared with using general-purpose processors, specifically designed chips for signal processing made operations much faster, allowing parallel and real-time signal-processing systems. An important DSP achievement of the 1980s was JPEG, which essentially relies on the DCT, and still is the international standard for still pictures compression. The success of JPEG inspired efforts to reach standards for moving images, which was achieved in the 1990s in the form of MPEG1 and MPEG2. Automated image-recognition found its way into both military and civil applications, as well as for Earth observation and monitoring. Nowadays, these applications involve petabyte and multimillion ventures.

The introduction of NNs played a decisive role in many applications. In the 1950s, Frank Rosenblatt introduced the *perceptron* (a simple linear classifier), and Bernard Widrow the *adaline* (adaptive linear filter). Nevertheless, neural nets were not extensively used until the 1980s: new and more efficient architectures and training algorithms, capability to implement networks in very-large-scale integrated circuits, and the belief that massive parallelism was needed for speech and image recognition.[1] Besides the famous multilayer perceptron (MLP), other important developments are

1 Interestingly, this is in current times a revived idea because of the concepts of "big data" and "deep learning."

worth mentioning: Hopfield's recurrent networks, radial basis function (RBF) networks, and Jordan's and Elman's dynamic and recurrent networks. NNs were implemented for automatic speech recognition, automatic target detection, biomedical engineering, and robotics.

1.3 Digital Signal Processing Meets Machine Learning

1.3.1 1990s: New Application Areas

The 1990s changed the rules with the effect of the Internet and PCs. More data to be transmitted, analyzed, and understood were present in our lives. On top of this, the continuing growth of the consumer electronics market impulsed DSP. New standards like MPEG1 and MPEG2 made efficient coding of audio signals widely used. Actually, the new era is "visual": image processing and digital photography became even more prominent branches of signal processing. New techniques for filtering and image enhancement and sharpening found application in many (in principle orthogonal) fields of science, like astronomy, ecology, or meteorology. As in the previous decade, space science introduced a challenging problem: the availability of multi- and hyperspectral images impulsed new algorithms for image compression, restoration, fusion, and object recognition with unprecedented accuracy and wide applicability.

New applications were now possible because of the new computer platforms and interfaces, the possibility to efficiently simulate systems, and the well-established mathematical and physical theories of the previous decades and centuries. People started to adopt DSP unconsciously when using voice-recognition software packages, accessing the Internet securely, compressing family photographs in JPEG, or trading in the stock markets using moving-average filters.

1.3.2 1990s: Neural Networks, Fuzzy Logic, and Genetic Optimization

NNs were originally developed for aircraft and automobile-engine control. They were also used in image restoration and, given its parallel nature, efficiently implemented on very-large-scale integrated architectures. Also exciting was the development and application of fractals, chaos, and wavelets. Fractal coding was extensively applied in image compression. Chaotic systems have been used to analyze and model complex systems in astronomy, biology, chemistry, and other sciences. Wavelets are a mathematical decomposition technique that can be cast as an extension of Fourier analysis, and intimately related to Gabor filters and time–frequency representations. Wavelets were considerably advanced in the mid-1980s and extensively developed by Daubechies and Mallat in the 1990s.

Other new departures for signal processing in the 1990s were related to fuzzy logic and genetic optimization. The so-called fuzzy algorithms use fuzzy logic, primarily developed by Zadeh, and they gave rise to a vast number of real-life applications. Genetic algorithms, based on laws of genetics and natural selection, also emerged. Since then they have been applied in different signal-processing areas. These are mere examples of the tight relation between DSP and computer science.

1.4 Recent Machine Learning in Digital Signal Processing

We are facing a totally new era in DSP. In this new scenario, the particular characteristics of signals and data are challenging the traditional signal-processing technologies. Signal and data streams are now massive, unreliable, unstructured, and barely fit standard statistical assumptions about the underlying system. The recent advances in interdisciplinary research are of paramount importance to develop new technologies able to deal with the new scenario. Powerful approaches have been designed for advanced signal processing, which can be implemented thanks to continuous advances in fast computing (which is becoming increasingly inexpensive) and algorithmic developments.

1.4.1 Traditional Signal Assumptions Are No Longer Valid

Standard signal-processing models have traditionally relied on the rather simplifying and strong assumptions of linearity, Gaussianity, stationarity, circularity, causality, and uniform sampling. These models provide mathematical tractability and simple and fast algorithms, but they also constrain their performance and applicability. Current approaches try to get rid of these approximations in a number of ways: by widely using models that are intrinsically nonlinear and nonparametric; by encoding the relations between the signal and noise (which are often modeled and no longer considered Gaussian independent and identically distributed (i.i.d.) noise); by using new approaches to treat the noncircularity and nonstationarity properties of signals; by learning in *anti-causal systems*, which is an important topic of control theory; and, in some situations, since the acquired signals and data streams are fundamentally unstructured, by not assuming uniform sampling of the representation domain.

It is also important to take into account the increasing diversity of data. For example, large and unstructured text and multimedia datasets stored in the Internet and the increasing use of social network media produce masses of unstructured heterogeneous data streams. Techniques for document classification, part-of-speech tagging, multimedia tagging or classification, together with massive data-processing techniques (known as "big data" techniques) relying on machine-learning theory try to get rid of unjustified assumptions about the data-generation mechanisms.

1.4.2 Encoding Prior Knowledge

Methods and algorithms are designed to be specific to the target application, and most of the times they incorporate accurate prior and physical knowledge about the processes generating the data. The issue is two-sided. Nowadays, there is a strong need of constraining model capacity with proper priors. Inclusion of prior knowledge in the machines for signal processing is strongly related to the issue of encoding invariances, and this often requires the design of specific regularizers that constrain the space of *possible* solutions to be confined in a *plausible* space. Experts in the application field provide such knowledge (e.g., in the form of physical, biological, or psychophysical models), while engineers and computer scientists design the algorithm in order to fulfill the specifications. For instance, current advances in graphical models allow us to learn more about structure from data (i.e., the dynamics and relationships of each variable and their interactions), and multitask learning permits the design of models that tackle the problem of learning a task as a composition of modular subtask problems.

1.4.3 Learning and Knowledge from Data

Machine learning is a powerful framework to reach the goal of processing a signal or data, turning it into information, and then trying to extract knowledge out of either new data or the learning machine itself. Understanding is much more important and difficult than fitting, and in machine learning we aim for this from empirical data. A naive example can be found in biological signal processing, where a learning machine can be trained from patients and control records, such as electrocardiograms (ECGs) and magnetic resonance imaging (MRI) spatio-temporal signals. The huge amount of data coming from the medical scanner needs to be processed to get rid of those features that are not likely to contain information. Knowledge is certainly acquired from the detection of a condition in a new patient, but there is also potentially important clinical knowledge in the analysis of the learning-machine parameters in order to unveil which characteristics or factors are actually relevant in order to detect a disease. For this fundamental goal, learning with hierarchical deep NNs has permitted increasingly complex and more abstract data representations. Similarly, cognitive information processing has allowed moving from low-level feature analysis to higher order data understanding. Finally, the field of causal inference and learning has irrupted with new refreshing algorithms to learn causal relations between variables.

The new era of DSP has an important constraint: the urgent need to deal with massive data streams. From images and videos to speech and text, new methods need to be designed. Halevy *et al.* (2009) raised the debate about how it becomes increasingly evident that machine learning achieves the most competitive results when confronted with massive datasets. Learning semantic representations of the data in such environments becomes a blessing rather than a curse. However, in order to deal with huge datasets, efficient automatic machines must be devised. Learning from massive data also poses strong concerns, as the space is never filled in and distributions reveal skewed and heavy tails.

1.4.4 From Machine Learning to Digital Signal Processing

Machine learning is a branch of computer science and artificial intelligence that enables computers to learn from data. Machine learning is intended to capture the necessary patterns in the observed data, such as accurately predicting the future or estimating hidden variables of new, previously unseen data. This property is known in general as *generalization*. Machine learning adequately fits the constraints and solution requirements posed by DSP problems: from computational efficiency, online adaptation, and learning with limited supervision, to their ability to combine heterogeneous information, to incorporate prior knowledge about the problem, or to interact with the user to achieve improved performance. Machine learning has been recognized as a very suitable technology in signal processing since the introduction of NNs. Since the 1980s, this particular model has been successfully exploited in many DSP applications, such as antennas, radar, sonar and speech processing, system identification and control, and time-series prediction (Camps-Valls and Bruzzone, 2009; Christodoulou and Georgiopoulos, 2000; Deng and Li, 2013; Vepa, 1993; Zhao and Principe, 2001).

The field of DSP revitalized in the 1990s with the advent of support vector machines (SVMs) in particular and of kernel methods in general (Schölkopf and Smola, 2002; Shawe-Taylor and Cristianini, 2004; Vapnik, 1995). The framework of kernel machines allowed the robust formulation of nonlinear versions of linear algorithms in a very simple way, such as the classical LMS (Liu *et al.*, 2008) or recursive least squares (Engel *et al.*, 2004) algorithms for adaptive filtering, Fisher's discriminants for signal classification and recognition, and kernel-based autoregressive and moving average (ARMA) models for system identification and time series prediction (Martínez-Ramón *et al.*, 2006). In the last decade, the fields of graphical models (Koller and Friedman, 2009), kernel methods (Shawe-Taylor and Cristianini, 2004), and Bayesian nonparametric inference (Lid Hjort *et al.*, 2010) have played an important role in modern signal processing. Not only have many signal-processing problems been tackled from a canonical machine-learning perspective, but the opposite direction has also been fruitful.

1.4.5 From Digital Signal Processing to Machine Learning

Machine learning is in constant cross-fertilization with signal processing, and thus the converse situation has also been satisfied in the last decade. New machine-learning developments have relied on achievements from the signal-processing community. Advances in signal processing and information processing have given rise to new machine-learning frameworks:

Sparsity-aware learning. This field of signal processing takes advantage of the property of sparseness or compressibility observed in many natural signals. This allows one to determine the entire signal from relatively few scarce measurements. Interestingly, this topic originated from the image and signal processing fields, and rapidly extended to other problems, such as mobile communications and seismic signal forecasting. The field of sparse-aware models has recently influenced other fields in machine learning, such as target detection in strong noise regimes, image coding and restoration, and optimization, to name just a few.

Information-theoretic learning. The field exploits fundamental concepts from information theory (e.g., entropy and divergences) estimated directly from the data to substitute the conventional statistical descriptors of variance and covariance. The field has encountered many applications in the adaptation of linear or nonlinear filters and also in unsupervised and supervised machine-learning applications. In the most recent years the framework has been interestingly related to the field of dependence estimation with kernels, and shown successful performance in kernel-based adaptive filter and feature extraction.

Adaptive filtering. The urgent need for nonlinear adaptive algorithms in particular communications applications and web recommendation tools to stream databases has boosted the interest in a number of areas, including sequential and active learning. In this field, the introduction of online kernel adaptive filters is remarkable. Sequential and adaptive online learning algorithms are a fundamental tool in signal processing, and intelligent learning systems, mainly since they entail constraints such as accuracy, algorithmic simplicity, robustness, low latency, and fast implementation. In addition, by defining an instantaneous information measure on observations, kernel adaptive filters are able to actively select training data in online learning scenarios. This active learning mechanism provides a principled framework for knowledge discovery, redundancy removal, and anomaly detection.

Machine-learning methods in general and kernel methods in particular provide an excellent framework to deal with the jungle of algorithms and applications. Kernel methods are not only attractive for many of the traditional DSP applications, such as pattern recognition, speech, audio, and video processing. Nowadays, as will be treated in this book, kernel methods are also one of the primary candidates for emerging applications such as brain–computer interfacing, satellite image processing, modeling markets, antenna and communication network design, multimodal data fusion and processing, behavior and emotion recognition from speech and videos, control, forecasting, spectrum analysis, and learning in complex environments such as social networks.

2

Introduction to Digital Signal Processing

Signal processing deals with the representation, transformation, and manipulation of signals and the information they contain. Typical examples include extracting the pure signals from a mixture observation (a field commonly known as *deconvolution*) or particular signal (frequency) components from noisy observations (generally known as *filtering*). Before the 1960s, the technology only permitted processing signals analogically and in continuous time. The rapid development of computers and digital processors, plus important theoretical advances such as the FFT, caused an important growth of DSP techniques. This chapter first outlines the basics of signal processing and then introduces the more advanced concepts of time–frequency and time–scale representations, as well as emerging fields of compressed sensing and multidimensional signal processing.

2.1 Outline of the Signal Processing Field

A crucial point of DSP is that signals are processed in *samples*, either in batch or online modes. In DSP, signals are represented as sequences of finite precision numbers and the processing is carried out using digital computing techniques. Classic problems in DSP involve *processing* an input signal to obtain another signal. Another important part consists on *interpreting* or *extracting information* from an input signal, where one is interested in characterizing it. Systems of this class typically involve several steps, starting from a digital procedure, like filtering or spectral density estimation, followed by a pattern recognition system producing a symbolic representation of the data. This symbolic representation may be the input to an expert system based on rules that gives the final interpretation of the original input signal. In machine learning we know this as a classifier or automatic detection algorithm. Classic problems also deal with processing of symbolic expressions. In these cases, signals and systems are represented as abstract objects and analyzed using basic theoretical tools such as convolution, filter structures, FFT, decimation, or interpolation. Classic problems found in the literature – and being present in most DSP courses (Oppenheim and Schafer, 1989; Proakis and Manolakis, 2006) – include (but are not limited to): (1) filter design using FIR, IIR, and adaptive filters; (2) autoregressive (AR) and moving-average (MA) time-invariant systems; (3) spectral analysis, mainly using the discrete Fourier

Digital Signal Processing with Kernel Methods, First Edition. José Luis Rojo-Álvarez, Manel Martínez-Ramón, Jordi Muñoz-Marí, and Gustau Camps-Valls.

transform (DFT); (4) time–frequency analysis, using the spectrogram, time–frequency distributions (TFDs), and wavelets; (5) decimation and interpolation techniques; and (6) deconvolution-based signal models.

2.1.1 Fundamentals on Signals and Systems

Basic Signals in Continuous and Discrete Time

In a first approach, a *signal* is generally (and very roughly) defined as a physical magnitude that changes with time, and it is traditional to distinguish between continuous-time signals (e.g., denoted $x(t)$) and discrete-time signals (e.g., denoted $x[n]$). Several specific signals are specially relevant because they can be used as building blocks of a vast majority of observations, in the sense that they form the *basis of a signal space*. In this setting, unit impulse functions such as the Dirac delta $\delta(t)$ and the Kronecker delta $\delta[n]$ are the most widely spread ones, and they are closely related to unit steps, either in continuous $u(t)$ or discrete cases $u[n]$, by means of the integral or the cumulative sum, respectively, of their unit impulse counterparts. Any signal fulfilling some integral convergence requirements can be expressed in terms of impulse signals:

$$x(t) = \int_{-\infty}^{+\infty} x(\tau)\delta(t-\tau)\,\mathrm{d}\tau \tag{2.1}$$

$$x[n] = \sum_{k=-\infty}^{+\infty} x[k]\delta[n-k] \tag{2.2}$$

for continuous- and discrete-time signals respectively. Note that this is essentially a time-domain representation, as far as impulse signals are perfectly well located in time. Another fundamental basis of signal spaces, which are intrinsic to the time domain, is given by "sinc" functions:

$$\mathrm{sinc}(t) = \frac{sin(\pi t)}{\pi t} \tag{2.3}$$

and

$$\mathrm{sinc}[n] = \begin{cases} \dfrac{\sin(Nn/2)}{\sin(Nn/2)} & n \neq 2\pi k \\ (-1)^{k(N-1)} & n = 2\pi k \end{cases} \tag{2.4}$$

These functions have to be understood in terms of their properties as ideal filters in the frequency domain. Finally, rectangular windows $p(t)$ and $p[n]$ are also useful signals for dealing with time-local properties of the signals.

Linear and Time-Invariant Systems

Systems are those physical or mathematical entities that transform signals into other signals, which in general we will denote as $y(t) = T\{x(t)\}$ and $y[n] = T\{x[n]\}$ when they act on continuous or discrete data streams. Mathematical properties of the systems are intimately related to their physical behavior. For instance, an output signal at a given time instant only depends on the input at that time instant in *systems with memory*. Also, an output signal at a time instant does not depend on the future of input or output signals in *causal systems*. Stable systems always yield a bounded output signal as their response to a bounded input signal, and linear systems fulfill additivity and scaling properties between input and output signals. Finally, time-invariant systems yield the same time-shifted response for any shifted input signal. For those systems which simultaneously fulfill time invariance and linearity, their response to the unit impulse (often denoted as $h(t)$, $h[n]$) can be readily obtained in the time domain by using Equation 2.1, which yields the continuous- and discrete-time convolution operators, respectively given by

$$y(t) = x(t) * h(t) = \int_{-\infty}^{+\infty} x(\tau)h(t-\tau)\,\mathrm{d}\tau \tag{2.5}$$

$$y[n] = x[n] * h[n] = \sum_{k=-\infty}^{+\infty} x[k]h[n-k], \tag{2.6}$$

where symbol $*$ is a common compact notation of the convolution operation, referred to continuous-time and discrete-time according to the context. Convolution stands for a shared method for analytical representation and practical working with linear and time-invariant (LTI) systems. A vast number of physical systems admit this fundamental representation, which makes LTI systems theory a fundamental framework for signal analysis.

Complex Signals and Complex Algebra

Complex representation of magnitudes in signal processing is often necessary depending on the nature of the signals to be processed. The most representative examples can be found in communications, where complex notation provides a natural expression of signals that make their manipulation easier and compact. The justification of the use of complex algebra arises from the fact that any band-pass signal centered at any given frequency admits a representation that can be decomposed into two orthogonal components. They are called in-phase and quadrature-phase components. The in-phase component is Hermitic around the central frequency, while the quadrature phase is anti-Hermitic. The combination of both components reconstructs any band-pass signal.

Usually, communication signals are treated at the receiver in their complex envelope form; that is, the central frequency is removed because it does not convey any information, thus leading to a signal which is complex in nature. While both components can be processed independently, usually, channel distortion produces in-phase and quadrature components that are linear combinations of the original ones. Then, the reception of the signal must include an equalization procedure; that is, a recovery of the original components from the received ones. Equalization is most easily formulated by the use of complex algebra.

Signal detection in communications or radio detection and ranging (radar for short) necessarily includes observation noise. Among all noise sources, thermal noise is the most prevalent one. Usually, thermal noise is assumed to be Gaussian and circularly symmetric in the complex plane. A complex signal $z(t)$ is said to have circular symmetry if $e^{\mathbb{i}\theta} z(t)$ has the same probability distribution as $z(t)$ for any value of t. In this case the least squares (LS) criterion is optimal. Actually, LS-criterion-based algorithms are straightforwardly formulated in complex algebra, and the noise is fully characterized by its variance in each complex component. In some situations, however, non-Gaussian noise components may appear, thus making LS suboptimal. In these situations, other criteria than the ones based on maximum margin must be formulated, where the use of complex numbers is still possible. Moreover, noise can still be Gaussian but noncircularly symmetric; thus, in general, three parameters are needed for its characterization, which are the variances of each component and their covariance. In order to properly apply an optimization criterion, Wirtinger algebra has been applied in Bouboulis and coworkers (Bouboulis and Theodoridis, 2010, 2011; Bouboulis *et al.*, 2012) and Ogunfunmi and Paul (2011).

Frequency-Domain Representations

Unit impulses, pulses, and sinc functions have a clear determination in time properties. However, the velocity of the variation of a signal cannot be easily established in these basic signals. The simplest signals for which their change velocity can be determined are sinusoids, $x(t) = A\sin(\omega_0 t + \phi_0)$ and $x[n] = A\sin[\omega_0 n + \phi_0]$, in terms of their angular frequency. Also, complex exponentials form a basis for the signal space, $x(t) = A\,e^{st}$, with $A, s \in \mathbb{C}$, or $x[n] = Az^n$, with $z \in \mathbb{C}$, and the subset of purely imaginary exponentials, $x(t) = e^{\mathbb{i}w_0 t}$ and $x[n] = e^{\mathbb{i}\omega_0 n}$ are straightforwardly related to sinusoids and makes operative their mathematical handling.

These signals, having their rate of change properties well determined, open the field to an alternative basis of signal spaces. For periodic signals, fulfilling $x(t) = x(t + T_0)$ or $x[n] = x[n + N_0]$, Fourier showed that they can be expressed in terms of imaginary exponentials:

$$x(t) = \sum_{k=-\infty}^{+\infty} a_k\, e^{\mathbb{i}k(2\pi/T_0)t} \tag{2.7}$$

$$x[n] = \sum_{k=0}^{N_0-1} a_k\, e^{\mathbb{i}k(2\pi/N_0)n} \tag{2.8}$$

$$a_k = \int_{-\infty}^{+\infty} x(t)\, e^{-\mathbb{i}k(2\pi/T_0)t}\, \mathrm{d}t \tag{2.9}$$

$$a_k = \sum_{k=0}^{N_0-1} x[n]\, e^{-\mathbb{i}k(2\pi/N_0)n}, \tag{2.10}$$

where exponential coefficients stand for the relative importance of each harmonic frequency in the periodic signal under analysis, and can be readily obtained by cross-comparison between the signal and each harmonic component. Therefore, the so-called Fourier series for periodic signals represents a set of comparisons between the periodic signal and the sinusoid harmonics, which gives a precise description of the variation rates present in a signal.

This precise description is also desirable for nonperiodic signals. This is given by the continuous- and discrete-time Fourier transforms (FTs), which are generalized from Equations 2.7 and 2.9 by limit concepts as follows:

$$X(\mathbb{i}\omega) = \mathcal{F}\{x(t)\} = \int_{-\infty}^{+\infty} x(t)\,\mathrm{e}^{-\mathbb{i}\omega t}\,\mathrm{d}t \tag{2.11}$$

$$X(\mathbb{i}\Omega) = \mathcal{F}\{x[n]\} = \sum_{n=-\infty}^{\infty} x[n]\,\mathrm{e}^{-\mathbb{i}\Omega n} \tag{2.12}$$

$$x(t) = \frac{1}{2\pi}\int_{-\infty}^{+\infty} X(\mathbb{i}\omega)\,\mathrm{e}^{\mathbb{i}\omega t}\,\mathrm{d}w \tag{2.13}$$

$$x[n] = \frac{1}{2\pi}\int_{-\pi}^{+\pi} X(j\Omega)\,\mathrm{e}^{\mathbb{i}\Omega n}\,\mathrm{d}\Omega, \tag{2.14}$$

where $\mathcal{F}\{x(t)\}$ and $\mathcal{F}\{x[n]\}$ denote the FT of a signal in the continuous-time and in the discrete-time domain, depending on the context. Among many implications for LTI system theory, the previous equations mean that the convolution operator in the time domain can be handled as a product-template in the frequency domain affecting the modulus and phase of the input signal, $Y(\mathbb{i}\omega) = H(\mathbb{i}\omega)X(\mathbb{i}\omega)$ and $Y(\mathbb{i}\Omega) = H(\mathbb{i}\Omega)X(\mathbb{i}\Omega)$. An additional extremely relevant consequence of using the frequency domain is the analytical availability of the modulation property, as far as signal product in the time domain is equivalent to the convolution operation in the frequency domain. For instance, given the signal $c(t) = x(t)\,\mathrm{e}^{\mathbb{i}\omega_0 t}$, its frequency expression $C(\mathbb{i}w)$ is a function of the frequency transform of the original signal $x(t)$ simply noted as $C(\mathbb{i}\omega) = X(\mathbb{i}\omega - \mathbb{i}\omega_0)$.

Sampling

A discrete-time signal can be obtained from a continuous-time signal, $x[n] = x(nt_s)$, which is known as a sampling process, and the most basic scheme is given by

$$x_s(t) = x(t)p_\delta(t) = x(t)\sum_{k=-\infty}^{\infty}\delta(t - kt_s) = \sum_{k=-\infty}^{\infty} x(kt_s)\delta(t - kt_s), \tag{2.15}$$

where t_s denotes the sampling period. Hence, processing of a continuous-time signal can be made in the discrete-time domain, which is closer to software implementations and more prone to flexibility. Recovering a continuous-time signal from a discrete-time one is made through an ideal filter whose impulse response is the sinc signal with an adequate bandwidth,

$$x(t) = \sum_{k=-\infty}^{\infty} x(nt_s)\mathrm{sinc}\left(\frac{t - nt_s}{t_s}\right), \tag{2.16}$$

which shows that the signal can be expressed as an expansion of shifted and scaled sinc functions. Recall that the spectrum of a discrete-time signal is periodic, and it can be related to its continuous time counterpart by

$$X(\mathbb{i}\Omega) = \sum_{k=-\infty}^{\infty} X(\mathbb{i}\omega - \mathbb{i}k\omega_s). \tag{2.17}$$

This makes a necessary assumption that *no aliasing* in the spectral replications in the discrete-time spectrum can be allowed. This is known informally as the Nyquist theorem. It was first formulated by Harry Nyquist in 1928, but proved by Shannon (1949). It is stated as follows:

Theorem 2.1.1 Let $x(t)$ be a continuous, square integrable signal whose FT $X(\omega)$ is defined in the interval $[-W, W]$. Then signal $x(t)$ admits a representation in terms of its sampled counterpart $x(n/f_s)$ as

$$x(t) = \sum_n x\left(\frac{n}{f_s}\right) \frac{\sin\left[\pi f_s\left(t - \frac{n}{f_s}\right)\right]}{\pi f_s\left(t - \frac{n}{f_s}\right)} = \sum_n x\left(\frac{n}{f_s}\right) \operatorname{sinc}\left[f_s\left(t - \frac{n}{f_s}\right)\right] \tag{2.18}$$

if and only if $f_s = 1/t_s \geq 2W$.

The theorem says that the continuous-time signal can be recovered as far as $f_s > 2W$ (Nyquist frequency), where W is given by the largest frequency in the spectrum with non-null Fourier component.

Energy-Defined and Power-Defined Signals

Signals are mathematically compared in terms of their properties. Among them, the *magnitude* of a signal is a key intuitive concept to determine whether or not a signal is larger than another. Two main kinds of signals can be described by the concepts of *power* and *energy* of a signal, which are defined as follows in the continuous and discrete cases:

$$P_x = \lim_{T\to+\infty} \frac{1}{2T} \int_{-T}^{T} |x(t)|^2 \,\mathrm{d}t \tag{2.19}$$

$$P_x = \lim_{N\to+\infty} \frac{1}{2N+1} \sum_{-N}^{N} |x[n]|^2 \tag{2.20}$$

$$E_x = \int_{-\infty}^{\infty} |x(t)|^2 \,\mathrm{d}t \tag{2.21}$$

$$E_x = \sum_{-\infty}^{\infty} |x[n]|^2. \tag{2.22}$$

These concepts have been inherited from the field of linear circuit theory, but open the field to algebraic comparisons in terms of dot products and projections onto signal Hilbert spaces. When convergence conditions are fulfilled, a signal is said to be energy defined (its energy is non-null and finite), or power defined (its power is non-null).

In addition, the power or the energy of a given signal can be expressed with precision in the transformed frequency domain, according to the so called energy spectral density (ESD) and power spectral density (PSD), which are given by

$$S_x(\omega) = |X(\omega)|^2 = \left|x(t)\,\mathrm{e}^{-\mathrm{i}\omega t}\,\mathrm{d}t\right|^2 \tag{2.23}$$

$$P_x(\omega) = \lim_{T\to\infty} E\left(|X_T(\omega)|^2\right), \tag{2.24}$$

where

$$X_T(\omega) = \frac{1}{2T}\int_{-T}^{T} x(t)\,e^{-\mathrm{i}\omega t}\,dt \tag{2.25}$$

for continuous-time signals. ESD is more appropriate for transients (time-duration limiting or evanescent signals), whereas PSD is more appropriate for stationary or periodic signals. A fundamental theorem in signal processing actually relates the energy/power preservation in both domains: Parseval's theorem indicates that the energy (or the power) of a signal can be interchangeably obtained from its time-domain expression or from its spectral density; that is:

$$E_x = \frac{1}{2\pi}\int_{-\infty}^{\infty} S_x(\omega)\,d\omega \tag{2.26}$$

$$P_x = \frac{1}{2\pi}\int_{-\infty}^{\infty} P_x(\omega)\,d\omega. \tag{2.27}$$

The theorem is intuitively telling us that, after all, the representation domain is not changing the intrinsic properties of the signal.

Autocorrelation and Spectrum

The *autocorrelation* concept is closely related to the PSD concept. If we define the autocorrelation of a signal $x(t)$ as

$$\gamma(\tau) = x(t) * x(-t) = \int_{-\infty}^{+\infty} x(\tau)x(t+\tau)\,d\tau, \tag{2.28}$$

then the PSD of $x(t)$ and its autocorrelation are a transform pair; that is, $S_{xx}(\omega) = \int_{-\infty}^{\infty}\gamma(\tau)\,e^{-\mathrm{i}\omega\tau}\,d\tau$. Hence, the autocorrelation conveys relevant information in the time and frequency of the signal.

Symmetric and Hermitian Signals

An additional set of properties of the signals is given by Equation *symmetry*. A real signal $x(t)$ is symmetric or even when $x(-t) = x(t)$. The function is antisymmetric or odd if $x(-t) = -x(t)$. A complex function $z(t)$ is said to be Hermitian if $z(-t) = z^*(t)$, where $*$ is the conjugate operator. If $z(-\nu) = -z^*(t)$, the function is said to be anti-Hermitian. Hermitian signals have important properties often exploited in signal processing.

Property 1 The real part of a Hermitian signal is symmetric and its imaginary part is antisymmetric. Let the complex signal $z(\nu) = z_R(\nu) + \mathrm{i}z_I(\nu)$, where $z_R(\nu)$ and $z_I(\nu)$ are its real and imaginary parts, be a Hermitian signal. Then

$$z(-t) = z_R(-t) + \mathrm{i}z_I(-t) = z^*(t) = z_R(t) - \mathrm{i}z_I(t) \tag{2.29}$$

Property 2 If a signal $x(t)$ is real, its FT $X(\mathbb{i}\omega)$ is Hermitic. Let $x(t)$ be a real signal. Its FT is defined as (see Equation 2.11)

$$X(\mathbb{i}\omega) = \int_{-\infty}^{+\infty} x(t)\,\mathrm{e}^{-\mathbb{i}\omega t}\,\mathrm{d}t. \tag{2.30}$$

Since $\mathrm{e}^{-\mathbb{i}wt} = \cos(wt) - \mathbb{i}\sin(wt)$, then

$$X(\mathbb{i}\omega) = \int_{-\infty}^{+\infty} x(t)\cos(\omega t) - \mathbb{i}\int_{-\infty}^{+\infty} x(t)\sin(\omega t)\,\mathrm{d}t. \tag{2.31}$$

Changing the sign of the argument w and knowing that functions cos and sin are respectively symmetric and antisymmetric gives

$$\begin{aligned} X(-\mathbb{i}\omega) &= \int_{-\infty}^{+\infty} x(t)\cos(-\omega t) - \mathbb{i}\int_{-\infty}^{+\infty} x(t)\sin(-\omega t)\,\mathrm{d}t \\ &= \int_{-\infty}^{+\infty} x(t)\cos(\omega t) + \mathbb{i}\int_{\infty}^{+\infty} x(t)\sin(-\omega t)\,\mathrm{d}t. \end{aligned} \tag{2.32}$$

Clearly, the real part of $X(\mathbb{i}\omega)$ is symmetric and its imaginary part is antisymmetric. Then, by Property 1, the function is Hermitian.

Property 3 If a real signal $x(t)$ is symmetric, its FT $X(\mathbb{i}\omega)$ is real and symmetric. If a function is symmetric, then $x(t) = x(-t)$; hence their FTs should be equal. The transform of $x(-t)$ is then equal to $X(\mathbb{i}\omega)$ and it can be expressed as

$$\begin{aligned} X(\mathbb{i}\omega) &= \int_{-\infty}^{+\infty} x(-t)\cos(\omega t)\,\mathrm{d}t - \mathbb{i}\int_{-\infty}^{+\infty} x(-t)\sin(\omega t)\,\mathrm{d}t \\ &= \int_{-\infty}^{+\infty} x(t)\cos(-\omega t)\,\mathrm{d}t - \mathbb{i}\int_{-\infty}^{+\infty} x(t)\sin(-\omega t)\,\mathrm{d}t \\ &= \int_{-\infty}^{+\infty} x(t)\cos(\omega t)\,\mathrm{d}t + \mathbb{i}\int_{-\infty}^{+\infty} x(t)\sin(\omega t)\,\mathrm{d}t \end{aligned} \tag{2.33}$$

Similarly, $X(\mathbb{i}\omega)$ is defined as $X(\mathbb{i}\omega) = \int_{-\infty}^{+\infty} x(t)\cos(\omega t) - \mathbb{i}\int_{-\infty}^{+\infty} x(t)\sin(\omega t)\,\mathrm{d}t$; hence

$$\mathbb{i}\int_{-\infty}^{+\infty} x(t)\sin(\omega t)\,\mathrm{d}t = 0, \tag{2.34}$$

which proves that the FT is real. It is straightforward to see that $\int_{-\infty}^{+\infty} x(t)\cos(\omega t)\,\mathrm{d}t$ is symmetric when $x(t)$ is symmetric.

Property 3 can be proven by simply stating that if signal $x(t)$ is symmetric, the product $x(t)\sin(\omega t)$ is antisymmetric, and then its integral between $-\infty$ and ∞ is zero.

2.1.2 Digital Filtering

In signal processing, a *filter* is a device that removes some unwanted components (or features) from an observation (a signal coming out of a natural or physical system). A broader and updated definition of a filter is "a system that operates on an input signal to produce an output signal according to some computational algorithm." The field of digital-filter theory builds upon extensive earlier work done in classic circuit theory, numerical analysis, and more recently sampled data systems. Actually, *digital filtering* is a vast field, also intimately related to engineering fields of system control and identification, statistical branches of signal interpolation and smoothing, as well as the machine-learning paradigm of regression and function approximation. This section reviews the fundamental basis of digital filter theory, and advanced concepts and applications are treated in further sections.

Digital Filters as Input–Output Systems in the Time Domain

Digital systems were originally introduced as discrete-time systems because signals were defined as a function of time. A digital system can be cast as a transformation T operating on an input signal $x[n]$ to produce an output signal $y[n] = T\{x[n]\}$. There are two main filters in DSP. The conventional MA filter, also known as a FIR filter, models the output as a linear combination of the previous inputs to the system and is defined according to

$$y[n] = \sum_{l=0}^{Q} b_l x[n-l]. \tag{2.35}$$

The AR filter, also known as an IIR filter, models the output as a linear combination of previous outputs of the system and is defined as

$$y[n] = \sum_{m=1}^{P} a_m y[n-m], \tag{2.36}$$

where a_k and b_k are the filter coefficients, and P and Q are the order (or tap delay) of the AR and MA filters respectively. The notation ARMA(P, Q) combines AR and MA processes, and refers to a model with P autoregressive terms and $Q+1$ moving-average terms. Note that we are restricted here to causal filters, meaning that future information cannot be used for modeling (i.e., $k < n$), also known as *nonanticipative* filters. Different classes of input–output filters (or systems) are defined by placing constraints on the properties of transformation $T(\cdot)$. Filters are considered stable if, for a bounded input signal, the filter returns a bounded output signal. Looking at the models, one may think of filters implemented by either convolution (FIR) or by recursion (IIR).

Filtering in the Frequency Domain

Filtering in the time domain is convenient when the information is encoded in the *shape* of the signal. Time-domain filtering is widely used for signal smoothing, DC removal, and waveform shaping, among many others. However, filters are very often designed in the frequency domain to remove some frequencies while keeping others of interest. Frequency-domain filters are used when the relevant information is contained not only in the signal amplitude with respect to time but in the frequency and phase of the signal components (sinusoids if the chosen basis is the Fourier basis). Designing filters in the frequency domain requires defining *specifications* in this domain (Proakis and Manolakis, 2006).

The frequency response (transfer or system function) of a filter (system or device) can be defined as the quantitative measure of the output versus the input signal spectral contents. It can also be cast as a measure of magnitude and phase of the output as a function of frequency. The frequency response is characterized by the magnitude of the system response, typically measured in decibels (dB) or as a decimal, and the phase, measured in radians or degrees, versus frequency in radians per second or hertz (Hz).[1] One typically represents both the Bode magnitude and phase plots and analyzes different features, such as cut-off frequencies, band-pass maximum allowed ripples, or band width. These are the characteristics that define the type of filter, such as low-pass, high-pass, band-pass, and band-stop filters (Proakis and Manolakis, 2006).

The representation of discrete signals using the z-transform is very useful to work in the frequency domain. Essentially, the z-transform of signal $x[n]$ is defined as

$$X(z) = \sum_{n=-\infty}^{\infty} x[n]z^{-n}, \tag{2.37}$$

where variable z is defined in a *region of convergence* such that $r_2 < |z| < r_1$. Note the relation with the FT if z is expressed in the polar domain, $z = r\,e^{\mathbb{i}\omega}$, where $r = |z|$ and $\omega = \angle z$. Actually, the FT is just the z-transform of the signal evaluated in the unit circle, provided that it lies within the region of convergence. Equivalently, the z-transform of the impulsive response $h[n]$ characterizing a filter is

$$H(z) = \sum_{n=-\infty}^{\infty} h[n]z^{-n}. \tag{2.38}$$

The frequency response of a system is thus typically expressed as a function of (polynomials in) z rather than the more natural variable Ω, which allows us to study the stability of the system easily and to design filters accordingly.

Finally, we should note that, if the system under investigation is *nonlinear*, applying purely linear frequency-domain analysis will not reveal all the system's characteristics. In such cases, and in order to overcome these limitations, generalized frequency-response functions and nonlinear output frequency-response functions have been defined that allow the user to analyze complex nonlinear dynamic effects (Pavlov *et al.*, 2007).

Filter Design and Implementation

The design of frequency-selective filters involves fitting (selecting) the coefficients of a causal FIR or IIR filter that tightly approximates the desired frequency-response

1 A Bode plot shows the transfer function of an LTI system versus frequency, plotted with a log-frequency axis.

specifications. Which type of filter to use (namely, FIR or IIR) depends on the nature of the problem and specifications. FIR filters exhibit a linear phase characteristic and stability, but their design typically requires high filter orders (sometimes referred to as "taps"), and thus their implementation is less simple than for an equivalent IIR filter. Contrarily, IIR filters show a nonlinear phase characteristic and are potentially unstable, but they can be implemented with lower filter orders, thus giving rise to simpler circuits. In practice, however, using one or the other ultimately depends on the specificities of the signal and on the filter phase characteristic. Speech signals, for example, can be processed in systems with nonlinear phase characteristic. When it is the frequency response that matters, it is better to use IIR digital filters, which have far lower order (Rabiner *et al.*, 1974).

The implementation of discrete-time systems such as FIR and IIR admits many approaches. Among them, FIR filters are typically implemented with direct-form, cascade-form, frequency-sampling, and lattice structures, while IIR systems adopt parallel and lattice–ladder as well. In addition to the filter design methods based on the analog-to-digital domain transformations, there are several methods directly defined in the discrete-time domain, such as the LS method, which is particularly useful for designing IIR filters and FIR Wiener filters. The literature of digital filters design is vast and overwhelming. As reference books, we refer the reader to the excellent old treaties in Kaiser (1972), Golden and Kaiser (1964), Helms (1967), Constantinides (1970), and Burrus and Parks (1970).

Besides selecting the structure and optimizing the weights, some other practical problems arise with the *hardware* real implementation of filters. The main issues concern the finite resolution for the numerical representation of the coefficients, that is, the quantization and round-off effects appearing in real implementations. These are nonnegligible problems that need to be addressed and have been the subject of intense research in the field (Parker and Girard, 1976).

Linear Prediction and Optimal Linear Filters

Designing filters for signal estimation is a recurrent problem in communications and control systems, among many other disciplines. Filter design has been approached from a statistical viewpoint as well. The most common approach consists of assuming a linear model and the minimization of the mean-square error, thus focusing only on second-order statistics of the stationary process (autocorrelation and cross-correlation). The approach leads to solving a set of linear equations, typically with the (computationally efficient) Levinson–Durbin and the Schür algorithms (Proakis and Manolakis, 2006).

The field of *optimal linear filter estimation* is very large, and many algorithms have been (and constantly are being) presented. The pioneering work was developed by Wiener in 1949 for filtering statistically stationary signals (Wiener, 1975). The Wiener filter estimates a desired (or target) random process by LTI filtering of an observed noisy process, assuming known stationary signal and noise spectra, and additive noise. The Wiener filter essentially minimizes the mean square error between the estimated random process and the desired process, and hence is known as a minimum mean square error (MMSE) estimator. The Wiener filter has had a crucial impact on signal and image-processing applications (Wiener, 1975).

Adaptive Filtering

Adaptive filtering (Haykin, 2001) is a central topic in signal processing. An adaptive filter is a filter structure (e.g., FIR, IIR) provided with an adaptive algorithm that tunes the transfer function typically driven by an error signal. Adaptive filters are widely applied in nonstationary environments because they can adapt their transfer function to match the changing parameters of the system generating the incoming data (Hayes 1996; Widrow *et al.* 1975). Adaptive filters have become ubiquitous in current DSP, mainly due to the increase in computational power and the need to process data streams. Adaptive filters are now routinely used in all communication applications for channel equalization, array beamforming or echo cancellation, and in other areas of signal processing such as image processing or medical equipment.

2.1.3 Spectral Analysis

The concept of *frequency* is omnipresent in any signal-processing application. Spectral representations give fundamental information valuable in many signal processing tasks in the experimental plane as well as in applications. From a practical point of view, *spectral analysis* is used in a large number of applications, the most obvious one being spectral visualization for research and development. But also, spectral analysis is used in voice recognition and synthesis, wireless digital telephony and data communications, medical imaging, or antenna array processing, just to mention some of the most common applications.

The field of spectral analysis is wide, and thus a great many techniques exist depending on the particular application at hand. These techniques can be decomposed into two main families, all of them under the umbrella of Fourier analysis. The first one is the set of classical techniques called *nonparametric spectral analysis*. These embrace a very wide set of different technologies, and are the most widely used nowadays. The second set of analysis methods are *parametric spectral analysis*, and the family can be subdivided into *time-series analysis methods* (using AR models) and *subspace methods*. When these methods first appeared in the literature, they were barely used due to the hardware limitations of the digital signal processors at those times. One of the most important novelties of parametric methods was their frequency resolution, which dramatically over-passed the nonparametric ones (Kay and Marple, 1981). Owing to algorithmic improvements and computational power of digital signal processors, their use has exponentially increased, and parametric methods have been incorporated in many real-life applications, ranging from voice processing to radar applications, to mention just a couple.

From a theoretical viewpoint, a spectral representation can be understood as the similarity of an arbitrary stationary signal to each one of a set of orthogonal base signals. Usually, complex exponentials are used since they are eigenfunctions of linear systems. In these cases, the Fourier analysis is used as a fundamental analysis tool, but, nevertheless, other basis functions can be used to decompose a signal into components (e.g., Hadamard, Walsh). For a discrete signal $x[n]$ which is absolutely summable – that is, $\sum_{-\infty}^{\infty} |x[n]|^2 < \infty$ – its DFT at a frequency Ω can be conceived as the *dot product* (measure of similarity) between the signal and a complex sinusoid

$\{e^{i\Omega n}\}_{n=-\infty}^{\infty}$ at this frequency; that is, $X(i\Omega) = \mathcal{F}\{x[n]\}$. We can write the spectrum of a signal as

$$S_x(i\Omega) = X(i\Omega)X^*(i\Omega). \tag{2.39}$$

It is obvious that in practice the number of observed samples available for the computation is limited. This estimated spectrum $\hat{S}(e^{i\Omega})$ can be modeled as an expectation of the product in time of the time-unlimited signal with a time-limited window $w[n]$ of N samples:

$$\hat{S}_x(i\Omega) = E\left(\sum_{n=0}^{N-1}\sum_{m=0}^{N-1} e^{-i\Omega n}x[n]x^*[m]\, e^{i\Omega m}\right). \tag{2.40}$$

Now, defining $\boldsymbol{R}_{xx}$ as the autocorrelation matrix of the signal, and the complex sinusoid vector $\boldsymbol{v}(\Omega) = [1, e^{i\Omega}, \ldots, e^{i\Omega(N-1)}]^{\mathrm{T}}$, we obtain an expression for the spectrum estimation with the form

$$\hat{S}(i\Omega) = \boldsymbol{v}^{\mathrm{H}}(\Omega)\boldsymbol{R}_{xx}\boldsymbol{v}(\Omega), \tag{2.41}$$

where ^{H} stands for the Hermitic transpose operator. Since in the frequency domain the product turns into a convolution, the estimated spectrum $\hat{S}(i\Omega)$ can be written as

$$\hat{S}(i\Omega) = S(i\Omega) * |W(i\Omega)|^2, \tag{2.42}$$

where $|W(i\Omega)|^2$ is the spectrum of the (implicit or explicit) time window $w[n]$ used when a finite set of observations is available. This window has a spectrum with a main lobe centered at the origin, and with side lobes. The width of the main lobe and the amplitude of the side lobes tend to zero as N increases. The estimated spectrum is then distorted by the temporal window by spreading it and reducing its resolution. Changing the shape of the window can be useful to reduce the side lobes at the expense of widening the main side lobe.

The signal will be also embedded in noise, whose random nature will corrupt the estimated spectrum. The spectrum variance due to noise will decrease with the number of spectra computed at different time windows and used to estimate the expectation of Equation 2.40. This procedure is known as the *Welch spectrum estimation* (Welch, 1967). Since the number of samples is limited, the procedure described will impose a trade-off between the number of available spectra and the length of their corresponding time signals. Less windows of longer lengths will produce spectral estimations with more resolution; and conversely, more windows of smaller lengths will produce estimations with less noise corruption.

An alternative spectral representation can be obtained by assuming that signal $x[n]$ under analysis is the result of convolving a white random process $e[n]$ with a linear AR system. Specifically, the signal model is

$$x[n] = \sum_{m=1}^{P} a_m x[n-m] + e[n]. \tag{2.43}$$

The spectrum of the signal is straightforwardly computed as

$$\hat{S}(\mathbb{i}\Omega) = \frac{\sigma_e^2}{|1 - \sum_{m=1}^{P} a_m e^{-\mathbb{i}\Omega m}|^2}, \tag{2.44}$$

where σ_e^2 is the variance of $e[n]$, also called the *innovation process*. Note that this expression includes the FT of the model coefficients. In order to estimate these coefficients, a simple LS algorithm can be applied to minimize the estimation error, though other methods also based on minimum mean square error (MMSE) have been proposed; for example, see Burg's method (Burg, 1967; Byrne and Fitzgerald, 1983). Spectral estimation with AR models has shown much better performance than any method based on signal windowing. Alternative strategies also exist to estimate an optimal length for the AR model (Akaike, 1974; Rissanen, 1978).

All these methods are based on an optimization criterion, which is usually the LS. Actually, LS leads to the ML estimator when the statistics of the signal are Gaussian. Indeed, the FT itself is derived from the LS optimization of the approximation of a time series with a linear combination of orthogonal complex exponentials. AR spectrum estimation is the most obvious form of spectral analysis that uses this criterion, but also the so-called minimum variance distortionless response (MVDR) (Van Trees, 2002) provides a theoretical result where one can see that the LS criterion is the ML estimate when the signal is buried under Gaussian noise. The minimum power distortionless response (MPDR) method (Capon, 1969) can be seen as a sample estimate of the MVDR, which is, in turn, an LS method. The most common subspace method is the so-called multiple signal classification (MUSIC) (Stoica and Sharman, 1990) method, for which a high number of variants exist. This method can be interpreted as a modification of the MPDR.

The MPDR can be formulated as follows. The spectrum of a signal $x[n]$ can be estimated by convolving the signal with a set of linear FIR filters with impulse response $w_k[n]$ that are centered at different frequencies $\Omega_k, k = 0, \dots, N-1$. The output of the filter will be $y[n] = \sum_{m=0}^{M} w[m]x[n-m]$. The power of the output filters can be computed as the following expectation:

$$\hat{S}(\mathbb{i}\Omega_k) = E\left(\sum_{m=0}^{N-1}\sum_{m'=0}^{N-1} w_k[m]x[n-m]x^*[n-m']w_k^*[m']\right) = \boldsymbol{w}_k^{\mathrm{H}}\boldsymbol{R}_{xx}\boldsymbol{w}_k, \tag{2.45}$$

where $\boldsymbol{w} = [w[0], \dots, w[N-1]]^{\mathrm{T}}$ is the coefficient vector for the filter centered at frequency Ω_k. The MPDR criterion consists of minimizing the power output while producing an undistorted response for a complex sinusoid with unit amplitude centered at frequency Ω_k. That is, we minimize

$$\boldsymbol{w}_k^{\mathrm{H}}\boldsymbol{R}_{xx}\boldsymbol{w}_k \tag{2.46}$$

subject to

$$\sum_{n=0}^{N-1} \mathrm{e}^{\mathbb{i}\Omega_k n} w_k[n] = 1. \tag{2.47}$$

A Lagrange optimization of this minimization problem for Equation 2.45 gives the following solution:

$$\hat{S}(i\Omega_k) = \frac{1}{v(\Omega_k)^H R_{xx}^{-1} v(\Omega_k)}. \tag{2.48}$$

When compared with Equation 2.41, it can be seen that the MPDR spectrum is the inverse of the FT of the inverse of the autocorrelation matrix. An intuitive explanation of this spectrum arises from the fact that in R_{xx}^{-1} the eigenvectors are the same as in the signal autocorrelation matrix but the eigenvalues are inverted. We assume that, in a high signal-to-noise ratio (SNR) regime, each sinusoid is represented by an eigenvector with a high associated eigenvalue and that the rest of eigenvectors represent the noise subspace, thus corresponding to low eigenvalues. In that case, the denominator of Equation 2.48 will take a low value when frequency Ω is equal to one of the sinusoids in the signal, which will correspond to the inverse of the corresponding eigenvalue. For other frequencies not corresponding to the sinusoids, values will be higher. Then, Equation 2.48 will show high values at the frequencies of the sinusoids. This shows a much better performance than Welch methods, and it is considered a *hyperresolution method* since it allows one to detect two sinusoids that are very close to each other, much beyond the DFT limit.

A step forward can be done if the signal eigenvectors are just removed from the autocorrelation matrix. This is exactly equivalent to changing the signal eigenvalues by zeros. One can also change the noise eigenvalues by ones. In that case, when the frequency of the denominator corresponds to one of the sinusoids, Equation 2.48 will tend to infinity. The procedure consists of expressing the inverted autocorrelation as a function of its eigenvectors, and eigenvalues as $R_{xx}^{-1} = Q\Lambda^{-1}Q^H$, where Q is the eigenvectors matrix and Λ is a diagonal matrix containing the corresponding eigenvalues. We assume that the inverse of the signal eigenvalues have negligible values. The following approximation is valid for spectrum estimation purposes:

$$R_{xx}^{-1} = Q\Lambda^{-1}Q^H = Q_s\Lambda_s^{-1}Q_s^H + Q_n\Lambda_n^{-1}Q_n^H \approx Q_n\Lambda_n^{-1}Q_n^H, \tag{2.49}$$

where in this context the subindices "s" and "n" respectively denote signal and noise eigenvectors and eigenvalues.

The MUSIC method can be simply understood as a modification of the MPDR method where R_{xx}^{-1} is approximated as in Equation 2.49, but with the noise eigenvectors changed by ones, leading to

$$\hat{S}(i\Omega) = \frac{1}{v(\Omega)^H Q_n Q_n^H v(\Omega)}. \tag{2.50}$$

Note that this expression corresponds exactly to computing the DFT norm of the noise eigenvectors, adding them together, and then computing the inverse of the result. The DFT norm will show zeros in the frequencies corresponding to the sinusoids, since all noise eigenvectors are mutually orthogonal. Then, its inverse will tend to infinity at these frequencies. MUSIC is another hyperresolution method that, with the price of increasing the computational burden with respect to the MVDR, improves its resolution performance.

The performance of a given spectral analysis method can be measured in terms of its *frequency resolution*; that is, the ability to discriminate between two signals having close frequencies. Resolution is then expressed in terms of the minimum frequency difference that can be measured. But there are other performance measurements that are important, such as the probability of signal detection and accuracy of frequency and amplitude estimation. Also important is the robustness against non-Gaussian noise and the minimum number of samples needed to provide one with a spectrum estimation that fits certain specifications over the rest of performance parameters. The LS criterion can impose a limitation in the two latter performance criteria, since it can produce poor results in the presence of non-Gaussian noise, and using it with no regularization strategies may result in severe overfitting to the noisy observations. This can be improved in many situations by replacing the LS criterion by alternative criteria forcing *sparsity* (simplicity) of the solution, as we will show using particular SVM-based formulations for spectral density estimation (Rojo-Álvarez *et al.*, 2003).

2.1.4 Deconvolution

Deconvolution can be defined as the process of recovering a signal $x[n]$ that has been transmitted through a medium that presents an LTI behavior and whose impulse response $h[n]$ is unknown. Since the original signal has been convolved with the impulse response, the convolution consists of solving the solution of the implicit equation

$$y[n] = h[n] * x[n], \tag{2.51}$$

where $*$ stands for the convolution operator and $y[n]$ is the observed signal. The earliest application of deconvolution was reflection seismology, introduced in 1950 by E. Robinson (Mendel 1986; Silvia and Robinson 1979). The particular problem by then was the study of the layered structure of the ground by the transmission of a signal and the recording and observation of its reflected versions in the different layers. In this application, the ground impulse response, called *reflectivity* in this context, is assumed to consist of a series of Dirac delta functions; that is, the original signal is reflected from the different layers and received at different time instants later. Hence, this impulse response is sparse in time. Since the spectrum of a single Dirac delta function is a constant, the spectrum of a train of deltas sufficiently far apart can be considered approximately constant, so the impulse response can be approximated by a constant too. Under these conditions, the spectrum of the received signal is an attenuated replica of the transmitted signal. The reflectivity is modeled as a series of Dirac delta functions whose amplitudes and time delays are estimated. Deconvolution has also been extensively used in image processing for those applications where due to optical effects the image has been distorted. In these cases, the optical effects can be conveniently modeled by a *spread function* from the study of the physical phenomenon that blurs the image. Similar techniques can be found in the field of interferometry, radio astronomy, and computational photography.

Usually, deconvolution methods use the LS criterion for estimating the Dirac delta coefficients or amplitudes, since it is a computationally affordable and theoretically simple approach. However, LS methods tend to give poor estimation performance in situations where the noise is non-Gaussian. Specifically, LS methods are sensitive to data outliers, defined as those samples that can be considered as having a low likelihood. For example, a sample drawn from a Gaussian distribution can be considered an outlier if its amplitude is higher than a given number of standard deviations. In addition, the number of available samples in some applications is small when compared with the number of parameters to be estimated. In those cases, the optimization problem may turn to be ill-posed, thus leading to solutions highly biased from the optimal solution due to the noise in the estimation of the inverse matrix of the method.

As in other signal processing problems, Tikhonov regularization (Tikhonov and Arsenin, 1977) has been applied to deconvolution aiming to alleviate the previously mentioned problems. Roughly speaking, Tikhonov regularization is a solution smoothing, meaning that the regularized solution is smoothed or simplified with respect to the nonregularized one; for instance, in terms of the inverse autocorrelation matrix eigenvalues as estimated by the LS criterion. A very usual form of Tikhonov regularization consists of including an additional term into the functional to be optimized, usually the ℓ_2 norm of the parameter vector, where, in general, the ℓ_p norm of a vector $\boldsymbol{x} = [x_1, \dots, x_N]^{\mathrm{T}}$ is defined as

$$\ell_p(\boldsymbol{x}) = \|\boldsymbol{x}\|_p := \left(\sum_{i=1}^{N} |x_i|^p \right)^{1/p} . \tag{2.52}$$

Different forms of Tikhonov regularization can be found in ridge regression, total LS, or covariance-shaping LS methods (Hoerl and Kennard, 1970; Van Huffel and Vandewalle, 1991). The minimization of this norm, resulting in a smoothed solution, can produce reasonable approximations in ill-posed problems. Nevertheless, forcing a minimization of the ℓ_2 norm of the parameters results in the solutions being less sparse.

Given that the ℓ_2 norm is not adequate in those cases where sparse solutions are needed, the ℓ_1 norm has instead been proposed in many studies. Nevertheless, ℓ_1 norm tends to drop variables correlated to one dominant signal; for example, in cases where instead of "isolated spikes" the structure of the function to be estimated is structured in "clusters of spikes." In these cases, the combination of ℓ_1 and ℓ_2 norms can give more realistic solutions.

Deconvolution problems have been addressed by making use of ML techniques in state-variable models (Kormylo and Mendel, 1983). A more general solution was presented by Mendel (1986), where the sparse signal presents a Bernoulli–Gauss distribution, and modeling the impulse response was done by assuming an ARMA model. Gaussian mixtures (GMs) have also been used in deconvolution, by assuming that the prior distribution of the signal is a mixture of a narrow and a broad Gaussian. Expectation–maximization (EM) algorithms can be used to estimate the posterior of these signals.

Like other signal-processing problems, performance of the deconvolution process degrades due to different causes. First, the algorithm can result in numerically ill-posed inversion problems for which regularization methods are required, as mentioned previously. Second, sometimes the knowledge about the noise nature is limited, so the LS criterion is typically adopted, thus yielding suboptimal results. Often, these solutions require inverse filtering modeled as ARMA models that typically fail when the signal to be estimated does not have minimum phase (thus leading to unstable solutions). In order to alleviate these inconveniences, SVM solutions have been proposed (Rojo-Álvarez *et al.*, 2008), since their optimization criterion is regularized and intrinsically robust to non-Gaussian noise sources, as will be seen in subsequent chapters.

2.1.5 Interpolation

Signal interpolation is an old, yet widely studied research area (Butzer and Stens 1992; Jerri 1977). Interpolation in the information and communication applications has its origins in sampling theory, and more specifically in the Whittaker–Shannon–Kotel'nikov equation, also known as *Shannon's sampling theorem* (García 2000; Shannon 1949). This theorem states, in simple terms, that a band-limited, noise-free signal can be reconstructed accurately from a uniformly sampled sequence of its values, as far as the sampling period is properly chosen according to the signal bandwidth. The nonuniform sampling of a band-limited signal is a nontrivial different situation, which can also be addressed when the average sampling period still fulfills Shannon's sampling theorem and the dispersion is moderate (finite). Nonuniform sampling is naturally present and used in a number of applications, such as approximation of geophysical potential fields, tomography, or synthetic aperture radar (Jackson *et al.*, 1991; Strohmer, 2000).

Given that noise is often present in real-world registered signals, the reconstruction of a band-limited and noise-corrupted signal from its nonuniformly sampled observations becomes an even a harder problem. According to Unser (2000), two strategies are mainly followed in the nonuniform interpolation literature: (1) considering shift-invariant spaces, with a similar theoretical background as for uniformly sampled time series; and (2) definition of a new signal basis (or new signal spaces) that can be better suited to the nonuniform structure of the problem. The first approach has been most widely studied, where the sinc function is used as an interpolation kernel (Yen, 1956). Although Yen's interpolator is theoretically optimal in the LS sense, ill-posing is often present when computing the interpolated values numerically. This is also due to the sensitivity of the inverse autocorrelation matrix estimation of noise, in a similar way to that introduced in the preceding section for deconvolution problems. To overcome this limitation, numerical regularization has also been widely used in the nonuniform interpolation literature. Alternatively, a number of iterative methods have been proposed, including *alternating mapping*, *projections onto convex sets*, and *conjugate gradient* (CG) (Strohmer 2000; Yeh and Stark 1990), which alleviate this algorithmic inconvenience in some specific cases. Others have used noniterative methods, such as *filter banks*, either to reconstruct the continuous time signal or to interpolate to uniformly spaced samples (Eldar and Oppenheim 1999; Unser 2000; Vaidyanathan 2001), but none of these methods is optimal in the LS sense, and thus many approximate forms of the Yen interpolator have been developed (Choi and Munson, 1998a,b). The previously mentioned methods have addressed the reconstruction of band-limited signals, but the practical question of whether a signal that is not strictly band limited can be recovered from its samples has emerged. In fact, a finite set of samples from a continuous-time function can be seen as a duration-limited discrete-time signal in practice, and then it cannot be band limited. In this case, the reconstruction of a signal from its samples depends on the *a priori* information that we have about the signal structure or the nature of the problem, and the classical sinc kernel has been replaced by more general kernels that are not necessarily band limited (Vaidyanathan, 2001). This issue has been mostly studied in problems of uniformly sampled time series. A vast amount of research on nonuniform interpolation has been also made in polynomial interpolation (Atkinson, 1978), which can be related theoretically to Information Theory fundamentals and ultimately to regression. Nevertheless, this literature usually focuses on fast methods for large amounts of available samples, such as image processing and finite elements applications.

The following main elements can be either implicitly or explicitly considered by signal interpolation algorithms: the kind of sampling (uniform or nonuniform), the noise (present or not in the signal model), the spectral content (band limited or not), and the use (or not) of numerical regularization to compensate noise sensitivity of the interpolation methods. However, and despite the huge amount of work developed to date, the search for new, efficient interpolation procedures is still an active research area, which especially remains relevant today for multidimensional problems.

2.1.6 System Identification

The field of *nonlinear system identification* has been studied for many years and is still an active research area (Billings 1980; Giannakis and Serpedin 2001; Kalouptsidis and Theodoridis 1993; Nelles 2000; Sjöberg *et al.* 1995). The main objective of system identification is to build mathematical models of *dynamical* systems from observed, possibly corrupted data (Ljung, 1999). Modeling real systems is important not only to analyze but also to predict and simulate their behavior. Actually, the field of system identification extends to other areas of engineering, from the design of system controllers, the optimization of processes and systems, as well as the diagnosis and fault detection of components and processes.

Processes and *systems* are interchangeable terms in this field, always referring to the *plant* under study. The processes and systems can be either *static* or *dynamic*, as well as *linear* or *nonlinear*. In general, system identification consists of trying to infer the functional relationship between a system input and output, based on observations of the in- and outgoing signals. For linear systems, identification follows a unified approach thanks to the *superposition* principle, which states that the output of a linear combination of input signals to a linear system is the same linear combination of the outputs of the system corresponding to the individual components[2] Nonlinear systems, however, do not satisfy the superposition principle, and there is not an equivalent canonical representation for all nonlinear systems. Consequently, the traditional approach to study nonlinear systems is to consider only one class of systems at a time, and to develop a (possibly parametric) description that fits this class and allows one to analyze it.

---

2 The response of an LTI system to a signal is characterized by the convolution of the signal with the impulse response of the system (see Section 2.1.1). Given a system with an impulse response $h[n]$ and two signals $x_1[n]$ and $x_2[n]$, then the response of the system to a linear combination of both signals is $y[n] = h[n] * (\alpha x_1[n] + \beta x_2[n]) = \alpha h[n] * x_1[n] + \beta h[n] * x_2[n]$ by virtue of the linearity of the convolution operator.

Classical Linear System Identification

The generalization of the Wiener filter theory to dynamical systems with random inputs was developed by Kalman in the 1960s in the context of state-space models. Discrete-time (and dynamical) linear filters can be nicely expressed in the state-space formulation (or Markovian) representation, by which two coupled equations are defined as

$$y[n] = Cx[n] + Du[n] \tag{2.53}$$

$$x[n+1] = Ax[n] + Bu[n], \tag{2.54}$$

where $\boldsymbol{x}[n]$ is the vector states at time n, $\boldsymbol{u}[n]$ is the input at time n, $\boldsymbol{y}[n]$ is the output at time n, $\boldsymbol{A}$ is the state-transition matrix and determines the dynamics of the system, $\boldsymbol{B}$ is the input-to-state transmission matrix, $\boldsymbol{C}$ is the state-to-output transmission matrix, and $\boldsymbol{D}$ is the input-to-output transmission matrix. The state-space representation is especially powerful for multi-input multi-output (MIMO) linear systems, time-varying linear systems, and has found many applications in the fields of system identification, economics, control systems, and communications (Ljung, 1999).

The KF is the most widely known instantiation of a state-space formulation. The KF uses a series of noisy measurements observed over time to produce estimates of unknown variables that tend to be more precise than those based on a single measurement alone. Essentially, the KF operates recursively on noisy input signals to produce a statistically optimal estimate of the underlying system state. Notationally, the KF assumes that the true state at time n evolves from the previous state at time $n-1$ following

$$\boldsymbol{x}[n] = \boldsymbol{F}[n]\boldsymbol{x}[n-1] + \boldsymbol{B}[n]\boldsymbol{u}[n] + \boldsymbol{w}[n], \tag{2.55}$$

where $\boldsymbol{F}[n]$ is the state transition model which is applied to the previous state $\boldsymbol{x}[n-1]$, $\boldsymbol{B}[n]$ is the control-input model applied to the control vector $\boldsymbol{u}[n]$, and $\boldsymbol{w}[n]$ is the process noise assumed to be drawn from a zero-mean multivariate normal distribution with covariance $\boldsymbol{Q}[n]$; that is, $\boldsymbol{w}[n] \sim \mathcal{N}(\boldsymbol{0}, \boldsymbol{Q}[n])$. The impact of KFs in the field of dynamical system identification has been enormous, and many specific applied fields, such as econometrics or weather forecasting, have benefited from it in recent decades. The main limitation of the estimation of the KF is that it is designed to represent a linear system, but many of the real systems are intrinsically nonlinear. A solution is to "linearize" the system first and then apply a KF, thus yielding the extended KF, which has long been the de facto standard for nonlinear state-space estimation, mainly due to its simplicity, robustness, and suitability for real-time implementations.

Classical Nonlinear System Identification: Volterra, Wiener, and Hammerstein

One of the earliest approaches to parametrize nonlinear systems was introduced by Volterra, who proposed to extend the standard convolutional description of linear systems by a series of polynomial integral operators H_p with increasing degree of nonlinearity (Volterra, 1887a,b,c). For discrete-time systems, this description becomes simply

$$y[n] = H_0 + \sum_{p=1}^{\infty} H_p(x[n]), \tag{2.56}$$

where $x[n]$ and $y[n]$ are the input and output signals. Assuming the nonlinear system described is causal and with finite memory, functionals H_p can be expanded in Volterra kernels that actually depend of the previous inputs until time n. These Volterra kernels can then be described as linear combinations of products between elements $x[n-m]$, $0 \leq m \leq M$, M being the memory of the system model. Then, Equation 2.56 is then described by the coefficients of these linear combinations (Alper, 1965).

Note that Volterra series extend Taylor series by allowing the represented system output to depend on past inputs. Interestingly, if the input signals are restricted to a suitable subset of the input function space, it can be shown that any continuous, nonlinear system can be uniformly approximated up to arbitrary accuracy by a Volterra series of finite order (Boyd and Chua 1985; Brilliant 1958; Fréchet 1910). Thanks to this approximation capability, Volterra series have become a well-studied subject (Franz and Schölkopf 2006; Giannakis and Serpedin 2001; Schetzen 1980) with applications in many fields.

Although the functional expansion of Volterra series provides an adequate representation for a large class of nonlinear systems, practical identification schemes based on this description often result in an excessive computational load. It is for this reason that several authors have considered the identification of specific configurations of nonlinear systems, notably cascade systems composed of linear subsystems with memory and continuous zero-memory nonlinear elements (Gardiner 1973; Narendra and Gallman 1966). When the output of a system depends nonlinearly on its inputs, it is possible to decompose the input–output relation as either a Wiener system (linear stage followed by a nonlinearity) or a Hammerstein system (nonlinearity stage followed by a linear system).

The Wiener system (Billings and Fakhouri, 1977) consists of a linear filter followed by a static memoryless nonlinearity. When an FIR filter with length L samples is chosen as the linear part, the system output for a given input signal $x[n]$ is obtained as

$$y[n] = f\left(\sum_{m=0}^{L-1} h[m]x[n-m]\right), \tag{2.57}$$

where $h[n]$ represents the impulse response of the linear filter and $f(\cdot)$ is the nonlinearity. The Wiener model is a much more simplified version of Wiener's original nonlinear system characterization (Wiener, 1958). However, despite its simplicity, it has been used successfully to describe a number of nonlinear systems appearing in practice. Common applications include biomedical engineering (Hunter and Korenberg, 1986; Westwick and Kearney, 1998), control systems (Billings and Fakhouri, 1982; Greblicki, 2004), digital satellite communications (Feher, 1983), digital magnetic recording (Sands and Cioffi, 1993), optical fiber communications (Kawakami Harrop Galvão *et al.*, 2007), and chemical processes (Pajunen, 1992).

The Hammerstein system (Billings and Fakhouri, 1979) consists of a static memoryless nonlinearity followed by a linear filter. When the linear part is represented by an FIR filter, the output of a Hammerstein system is obtained as

$$y[n] = \sum_{m=0}^{L-1} h[m] f(x[n-m]). \tag{2.58}$$

Hammerstein systems are encountered, for instance, in electrical drives (Balestrino *et al.*, 2001), acoustic echo cancellation (Ngia and Sjobert, 1998), heat exchangers, and biomedical modeling (Westwick and Kearney, 2000). Wiener and Hammerstein systems form particular types of block-oriented structures (Chen, 1995; Giannakis and Serpedin, 2001). Other popular cascade models include the so-called sandwich models that combine more than two blocks; for instance, the Hammerstein–Wiener model, which consists of a regular Hammerstein system followed by an additional nonlinearity.

Modern Nonlinear System Identification

The area of nonlinear system identification is huge and any attempt to summarize the contributions and techniques is certainly futile (Ljung, 2008). The previous building blocks have generated a plethora of identification approaches for Wiener and Hammerstein systems since the late seventies. Most of them are *supervised* techniques, although in the last decade a number of blind, *unsupervised* identification methods have been proposed as well.

Supervised Nonlinear System Identification

Following the excellent book by Nelles (2000), one may split the field according to the static or dynamic nature of the models. The family of nonlinear static model architectures comprises simple linear models and parametric polynomial models, which advantageously are linear in the parameters but limited in terms of generalization capabilities and exhibit poor behavior in high-dimensional scenarios as well. One also finds look-up tables (LUTs), which essentially dominate in industrial applications, being simple and fast to implement in microcontrollers. The LUT approach has as the main feature that the parameters do not necessarily come from any regression method, but are typically determined directly. The main problems of LUTs come with even a low number of inputs. A more robust and efficient family of nonlinear static models are NNs, whose most successful models are the classical MLP (typically working with RBF nodes, hence called RBF networks), and neuro-fuzzy models which may shed light on the problem due to the implementation of tunable *membership functions*. Additionally, local linear (Takagi–Sugeno) neuro-fuzzy models have been very popular in recent decades, especially because of the efficient training, interpretability of the model, and relative robustness in high-dimensional problems.

On the other hand, the family of nonlinear dynamic models builds upon the previous family of static models. Therefore, one also encounters approaches based on polynomials, NNs, fuzzy model architectures, and linear neuro-fuzzy models. The field of *dynamic NNs* is a distinct one: these are NN architectures encoding the dynamics internally, and have been very popular for speech processing, natural language processing, control, and biomedical signal applications. The most successful architectures involve extensions of the MLP following two main philosophies. The first one considers including additional recurrent layers that process the predictions to feed either the input (Elman's net) or the hidden (Jordan's net) layer. The second approach considers replacing the standard scalar weights in the MLP by vectors and multiplications by dot products. If an FIR filter is included one speaks about the FIR NN that processes the temporal information internally in the network synapses, not as a particular ad hoc embedding at the input layer. Other more sophisticated filters can replace the weights, such as gamma filters that introduce recursive connections locally. Interestingly, there are extensions of the back-propagation algorithm for these architectures in which error terms are symmetrically filtered backward through the network.

Including the System Structure

Traditional supervised methods do not make any assumption about the system structure. Nonlinear identification was tackled by considering nonlinear structures such as MLPs (Erdogmus *et al.*, 2001a), recurrent NNs (Kechriotis *et al.*, 1994), or piecewise linear networks (Adali and Liu, 1997). Other black-box identification methods include techniques based on orthogonal LS expansion (Chen *et al.*, 1989; Korenberg, 1989) and separable LS (Bruls *et al.*, 1999). In order to improve the identification results, a number of algorithms were introduced that exploit the Wiener or Hammerstein structure explicitly. This can be done, for instance, by an identification scheme that mimics the unknown system and estimates its parameters iteratively (Billings and Fakhouri 1977, 1982; Dempsey and Westwick 2004; Greblicki 1997, 2004; Haykin 2001; Hunter and Korenberg 1986; Narendra and Gallman 1966; Westwick and Kearney 2000; Wigren 1994). Another approach that exploits the system structure consists of a two-step procedure that consecutively estimates the linear part and the nonlinearity of the Wiener or Hammerstein systems. Most of the proposed two-step techniques are based on predefined test signals (Bai, 1998; Pawlak *et al.*, 2007; Wang *et al.*, 2007).

A different proposition is found in Aschbacher and Rupp (2005), where both blocks of the nonlinear system are estimated simultaneously through a coupled regression on the unknown intermediate signal. A generalization of this technique for robust identification based on nonlinear canonical correlation analysis has also been presented (Van Vaerenbergh and Santamaría, 2006; Van Vaerenbergh *et al.*, 2008). Extensions to the standard system identification settings have also been proposed; for instance, to account for complex signals (Cousseau *et al.*, 2007) and MIMO Wiener or Hammerstein systems (Goethals *et al.*, 2005a).

Unsupervised Nonlinear System Identification

Although all the aforementioned techniques are supervised approaches (i.e., input and output signals are known during estimation), there have also recently been a few attempts to identify Wiener and Hammerstein systems blindly. Most of these techniques make certain assumptions about the input signal; for instance, requiring it to be Gaussian and white (Gómez and Baeyens, 2007; Vanbeylen *et al.*, 2008). A less restrictive method in Taleb *et al.* (2001) only assumed that the input signal was i.i.d., and the resulting technique aimed to recover the input signal by minimizing the mutual information of the inversion system output.

2.1.7 Blind Source Separation

Blind source separation (BSS) is an important problem in signal processing. The topic is recurrent in many applications, including communications, speech processing, hyperspectral image analysis, and biomedical signal processing. In general, the goal of BSS is to recover a number of source signals from their observed linear or nonlinear mixtures (Cardoso 1998; Comon *et al.* 1991). Three main problem settings can be identified, depending on the type of the mixing process, the number of sources n_s, and the number of available mixtures m.

Standard Problem Setting

The most basic scenario of BSS assumes a simple linear model in which the measurement random vector $\boldsymbol{y} \in \mathbb{R}^m$ can be described as an instantaneous mixture, $\boldsymbol{y} = \boldsymbol{A}\boldsymbol{s} + \boldsymbol{n}$, where $\boldsymbol{s} \in \mathbb{R}^{n_s}$ is a zero-mean independent random vector representing a statistically independent source, $\boldsymbol{A} \in \mathbb{R}^{n_s \times m}$ is the unknown mixing matrix, and $\boldsymbol{n} \in \mathbb{R}^m$ is an independent random vector representing sensor noise.

Many techniques have been developed for the case where as many mixtures as unknown sources are available ($m = n_s$). Most of them are based on independent component analysis (ICA) (Comon, 1994; Hyvärinen *et al.*, 2001), a statistical technique whose goal is to represent a set of random variables as linear functions of statistically independent components. In the absence of noise, this reduces to the problem of estimating the *unmixing matrix* $\boldsymbol{W} = \boldsymbol{A}^{-1}$, which allows one to retrieve the estimated sources as

$$\hat{\boldsymbol{x}} = \boldsymbol{W}\boldsymbol{y} = \boldsymbol{W}\boldsymbol{A}\boldsymbol{s}. \tag{2.59}$$

The unmixing matrix $\boldsymbol{W}$ is found by minimizing the dependence between the elements of the transformed vector $\hat{\boldsymbol{x}}$. In order to estimate this matrix, a suitable dependence measure (or contrast function) must be defined and minimized with respect to $\boldsymbol{W}$.

Most ICA contrast functions can be derived using the ML principle. For this problem setting, an impressive amount of algorithms have been proposed in the literature. Note that these methods only exploit the spatial structure of the mixtures, which is a suitable approach for mixtures of i.i.d. sources. If the sources have temporal structure, this information can also be exploited; for instance, see Molgedey and Schuster (1994). We will briefly review some of the most interesting techniques.

Among them, the most relevant ones are: (1) the infomax principle/algorithm (Bell and Sejnowski, 1995), which tries to maximize the log-likehood function of the mixing matrix, and it is simple and computationally efficient; (2) the joint approximate diagonalization of eigenmatrices algorithm (Cardoso and Souloumiac, 1993), which is based on higher order statistics, whose computation makes it unsuitable in large-scale scenarios; (3) the FastICA algorithm (Hyvärinen and Oja, 1997), which uses a fast "fixed-point" iterative scheme to maximize the non-Gaussianity of the estimated sources, and it is fast and efficient, but its solution is too flexible and dependent on the chosen nonlinearity; (4) the kernel ICA (KICA) (Bach and Jordan, 2005b), which uses a kernel canonical correlation analysis based contrast function to obtain the unmixing matrix $\boldsymbol{W}$. KICA outperforms the rest in accuracy and robustness, but it comes at the price of higher computational burden.

Underdetermined Blind Source Separation

The previous methods dealt with the standard BSS problem, where the number of mixtures and sources are the same. If more mixtures than sources are available ($m > n_s$), the redundancy in information can be used to extend existing techniques to achieve additional noise reduction (Joho *et al.*, 2000). If less mixtures than sources are available ($m < n_s$), a more complex scenario arises. This so-called underdetermined BSS problem can only be solved if one relies on a priori information about the sources. Two-step procedures are typically deployed: first the mixing matrix is estimated, often based on geometric properties, and then the original sources are recovered.

An extreme case of underdetermined BSS problems occurs when only one mixture is available. In audio applications, this scenario is known as the single-microphone source separation problem (Bach and Jordan 2004; Roweis and Saul 2000). In more general problems, it is often assumed that the data can be transformed to a domain in which the sources are sparse, meaning that they equal zero most of the time. For instance, audio separation algorithms typically work in the time–frequency domain, where overlap of the sources is likely to be much smaller than in the time domain (Belouchrani and Amin, 1998).

Under a geometrical viewpoint, several algorithms have been proposed to separate sparse sources in an underdetermined problem setting (Bofill and Zibulevsky 2001; Lee *et al.* 1999; Luengo *et al.* 2005; Vielva *et al.* 2001). In a first stage, these methods estimate the mixing matrix $\boldsymbol{A}$ by identifying the main vectors to which the samples in the scatter plot are aligned. A large number of estimators have been proposed for this purpose, among them a Parzen windowing-based method (Erdogmus *et al.*, 2001b), a line spectrum estimation method (Vielva *et al.*, 2002), and a kernel principal component analysis (KPCA) based technique (Desobry and Févotte, 2006). Once $\boldsymbol{A}$ is known, the original source samples are estimated. The most straightforward method to do this is by using the pseudo-inverse of $\boldsymbol{A}$. Better estimates can be obtained by the shortest path (ℓ_1-norm) method introduced in Bofill and Zibulevsky (2001) or the *maximum a posteriori* (MAP) estimator described in Vielva *et al.* (2001).

Post-nonlinear Blind Source Separation

A considerable amount of research has also been done on the so-called post-nonlinear BSS problem, in which the sources are first mixed linearly and then transformed nonlinearly, hence boiling down to a sort of Hammerstein model. In this scenario, the m sensors that measure the mixtures reveal some kind of saturation effect or another nonlinearity, which results in the extension of the basic model to a post-nonlinear mixture model:

$$\boldsymbol{x} = f(\boldsymbol{A}\boldsymbol{s}) + \boldsymbol{n}, \tag{2.60}$$

where $f(\cdot)$ returns an $m \times 1$ vector of nonlinear pointwise functions and $\boldsymbol{x} \in \mathbb{R}^m$ is the measurement random vector. In this model, each component $f_i(\cdot)$ of $f(\cdot)$ affects only one mixture so that there are no cross-nonlinearities. In the underdetermined case ($m < n$), the addition of the post-nonlinear transformations introduces an additional difficulty for the estimation of the mixing matrix. Therefore, it is not sufficient to apply a clustering technique on the angle histogram.

For an equal number of mixtures and sources ($m = n$), some algorithms have been proposed (Babaie-Zadeh *et al.* 2002; Solazzi *et al.* 2001; Taleb and Jutten 1999; Tan and Wang 2001) that treat the post-nonlinear problem. However, these algorithms cannot deal with the more restricted problem of underdetermined post-nonlinear BSS. An underdetermined algorithm was successfully proposed in Theis and Amari (2004), which requires the number of active sources at each instant to be lower than the total number of available mixtures m and assumes noiseless mixtures.

2.2 From Time–Frequency to Compressed Sensing

2.2.1 Time–Frequency Distributions

TFDs aim to represent the energy (or intensity) of a signal in time and frequency simultaneously. Standard spectral analysis (i.e., Fourier decomposition) gives us the signal components (frequencies) and their relative intensity, but it does not inform us about the specific time in which those frequencies dominate the signal. If we want to analyze the spectra of a signal that changes over time, a straightforward way to do so is to obtain its spectrum in time intervals of duration t. This simple strategy is valid if the signal changes at a rate slower than t, otherwise we have to decrease the time interval accordingly. This is the idea behind the so-called *short-time FT* (STFT), or *spectrogram*, which nowadays is still a standard and powerful method to describe time-varying signals. The spectrogram was initially developed during the 1940s to analyze human speech. Despite its widespread use, it has two main drawbacks. First, some signals can change so quickly that it makes it almost impossible to find a short-time window for which the signal can be considered stationary; that is, the *time resolution* is bounded. Second, shortening the time window reduces the *frequency resolution*. These two problems are tied in such a way that we have a trade-off between time and spectral resolutions for any signal representation: improving one resolution decreases the resolution of the other. Section 2.2.2 discusses the intuition behind this trade-off and the relation to the Heisenberg principle in quantum physics.

A time distribution represents the intensity of a signal per unit time at time t. Similarly, a frequency distribution represents the intensity of a signal per unit frequency at frequency ω. Therefore, a straightforward definition of a TFD is the one representing the energy or intensity of a signal per unit time and per unit frequency. Therefore, a joint distribution $P(t, \omega)$ describes the intensity at time t and frequency ω, and $P(t, \omega)\Delta t \Delta \omega$ is the fractional energy in the time–frequency interval $\Delta t \Delta \omega$ at (t, ω). Ideally, a TFD should fulfill the following properties:

Property 1 $|x(t)|^2 = \int P(t, \omega)\, \mathrm{d}\omega$

Property 2 $|X(\omega)|^2 = \int P(t, \omega)\, \mathrm{d}t$

Property 3 $E = \int P(t, \omega)\, \mathrm{d}\omega\, \mathrm{d}t$

Property 1 states that adding up the energy distribution for all frequencies at a particular time should give the instantaneous energy at that time, $|x(t)|^2$. On the other hand, property 2 states that adding up over all times at a particular frequency should give the energy density spectrum, $|X(\omega)|^2$. Finally, integrating both in frequency and time gives the total energy E of the signal (property 3).

From the theoretical definition of what a TFD should be, the question is whether there is a practical TFD that can fulfill the aforementioned properties and whether it would be a true density distribution. As it turns out that this is not actually possible, the question is then to find reasonable approaches to theoretical TFDs, according to different specifications for different problems.

In the following, we will use $P(t, \omega)$ for denoting a theoretical TFD of $x(t)$, but we will use instead $X(t, \omega)$ for denoting the different definitions of the existing and specific time–frequency transforms.

As we mentioned before, a simple yet powerful way to deal with signals changing both in time and frequency is the STFT. In the continuous case, the time function to be transformed is multiplied by a window function, and then the FT is taken as the window moves along the time axis:

$$X(t, \omega) = \int x(\tau) w(\tau - t)\, \mathrm{e}^{-\mathrm{i}\omega\tau}\, \mathrm{d}\tau, \tag{2.61}$$

where $w(t)$ is a window function, such as Hamming, Hanning, Han, or Gaussian windows. The spectrogram is obtained by taking the squared magnitude; that is: $|X(t, \Omega)|^2$. The discrete case can be expressed as

$$X(n, \Omega) = \sum_{n=-\infty}^{\infty} x[k] w[k - n]\, \mathrm{e}^{-\mathrm{i}\Omega k}, \tag{2.62}$$

where in this case n is discrete, but Ω is still a continuous variable.

The first studies that considered the question of a joint distribution function in time–frequency were presented by Gabor (1946) and Ville (1948). The Gabor transform, although not a TFD, is still a special case of an STFT where the window is a Gaussian function; that is:

$$X(t, \omega) = \int x(\tau)\, \mathrm{e}^{-\pi(\tau - t)^2}\, \mathrm{e}^{-\mathrm{i}\omega\tau}\, \mathrm{d}\tau, \tag{2.63}$$

where we can easily identify the window function as $w = \mathrm{e}^{-\pi(\tau - t)^2}$.

On the other hand, Ville obtained a TFD from the distribution presented in Wigner (1932) 16 years before in the field of quantum statistical mechanics. The *Wigner–Ville distribution*, often referred to only as the Wigner distribution, is given by

$$X(t, \omega) = \frac{1}{2\pi} \int x\left(t + \frac{1}{2}\tau\right) x^*\left(t - \frac{1}{2}\tau\right)\, \mathrm{e}^{-\mathrm{i}\omega\tau}\, \mathrm{d}\tau. \tag{2.64}$$

The *Wigner distribution* satisfies the marginals property; that is, we can obtain the instantaneous energy in a given time instant or for a given frequency, integrating for all frequencies or for all time instants respectively. However, the Wigner distribution is a nonlinear transform. When the input signal has more than one component, a cross-term appears. Another problem of the Wigner–Ville distribution is that it is not always positive for an arbitrary signal, which makes it difficult (or even impossible) to obtain a physical interpretation in some cases. In contrast, the STFT is always positive, although it does not satisfy the marginals property.

Another interesting particularity of the Wigner distribution is that, in general, it can have nonzero values even when there is no signal. A simple example of a signal that exhibits this behavior is one which has nonzero values initially, then is zero in the next time interval, and then is again nonzero in the next time interval. The Wigner TFD for such a signal shows nonzero values in the time–frequency representation just whenever the signal is zero. This is caused by the cross-terms mentioned before.

Another well-known TFD was devised by Kirkwood (1933) a year after Wigner proposed his, also in the field of quantum mechanics. However, this TFD is known as the *Rihaczek distribution* (Rihaczek, 1968), derived from the Kirkwood distribution taking into account physical considerations:

$$X(t,\omega) = \frac{1}{\sqrt{2\pi}} x(t) X^*(\omega)\, \mathrm{e}^{-\mathrm{i}\omega t}. \tag{2.65}$$

One of the reasons the Rihaczek distribution is less popular than the Wigner distribution is because it is complex and thus it can hardly represent a physical process. Nevertheless, the Wigner distribution has its physical issues too, giving in some cases negative values for the energy. Moreover, the real part of the Rihaczek distribution is also a TFD, called the *Margenau–Hill distribution* (Margenau and Hill, 1961), that satisfies the marginals property.

Page (1952) derived another famous TFD, introducing the concept of a *running spectrum,* based on the assumption that we have access to signal samples until time t and hence no information is available about the future, $\tau \geq t$. Note that this is in contrast to other distributions where the integration is done covering the full time axis. Taking the FT of such a signal and deriving it with regard to time t gives the *Page distribution,* expressed as:

$$X(t,\omega) = \frac{\partial}{\partial t} \left| \frac{1}{\sqrt{2\pi}} \int_{-\infty}^{t} x(\tau)\, \mathrm{e}^{-\mathrm{i}\omega\tau}\, \mathrm{d}\tau \right|^2 . \tag{2.66}$$

Interestingly, the method used by Page (1952) to derive this distribution was employed extensively to obtain other TFDs.

All the distributions we have seen here are valid TFDs. There is none that may be said to be better than another. All of them satisfy the marginals, and each one has interesting and useful properties. Choosing one or another depends on the specific problem at hand. By the time these TFDs were discovered and presented, there was the unanswered question of how many different TFDs existed, and what properties they might have. It was Cohen (1966) who showed that an infinite number of TFDs can be obtained from the following general equation:

$$X(t,\omega) = \frac{1}{4\pi^2} \int\int\int \mathrm{e}^{-\mathrm{i}\theta t - \mathrm{i}\tau\omega + \mathrm{i}\theta u} \phi(\theta,\tau) x^*\left(u - \frac{1}{2}\tau\right) x\left(u + \frac{1}{2}\tau\right) \mathrm{d}u\, \mathrm{d}\tau\, \mathrm{d}\theta, \tag{2.67}$$

where $\phi(\theta, \tau)$ is a general function called the *kernel*. Choosing different kernels leads to different distributions. For instance, fixing $\phi(\theta, \tau) = 1$ yields the Wigner distribution, setting $\phi(\theta, \tau) = e^{i\theta\tau/2}$ gives the Rihaczek distribution, and if $\phi(\theta, \tau) = e^{i\theta|\tau|/2}$ the Page distribution emerges. This procedure to generate all distributions has many advantages. Most notably, it allows one to compare the properties of the different distributions. It is also worth noting that these properties can be obtained directly from the kernel definition.

In order to illustrate the usefulness of having a unified way to derive TFDs based on a kernel function, we will finish this section showing the *Choi–Williams distribution*. Choi and Willians (1989) addressed the problems of the Wigner distribution related to the cross-terms. Starting from considering a signal made up of components, Choi and Willians (1989) used the Cohen equation to obtain an expression where the self- and cross-terms appeared in a separate form. They realized that by choosing the kernel wisely it was possible to minimize the cross-terms, retaining at the same time most parts of the self-terms. The kernel they engineered was $\phi(\theta, \tau) = e^{-\theta^2\tau^2/\sigma}$, where σ is a free parameter that controls the ability of the Choi–Williams distribution to suppress the cross-terms. With this kernel, the obtained TFD is

$$X(t, \omega) = \frac{1}{4\pi^{3/2}} \int\int \frac{1}{\sqrt{\tau^2/\sigma}} e^{-[(u-t)^2/(4\tau^2/\sigma)]-i\omega\tau} x^*\left(u - \frac{1}{2}\tau\right) x\left(u + \frac{1}{2}\tau\right) du\, d\tau, \tag{2.68}$$

which is the Choi–Williams distribution, which greatly alleviates the cross-terms problem of the Wigner distribution.

2.2.2 Wavelet Transforms

Time–frequency representations of nonstationary signals are based on the strategy of taking windows of the signal and analyzing them separately. That way, under the assumption that the signal is almost stationary inside each window, one can observe its frequency properties as an approximation. Nevertheless, this approach will be more or less effective depending on the length of the window with respect to the properties of the signal. If it is highly nonstationary, frequency properties may only be observed if the window is small. But the frequency effect of the windowing is the convolution of the FT of the signal with the transform of the window. If the window is narrow in time, its FT will be wide, thus blurring the frequency components of the signal that we want to analyze. This is, straightforwardly, a form of Heisenberg's principle, which states, in this case, that it is not possible to know at the same time the frequency and the exact time where a signal has been measured. This indicates that a time–frequency representation of a signal is eventually not possible. A trade-off is then imposed by nature between the accuracy of a measure in its frequency and in its time.

Although this limitation is a fundamental rule, there are ways to improve the joint representation of a signal. Roughly speaking, it is possible to use fully scalable modulated windows to analyze the properties of a signal. The idea is to use different versions of the window in different scales, thus obtaining a multiscale representation of the signal. The result will be a representation with different levels of detail, from a more general representation, covering longer time intervals, to a more detailed representations, with shorter intervals. This allows an analysis that goes from global to local signal representations. This is in fact the underlying idea of the so-called *wavelet representations*.

A *wavelet* is a time-limited wave whose limits continuously tend to zero when time tends to minus and plus infinity. The basic properties of the wavelets can be summarized as follows (Sheng, 1996): if a transform is used to construct a time–frequency, or more generally speaking, a time–scale representation, it should be possible to reconstruct the original signal from its representation.

In DSP, only discrete wavelets are considered. Nevertheless, the fastest way to describe the properties of wavelets is to first describe the continuous-time wavelet transform and their properties. This paves the way to later introduce then the discrete-time wavelet transform, which is constructed as a discretized version of a continuous one, thus keeping its properties.

Continuous-Time Wavelet Transform Analysis

The wavelet transform of a time function $x(t)$ can be described as the linear operation

$$X(s,\tau) = \int_t x(t)\phi^*_{s,\tau}(t)\,\mathrm{d}t, \tag{2.69}$$

where $\phi_{s,\tau}(t)$ is the *wavelet function*. This function has two parameters. The first one is the scale s, which determines the width of the function in the time domain. The second one is the delay τ, and determines the time instant where the signal is centered with respect to the time origin. Then, a wavelet transform is obtained for each pair (s,τ), and hence this is a 2D transform. The continuous wavelet set is constructed from a basic function called a *mother wavelet* through its scaling and translation. That is, $\phi_{s,\tau}(t) = s^{-1/2}\phi(\frac{t-\tau}{s})$. Factor $s^{-1/2}$ is added to keep the energy of the wavelets constant with the scale factor s. The wavelet transform is the *convolution of the wavelet with the signal under analysis*:

$$X(s,\tau) = \int_t x(t)\phi^*\left(\frac{t-\tau}{s}\right)\,\mathrm{d}t. \tag{2.70}$$

This definition, translated into the discrete domain, leads to the idea of *filter bank analysis*. For a wavelet to be valid, the following inverse transform must be satisfied:

$$x(t) = \int_{s,\tau} X(s,\tau)\phi_{s,\tau}(t)\,\mathrm{d}s\,\mathrm{d}\tau. \tag{2.71}$$

An example of continuous wavelet is the so-called *Mexican hat wavelet*, whose expression and FT are respectively:

$$\phi(t) = (1-t^2)\,\mathrm{e}^{-(t^2/2)} \tag{2.72}$$

$$\Phi(\omega) = \omega^2\,\mathrm{e}^{-(\omega^2/2)}. \tag{2.73}$$

Note that the function in time has a maximum in the origin, and it vanishes with time tending to infinity. Its spectrum is band-pass. This function is the second derivative of the Gaussian function, which is also an admissible wavelet.

Discrete-Time Wavelet Transform Analysis

Under certain restrictions, the original signal can be recovered even if the wavelet functions are discretized in delay τ and scale s. In particular, wavelet functions are discretized by taking $s_j = s_0^j$ and $\tau_j = ks_0^j\tau_0$, where k and j are integers and usually $\tau_0 = 1$ and $s_0 = 2$.

A discrete time signal $x[n]$ can be represented as a function of wavelets as

$$x[n] = c_\phi \sum_k \sum_j a_{j,k}\phi_{j,k}(n), \tag{2.74}$$

where $a_{j,k} = \sum_n x[n]\phi_{j,k}[n]$, and c_ϕ is an amplitude factor.

The discrete wavelet transform analyzes the signal using two different functions called scaling functions $l[n]$ (low-pass filters) and wavelet functions $h[n]$ (high-pass filters). The decomposition of the signal is simply performed by successive decomposition of the signal in high-pass and low-pass signals. An example is the *subband coding algorithm*. In that algorithm, the signal is convolved with a scaling and a wavelet function acting as two half-band filters. After the filtering, the resulting signals can be downsampled by two since they have half the bandwidth. This downsampling is what actually produces the scaling of the wavelet functions. The low-pass signal is filtered again to produce two new subbands, downsampled again, and the process is iterated until, due to the downsampling, just one sample is kept at the output of the filters. The step j of the procedure can be summarized as follows:

$$y_{\mathrm{low},j}[k] = \sum_n y_{\mathrm{low},j-1}[n]l[2k-n] \tag{2.75}$$

$$y_{\mathrm{high},j}[k] = \sum_n y_{\mathrm{low},j-1}[n]h[2k-n]. \tag{2.76}$$

Note that the total number of samples is not increased with respect to the original signal. The highest band will contain $W/2$, W being the bandwidth of the original signal, and the subsequent bands will have a decreasing bandwidth $W/2^j$.

This representation of the signal has a major advantage with respect to the FT, and it is that the time localization of the signal is never lost. The most important frequencies in the signal will be detected by a higher amplitude in the corresponding subband. Higher frequency subbands will have a higher time resolution, but a lower frequency resolution, since these subbands have a higher bandwidth. Lower bands will have an increasing frequency resolution, and thus a decreasing time resolution. Note that the signal can be compressed by discarding those bands whose energy is negligible (typically via thresholding), thus resulting in a compressed signal that can be further reconstructed using a procedure inverse to the one in Equations 2.75 and 2.76.

Applications

Applications of wavelet transforms in communications are diverse. Multiresolution analysis is useful to decompose the signal of interest in meaningful components. An example of multiresolution analysis can be found in the detection of transient effects in power lines. Such analysis is able to distinguish transient events, present in higher frequencies at much lower power, from the baseline low-pass high-power signal. This helps to characterize the occurrence instants and the nature of the transients (Wilkinson and Cox, 1996). In communications, wavelets have been used to detect weak signals in noise (Abbate *et al.*, 1997). Wavelets have also been used in biomedical signals in order to characterize ECG waves for diagnostics (Huang and Wang, 2009).

Relevant applications can also be found in audio compression, where the filter functions approximate the critical bands of the human hearing system. After a subband decomposition, only bands with relevant energy are preserved, and the rest are discarded. This provides reasonable rates of compression without a noticeable distortion, or high compression rates with lower – still reasonable – quality levels (Sinha and Tewfik, 1993). 2D wavelet transforms are also used for image and video compression (Ohm, 2005). These compression strategies are present in the JPEG2000 standards. Wavelets have also been proposed for texture classification, denoising, system identification, speech recognition, and many other applications in signal processing and related areas (e.g., Randen and Husoy, 1999).

2.2.3 Sparsity, Compressed Sensing, and Dictionary Learning

The last decade has witnessed a growing interest in the field of *sparse learning*, mostly revived by the achievements of *compressed sensing* (Candes *et al.*, 2006) and *dictionary learning* (Tropp and Wright, 2010). These two fields are at the intersection of signal processing and machine learning, and have revolutionized signal processing and many application domains dealing with sensory data. The reason for that is in the core of signal processing: both approaches try to find signal descriptions and feature representation spaces in terms of a few, yet fundamental, components. Let us briefly review the main elements in this field.

Sparse Representations

As we have seen in the previous section, an adequate representation of the signal is crucial to obtain good results. The wavelet domain is an adequate representation domain for natural signals (like audio and images) because it captures the essence of the underlying process generating the data with minimal redundancy (i.e., high compact representation). The representation domain is called *sparse* because only a few coefficients (much lower than the signal dimensionality) are needed to represent the data accurately. *Sparsity* is an attribute that is met in many natural signals, since nature tends to be parsimonious and optimal in encoding information. Current models of the brain suggest that the different hierarchically layered structures in the visual, olfactory, and auditory signal pathways pursue lower-to-higher abstract (more independent and sparse) representations.

Signal processing has not been blind to this description of living sensory systems. Many applications have benefited from sparse representations of the signal. For example, one may find the case of Internet telephony and acoustic and network environments

(Arenas-García and Figueiras-Vidal 2009; Naylor *et al.* 2006), where the echo path, represented by a vector comprising the values of the impulse response samples, is sparse. A similar situation is found in wireless communication systems involving multipath channels. More recently, sparsity has been exploited in channel estimation for multicarrier systems, both for single antenna ands for MIMO systems (Eiwen *et al.*, 2010a,b). Sparsity in multipath communication systems has been extensively treated as well (Bajwa *et al.*, 2010). But perhaps the most widely known example of exploiting sparse signal representations has to do with signal compression: the DFT, the DCT, and the discrete wavelet transform (DWT) are the most widely known and successful examples. While the DCT is the basis for the familiar JPEG algorithm, the DWT is in the core of the more recent JPEG-2000 coding algorithm (Gersho and Gray, 1991).

Notationally, a transform $\boldsymbol{\Phi}$ is simply an operation that projects the acquired signal $\boldsymbol{y} \in \mathbb{R}^d$ into new coordinate axes, thus giving rise to a new signal representation $\boldsymbol{x} = \boldsymbol{\Phi}^H \boldsymbol{y}$ (known as the *analysis equation*). Typically, such a transform is required to be orthonormal; that is, $\boldsymbol{\Phi}^{\mathrm{H}}\boldsymbol{\Phi} = \boldsymbol{I}$. In the case of orthonormal transforms, the inverse of $\boldsymbol{x}$ reduces to $\boldsymbol{y} = \boldsymbol{\Phi}\boldsymbol{x}$ (known as the *synthesis equation*). The new representation is called sparse (and the signal is termed *compressible*) if one can throw away a large number of components $n - m$ of $\boldsymbol{x}$, for $m \ll n$, in such a way that a good representation of $\boldsymbol{x}$ can still be recovered with minimal error by inverting the largest m components of $\boldsymbol{x}$ only. More formally, the signal $\boldsymbol{x}$ is called m-sparse if it is a linear combination of only m basis vectors; that is, only m coefficients in $\boldsymbol{y}$ are nonzero.

Dictionary Learning

We have identified common transforms for $\boldsymbol{\Phi}$ such as the DWT. These transforms lead to compact signal representations, but they involve fixed operations that might be not adapted to the signal characteristics. The field of *dictionary learning* deals with the more ambitious goal of inferring the transform directly from the data. Therefore, the goal of attaining a sparse representation boils down to simply expressing $\boldsymbol{y}$ of dimension d as a linear combination of a smaller number of signals taken from a database called a *dictionary*. The columns of the DWT compose a dictionary, which in the case of an orthonormal transform is a *complete dictionary*. Elements in the dictionary are typically unit norm functions called *atoms*. Let us denote in this context the dictionary as $\mathcal{D}$ and the atoms as ϕ_k, $k = 1, \ldots, N$, where N is the size of the dictionary. When we have more atoms in the dictionary than signal samples (i.e., $N \gg d$), the dictionary is called *overcomplete*.

The signal is represented as a linear combination of atoms in the dictionary as before, $\boldsymbol{y} = \boldsymbol{\Phi}\boldsymbol{x}$. The goal in dictionary learning is to estimate both the signal $\boldsymbol{x}$ and the dictionary $\boldsymbol{\Phi}$ with as few measurements as possible. The dictionary needs to be large enough (representative, expressive), and hence one typically works with overcomplete dictionaries. This poses a difficult problem, since the atoms often become linearly dependent, which gives rise to undetermined systems of equations and hence nonunique solutions for $\boldsymbol{x}$. It is here precisely when imposing sparsity in the solution alleviates the problem: we essentially aim to find a sparse linear expansion with an error $\boldsymbol{e}$ bounded to be lower than a threshold parameter ε. A possible formulation of the problem is

$$\min_{\boldsymbol{\Phi},\boldsymbol{x}} \|\boldsymbol{x}\|_0 \tag{2.77}$$

subject to

$$\boldsymbol{y} = \boldsymbol{\Phi x} + \boldsymbol{e} \quad \text{and} \quad \|e\|_2^2 < \varepsilon, \tag{2.78}$$

where $\|\cdot\|_0$ is the ℓ_0-norm. This problem, however, is NP-hard and cannot be solved exactly. Several approximations exist, either based on greedy algorithms, such as the matching pursuit (Mallat and Zhang, 1993) or the orthogonal matching pursuit (Tropp, 2004), or convex relaxation methods such as the basis pursuit denoising (Chen *et al.*, 1998) or the least absolute shrinkage and selection operator (LASSO) (Tibshirani, 1994), which solve

$$\min_{\boldsymbol{\Phi},\boldsymbol{x}} \left\{ \|\boldsymbol{y} - \boldsymbol{\Phi x}\|_2^2 + \lambda \|\boldsymbol{x}\|_1 \right\}, \tag{2.79}$$

where λ is a trade-off parameter controlling the sparseness versus the accuracy of the solution. This convex relaxation replaced the nonconvex ℓ_0-norm with the ℓ_1-norm that has surprisingly good properties.

There are alternative algorithms for solving this problem, and a good review can be found in Tropp and Wright (2010). Tosic and Frossard (2011) pointed out that three main research directions have been followed in methods for dictionary learning: (1) *probabilistic algorithms*, such as those based on ML (Olshausen and Fieldt, 1997), MAP (Kreutz-Delgado *et al.*, 2003), and Kulback–Leibler divergences, as proxies to the ℓ_1-norm (Bradley and Bagnell, 2008); (2) *vector quantization algorithms*, such as the Schmid–Suageon method for video coding, or the kernel singular value decomposition (SVD) algorithm (Aharon *et al.*, 2006); (3) *structure-preserving algorithms* that encode some form of prior knowledge during learning, which are found especially useful for audio (Yaghoobi *et al.*, 2009) and image applications (Olshausen *et al.*, 2003).

On the other hand, supervised dictionary learning is a vast field with many developments, and can be seen as another form of encoding explicit knowledge. The typical formulations include additional terms in charge of encouraging class separability, such as

$$\min_{\boldsymbol{\Phi},\boldsymbol{x}} \left\{ \|\boldsymbol{y} - \boldsymbol{\Phi x}\|_2^2 + \lambda \mathcal{L}(\boldsymbol{\Phi}, \boldsymbol{x}) \right\}, \tag{2.80}$$

where $\mathcal{L}(\boldsymbol{\Phi}, \boldsymbol{x})$ measures class separability when data are represented by the dictionary, or

$$\min_{\boldsymbol{\Phi},\boldsymbol{x}} \left\{ \|\boldsymbol{y} - \boldsymbol{\Phi x}\|_2^2 + \lambda_1 \|\boldsymbol{x}\|_1 + \lambda_2 C(\boldsymbol{\Phi}, \boldsymbol{x}, \boldsymbol{\theta}) \right\}, \tag{2.81}$$

where $C(\boldsymbol{\Phi}, \boldsymbol{x}, \boldsymbol{\theta})$ is a classification term that depends on the dictionary and a particular statistical classifier (e.g., an SVM) parametrized with weights $\boldsymbol{\theta}$.

Learning dictionaries and compact signal representations have been found very useful in image and audio applications. Despite the great advances in the last decade, some open problems still remain, such as learning in distorted, unevenly sensed and noisy data, or, perhaps more importantly, learning dictionaries capable of dealing with nonlinearities.

Compressive Sensing

Sparse representations are the foundation of *transform coding*, and the core of many technological developments like digital cameras. After acquisition of an image $\boldsymbol{y}$, a transform $\boldsymbol{\Phi}$ is applied, and only the positions and values of the largest coefficients of $\boldsymbol{x}$ are encoded, stored, or transmitted. Image compression is simply this. Unfortunately, this procedure is obviously suboptimal because: (1) even for a small value of m, the number of sampling points may be large; (2) one has to compute all the n coefficients while most of them will be later discarded; and (3) encoding implies storing not only the most important coefficients but also their locations, which is obviously a wasteful strategy. The brain does not work like that. Compressive sensing (CS) addresses these inefficiencies by directly acquiring a compressed signal representation without going through the intermediate stage of acquiring n samples (Candes and Wakin, 2008; Candes *et al.*, 2006; Donoho, 2006). The main achievement of CS is that the number of measurements needed is based on the sparsity of the signal, which permits signal recovery from far fewer measurements than those required by the traditional Shannon/Nyquist sampling theorem.

In CS we assume that a one-dimensional (1D) discrete-time signal $\boldsymbol{x} \in \mathbb{R}^d$ can be expressed as a linear combination of basis functions ϕ_k, $k = 1, \ldots, d$ which, in matrix notation, is given by $\boldsymbol{x} = \boldsymbol{\Psi s}$. Here, the $d \times d$ dictionary matrix $\boldsymbol{\Psi}$ is assumed orthonormal and describes the domain where $\boldsymbol{x}$ accepts a sparse representation. Accordingly, we assume that the signal $\boldsymbol{x}$ is k-sparse. Let us now consider a general linear measurement process that computes dot products between acquisitions, and a second set of basis vectors ψ_j, $j = 1, \ldots, m$, for $m < d$, that are collectively grouped in the matrix $\boldsymbol{\Psi}$. Therefore, the m measurement vector can be expressed as

$$\boldsymbol{y} = \boldsymbol{\Phi x} = \boldsymbol{\Phi \Psi s} = \boldsymbol{\Theta s}, \tag{2.82}$$

where $\boldsymbol{\Theta}$ is an $m \times d$ matrix. Here, the measurement process is not adaptive; that is, $\boldsymbol{\Phi}$ is fixed and does not depend on the signal $\boldsymbol{x}$. The problem in compressed sensing involves the design of: (1) a stable measurement matrix $\boldsymbol{\Phi}$ such that the dimensionality reduction from $\boldsymbol{x} \in \mathbb{R}^d$ to $\boldsymbol{y} \in \mathbb{R}^m$ does not hurt the k-sparse representation of the signal; and (2) a reconstruction algorithm to recover $\boldsymbol{x}$ from only $m \approx k$ measurements of $\boldsymbol{y}$. For the first step, one can achieve stability by simply selecting $\boldsymbol{\Phi}$ as a random matrix.[3] For the second step, one may resort to the techniques presented before for dictionary learning. Note that, here, the signal reconstruction algorithm must take the m measurements in the vector $\boldsymbol{y}$, the random measurement matrix $\boldsymbol{\Phi}$ (or the random seed that generated it), and the basis $\boldsymbol{\Psi}$, and reconstruct the signal $\boldsymbol{x}$ or, equivalently, its sparse coefficient vector $\boldsymbol{s}$. The problem can be expressed as

$$\min_{\boldsymbol{s}} \|\boldsymbol{s}\|_0 \ \text{s.t.} \ \boldsymbol{y} = \boldsymbol{\Theta s}, \tag{2.83}$$

3 The *restricted isometry property* and a related condition called incoherence require that the rows of $\boldsymbol{\Phi}$ cannot sparsely represent the columns of $\boldsymbol{\Psi}$ and vice versa. The selection of a random Gaussian matrix for $\boldsymbol{\Phi}$ has two nice properties: it is incoherent with the basis $\boldsymbol{\Psi} = \boldsymbol{I}$ of delta spikes, and it is universal in the sense that $\boldsymbol{\Theta}$ will be i.i.d. Gaussian, thus holding the restricted isometry property.

for which an equivalent optimization based on the ℓ_1-norm is

$$\min_{s} \left\{ \|y - \Theta s\|_2^2 + \lambda \|s\|_1 \right\}, \tag{2.84}$$

which typically yields good results and can recover k-sparse signals.

It is important to note that x needs not be stored and can be obtained anytime once s is retrieved. Additionally, measurements y can be obtained directly from an analog signal $x(t)$, even before obtaining its sampled discrete version x. Intuitively, one can see CS as a technique that tackles sampling and compression simultaneously. These properties have been used to develop efficient cameras and MRI systems that take advantage of signal characteristics and smart sampling designs.

Finally, we should mention that, similar to dictionary learning, the k-sparse concept in CS has also recently been extended to *nonlinear* k-sparse by means of kernel methods (Gangeh *et al.* 2013; Gao *et al.* 2010; Nguyen *et al.* 2012; Qi and Hughes 2011). Natural audio and images lie in particular low-dimensional manifolds and exhibit this sort of nonlinear sparsity, which is not taken into account with previous linear algorithms.

2.3 Multidimensional Signals and Systems

Multidimensional signal processing refers to the application of DSP techniques to signals that can be represented in a space of two or more dimensions. The most straightforward examples of multidimensional signals are images and videos, but there are many other signals generated from physical systems that can be described in high-dimensional spaces. Surprisingly, classical DSP has not been naturally extended in a natural way. In many applications the operations of mathematical morphology are more useful than the convolution operations employed in signal processing because the morphological operators are related directly to shape (Haralick *et al.*, 1987). Other types of techniques based in part on heuristic considerations like local scale-invariant feature extraction appear as natural extensions of classical 1D signal-processing techniques (Lowe, 1999).

When moving to multidimensional signal processing, a modern approach is taken from the point of view of statistical (machine) learning, but we should not forget the solid signal-processing concepts applied to this topic. Machine learning has also been used for image and video processing, and actually dominates the particular field of computer vision. That said, classical DSP is also used in image processing by extending the definitions and tools to two dimensions for many tasks, from object recognition to denoising. In addition, multidimensional processing is not just exclusive to 2D imaging, with it also having applications in other kinds of multidimensional signals (Dudgeon and Merserau, 1984).

Medical imaging can be seen as a source of 3D or four-dimensional (4D) signals. For example, computer-aided tomography scan images are 3D images, while functional MRI of the brain (Friston *et al.*, 1995) is a 4D signal in nature. The processing of such signals includes classical DSP techniques extended to the multidimensional field. Other fields where one can find multidimensional signals include antenna array processing (Van Trees, 2002). Planar arrays, consisting of a layer of radio sensors or antennas placed in a 2D structure, are a source of multidimensional signals. In a basic framework, a snapshot or sample of the excitation current produced by an incoming radioelectric wave constitutes itself a 2D signal susceptible to being processed by 2D DSP in order to obtain information about the incoming wave. This information can be its angle of arrival, the symbols carried by this signal in a communications application, or even just the power of that signal in, for example, a radiometry application. Many times, this process is carried out in the frequency domain of the signal.

When dealing with dispersive channels, the process turns out to be 3D and often time–space processing strategies are used (Verdú, 2003). Sometimes, the application is not targeted to the observation of a single or a few arbitrary discrete directions of arrival, but it focuses on the reconstruction of radar images of surfaces. An example of this is synthetic aperture array techniques (Van Trees, 2002), where using conventional time–frequency DSP techniques, high-resolution radar images of remote areas can be reconstructed. In summary, multidimensional DSP is a framework that inherits concepts and elements from DSP in order to implement them in applications such as image processing, radar, communications, or biomedical imaging, while still borrowing others from machine learning and statistics.

2.3.1 Multidimensional Signals

Fundamental multidimensional DSP is constructed by the extension of the elements of DSP to more than one dimension. The first and more straightforward extension is the definition of a 2D signal. A 2D signal or *sequence* can be represented as a function ordered in pairs of integers:

$$x[\boldsymbol{n}] = x[n_1, n_2]. \tag{2.85}$$

The definition of special sequences follows straightforwardly. For example, one can say that a sequence is separable if $x[n_1, n_2] = x_1[n_1]x_2[n_2]$. A sequence is periodic if it satisfies the properties $x[n_1 + N_1, n_2]x[n_1, n_2]$ and $x[n_1, n_2 + N_2]x[n_1, n_2]$, with N_1 and N_2 being the periods of the sequence. In general, a multidimensional sequence can be expressed as $x[\boldsymbol{n}]$, where $\boldsymbol{n} = [n_1, \dots, n_K]^{\mathrm{T}}$.

We know that a fundamental operator in DSP is convolution. The convolution operator can be defined in a multidimensional space as well, and appears recurrently in machine learning in general and kernel methods in particular, as we will see in the book. Given two multidimensional sequences $x_1[\boldsymbol{n}]$ and $x_2[\boldsymbol{n}]$, its convolution is expressed as

$$x_1[\boldsymbol{n}] * x_2[\boldsymbol{n}] = \sum_{\boldsymbol{k}} x_1[\boldsymbol{k}]x_2[\boldsymbol{n} - \boldsymbol{k}], \tag{2.86}$$

where $\sum_{\boldsymbol{k}} = \sum_{k_1=-\infty}^{\infty} \cdots \sum_{k_K=-\infty}^{\infty}$.

Multidimensional Sampling

Multidimensional signals can be uniformly sampled in an arbitrary way, depending on the directions of the sampling. Usually, images are signals sampled in a rectangular pattern. In a 2D space, positions t_u, t_v of the samples can be represented with the following pair of dot products:

$$\begin{aligned} t_u &= \boldsymbol{u}^{\mathrm{T}}[n_1, 0]^{\mathrm{T}} = dn_1, \\ t_v &= \boldsymbol{v}^{\mathrm{T}}[0, n_2]^{\mathrm{T}} = dn_2, \end{aligned} \tag{2.87}$$

where $\boldsymbol{u} = [d, 0]^{\mathrm{T}}$ and $\boldsymbol{v} = [0, d]^{\mathrm{T}}$ are a vector basis and d is the sampling distance. A more general type of sampling can be performed by using an arbitrary, non-necessarily orthogonal vector basis $\boldsymbol{u} = [u_1, u_2]^{\mathrm{T}}$ and $\boldsymbol{v} = [v_1, v_2]^{\mathrm{T}}$. Then, the expression becomes

$$
\begin{aligned}
t_u &= \boldsymbol{u}^{\mathrm{T}}[n_1, n_2]^{\mathrm{T}}, \\
t_v &= \boldsymbol{v}^{\mathrm{T}}[n_1, n_2]^{\mathrm{T}}.
\end{aligned}
\tag{2.88}
$$

This sampling procedure can be obviously extended to any number of dimensions.

DFT of a Multidimensional Signal

The Fourier transform of a multidimensional signal is, as its unidimensional counterpart, a projection of the signal onto a basis of complex sinusoid functions. In order to find an expression for a multidimensional sequence, it is only necessary to follow the same procedure as for unidimensional signals. Consider that $x[\boldsymbol{n}]$ is a linear multidimensional impulse response of a linear multidimensional system. Since the complex exponential signals are the eigenfunctions of linear systems, the convolution of such a signal with any linear system will produce a new complex sinusoid with the same frequency, but with different amplitude and phase. The behavior of the amplitude and phase with respect to the frequency is equivalent to the FT. Since the impulse response fully characterizes the system, so will its FT. A complex multidimensional sinusoid can be expressed as

$$
\boldsymbol{v}(\boldsymbol{\Omega}) = \boldsymbol{v}(\Omega_1 n_1, \dots, \Omega_K n_K) = \exp\left(\mathbb{i} \sum_k \Omega_k n_k\right) = \exp(\mathbb{i}\boldsymbol{\Omega}^{\mathrm{T}}\boldsymbol{n})^{\mathrm{T}}, \tag{2.89}
$$

where discrete frequencies can be expressed as $\Omega_k = 2\pi/N_k$, with N_k being each one of the periods of the multidimensional signal. The convolution of both sequences is

$$
\begin{aligned}
x[\boldsymbol{n}] * \exp(\mathbb{i}\boldsymbol{\Omega}^{\mathrm{T}}\boldsymbol{n}) &= \sum_{\boldsymbol{k}} \exp(\mathbb{i}(\boldsymbol{\Omega}^{\mathrm{T}}\boldsymbol{n} - \boldsymbol{k}))x[\boldsymbol{k}], \\
&= \exp(\mathbb{i}\boldsymbol{\Omega}^{\mathrm{T}}\boldsymbol{n}) \sum_{\boldsymbol{k}} \exp(-\mathbb{i}\boldsymbol{\Omega}^{\mathrm{T}}\boldsymbol{k})x[\boldsymbol{k}].
\end{aligned}
\tag{2.90}
$$

The term $\exp(\mathbb{i}\boldsymbol{\Omega}^{\mathrm{T}}\boldsymbol{n})$ in the right side of Equation 2.90 shows that the output is indeed a complex sinusoid with the same frequency as the input. The second term is the complex amplitude response of the system, and equals its FT, which is simply the dot product between the multidimensional sequence and the exponential sequence. Therefore:

$$
\mathcal{F}(x[\boldsymbol{n}]) = X(\boldsymbol{\Omega}) = \sum_{\boldsymbol{k}} \exp(-\mathbb{i}\boldsymbol{\Omega}^{\mathrm{T}}\boldsymbol{k})x[\boldsymbol{k}]. \tag{2.91}
$$

The DFT has efficient ways of computation based on the well-known *butterfly algorithm* for efficient computation of the unidimensional DFT.

2.3.2 Multidimensional Systems

FIR filters are widely used in multidimensional DSP because they are relatively easy to design and, as their unidimensional counterparts, they are inherently stable (Hertz and Zeheb, 1984). Nevertheless, multidimensional FIR filter design involves a much higher computational burden and design constraints that need to be taken into account in their operation and design. Since the FIR can always be designed, the most straightforward way to use an FIR filter is the direct convolution of Equation 2.86 of this response with the signal to be processed. That is, if $h[\boldsymbol{n}]$ and $x[\boldsymbol{n}]$ are an FIR and a signal, the filtering operation is simply

$$y[\boldsymbol{n}] = \sum_{\boldsymbol{k}} x[\boldsymbol{k}]h[\boldsymbol{n} - \boldsymbol{k}]. \tag{2.92}$$

Nevertheless, if the filter lengths in each dimension are $N_1, N_2 \dots, N_K$, each output sample needs $\mathcal{O}(N_1 \cdot N_2 \cdots N_K)$ multiplications and additions, though this number can be further reduced when the filter presents symmetries. A reduced computational burden can be achieved if the FT of the filter is used for the output computation. In that case, provided that the dimensions of the input signal are $M_1, M_2, \dots, M_K$, the computational burden per sample is $2\Pi_k N_k(1 + \log_2 N_k)/\Pi N_k$. This is advantageous when the filter order is high, but the whole signal needs to be stored, which can challenge the capability of certain processing structures. Other strategies, such as the block convolution, allow processing the signal in smaller (eventually disjoint) blocks, thus saving internal memory. This can turn an infeasible filter operation into an affordable one.

Regarding the design of multidimensional FIR filters, the use of standard windows of low-pass filtering (or other forms) is widely extended in image processing. In certain situations, ad hoc filters need to be designed for specific applications. Then, the MMSE criterion can be applied. In addition, *approximation theory* extended to the multidimensional domain is often used (Bose and Basu, 1978).

A more compact, and hence less computationally heavy, approach consists of using IIR filters for multidimensional signal processing (Gorinevsky and Boyd, 2006; Madanayake *et al.*, 2013, 2015). Here, the nature of the impulse response makes direct or indirect use of a recursive structure mandatory for the process. Basically, the implementation of an IIR filter translates into the implementation of a multidimensional finite difference equation of the form

$$\sum_{m_1,\dots,m_K} a_{m_1\dots m_K} y[n_1 - m_1 \dots n_k - m_k] = \sum_{p-1,\dots,p_K} b_{p_1\dots p_K} x[n_1 - p_1 \dots n_K - p_k].$$

Note that the output is multidimensional. The recursive computation of this difference equation is possible by means of just four multidimensional matrices containing coefficients a and b and for input and output sequences. Nevertheless, it is important to notice that, in order to compute an output, the rest of them must be known. This implies that some samples must be computed before others. If a computation order cannot be established, the filter cannot be implemented recursively. As for FIR filters, implementations in the frequency domain exist for IIR filters through the use of a multidimensional extension of the z-transform, which in turn allows us to study stability as well.

2.4 Spectral Analysis on Manifolds

An interesting branch of multidimensional signals and systems analysis, different from the previously described approaches, is that of spectral analysis on manifolds. The previously described discrete Fourier analysis was defined to transform uniformly sampled signals from the temporal domain to the frequency domain. However, this transformation cannot be directly applied to 3D discrete triangle meshes because they are irregularly sampled; that is, they have different resolutions at different places depending on the required details in each place. For that reason, a large number of spectral mesh processing methods have been proposed for a variety of computer graphics applications, such as clustering, mesh parameterization, mesh compression, mesh smoothing, mesh segmentation, remeshing, surfaces reconstruction, texture mapping, or watermarking, among others (Sorkine, 2005; Zhang *et al.*, 2010). In this section we provide an introduction to these kinds of techniques. Some background on graph theory is assumed. The interested reader will find other kernel approaches based on graphs in Chapters 8, 10, and 12.

2.4.1 Theoretical Fundamentals

A mesh is a piecewise-linear representation of a smooth surface with arbitrary topology, and it is defined by a set of vertices V, edges E, and faces F (see Figure 2.1). Each vertex $v_i \in V$ has an associated position in the 3D space $v_i = [x_i, y_i, z_i]$, and each edge and face connect two and three vertices, respectively. The vertex positions and the faces capture the geometry of the surface, and the connectivity between vertices (edges) captures the topology of the surface. Here, we consider a mesh as a manifold (2-manifold embedded in a 3D space), which obeys the following two rules: (1) every edge is adjacent to one or

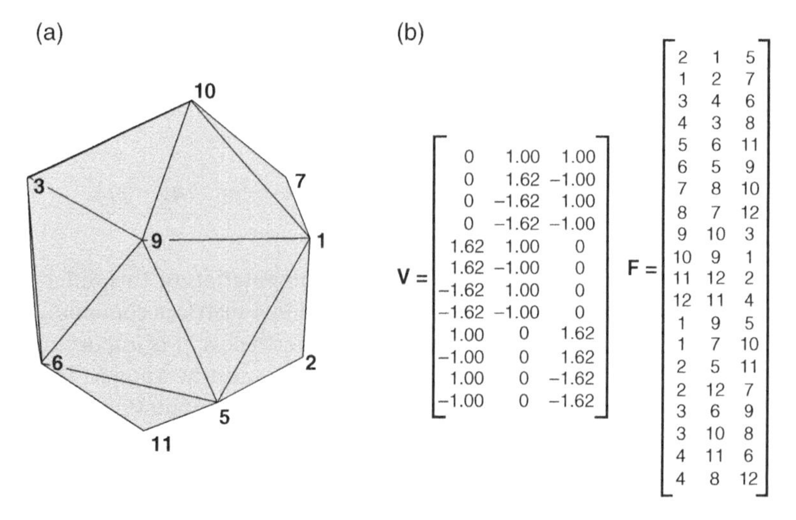

Figure 2.1 Icosahedron mesh (a) and its vertex and face matrices (b).

two faces; and (2) the faces around every vertex form a closed or open fan. In addition, the manifold mesh fulfills the Euler–Poincaré formula for closed polygonal meshes: $V - E + F = \chi$, where $\chi = 2(1 - G)$ and G is the genus of the surface (i.e., the number of holes).

Current developments in spectral mesh processing are based on the work published by (Taubin, 1995), which showed that the well-known FT of a signal can be seen as the decomposition of this signal into a linear combination of Laplacian operator eigenvectors. Hence, most of the spectral methods for mesh processing define the most appropriate Laplacian operator for their spectral application (Sorkine, 2005; Zhang *et al.*, 2010).

The Laplacian operator of a twice-differentiable real-valued function $\boldsymbol{f}$ is defined by the divergence of its gradient as

$$\Delta \boldsymbol{f} = \mathrm{div}(\mathrm{grad}\boldsymbol{f}) = \nabla \cdot \nabla \boldsymbol{f}. \tag{2.93}$$

When functions are defined over a Riemannian or pseudo-Riemannian manifold M with metric g – for example, the coordinate function given by the vertex position (x_i, y_i, z_i) – the Laplace operator is known as the Laplace–Beltrami operator and is defined as

$$\Delta_M \boldsymbol{f} = \mathrm{div}_{\mathrm{M}}(\mathrm{grad}_{\mathrm{M}}\, \boldsymbol{f}). \tag{2.94}$$

The definition of the Laplacian can also be extended to k-forms, and then it is called the Laplace–de Rham operator (Rosenberg, 1997). This operator is defined as $\Delta = \delta d + d\delta$, where d is the exterior derivative on k-forms and δ is the codifferential, acting on $k-1$-forms. In the Laplace–de Rham operator for functions on a manifold (0-forms) it can be seen that to $d\delta$ vanishes, and the result $\Delta = \delta d$ is the actual definition of the Laplace–Beltrami operator.

In general, the procedure for mesh spectral analysis follows the following three stages: (1) compute a Laplacian matrix, normally a sparse matrix; (2) compute the eigendecomposition of this Laplacian matrix (i.e., eigenvalues and eigenvectors); (3) process these eigenvectors and eigenvalues for the specific purpose of the application. A wide range of Laplacian operators, namely combinational graph Laplacian and geometric mesh Laplacian, have been proposed with the aim of finding a Laplacian operator which fulfills all the desired properties in a discrete Laplacian, such as symmetry, orthogonal eigenvectors, positive semi-definiteness, or zero row sum. Readers are referred to Wardetzky *et al.* (2007) for a detailed explanation of these properties and their compliance in different Laplace operators. The combinational graph Laplacians only take into account the graph of the mesh (vertices and connectivity), whereas the geometry information is not included explicitly. Hence, two different embeddings of the same graph yield the same eigenfunctions, and two different meshings of the same object yield different eigenfunctions (Vallet and Lévy, 2008).

While the graph Laplacian can be seen as a discrete analogue of the Laplace–Beltrami operator, the geometric mesh Laplacian is the direct discretization of the Laplace–Beltrami operator, which includes both the topology and geometry information about the mesh (face angles and edge lengths). Dyer *et al.* (2007) showed a higher robustness to changes in sampling density and connectivity of the geometric Laplacian than the graph Laplacian. For both Laplacian operators, the Laplacian matrix can be written in a local form as

$$\boldsymbol{\Delta f}(\boldsymbol{v}_i) = \frac{1}{b_i} \sum_{\boldsymbol{v}_j \in N(\boldsymbol{v}_i)} w_{ij}(\boldsymbol{f}(\boldsymbol{v}_i) - \boldsymbol{f}(\boldsymbol{v}_j)), \tag{2.95}$$

where $\boldsymbol{f}(\boldsymbol{v}_i)$ is the value of the function $\boldsymbol{f}$ in the position of the vertex v_i, $N(v_i)$ is the one-ring first-order neighbor of $\boldsymbol{v}_i$, w_{ij} is the symmetric weight ($w_{ij} = w_{ji}$), and b_i is a positive value.

Once the Laplacian $\boldsymbol{\Delta}$ is defined, its eigendecomposition on a manifold satisfies

$$-\boldsymbol{\Delta H}^k = \lambda_k \boldsymbol{H}^k \tag{2.96}$$

where $\boldsymbol{H}^k$ are the eigenvectors, which define a basis for functions over the surface; and λ_k are the eigenvalues, which represent the mesh spectrum (frequencies). Note that the negative sign is required to have positive eigenvalues. The basis functions are ordered according to the increasing eigenvalues; hence, lower eigenvectors correspond to a basis function associated with lower frequencies. Applying this concept to Fourier analysis, we can prove that the complex exponential of the DFT, $\mathrm{e}^{\mathrm{i}\Omega x}$, is an eigenfunction (orthonormal basis) of the Laplace operator:

$$\Delta \mathrm{e}^{\mathrm{i}\Omega x} = -\frac{\mathrm{d}^2}{\mathrm{d}x^2}\, \mathrm{e}^{\mathrm{i}\Omega x} = (\Omega)^2\, \mathrm{e}^{\mathrm{i}\Omega x} \tag{2.97}$$

Note that the set of basis functions in the DFT are fixed; meanwhile, the eigenvectors of the Laplacian operator could change based on the mesh geometry and connectivity. As in Fourier analysis, the spectral coefficients can be obtained by projecting the vertices on the H^k basis as

$$\boldsymbol{a}_k = \sum_{i=1}^{n} \boldsymbol{v}_i \boldsymbol{H}_i^k \tag{2.98}$$

where n is the number of vertices, and the reconstruction of the original mesh can be obtained as

$$\hat{\boldsymbol{v}}_i = \sum_{k=1}^{m} \boldsymbol{a}_k \boldsymbol{H}_i^k \tag{2.99}$$

where m is the number of coefficients used for reconstruction. As lower eigenvalues (and their associated eigenvectors) correspond to the general shape of the mesh (lower frequencies), a low-pass-filtered shape of the mesh is obtained for $m \ll n$. The original mesh is reconstructed when $m = n$.

2.4.2 Laplacian Matrices

Among the wide variety of combinational graph Laplacians and geometric graph Laplacians, we present here a representative set of examples. For a more detailed survey of combinational and geometric Laplacians, readers are referred to Sorkine (2005) and Zhang *et al.* (2010). The three well-known combinational mesh Laplacians are the graph Laplacian (also known as Kirchoff operator), the Tutte Laplacian, and the normalized graph Laplacian (also known as the graph Laplacian).

The graph Laplacian matrix $\boldsymbol{L}$ is obtained from the adjacent matrix $\boldsymbol{W}$ and the degree matrix $\boldsymbol{D}$ as $\boldsymbol{L} = \boldsymbol{D} - \boldsymbol{W}$. Both $\boldsymbol{W}$ and $\boldsymbol{D}$ are $n \times n$ matrices defined as

$$\boldsymbol{W}_{ij} = \begin{cases} 1 & \text{if } (i,j) \in E \\ 0 & \text{otherwise} \end{cases} \tag{2.100}$$

$$\boldsymbol{D}_{ij} = \begin{cases} |N(\boldsymbol{v}_i)| & \text{if } i = j \\ 0 & \text{otherwise} \end{cases} \tag{2.101}$$

where $|N(\boldsymbol{v}_i)|$ is the number of first-order neighbors of $\boldsymbol{v}_i$. Here, $b_i = N(\boldsymbol{v}_i)$, and $w_{ij} = 1$ in Equation 2.95.

The Tutte Laplacian $\boldsymbol{T}$ is defined as $\boldsymbol{T} = \boldsymbol{D}^{-1}\boldsymbol{L} = \boldsymbol{I} - \boldsymbol{D}^{-1}\boldsymbol{W}$ and it was used by Taubin (1995) for surface smoothing (as a linear low-pass filter of the natural frequencies of the surface). In Equation 2.95, $b_i = -1/N(\boldsymbol{v}_i)$, and $w_{ij} = 1$ for the Tutte Laplacian. Given that $\boldsymbol{T}$ is not a symmetric matrix, several variations of this Laplacian have been proposed in order to symmetrize $\boldsymbol{T}$, such as $\boldsymbol{T}^* = (\boldsymbol{T} + \boldsymbol{T}')/2$ (Lévy, 2006) or $\boldsymbol{T}^{**} = \boldsymbol{T}' \cdot \boldsymbol{T}$ (Zhang *et al.*, 2004).

Finally, the normalized graph Laplacian is defined as $\boldsymbol{Q} = \boldsymbol{D}^{-1/2}\boldsymbol{L}\boldsymbol{D}^{-1/2}$, where $b_i = -1/(N(\boldsymbol{v}_i)N(\boldsymbol{v}_j))^1/2$, and $w_{ij} = 1$ in Equation 2.95. $\boldsymbol{Q}$ and $\boldsymbol{T}$ have the same spectrum due to $\boldsymbol{T} = \boldsymbol{D}^{-1/2}\boldsymbol{Q}\boldsymbol{D}^{-1/2}$, and although $\boldsymbol{Q}$ is not a Laplacian in Equation 2.95, its symmetry can help to calculate the eigenvectors of $\boldsymbol{T}$.

The geometric mesh Laplacians use cotangent weights in Equation 2.95 to compute the Laplacian:

$$w_{ij} = \frac{\cot(\beta_{ij}) + \cot(\alpha_{ij})}{2}, \tag{2.102}$$

where $b_i = 1$ and β_{ij} and α_{ij} are the opposite angles to the edge between $\boldsymbol{v}_i$ and $\boldsymbol{v}_j$, as is shown in Figure 2.2. Pinkall and Polthier (1993) presented this first cotangent scheme for discrete minimal surfaces. In a different way, Meyer *et al.* (2003) also obtained the same cotangent weights but the factor $b_i = |A_i|$, where $|A_i|$ is the area of the barycell A_i which is formed by connecting the barycenters of the triangles adjacent to $\boldsymbol{v}_i$ with the midpoints of the edges incident to the triangles. Given that the cotangent can give problems when the angle is close to π radians, Floater (2003) proposed an alternative scheme called mean-value coordinates and defined as

$$w_{ij} = \frac{\tan(\theta_{ij}) + \tan(\gamma_{ij})}{||\boldsymbol{v}_i - \boldsymbol{v}_j||}, \tag{2.103}$$

where θ_{ij} and γ_{ij} are the angles shown in Figure 2.2. The finite-element method also provides a new formulation of the geometric Laplacian by redefining Equation 2.96 as (Vallet and Lévy, 2007)

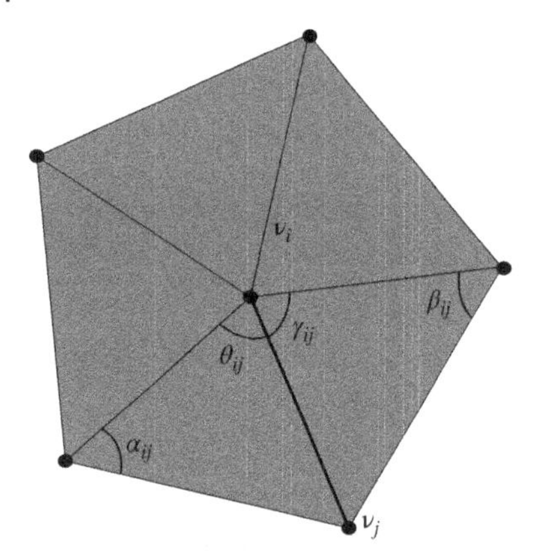

Figure 2.2 Angles in the cotangent weights.

$$-\boldsymbol{Q}\boldsymbol{h}^k = \lambda_k \boldsymbol{Q}\boldsymbol{h}^k \tag{2.104}$$

where $\boldsymbol{h}^k = [\boldsymbol{H}_1^k, \boldsymbol{H}_2^k, \dots, \boldsymbol{H}_n^k]$; $\boldsymbol{Q}$ is the stiffness matrix and $\boldsymbol{B}$ is the mass matrix, defined as follows:

$$\boldsymbol{Q}_{ij} = \begin{cases} \frac{\cot(\beta_{ij})+\cot(\alpha_{ij})}{2} & \text{if } (i,j) \in E \\ -\sum_j Q_{ij} & \text{if } i = j \\ 0 & \text{otherwise} \end{cases} \tag{2.105}$$

$$\boldsymbol{B}_{ij} = \begin{cases} \frac{|t_\alpha|+|t_\beta|}{12} & \text{if } i \neq j \\ \frac{\sum_{j\in N(\boldsymbol{v}_i)} |t_j|}{6} & \text{if } i = j \\ 0 & \text{otherwise} \end{cases} \tag{2.106}$$

where $|t_\alpha|$ and $|t_\beta|$ are the areas of the faces which share the edge (i, j), and $|t_j|$ is the area of jth one-ring first-order neighbor of $\boldsymbol{v}_i$. The previous formulation can be simplified as $-\boldsymbol{D}^{-1}\boldsymbol{Q}\boldsymbol{h}^k = \lambda^k \boldsymbol{h}^k$, where $\boldsymbol{D}$ is a diagonal matrix defined as

$$\boldsymbol{D}_{ii} = \sum_j \boldsymbol{B}_{ij} = \frac{\sum_{j\in N(\boldsymbol{v}_i)} |t_j|}{3}. \tag{2.107}$$

Finally, the most recent geometric Laplacian definition has been derived by using discrete exterior calculus, a new language of differential geometry and mathematical physics (Desbrun *et al.*, 2005; Grinspun *et al.*, 2006). Vallet and Lévy (2008) proposed a symmetrized Laplace–de Rham operator for mesh filtering by using discrete exterior calculus:

$$\Delta_{ij} = \begin{cases} -\frac{\cot(\beta_{ij})+\cot(\alpha_{ij})}{\sqrt{|\mathbf{v}_i||\mathbf{v}_j|}} & \text{if } i \neq j \\ -\sum_{k\in N(\mathbf{v}_i)} \Delta_{ik} & \text{if } i = j \\ 0 & \text{otherwise} \end{cases} \tag{2.108}$$

where $|\mathbf{v}_i|$ is the area of the Voronoi region of the vertex $\mathbf{v}_i$ in its one-ring neighbor.

2.5 Tutorials and Application Examples

In this section we present a selected set of illustrative examples to review the basic concepts of signal processing and standard techniques. In particular, we will introduce:

1) Examples of real and complex signals along with their basic graphical representations.
2) Basic concepts on convolution, FT, and spectrum for discrete-time signals.
3) Continuous-time signals concepts which are highlighted by using symbolic representations, and a simple case study analyzed for the reconstruction via Fourier series of a synthetic squared periodic wave.
4) Filtering concepts by using some simple synthetic signals and one real cardiac recording.
5) Nonparametric and parametric spectrum estimation, which will be compared for real sinusoidal signals buried in noise, for highlighting some of the advantages and drawbacks of each method in practice.
6) Standard source separation techniques and wavelet analysis, which will be briefly presented in a couple of case studies.
7) Finally, spectral analysis on meshes by using manifold harmonic analysis is studied with a toy example application with known solution to highlight the main elements of this kind of harmonic analysis in a multidimensional and nonconventional subsequent application for cardiac meshes analysis.

2.5.1 Real and Complex Signal Processing and Representations

Complex signals are a convenient representation in signal processing and communications, where the information is transmitted using both the amplitude and the phase of the signal. In this first example, a Hanning pulse is convolved with a train of delta functions with positive or negative values representing one- or zero-valued bits. The first panel of Figure 2.3 shows the pulse represented with 128 samples, where its FT (which is real) is represented in the second panel. The third panel represents the pulse modulation, which is a train of three pulses carrying bits 1, 1, and 0. The FT of this train of pulses is seen in the fourth panel of Figure 2.3. Since the train of pulses is a real signal, its FT has a symmetric real part and an antisymmetric imaginary part. The code generating this figure is in Listing 2.1. Note that the seed initialization of the first line can be changed to generate other random realizations.

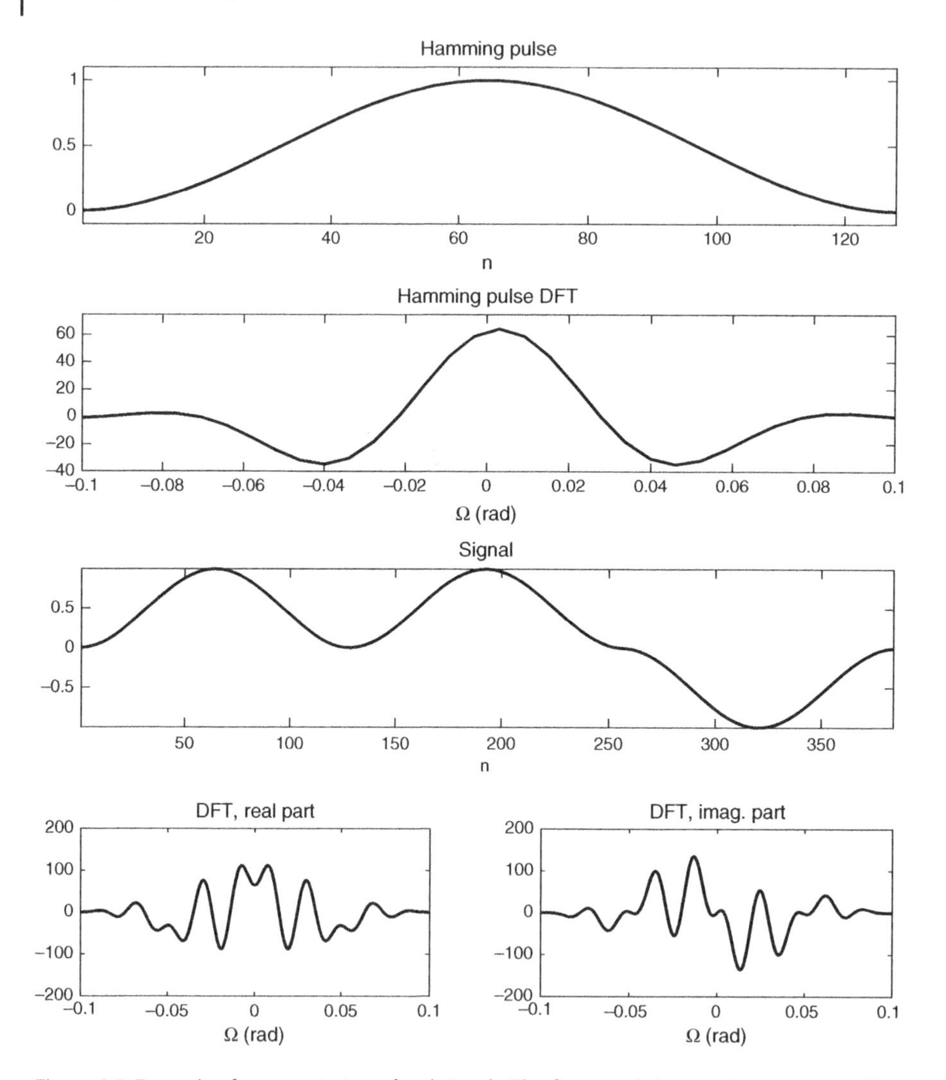

Figure 2.3 Example of representation of real signals. The first panel shows a Hanning pulse. The second panel shows the discrete-time FT of the pulse. In the third panel, a binary modulation is shown, consisting of a train of pulses multiplied by +1 or −1 to represent bits 1 and 0. The fourth panel shows the real (left) and imaginary (right) parts of the FT of the signal in the third panel. Since this signal is real, its transform is Hermitian.

```
randn('seed',2)
x = hanning(128);
figure(1), subplot(411), plot(x)
axis([1 128,-0.1,1.1]), xlabel('n'), title('Hamming pulse')
subplot(412)
plot(linspace(-pi,pi,1024), real(fftshift(fft(x,1024))))
axis([-0.1,0.1,-40,75]), xlabel('\Omega (rad)'), title('Hamming pulse ...
    DFT')
s = sign(randn(1,3));
y = kron(s',x);
subplot(413), plot(y), xlabel('n'), title('Signal'), axis tight
```

```
subplot(427),
plot(linspace(-pi,pi,65535), real(fftshift(fft(y,65535))))
axis([-0.1,0.1,-200,200]), xlabel('\Omega (rad)'), title('DFT, real part')
subplot(428)
plot(linspace(-pi,pi,65535), imag(fftshift(fft(y,65535))))
axis([-0.1,0.1,-200,200]), xlabel('\Omega (rad)')
```

Listing 2.1 DFT example: real and complex counterparts (dft.m).

This signal is modulated in signal $z[n]$, as seen in Figure 2.4, first panel. Here, the pulse train $y[n]$ is multiplied by a sinusoid $\cos(\Omega_0 n) = 0.5\exp(\mathbb{i}\Omega_0 n) + 0.5\mathbb{i}\exp(\mathbb{i}\Omega_0 n)$. This panel shows the modulated sinusoid, where it can be clearly seen that the information of the symbols is carried in the phase of the signal. Indeed, the third pulse is shifted π radians with respect to the first and second ones.

Multiplying a signal in time by a complex exponential $\mathrm{e}^{\mathbb{i}\Omega_0 n}$ is equivalent to shifting its FT in Ω_0 radians. Since the sinusoid is composed of two complex exponentials of positive and negative sign, the modulation appears in frequency as a pair of signals shifted to the positive and negative parts of the frequency axis (second panel). The spectra are centered around $\pm\Omega_0$. Note that both the pulse train and the modulation are Hermitian with respect to the origin and with respect to $\pm\Omega$.

Such modulations are called double side-band modulations, since both the subbands at the right and left sides of $\pm\Omega_0$ carry the same information. This property is exploited to recover the modulating pulse. Both sides of the spectrum contain a signal which is simply a shifted version of the original spectrum. In order to recover the signal, we only need to remove the negative part and shift the positive part Ω radians to the left.

Signal $Z(\mathbb{i}\Omega)$ in the second panel of Figure 2.4 can be analytically represented as

$$Z(\mathbb{i}\Omega) = \frac{1}{2}Y(\mathbb{i}\Omega - \mathbb{i}\Omega_0) + \frac{1}{2}Y(\mathbb{i}\Omega + \mathbb{i}\Omega_0). \tag{2.109}$$

Hence, signal $\hat{Z}(\mathbb{i}\Omega)$ of the third panel is simply $\hat{Z}(\mathbb{i}\Omega) = Z(\mathbb{i}\Omega - \mathbb{i}\Omega_0)$, which in time is equal to the complex signal $\hat{z}[n] = \exp(\mathbb{i}\Omega_0 n)$. In order to recover the original signal, we simply multiply this complex signal by a complex exponential of frequency $-\Omega_0$. The FT and time versions of the recovered signal are shown in the fifth panel of of Figure 2.4. The corresponding code of the experiment is in Listing 2.2. Note in particular that the modulation is carried out by using a carrier of $2\pi/16$ rad/s.

```
figure(2)
p = cos(2*pi/16.*(0:length(y)-1)); % Carrier
z = p'.*y;
subplot(511), plot(z)
axis tight, xlabel('n'), title('DSB modulation')
subplot(523)
plot(linspace(-pi,pi,65536),real(fftshift(fft(z,65536))))
axis([-0.8 0.8 -100 100]), xlabel('\Omega (rad)'), title('DFT, real part')
subplot(524)
plot(linspace(-pi,pi,65536),imag(fftshift(fft(z,65536))))
axis([-0.8 0.8 -100 100]), xlabel('\Omega (rad)'), title('DFT, imag. ...
    part')
z2 = z;
Z2 = fftshift(fft(z,65536));
Z2(1:end/2) = 0;
```

```
subplot(525), plot(linspace(-pi,pi,65536),real(Z2))
axis([-0.8 0.8 -100 100])
xlabel('\Omega (rad)'), ylabel('Negative part removal')
subplot(526), plot(linspace(-pi,pi,65536),imag(Z2))
axis([-0.8 0.8 -100 100])
xlabel('\Omega (rad)'), ylabel('Negative part removal')
fc = 6553;
Z3 = [2*Z2(4097:end) ; zeros(4096,1)];
subplot(527), plot(linspace(-pi,pi,65536),real(Z3))
axis([-0.1 0.1 -200 200])
xlabel('\Omega (rad)'), ylabel('Frequency shift')
subplot(5,2,8), plot(linspace(-pi,pi,65536),imag(Z3))
axis([-0.1 0.1 -200 200])
xlabel('\Omega (rad)'), ylabel('Frequency shift')
subplot(515), plot(real(ifft(fftshift(Z3))))
axis([1 length(y) -1 1]), title('Recovered signal'), xlabel('n')
```

Listing 2.2 DFT example: real and complex counterparts (dsb.m).

This example modulates a signal in a double side band, which carries redundancy in its spectrum, in exchange for very inexpensive forms of demodulation. Nevertheless, it is possible to process the signal using just half the bandwidth since, as said before, both sides of the spectrum around the central frequency Ω carry the same information. Then, it is possible to just transmit one of the sides of the spectrum and then reconstruct the whole spectrum in the receiver by forming the symmetric and antisymmetric counterparts parts of the real and imaginary components of the transmitted signal.

This is depicted in Figures 2.5 and 2.6. The second panel of Figure 2.5 shows the FT of the signal where the negative part has been removed, since it carries exactly the same information as the positive one. The idea is to transmit this signal, which requires only half of the bandwidth. This signal is complex, since the FT is neither symmetric nor antisymmetric, and the real part of this signal in time can be readily shown to be identical to the original signal. Real component $y_R[n]$ and imaginary component $y_I[n]$ are presented in the third and fourth panels of the figure. The reader can use the code in Listing 2.3 to reproduce this part of the experiment.

```
figure(3)
y = kron(s',x);
subplot(411), plot(y), title('Signal'), xlabel('n'), axis tight
Y = fftshift(fft(y,65535));
Y(1:end/2) = 0;
subplot(423), plot(linspace(-pi,pi,65535),real(Y))
axis([-0.1,0.1,-200,200])
ylabel('Negative spectrum removal'), xlabel('\Omega (rad )')
subplot(424), plot(linspace(-pi,pi,65535),imag(Y))
axis([-0.1,0.1,-200,200])
xlabel('\Omega (rad)'), ylabel('Negative spectrum removal')
yssb = ifft(fftshift(Y));
yssb = yssb(1:length(y));
subplot(413), plot(real(yssb)), title('Baseband SSB, real part')
xlabel('n'), axis tight
subplot(414), plot(imag(yssb))
axis tight, title('Baseband SSB, imag. part'), xlabel('n')
```

Listing 2.3 DFT example: transmitting just one side of the spectrum (ssb.m).

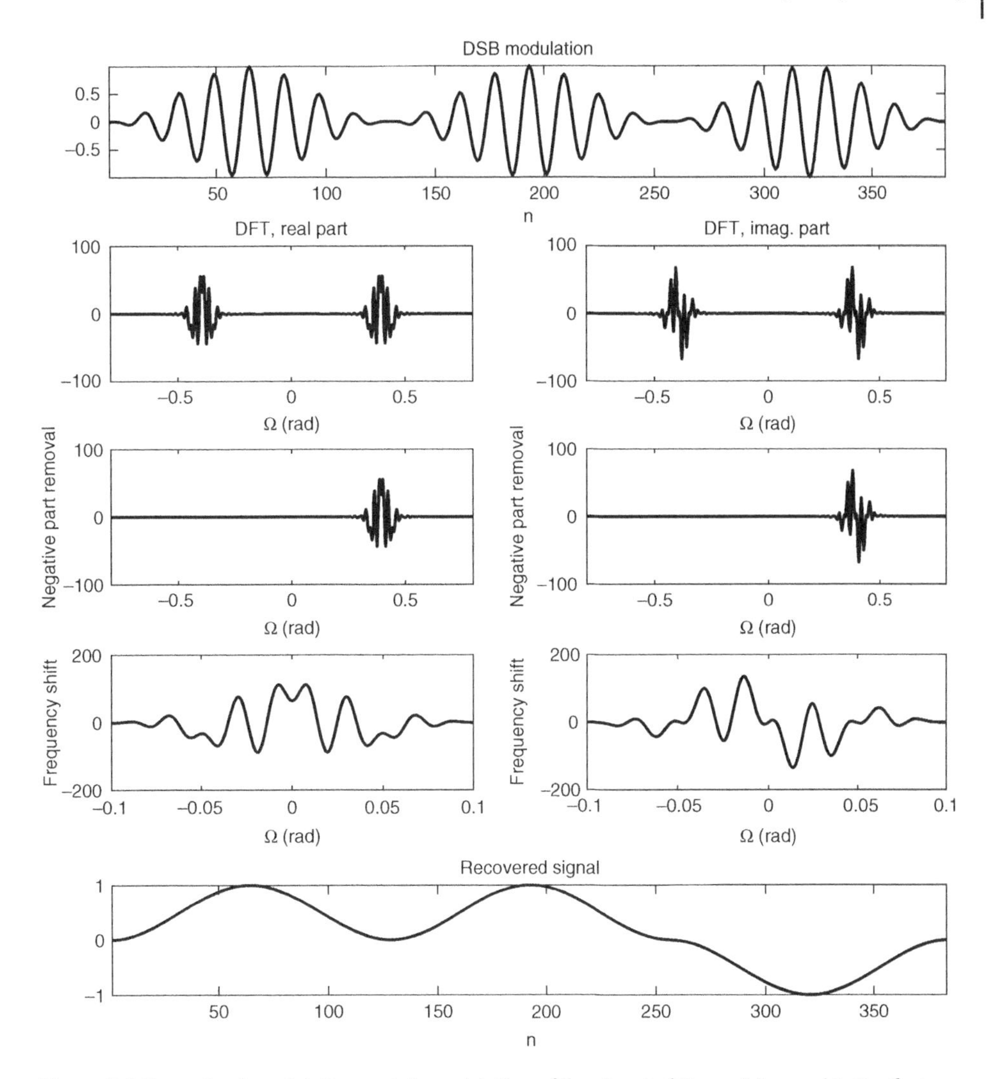

Figure 2.4 Example of modulation and demodulation of the signal of Figure 2.3, panel 3. The first panel shows the modulation of the signal using a sinusoid function. The corresponding FT can be seen in the second panel. Since the signal is real, its FT is Hermitian. In the third panel, the negative components of the frequency-domain panel show the real and imaginary parts of the FT of the signal are removed, resulting in a complex version of the signal in time. The pulse train is recovered by simply shifting the signal in time toward the origin, which produces a Hermitian FT, and hence a real signal in time (fourth and fifth panels).

Figure 2.6 shows the modulated signal. The modulation process consists of embedding both the real and imaginary parts of the spectrum in a real signal. To this end, we simply construct a signal symmetric in the frequency domain with the FT of $y_R[n]$ and an antisymmetric one with the FT of $y_I[n]$, as shown in the second panel of the figure. These will constitute the real and imaginary parts of the Hermitian FT of the modulated signal, which is real. This is equivalent to constructing a modulated signal of the form

$$z[n] = y_R[n]\cos(\Omega n) - y_I[n]\sin(\Omega n), \tag{2.110}$$

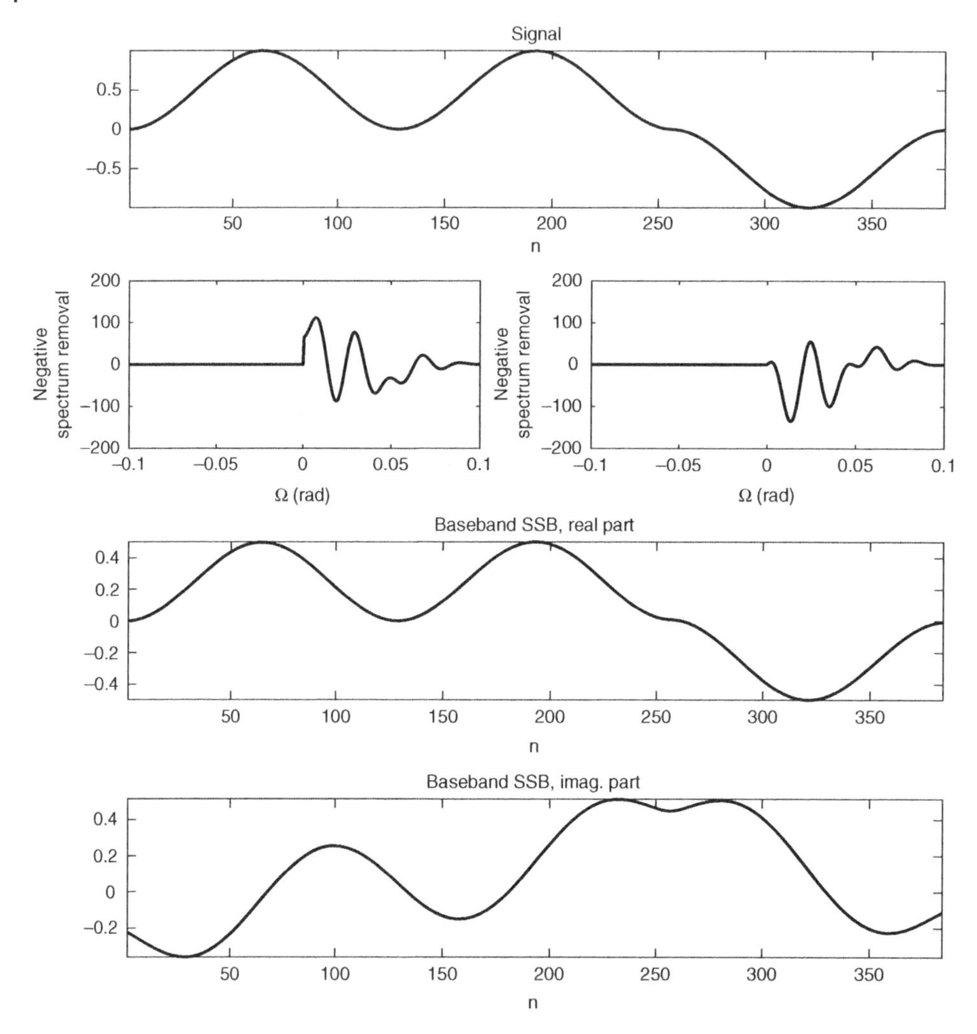

Figure 2.5 Example of single side-band modulation and demodulation of the signal of Figure 2.3, third panel. The original signal is in the first panel. Subsequent panels show the FT of the signal with the negative part removed. This results in a complex signal whose real part is identical to the original one (third panel), but which has an imaginary part (shown in fourth panel).

whose proof is proposed as an exercise. The demodulation is performed in an identical way as in the modulation of Figure 2.4, since the real part of the single side-band signal, once shifted back to the origin (Figure 2.6, third panel), is identical to the original pulse train. Finally, the part of the experiment depicted in Figure 2.6 can be reproduced with the code in Listing 2.4.

```
figure(4)
p = cos(2*pi*4096/65536.*(0:length(y)-1));
q = sin(2*pi*4096/65536.*(0:length(y)-1));
z = p'.*real(yssb)-q'.*imag(yssb);
subplot(511), plot(z), title('Modlulated SSB signal'), axis tight
```

```
subplot(523), plot(linspace(-pi,pi,65536),real(fftshift(fft(z,65536))))
axis([-0.8 0.8 -100 100]), xlabel('\Omega (rad)'), title('DFT, real part')
subplot(524), plot(linspace(-pi,pi,65536),imag(fftshift(fft(z,65536))))
axis([-0.8 0.8 -100 100]), xlabel('\Omega (rad)'), title('DFT, imag. ...
    part')
z2 = z;
Z2 = fftshift(fft(z,65536));
Z2(1:end/2+1) = 0;
subplot(525), plot(linspace(-pi,pi,65536),real(Z2))
axis([-0.8 0.8 -100 100])
ylabel('Negative spectrum removal'), xlabel('\Omega (rad )')
subplot(526), plot(linspace(-pi,pi,65536),imag(Z2))
axis([-0.8 0.8 -100 100])
ylabel('Negative spectrum removal'), xlabel('\Omega (rad )')
fc = 6553;
Z3 = [2*Z2(4097:end);zeros(4096,1)];
subplot(527), plot(linspace(-pi,pi,65536),real(Z3))
axis([-0.1 0.1 -200 200])
ylabel('Frequency shift'), xlabel('\Omega (rad )')
subplot(5,2,8), plot(linspace(-pi,pi,65536),imag(Z3))
axis([-0.1 0.1 -200 200])
ylabel('Frequency shift'), xlabel('\Omega (rad )')
subplot(515), plot(real(ifft(fftshift(2*Z3)))), axis([1 length(y) -1 1])
title('Recovered signal, real part'), xlabel('n')
```

Listing 2.4 DFT example: modulated signal (ssb2.m).

2.5.2 Convolution, Fourier Transform, and Spectrum

Let $x[n]$ be a discrete time signal with samples defined between $n = 0$ and $n = 100$ as follows:

$$x[n] = \begin{cases} n/100 & \text{if } 0 \leq n \leq 100 \\ 0 & \text{otherwise} \end{cases} \tag{2.111}$$

and $h[n]$ the impulse response of a given discrete time signal, with the form of a pulse with 150 samples length:

$$h[n] = \begin{cases} 1 & \text{if } 0 \leq n \leq 150 \\ 0 & \text{otherwise} \end{cases} \tag{2.112}$$

The response $y[n]$ of the system $h[n]$ to the signal $x[n]$ can be computed as the convolution of both signals as in Equation 2.5:

$$y[n] = \sum_{k=-\infty}^{\infty} h[k]x[n-k] \tag{2.113}$$

By inspection, it can be seen that there are five different cases of this equation (see Figure 2.7). The first one is when $n \leq 0$. In this case, if $k \geq 0$, then $x[n-k] = 0$, and for $k \leq 0$ then $h[k] = 0$, so the convolution is zero for this interval of n, as shown in

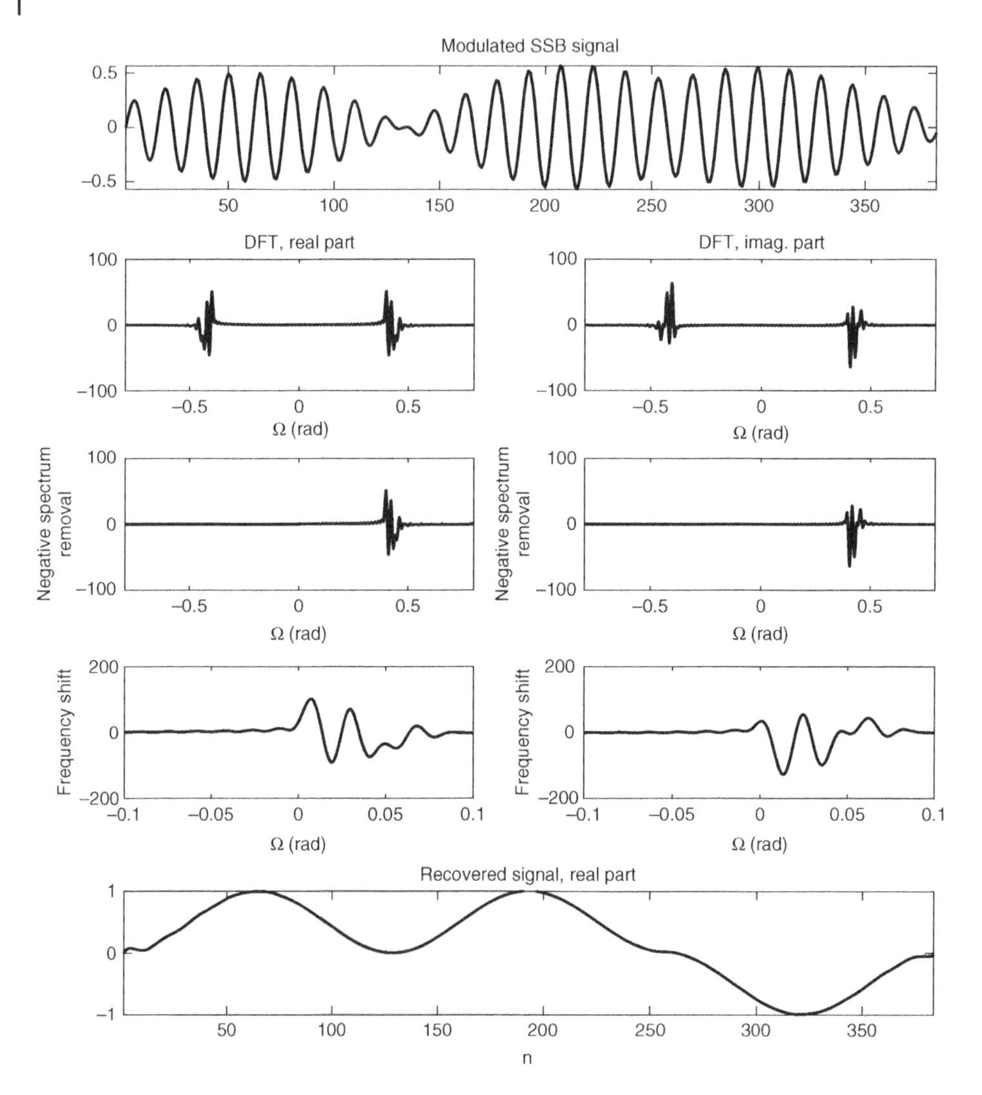

Figure 2.6 Example of modulation and demodulation of the signal of Figure 2.3, panel 3. The first panel shows the modulation, which contains a shifted version of one side of the base-band frequency components of the pulse train (second panel). The demodulation consists of discarding the negative frequency components (third panel) plus a frequency shifting back to the origin (fourth panel). The real part of the signal is identical to the original pulse train.

the lower panel of the figure. The second case corresponds to the interval $0 \leq n \leq 100$. In this case, $x[n-k]$ is nonzero for $0 \leq k < n$ and the convolution corresponds to the integral of a triangle with base equal to n and height equal to $n/100$, and whose value is given by $n^2/200$. The third case corresponds to the interval $100 \leq n \leq 150$. There, the convolution for all values of n is equal to the area of a triangle of unit height and base 100, which is 50. The fourth for n between 150 and 250 corresponds to the difference between a triangle of area 50 and the triangle defined between $k = 150$ and $k = n$. This triangle has a base equal to $n - 150$, and a height equal to $(n - 150)/100$. Then the

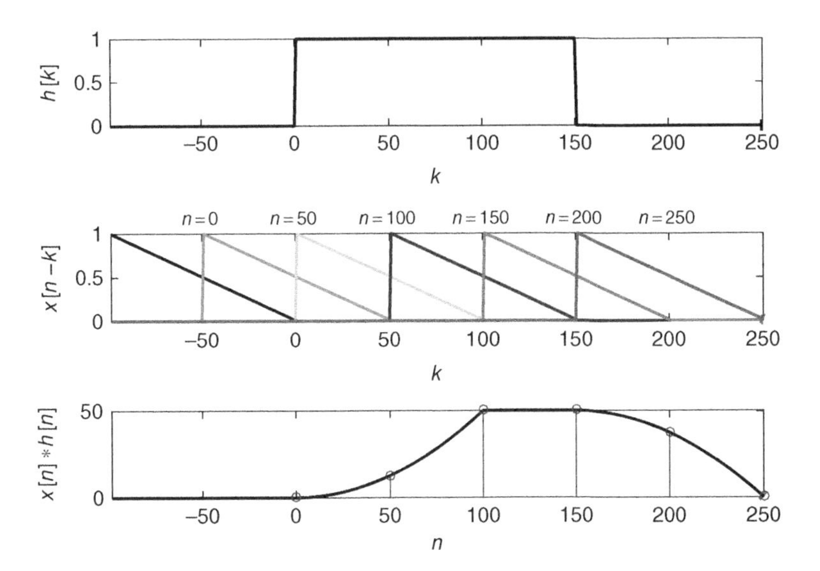

Figure 2.7 Convolution of signal $x[n]$ with system $h[n]$. The impulse response of the system $h[k]$ is shown in the upper panel. The middle panel shows $x[n-k]$ for different values of n. The lower panel shows the result of the convolution. The circles show the values of the convolution for the values of n shown in the middle panel.

convolution is $50 - [(n-150)^2/200]$. The fifth case corresponds to $k \leq 250$, where the convolution is zero.

The convolution can be written as

$$y[n] = \begin{cases} 0 & \text{if } n \leq 0 \\ \frac{n^2}{200} & \text{if } 0 \leq n \leq 100 \\ 50 & \text{if } 100 \leq n \leq 150 \\ 50 - \frac{(n-150)^2}{200} & \text{if } 150 \leq n \leq 250 \\ 0 & \text{otherwise.} \end{cases} \tag{2.114}$$

The corresponding MATLAB code is given in Listing 2.5.

```
h = [zeros(1,100), ones(1,150) zeros(1,100)]; % h[n]
x = [1:100]/100; % x[n]
xa = [fliplr(x),zeros(1,250)];                 % x[-n]
xb = [zeros(1,50),fliplr(x),zeros(1,200)];   % x[50-n]
xc = [zeros(1,100),fliplr(x),zeros(1,150)];% x[100-n]
xd = [zeros(1,150),fliplr(x),zeros(1,100)];% x[150-n]
xe = [zeros(1,200),fliplr(x),zeros(1,50)];   % x[200-n]
xf = [zeros(1,250),fliplr(x)];                 % x[250-n]
% Represent impulse response
figure(1), subplot(311), plot([-99:250],h)
% Represent x[t-n]
subplot(312), plot([-99:250],xa), hold all
plot([-99:250],xb), plot([-99:250],xc)
plot([-99:250],xd), plot([-99:250],xe)
```

```
plot([-99:250],xf)
% Convolution
y = [zeros(1,99) conv(h,x,'valid')];
% Represent convolution
subplot(313), plot([-99:250],y)
hold all, stem(0:50:250,y(100:50:end))
```

Listing 2.5 Convolution example (convolution.m).

Consider now a pulse $x[n]$ of length $2N+1$ centered around the origin. Its FT can be computed as in Equation 2.12, where the domain is $-N \leq n \leq N$; that is:

$$X(\mathbb{i}\Omega) = \mathcal{F}\{x[n]\} = \sum_{n=-N}^{N} \mathrm{e}^{-\mathbb{i}\Omega n}. \tag{2.115}$$

The series can be split into the positive and the negative parts, and their values can be computed using the result $\sum_{n=1}^{N} r^n = (r - r^{N+1})/(1-r)$ with $r = \mathrm{e}^{\mathbb{i}\Omega n}$ and $\mathrm{e}^{-\mathbb{i}\Omega n}$ respectively:

$$\begin{aligned}
X(\mathbb{i}\Omega) &= \sum_{n=1}^{N} \mathrm{e}^{-\mathbb{i}\Omega n} + \sum_{n=1}^{N} \mathrm{e}^{\mathbb{i}\Omega n} + 1 \\
&= \frac{\mathrm{e}^{-\mathbb{i}\Omega} - \mathrm{e}^{-\mathbb{i}(N+1)\Omega}}{1 - \mathrm{e}^{-\mathbb{i}\Omega}} + \frac{\mathrm{e}^{\mathbb{i}\Omega} - \mathrm{e}^{\mathbb{i}(N+1)\Omega}}{1 - \mathrm{e}^{\mathbb{i}\Omega}} + 1 \\
&= \frac{\mathrm{e}^{-\mathbb{i}(\Omega/2)} - \mathrm{e}^{-\mathbb{i}[N+(1/2)]\Omega}}{\mathrm{e}^{\mathbb{i}(\Omega/2)} - \mathrm{e}^{-\mathbb{i}(\Omega/2)}} + \frac{-\mathrm{e}^{\mathbb{i}(\Omega/2)} + \mathrm{e}^{\mathbb{i}[N+(1/2)]\Omega}}{-\mathrm{e}^{-\mathbb{i}(\Omega/2)} + \mathrm{e}^{\mathbb{i}(\Omega/2)}} + 1 \\
&= \frac{\mathrm{e}^{\mathbb{i}[N+(1/2)]\Omega} - \mathrm{e}^{-\mathbb{i}[N+(1/2)]\Omega}}{\mathrm{e}^{\mathbb{i}(\Omega/2)} - \mathrm{e}^{-\mathbb{i}(\Omega/2)}} = \frac{\sin[N+(1/2)]\Omega}{\sin(\Omega/2)}.
\end{aligned} \tag{2.116}$$

Note that since the signal in time is symmetric, then its FT is real and symmetric (see Property 3). As can be seen in the equation, there is an indetermination at the origin. The limit of the function when $x \to 0$ is $2[N+(1/2)]$. Note also that the signal is periodic of period 2π. Its representation for $N = 5$ (see Figure 2.8) can be obtained with the code snippet in Listing 2.6.

```
N = 5;
X = sin((N+0.5)*Omega)./sin(Omega/2); % Theoretical sinc
X(end/2) = 2*(N+0.5); % Remove the indetermination at the origin
subplot(211), plot(Omega,X)
% Signal in time domain
x2 = [ones(1,N+1) zeros(1,256-2*N-1), ones(1,N)];
% Its transform
X2 = fftshift(fft(x2));
subplot(212), plot(Omega,X2)
```

Listing 2.6 Example on FT (fftshiftdemo.m).

Assume now a periodic signal $\tilde{x}[n]$ where each period is a version of $x[n]$ delayed lM samples; that is:

$$\tilde{x}[n] = \sum_{l=-\infty}^{\infty} x[n - lM]. \tag{2.117}$$

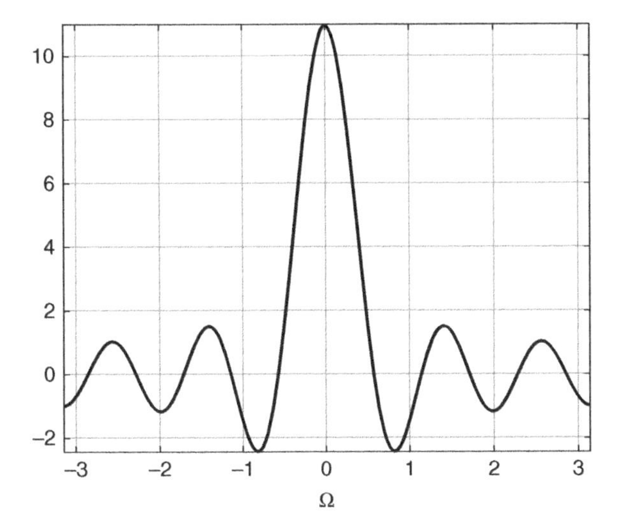

Figure 2.8 FT of a pulse with 256 samples length and five samples width.

By virtue of the convolution and Kronecker delta properties, it can be expressed as

$$\tilde{x}[n] = x[n] * \sum_{l=-\infty}^{\infty} \delta[n - lM]. \tag{2.118}$$

Since the FT of an infinite sequence of deltas can be written as an infinite sequence of deltas (show it as proposed exercise), then the FT of $\tilde{x}[n]$ can be expressed as

$$\tilde{X}(\mathbb{i}\Omega) = X(\mathbb{i}\Omega) \sum_{l=-\infty}^{\infty} \delta\left[\Omega - \frac{2\pi}{lM}\right]. \tag{2.119}$$

This can be interpreted as function $X(\mathbb{i}\Omega)$ being sampled at frequencies $2\pi/lM$. A discrete representation of this function is

$$\tilde{X}(\mathbb{i}\Omega) = X\left(\mathbb{i}\frac{2\pi}{lM}\right), \tag{2.120}$$

which is known as the DFT). For a periodic version of period M of the pulse with width $2N + 1$, the corresponding DFT is

$$X\left(\mathbb{i}\frac{2l\pi}{M}\right) = \frac{\sin[N + (1/2)](2l\pi/M)}{\sin(l\pi/M)}, \tag{2.121}$$

which becomes a single delta at the origin when $N = 2M + 1$.

2.5.3 Continuous-Time Signals and Systems

The MATLAB Symbolic toolbox allows one to work with functions and signals standing for continuous time. A symbolic variable is explicitly declared, and the result of scalar and symbolic variables is another symbolic variable. In Listing 2.7, we see, for instance, how to graphically represent symbolic variables and how to obtain first- and higher-order derivatives easily.

```
syms t
x = sin(2*pi*t);
y = exp(-t/10);
z = x*y;
whos

figure(1), clc,
subplot(3,1,1); ezplot(x,[-10 10]);
subplot(3,1,2); ezplot(y,[-10 10]);
subplot(3,1,3); ezplot(z,[-10 10]);

dx = diff(x);
ezplot(x,[-1 1]);  hold on
ezplot(dx,[-1 1]); hold off
grid on
d2x = diff(x,2)
d3x = diff(x,3)
```

Listing 2.7 Definition and plot of symbolic functions (symbolic.m).

Symbolic expressions are not always easy to read; for instance, consider the signal

$$s(t) = \ln \sqrt{\frac{1+\cos(t)}{1-\cos(t)}}. \tag{2.122}$$

If we calculate and show its derivative, then the result is not readily read. We can use commands `simple` and `pretty` for simplifying and for providing with a readable text form respectively, as seen in Listing 2.8.

```
s = log(sqrt((1+cos(t))/(1-cos(t))))
ds = diff(s)

pretty(ds)
simple(ds)
pretty(simple(ds))
```

Listing 2.8 Example of symbolic simplifications (symbolic2.m).

The integration operator is provided by command `int(Integrand,VarInd,LimInf,LimSup)`, where `Integrand` is the symbolic signal to integrate; `VarInd` is the integration independent variable; `LimInf` and `LimSup` are the limits. Consider, for instance, the simple example

$$\int t \arctan t \, \mathrm{d}t, \tag{2.123}$$

or this, another simple example:

$$\int_0^{\infty} \mathrm{e}^{-3t} \, \mathrm{d}t \tag{2.124}$$

Both are obtained immediately in Listing 2.9.

```
% One example
f = t * atan(t);
F = int(f,t);
pretty(F)

% Another example
g = exp(-3*t);
G = int(g,t,0,Inf);
pretty(F)
```

Listing 2.9 Simple examples of integration (symbolic3.m).

The independent time variable can be readily substituted for signal sampling by means of `subs(x,Expres1,Expres2)` which for expression `x` substitutes the expression `Expres1` by `Expres2` in every case, such as in Listing 2.10.

```
x = 1/t; x2 = subs(x,t,t-2); x3 = subs(x,t,-t);
figure(1), subplot(2,1,1);
ezplot(x); subplot(2,1,2); ezplot(x2);
figure(2), subplot(2,1,1);
ezplot(x); subplot(2,1,2); ezplot(x3);
```

Listing 2.10 Example of substituting the time variable for sampling (symbolic4.m).

As an example, you are invited to write a function generating the following continuous-time signal:

$$x(t) = \mathrm{e}^{-\alpha t}\sin(\omega_0 t)u(t) \tag{2.125}$$

given input parameters for the exponent and for the frequency, and recalling that $u(t)$ is obtained by MATLAB function `heaviside(t)`. Check the effect of the independent variable transformation for $y(t) = x(t-3)$ and $z(t) = y(2t)$, compared with $r(t) = x(2t)$ and $v(t) = r(t-3)$.

As another example, we can visualize the well-known effect of approximating a squared periodic wave with its Fourier series with increasing order. Note the resulting effect of 50 terms with a unit amplitude and period wave, by completing the code in Listing 2.11, where you can analyze the convergence issues problems that are visible through the development.

```
syms k a_k a_kbis serie t
T = 2;
serie = 0;
for k = 1:50
    a_k = 1/T*(int(...,0,1) + int(...,1,2));
    a_kbis = ...;
    serie = serie + a_k * exp(j*2*pi/T*k*t) + ...
                    a_kbis * exp(-j*2*pi/T*k*t);
    serie = simple(serie); % Note the effect of simplifying

    if rem(k,5) == 0;
        ezplot(serie,[-4 4]); legend(num2str(k)); pause
    end
end
```

Listing 2.11 Example to complete: Fourier series (symbolic5.m).

An additional example can be used for providing the continuous-time convolution of elementary theoretical signals; for instance, if we define the function in Listing 2.12, the reader can check it working for the signals

$$x(t) = \mathrm{e}^{-2t}u(t) \tag{2.126}$$

$$v(t) = \mathrm{e}^{-4|t|} \tag{2.127}$$

$$w(t) = (1 - |t|)(u(t+1) - u(t-1)) \tag{2.128}$$

in the following continuous-time convolutions:

$$y_{\mathrm{a}}(t) = x(t) * v(t) \tag{2.129}$$

$$y_{\mathrm{b}}(t) = x(t) * w(t) \tag{2.130}$$

$$y_{\mathrm{c}}(t) = y_{\mathrm{b}}(t) * \delta(t-3) \tag{2.131}$$

```
function y = symbolic6(x,h)

% Example of use on command window:
% syms t
% x = exp(-3*t) .* heaviside(t);
% h = heaviside(t) - heaviside(t-3);
% y = symbolic6(x,h)

syms t tau
x = subs(x,t,tau);
h = subs(h,t,t-tau);
y = int( x.*h, tau, -Inf, Inf);
```

Listing 2.12 Example of symbolic convolutions (symbolic6.m).

Finally, we can check some simple examples for the continuous-time FT with the function in Listing 2.13.

```
function symbolic7

syms t w
x = heaviside(t+2) - heaviside(t-2);
X = fourier(x,t,w);
ezplot(X, [-2*pi,2*pi]), axis tight, grid on
```

Listing 2.13 Example of symbolic FT calculation (symbolic7.m).

Note, however, that symbolic calculations can be limited, even for simple signals involved, and will not always return a nice and explicit result.

2.5.4 Filtering Cardiac Signals

Here, we show some filtering examples in synthetic and in cardiac signals. First, the effect of including additional terms on Fourier series in a cardiac signal is explored. Cardiac signals are not exactly periodic, but rather small changes are present on a beat basis. For this part, you can use the code in Listing 2.14. As seen in Figure 2.9a, we can check

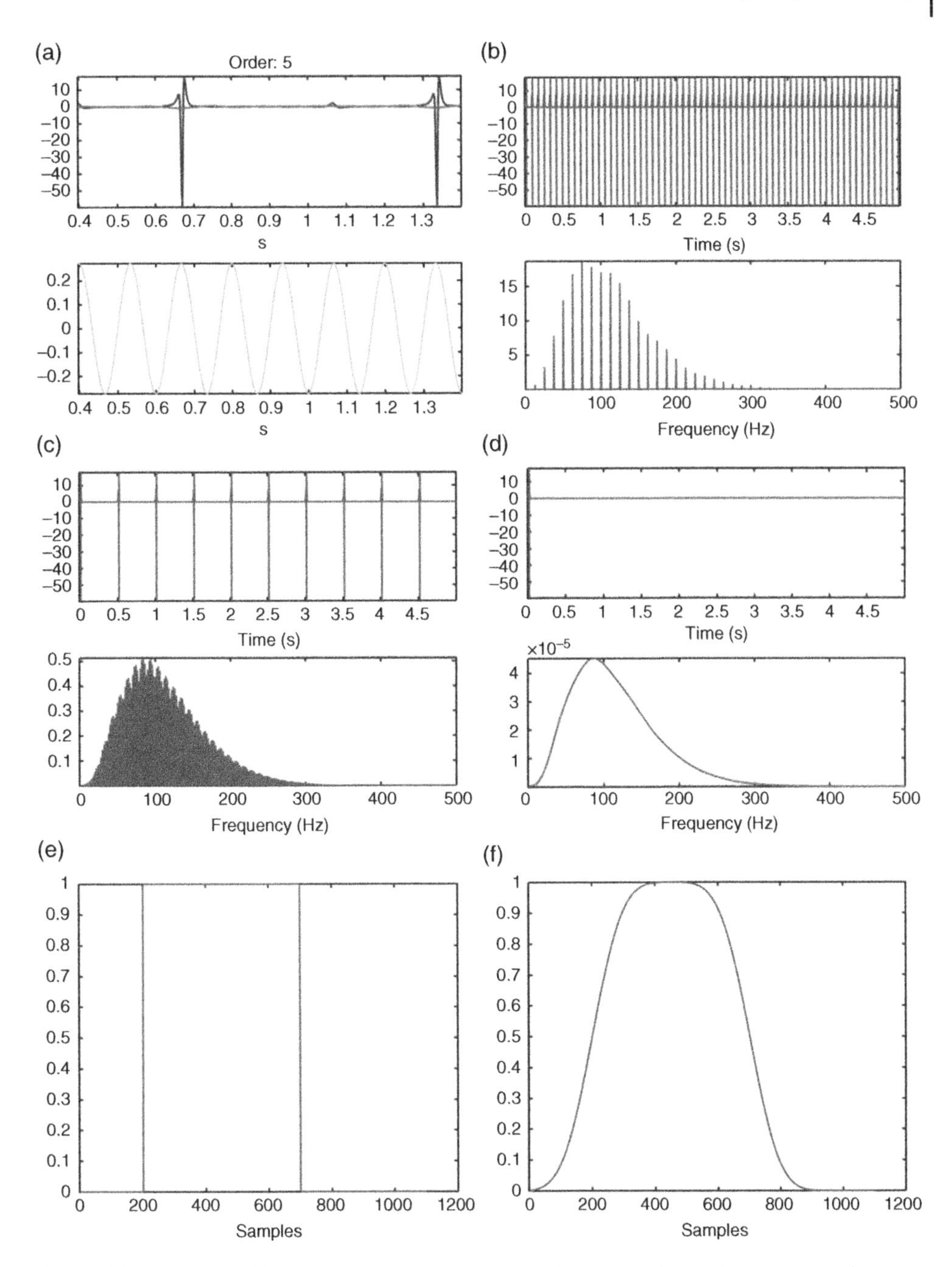

Figure 2.9 Examples for filtering (I). (a) Demo for Fourier series on a cardiac and near-to-periodic signal. (b–d) Demo for taking the infinite limit on a periodic signal and its effect on the Fourier coefficients, leading to the FT. (e,f) Simple examples for low-pass and high-pass filtering on a pulse signal.

that the ventricular activity, given by the most energetic waveform in a given cardiac cycle, is not sufficiently represented until a large enough number of terms are included.

```
load EGM_bipolar
SeriesFourier2(EGM(2,:));
```

Listing 2.14 Example of Fourier series for a cardiac signal (cardiac1.m).

Second, we review the well-known mathematical foundation for introducing the FT from Fourier series. For this purpose, you can use the code in Listing 2.15. In this example, we artificially build a periodic signal by repeating a single cardiac cycle with changing period on each step. As seen in Figure 2.9b–d, a longer signal period accounts for Fourier coefficients on a frequency basis which are getting closer, with an extreme case limit of infinite period corresponding to an aperiodic signal represented by its Fourier coefficients envelope.

```
egm = EGM(2,340:380);
fs = 1e3; ts = 1/fs;
t = (0:5000-1) * ts;
for ciclo = [80 160 500 1000 2000 5000];
    x = zeros(1,5000);
    x(1:ciclo:end) = 1;
    v = filter(egm,1,x);
    subplot(211), plot(t,v),xlabel('time (s)'), axis tight
    subplot(212), [P,f] = pwelch(v,[],length(v)/2,[],fs);
    plot(f,P), axis tight, xlabel('frequency (Hz)'), pause
end
```

Listing 2.15 Example of Fourier series becoming an FT in the limit for a cardiac signal (cardiac2.m).

Third, we examine a simple filtering case for an artificial signal, given Listing 2.16. As seen in Figure 2.9e and f and in Figure 2.10a, the different effects of low-pass, high-pass, and notch filtering can be readily observed.

```
fs = 1e2;
h = fir1(256,1e-9/fs);
x = [zeros(1,2*fs),ones(1,5*fs),zeros(1,4*fs)];
y1 = filtfilt(h,1,x);
[H,f] = freqz(h,1,1024,fs);
plot(f,abs(H));
h2=fir1(128,30/fs,'high');
y2 = filtfilt(h2,1,x);
v1 = exp(1j*pi*50/fs);
v = [v1 conj(v1)]; h3=poly(v);
H = freqz(h3);  h3 = h3/max(abs(H));
n = .1*cos(2*pi*50/200*(1:length(x)));
xx = x+n;
y3 = filtfilt(h3,1,xx);
```

Listing 2.16 Example of filtering synthetic signals (cardiac3.m).

Finally, Figure 2.10b–l shows the effect of low-pass filtering an intracardiac signal for trying to remove its noise, which has been generated with Listing 2.17. The reader is encouraged to modify the text and obtain the results for different cut-off frequencies (in the figure, 75 and 30 Hz have been used).

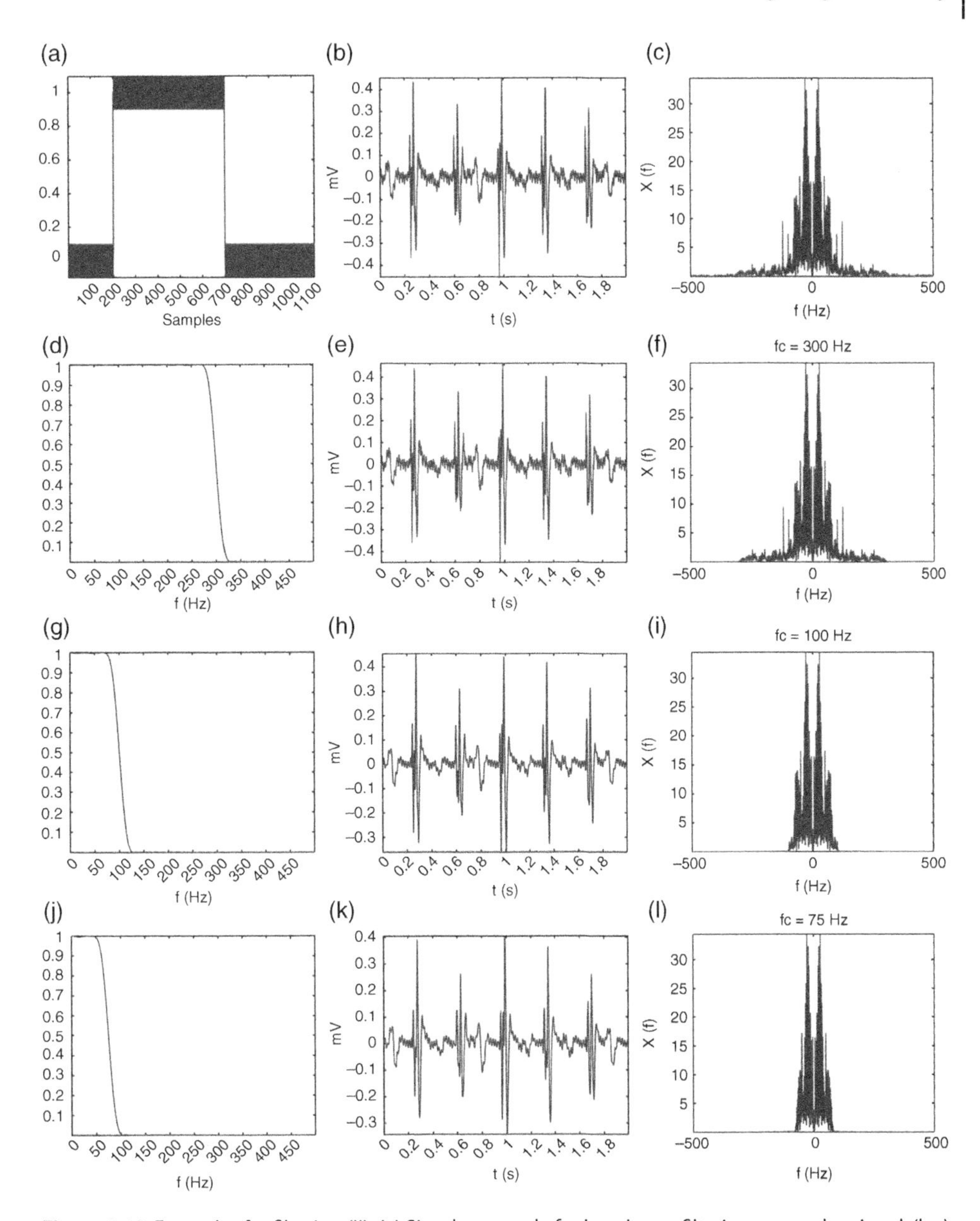

Figure 2.10 Examples for filtering (II). (a) Simple example for band-pass filtering on a pulse signal. (b,c) Examples for filtering intracardiac recorded signals, cardiac signal, and its FT. (d–f) Filtered signal, frequency response of the filter, and resulting spectrum for cutting frequency $f_c = 300$ Hz. (g–l) Same as previous for $f_c = 100$ Hz (g–i) and $f_c = 75$ Hz (j–l).

```
load filtradoegm
fs = 1e3; ts = 1/fs;
N = length(x1);
t = ts*(0:N-1);
figure(1), plot(t,x1); axis tight,
```

```
xlabel('t (s)'), ylabel('mV');
X1 = abs(fftshift(fft(x1)));
f=linspace(-fs/2,fs/2,length(X1));
figure(2), plot(f,X1), axis tight,
xlabel('f (Hz)'), ylabel('X(f)');
% Filter at 300 Hz
h1 = fir1(64+1,2*300/fs);
[H1,f1] = freqz(h1,1,1024,fs);
y1 = filtfilt(h1,1,x1);
figure(3), plot(t,y1);axis tight,
xlabel('t (s)'), ylabel('mV');
figure(4), plot(f1,abs(H1))
axis tight, xlabel('f(Hz)');
Y1 = abs(fftshift(fft(y1)));
figure(5), plot(f,Y1), axis tight,
xlabel('f (Hz)'), ylabel('X(f)');
title('fc = 300 Hz')
```

Listing 2.17 Example of filtering cardiac signals (cardiac4.m).

Note that for an extremely low-pass cutting frequency the noise is removed, but also the morphology of the ventricular activation (sharp peaks) is severely modified, which can have an impact on the diagnosis when it is based on morphological information. The reader is encouraged to represent the FT of the cardiac signal, see its spectrum, and build two simple filters, one as a low-pass and another as a notch filter, for removing the noise while respecting as much as possible the signal morphology.

2.5.5 Nonparametric Spectrum Estimation

This section compares the different nonparametric spectrum estimation approaches presented in Section 2.1.3. The first approach is based on the expectation estimation of the FT of a window of the signal (Welch periodogram); see Equation 2.41. As will become apparent in the experiments, this is a low-resolution procedure, and it will depend on the length of the window. The resolution increases only when the window width increases, which makes this solution impractical where high resolution is needed. The second and third approaches are based on the so-called MVDR algorithm in Equation 2.45, the third one being an approximation based on the eigenvectors of the data autocorrelation matrix in Equation 2.50.

Assume a signal $x[n]$, $1 \leq n \leq N$ with $N = 1024$, containing three sinusoids of different frequencies: $\Omega_1 = 2\pi/10$, $\Omega_2 = 2\pi/20$, and $\Omega_3 = 2\pi/30$. The signal is corrupted with additive white Gaussian noise with standard deviation $\sigma = 0.4$. The code to produce the signal is given in Listing 2.18.

```
N = 1024;
t = 0:N-1;
w1 = 0.3*pi; w2 = 0.7*pi; w3 = 0.8*pi; % Frequencies
x = sin(w1*t) + sin(w2*t) + sin(w3*t); % Signal
x = x + 0.4 * randn(size(x));          % Noise added
```

Listing 2.18 Nonparametric spectrum estimation example: signal with noise (spectrum1.m).

The Welch spectrum is estimated by computing the DFT of all the windows of the signal containing samples from $x[n]$ to $x[n+T-1]$, $1 \leq n \leq N-T+1$. The spectrum of each window is computed as the squared norm of the DFT at each frequency, and

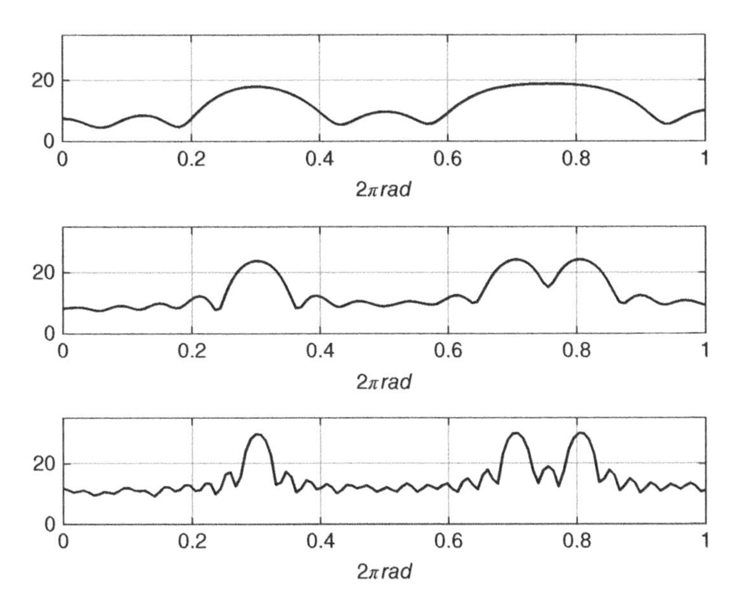

Figure 2.11 Welch spectrum of a signal containing signals of frequencies 0.2π, 0.7π, and 0.8π radians, using windows of 16, 32, and 64 samples.

then all the spectra are averaged. The code in Listing 2.19 performs these operations for a window length of 32 samples.

```
T = 32; % Window length
x2 = buffer(x,T,T-1,'nodelay'); % Windowed signal vectors
X = abs(fft(x2,256)).^2; % FFT
X = mean(X,2); % Expectation
figure(1), plot(linspace(0,1,128),10*log10(X(1:end/2)))
```

Listing 2.19 Nonparametric spectrum estimation example: windowing and averaging the spectra (spectrum2.m).

Function `buffer(T,T-K)` constructs a matrix whose columns contain windows of length T, each one with a delay of K samples from the previous one. Figure 2.11 depicts the Welch spectra for windows of 16, 32, and 64 samples. As a result, when the window is of 16 samples, the algorithm cannot resolve the two closest frequencies.

The MVDR spectrum is computed by using the autocorrelation of the windowed signals; that is, computing the sample expectation:

$$\hat{\boldsymbol{R}}_{xx} = \frac{1}{N} \sum_{n=1}^{N-T+1} \boldsymbol{x}[n]\boldsymbol{x}^{\mathrm{H}}[n] \tag{2.132}$$

where $\boldsymbol{x}[n] = [x[n] \cdots x[n+T-1]]^{\mathrm{T}}$. The spectrum at a given frequency Ω is computed as the inverse of dot product $\boldsymbol{v}(\Omega)^{\mathrm{H}}\hat{\boldsymbol{R}}_{xx}^{-1}\boldsymbol{v}(\Omega)$. Provided that $\boldsymbol{R}_{xx}$ is Hermitian (i.e., $\boldsymbol{R}_{xx} = \boldsymbol{R}_{xx}^{\mathrm{H}}$), it can be proven (exercise) that the product is simply the sum of the DFT of all the rows or the columns of the autocorrelation matrix. Then, a simple code that produces the spectrum is given in Listing 2.20. A representation of the MVDR spectrum for three different window lengths is shown in Figure 2.12.

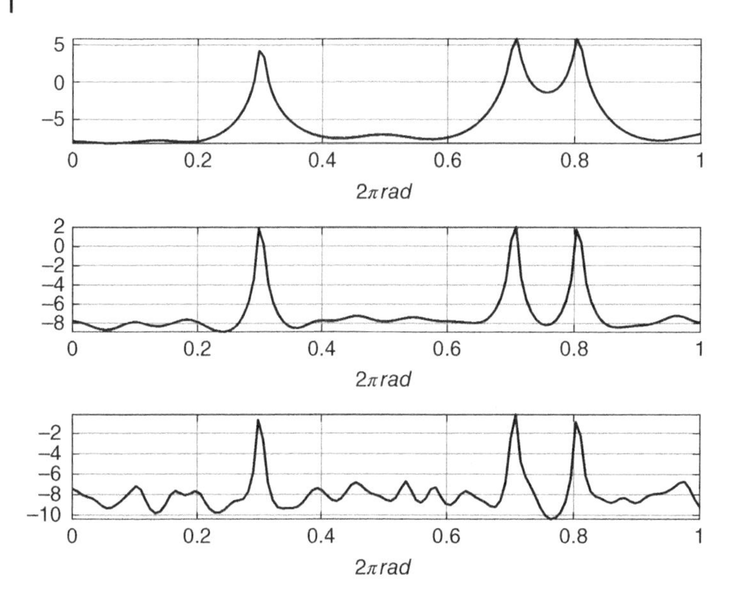

Figure 2.12 MVDR spectrum of a signal containing signals of frequencies 0.2π, 0.7π and 0.8π radians, using windows of 16, 32, and 64 samples.

```
x2 = buffer(x,T,T-1,'nodelay'); % Windowed signal vectors
R = x2 * x2' / size(x2,2);
X = 1./sum(abs(fft(pinv(R),256)),2);
plot(linspace(0,1,128),10*log10(X(1:end/2)))
```

Listing 2.20 Example of MVDR spectrum estimation (spectrum3.m).

Similarly, the MUSIC spectrum is estimated by computing the inverse of the sum of the DFTs of all columns of matrix $\boldsymbol{Q}_\mathrm{n}\boldsymbol{Q}_\mathrm{n}^\mathrm{H}$. The noise eigenvectors in matrix $\boldsymbol{Q}_\mathrm{n}$ are found by excluding the eigenvectors with highest eigenvalues. The signal of the experiment has three real sinusoids. A real sinusoidal is a linear combination of two complex exponentials, since

$$\mathrm{e}^{\mathbb{i}\Omega n} = \cos(\Omega n) + \mathbb{i}\sin(\Omega n), \tag{2.133}$$

so the cosine function is expressed as $(1/2)\exp(\mathbb{i}\Omega n) + (1/2)\exp(-\mathbb{i}\Omega n)$, and similarly for the sine function. Hence, the three signals are represented by six complex exponentials, or, in other words, six eigenvectors. Then, the noise eigenvectors are obtained here by removing the six eigenvectors with higher eigenvalues. The corresponding simple code is given in Listing 2.21, and Figure 2.13 shows the MUSIC spectrum for three different window lengths.

```
x2 = buffer(x,T,T-1,'nodelay'); % Windowed signal vectors
R = x2 * x2' / 1024;
[Q,L] = eig(R);
X = 1./sum(abs(fft(Q(:,1:end-6) * Q(:,1:end-6)',256)),2);
plot(linspace(0,1,128),10*log10(X(1:end/2)))
```

Listing 2.21 Example of MUSIC spectrum estimation (spectrum4.m).

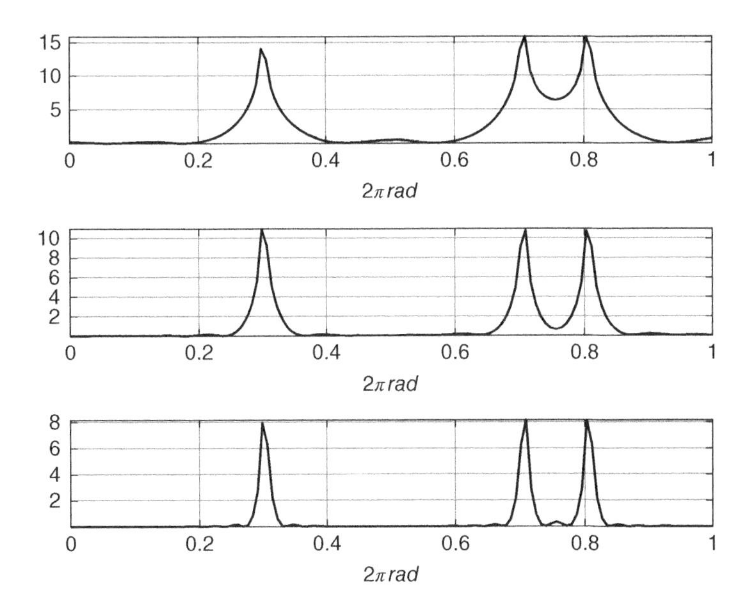

Figure 2.13 MUSIC spectrum of a signal containing signals of frequencies 0.2π, 0.7π and 0.8π radians, using windows of 16, 32, and 64 samples.

2.5.6 Parametric Spectrum Estimation

An example of parametric estimation is the previously described AR method of Equation 2.44. The goal here is to construct an AR model of the signal whose parameters are those of the AR model in Equation 2.43. The model can be constructed by minimizing a cost function over the error:

$$e[n] = x[n] - \sum_{m=1}^{P} a_m x[n-m]. \tag{2.134}$$

If the MMSE criterion is chosen, and defining $\boldsymbol{e} := [e_1, \ldots, e_N]^{\mathrm{T}}$, then the cost function is

$$\mathbb{E}[\|\boldsymbol{e}\|^2] = \mathbb{E}\left[\left|x[n] - \sum_{m=1}^{P} a_m x[n-m]\right|^2\right]. \tag{2.135}$$

Assuming that the signal is stationary, the expectation can be approximated by a sample average as

$$\mathbb{E}[\|\boldsymbol{e}\|^2] \approx \sum_{n=P+1}^{N} \left|x[n] - \sum_{k=1}^{P} a_k x[n-k]\right|^2. \tag{2.136}$$

We can write the following set of linear equations:

$$\begin{pmatrix} x[P+1] \\ x[p+2] \\ \vdots \\ x[N] \end{pmatrix} - \begin{pmatrix} a_1 \\ a_2 \\ \vdots \\ a_p \end{pmatrix}^{\mathrm{T}} \begin{pmatrix} x[1] & \cdots & x[N-P] \\ x[2] & \cdots & x[N-P+1] \\ \vdots & \cdots & \vdots \\ x[P] & \cdots & x[N-1] \end{pmatrix} = \begin{pmatrix} e[1] \\ e[2] \\ \vdots \\ e[N] \end{pmatrix} \tag{2.137}$$

that can be expressed in matrix form as

$$\boldsymbol{x} - \boldsymbol{a}^{\mathrm{T}}\boldsymbol{X} = \boldsymbol{e}. \tag{2.138}$$

The minimization of Equation 2.136 is equivalent to the minimization of the norm of vector $\boldsymbol{e}$ with respect to $\boldsymbol{a}$, which leads to the solution

$$\boldsymbol{a} = (\boldsymbol{X}\boldsymbol{X}^{\mathrm{T}})^{-1}\boldsymbol{X}\boldsymbol{x}, \tag{2.139}$$

whose proof is left as an exercise for the reader.

Nevertheless, this solution is not convenient since it does not guarantee that the poles in Equation 2.43 are inside the unit circle. Thus, the corresponding estimator is not necessarily stable. A solution that minimizes the mean square error (MSE) with the constraint of providing a stable solution for the AR estimator is the Burg algorithm (Marple, 1987), which is based on a recursive computation of the model parameters.

In the following example, we use a signal containing a fragment of the "Etude for guitar No. 19 in E minor" by Spanish guitar composer Francisco de Aís Tárrega i Eixea (Castellón de la Plana, Spain, 1852 – Barcelona, Spain, 1909). The example shows the spectrum computed with the estimation provided in Equation 2.139. The spectrum described in Section 2.5.5 can be computed as follows. We first construct a matrix $\boldsymbol{x}$ whose column vectors take the form

$$\begin{aligned} \boldsymbol{x}[0] &= [x[1], \dots, x[P]]^{\mathrm{T}} \\ \boldsymbol{x}[1] &= [x[P-W], \dots, x[2P-W]]^{\mathrm{T}} \\ &\cdots \\ \boldsymbol{x}[k] &= [x[kP-kW], \dots, x[2kP-W]]^{\mathrm{T}} \end{aligned}$$

Therefore, the matrix contains windows of the signal of length P, and there is an overlap W between consecutive windows. Each column is multiplied by a window. In our example we choose rectangular, triangular, Hamming, and Blackman windows. The windowing helps to reduce the side lobes of the FFT. The code can be written as seen in Listing 2.22.

```
function param1
% Load the samples, contained in variable 'data'
% with frequency fs=11025 samples/second
load guitar; y=data;

%% Variable setup
ts = 1/fs;           % sampling period
N = length(y);       % signal length
```

```
tw = 0.025;          % window length in seconds
to = 0.0125;         % window overlap in seconds
nw = floor(tw / ts)+1;   % window length in samples
no = floor(to / ts)+1;   % window overlap in samples
res1 = fs / N;       % Signal resolution
res2 = fs / nw;      % Window resolution
Nt   = floor(N/no); % Time length of the spectrogram
% in samples
Nw   = nw/2;          % Frequency length onf the spectrogram
% in samples
t = no/fs*(0:Nt-1); % time axis for the spectrum
tt = 1/fs*(0:N-1);  % time axis for the time representation
f = fs*(0:Nw-1)/2/Nw; %frequency axis

%% Signal representation
figure(1), plot(tt,y);
axis tight, xlabel('t(s)'), ylabel('mV')

%% Windows used for the representation
wwT = repmat(triang(nw),1,Nt);
wwH = repmat(hamming(nw),1,Nt);
wwB = repmat(blackman(nw),1,Nt);

%% Signal buffering and windowing
s = buffer(y,nw,no,'nodelay');
sT = s .* wwT;
sH = s .* wwH;
sB = s .* wwB;

%% Spectrogram computation
S = spectrogr(s);
ST = spectrogr(sT);
SH = spectrogr(sH);
SB = spectrogr(sB);

%% Representation
figure(2), subplot(211), imagesc(t,f,S);
xlabel('t(s)'), ylabel('dBm')
title('Spectrogram with a rectangular window')
subplot(212), imagesc(t,f,ST);
xlabel('t(s)'), ylabel('dBm')
title('Spectrogram with a triangular window')
figure(3), subplot(211), imagesc(t,f,SH);
xlabel('t(s)'), ylabel('dBm')
title('Spectrogram with a Hamming window')
subplot(212), imagesc(t,f,SB);
xlabel('t(s)'), ylabel('dBm')
title('Spectrogram with a Blackman window')

%% Auxiliary function
function S = spectrogr(s,Nw)
Nw = size(s,1) / 2;
S = 10 * log10(abs(fftshift(fft(s))));
S = S(Nw+1:2*Nw,:);
```

Listing 2.22 Signal windowing (param1.m).

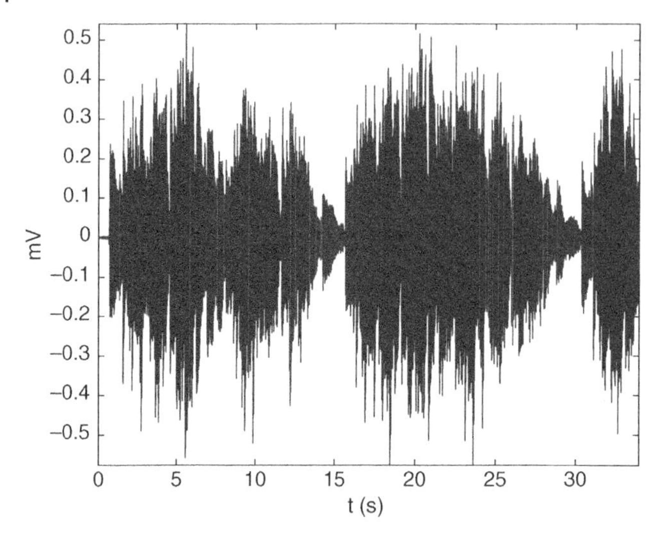

Figure 2.14 Time representation of the analyzed signal.

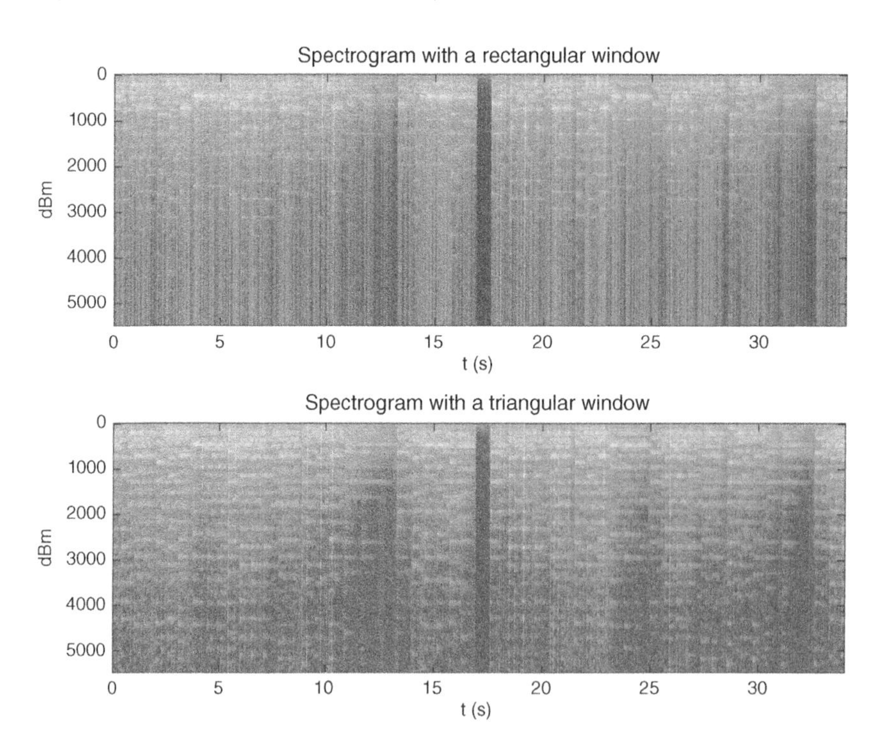

Figure 2.15 Upper panel: spectrogram of the signal with a rectangular windowing. Lower panel: spectrogram with a triangular window.

The signal has a sampling frequency of 11 025 samples per second, and thus the spectrum representation ranges from 0 to 55 125 Hz. Figure 2.14 shows the time representation of the signal. Figures 2.15 and 2.16 show the spectrograms of the signal with different windows. In these figures, the peaks correspond to the pulsation of

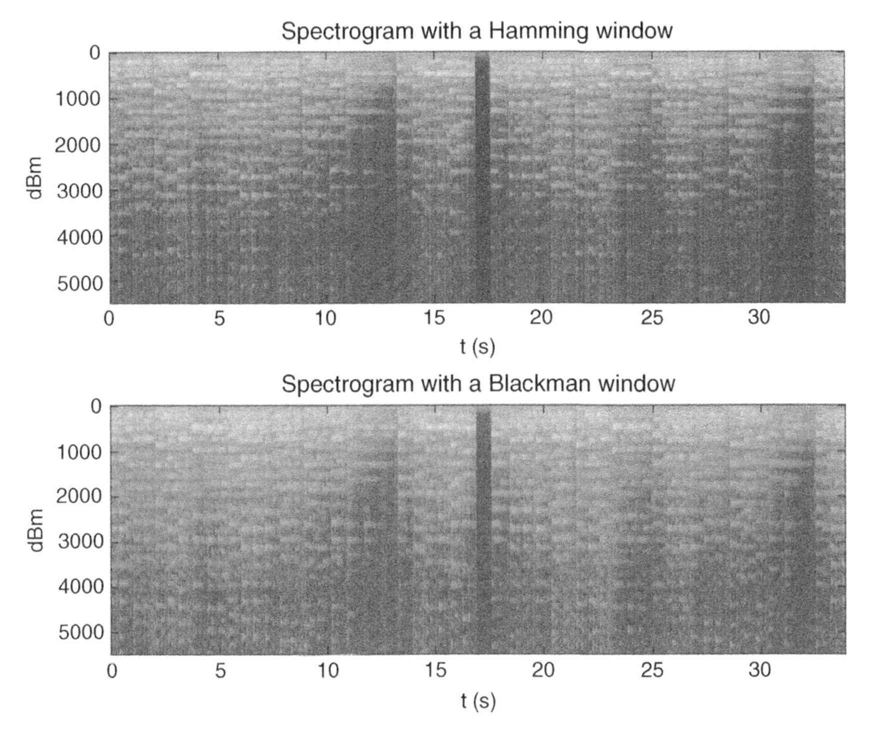

Figure 2.16 Upper panel: spectrogram of the signal with a Hamming windowing. Lower panel: spectrogram with a Blackman window.

the different notes of the music in the guitar. The rectangular window exhibits better frequency resolution, while the rest of the windows exhibit lower resolution, but the peaks can actually be better seen thanks to the reduction of the secondary lobes.

2.5.7 Source Separation

The following example uses the well-known FastICA algorithm for blind source separation. The example is reproduced from the one in Murphy (2012), but using a simplified algorithm for demonstration purposes.

The simulated scenario consists of a set of four sources, one of them being white noise. The sources are supposed to be in different places of the space and they are recorded by four sensors. The recorded signals are simulated by a 4×4 mixing matrix, given by

$$\boldsymbol{W} = \begin{pmatrix} 1 & 1 & 0.5 & 0.2 \\ 0.2 & 1 & 1 & 0.5 \\ 0.2 & 1 & 1.4 & 0.4 \\ 1 & 0.3 & 0.5 & 0.5 \end{pmatrix} \tag{2.140}$$

The observed data are then

$$\boldsymbol{x}[n] = \boldsymbol{W}\boldsymbol{z}[n] \tag{2.141}$$

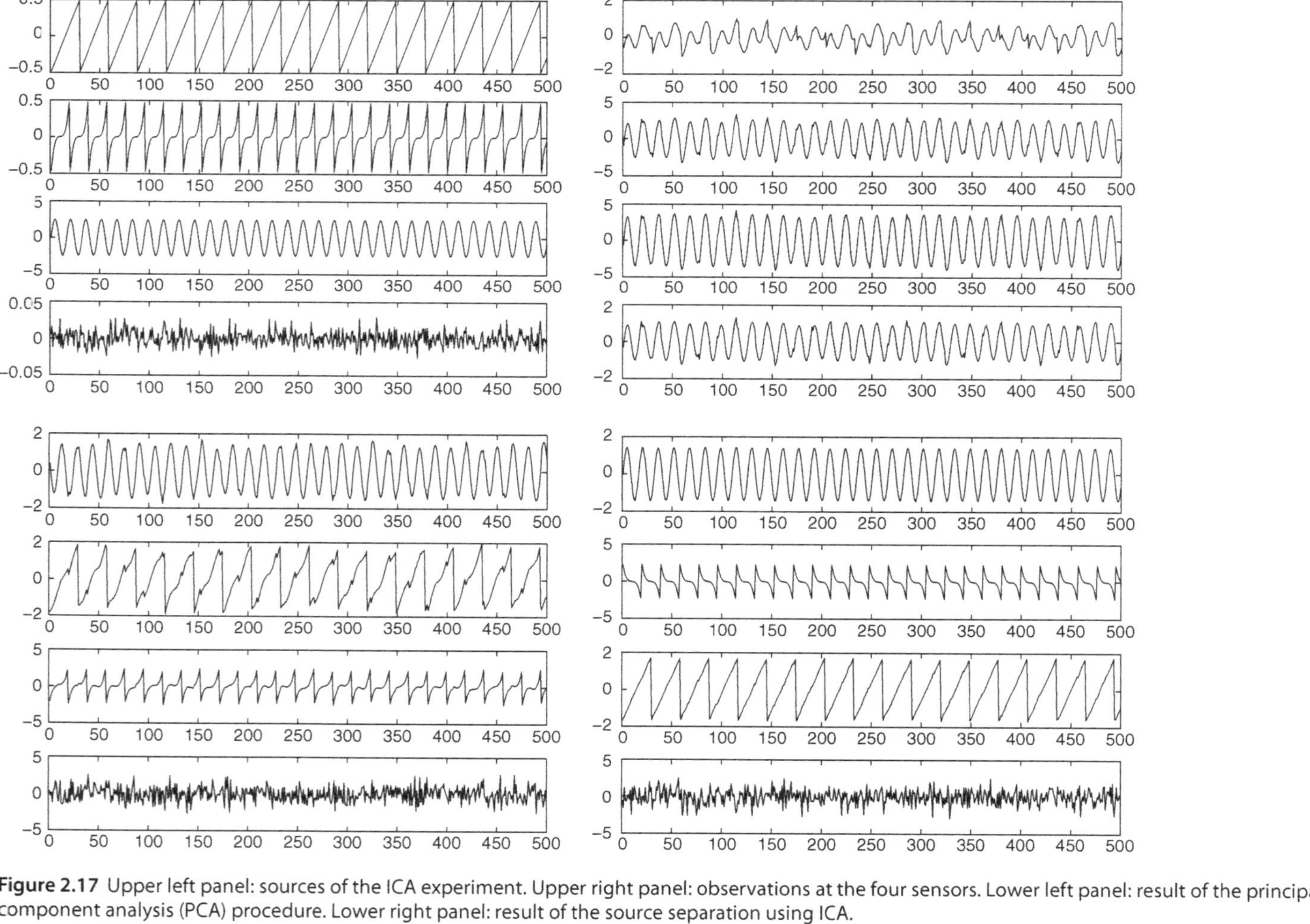

Figure 2.17 Upper left panel: sources of the ICA experiment. Upper right panel: observations at the four sensors. Lower left panel: result of the principal component analysis (PCA) procedure. Lower right panel: result of the source separation using ICA.

where $\boldsymbol{z}[n]$ is the vector of samples of the four sources. Then, the source separation task consists of the estimation of matrix $\boldsymbol{V} = \boldsymbol{W}^{-1}$. The corresponding MATLAB code for the data generation is given in Listing 2.23. Figure 2.17 (upper left panel) shows the data sources $z_i[n]$, and the upper right panel shows the mixed signals $x_i[n]$ received by the sensors.

```
x  = (-14:14) / 29;
z1 = kron(ones(1,18),x); z1 = z1(1:500);
x  = (-9:9) / 25;
z2 = kron(ones(1,28),x.^3)/0.1; z2 = z2(1:500);
z3 = 5 * sin(2*pi*32*(0:499)/500)/2;
z4 = 0.01 * randn(1,500);
W  = [1 1 0.5 0.2;0.2 1 1 0.5;0.2 1 1.4 0.4; 1 0.3 0.5 0.5];
X  = W' * [z1;z2;z3;z4];
```

Listing 2.23 Data generation for BSS example (bss1.m).

The ICA approach assumes that the data is non-Gaussian, so the observations have to be be whitened using a PCA. After that, the covariance of the observations is given by

$$\mathrm{Cov}[\boldsymbol{x}] = \boldsymbol{W}\boldsymbol{W}^{\mathrm{T}}. \tag{2.142}$$

The ICA procedure consists of computing the log likelihood of matrix $\boldsymbol{V}$ according to a set of prior probability distribution models for the source signals, and then maximizing them. The FastICA approach assumes that vectors in matrix $\boldsymbol{V}$ are orthonormal and adds it as a constraint of the optimization. It iteratively maximizes the log likelihood by using its gradient and Hessian. The code used in the experiments is given in Listing 2.24. The code iteratively updates the values of the vectors and forces their orthonormality by using a Gram–Schmidt orthonormalization. Figure 2.17 shows the results. The left lower panel shows the PCA approach for source separation and the left lower one shows the result of the ICA algorithm.

```
function [V Z iter L] = bss2(X,epsilon)
[Dim Ndata] = size(X);
V = randn(Dim);
L = [];
for j = 1:Dim
    current = V(:,j);
    last = zeros(Dim,1);
    iter_max = 40;
    iter(j) = 0;
    while ~(abs(current-last)<epsilon)
        last = V(:,j);
        V(:,j) = 1 / Ndata * X * tanh(V(:,j)'*X)' - 1 / Ndata *
            sum((1-tanh(V(:,j)'*X).^2)) * V(:,j);
        V(:,1:j) = gramsmith(V(:,1:j));
        current = V(:,j);
        P = -log(cosh(V'*X));
        L = [L 1/Ndata*sum(P(:))];
        iter(j) = iter(j)+1;
        if iter(j) > iter_max
            break
```

```
            end
        end
end
Z = V' * X;
function V = gramsmith(V)
for i = 1:size(V,2)
        V(:,i) = V(:,i) - V(:,1:i-1) * V(:,1:i-1)' * V(:,i);
        V(:,i) = V(:,i) / norm(V(:,i));
end
```

Listing 2.24 BSS example with iterative simple algorithm (bss2.m).

2.5.8 Time–Frequency Representations and Wavelets

As stated in Section 2.2.2, the DWT decomposes the signal using a cascade of low-pass and high-pass filters. Wavelets have revolutionized the field of signal processing, image processing, and computer vision in recent decades. Many problems exploit these data representations to analyze and process signals of any kind, from time series to images and videos. Being an invertible transformation, one can do smart operations in the transformed domain and invert back the result to the original domain.

In what follows, we will just illustrate this as a naive suppression of signal/image features, but one could actually combine components, downweight others, and think of many alternative analyses. Actually, defining appropriate operations in wavelet representations is a whole field in its own right. We can currently find in the literature lots of algorithms for image restoration, image compression, image fusion, and object detection and recognition.

Wavelets for Signal Decomposition

Let us start with a very simple example of 1D signal decomposition. One example of how this can be used is in the decomposition of the observed (and possibly noise-corrupted) signal into its signal and noise components. We will illustrate the use of (time–frequency) representations, and in particular of wavelets, in both signal and image processing.

Let us assume first we have a simple observed signal composed of a low-frequency sinusoid buried on high-frequency Gaussian noise of low variance, $\sigma_\mathrm{n}^2 = 0.25$. Such a signal can be generated and represented with the code in Listing 2.25. We can use the DWT with only two scales of decomposition to separate high- and low-frequency components, as seen in the code, where `cA` and `cD` are the low and high DWT frequency coefficients respectively. Now, we can use these coefficients to perform a full or a partial reconstruction of the signal, since we are using an orthogonal (hence invertible) wavelet transform. Finally, we represent and compare the three reconstructions.

```
% Signal generation
N = 1000;
n = linspace(0,pi,N);
s = cos(20.*n) + 0.5.*rand(1,N);
figure(1), clf, subplot(221),
plot(n,s), grid on, axis tight
title('Original signal')
% Two scales decomposition
[cA,cD] = dwt(s,'db2');
```

```
ssf = idwt(cA,cD,'db2');                   % Full reconstruction
ssl = idwt(cA,zeros(1,N/2+1),'db2');       % Inverse using LF
ssh = idwt(zeros(1,N/2+1),cD,'db2');       % Inverse using HF
% Representations
subplot(222), plot(n,ssf)
grid on, axis tight, title('Reconstructed signal')
subplot(223), plot(n,ssl)
grid on, axis tight, title('Low freq. reconstruction')
subplot(224), plot(n,ssh)
grid on, axis tight, title('High freq. reconstruction')
```

Listing 2.25 Simple example for DWT signal decomposition (wdt1.m).

The results are shown in Figure 2.18, where the full observation, its wavelet decomposition in low- and high-frequency components, and the reconstructed signals are represented.

Wavelets for Image Decomposition

Let us now illustrate the use of wavelets in a 2D example, so that the signal is now a grayscale image. Images and time series are both structured domains with clear correlations (either in time or space) among nearby samples. Such correlations may

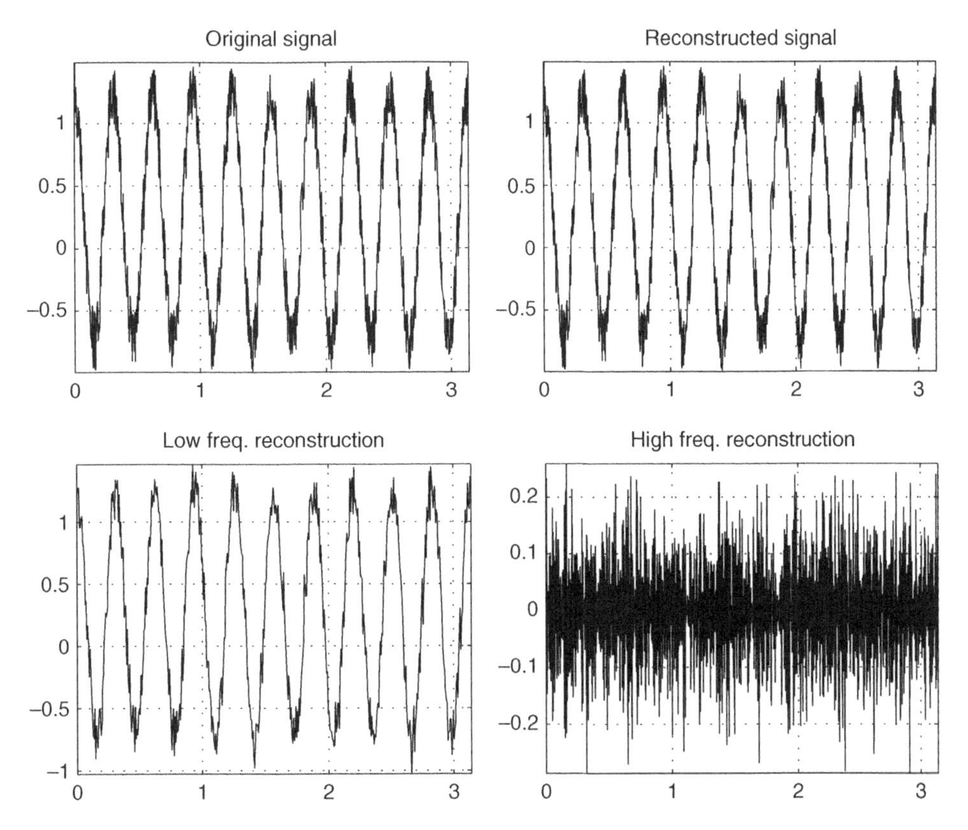

Figure 2.18 Full, low-only and high-only frequency reconstructions of a low-frequency sinusoid signal contaminated with high-frequency noise using DWT and inverse DWT.

Figure 2.19 Illustration of the use of wavelet-based image decomposition (left) and the reconstruction error when the horizontal component is removed (right).

occur at particular (frequency) scales, but in the case of images the correlations also arise at *spatial orientations*. Therefore, a wavelet decomposition of an image is an alternative representation composed of a number of scales and orientations. Typically, only three orientations are considered (horizontal, vertical, and diagonal), but there is a plethora of wavelet transformations accounting for more angular scales.

In this illustrative case, we just plan to take an image and represent the wavelet coefficients, to do a simple operation there by suppressing some scales, and to invert back the result, as seen in Figure 2.19. The wavelet decomposition of the selected standard image of `barbara` reveals relevant information about the image structure by highlighting horizontal, vertical, or diagonal edges in the corresponding scale only. Actually, by setting the horizontal coefficients to zero and inverting back, the reconstruction error concentrates in such image features (i.e., horizontal edges are affected by large errors). The MATLAB code to reproduce this simple example of image decomposition is given in Listing 2.26.

```
load barbara512
% Single-level decomposition of X of a particular wavelet
wname = 'db1'; % wname = 'sym4';
[CA,CH,CV,CD] = dwt2(X,wname);
decompo = [ CA/max(max(CA)), CH/max(max(CH));
            CV/max(max(CV)), CD/max(max(CD)) ];
figure, imshow(decompo,[])
% Let's remove the horizontal coefficients and invert the representation
CH = zeros(size(CH));
Xr = idwt2(CA,CH,CV,CD,wname);
% Plot the reconstruction error
figure, imshow(abs(X-Xr),[]);
```

Listing 2.26 Example of DTW for image decomposition (wdt2.m).

Figure 2.20 Illustration of the use of wavelet decomposition for image fusion.

Wavelets for Image Fusion

Another example of the use of wavelet decomposition is the field of image fusion, in which two (distorted) acquisitions of the same scene are decomposed, only the interesting components are smartly combined in the (wavelet) representation space, and then inverted back to the image representation space.

Figure 2.20 shows an example of this process of image fusion with wavelets. Essentially, we merge the two images from wavelet decompositions at level 5 by taking the maximum of absolute value of the coefficients, for both approximations and details components. The code reproducing this is given in Listing 2.27.

```
% Load two fuzzy versions of an original image
load checkboard X1 X2
% Merge two images from wavelet decompositions at level 5
% using sym4 by taking the maximum of absolute value of the
% coefficients for both approximations and details
XFUS = wfusimg(X1,X2,'sym4',5,'max','max');
% Plot original and synthesized images
figure, imshow(X1), axis square off, title('Catherine 1')
figure, imshow(X2), axis square off, title('Catherine 2')
figure, imshow(XFUS), axis square off, title('Synthesized image')
```

Listing 2.27 Example of DWT for image fusion (wdt3.m).

2.5.9 Examples for Spectral Analysis on Manifolds

In this section, the spectral analysis of a mesh is presented by using a combinational graph Laplacian and a geometric mesh Laplacian. First, an example of the graph Laplacian shows the transformation from the spatial domain to the frequency domain and the transformation back to the spatial domain. In addition, a mesh deformation is done by using this frequency transformation. The second example uses a geometric Laplacian to filter a noisy mesh. Finally, the graph Laplacian and the geometric Laplacian are compared by computing the mean absolute error (MAE) during the reconstruction process.

Let the 2-manifold triangular mesh be $M = (V, F, E)$; the graph Laplacian of this mesh can be computed by

$$\boldsymbol{L}_{ij} = \begin{cases} d_i & \text{if } i = j \\ -1 & \text{if } (i,j) \text{ is an edge} \\ 0 & \text{otherwise,} \end{cases} \tag{2.143}$$

where $d_i = |N(\boldsymbol{v}_i)|$ is the number of vertices in a one-ring first-order neighbor. Here, we use a tear-shaped mesh of 482 vertices. The reader can access a variety of meshes in Stanford University (2016a,b). The eigendecomposition of the graph Lagrangian $\boldsymbol{L}$ provides the eigenvectors (basis) and the eigenvalues (spectrum) of the mesh. Lower eigenvalues (lower frequencies) correspond to first basis components.

The code for the spectral analysis is given in Listing 2.28, and the code for graph_Laplacian.m is in Listing 2.29.

Figure 2.21 shows the basis (eigenvectors) projected on the surface of the mesh. Note that the first eigenvector is constant and the variation of the basis (frequency) increases when the a higher basis is considered. As was defined in Section 2.4, the spectral coefficients can be obtained by using Equation 2.98 and the reconstruction of the original mesh with Equation 2.99.

```
% Load the vertex and faces matrices
load vertex;     load faces;
n = length(vertex);

% The Laplacian is computed
L = graph_Laplacian(n,faces);

% Eigendecompositon of K:
[H,A]  = eig(L);

% Plot the eigenvectors on the mesh surface
figure;
selec_basis =  [1:5 20 50 100 200 400];
for i=1:length(selec_basis)
    subplot(2,5,i);
    h = patch('vertices',vertex,'faces',faces,...
        'FaceVertexCData', H(:,selec_basis(i)),  ...
    'FaceColor','interp');
    title(['H_{' num2str(selec_basis(i))  '}'],'FontSize', 20)
    caxis([-0.05 0.05]);
    axis([20    60    -30    60])
    lighting phong;     camlight('right');     camproj('perspective');
    shading interp;     axis off;              colormap('jet')
end

% Spectral coefficients
a = H'*vertex;
figure;
plot(a,'Linewidth',2);  axis tight; title('Spectral coefficients');
legend('a_x coefficients','a_y coefficients','a_z coefficients');

% Mesh reconstruction
figure; colormap gray(256);
selec_basis =  [1:5 50 200 482];
for i=1:length(selec_basis)
    subplot(2,4,i);
    x_hat  = H(:,1:selec_basis(i))*a(1:selec_basis(i),:);
    if selec_basis(i)>2
        patch('vertices',x_hat,'faces',faces,  ...
        'FaceVertexCData',[0.5 0.5 0.5],'FaceColor','flat');
        title(['H_1 - H_{' num2str(selec_basis(i))  '}'])
```

```
        axis([20   60   -30    60]);    axis off;
        lighting phong;     camlight('right');   camproj('perspective');
    else
        plot3(x_hat(:,1),x_hat(:,2),x_hat(:,3),'k.-');
        title('H_1','FontSize', 20);    axis off;
    end
end
```

Listing 2.28 Basis projected on the surface of the mesh (MHexamples1.m).

```
function L = graph_Laplacian(n,faces)

% Firstly, the edges and their indices are obtained from the face matrix
edges = [faces(:,1) faces(:,2); faces(:,1) faces(:,3);
         faces(:,2) faces(:,3); faces(:,2) faces(:,1);
         faces(:,3) faces(:,1); faces(:,3) faces(:,2)];
indices = (edges(:,2)-1)*n + edges(:,1);

% Adjacency matrix
W = zeros(n,n);
W(indices) = 1;

% Degree matrix
D = diag(sum(W,2));

% Graph Laplacian
L = D - W;
```

Listing 2.29 Graph Laplacian code (graph_Laplacian.m).

Figure 2.22 shows the process of reconstruction when an increased number of components are included in the reconstruction. Note that all vertex positions are

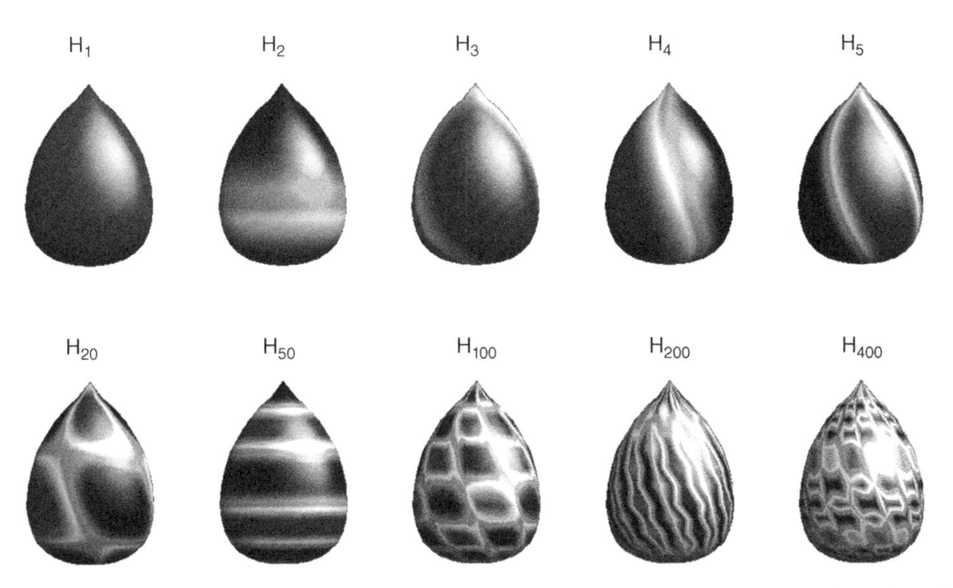

Figure 2.21 Eigenvectors (basis) of the tear-shaped mesh graph Laplacian projected on the surface of this mesh.

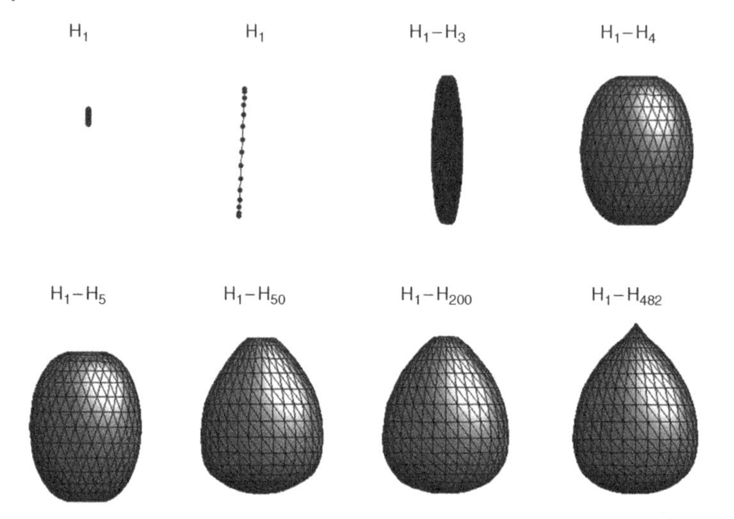

Figure 2.22 Reconstruction of the tear-shaped mesh from the eigendecomposition of the graph Laplacian.

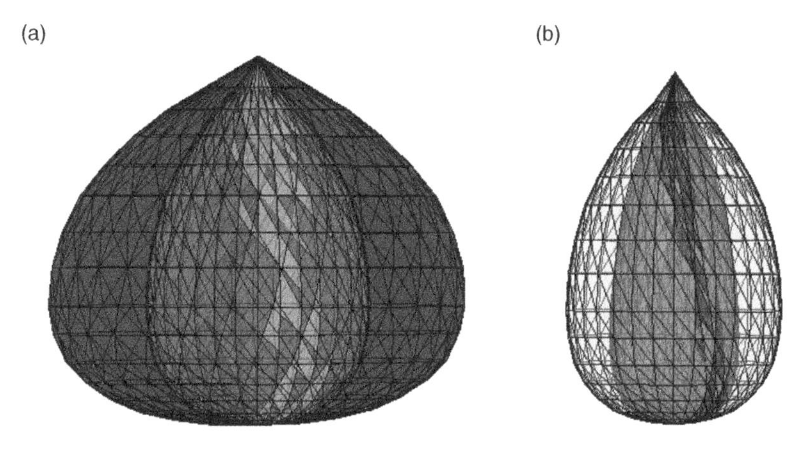

Figure 2.23 The green mesh is the original tear-shaped mesh and the red mesh is the modified one by increasing the fourth spectral component (a) or decreasing the fourth spectral component (b).

concentrated in small area when only the first component, which corresponds to the constant eigenvector, is considered for the reconstruction. If one more component (two components in total) is included in the reconstruction, the vertex positions are distributed along a line. For the first three components, the vertices form a plane, and finally the 3D mesh is obtained when four components are used. When more components are included in the reconstruction, a more detailed mesh is obtained; hence, the last components with no additional information or noise could be discarded for a compression or denoising application.

As with filtering in signal processing, it is also possible to modify the geometry of the mesh by applying a filter in the frequency domain. For example, Figure 2.23 shows

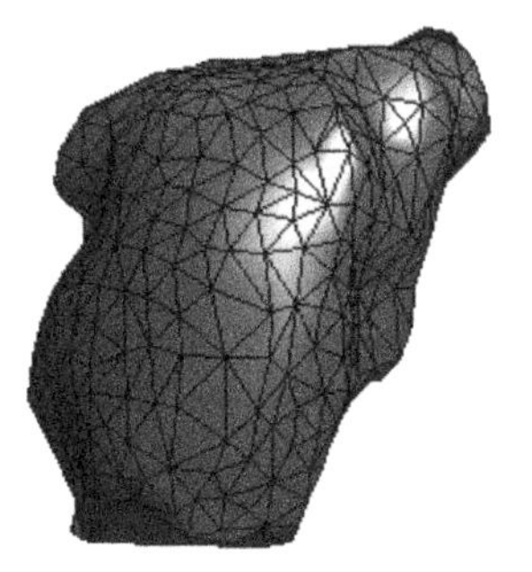
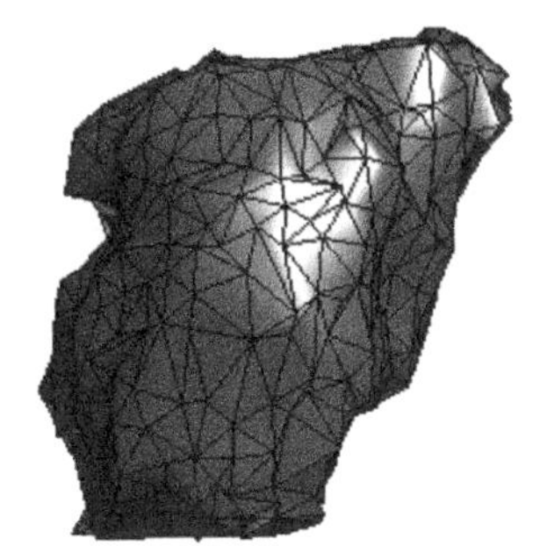

Figure 2.24 Left atrial mesh and its noisy version.

the modification of the fourth spectral component (equivalent to a notch filter in the frequency associated with the fourth component) by increasing it with a value of 2 (Figure 2.23a), or decreasing with a value of 0.5 (Figure 2.23b). These modifications respectively widen or narrow the tear-shape mesh. The additional code for the mesh modification is given in Listing 2.30.

```
% Filter
filtcoef = ones(size(vertex));
filtcoef(4,:)=2;

% Reconstructed mesh
x_hat_n = H*(filtcoef.*a);
```

Listing 2.30 Code for the mesh modification (MHexamples2.m).

As pointed out in the previous example, the mesh compression can be obtained by just taking the first components which provide the shape of the mesh. This compression can also be considered as a low-pass filtering procedure. Now, we explore further this filtering feature by using a more complex mesh. The left mesh of Figure 2.24 shows the aforementioned mesh, which was obtained by segmenting the computed tomography image of a left atrium. Also shown is its noisy version (Gaussian noise) in the right part of the figure. This mesh had hundreds of thousands of vertices; however, it was decimated to 500 vertices in order to reduce the computational cost of the eigendecomposition algorithm.

For this filtering example, we use the second type of Laplacian, the geometric Laplacian $\boldsymbol{L}$, which is defined as

$$L_{ij} = \begin{cases} -\sum_{\boldsymbol{v}_j \in N(\boldsymbol{v}_i)} w_{ij} & \text{if } i = j \\ w_{ij} & \text{if } (i,j) \text{ is an edge} \\ 0 & \text{otherwise,} \end{cases} \tag{2.144}$$

where w_{ij} was defined in Equation 2.102. The eigendecomposition of $\boldsymbol{L}$, the spectral coefficients, and the mesh reconstruction are obtained as in the previous example. Figure 2.25 shows the evolution of the reconstruction for the noisy left atrial mesh. When 100 components have been included in the reconstruction, the atrial mesh has

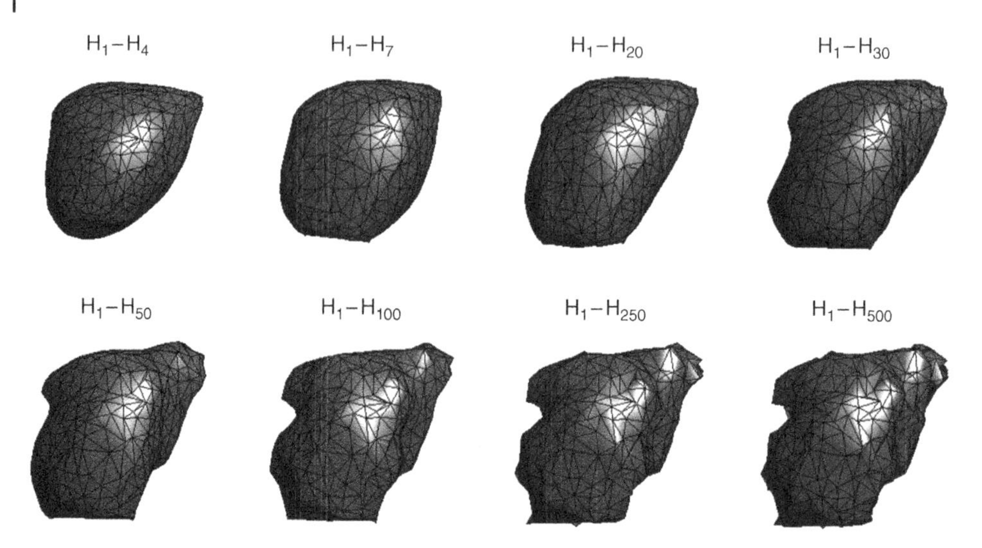

Figure 2.25 Reconstruction process of the noisy left atrial mesh.

its complete shape but without the additional noise. Hence, a low-pass filter of the mesh is computed by just reconstructing the mesh with the first few components. The code for the computation of the geometric mesh Laplacian is given in Listing 2.31.

```
function L = geometric_Laplacian(vertex,faces)

n = length(vertex);
L=zeros(n,n);

for i=1:n
    vi=vertex(i,:);

    % Search the faces which share the vertex vi
    ind_vi=[find(faces(:,1)==i); find(faces(:,2)==i); ...
        find(faces(:,3)==i)];
    faces_vi = faces(ind_vi,:);
    % vi neighborgs
    Ni=setdiff(faces_vi,i);

    % For each neighborg of vi
    for j=1:length(Ni)
        vj=vertex(Ni(j),:);
        % Search the two faces shared by the edge vi-vj
        ind_vj=[find(faces_vi(:,1)==Ni(j)); ...
            find(faces_vi(:,2)==Ni(j)); find(faces_vi(:,3)==Ni(j))];
        % Search the third vertex in the face
        for k = 1: length(ind_vj)
            ind_vk=setdiff(faces_vi(ind_vj(k),:),[i Ni(j)]);
            % Compute the angle opposite to the edge vi-vj
            va = vi - vertex(ind_vk,:);
            vb = vj - vertex(ind_vk,:);
```

```
            % Angles alpha and beta
            opp_angle = acos(dot(va,vb)/(norm(va)*norm(vb)));
            L(i,Ni(j))= L(i,Ni(j)) + cot(opp_angle);
            L(Ni(j),i)= L(Ni(j),i) + cot(opp_angle);
        end
        L(Ni(j),i)  = L(Ni(j),i)/2;
        L(i,Ni(j))  = L(i,Ni(j))/2;
    end
end

L= diag(sum(L,2)) - L;
```

Listing 2.31 Code for geometric mesh Laplacian computation (geometric_Laplacian.m).

Finally, as Dyer *et al.* (2007) remarked that the robustness of the geometric Laplacian is greater than the graph Laplacian, we also consider the graph Laplacian in the last example in order to compare the quality of the reconstruction with both Laplacians. For this purpose, we use the MAE as a merit figure of comparison. The code for the comparison is in Listing 2.32.

```
% Load the vertex and faces matrices
load vertex;
load faces;

n = length(vertex);
L = geometric_Laplacian(vertex,faces);
K = graph_Laplacian(n,faces);

% Eigendecompositon of L and K:
[H_K,A_K] = eig(K);
[H_L,A_L] = eig(L);

% Spectral coefficients
a_K = H_K'*vertex;
a_L = H_L'*vertex;

% Mesh reconstruction error and mean absolute error
for i=1:n
    x_hat_K = H_K(:,1:i)*a_K(1:i,:);
    MAE_K(i) = mean(abs(vertex(:)-x_hat_K(:)));
    x_hat_L = H_L(:,1:i)*a_L(1:i,:);
    MAE_L(i) = mean(abs(vertex(:)-x_hat_L(:)));
end

% Plot the error figure
figure; set(gcf,'color','w');    set(gca,'FontSize',20);
plot(MAE_K,'LineWidth',2);
hold on; plot(MAE_L,'r','LineWidth',2); axis tight;
xlabel('Components');    ylabel('Mean absolute error')
legend('Graph Laplacian','Geometric Laplacian')
```

Listing 2.32 Comparisons in terms of the MAE (MHexamples3.m).

Figure 2.26 shows the MAE for the mesh reconstruction using the graph Laplacian and the geometric Laplacian. During all the reconstruction process the quality of the reconstructed mesh (less error) is superior for the geometric Laplacian than the

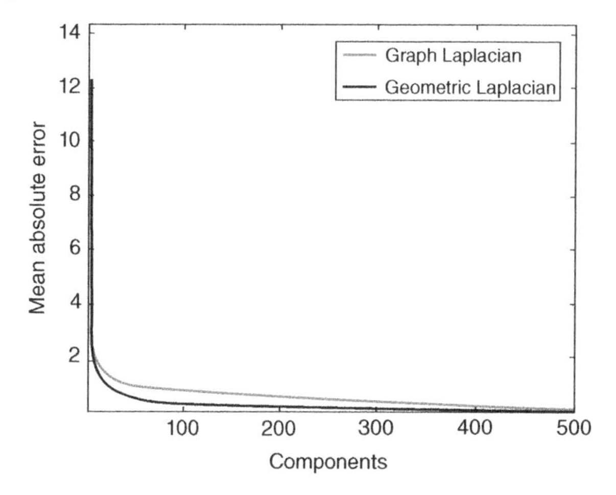

Figure 2.26 MAE for the mesh reconstruction using the graph Laplacian and the geometric Laplacian.

graph Laplacian. Note that the highest error is obtained for the first components because the mesh obtained a 3D shape when four component are included in the reconstruction.

2.6 Questions and Problems

Exercise 2.6.1 Explain in a physical sense and with your own words why we need to use complex algebra for defining the FT.

Exercise 2.6.2 Does the PSD contain all the relevant information of a signal? Is the same response true for a stochastic process?

Exercise 2.6.3 Can you describe three effects of windowing a stochastic process on its PSD?

Exercise 2.6.4 Write the equations of deconvolution with the MMSE criterion, and then repeat this exercise after adding a term with an ℓ_2 regularization term of model parameters. Whath are the main differences between both equations?

Exercise 2.6.5 Program a simple example for your equations in Exercise 2.6.4, where the signal to be estimated is a sparse signal (only some few samples are nonzero). What is the effect of the presence or absence of regularization on the estimated amplitude for the samples?

Exercise 2.6.6 Write the equations of a regular sampling interpolation with sinc signals using the MMSE criterion, and then repeat the derivation adding an ℓ_2 regularization term. Program a simple example where you can see the effect of interpolating a signal, and see the effect of regularization in the same signal when sampled at 10 times the sampling period used for the observed samples.

Exercise 2.6.7 Can you explain qualitatively the differences between the Gabor transform, the Wigner–Ville distribution, and the Margenau–Hill distribution?

Exercise 2.6.8 When would it be better to use the CWT and when the DWT?

Exercise 2.6.9 Can you explain qualitatively what is the physical meaning of the Laplacian operator in the manifold harmonics algorithm? Can you explain the meaning of the eigenvectors and eigenvalues in simple geometrical terms?

3

Signal Processing Models

3.1 Introduction

Classical taxonomies on signal processing and analysis often distinguish between *estimation* and *detection* problems, which are often known in machine learning as *regression* and *classification* respectively. Up to four main data models in *statistical Learning* were pointed out further by Cherkassky and Mulier (1998); namely, clustering, density estimation, classification, and regression. Models in these paradigms can be adjusted (tuned, trimmed, fitted) to the available datasets by following different inductive principles, including MMSE and ML criteria. On the other hand, *time series analysis* has provided a wide variety of paradigms for parametric modeling to be adjusted in terms of their time properties, which can be summarized in the autocorrelation and cross-correlation statistical descriptions. The good news with SVM and kernel methods is that they provide at least two main advantageous elements, such as the kernel trick and the single solution inductive principle. These advantages allowed the kernel methods community to revisit old, well-established data models successfully. However, the existing richness of data models in time series analysis, statistics, and many other data processing fields has not always been taken into account, so there is still room for improving experimental performance and more solid theoretical analysis.

Signal estimation, regression, and function approximation are old, largely studied problems in signal processing, statistics, and machine learning. The problem boils down to optimizing a loss (cost, energy) function over a class of functions. The problem of building function approximation models is interesting and highly relevant in DSP. A model that can interpolate and extrapolate well is useful to identify a system that generated the observations and to generalize (give predictions) to unseen data. Signal estimation is ubiquitous in many signal processing settings, such as system identification, time-series prediction, channel equalization, and signal denoising, just to name a few.

A major goal of this book is to show how to adapt a good existing algorithm given the specificities of the data and a proper signal model, and how to do this via kernel methods. This algorithmic ductility and malleability of kernel methods has already yielded an unprecedented number of mathematical models to be adapted and delivered, rediscovered, or brought to the fore during the last decade. But kernel methods also bring new opportunities to develop new algorithms for dealing with high-dimensional data from 1D signal principles and models. Therefore, the main purpose of this chapter and Chapter 4 is to pay attention to a number of signal models, which will be widely

Digital Signal Processing with Kernel Methods, First Edition. José Luis Rojo-Álvarez, Manel Martínez-Ramón, Jordi Muñoz-Marí, and Gustau Camps-Valls.

used throughout the book. We will make emphasis on understanding, in a simple way, those paradigms that take into account temporal relationships or correlations (*signal processing models*), and those paradigms that do not make any assumption on the correlation properties of the data (*data processing models*). We will see that the latter can be seen as just instances of the former in their multidimensional versions. This observation holds when we replace the need of accounting for multidimensional correlations with random sampling of the multidimensional spaces.

This chapter conveys the ideas from DSP to be clearly kept in mind when working on kernel-based signal processing algorithms. Basic concepts of signal Hilbert spaces, noise, and optimization are presented. Attention is paid to well-known signal models in terms of signal expansions in orthogonal and nonorthogonal signal spaces; that is, nonparametric spectral estimation (sinusoidal expansions), autoregressive exogenous (ARX) and γ-filter system identification (lagged signals expansions), sinc interpolation and deconvolution (impulse response expansions), and array processing (subspace and other expansions). Other fundamental concepts in DSP for time series are introduced, specifically from adaptive filtering concepts, noise concepts, and complex signals, which will be used throughout the rest of the book. A set of synthetic and real examples is presented for clarifying these concepts from simple MATLAB scripts.

3.2 Vector Spaces, Basis, and Signal Models

We start the chapter with a brief overview of vector spaces and basis. This will allow us to express signals and signal models in vector (and more generally in matrix) form. By doing so we inherit all the powerful and well-established tools of linear algebra, which in turn makes it easy to build algorithms and derive mathematical proofs. Another advantage of expressing signals in vector/matrix form is that standard software packages, such as R and MATLAB, provide extensive tools to work with them.

3.2.1 Basic Operations for Vectors

Let us start by defining a *point*[1] in a vector space $\mathcal{X}$. We express a point in $\mathbb{R}^N$ as a (column) vector of the form

$$\boldsymbol{x} = [x_1, x_2, \dots, x_N]^{\mathrm{T}},$$

where x_i, $i = 1, \dots, N$, is an element of that vector, and the symbol T indicates the transpose operation. Note that by definition our points are *column vectors*. We will distinguish the notation for a vector and for its components by using boldfaced fonts for vectors; that is, $\boldsymbol{x}$.

Definition 3.2.1 (Sum of two vectors) The sum of two vectors $\boldsymbol{x}, \boldsymbol{y} \in \mathcal{X}$, is given by the sum of their individual components: $\boldsymbol{x} + \boldsymbol{y} = [x_1 + y_1, x_2 + y_2, \dots, x_N + y_N]^{\mathrm{T}}$, where x_i is the ith component of vector $\boldsymbol{x}$.

1 Throughout the book we will use the terms "sample," "example," "pattern," or "instance" interchangeably to denote a *point* in a geometrical space interchangeably.

Definition 3.2.2 (Inner product) The inner (scalar, or dot) product of two vectors $\boldsymbol{x}, \boldsymbol{y} \in \mathcal{X}$ is defined as the sum of the products of all elements of the vectors, and is denoted as $\langle \boldsymbol{x}, \boldsymbol{y} \rangle_{\mathcal{X}} = \sum_{k=1}^{N} x_k y_k$. Very often one omits the subscript $\mathcal{X}$ indicating the space where the inner product occurs, being obvious from the context, and simply denote $\langle \boldsymbol{x}, \boldsymbol{y} \rangle$.

Given two vectors $\boldsymbol{x}$ and $\lambda \boldsymbol{y}$, its difference vector $\boldsymbol{x} - \lambda \boldsymbol{y}$ has a positive norm for all $\lambda \in \mathbb{R}$:

$$\|\boldsymbol{x} - \lambda \boldsymbol{y}\|^2 = \langle \boldsymbol{x} - \lambda \boldsymbol{y}, \boldsymbol{x} - \lambda \boldsymbol{y} \rangle \geq 0. \tag{3.1}$$

Now assume that the difference vector $\boldsymbol{x} - \lambda \boldsymbol{y}$ is orthogonal to $\boldsymbol{y}$. In this case, it is straightforward to prove that $\langle \boldsymbol{x} - \lambda \boldsymbol{y}, \boldsymbol{y} \rangle = \langle \boldsymbol{x}, \boldsymbol{y} \rangle - \lambda \|\boldsymbol{y}\|^2 = 0$, and thus necessarily $\lambda = \langle \boldsymbol{x}, \boldsymbol{y} \rangle / \|\boldsymbol{y}\|^2$, and the Cauchy–Schwartz inequality follows:

$$\langle \boldsymbol{x}, \boldsymbol{y} \rangle \leq \|\boldsymbol{x}\| \|\boldsymbol{y}\|. \tag{3.2}$$

Also, $\lambda \boldsymbol{y}$ is the projection of $\boldsymbol{x}$ over $\boldsymbol{y}$, and then $\lambda \|\boldsymbol{y}\| = \|\boldsymbol{x}\| \cos \theta$, where θ is the angle between these vectors. In that case, with the previous value of λ, it is straightforward to show that $\langle \boldsymbol{x}, \boldsymbol{y} \rangle = \|\boldsymbol{x}\| \|\boldsymbol{y}\| \cos \theta$. This second definition shows clearly that the inner product is related to the concept of *similarity* between vectors, and therefore is of crucial importance in statistics, and in signal processing in particular, because filtering, correlation, and classification are ultimately concerned about estimating similarities and finding patterns in the data.

Definition 3.2.3 (ℓ_q-norm) The ℓ_q-norm of a vector $\boldsymbol{x}$ is denoted as $\|\boldsymbol{x}\|_q = (\sum_{i=1}^{N} x_i^q)^{1/q}$. Usually the ℓ_2 norm is employed (called simply *norm* of a vector) and the subindex is omitted, being $\|\boldsymbol{x}\|_2 = \|\boldsymbol{x}\| = \langle \boldsymbol{x}, \boldsymbol{x} \rangle^{1/2}$.

Definition 3.2.4 (Euclidean distance) Given the ℓ_2 norm, the Euclidean distance between two vectors $\boldsymbol{x}$ and $\boldsymbol{y}$ is defined as

$$d(\boldsymbol{x}, \boldsymbol{y}) = \|\boldsymbol{x} - \boldsymbol{y}\|_2 = \sqrt{\sum_{k=1}^{N} (x_k - y_k)^2}.$$

Definition 3.2.5 (Linear independence) A set of M vectors $\boldsymbol{x}_i \in \mathbb{R}^N$, $i = 1, \ldots, M$, are linearly independent if and only if $\sum_{i=0}^{M} b_i \boldsymbol{x}_i = 0 \Leftrightarrow b_i = 0 \; \forall i$. Linear independence is a very important property related to the definition of basis.

Definition 3.2.6 (Basis) A basis is any set of vectors, $\boldsymbol{x}_i$, $i = 1, \ldots, M$, $M \geq N$, from which we can express any vector in $\mathbb{R}^N$. In other words, for any vector $\boldsymbol{y} \in \mathbb{R}^N$ we can find a set of M coefficients, $a_i \in \mathbb{R}$ so $\boldsymbol{y}$ can be expressed as

$$\boldsymbol{y} = \sum_{i=1}^{M} a_i \boldsymbol{x}_i$$

A set of vectors $\boldsymbol{x}_i$ being a basis contains at least N linear independent vectors. Among all the possible bases, we are usually interested in *orthonormal* bases, which are those formed with orthogonal vectors of norm 1:

$$\langle \boldsymbol{x}_i, \boldsymbol{x}_j \rangle = \delta_{i,j} = \begin{cases} 1 & i = j \\ 0 & i \neq j \end{cases} \tag{3.3}$$

for $i = 1, \dots, M$, where $\delta_{i,j}$ is the Kronecker function.

An important and well-known basis in $\mathbb{R}^N$ is the *canonical basis*. Vectors $\boldsymbol{x}_i$ of this basis have their kth elements of the form $\boldsymbol{x}_{i,k} = \delta_{i,k}$. Another relevant basis is the *Fourier basis*, consisting of functions

$$\boldsymbol{w}_i = \left[1, \frac{1}{N} \mathrm{e}^{-\mathbb{i}\frac{2\pi i}{N}}, \dots, \frac{1}{N} \mathrm{e}^{-\mathbb{i}\frac{2\pi i(N-1)}{N}}\right]^{\mathrm{T}},$$

whose components are $w_{i,k} = \mathrm{e}^{-\mathbb{i}(2\pi ik/N)}$, where $k = 0, \dots, N-1$. The dot product between two vectors is

$$\langle \boldsymbol{w}_l, \boldsymbol{w}_m \rangle = \frac{1}{N^2} \sum_{k=0}^{N-1} \mathrm{e}^{-\mathbb{i}(2\pi lk/N)} \, \mathrm{e}^{-\mathbb{i}(2\pi mk/N)} = \frac{1}{N^2} \sum_{k=0}^{N-1} \mathrm{e}^{-\mathbb{i}(2\pi(l-m)k/N)}, \tag{3.4}$$

where we want to note that if $l = m$ then the argument of the summation is 1, and thus the dot product is equal to $1/N$. Otherwise, since the exponential function is a zero mean function, the dot product is zero.

3.2.2 Vector Spaces

In the following definitions, $\boldsymbol{x}$, $\boldsymbol{y}$, and $\boldsymbol{z}$ are a set of vectors $V \in \mathbb{R}^N$, and a is any complex number, $a \in \mathbb{C}$.

Definition 3.2.7 (Vector space) A vector space, $\mathcal{H}(V, S)$, is formed by a set of vectors $V \in \mathbb{R}^N$ and a set of scalars $S \in \mathbb{C}$ fulfilling the following properties.

- Commutative: $\boldsymbol{x} + \boldsymbol{y} = \boldsymbol{y} + \boldsymbol{x}$.
- Associative: $(\boldsymbol{x} + \boldsymbol{y}) + \boldsymbol{z} = \boldsymbol{x} + (\boldsymbol{y} + \boldsymbol{z})$.
- Distributive with respect to scalar multiplication:

$$\begin{aligned} a(\boldsymbol{x} + \boldsymbol{y}) &= a\boldsymbol{x} + a\boldsymbol{y} \\ (a + b)\boldsymbol{x} &= a\boldsymbol{x} + b\boldsymbol{x} \\ a(b\boldsymbol{x}) &= (ab)\boldsymbol{x}. \end{aligned}$$

- Existence of null vector in V for the sum operation: $\boldsymbol{x} + \boldsymbol{0} = \boldsymbol{x}$, where the null vector $\boldsymbol{0} = [0, \dots, 0]^{\mathrm{T}}$ has the same size as $\boldsymbol{x}$.
- Existence of the inverse of a vector: $\boldsymbol{x} + (-\boldsymbol{x}) = \boldsymbol{0}$.
- Existence of an identity element for scalar multiplication: $1 \cdot \boldsymbol{x} = \boldsymbol{x} \cdot 1 = \boldsymbol{x}$.

Definition 3.2.8 (Inner product space) An inner product space is a vector space equipped with an inner product. The inner product allows one to define the notion of distance between vectors in a given vector space. An inner product for a vector space is defined as a function $f : V \times V \rightarrow S$ satisfying the following properties.

- Distributive with respect vector addition: $\langle \boldsymbol{x}+\boldsymbol{y}, \boldsymbol{z}\rangle = \langle \boldsymbol{x}, \boldsymbol{z}\rangle + \langle \boldsymbol{y}, \boldsymbol{z}\rangle$.
- Scaling with respect to scalar multiplication:

$$\begin{aligned} \langle \boldsymbol{x}, a\boldsymbol{y}\rangle &= a\langle \boldsymbol{x}, \boldsymbol{y}\rangle \\ \langle a\boldsymbol{x}, \boldsymbol{y}\rangle &= a^*\langle \boldsymbol{x}, \boldsymbol{y}\rangle. \end{aligned}$$

- Commutative: $\langle \boldsymbol{x}, \boldsymbol{y}\rangle = \langle \boldsymbol{y}, \boldsymbol{x}\rangle^*$.
- Self-product is positive: $\langle \boldsymbol{x}, \boldsymbol{x}\rangle \geq 0$, and $\langle \boldsymbol{x}, \boldsymbol{x}\rangle = 0 \Leftrightarrow \boldsymbol{x} = 0$.

From the definition of the inner product, the following properties arise:

- Orthogonality. Two nonzero vectors $\boldsymbol{x}$ and $\boldsymbol{y}$ are orthogonal, denoted as $\boldsymbol{x} \perp \boldsymbol{y}$, if and only if $\langle \boldsymbol{x}, \boldsymbol{y}\rangle = 0$.
- The ℓ_2 norm of a vector is defined as $||\boldsymbol{x}|| = \langle \boldsymbol{x}, \boldsymbol{x}\rangle^{1/2}$.
- The norm satisfies the *Cauchy–Schwartz* inequality: $||\langle \boldsymbol{x}, \boldsymbol{y}\rangle|| \leq ||\boldsymbol{x}|| \cdot ||\boldsymbol{y}||$, and $||\langle \boldsymbol{x}, \boldsymbol{y}\rangle|| = ||\boldsymbol{x}|| \cdot ||\boldsymbol{y}|| \Leftrightarrow \boldsymbol{x} = a\boldsymbol{y}, a \in \mathbb{R}$.
- Triangle inequality: $||\boldsymbol{x}+\boldsymbol{y}|| \leq ||\boldsymbol{x}|| + ||\boldsymbol{y}||$, and $||\boldsymbol{x}+\boldsymbol{y}|| = ||\boldsymbol{x}|| + ||\boldsymbol{y}|| \Leftrightarrow \boldsymbol{x} = a\boldsymbol{y}, a \in \mathbb{R}$.
- Pythagorean theorem: for orthogonal vectors, $\boldsymbol{x} \perp \boldsymbol{y}$, $||\boldsymbol{x}+\boldsymbol{y}||^2 = ||\boldsymbol{x}||^2 + ||\boldsymbol{y}||^2$.

3.2.3 Hilbert Spaces

Hilbert spaces are of crucial interest in signal processing because they allow us to deal with signals of infinite length. Note that given $\mathbb{C}^N$, letting N go to infinity to obtain $\mathbb{C}^\infty$ in a standard Euclidean space will raise problems, even with trivial signals such as $x_i = 1$ for all i, because operations like the inner product (and consequently the norm) will be impossible to compute. The proper way to generalize $\mathbb{C}^N$ to an infinite number of dimensions is through Hilbert spaces. Hilbert spaces impose a set of constraints that ensure that divergence problems will not arise, and that it would be possible to perform all the required operations, like the inner product. Essentially, the constraints imposed in Hilbert spaces ensure that infinite signals embedded into them have finite energy.

We now have the fundamental bricks to define a Hilbert space: a vector space and an inner product space. A Hilbert space is an inner product space $(\mathcal{H}, V)$ that fulfills the property of *completeness*. Completeness is the property by which any convergent sequence of vectors in V has a limit in V. An example of a Hilbert space is the vector space $\mathbb{C}^N$ with a sum operation and the inner product defined as $\langle \boldsymbol{x}, \boldsymbol{y}\rangle = \sum_{i=1}^N x_i^* y_i$, where superscript $*$ is the conjugate operation.

Definition 3.2.9 (Subspace) A subspace is a *closed* region of a Hilbert space, meaning that all vector operations in that subspace remain inside it. A subspace has its own bases describing it, just as the vector space it belongs to. Formally, given a Hilbert space $\mathcal{H}(V, S)$, $S \in \mathbb{C}$, a subspace is a subset $P \subseteq V$ that satisfies that any linear combination of vectors of that subspace also belong to the subspace. Therefore, given $\boldsymbol{x}, \boldsymbol{y} \in P$, for any pair of scalars $a, b \in \mathbb{C}$, vector $a\boldsymbol{x} + a\boldsymbol{y}$ belongs to P.

A set $\boldsymbol{x}_i$, $i = 1, \dots, M$, of M vectors from a subspace P is a basis for that subspace if they are linearly independent. Also note that any vector in P can be expressed as a linear combination of the elements of the set; that is, $\forall\, \boldsymbol{y} \in P$:

$$\boldsymbol{y} = \sum_{i=1}^{M} a_i \boldsymbol{x}_i, \quad a_i \in S, \tag{3.5}$$

which is called the *Fourier analysis equation*. If all vectors in a base of P are orthonormal, then they form an orthonormal base for P, and they satisfy that $\langle \boldsymbol{x}_i, \boldsymbol{x}_j \rangle = \delta_{i,j}, 0 \leq i,j \leq M$. Actually, an interesting property of orthonormal bases is that the coefficients in the Fourier analysis Equation 3.5 can be retrieved from

$$a_i = \langle \boldsymbol{x}_i, \boldsymbol{y} \rangle, \tag{3.6}$$

which are called *Fourier coefficients*, although this term is most often used to refer to the particular base where these coefficients have the form $a_k = \mathrm{e}^{-\mathbb{i}\omega k}$. On the other hand, Equation 3.6 is known as the *synthesis formula*. Another important property of orthonormal bases is the *Parseval identity*, $\|\boldsymbol{y}\|^2 = \sum_{i=1}^{M} |\langle \boldsymbol{x}_i, \boldsymbol{y} \rangle|^2$. From a physical point of view, the norm of a vector is related to the energy, and the Parseval identity is an expression of energy conservation, irrespective of the orthogonal base used for representation.

3.2.4 Signal Models

The SVM framework for DSP that will be presented in Chapter 5 uses several basic tools and procedures. We will need to define a general signal model equation for considering a time-series structure in our observed data, consisting of an expansion in terms of a set of signals spanning a Hilbert signal subspace and a set of model coefficients to be estimated. Then, we next define a general signal model for considering a time-series structure in our observed data, consisting of an expansion in terms of a set of signals spanning a Hilbert signal subspace, and a set of model coefficients to be estimated with some criteria.

Definition 3.2.10 (General signal model) Given a time series $y[n]$, we use the samples set $\{y_n\}$ consisting of $N + 1$ observations of $y[n]$, with $n = 0, \dots, N$, which will also be denoted in vector form as $\boldsymbol{y} = [y[0], y[1], \dots, y[N]]^{\mathrm{T}}$. An expansion for approximating the set of samples for this signal can be built with a set of signals $\{s_n^{(k)}\}$, with $k = 0, \dots, K$, spanning a projection Hilbert signal subspace. This expansion is given by

$$y_n = \sum_{k=0}^{K} z_k s_n^{(k)} + e_n, \tag{3.7}$$

where z_k are the expansion coefficients, to be estimated according to some adequate criterion.

Importantly, before continuing, let us note that after this point we will be using $y[n]$ for denoting the infinite time series, y_n for denoting the set of $N + 1$ time samples, and $\boldsymbol{y}$ when vector notation is more convenient for the same $N+1$ samples. The set $\{s_n^{(k)}\}$ are the *explanatory signals*, which are selected in this signal model to include the a priori knowledge about the time-series structure of the observations, and e_n stands for the error approximation signal, or error approximation in each sample; that is, $y_n = \hat{y}_n + e_n$. The estimation of the expansion coefficients has to be done by optimizing some convenient criterion, as seen next.

In the following section, we review several signal models and problems and write them with this form as a signal model equation specified for the DSP problems which will be addressed in Chapter 5. After we have defined this general signal model, we need a criterion to find its parameters optimizing some criterion. We want to note the following points at this moment:

1) It can be assumed, without loss of generality, that estimation and regression are different names for the same kind of problem within different knowledge fields. Throughout this book, sometimes we refer to one and sometimes to the other, trying to emphasize the use of regression for the familiar estimation from samples of a continuous output.
2) Moreover, in this and in following chapters, we will use a number of signal models that correspond to estimation problem statements based on time samples, and also (linear and nonlinear) regression problem statements.

For these reasons, we follow the approach of describing the general signal model in Definition 3.2.10, and the general regression model in Definition 4.6.1. We also define a general form for the optimization functional of estimation problems, in terms of a regression model, so that it can be valid for both signal models and regression models in subsequent chapters.

Definition 3.2.11 (Optimization functional) Given a set of observations y_i and a corresponding set of explanatory variables $\boldsymbol{x}_i$, an optimization functional is used to estimate model coefficients $\boldsymbol{w}$. The functional consists of the combination of two terms; namely, a loss term (sometimes referred as to cost, risk, or energy function) and a regularization term. The framework of statistical learning theory (Vapnik, 1995) calls these two terms the empirical risk $\mathcal{L}_{\text{emp}}$ and the structural risk $\mathcal{L}_{\text{str}}$ respectively, and it seeks for predictive functions $f(\boldsymbol{x}, \boldsymbol{w})$, parametrized by a set of coefficients $\boldsymbol{w}$, such that we jointly optimize (minimize) both terms; that is:

$$\mathcal{L}_{\text{reg}} = \mathcal{L}_{\text{emp}} + \lambda \mathcal{L}_{\text{str}} = \sum_{i=0}^{N} V(\boldsymbol{x}_i, y_i, f(\boldsymbol{x}_i, \boldsymbol{w})) + \lambda\, \Omega(f), \tag{3.8}$$

where V is a loss/cost/risk term acting on the N available examples for learning, λ is a trade-off parameter between the cost and the regularization terms, and $\Omega(f)$ is defined in the space of functions, $f \in \mathcal{H}$.

Note that the previous functional can be readily particularized for the general signal model using coefficient vector $\boldsymbol{z}$ and the signal observations $s_i^{(k)}$ to obtain prediction functions $f(s_i^{(k)}, \boldsymbol{z})$. This leads to minimizing a quite similar functional given by

$$\mathcal{L}_{\text{reg}} = \mathcal{L}_{\text{emp}} + \lambda \mathcal{L}_{\text{str}} = \sum_{i=0}^{N} V(s_i^{(k)}, y_i, f(s_i^{(k)}, z)) + \lambda\, \Omega(f). \tag{3.9}$$

In order to ensure unique solutions, many algorithms use strictly convex loss functions. The regularizer $\Omega(f)$ limits the capacity of the classifier to minimize the empirical risk $\mathcal{L}_{\text{emp}}$ and favors smooth functions f such that close enough examples should provide close-by predictions. Different losses and regularizers can be adopted for solving the problem, involving completely different families of models and solutions.

Therefore, a general problem involving time-series modeling consists first in looking for an advantageous set of explanatory signals (hence giving the signal expansion), and then estimating the coefficients with some adequate criterion for the residuals and for these coefficients. In what follows, we will pay attention to some signal models representing relevant DSP problems:

- In *nonparametric spectral estimation*, the signal model hypothesis is the sum of a set of sinusoidal signals. The observed signal is expressed as a set of sinusoidal waveforms acting as explanatory signals within a grid of possible oscillation frequencies, with unknown amplitudes and phases to be determined from the data.
- In *parametric system identification and time series prediction*, a difference equation signal model is hypothesized. The observed signal is built by using explanatory signals, which are delayed versions of the observed signal and in system identification problems by delayed versions of an *exogenous* signal.
- In *system identification with γ-filtering*, a particular case of parametric system identification is defined. Here, the signal model can be tuned by means of a single parameter related to memory, while ensuring signal stability in the exogenous signal.
- In *sinc interpolation*, a band-limited signal model is hypothesized, and the explanatory signals are delayed sinc functions. This is known to be a very *ill-posed* DSP problem, especially for nonuniform sampling and noisy conditions.
- In *sparse deconvolution*, a convolutional signal model is hypothesized, and the explanatory signals are the delayed versions of the impulse response from a previously known LTI system.
- Finally, in *array processing*, a complex-valued, spatio-temporal signal model is needed to configure the properties of an array of antennas in several signal processing applications.

All of the aforementioned signal models have a signal structure that is better analyzed by taking into account their correlation information. In Chapter 5 we will place them in a framework that will allow us to indicate their differences and commonalities.

3.2.5 Complex Signal Models

Complex signal models are advantageous in those situations where the signal to be represented is a band-pass signal around a given frequency. Assume an arbitrary real-valued discrete-time signal $y[n]$. It is immediate to prove that its FT $Y(\Omega)$ is Hermitic around the origin; that is, its real part is symmetric with respect to the vertical axis, and its imaginary part is antisymmetric with respect to the coordinates origin.

$$\text{Re}\{Y(\Omega)\} = \sum_{n=-\infty}^{\infty} y[n]\cos(\Omega n), \tag{3.10}$$

$$\text{Im}\{Y(\Omega)\} = -\sum_{n=-\infty}^{\infty} y[n]\sin(\Omega n), \tag{3.11}$$

Actually, from which follows that, when signal $y[n]$ is real, its FT is Hermitic. Now assume that the signal is band-pass; that is, that its FT is different from zero between two given frequencies ω_1 and ω_2, and hence it is also nonzero between $-\omega_2$ and $-\omega_1$ since it is real, and thus Hermitic around the origin. For this reason, both positive and negative intervals will carry a signal with the same information.

Let us now assume an arbitrary central frequency ω_0 defined by convenience as $\omega_1 \leq \omega_0 \leq \omega_2$. A base-band representation $\tilde{y}[n]$ of the signal that carries all the information can be used by representing just the positive part of the signal shifted to the origin a quantity equal to the central frequency. The original FT of the signal can then be represented as

$$Y(\Omega) = \tilde{Y}(\Omega)\,\text{e}^{\text{i}\omega_0} + \tilde{Y}(-\Omega)\,\text{e}^{-\text{i}\omega_0}. \tag{3.12}$$

It is easy to prove that this signal is Hermitic, and hence its inverse FT is real. Owing to the above representation, $\tilde{Y}(\Omega)$ is often called the complex envelope of $Y(\Omega)$. Nevertheless, $\tilde{Y}(\Omega)$ does not need to be Hermitic, so its inverse FT is not necessarily real, but instead it is complex in general. The inverse FT of $Y(\Omega)$ is

$$y[n] = \text{e}^{\text{i}\omega_0}\int_{-\infty}^{\infty} \tilde{Y}(\Omega)\,\text{e}^{-\text{i}\omega n}\,\text{d}\Omega + \text{e}^{-\text{i}\omega_0}\int_{-\infty}^{\infty} \tilde{Y}(-\Omega)\text{e}^{-\text{i}\Omega n}\,\text{d}\Omega. \tag{3.13}$$

The second integral on the right side of the equation corresponds to the inverse FT of $\tilde{y}^*[n]$. This leads to

$$y[n] = \text{e}^{\text{i}\omega_0}\tilde{y}[n] + \text{e}^{-\text{i}\omega_0}\tilde{y}^*[n] = \text{Re}\{\tilde{y}[n]\}\cos(\Omega n) - \text{Im}\{\tilde{y}[n]\}\sin(\Omega n). \tag{3.14}$$

The real and imaginary parts $\text{Re}\{\tilde{y}[n]\}$ and $\text{Im}\{\tilde{y}[n]\}$ of complex envelope $\tilde{y}[n]$ are often called in-phase and quadrature-phase signals, and denoted as $y^{\text{I}}[n]$ and $y^{\text{Q}}[n]$. Equation 3.14 can be rewritten as

$$y[n] = \text{Re}\{\tilde{y}[n]\,\text{e}^{\text{i}\omega_0 n}\} = y^{\text{I}}[n]\cos(\Omega n) - y^{\text{Q}}[n]\sin(\Omega n) \tag{3.15}$$

and the envelope is expressed as $\tilde{y}[n] = y^{\text{I}}[n] + \text{i}y^{\text{Q}}[n]$.

3.2.6 Standard Noise Models in DSP

In Section 3.2.5 we revised the main aspects of the signal models encountered in DSP. While the vast majority of the problems in DSP are tackled by fitting a signal model that captures the relations between the observed data, it is imperative to describe the behavior of the noise or disturbances in the acquired data. Remember that a key for the estimation of parameters is to define a proper loss function, and the residuals

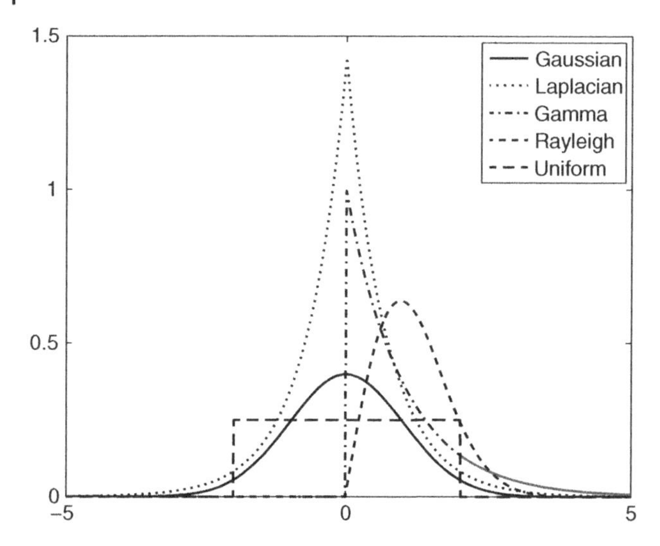

Figure 3.1 Examples of *pdfs* for widely used noise models.

minimization is one of its components. On the one hand, it is typically assumed that the observations, called now $x[n]$, come from the direct sum of a signal $s[n]$ and a noise component $e[n]$; thus, $x[n] = s[n] + e[n]$. In this case, one says that an *additive noise model* was assumed. While this additive assumption is very convenient to formalize the problems (either function approximation, deconvolution, or unmixing), it is widely acknowledged to be somewhat limited. Multiplicative noise given by $x[n] = s[n]e[n]$ is a better assumption in some application fields (e.g., in radar signal processing). In such cases, a simple logarithmic transformation of the observations returns the problem to the additive setting; that is, $\log x[n] = \log s[n] + \log e[n]$.

On the other hand, even though many noise sources in DSP problems can be modeled to fairly follow a Gaussian model (e.g., thermal noise in acquisition systems), many systems are governed by *non-Gaussian* and *colored* (noise power is frequency dependent) noise sources. Finally, we should introduce the assumption of i.i.d. observations being ubiquitous in statistics and signal processing.

Common noise distributions used in signal processing are the Gaussian, the Laplacian, or the uniform. Gaussian noise provides a good noise model in many signal and image systems affected by *thermal noise*. The probability density function (*pdf*) in this case is

$$p_e(e|\sigma) = \frac{1}{\pi\sigma^2}\exp(-e^2/\sigma^2), \tag{3.16}$$

where e refers to the noise and σ accounts for the standard deviation (see Figure 3.1). The Gaussian distribution has an important property: to estimate the mean of a stationary Gaussian random variable, one cannot do any better than the linear average. This makes Gaussian noise a worst-case scenario for nonlinear restoration filters, in the sense that the improvement over linear filters is least for Gaussian noise. To improve on linear filtering results, nonlinear filters can exploit only the non-Gaussianity of the signal distribution.

A second important *pdf* is the Laplacian noise model, also called the bi-exponential, which follows the *pdf* given by

$$p_e(e|\sigma) = \frac{1}{\sqrt{2}\sigma} \exp(-\sqrt{2}|e|/\sigma), \tag{3.17}$$

where e refers to the noise, and σ accounts for the standard deviation (see Figure 3.1). The bi-exponential distribution is particularly useful for modeling events that occur randomly over time. Nonlinear estimators can provide a much more accurate estimate of the mean of a stationary Laplacian random variable than the linear average.

Other common *pdfs* in signal processing applications are, first, the Gamma distribution (see Figure 3.1), modeled by

$$p_e(e|a,b) = \frac{e^{a-1}}{b^a \Gamma(a)} \exp(-e/b), \tag{3.18}$$

where $\Gamma(\cdot)$ is the Gamma function, and a and b are parameters controlling the shape and scale of the *pdf*; and second, the Rayleigh distribution (see Figure 3.1), expressed as

$$p_e(e|c) = \frac{e}{c^2} \exp(-e^2/2c^2), \tag{3.19}$$

where c is a scale parameter. The Gamma distribution is present, for instance, in reliability models of lifetimes for some natural processes, while the Rayleigh (along with Nakagami and Rician distributions) commonly appears in communications theory to model scattered signals that reach a receiver by multiple paths (depending on the density of the scatter, the signal will display different fading characteristics).

Finally, we mention the case of *uniform noise models*. Even though they are not often encountered in real-world systems, they still provide a useful comparison with Gaussian noise sources. The linear average is comparatively a poor estimator for the mean of a uniform distribution, and this implies that nonlinear filters should be better in removing uniform noise than Gaussian noise. Figure 3.1 illustrates the aforementioned *pdf* for zero-mean, unit-variance noise.

3.2.7 The Role of the Cost Function

Many of the inference problems that we will see throughout this book will require explicit or implicit assumptions about both the signal and the noise distributions. The standard assumption about the signal is that the observations are i.i.d., which we actually point out as one of the most important problems of the standard machine learning approximation to signal processing problems. When talking about the noise, one typically observes that the approaches rely on two important assumptions: first, the independence between the signal and the noise counterparts; and second, the implicit assumption of a noise model through the minimization of a particular cost (or loss) function.

On the one hand, the SNR is defined as the ratio of signal power to the noise power. This concept has been extraordinarily useful in signal processing for decades, since it allows us to quantify the quality and robustness of systems and models.

Noise reduction is an issue in signal processing, typically tackled by controlling the acquisition environment. Alternatively, one can filter the signal by looking at the noise characteristics. As we will see, filter design and regularization are intimately related topics, since maximizing the SNR can be cast as a way to seek for smooth solutions preserving signal characteristics and discarding features mostly influenced by the noise. In this context, one typically looks for transformations of the observed signal such that the SNR is maximized, or alternatively for transformations that minimize the fraction of noise. This issue will be treated in detail in Chapter 12 by introducing kernel transformations suited to cope with signal-to-noise correlations.

On the other hand, one admits that most of the signal processing problems can be cast as *estimation* problems, known in statistics as *curve fitting* or *function approximation*, and in machine learning simply as *regression*. Regression models involve the following variables: the unknown parameters (weights) $\boldsymbol{w}$, the independent variables (input variables, features) $\boldsymbol{x}$, and the dependent variable (output, target, observation) to be approximated y. A model can be defined as a function f that relates y with $\boldsymbol{x}$; that is, $y = f(\boldsymbol{x}, \boldsymbol{w}) + r$, where $r = y - f(\boldsymbol{x}, \boldsymbol{w})$ are called residuals, discrepancies, or simply errors.[2] The question here reduces to choosing a good set of weights that allows the signal model to perform well in unseen (test) data. Many techniques are available in the literature, but mostly involve selecting a quality (performance, objective) criterion to be optimized. Taking into account Definition 3.2.11, the first task is to select a cost/loss/risk/energy function. The most common cost functions are depicted in Figure 3.2.

Once the loss is selected, the selection of the parameters and the model weights is the second crucial issue. A standard approach relies on the ML estimation. In statistics in general and in signal processing in particular, ML plays a crucial role as perhaps

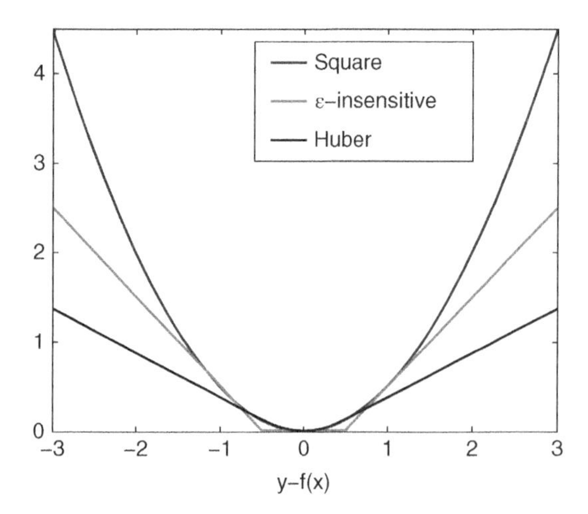

Figure 3.2 Standard cost functions (losses) in signal processing and in machine learning.

---

2 Note that the observation noise is different from the residuals, as far as the former refers to the error in the abstract mathematical signal or data model, whereas the latter refers to the approximation errors after the model has been fit to the data or observations.

Table 3.1 Common cost functions and their assumed noise models.

Loss/cost	$\mathcal{L}_{\text{emp}}(y, f(\boldsymbol{x}, \boldsymbol{w}))$	Noise model, $p(e)$
Squared loss	$0.5(y - f(\boldsymbol{x}, \boldsymbol{w}))^2$	$\frac{1}{\sqrt{2\pi}} \exp(-e^2/2)$
Absolute loss, Laplacian	$\lvert y - f(\boldsymbol{x}, \boldsymbol{w})\rvert$	$\frac{1}{2} \exp(-\lvert e\rvert)$
ε-insensitive loss	$(\lvert y - f(\boldsymbol{x}, \boldsymbol{w})\rvert - \varepsilon)_+$	$\frac{1}{\sqrt{2(1+\varepsilon)}} \exp(-\lvert e\rvert_\varepsilon)$
Huber loss	$\begin{cases} \frac{1}{2}e^2 & \lvert e\rvert \leq \delta \\ \delta(\lvert e\rvert - \frac{\delta}{2}) & \text{otherwise} \end{cases}$	$\begin{cases} \exp(-e^2/2) & \lvert e\rvert \leq \delta \\ \exp(\delta/2 - \lvert e\rvert) & \text{otherwise} \end{cases}$

one of the most widely used methods for estimating the parameters of a statistical model. Roughly speaking, ML reduces to finding the most likely function (model) that generated the data, and hence this criterion tries to maximize

$$\begin{aligned} \max_{\boldsymbol{w}}\{p(\boldsymbol{x}, y|\boldsymbol{w})\} &= \max_{\boldsymbol{w}} \left\{ \prod_i p(\boldsymbol{x}_i, y_i|\boldsymbol{w}) \right\} = \max_{\boldsymbol{w}} \left\{ \prod_i p(y_i|\boldsymbol{x}_i, \boldsymbol{w}) p(\boldsymbol{x}_i) \right\} \\ &= \max_{\boldsymbol{w}} \left\{ \prod_i p(y_i|\boldsymbol{x}_i, \boldsymbol{w}) \right\} \approx \max_{\boldsymbol{w}} \left\{ -\log(p(\boldsymbol{x}, y|\boldsymbol{w})) \right\} \end{aligned} \tag{3.20}$$

When a particular noise model is used in this optimization problem, a likelihood function is obtained, thus giving rise to a particular noise model. Table 3.1 summarizes the main cost functions and the associated noise models that are implicitly assumed. Detailed derivation of the ML relationship between the cost function and the noise models can be found in Smola and Schölkopf (2004).

3.2.8 The Role of the Regularizer

The minimization of a particular function yields a model that can be used to make predictions by using new incoming data. However, the problem of generalization arises here. Generalization is the capability of a method to extrapolate to unseen situations; that is, the function f should accurately predict the label $y_* \in \mathcal{Y}$ for a new input example $\boldsymbol{x}_* \in \mathcal{X}$, i.e. $\hat{y}_* = f(\boldsymbol{x}_*, \boldsymbol{w})$. Generalization has recurrently appeared in the statistics literature for decades under the names of *bias–variance dilemma*, *capacity*, and *complexity control*.

The underlying idea in *regularization* is to constrain the flexibility of a class of functions in order to avoid overfitting to the training data. For example, it is well known that if one aims to fit five points with a polynomial of order 10, the curve will pass through all of them, being a too-fitted and too-wavy function unable to estimate well other unseen points. Thus, the class of polynomial functions is a flexible one, and one has to control it, for example, selecting lower orders or constraining coefficients variance. We will come back later to this issue, which is fundamental in kernel machines. Actually, regularization is one of the keys to prevent overfitting and to simplify the solution. The idea is essentially to explicitly (or implicitly) penalize the excess of model capacity based on the number of effective parameters. Another perspective to the issue of regularization is provided by Bayesian inference, as far as Bayesian methods use a *prior probability* that gives lower probability to more complex models. Either way, selecting an appropriate regularizer is key in statistical inference. According to Definition 3.2.11, selecting regularizer $\Omega(f)$ is our second task to fully define the optimization functional.

Regularizer $\Omega(f)$ plays the role of encoding prior knowledge about the signal (and possibly noise) characteristics. Very often, the regularizer acts on model weights, imposing a particular (expected or desired) characteristic. Defining the regularizer typically reduces to acting on model weights norms. Recall that the q-norm is defined as $\|\boldsymbol{w}\|_q = (\sum_i |w_i|^q)^{1/q}$ and the zero-norm $\|\boldsymbol{w}\|_0$ reduces to compute the sum of all nonzero entries in the vector $\boldsymbol{w}$. For example, one may wish to limit their variance (to prevent unstable predictions in unseen data) or the number of active coefficients (to achieve simple, sparse, compact models). The former strategy uses an ℓ_2-norm of model weights (or coefficients), while the latter would aim to use an ℓ_0-norm which essentially accounts for the number of nonzero components in the vector $\boldsymbol{w}$. In this context, *Tikhonov regularization* is the most well-known and widely used regularizer in machine learning and statistics.

Definition 3.2.12 (Tikhonov regularization) Given a linear function f parametrized by $\boldsymbol{w}$, the Tikhonov regularizer (Tikhonov and Arsenin, 1977) constrains some form of energy of the function $\Omega(f) = \|f\|_{\boldsymbol{H}}^2 = \boldsymbol{w}^{\mathrm{T}}\boldsymbol{H}\boldsymbol{w}$. Matrix $\boldsymbol{H}$ is known as the regularization matrix, and in many approaches the identity matrix is chosen in order to force minimum norm solutions.

Many models have actually used ℓ_0 and ℓ_1 norms as alternatives to the more amenable and tractable ℓ_2-norm. The rationale behind this choice is that $\Omega = \|\boldsymbol{w}\|_1$ promotes sparsity of the solution and constitutes a relaxation of the more computationally demanding problem induced by using the ℓ_0-norm. A geometrical interpretation of the most commonly used regularizers is depicted in Figure 3.3. It can be shown that: (1) for $q < 1$, the regularizer is not convex and hence the corresponding function is not a true norm; (2) for $q \geq 1$, large values become the dominant ones and the optimization algorithm will penalize them to reduce the error, so that the overall cost is reduced. Interestingly, it can be shown that: (1) the regularizer is not convex for values $q < 1$; (2) the value $q = 1$ is the smallest one for which convexity is retained; (3) for large values of $q > 1$, the contribution of small values of the regularization function to the respective norm becomes insignificant. Actually, the extreme case is when one considers the ℓ_0-norm, because even a small increase of a component from zero makes its contribution to the norm very large, so making an element nonzero has

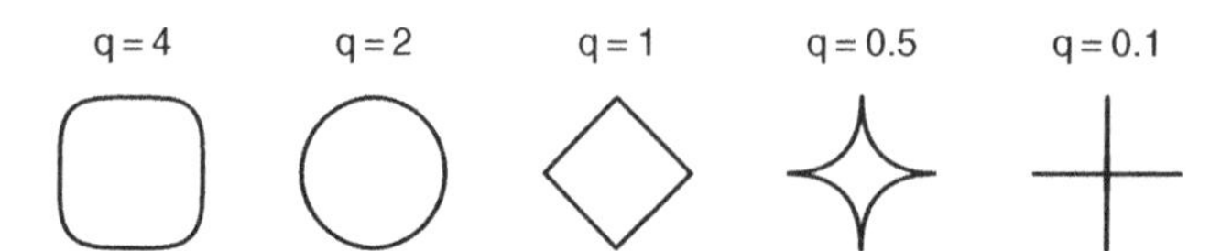

Figure 3.3 Geometrical interpretation of the most commonly used regularizers. The isocurves for different values of $q = \{4, 2, 1, 0.5, 0.1\}$ in the 2D space. Note that for the ℓ_0-norm, the respective values cover the two axes with the exception of the point (0, 0). For the ℓ_1-norm the isovalue curve is a rhombus, and for the ℓ_2-norm (Euclidean) norm it is a circle.

Table 3.2 Combination of cost functions and regularizers result in different forms of statistical inference models.

Model	Loss	Regularizer
Akaike information criterion/Bayesian information criterion (Akaike, 1974)	$(y-f(\boldsymbol{x},\boldsymbol{w}))^2$	$\|\boldsymbol{w}\|_0$
Ridge regression, kriging, kernel ridge regression/-Gaussian process (Rasmussen and Williams, 2006)	$(y-f(\boldsymbol{x},\boldsymbol{w}))^2$	$\|\boldsymbol{w}\|_2$
Support vector regression (Smola and Schölkopf, 2004)	$\lvert y-f(\boldsymbol{x},\boldsymbol{w})\rvert_\varepsilon$	$\|\boldsymbol{w}\|_2$
LASSO (Tibshirani, 1994)	$(y-f(\boldsymbol{x},\boldsymbol{w}))^2$	$\|\boldsymbol{w}\|_1$
Basis pursuit denoising (Chen *et al.*, 1998)	$(y-f(\boldsymbol{x},\boldsymbol{w}))^2$	$\|\boldsymbol{w}\|_1$
Least absolute deviations (LADs), errors (LAEs), residuals (LARs) (Bloomfield and Steiger, 1980)	$\lvert y-f(\boldsymbol{x},\boldsymbol{w})\rvert$	$\|\boldsymbol{w}\|_1$

a dramatic impact on the solution. Note that different losses and regularizers can be adopted for solving an inference problem. Each combination will lead to a completely different family of models and solutions (see Table 3.2 and references therein).

3.3 Digital Signal Processing Models

The basic assumption in the vast majority of statistical inference problems is that the observations are i.i.d. This assumption of independence between samples is not fulfilled in DSP problems in general, and in time-series analysis in particular; instead, statistical similarity among samples plays a key role in their analysis. Algorithms that do not take into account temporal (or spatial) dependencies are ignoring relevant structures of the analyzed signals, such as their autocorrelation or their cross-correlation functions.

The list of examples of this is very long. As a reminder, the following are some examples. The acquired signals in many system identification and control problems give rise to strong temporal correlations and fast decaying spectra. Speech signals are usually coded in digital telephony and media, profiting from the temporal correlation in small signal windows of some milliseconds in which stationarity can be assumed and providing good voice signal representation by just using a few modeling coefficients. Audio coding techniques for high-fidelity daily applications greatly exploit the temporal redundancy of the signals. Biomedical signals, such as ECGs or electroencephalograms (EEGs), usually exhibit well-defined spectral characteristics, which are intimately related to the autocorrelation properties, and allow the use from simple DSP techniques (such as filtering) to sophisticated analysis (such as power law representations). Also, current models of natural images assume they are spatially smooth signals, as far as the joint *pdf* of the luminance samples is highly uniform, the covariance matrix is highly nondiagonal, and the autocorrelation functions are broad and have generally a $1/f$ band-limited spectrum. And in the case of color images, the correlation between the tri-stimulus values of the natural colors is typically high.

Standard machine learning and signal processing approaches typically consider signals as a collection of *independent* samples, mainly because the estimation of the underlying *pdf* is very difficult, and the random sampling of the input space is a convenient (if not the only) assumption allowing to aboard this problem. This, however, can sometimes be alternatively alleviated by introducing prior knowledge on the signal and noise structure, and by imposing suitable signal models (as we will see throughout this and the next chapters).

Several signal models have been paid attention in signal processing and information theory, whose signal structure is better analyzed by taking into account their particular features. This section builds upon Definition 3.2.10 of the *general signal model* and introduces specific signal models for the most common problems in DSP; namely, nonparametric spectral estimation, system identification, interpolation, deconvolution, and array processing. These signal models will be used in Chapter 5 for building specific algorithms for particular applications in DSP.

3.3.1 Sinusoidal Signal Models

Nonparametric spectral analysis is usually based on the adjustment of the amplitudes, frequencies, and phases of a set of sinusoidal signals, so that their linear combination minimizes a given optimization criterion. In general, this adjustment is a hard problem with *local minima*; thus, a simplified solution consists of the optimization of the amplitudes and phases of a set of orthogonal sinusoidal signals in a grid of previously specified oscillation frequencies. This is the basis of the classical *nonparametric spectral analysis.*

When the signal to be spectrally analyzed is uniformly sampled, the LS criterion yields FT-based methods, such as the *Welch periodogram* and the *Blackman–Tukey correlogram* (Marple, 1987), which establish a bias–variance trade-off and allow one to adjust the frequency resolution in terms of the application requirements. These estimators are based on the FT representation of the observed time series and on its estimated autocorrelation function respectively. However, when the signal is nonuniformly (unevenly) sampled in time, the basis of the signal Hilbert space is chosen such that their in-phase and quadrature-phase components are orthogonal at the uneven sampling times, which leads to the *Lomb periodogram* (Lomb, 1976). Methods based on the FT are computationally lighter than those based on LS. However, the latter are optimal for the case of Gaussian noise, but more sensitive to the presence of outliers, which is often encountered in communications applications.

The following is a signal model formulation for sinusoidal representation of a given signal in the presence of noise. This signal model is widely known and was analyzed in detail in Rojo-Álvarez *et al.* (2014).

Definition 3.3.1 (Sinusoidal signal model) Given a set of observations $\{y_n\}$, which is known to present a spectral structure, its signal model can be stated as

$$\begin{aligned}\hat{y}_n &= \sum_{k=0}^{K} z_k s_n^{(k)} = A_0 + \sum_{k=1}^{K} A_k \cos(k\omega_0 t_n + \phi_k) \\ &= \sum_{k=1}^{K} (c_k \cos(k\omega_0 t_n) + d_k \sin(k\omega_0 t_n)),\end{aligned} \tag{3.21}$$

where angular frequencies are assumed to be previously known or fixed in a regular grid with spacing ω_0; A_0 is the mean value; A_k and phi_k are the amplitudes and phases of the kth components, and $c_k = A_k \cos(\phi_k)$ and $d_k = A_k \sin(\phi_k)$ are the in-phase and in-quadrature model coefficients, respectively; and $\{t_n\}$ are the (possibly unevenly separated) sampling time instants.

Note that the sinusoidal signal model corresponds to the general signal model in Definition 3.2.10 for z_k given by c_k and d_k, and for $\{s_n^{(k)}\}$ given by $\sin(k\omega_0 t_n)$ and by $\cos(k\omega_0 t_n)$. Note that this signal model also accounts for the spectral analysis of continuous-time unevenly sampled time series.

3.3.2 System Identification Signal Models

A common problem in DSP is to model a functional relationship between two simultaneously recorded discrete-time processes (Ljung, 1999). When this relationship is linear and time invariant, it can be addressed by using an ARMA difference equation, and when a simultaneously observed signal $\{x_n\}$ is available, called an exogenous signal, the parametric model is called an ARX signal model for system identification. Essentially, in these cases of (parametric) system identification and time series prediction, the signal model is driven by one or several difference equations, and the explanatory signals are delayed versions of the same observed signal, and possibly (for system identification) by delayed versions of an exogenous signal.

In this section we introduce two signal models for system identification that will be used in the following chapters when dealing with kernel functions; namely, the ARX signal model and a particular instantiation of an ARMA filter called the γ-filter. The section also presents the state-space models and signal models which are used for recursion in many DSP applications.

Real and Complex Autoregressive Exogenous Signal Models

The signal model for ARX time series can be stated as follows.

Definition 3.3.2 (ARX signal model) Given a set of output observations $\{y_n\}$ from a system, and its simultaneously observed input signal $\{x_n\}$, an ARX signal model can be stated between them in terms of a parametric model described by an ARMA difference equation, given by delayed versions of both processes:

$$\hat{y}_n = \sum_{k=0}^{K} a_k s_n^{(k)} = \sum_{m=1}^{M} a_m y_{n-m} + \sum_{l=0}^{L} b_l x_{n-l}, \tag{3.22}$$

where $\{x_n\}$ is the exogenous signal; a_m and b_l are the AR and the exogenous (X) model coefficients respectively, and the system identification is an ARX signal model.

Note that the ARX system identification model is the general signal model in Definition 3.2.10 for z_k given by a_m and b_l, and $s_n^{(k)}$ given by y_{n-m} and x_{n-l}.

An important field of application of DSP can be found in digital communications problems, such as *array processing, equalization, channel estimation,* and *multiuser detection*. In these applications, the complex envelope of modulated signals is commonly used. For instance, many important communications systems, such as

Universal Mobile Telecom Service, perform modulation through quadrature schemes, like *quadrature-phase shift keying* (QPSK). Other modulation schemes (for instance, Digital Video and Audio Broadcasting) use *quadrature amplitude modulations* (QAMs). In all these cases, receivers need to deal with complex signals. The statement of complex-valued ARX signal models is also remarkable in the field of DSP, especially for communication applications dealing with many channels and complex constellations.

System Identification with Constrained Recursivity

Identification by means of parametric structures using AR substructures may eventually lead to unstable solutions if no constraints regarding the allocation of the system poles is considered, which is not easy to design. A reasonable solution to force stability consists of using structures that are intrinsically stable. This is the case of the γ-filter structure introduced in Principe *et al.* (1993). The filter is a parsimonious structure that has been used for echo cancellation (Palkar and Principe, 1994), time-series prediction (Kuo *et al.*, 1994), system identification (Principe *et al.*, 1993), and noise reduction (Kuo and Principe, 1994). This structure provides stable models, while allowing the measurement of the memory depth of a model (i.e., how much past information the model can retain) in a very simple and intuitive way. As seen in Figure 3.4, the γ-filter structure is defined by using the difference equations of the linear ARMA model of a discrete time series y_n as a function of a given input sequence x_n as follows.

Definition 3.3.3 (**γ-filter signal model**) Given an observed signal $\{y_n\}$ and a simultaneously observed exogenous signal $\{x_n\}$, a parametric signal model can be stated between them by a γ-structure given by

$$y_n = \sum_{p=0}^{P} w_p x_n^p + e_n \tag{3.23}$$

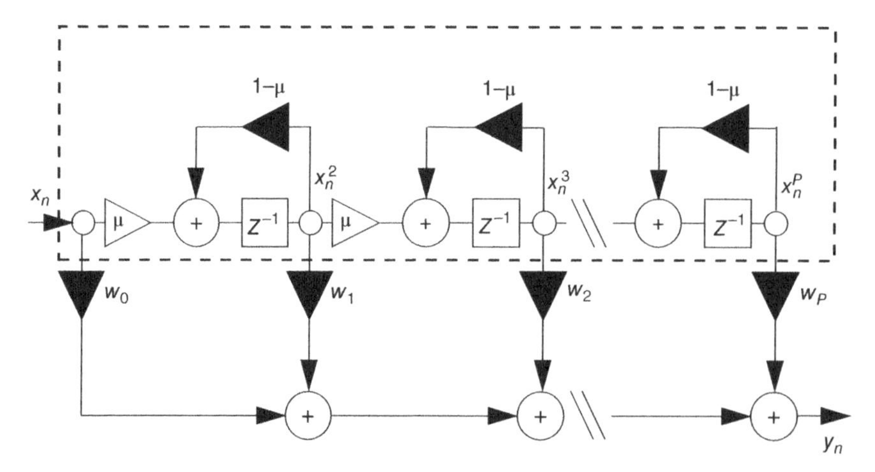

Figure 3.4 The γ structure. The γ filter can be regarded as a cascade of IIR filters where loops are kept local and loop gain is constant.

$$x_n^p = \begin{cases} x_n, & p = 0 \\ (1-\mu)x_{n-1}^p + \mu x_{n-1}^{p-1}, & p = 1, \dots, p \end{cases} \tag{3.24}$$

where μ is a free parameter of the model and w_p are the model coefficients.

The structure is stable for $0 < \mu < 1$ if the transfer function is low pass, and for $1 < \mu < 2$ if it is high pass. For $\mu = 1$ the recursion vanishes and the filter has the Widrow adaline structure. Since P is the number of samples, note that $M = P/\mu$ is an approximate measure of memory depth (Principe *et al.*, 1993). Other IIR structures with restricted recursivity have been developed, such as Laguerre- and gamma-II-type filter structures (De Vries and Principe, 1992), whose signal models can be trivially derived in a similar way.

State-Space Models and Signal Models for Recursion

A generalization of the Wiener filter theory to dynamical systems with random inputs was developed by Kalman in the 1960s in the context of state-space models (Kalman, 1960). Discrete-time (and dynamical) linear filters can be nicely expressed in the state-space formulation (or Markovian) representation, by which two coupled equations are defined as seen in Chapter 2. State-space models give rise to the important property of recursivity in the system equations. Given a set of samples $\{x_n, y_n\}_{n=1}^N$, an arbitrary (Markovian) signal model representation of recursion between input–output time-series pairs can be defined as

$$\begin{aligned} y_n &= f(\langle w_i, x_n^i \rangle | \theta_f) + e_n^x \\ x_n^i &= g(x_{n-k}^i, y_{n-1} | \theta_g) + e_n^y, \quad \forall k > 1, \ l \geq 0 \end{aligned} \tag{3.25}$$

where x_n^i is the signal at the input of the ith filter tap at time n, y_{n-l} is the previous output at time $n - l$, e_x is the state noise, and e_y is the measurement noise. Here, f and g are linear functions parametrized by model parameters w_i and hyperparameters θ_f and θ_g.

The KF is the most widely known instantiation of a state-space formulation. Also note that, when a linear model is imposed for $f(\cdot)$ and $g(\cdot)$, the AR and MA filter structures for time-series prediction readily emerge from Equation 3.25, since a conventional MA filter structure is given by

$$y_n = \sum_{l=0}^{L} b_l x_{n-l} + e_n, \tag{3.26}$$

and a conventional AR filter structure is defined by

$$y_n = \sum_{m=1}^{M} a_m y_{n-m} + e_n. \tag{3.27}$$

The state-space representation is especially powerful for MIMO linear systems, as well as for time-varying linear systems, and it has found many applications in the fields of system identification, economics, control systems, and communications (Ljung, 1999).

3.3.3 Sinc Interpolation Models

In *sinc interpolation*, a band-limited signal model is hypothesized, and the explanatory signals are delayed sincs. The *sinc* kernel provides the perfect reconstruction of an evenly sampled noise-free signal (Oppenheim and Schafer, 1989). In the presence of noise, the sinc reconstruction of a possibly nonuniformly sampled time series is an ill-posed problem (Choi and Munson, 1998a; Yen, 1956; Ying and Munson, 2000). The main elements that are considered by signal interpolation algorithms in the literature are the type of sampling (uniform or nonuniform), the noise (present or not), the spectral content (band-limited or not), and the numerical regularization adopted (none, Tikhonov, or other forms). Despite of the great amount of work developed to date in this setting, the search for new efficient interpolation procedures is still an active research area. Accordingly, we propose to use the following signal model for sinc interpolation in DSP.

Definition 3.3.4 (Sinc kernel signal model) Let $x(t)$ be a band-limited, possibly Gaussian noise-corrupted, signal and be $\{x_k = x(t_k), k = 0, \dots, N\}$, a set of $N+1$ nonuniformly sampled observations. The sinc interpolation problem consists of finding an approximating function $\hat{x}(t)$ fitting the data, $\hat{x}(t) = \sum_{k=0}^{N} f_k \text{sinc}(\sigma_0(t - t_k))$. The previous continuous-time model, after nonuniform sampling, is expressed as the following discrete-time model:

$$x_n = \hat{x}_n + e_n = \sum_{k=0}^{N} f_k \text{sinc}(\sigma_0(t_n - t_k)) + e_n, \tag{3.28}$$

where $\text{sinc}(t) = \frac{\sin(t)}{t}$, and $\sigma_0 = \frac{\pi}{T_0}$ is the sinc units bandwidth.

Therefore, we are using an expansion of *sinc* kernels for interpolation of the observed signal. The sinc kernel interpolation signal model straightforwardly corresponds to the general signal model in Definition 3.2.10 for explanatory signals $s_n^{(k)}$ corresponding to $\text{sinc}(\sigma_0(t_n - t_k))$. An optimal band-limited interpolation algorithm, in the MMSE sense, was first proposed in Yen (1956). As explained next, note that in Yen's algorithm we have as many free parameters as observations, so that this is an ill-posed problem. It can be checked that in the presence of (even a low level of) noise, some of the coefficient estimations grow dramatically, leading to huge interpolation errors far away from the observed samples. To overcome this limitation, the regularization of the quadratic loss was also proposed in the same work.

In this section, two nonuniform interpolation algorithms are presented; namely, the Wiener filter (Kay, 1993) and the Yen regularized interpolator (Kakazu and Munson 1989; Yen 1956). Although many more interpolation methods have already been proposed in the literature, we limit ourselves to these cases for two reasons: (1) they are representative cases of optimal algorithms (with a different inductive principle for each case); and (2) they have a straightforward spectral interpretation, which allows an interesting comparison with the algorithms proposed in later chapters.

Wiener Filter for Nonuniform Interpolation

For defining the Wiener filter in nonuniform interpolation, the following signal model and slightly different notation are followed. Let $x(t)$ be a continuous-time signal with finite energy, consisting of a possibly band-limited signal $z(t)$, which can be seen as a realization of a random process, corrupted with additive noise $e(t)$; that is, $x(t) = z(t)+e(t)$, where the noise is modeled as a zero-mean *wide sense stationary* process. This signal has been observed on a set of N unevenly spaced time instants, $\{t_n\}, n = 1, \dots, N$, obtaining the set of observations $\boldsymbol{x} = [x(t_1), \dots, x(t_n), \dots, x(t_N)]^{\mathrm{T}}$. Then, the nonuniform interpolation problem consists of finding a continuous-time signal $\hat{z}(t)$ that approximates the noise-free interpolated signal in a set of K time instants, $\{t'_k, k = 1, \dots, K\}$.

As described by Kay (1993), a Bayesian approach to solve this problem amounts to the Wiener filter. Assuming that $z(t)$ is zero mean, the linear estimator is given by

$$\hat{z}(t'_k) = \boldsymbol{a}_k^{\mathrm{T}}\boldsymbol{x} \qquad \text{for } k = 1, \dots, K, \tag{3.29}$$

where $\boldsymbol{a}_k$ is the Wiener filter estimator for the signal in time instant t_k. The scalar MMSE estimator is obtained when $\boldsymbol{a}_k$ is chosen to minimize the MSE and takes the following form (Kay, 1993):,

$$\hat{z}(t'_k) = \boldsymbol{r}_{z_k}^{\mathrm{T}}\boldsymbol{C}_{xx}^{-1}\boldsymbol{x} \qquad \text{for } k = 1, \dots, K \tag{3.30}$$

Vector $\boldsymbol{r}_{z_k}$ contains the cross-covariance values between the observed signal and the signal interpolated at time t'_k; that is:

$$\boldsymbol{r}_{z_k} = [r_{zz}(t'_k - t_1), \dots, r_{zz}(t'_k - t_N)]^{\mathrm{T}}, \tag{3.31}$$

where $r_{zz}(\tau)$ is the autocorrelation of the *noise-free* signal for a time shift τ. Note that $\boldsymbol{C}_{xx}$ is the covariance matrix of the observations and, assuming wide sense stationary data with zero mean, it is computed as

$$\boldsymbol{C}_{xx} = \boldsymbol{R}_{zz} + \boldsymbol{R}_{nn}, \tag{3.32}$$

where $\boldsymbol{R}_{zz}$ is the autocovariance matrix of the signal with component i, j given by

$$\boldsymbol{R}_{zz}(i,j) = r_{zz}(t_i - t_j), \tag{3.33}$$

and $\boldsymbol{R}_{nn}$ is the noise covariance matrix. For the i.i.d. case, $\boldsymbol{R}_{nn} = \sigma_n^2\boldsymbol{I}_N$, with σ_n^2 the noise power and $\boldsymbol{I}_N$ the identity matrix of size $N \times N$. Thus, $\hat{z}(t'_k)$ is given by

$$\hat{z}(t'_k) = ((\boldsymbol{R}_{zz} + \sigma_n^2\boldsymbol{I}_N)^{-1}\boldsymbol{r}_{z_k})^{\mathrm{T}}\boldsymbol{x}. \tag{3.34}$$

Although the solution in Equation 3.34 is optimal in the MMSE sense, two main drawbacks can arise when using it in real practice:

1) It implies the inversion of a matrix that can be almost singular, so the problem can become numerically ill-posed.
2) The knowledge of the autocorrelation of the signal $r_{zz}(\tau)$ at every $\tau = t_i - t_j$ is needed, so it must be estimated from the observed samples, which is not an easy task in general.

A frequency-domain analysis can be done from this Wiener filter for interpolation. The solution of the MMSE estimator given by Equation 3.30 can be seen as the convolution of the observations with a filter with impulse response $h_W^{(k)}[n] = a[k-n]$.

For a finite number of nonuniform samples, the solution cannot be converted into a time-invariant filter since it depends on the k index, which is significantly different with the uniform-sampling case. However, in order to provide a simple spectral interpretation of the interpolator we assume that $N \to \infty$ and then Equation 3.34 can be approximated as the convolution of the observations with a time-invariant filter with response $h_W[n]$, which does not depend on the time index k (Kay, 1993); that is:

$$\hat{z}(t'_k) = \sum_{n=-\infty}^{\infty} h_W[n]x(t_k - t_n). \tag{3.35}$$

In this case, the coefficients of the filter $h_W[n]$ can be computed by using the Wiener–Hopf equations, also known as the *normal equations*. By applying the FT to these equations, the transfer function of the filter is finally obtained:

$$H_W(\Omega) = \frac{P_{zz}(\Omega)}{P_{zz}(\Omega) + P_{ee}(\Omega)} = \frac{\eta(\Omega)}{\eta(\Omega)+1}, \tag{3.36}$$

where $P_{zz}(\Omega)$ and $P_{ee}(\Omega)$ are the PSD of the original signal and the noise respectively, and

$$\eta(\Omega) = \frac{P_{zz}(\Omega)}{P_{ee}(\Omega)} \tag{3.37}$$

represents the local SNR in frequency Ω. Obviously, $0 \leq H_W(\Omega) \leq 1$, tending to unity (to zero) in spectral bands with high (low) SNR. Hence, the Wiener filter enhances (attenuates) the signal in those bands with high (low) SNR, and the autocorrelation of the process to be interpolated is a natural indicator of the relevance of each spectral band in terms of SNR.

Yen Regularized Interpolator

Inspired by Shannon's sampling theorem, a priori information can be used for band-limited signal interpolation by means of a sinc kernel. In this case, the signal is modeled with a sinc kernel expansion as

$$x(t'_k) = z(t'_k) + e(t'_k) = \boldsymbol{f}^{\mathrm{T}}\boldsymbol{s}_k + e(t'_k), \tag{3.38}$$

for $k = 1, \dots, K$, with $\boldsymbol{s}_k$ being an $N \times 1$ column vector with components

$$s_k(i) = \mathrm{sinc}(\sigma_0(t'_k - t_n)), \tag{3.39}$$

where $\mathrm{sinc}(t) = \sin(t)/t$ and parameter $\sigma_0 = \pi/T_0$ is the sinc function bandwidth. Then, the interpolator can be stated as

$$\hat{z}(t'_k) = \boldsymbol{f}^{\mathrm{T}}\boldsymbol{s}_k \qquad \text{for } k = 1, \dots, K. \tag{3.40}$$

When LS strategy is used to estimate $\boldsymbol{f}$, the Yen solution is obtained; see Theorem IV in Yen (1956). If a regularization term is used to prevent numerical ill-posing, $\boldsymbol{f}$ is obtained by minimizing

$$\mathcal{L}_{\mathrm{reg}} = \|\boldsymbol{x} - \boldsymbol{S}\boldsymbol{f}\|^2 + \lambda\|\boldsymbol{f}\|^2, \tag{3.41}$$

where $\boldsymbol{S}$ is a square matrix with elements $\boldsymbol{S}(n,k) = \mathrm{sinc}(\sigma_0(t_n - t'_k))$, and λ tunes the trade-off between solution smoothness and the errors in the observed data. In this case, $\boldsymbol{f}$ is given by

$$\boldsymbol{f} = (\boldsymbol{S}^2 + \lambda\mathbf{I}_N)^{-1}\boldsymbol{S}\boldsymbol{x}. \tag{3.42}$$

In this case, the use of the regularization term leads to solutions that are suboptimal in the MSE sense.

Similar to the previous Wiener interpolation, the frequency-domain analysis can be established here as follows. An asymptotic analysis similar to the one presented for the Wiener filter can be done based on Equations 3.42 and 3.40. Using a continuous-time equivalent model for the interpolation algorithm (see Rojo-Álvarez *et al.* (2007) for further details), the interpolation algorithm can be interpreted as a filtering process over the input signal; that is:

$$\hat{z}(t) = h_Y(t) * x(t), \tag{3.43}$$

where $*$ again denotes the convolution operator. Now, the transfer function of the filter is given by

$$H_Y(\Omega) = \frac{P_{ss}(\Omega)}{P_{ss}(\Omega) + \lambda}, \tag{3.44}$$

where $P_{ss}(\Omega)$ is the PSD of $\mathrm{sinc}(\sigma_0 t)$, and since this one is deterministic, $P_{ss}(\Omega) = |S(\Omega)|^2$, with $S(\Omega)$ the FT of the sinc function. This is just a rectangular pulse of width σ_0. Note now that $H_Y(\Omega)$ takes the value $1/(1 + \lambda)$ inside the pass band of $P_{ss}(\Omega)$ and 0 outside. Therefore, if σ_0 is equal to the signal bandwidth, the filter attenuates the noise outside the signal band and does not affect the components inside the band. A comparison between Equations 3.44 and 3.36 reveals that both interpolators can be interpreted as filters in the frequency domain, but in the case of Yen's algorithm the local SNR $\eta(\Omega)$ is simply approximated by the sinc kernel PSD, $P_{ss}(\Omega)$.

Remarks

It can be seen that both Yen and Wiener filter algorithms use the LS (regularized for Yen method) criterion. However, the Wiener algorithm is linear with the observations, and it does not assume any a priori decomposition of the signal in terms of building functions.

Instead, it relies on the knowledge of the autocorrelation function, which can be hard to estimate in a number of applications. Alternatively, Yen's algorithm is nonlinear with respect to the observations and assumes an a priori model based on sinc kernels. Hence, the knowledge of the signal autocorrelation is not needed. As seen in later chapters, the SVM interpolation uses a different optimization criterion, which is the structural risk minimization (SRM), and its solution is nonlinear with respect to the observations since it assumes a signal decomposition in terms of a given Mercer kernel.

3.3.4 Sparse Deconvolution

In *sparse deconvolution*, the signal model is given by a convolutional combination of signals, and the explanatory signals are the delayed and scaled versions of the impulse response from a previously known LTI system. More specifically, the sparse deconvolution problem consists on the estimation of an unknown sparse sequence which has been convolved with a time series (impulse response of the known system) in the presence of noise. The non-null samples of the sparse series contain relevant information about the underlying physical phenomenon in a number of applications.

One of the main fields for sparse deconvolution is seismic signal analysis, where the users send a known waveform through the ground and then collect the echoes produced in the interfaces between two layers of different materials. The returning signal is the convolution of a waveform h_n and a sparse time series x_n (ideally, a sparse train of Kronecker delta functions) that contains information about the depth of different layers.

Definition 3.3.5 (Sparse deconvolution signal model) Let $\{y_n\}$ be a discrete-time signal given by $N+1$ observed samples of a time series, which is the result of the convolution between an unknown sparse signal $\{x_n\}$, to be estimated, and a known time series $\{h_n\}$ (with $K+1$ duration). Then, the following convolutional signal model can be stated for these signals:

$$\hat{y}_n = \hat{x}_n * h_n = \sum_{j=0}^{K} \hat{x}_j h_{n-j} \tag{3.45}$$

where $*$ denotes the discrete-time convolution operator, and $\hat{x}_n$ is the estimation of the unknown input signal.

The sparse deconvolution signal model corresponds to the general signal model in Definition 3.2.10 for model coefficients a_k given by $\hat{x}_k$ and explanatory signals $s_n^{(k)}$ given by h_{n-k}.

The performance of sparse deconvolution algorithms can be degraded by several causes. First, they can result in ill-posed problems (Santamaría-Caballero *et al.*, 1996), and regularization methods are often required. Also, when the noise is non-Gaussian, either LS or ML deconvolution can yield suboptimal solutions (Kassam and Poor, 1985). Finally, if h_n has nonminimum phase, some sparse deconvolution algorithms built by using inverse filtering become unstable.

3.3.5 Array Processing

An antenna array is a group of ideally identical electromagnetic radiator elements placed in different positions of the space. This way, an electromagnetic flat wave illuminating the array produces currents that have different amplitudes and phases depending of the direction of arrival (DOA) of the wave and of the position of each radiating element. The discrete time signals collected from the array elements can be seen as a time and space discrete process.

The fundamental property of an array is that it is able to spatially process an incoming signal, thus being able to discriminate among various signals depending on their DOA. Applications of antenna array processing range from radar systems (which minimize the mechanical components by electronic positioning of the array radiation beam), to communication systems (in which the system capacity is increased by the use of spatial diversity), and to radio astronomy imaging systems, among many others.

The simplest system model in array processing consists of a linear array of $K + 1$ elements equally spaced a distance d, whose output is a time series of vector samples $\boldsymbol{x}_n = [x_n^0, \dots, x_n^K]^T$ or snapshots (Van Trees, 2002). In order to obtain an expression of such a signal, we first model the output signal from transmitter l, received by the sensor located at the right end of Figure 3.5, which can be written as

$$\tilde{x}_n^l = b_n^{l,I} \cos(2\pi f_c t + \phi_l) - b_n^{l,Q} \sin(2\pi f_c t + \phi_l), \tag{3.46}$$

where $b^{l,I}$ and $b^{l,Q}$ are the in-phase and quadrature phase of the transmitted modulation, and f_c is the carrier frequency of the radio signal. This expression can be written as

$$\tilde{x}_n^l = \mathrm{Re}\{b_n^l \, e^{i(2\pi f_c t + \phi_l)}\}, \tag{3.47}$$

where $b_n^l = b_n^{l,I} + \mathbb{i} b_n^{l,Q}$. The signal will illuminate the rest of elements with a time delay that will depend on the array spacing and the illumination angle. In particular, the time delay between two consecutive elements will be $\tau_l = (d/c)\sin(\theta_l)$, c being the light

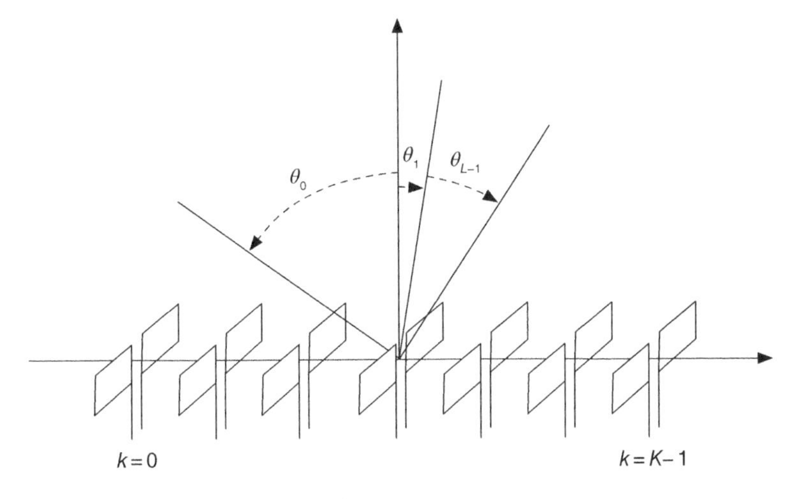

Figure 3.5 Array of K antennas and L transmitters.

speed. The delay will be zero if the element is in the broadside direction ($\theta_l = 0$) and maximum if the direction is the endfire ($\theta = \pm 90°$). Then, knowing that the phase delay between elements will be $\varphi_l = 2\pi f_c \tau$ and that $c = \lambda f_c$, with λ being the wavelength, the vector of amplitudes measured in the array produced by signal $x_l[n]$ can be written

$$\tilde{x}_n^l = \text{Re}\{b_n^l \, e^{i(2\pi f_c t + \phi_l)}\}\text{Re}\{[1, e^{i2\pi(d/\lambda)\sin(\theta_l)}, \dots, e^{i2\pi[(K-1)d]/\lambda \sin(\theta_l)}]^T\}. \tag{3.48}$$

Usually, these signals are represented by their complex low-pass equivalent signals $x_l[n]$. Indeed, the radiofrequency signal can be removed from the expression since it is common to all signals. Also, we assume that all phase delays ϕ_l are zero for simplicity (see Equation 3.15). Then, a given source with constant amplitude, whose DOA and wavelength are θ_l and λ respectively, yields the following array output (so-called *steering vector*):

$$\boldsymbol{v}_l = (1, e^{i2\pi(d/\lambda)\sin(\theta_l)}, \dots, e^{i2\pi[(K-1)d]/\lambda \sin(\theta_l)})^T. \tag{3.49}$$

If L transmitters are present, the snapshot can be represented as

$$\boldsymbol{x}_n = \boldsymbol{E}\boldsymbol{b}_n + \boldsymbol{e}_n, \tag{3.50}$$

where $\boldsymbol{E}$ is a matrix containing all steering vectors of the L transmitters, $\boldsymbol{b}_n$ is a column vector containing (complex-valued) symbols transmitted by all users, and $\boldsymbol{e}_n$ is the thermal noise present at the output of each antenna. The application consists of the estimation of the symbols received from transmitted 0 (or desired user) in the presence of noise and $L-1$ interfering transmitters.

Definition 3.3.6 (Array processing signal model) Let $\{y_n\}$ be a discrete time signal given by $N-1$ symbols transmitted by user 0, and be x_n^k, $0 \le k \le K$, a set of discrete-time processes representing time samples of the measured current at each of the array elements. The array processing problem consists of estimating y_n as

$$\hat{s}_n = y_n = \sum_{k=0}^{K-1} w_k x_n^k. \tag{3.51}$$

The problem is called *array processing with temporal reference* estimation when a set of transmitted signals is previously observed for training purposes. Whenever the DOA of the desired user is known, the problem is an *array processing with spatial reference* estimation. Note that this signal model agrees with the general signal model in Definition 3.2.10 for s_n^k given by x_n^k.

3.4 Tutorials and Application Examples

We next include a set of examples for highlighting the concepts in the preceding sections:

1) First, we work on noise examples from different applications; namely, for images, for communication systems, and for biomedical signals.

2) Some basic examples on system identification models are presented on synthetic data for fitting ARX models, and nonlinear system identification is also included as an example by using Volterra models.
3) Sinusoidal signal models adjustment from an LS criterion are used in simple synthetic data and in heart rate variability (HRV) signals.
4) Sinc interpolation models are addressed by using an LS criterion as well as simple methods in the literature.
5) Examples of sparse deconvolution are scrutinized for synthetic data.
6) Finally, array antenna processing shows an example of handling complex algebra data structures.

3.4.1 Examples of Noise Models

A probabilistic description of noise provides a solid mathematical framework to cope with many noise sources. However, many signal processing problems exhibit signals contaminated with specific noise distributions which cannot be easily modeled probabilistically. Very often we find that Gaussian noise is assumed in DSP models because of lack of knowledge or for ease of treatment. Nevertheless, this assumption may be far from being realistic and lead to suboptimal results and even meaningless conclusions. Correctly characterizing the noise in our observations is a relevant first step in DSP applications.

In this section we first visualize and review standard noise sources found in images, as this gives an intuitive idea of the nature and properties of some noise distributions. For instance, accounting for particular characteristics of the noise, such as revealing heavy tails, autocorrelation structures, or asymmetry, can make the difference in accuracy and robustness of the models. Then, we present some examples of noise in signals from communication systems, where Gaussian and non-Gaussian distributions are well characterized in current telecommunications. Finally, we show some examples of the different nature of the noise that can be present in physiological signals, specifically in ECG signals stored in long-term monitoring Holter recordings.

Noise in Images

Images are affected by many kinds of noise and distortions, depending on the acquisition process and the imaging system, but also due to artifacts occurring in the image transmission and storage. While academically most of the denoising (sometimes referred to as restoration) algorithms work under the assumption of Gaussian uncorrelated noise, this is actually far from being realistic. Images can be affected by missed pixels due to errors in the acquisition, flickering pixels, shifts and optical image aberrations, vertical striping in some satellite imaging sensors and thermal cameras, spatial–spectral distortion produced by image compression, fast-fading noise due to passing the image through a communication transmission channel, or speckle and coherent noise in ultrasound images and radar imagery. Figure 3.6 illustrates the explicit appearance of the noise in the familiar image of Barbara.

It is important to note that dealing with these heterogeneous noise sources constitutes a very complex problem for many denoising algorithms because the noise *pdf* is known (or can be fairly well estimated) for the Gaussian noise only. This compromises the use of parametric models for doing denoising, and in many cases nonparametric algorithms

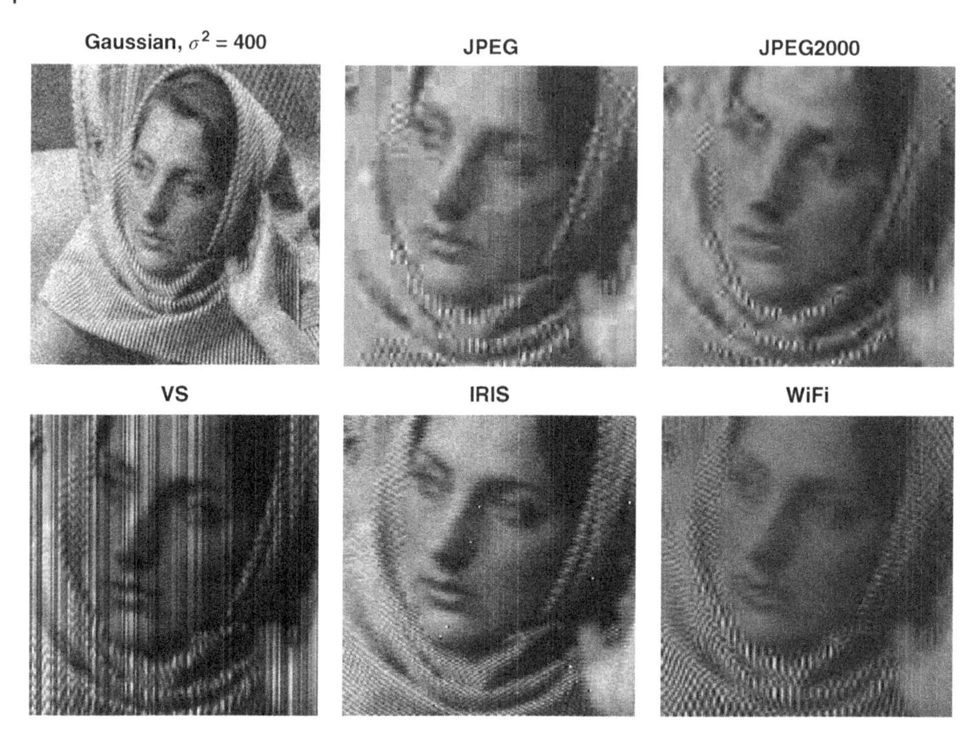

Figure 3.6 Illustration of different noise sources in a common standard image: Gaussian white noise, JPEG and JPEG2000 quantization noise, vertical striping acquisition noise, a simulated infrared imaging system (IRIS) noise as a mixture of uncorrelated Gaussian noise with paired salt-and-pepper noise and lines interleaving, and WiFi fast-fading transmission noise.

typically excel (Laparra *et al.*, 2010). Actually, a successful class of image denoising methods is based on Bayesian approaches working in wavelet representations. The performance of these methods improves when information a priori for the local frequency coefficients is explicitly included. However, in these techniques, analytical estimates can be obtained only for particular combinations of analytical models of signal and noise, thus precluding its straightforward extension to deal with other arbitrary noise sources.

We here recommend the reader interested in image processing to consult the MATLAB toolboxes `ViStaCoRe` referred to on this book web page. The toolbox module on image restoration contains implementations of classical regularization methods, Bayesian methods, and an SVM regression method applied in wavelet domains. The toolbox includes other schemes based on regularization functionals that also take into account the relations among local frequency coefficients (Gutiérrez *et al.*, 2006). Two families of methods are included therein. On the one hand, the toolbox contains techniques for denoising and deconvolution by regularization in local-Fourier domains, such as the second derivative regularization functional, AR models (Banham and Katsaggelos, 1997), and regularization by perceptually based nonlinear divisive normalization functionals (Gutiérrez *et al.*, 2006). On the other hand, the second family of methods considers denoising in wavelet domains, such as classical thresholding techniques (Donoho and Johnstone, 1995), Bayesian denoising assuming Gaussian noise and signal with Gaussian marginals in the wavelet domain (Figueiredo and Nowak, 2001), and mutual information kernel-based regularization in the steerable wavelet domain (Laparra *et al.*, 2010). The toolbox allows one to study the nature of different noise sources in the particular case of images since it includes image degradation routines to generate controlled image blurring plus white (or tunable colored) Gaussian noise.

Noise in Communication Systems

In a typical communications system, the signal demodulated by the receiver can be expressed as a train of pulses of the form

$$p(t) = \text{sinc}(t - T) = \frac{\sin[\pi(t - T)/T]}{\pi(t - T)/t}. \tag{3.52}$$

The pulse is represented in Figure 3.7. This is known as a *Nyquist pulse* (Proakis and Salehi, 2004), a pulse with all samples equal to zero except at instant T, where the sampling frequency $f_s = 1/T$. Actually, it is easy to show that $\lim_{t\to T} \text{sinc}(t - T) = 1$ and $\lim_{t\to nT} \text{sinc}(t - T) = 1$ if $n \neq 1$. Then, when a train of pulses $\text{sinc}(t - mT)$ is modulated as

$$x(t) = \sum_{m=0}^{N} a_m \text{sinc}(t - mT), \tag{3.53}$$

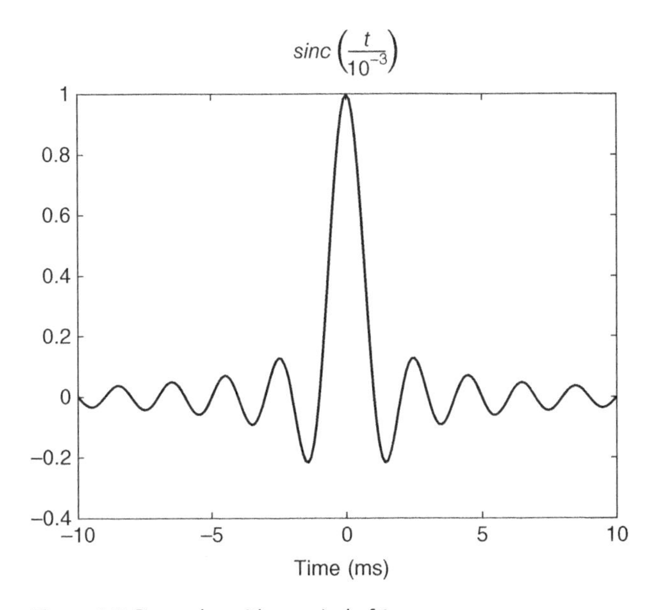

Figure 3.7 Sinc pulse with a period of 1 ms.

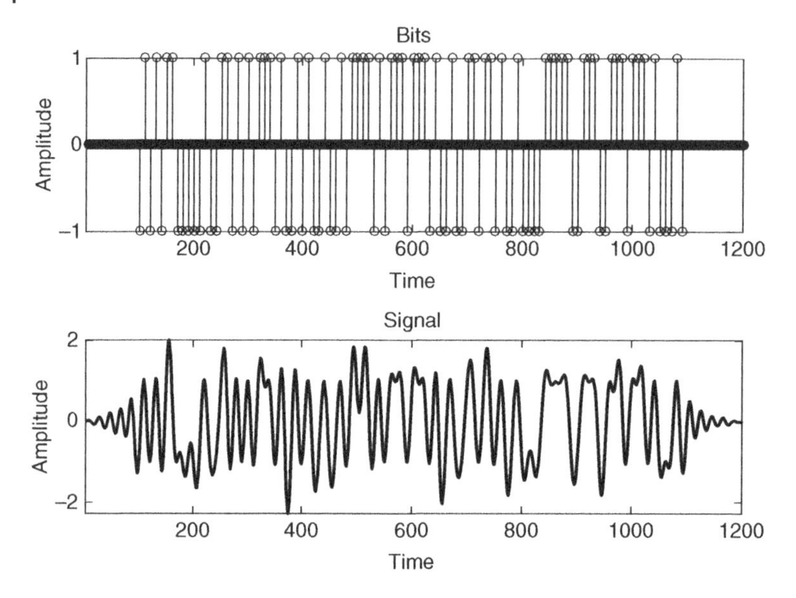

Figure 3.8 Upper panel: sequence of binary symbols. Lower panel: train of sinc pulses modulated by the symbol sequence of the upper panel.

and it is sampled at a frequency $f_s = 1/T$, the resulting discrete signal is

$$x_n = x(nT) = \sum_{n=0}^{N} a_m \delta_{n-m}, \tag{3.54}$$

where δ_n is the Kronecker delta, thus recovering each one of the transmitted symbols without intersymbol interference. That is, at each instant nT a symbol a_n is recovered. This pulse has a bandwidth equal to $1/2T$, so it can be transmitted through a channel having the same bandwidth without distortion.

Figure 3.8 depicts a sequence of binary signals that modulate the sinc pulse as in Equation 3.53. Figure 3.9 shows the detailed modulation process. The thick line represents the modulated signal, which is a composition of each one of the pulses delayed a time nT, for $0 \leq n \leq 5$. Each pulse is multiplied by a binary symbol of value ± 1. The squares represent the sampling instants, where it is clear that the pulse sequence has a value equivalent to the corresponding bit.

Figure 3.10 represents the so-called *eye diagram*, which is a representation of the superposition of each interval $(n-1)T \leq t \leq (n+1)T$. The center instant of the graph contains all the sampling instants of the signal. It is easy to see that at instants nT the signal is either +1 or −1, where the zero crosses correspond to the initial and final tails of the signal.

In the conditions of the aforementioned model, the signal can be recovered in the receiver without error. The received signals are nevertheless corrupted by additive noise of power σ_n^2. The noise is usually Gaussian in nature, and it is the main source or detection error. If the noise has a power spectral density N_0 W/Hz, then the received samples will contain a noise of power $\sigma_n^2 = N_0/2T$. Hence, the noise power will increase with the symbol rate. Thus, noise poses a limit in the symbol rate of a communication

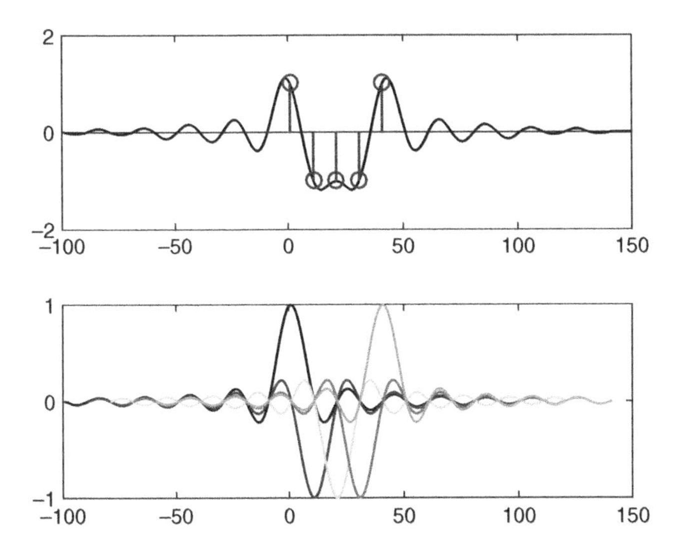

Figure 3.9 Detail of a train of pulses modulated by five binary symbols. The signals represented with thin lines in the lower panel are each one of the pulses whose linear combination is the modulated signal, represented in the upper panel with a thick line. Each circle corresponds to the amplitude of the signal at sampling instants nT.

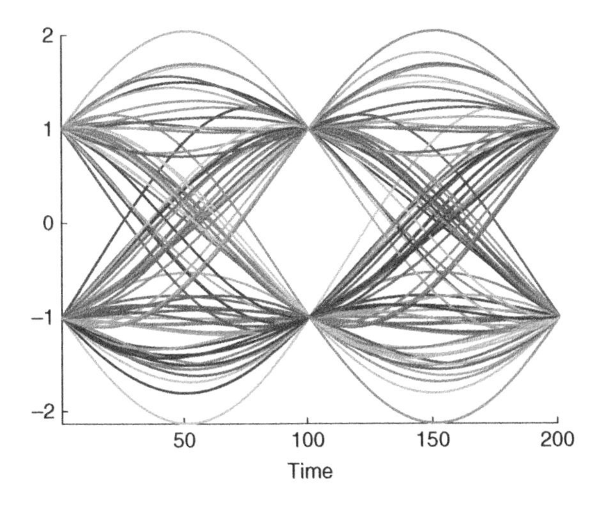

Figure 3.10 Eye diagram of a pulse train modulated by a binary sequence of symbols.

system. As an example, assume a 64 QAM transmission (e.g., see Proakis and Salehi (2004)); this is a signal carrying $M^{0.5} = 8$ possible symbols in the in-phase and quadrature-phase components (see Section 3.2.5 for a more detailed description of phase and quadrature representations of a signal). The signal corrupted by additive Gaussian noise is expressed as

$$x(t) = \sum_{m=0}^{N} a_{m,\mathrm{I}}\mathrm{sinc}(t - mT) + \mathrm{i} \sum_{m=0}^{N} a_{m,\mathrm{Q}}\mathrm{sinc}(t - mT). \tag{3.55}$$

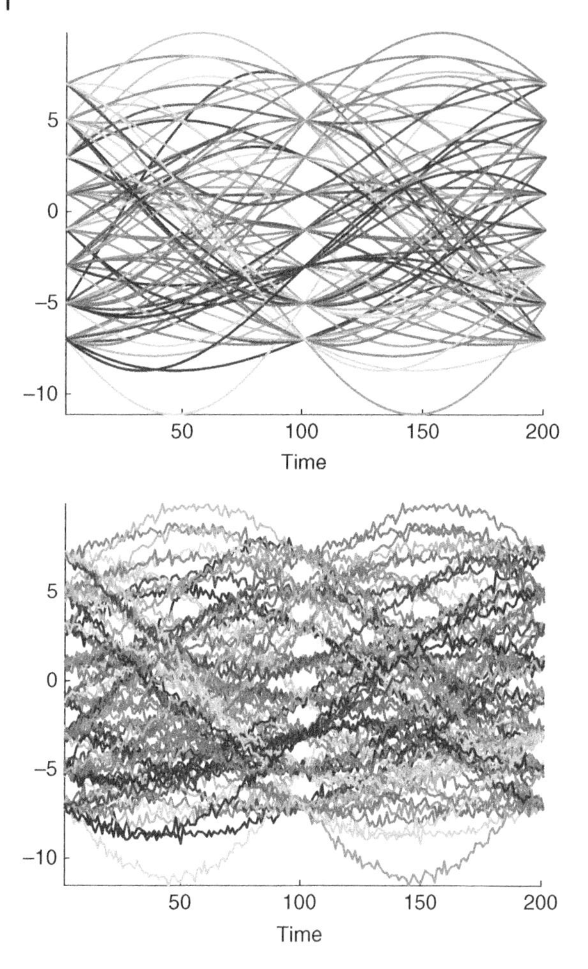

Figure 3.11 Representation of the in-phase or quadrature-phase eye diagram of a 64 QAM. Upper panel shows the noise-free diagram, where it can be seen that all symbols cross a specific level during the sampling instant. Lower panel shows the same signal, but corrupted by a white noise of power $\sigma_n^2 = 0.05$ W.

When sampled at a frequency $f_s = 1/T$, the resulting discrete signal is

$$x_n = x(nT) = \sum_{m=0}^{N} (a_{m,\mathrm{I}}\delta_{n-m} + \mathrm{i}a_{m,\mathrm{Q}}\delta_{n-m}). \tag{3.56}$$

In the presence of noise, the eye diagram of the signal will not cross the symbol amplitudes at the sampling instant, but an amplitude error will be measured. Figure 3.11 shows the eye diagrams of one of the phases of 1000 periods of the signal (in phase or quadrature phase) in a noise-free environment and in the presence of noise. The recovered symbols, once sampled, can be represented in the complex plane. This representation is called a signal constellation, and gives a visual representation of the noise corruption. Figure 3.12 shows the constellations of the signals of Figure 3.11.

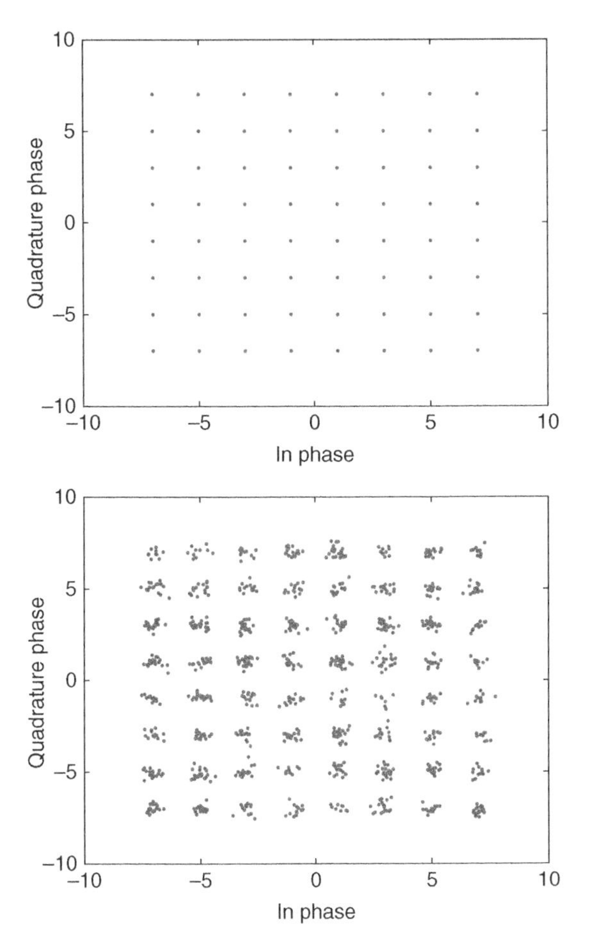

Figure 3.12 Constellation of a 64 QAM. Upper panel shows the noise-free constellation, where it can be seen that all symbols can be detected without error. Lower panel shows the constellation of the QAM, corrupted by a white noise of power $\sigma_n^2 = 0.05$ W. Here, the signal detection is subject to an error rate that increases with the noise power.

Noise in Biomedical Signals

A white Gaussian noise distribution is a reasonable assumption for many DSP scenarios, but it is not always the best choice for physiological and medical signals. A good example can be found in ECG noise. Holter monitoring systems are used for long-term recording of the cardiac activity in patients with suspect of underlying arrhythmia which cannot be elucidated with the usual 10 s of signals during consultation with the cardiologist. In this situation, an ambulatory device is used by the patient over about 24 h, hence increasing the amount of information about the heart rhythm and behavior (Lee *et al.*, 2012). The daily activity, in combination with the adherence of the patches to the patient's skin, are sources of very different kinds of noise.

Figure 3.13 shows several examples of noise in ECG recordings during Holter monitoring. One of the noise sources most often present is *baseline wander*, which is a low-frequency noise with changing bandwidth, often due to patient activity or movement, respiration, and changes in the time scale of several seconds. A different kind of noise or artifacts is due to the movement or slipping of the ECG electrode patches on the skin. This exhibits some nonphysiological patterns which are not easy to characterize and to eliminate from the recording. Also, the ECG quantization is usually done with 12 or 16 bits in the analog to digital conversion, which is a quite generous rate compared with other physiological signals. However, when the ECG amplitude becomes extremely small (maybe due to a change in the impedance or contact conditions), the low-amplitude recording can be noticeably affected by the quantization noise, which is characterized by a step-like appearance of the signal, and has a distorting impact on the ECG morphology and clinical measurements.

We already discussed in Section 2.5.4 that the presence of high-frequency noise can be distorting, especially due to the network interference at 50 or 60 Hz (and maybe several of their harmonics). This is a well-localized noise that can often be cancelled with notch-filtering, which is implemented in all the cardiac signals processing systems. However, another high-frequency noise of different nature is due to the interference of the muscle activity, which typically lasts some seconds. Muscle noise could be easily confounded

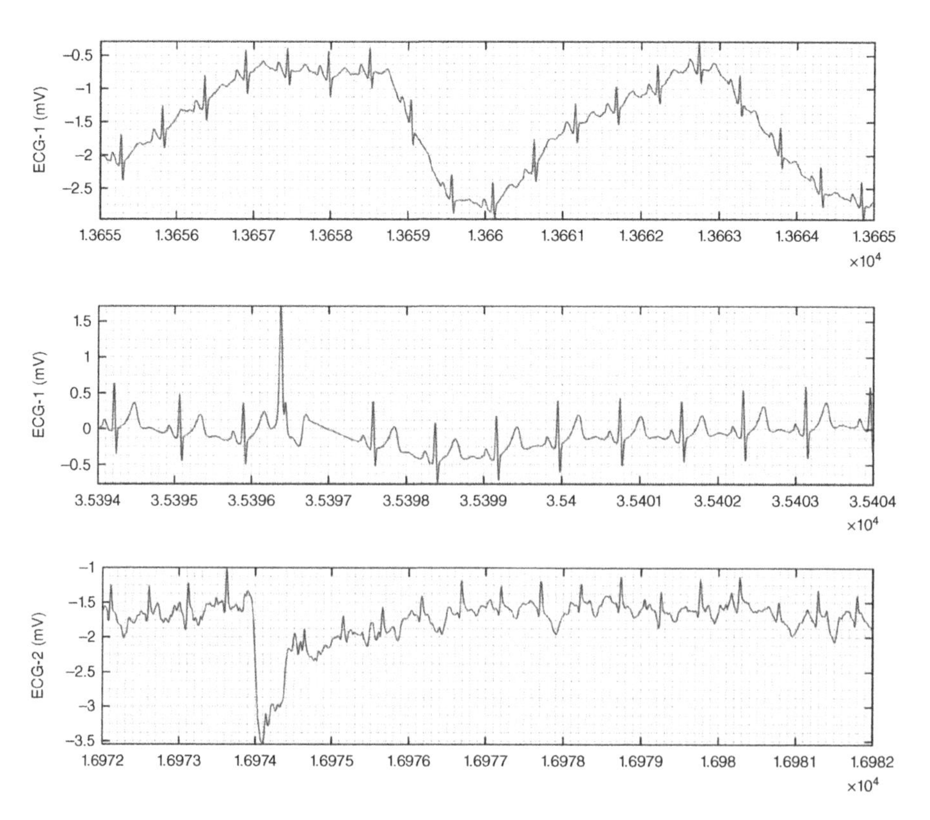

Figure 3.13 Different kinds of noise in ECG recorded during Holter monitoring, from top to bottom: baseline wander; probably lack of electrode contact; step in the middle; some other unknown source of distortion, affecting the ECG morphology; strong quantification noise; and weak quantification noise. Time in seconds from short segments of long-term monitoring recordings.

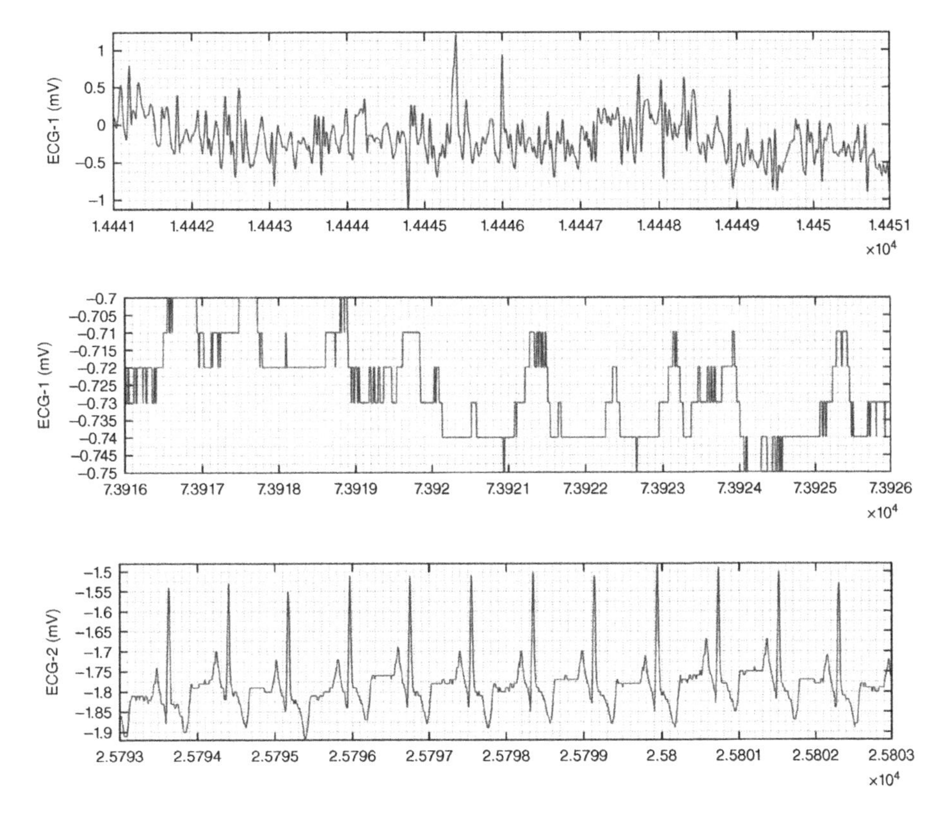

Figure 3.13 (Continued)

with network noise in the time domain. Nevertheless, its observation in the frequency domain shows that muscle interference cannot be readily filtered out, as far as it may consist of a broadband activity, strongly overlapped with the ECG band of interest. This can be seen in Figure 3.14, which has been made using the code in Listing 3.1.

```
%% Load data examples
load Noisy_ecg.mat
fs = 600;    % We know the sampling frequency
ts = 1/fs; n = length(clean_ecg); t = ts*(0:n-1);
% Time axis to plot
tmax = 20; % [s]
ind = find(t<tmax);
%% Plots in time
figure(1),
subplot(311), plot(t(ind),clean_ecg(ind)); ylabel('mV'),
subplot(312), plot(t(ind),noise(ind)); ylabel('mV'),
subplot(313), plot(t(ind),ecg_signal(ind));
ylabel('mV'), xlabel('t (s)')
%% Plots in frequency
X1 = abs(fftshift(fft(clean_ecg(ind)-mean(clean_ecg(ind)))));
X2 = abs(fftshift(fft(noise(ind)-mean(noise(ind)))));
X3 = abs(fftshift(fft(ecg_signal(ind)-mean(ecg_signal(ind)))));
```

```
f = linspace(-fs/2,fs/2,length(X1));
figure(2),
subplot(311), plot(f,X1); ylabel('FFT'),
axis tight, ax = axis; ax(1:2)=[-100 100]; axis(ax);
subplot(312), plot(f,X2); ylabel('FFT'),
axis(ax); subplot(313), plot(f,X3); ylabel('FFT'), xlabel('f (Hz)')
%% Try to filter out ...
filtord = 256+1;      % Try other orders
fc = 50;              % Try e.g. cutting freq 80 and 50
b = fir1(filtord,2*fc/fs);
[H,f] = freqz(b,1,1024,fs);  % Check and plot the design is OK
figure(3),
subplot(211), stem(b); axis tight; xlabel('n');
subplot(212), plot(f,abs(H)); axis tight; xlabel('f (Hz)');
%% ... but check for distortion in the residuals
ecg_filtered = filtfilt(b,1,ecg_signal);
resid = ecg_signal - ecg_filtered;
% Plot results
figure(4)
subplot(211), plot(t(ind),ecg_filtered(ind)), ylabel('mV'); axis tight
subplot(212), plot(t(ind),resid(ind)), ylabel('mV'); axis tight; ...
    xlabel('t(s)');
```

Listing 3.1 Frequency analysis and filtering of ECG signals (ecg1.m).

It is worth noting that *filtering* can be useful in many cases, but not always. If we do not filter with caution, we could be distorting the ECG morphology, which is highly valuable from a clinical point of view. In Figure 3.15, we can see the effect of filtering on several of the preceding example signals. One of the cautions to decide the suitability of the filter is to plot the residuals. Residuals showing visible peaks coming from the ECG peaks suggest that we have distorted the ECG signal in the QRS complex. These are the prominent and narrow visible peaks corresponding to the ventricular depolarization. Then, many clinically meaningful measurements made on the QRS complex, such as its time duration, will be likely distorted. The reader can analyze the code for generating the previous examples, and is encouraged to change the filter settings in order to see the effect on the signal distortion and on the noise distribution.

3.4.2 Autoregressive Exogenous System Identification Models

Let us now deal with system identification problems. In particular, we will work with the following test system to be identified:

$$y_n = 0.03y_{n-1} - 0.01y_{n-2} + 3x_n - 0.5x_{n-1} + 0.2x_{n-2}. \tag{3.57}$$

This system is used here because of the different orders of magnitude of the samples in its impulse response. In order to identify this process we will use as input a white Gaussian noise discrete-time process with unit variance, denoted as $x_n \sim \mathcal{N}(0,1)$. To make the problem harder and more interesting, we will corrupt the corresponding output by an additive, small-variance random process $e_n \sim \mathcal{N}(0, 0.1)$. This leads to an observed process $o_n = y_n + e_n$. The number of observed samples is $N = 100$. We use an

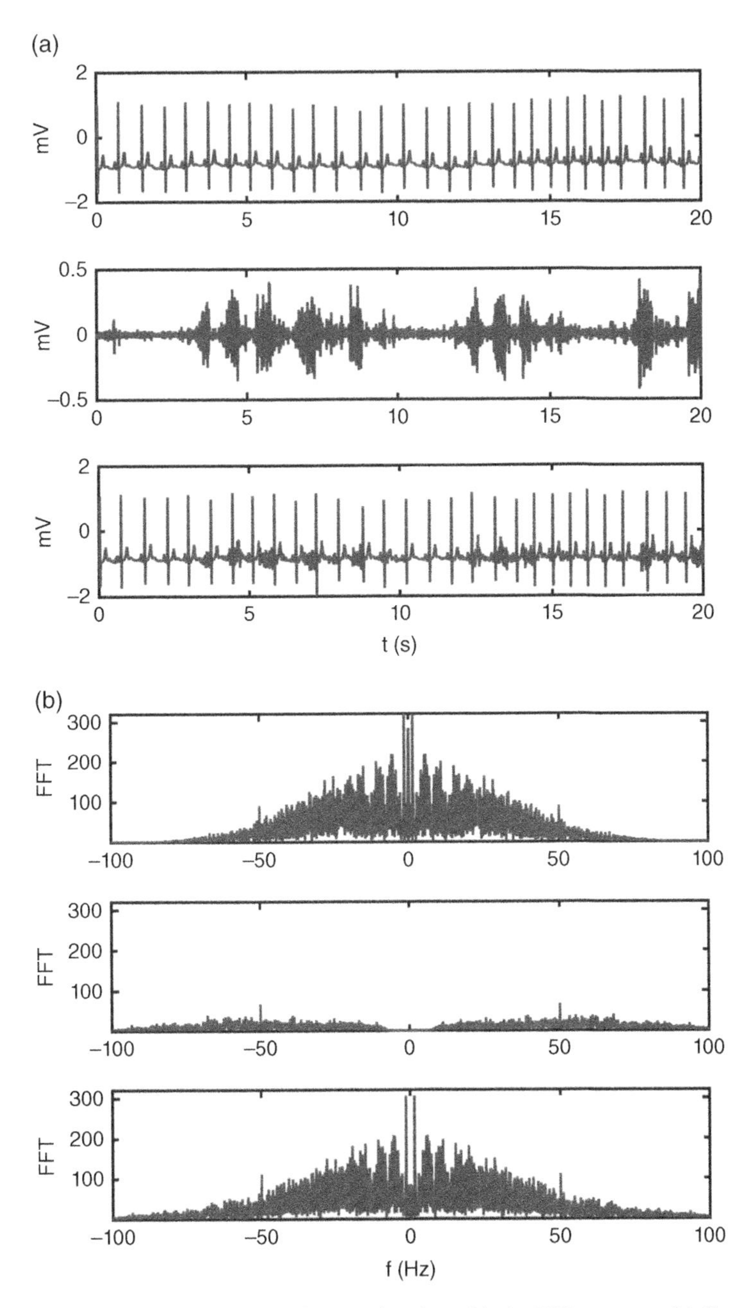

Figure 3.14 Muscle noise and its overlapping with the ECG spectrum. (a) Clean ECG, muscle noise, and noisy ECG. (b) Spectrum for each of the aforementioned signals.

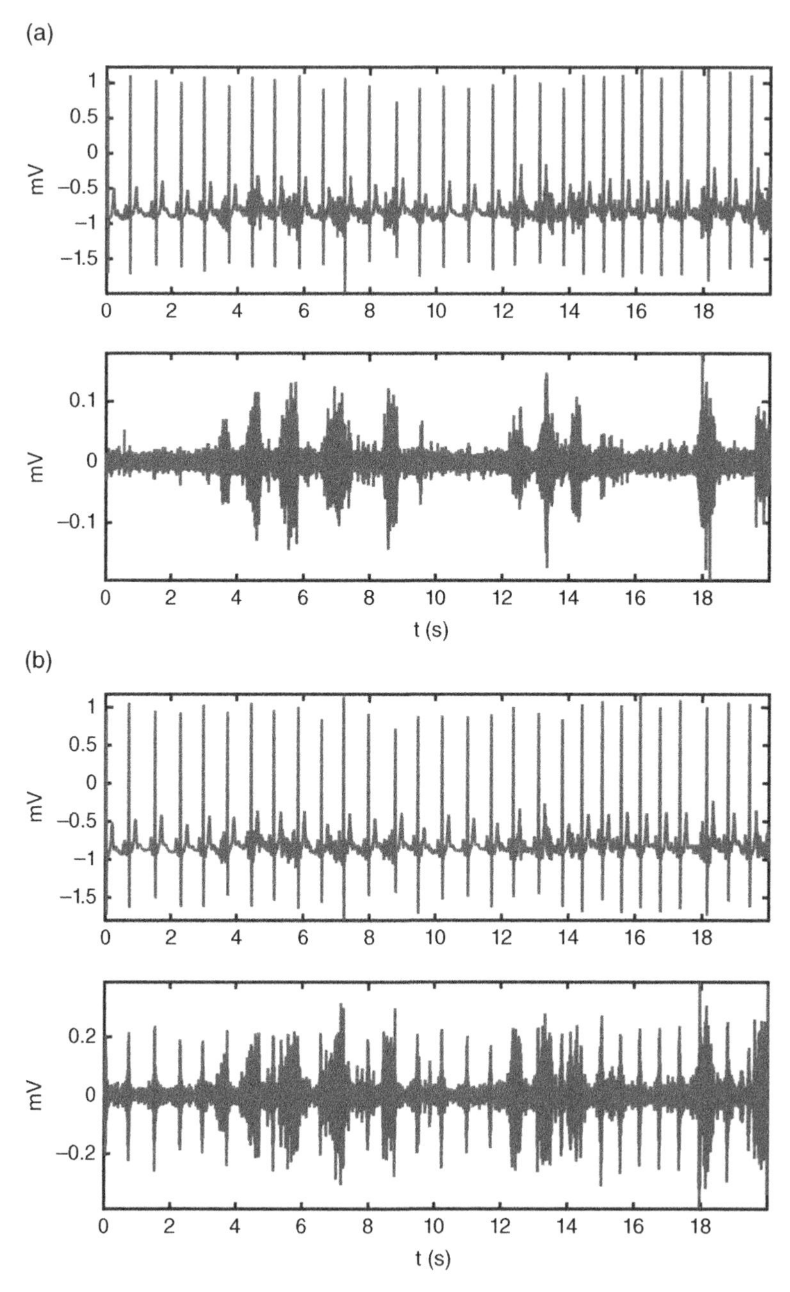

Figure 3.15 Muscle noise and its overlapping with the ECG spectrum. ECG noise filtering (upper panel) and the residuals after filtering (lower panel), for (a) $f_s = 80$ Hz and (b) $f_s = 50$ Hz.

ARX model with LS to identify the system, and we assume that we know the order of the input system. The sample generation code is given in Listing 3.2, and the MATLAB implementation with the `arx` function is given in Listing 3.3.

```
N = 100;
x = randn(N,1);
b = [3 -0.5  0.2];
a = [1 -0.03 0.01];
y = filter(b,a,x);
% ... plus noise ...
yy = y + sqrt(.1) * randn(size(y));
% Filter impulse response
h = impz(b,a,10);
```

Listing 3.2 Sample generation from a system and impulse response (sysid1.m).

```
% Assume we know the order of the system to be identified
p = 2; q = 3; m = arx([yy(:),x(:)],[p,q,0]); [a_est,b_est] = polydata(m);
% Instead of comparing the obtained coefficients, we will compare the ...
   response of the estimated system in terms of the impulse and ...
   frequency responses
fvtool(b,a,b_est,a_est);
```

Listing 3.3 System identification via ARX and frequency analysis (sysid2.m).

Figure 3.16 shows that the method is able to successfully simulate (and thus to identify) the system, despite the Gaussian noise present in the output signal. However, the LS method is very sensitive to outliers. To illustrate this point, let us add some impulsive noise to the output signal. The impulsive noise is generated as a sparse sequence with 30% of randomly placed, high-amplitude samples, given by $\pm 10+\mathcal{U}(0,1)$, where $\mathcal{U}(a,b)$ denotes the uniform distribution in interval $[a,b]$, and the remaining are zero samples. This noise sequence is denoted by j_n. The observations consists of input x_n and the observed output plus impulsive noise; that is, $o_n + \sigma_w j_n$. Values of σ_w are in the range from −18 to 0 dB. The code is given in Listing 3.4, and Figure 3.17 shows the results. As we can see, the identification is now very poor. We will see in the next chapters how kernel methods can successfully deal with this kind of noise.

```
% Jitter generation
w = ceil(N*rand(30,1));
w = unique(w); % indexes for jitter
jit = rand(size(w)) - .5;
jit = (sign(jit)*10)+jit;
% Mixing
yy(w) = jit;
% Now we repeat the same procedure to model the system ...
p = 2; q = 3;
m = arx([yy(:),x(:)],[p,q,0]);
[a_est,b_est] = polydata(m);
% Compare results using fvtool
fvtool(b,a,b_est,a_est);
```

Listing 3.4 Identification of a system with ARX under jitter noise (sysid3.m).

(a)

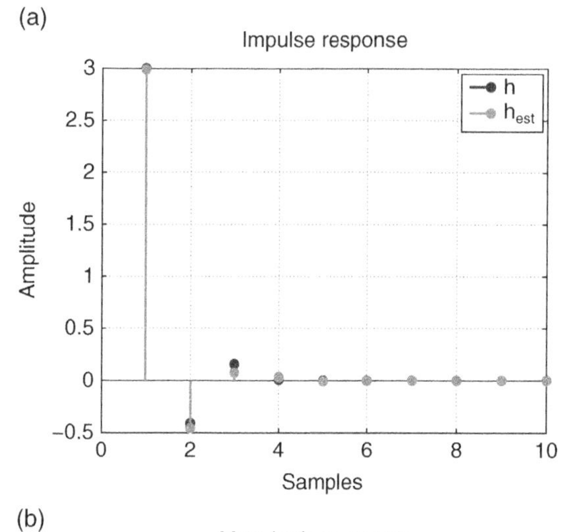

(b)

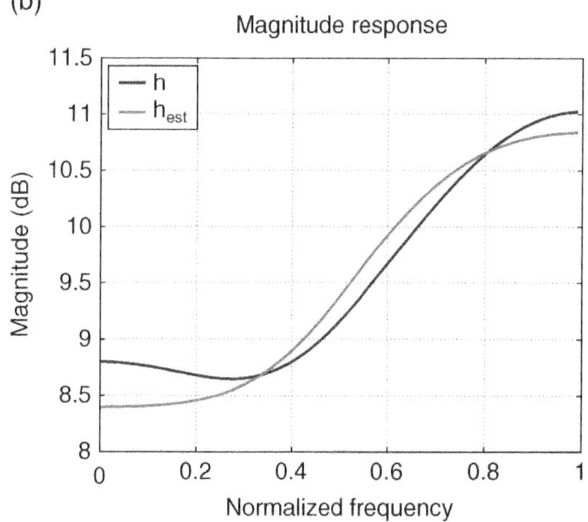

(c)

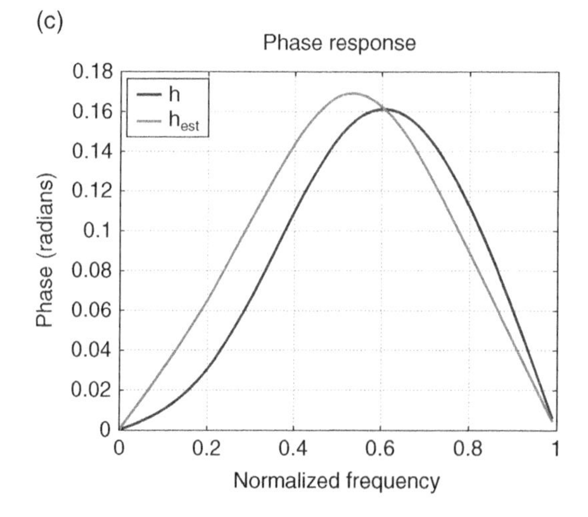

Figure 3.16 Comparison between the original and the estimated systems in terms of the impulse response (top), the magnitude response (middle), and the phase response (bottom).

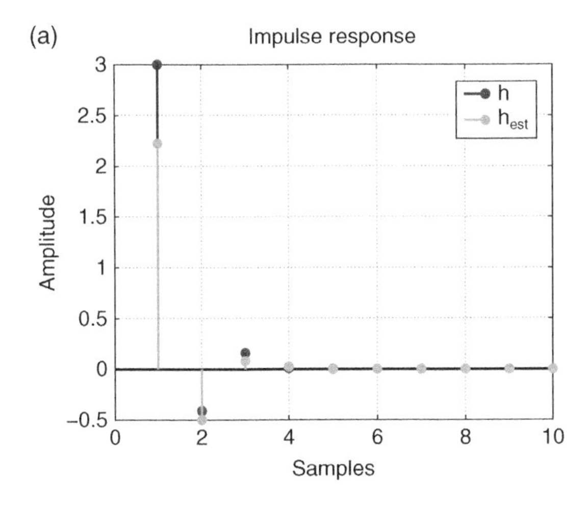

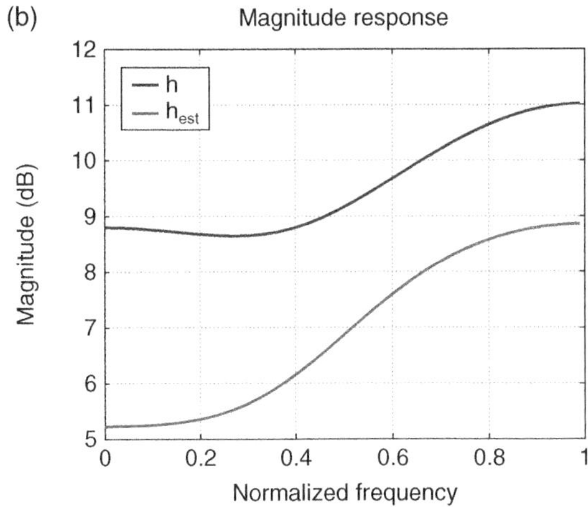

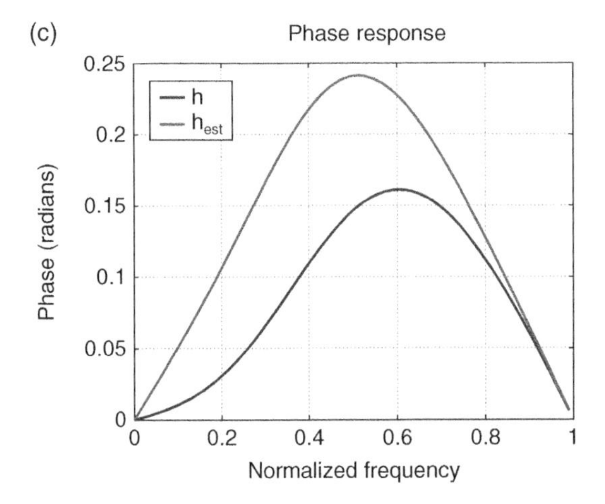

Figure 3.17 Comparison between the original and the estimated systems in terms of the impulse response (top), the magnitude response (middle), and the phase response (bottom).

3.4.3 Nonlinear System Identification Using Volterra Models

A system that shows a nonlinear behavior can be identified by constructing a polynomial function of the output with a Volterra model. Assume a signal x_n passing through a system as

$$y_n = f(x_n, \dots, x_{n-K}), \tag{3.58}$$

where f is a nonlinear function. A model to identify this signal can be expressed as

$$\begin{aligned} y_n = b_0 &+ \sum_{k=1}^{D} b_k x_{n-k} + \sum_{k_1=1}^{D}\sum_{k_2=1}^{D} b_{k_1,k_2} x_{n-k_1} x_{n-k_2} \\ &+ \sum_{k_1=1}^{D}\sum_{k_2=1}^{D}\sum_{k_3=1}^{D} b_{k_1,k_2,k_3} x_{n-k_1} x_{n-k_2} x_{n-k_3} + \cdots, \end{aligned} \tag{3.59}$$

which contains all possible monomials $x_{n-k_1} x_{n-k_2} x_{n-k_3} \cdots$ up to order P among all sample signals x_n to x_n. Model coefficients $b_{k_1,k_2,\dots}$ of the monomials can be adjusted by a simple MMSE strategy. The algorithm consists of applying a nonlinear transformation $\boldsymbol{\phi}(\boldsymbol{x}_n)$ to vector $\boldsymbol{x}_n = [x_n, \dots, x_{n-D}]^{\mathrm{T}}$ of the form

$$\boldsymbol{\phi}(\boldsymbol{x}_n) = \begin{pmatrix} \mathbf{1} \\ \boldsymbol{x}_n \\ \vdots \\ x_{n-k_1} x_{n-k_2} \\ \vdots \\ x_{n-k_1} x_{n-k_2} x_{n-k_3} \\ \vdots \end{pmatrix} \tag{3.60}$$

that contains all the monomials, and a weight vector of the form

$$\boldsymbol{w} = \begin{pmatrix} b_0 \\ \vdots \\ b_k \\ \vdots \\ b_{k_1,k_2} \\ \vdots \\ b_{k_1,k_2,k_3} \\ \vdots \end{pmatrix} \tag{3.61}$$

such that the model can be written in the compact form $y_n = \boldsymbol{w}^{\mathrm{T}} \boldsymbol{\phi}(\boldsymbol{x}_n)$.

Clearly, a simple way to find a solution of these parameters is

$$\boldsymbol{w} = (\boldsymbol{\Phi}^{\mathrm{T}} \boldsymbol{\Phi})^{-1} \boldsymbol{\Phi} \boldsymbol{y}, \tag{3.62}$$

where $\boldsymbol{\Phi}$ is a matrix containing all training vectors $\boldsymbol{\phi}(\boldsymbol{x}_n)$, and $\boldsymbol{y}$ is a vector containing all training observations y_n. We should note here that, while this Volterra approach considers the explicit calculation of matrix $\boldsymbol{\Phi}$, we will see in subsequent chapters that this is not really necessary, and we can work by using kernel functions that implicitly approximate this matrix.

```
function volterra1
for P = 1:2:7 % From order 1 to 7
    [b,a] = butter(4,0.005); randn('seed', 0);
    D = 5; % Dimension of the input
    C = volt(D,P); % Compute the volterrra monomial indexes
    % System, training data
    m = randn(1,1000);
    xm = filter(b,a,m)+fliplr(filter(b,a,m));
    x = 10*sin(0.1*(1:1000)).*xm; % Modulated signal
    y = system(x); % Apply distortion
    % System, test data
    m = randn(1,1000);
    xm = filter(b,a,m)+fliplr(filter(b,a,m));
    xt = 10*sin(0.1*(1:1000)).*xm;
    yt = system(xt);
    % Train
    X = buffer(x,D,D-1,'nodelay');
    XX = volterraMatrix(X,C);
    XX = [XX ; ones(1,size(XX,2))]; % Add a constant for the bias
    % Compute the MMSE coefficients
    w = pinv(XX*XX')*XX*y(1:size(XX,2));
    % Test
    Xtst = buffer(xt,D,D-1,'nodelay');
    XXtst = volterraMatrix(Xtst,C);
    XXtst = [XXtst ; ones(1,size(XXtst,2))];
    Ytst_ = XXtst'*w;
    % Represent the signal
    hold off, plot(Ytst_),hold all, plot(yt), drawnow
    pause
end

%% Callback functions

% Compute the indexes of the monomials
function C = volt(D,P)
x = (1:D)';
for i = 1:P-1
    x = coefficients(x);
    C.(['x' num2str(i+1)]) = x;
end
C.P = P; % Store the order

% Compute the next set of indices from indices x
function z = coefficients(x)
z = [];
d = max(x(:));
for i = 1:size(x,1)
    for k = x(i,end):d
        z = [z; [x(i,:) k]];
    end
end
```

```
% Compute all monomials of vectors in X
function XX = volterraMatrix(X,C)
XX = X;
P = C.P;
for i = 2:P
    aux = 1;
    idx = C.(['x' num2str(i)]);
    for j = 1:size(idx,2)
        aux = aux.* X(idx(:,i),:);
    end
    XX = [XX ; aux];
end

% Nonlinear system that filters and compresses the signal
function y = system(x)
a = [0.1 0.1 0.43 0.9 0.81]';
X = buffer(x,5,4,'nodelay');
z = X'*a; y = tansig(0.7*z);
```

Listing 3.5 Example of Volterra identification (volterra1.m).

An implementation of the algorithm can be found in Listing 3.5, where a modulated sinusoidal is linearly filtered and processed through a hyperbolic tangent function that simulates a signal distortion by saturation. The system is identified by a Volterra expansion of order P. The input signal is buffered in a vector of dimension $D = 5$. Function `volt` computes the indexes corresponding to all Volterra monomials and then function `volterraMatrix` computes all these monomials to construct a matrix containing the Volterra expansion of all input vectors. The script first synthesizes a training dataset to adjust the parameters, and then the estimator is tested with a second dataset. The system identification method has been tested for different degrees of the Volterra model from $P = 1$ (linear identification) to $P = 7$ (higher order relations).

Figure 3.18 shows the result. The first panel shows the system output and the output simulated through linear identification, which presents high differences with respect to the original, due to its inability to identify the nonlinearities. The next panels show the approximation with orders $P = 3$, 5, and 7 increasing the approximation quality.

3.4.4 Sinusoidal Signal Models

In this subsection we present some examples of *spectral estimation* in nonuniform sampling conditions with the sinusoidal signal model. Let us start by building the LS solution. Using matrix notation in Equation 3.21, we can state the problem as that of minimizing the quadratic error, and thus the minimization functional can be written as

$$\mathcal{L}(\boldsymbol{b}, \boldsymbol{c}) = \|\boldsymbol{e}\|^2 = \|\boldsymbol{y} - \boldsymbol{M}\boldsymbol{b} - \boldsymbol{N}\boldsymbol{c}\|^2. \tag{3.63}$$

We encourage the reader to obtain the gradient with respect to the unknowns and to make it equal to zero in order to obtain the solution. After doing so, the MATLAB function in Listing 3.6 can be used.

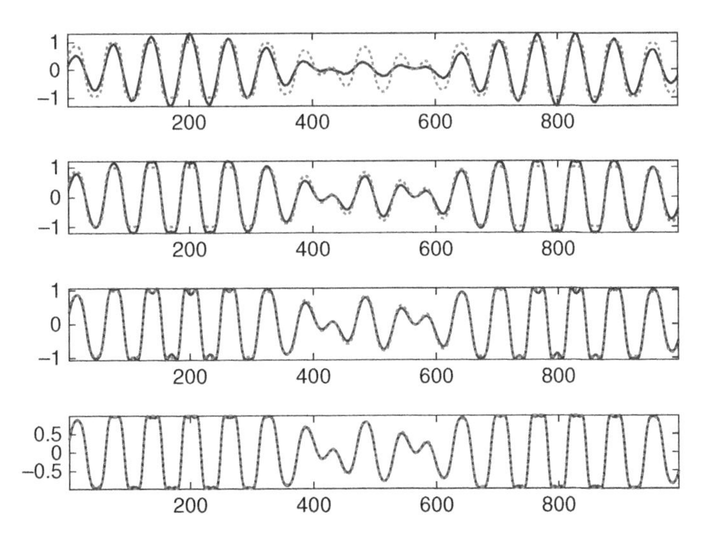

Figure 3.18 Amplitude-modulated sinusoidal processed by a linear system and by a Volterra plant identification system. The first panel shows the system output (dots) and the output simulated through linear identification (continuous line), which presents high differences with respect to the original, due to its inability to identify the nonlinearities. The next panels show the approximation with orders $P = 3, 5$, and 7.

```
function [B,C,A,Phi,y_test] = sinusoidals(y_train,t_train,t_test,w)
% Data
n = length(t_train); m = length(w);
M = zeros(m,n);    % Data matrix
N = zeros(m,n);
for i = 1:m;
    M(i,:) = cos(w(i)*t_train');
    N(i,:) = -sin(w(i)*t_train');
end
MN = [M',N'];
gam = 1e-5;
D = inv(MN'*MN + gam*eye(2*m)) * (MN'*y_train);
% Solution in polar coordinates
B = D(1:m);
C = D(m+1:end);
% Solution in cartesian coordinates
A = sqrt(B.^2+C.^2);
Phi = atan(C./B);
% Test estimation
y_test = zeros(size(t_test));
for i = 1:m
    y_test = y_test + A(i)*cos(w(i)*t_test+Phi(i));
end
```

Listing 3.6 LS optimization of a sinusoidal signal model (sinusoidals.m).

Note that this function requires a vector of frequency hypothesis, as well as the pairs of time sampling and measured samples. It also supports a test set of time samples, which usually consist of a grid of equally sampled time instants, to check the interpolation achieved by the method. One should note also that a regularization term is included, which controls the smoothness of the solution. The code in Listing 3.7 can be executed for testing these results. Note that, in this case, we are using a regularization value $\gamma = 10^{-3}$.

```
%% Example of synthetic data
ntr = 100; nsecs = 20;
t_train = nsecs*rand(ntr,1);
fdata = 0.3;
Ampdata = 3;
phidata = 0.5;
% Generate data
y_train = Ampdata*cos(2*pi*fdata*t_train + phidata);
% Hypothesis frequencies
f = 0:.1:15;
f = f(2:end);    % zero mean
%f = [0.1 0.2 0.3 0.4 0.5];
ntest = 1000;
t_test = linspace(0,nsecs,ntest);
% Least squares optimization of the sinusoidal model
[B,C,A,Phi,y_test] = sinusoidals(y_train,t_train,t_test,2*pi*f);
% Plot results
figure(1), plot(t_train,y_train,'.r'); hold on; plot(t_test,y_test,'b'); ...
    hold off;
xlabel('t (s)'); ylabel('Estimated and training signal')
figure(2), subplot(211), plot(f,A); grid on; xlabel('f (Hz)'); ylabel('A')
subplot(212), plot(f,Phi); grid on; xlabel('f (Hz)'); ylabel('\phi');
```

Listing 3.7 Testing a sinusoidal-based interpolator (sinusoidals2.m).

As shown in Figure 3.19a and b, the time reconstruction is loose; however, the amplitude is detected to be higher in the sinusoid frequency. Nevertheless, there is noticeable spectral noise due to the nonuniform sampling and to the grid of hypothesis frequencies. Better results now can be obtained in this trivial example by using a more specific frequency grid; for instance, `f = [0.1 0.2 0.3 0.4 0.5]` instead of the detailed grid yields the solution in Figure 3.19c and d, so that the amplitudes and phases are now more accurately estimated, as well as the reconstructed signal.

The reader is encouraged to check and explain the effect of changing the regularization parameter, as well as to use a frequency grid with values close to (but not exactly) the actual frequency of the sinusoidal data. Also, the effect of the presence of noise in the data can be readily scrutinized. All these checks will give an idea on the sensitivity of the method to the selection of the free parameters.

Another example is given by the analysis of an HRV signal. In this case, we use an example of an RR signal (sequence of times between consecutive beats) and extract 2000 time samples. Note that the beat sequence itself is used for the time instants when each RR interval is measured, hence giving a temporal basis in seconds. This is usually known as instantaneous cycle representation of the HRV, whereas the consideration of the discrete time series for the consecutive RR intervals, just in terms of their ordinal index, is called a *tachogram*. Hence, the instantaneous cycle has a clearer sense of actual time when the beats occur, and the corresponding spectral representations will be in a basis of hertz, whereas the tachogram is usually represented in terms of discrete frequency (often known as beatquency domain in the HRV literature).

```
% Example for HRV signal
load ejemploHRV RR
auxRR = RR(500:2500); auxt = cumsum(auxRR);
auxRR = auxRR/1000; auxt = auxt/1000;        % in seconds
% Plot
figure(1), clf
subplot(311), plot(auxRR), grid, hold on, axis tight
subplot(312), plot(auxt,auxRR), grid, hold on, axis tight
%% Remove ectopic beats
aux = find(abs(diff(auxRR))>0.1);
aux = unique([aux,aux+1]);
aux = setdiff(1:length(auxRR), aux);
auxt = auxt(aux);
auxRR = auxRR(aux);
subplot(313), plot(auxt,auxRR); grid, hold on, axis tight
%% Plot spectra
X = abs(fftshift(fft(auxRR-mean(auxRR))));
f = linspace(-0.5, 0.5, length(X));
% Plot results
figure(2), subplot(211), plot(f,X,'b'); hold on; ejes = axis; ejes(1)=0; ...
    axis(ejes); xlabel('f (tachogram)');
% Estimation
t_test = 0:.25:length(auxt);
ff = linspace(0,0.5,400);
[B,C,A,Phi,y_test] = sinusoidalLS(auxRR-mean(auxRR),auxt,t_test,2*pi*ff);
subplot(212), plot(ff,A), axis tight, xlabel('f (Hz)'), ...
    ylabel('Estimated amplitude');
% Plot results
figure(3),
subplot(211), plot(auxt,auxRR-mean(auxRR)), grid, axis tight, xlabel('t ...
    (s)'), ylabel('RR - mean (s)')
subplot(212), plot(t_test,y_test), grid, axis tight, xlabel('t (s)'), ...
    ylabel('RR - mean (s)')
```

Listing 3.8 Example of an HRV signal analysis (hrv1.m).

The code in Listing 3.8 is used to perform a set of preprocessing transformations on the RR sequence and to make a simple analysis of such an HRV signal in terms of tachogram and sinusoidal interpolation. First, the time basis is recovered, in seconds. Second, possible ectopic beats are removed, by just establishing a threshold in the difference of consecutive beats. Note that this is a rough step for this example, and the HRV signal will be strongly sensitive to this stage and to the threshold choice. The spectra are represented in terms of both instantaneous cycle and beatquency. Then, we use the LS criterion to perform an estimate of the spectrum with a hypothesis grid for the frequencies in the signal. Note that the zero frequency is removed from the frequency grid, but also the mean is removed from the observations, as far as it has not been included in our LS equations. The LS solution for the amplitudes is represented in terms of frequency (physical units or hertz). Figure 3.20 shows the results of the code. The reader is encouraged to determine which error is taking place in the estimation with the LS criterion and to propose a modification in the code in order to to obtain a better estimate.

3.4.5 Sinc-based Interpolation

In this section we evaluate the performance of five standard interpolation techniques. We will follow the steps in the toolbox `simpleInterp` available on this book's webpage. Details on the toolbox structure and main routines can also be found in Appendix 3.A. The reader is encouraged to spend some time mastering this, as it will be used in subsequent chapters for dealing with many of the signal problems in the framework in a systematic way. Note that in order to run experiments with `simpleInterp`, we will need for each problem: (1) an experiment configuration file; (2) a data load or generation function; (3) the algorithm function to be used; and (4) some code to visualize the results. We will go through all these steps in each of the examples using this toolbox.

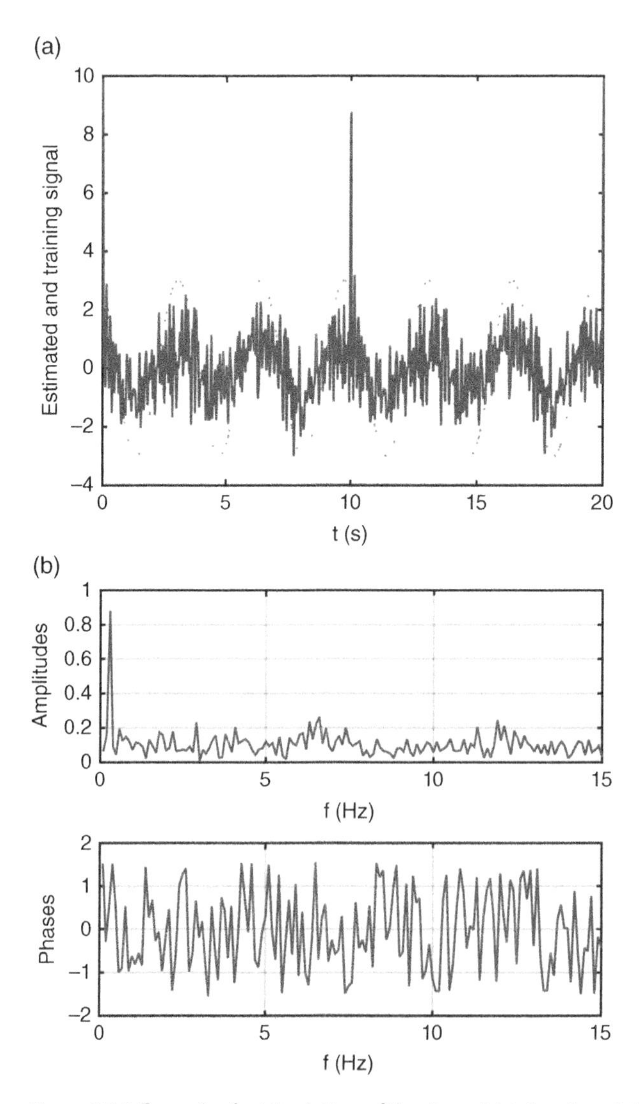

Figure 3.19 Examples for LS solution of the sinusoidal signal model: (a) training set of samples and solution for regularized inversion matrix; (b) estimated amplitudes and phases; (c,d) the same for a more reduced set of frequency hypothesis.

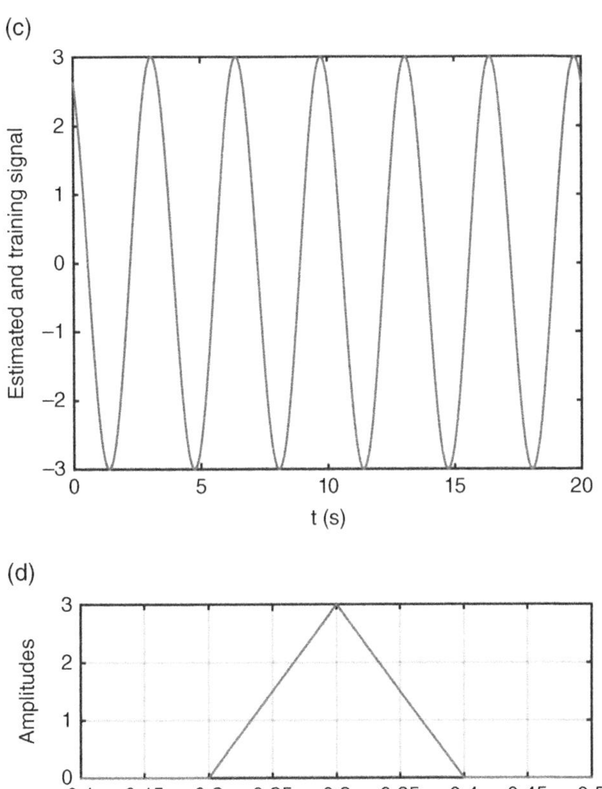

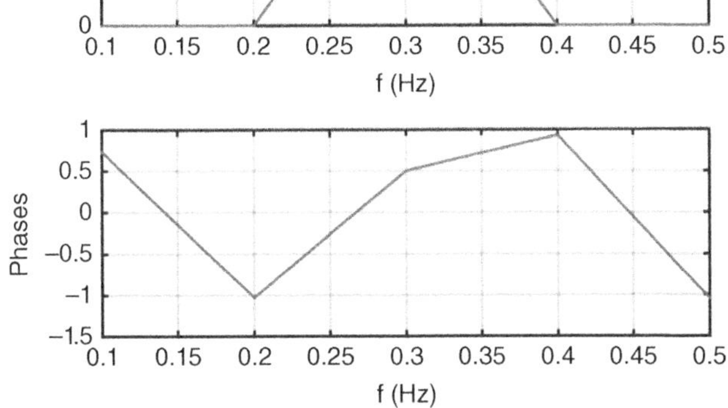

Figure 3.19 (Continued)

Interpolation Algorithms

Many nonoptimal interpolation algorithms have been proposed in the literature, which were benchmarked in Rojo-Álvarez *et al.* (2007). In this example, we include a detailed explanation of how to perform the comparison among methods. The Jacobian weighting (Choi and Munson, 1998a) uses the following direct interpolation equation:

$$x(t) = x^{\mathrm{J}}(t) + e(t) = \sum_{k=0}^{N} f_k x_k \mathrm{sinc}(\sigma_0(t - t_k)) + e(t), \tag{3.64}$$

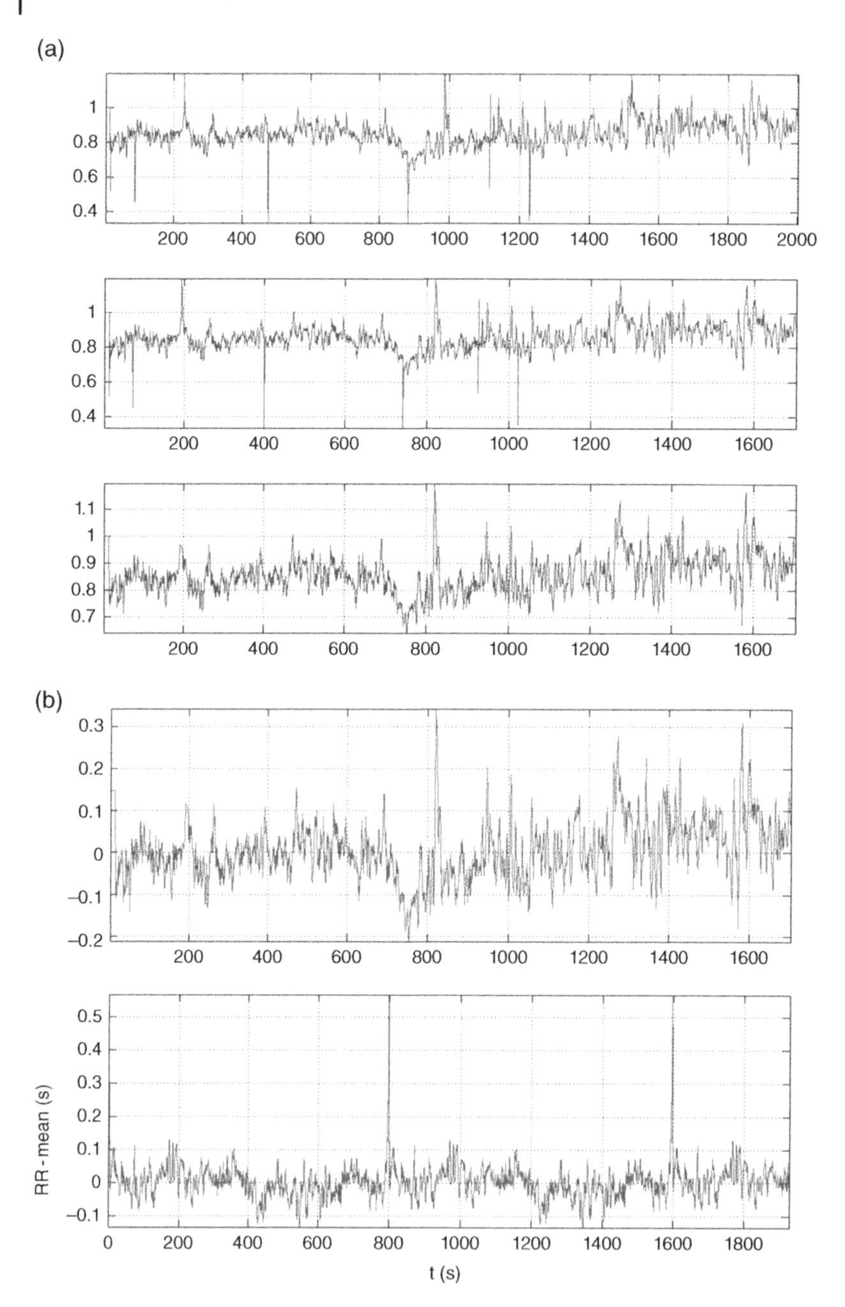

Figure 3.20 Examples for LS solution of the HRV estimation with the sinusoidal signal model. (a) Original signal in terms of tachogram, of instantaneous cycle, and without ectopic beats. (b) Original and LS reconstructed. (c) Estimated spectrum with FT, and estimated amplitudes with LS.

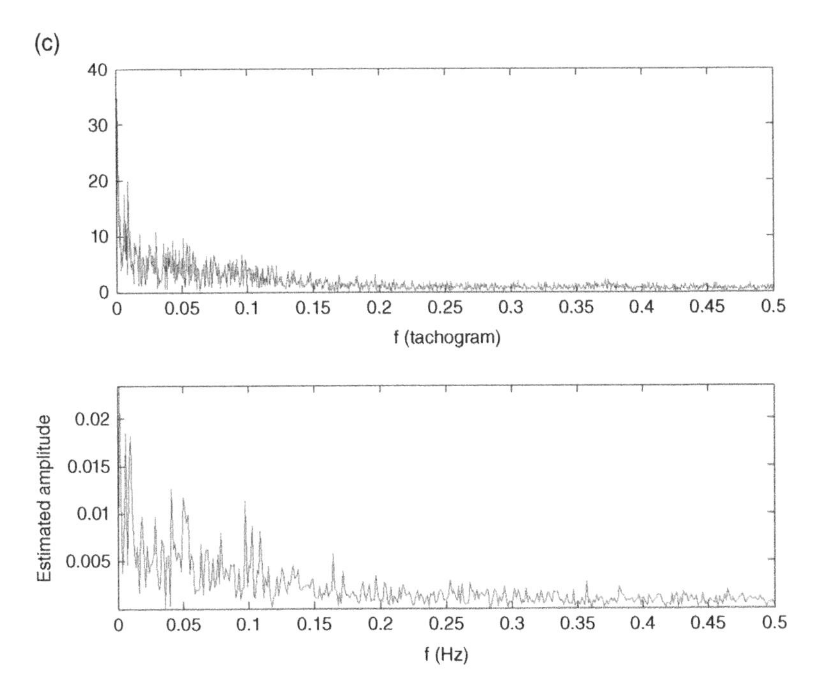

Figure 3.20 (Continued)

where coefficients f_i are chosen to be the sample spacings; that is, $f_k = t_{k+1} - t_k$. This corresponds to a Riemann sum approximation to the following integral:

$$x(t) = \int_{-\infty}^{+\infty} x(\tau)\mathrm{sinc}(\sigma_0(t-\tau))\,\mathrm{d}\tau, \tag{3.65}$$

where we are assuming that σ_0 is the true bandwidth of $x(t)$. This algorithm sometimes has poor performance in terms of interpolation accuracy, but it requires an extremely reduced computation burden, which can provide a nice trade-off in problems with a large number of available observations.

Another suboptimal yet rather improved approach was proposed by Choi and Munson (1998b), where a generalization of the sinc kernel interpolator was presented. The model relies on a *minimax optimality criterion* as an approximate design strategy, and it yields the following expression for the coefficients:

$$f_k = \frac{\pi}{\sigma_0}\left(\sum_{k=0}^{N} \mathrm{sinc}^2(\sigma_0(t_n - t_k))\right)^{-1}. \tag{3.66}$$

Both the performance and the computational burden of this approach are between the Yen and Jacobian sinc kernel interpolators. However, all these approaches exhibit some limitations, such as poor performance in low SNR scenarios, or in the presence of non-Gaussian noise (in this last case, as a direct consequence of the use of a quadratic loss function).

In this example, the following signal interpolators are considered:

1) Yen's interpolator without regularization (Y1).

```
function [y,coefs] = Y1(tk,yk,t,T0)
nk = length(tk);
S = sinc((repmat(tk,1,nk) - repmat(tk',nk,1))/T0)/T0;
B = pinv(S);
[y,coefs] = interpSinc(tk,yk,t,T0,B);
```

Listing 3.9 Yen's interpolator without regularization – Y1 algorithm (Y1.m).

2) Yen's interpolator with regularization (Y2).

```
function [y,coefs] = Y2(tk,yk,t,T0,gamma)
nk = length(tk);
S = sinc((repmat(tk,1,nk) - repmat(tk',nk,1))/T0)/T0;
B = (S'*S + gamma*eye(size(S))) \ S';
[y,coefs] = interpSinc(tk,yk,t,T0,B);
```

Listing 3.10 Yen's interpolator with regularization – Y2 algorithm (Y2.m).

3) Sinc interpolator with uniform weighting (S1).

```
function [y,coefs] = S1(tk,yk,t,T0)
B = 1; [y,coefs] = interpSinc(tk,yk,t,T0,B);
```

Listing 3.11 Sinc interpolator with uniform weighting – S1 algorithm (S1.m).

4) Sinc interpolator with Jacobian weighting (S2).

```
function [y,coefs] = S2(tk,yk,t,T0)
nk = length(tk);
tkdesp = zeros(nk,1);
tkdesp(2:nk) = tk(1:nk-1);
subtr = tk(2:end) - tkdesp(2:end);
subtr = [subtr(:); mean(subtr(1:end-1))];
B = diag(subtr,0);
[y,coefs] = interpSinc(tk,yk,t,T0,B);
```

Listing 3.12 Sinc interpolator with Jacobian weighting – S2 algorithm (S2.m).

5) Sinc interpolator with minimax weighting (S3).

```
function [y,coefs] = S3(tk,yk,t,T0)
nk = length(tk);
S = sinc((repmat(tk,1,nk) - repmat(tk',nk,1))/T0);
b = T0./sum(S.^2);
B = diag(b,0);
[y,coefs] = interpSinc(tk,yk,t,T0,B);
```

Listing 3.13 Sinc interpolator with minimax weighting – S3 algorithm (S3.m).

All these methods use the `interpSinc` function, which obtains a `y` reconstruction from the `yk` signal in a `t` nonuniform interpolation grid from the sinc base `B`. Listing 3.14 shows the code for the common part of all of these interpolation algorithms.

```
function [y,Ak] = interpSinc(tk,yk,t,T0,B)
Ak = B * yk;         % Coefficients
y=zeros(size(t)); % Reconstruction
```

```
for i=1:length(Ak)
    y = y + Ak(i)*(sinc((t-tk(i))/T0)/T0);
end
```

Listing 3.14 Common part of interpolation algorithms (interpSinc.m).

Some of these methods have free parameters and they have to be tuned. In the present set of experiments, we assume that a test set is available, which is not a usual situation, though it can be addressed when dealing with synthetic data. Note that, in this case, a grid search procedure is used by means of the `gridSearch.m` function.

Signals Generation and Settings

In order to benchmark all previous methods, simulations with known solutions were conducted. Our experimental setup is adapted from Choi and Munson (1998b), where a set of signals with stochastic band-limited spectra were generated. Nevertheless, here we used a signal with deterministic band-limited spectra instead. The signals were constructed by adding two squared sincs, one of them being a smaller and amplitude-modulated version of the other:

$$x(t) = \mathrm{sinc}^2(\pi t)\left[1 + \frac{1}{2}\sin(2\pi f t)\right] + e(t), \tag{3.67}$$

where $f = 0.4$ Hz and $e(t)$ is additive noise. Note that, in spite of the simplicity of Equation 3.67, it cannot be trivially adjusted by a weighted combination of sinc basis.

A set of L samples can be used with averaged sampling interval T s. The sampling instants are obtained by adding uniform noise, in the range $[-0.1T, 0.1T]$, to equally spaced time points $\{t_k = kT\}_{k=1}^{L}$. Different values of L can be taken, changing accordingly averaged sampling interval T; that is, when $L = m \times 32$ samples are considered, the averaged sampling interval is changed to $T = 0.5/m$ s, with $m = 1, 2, 3, 4$. Different SNRs were explored (no noise, 40 dB, 30 dB, 20 dB, and 10 dB). Sampling intervals falling outside $[0, LT]$ are wrapped inside. The code for generating the signal is given in Listing 3.15, and the `interpSignal` function was implemented as in Listing 3.16.

```
function data = genDataInterp(conf)
L = conf.data.L; LT = L*conf.data.T;
% Uniform sampling generation to training and reconstruction
Xtrain = linspace(1, LT, L); Xtrain = Xtrain(:);
Xtest = linspace(1, LT, L*16); Xtest = Xtest(:);
% Nonuniform sampling generation
Xtrain(2:L-1) = Xtrain(2:L-1)+ 2*conf.u .* (rand(L-2,1)-0.5);
% Avoiding samples out of the signal limits
Xtrain(Xtrain>LT) = 2*LT-Xtrain(Xtrain>LT);
Xtrain(Xtrain<1)  = 2-Xtrain(Xtrain<1);
% Signal generation
[Ytrain, Ytest] = interpSignal(conf,Xtrain,Xtest);
potY = mean(Ytest.^2);
N = sqrt(potY./(10.^(conf.SNR./10)));
Ytrain = Ytrain + N*randn(size(Ytrain));
% Return the data structure
data.Xtrain=Xtrain; data.Xtest=Xtest; data.Ytrain=Ytrain; data.Ytest ...
    =Ytest;
```

Listing 3.15 Data generation for interpolation examples (genDataInterp.m).

```
function [varargout] = interpSignal(conf,varargin)
cf = conf.data;
varargout = cell(nargout,1);
for m = 1:nargout,
    t = varargin{m};
    y = zeros(size(t));
    x = ((t-cf.mu)./cf.sigma);
    switch cf.FNAME
        case 'MSSF'
            y = cos(2*pi*cf.f.*t) .* ...
                (1/(cf.sigma*sqrt(2*pi))).*exp((-1/2).*(x.^2));
        case 'DMGF'
            sigma=3; f1=.75; f2=0.25;
            y = (cos(2*pi*f1.*t) + cos(2*pi*f2.*t)) .* ...
                (1/(sigma*sqrt(2*pi))).*exp((-1/2).*((x*cf.
                                        sigma/sigma).^2));
        case 'gauss'
            y = (1/(cf.sigma*sqrt(2*pi))).*exp((-1/2).*(x.^2));
        case 'sinc2'
            y = sinc(t/(cf.T0*2)).^2;
        case 'doblesinc2'
            t0 = 6*cf.T0;
            w = 2/(3*cf.T0);
            tc = t - (t(1)+t(end))/2;
            y = sinc(tc/t0).^2 + cf.k*(sinc(tc/t0).^2) .* sin(w*tc);
        case '50sincs'
            A = rand(50,1);
            mu = (rand(50,1)+cf.DESP).*cf.L*cf.T;
            for n = 1:50,
                sinc_aux = A(n) .* sinc((t-mu(n))/cf.T0);
                y = y + sinc_aux;
            end
    end
    varargout{m} = y;
end
```

Listing 3.16 Signal generation part for interpolation examples (interpSignal.m).

We generated 100 realizations for each set of experiments. The performance of the interpolators was measured on a validation set given by a noise-free, uniformly sampled version of the output signal (sampling interval $T/16$). The S/E ratio is computed in decibels in the training set as

$$\left(\frac{S}{E}\right)_{\text{dB}} = 10\log_{10}\left(\frac{\mathbb{E}[(x_n^N)^2]}{\mathbb{E}[e_n^2]}\right). \tag{3.68}$$

Means and standard deviations of S/E were averaged over the 100 realizations. The code for calculating this signal-to-error ratio is as simple as seen in Listing 3.17.

```
function value = SE(ypred,y)
signal = sum(y.^2);
error = sum((ypred - y).^2);
value = signal/error;
```

Listing 3.17 Signal-to-error ratio (SE.m).

Table 3.3 *S/E* ratios (mean ± std) for Gaussian noise.

Method	*S/E* ratio				
	No noise	40 dB	30 dB	20 dB	10 dB
Y1	47.5 ± 4.1	38.7 ± 1.7	29.6 ± 1.3	19.9 ± 1.2	9.9 ± 1.1
Y2	**53.4 ± 1.9**	**39.5 ± 1.4**	**29.8 ± 1.3**	**20.1 ± 1.2**	**10.6 ± 1.2**
S1	−0.5 ± 0.7	−0.5 ± 0.7	−0.5 ± 0.7	−0.6 ± 0.8	−1.3 ± 1.0
S2	15.9 ± 3.0	15.9 ± 3.0	15.7 ± 2.9	14.5 ± 2.2	9.8 ± 1.3
S3	16.9 ± 2.9	16.9 ± 2.9	16.7 ± 2.8	15.4 ± 2.2	10.2 ± 1.3

All the configuration parameters defining these experiments are fixed in the configuration file `config_interp.m` available on the book webpage.

Experimental Results

Table 3.3 shows the performance of the algorithms in the presence of additive and Gaussian noise, as a function of SNR. The poorest performance was noticeably exhibited by S1, and some improvement was observed with S2 and S3. Y1 yielded good performance only for low noise levels, whereas Y2 showed good performance for all the noise levels, which was in agreement with its theoretical optimality for the Gaussian noise.

The top panels in Figure 3.21 show a representative example of a modulated sinc signal with SNR = 20 dB. The interpolation in the time domain was shown to provide better approximation to the validation signal for Y2 (theoretical optimum). These results will be compared with the SVM algorithm implementations in Chapters 5 and 7. The code in Listing 3.18 can be used to plot the results, in the time and in the frequency domain.

```
function solution_summary = display_results_interp(solution,conf)
% Initials
algs = fields(solution);
nalgs = length(algs);
Xtest = solution.(algs{1})(end).Xtest;
fm = 1/conf.data.T;
f = linspace(-fm/2,fm/2,length(Xtest))';
Ytf = abs(fftshift(fft(solution.(algs{1})(end).Ytest)));
options = {'FontSize',20,'Interpreter','Latex'};
options2 = {'FontSize',18,'FontName','Times New Roman'};
% Figures
figure
for ia = 1:nalgs
    subplot(3,nalgs,ia)
    plot(solution.(algs{ia})(end).Xtrain,solution.(algs{ia})
               (end).Ytrain,'ok'); hold on
    plot(Xtest,solution.(algs{ia})(end).Ytest);
    plot(Xtest,solution.(algs{ia})(end).Ytestpred,'r')
    title(strrep(algs{ia},'_','-'),options{:});
    if ia==1; H=legend('$y_{train}$','$y_{test}$','$y_{pred}$');
        set(H,options{:}); end
    axis tight; set(gca,options2{:})
    subplot(3,nalgs,nalgs+ia)
```

```
    stem(solution.(algs{ia})(end).coefs);
    axis tight; set(gca,options2{:})
    subplot(3,nalgs,2*nalgs+ia)
    plot(f, Ytf, 'b'); hold on;
    plot(f,abs(fftshift(fft(solution.(algs{ia})(end)
                .Ytestpred))),'r');
    xlabel('f',options{:}); xlim([0, .2]); set(gca,options2{:})
    if ia == 1; H = legend('Test','Pred.'); set(H,options{:}); end
end
suptitle(['SNR=',num2str(conf.SNR),' - u=',num2str(conf.u)])
% Performance summary
solution_summary = results_summary(solution,conf);
```

Listing 3.18 Displaying interpolation results (display_results_interp.m).

3.4.6 Sparse Deconvolution

Let y_n be a discrete-time signal that contains in its lags $0 \leq n \leq N$ a set of $N+1$ observed samples of a time series obtained as the convolution between an unknown sparse signal x_n, whose samples in the lags $0 \leq n \leq M$ we want to estimate, and a time series h_n, with known samples in lags $0 \leq n \leq Q-1$. Samples of y_n, x_n, and h_n are treated as null outside the observation intervals, and $N = M + Q + 1$. Then, the following signal model can be used:

$$y_n = \hat{x}_n * h_n + e_n = \sum_{j=0}^{M} \hat{x}_j h_{n-j} + e_n, \tag{3.69}$$

where $*$ denotes the discrete-time convolution operator, $\hat{x}_n$ is the estimation of the unknown input signal, and e_n is the noise. Equation 3.69 is required to be fulfilled for lags $n = 0, \ldots, N$ of observed signal y_n.

The solution is usually regularized (Tikhonov and Arsenin, 1977) by minimizing the qth power of the q-norm of estimate $\hat{x}_n$, which promotes smooth solutions. The regularized functional to be minimized can be written as

$$J_{\mathrm{D}} = \|\boldsymbol{e}\|_p^p + \lambda \|\hat{\boldsymbol{x}}\|_q^q \tag{3.70}$$

where vector notation has been introduced for the observed set of samples and the errors, and parameter λ tunes the trade-off between model complexity and the minimization of estimation errors.[3] It is worth noting again that different norms can be adopted, hence involving different families of models and solutions. For instance, setting $p = 2$ and $\lambda = 0$ yields the LS criterion, whereas setting $p = 2$, $\lambda \neq 0$, and $q = 2$ yields the well-known Tikhonov–Miller regularized criterion (Tikhonov and Arsenin, 1977). Also, for $p = 1$ and $\lambda = 0$ we obtain the ℓ_1 criterion, and setting $p = 1$, $\lambda \neq 0$, and $q = 1$ provides the ℓ_1-penalized method (O'Brien *et al.*, 1994).

From the point of view of ML deconvolution (Kormylo and Mendel 1982; Mendel and Burrus 1990) or GM (Santamaría-Caballero *et al.*, 1996), sparse deconvolution algorithms can be written as the minimization of the (power of the) ℓ_p norm of the

3 Note that, for notation simplicity, sometimes we refer to e_n as the noise and at other times as the residuals, although it is well known that, in general, residuals can be seen as an approximation $\hat{e}_n$ to the noise e_n.

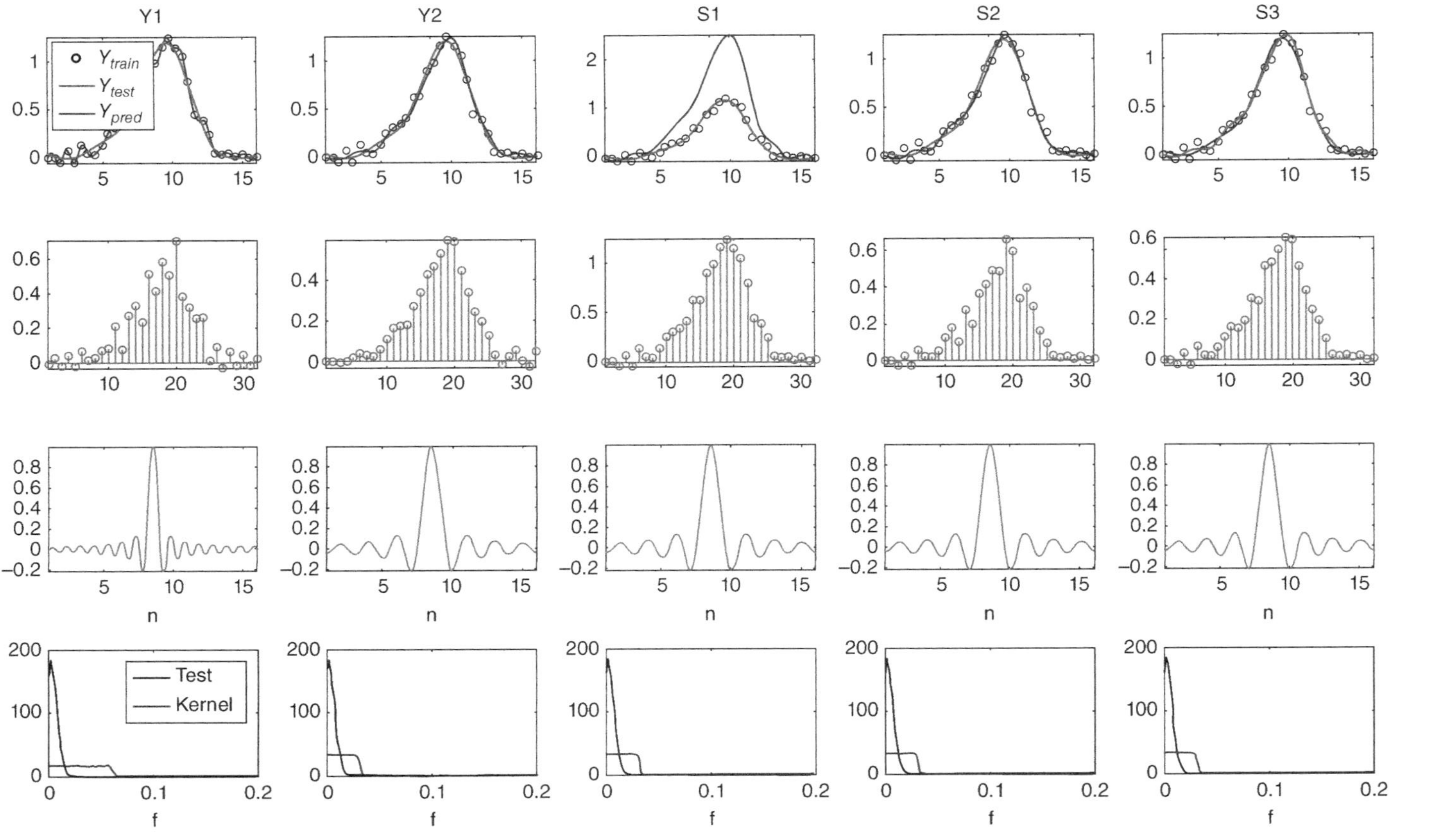

Figure 3.21 Examples of interpolation in the time domain. The second row shows the coefficients obtained to construct the kernel of each interpolation algorithm. The third and fourth rows display the kernel of each algorithm in the time and frequency domains respectively for Gaussian noise (SNR = 20 dB, $L = 32$ samples).

residuals minus a regularization term which consists of the log-likelihood of the sparse time series for an appropriate statistical model; that is:

$$J_{\mathrm{ML}} = \|\boldsymbol{e}\|_p^p - \lambda \log l(\hat{\boldsymbol{x}}), \tag{3.71}$$

where $l(\hat{\boldsymbol{x}})$ denotes the likelihood. White Laplacian statistics lead to the ℓ_1-penalized criterion, whereas non-Gaussian input signal statistics have been proposed, such as Bernoulli–Gaussian (BG) and GM distributions (Mendel and Burrus, 1990; Santamaría-Caballero *et al.*, 1996). If the prior distribution of the sparse signal can be approximated with a mixture of two zero-mean Gaussians, one of them being narrow (small variance) and the other being broad (larger variance) (Santamaría-Caballero *et al.*, 1996), then the functional to be minimized in GM deconvolution is

$$J_{\mathrm{GM}} = \|\boldsymbol{e}\|^2 - \lambda_{\mathrm{GM}} \sum_{n=0}^{N} \log \sum_{j=1}^{2} \pi_j p_j(\hat{x}_n), \tag{3.72}$$

where π_j are the prior probabilities of each Gaussian component, and

$$p_j(\hat{x}_n) = (\sigma_j \sqrt{2\pi})^{-1} \exp\left(-\frac{\hat{x}_n^2}{2\sigma_j^2}\right) \tag{3.73}$$

are their probability densities. These functionals contain a term that is intended to minimize the training error of the solution plus a regularization term from a Tikhonov point of view. As stated by Girosi *et al.* (1995) and Poggio and Smale (2003), these stabilizer terms used in Equations 3.70–3.72 are also seen as *smoothers*, as lower values of these functions produce smoother solutions.

Deconvolution Algorithms for Benchmarking

We choose ℓ_1-penalized minimizing Equation 3.70 and GM minimizing Equation 3.72 as algorithms for benchmarking sparse deconvolution, and following the experiments in Rojo-Álvarez *et al.* (2008). MATLAB codes for these two methods are given in Listings 3.19 and 3.20.

```
function [predict,coefs] = deconv_L1(y,h,u,lambda)
% Init
N       = length(y); coefx = ones(N,1); coefe = ones(N,1); lb      = ...
    zeros(1,4*N);
% Construct kernel matrix
Haux = convmtx(h,N);
H    = Haux(1:N,:);
A    = [H -H eye(N) -eye(N)];
% Deconvolution
options = optimset('Display','off');
fo      = [lambda*coefx; lambda*coefx ; coefe ; coefe];
x       = linprog(fo,[],[],A,y,lb,[],[],options);
predict = x(1:N)-x(N+1:2*N);
coefs   = filter(h,1,predict);
if ~isempty(u); aux=predict; predict=coefs; coefs=aux; end
```

Listing 3.19 ℓ_1-regularized deconvolution algorithm (deconv_L1.m).

```
function [predict,coefs] = deconv_GM(z,h,u,gamma,alfa)

% Initials
pr    = [0.5 0.5];
sn    = [0.5 1]*mean(z.^2);
m     = [1000 50];
mu    = 1e-2;
N     = length(z);
x     = zeros(N,1);
% Construct kernel matrix
Haux = convmtx(h,length(x));
H    = Haux(1:length(x),:);
% Initial norm-2 loop
for i = 1:m(1)
   x = x+2*mu*H'*(z-H*x);
end
% Main loop
for i = 1:m(2)
   % Posteriori
   p(:,1) = gauss(0,sn(2),x);
   p(:,2) = gauss(0,sn(1),x);
   pp     = (pr(2)*p(:,1)+pr(1)*(p(:,2)));
   r(:,1) = pr(2)*p(:,1)./pp;
   r(:,2) = pr(1)*p(:,2)./pp;
   % q caltulation
   q      = x.*(r(:,1)/sn(2)+r(:,2)/sn(1));
   % Intermediate steps
   c      = 2*H'*(H*x-z)+alfa*q;
   A      = H'*H;
   lam    = max(eig(A,'nobalance'));
   v      = c.*x;
   d      = zeros(length(x),1);
   d(v<0) = 1/(lam+alfa/(2*sn(1)));
   d(v>0) = min(1/(lam+alfa/(2*sn(1))), x(v>0)./c(v>0));
   D      = diag(d);
   % Main step
   x      = x-D*c;
   x      = x*(max(z))/max(abs(x));
   % Updating posteriori
   p(:,1) = gauss(0,sn(2),x);
   p(:,2) = gauss(0,sn(1),x);
   pp     = (pr(2)*p(:,1)+pr(1)*(p(:,2)));
   r(:,1) = pr(2)*p(:,1)./pp;
   r(:,2) = pr(1)*p(:,2)./pp;
   % Updating parameters
   sn(2)  = gamma*sn(2) + (1-gamma)*sum(x.^2.*r(:,1))/sum(r(:,1));
   sn(1)  = gamma*sn(1) + (1-gamma)*sum(x.^2.*r(:,2))/sum(r(:,1));
   pr(2)  = gamma*pr(2) + (1-gamma)*mean(r(:,1));
   pr(1)  = gamma*pr(1) + (1-gamma)*mean(r(:,2));
end
predict = x;
coefs = filter(h,1,x);

if ~isempty(u); predict=coefs; coefs=x; end
```

Listing 3.20 GM deconvolution algorithm (deconv_GM.m).

Signals Generation and Experimental Setup

A deterministic sparse signal with 128 samples and five nonzero values ($x_{20} = 8$, $x_{25} = 6.845$, $x_{47} = -5.4$, $x_{71} = 4$, and $x_{95} = -3.6$) (Figueiras-Vidal *et al.*, 1990) was used for simulations. The other time series consists of the first 15 samples of the impulse response of

$$H(z) = \frac{z - 0.6}{z^2 - 0.414z + 0.64}. \tag{3.74}$$

A noise sequence was added, with a variance corresponding to SNR from 4 to 20 dB. Performance was studied for three different kinds of additive noise; namely, Gaussian, Laplacian, and uniform. The MATLAB code in Listing 3.21 can be used to generate the sparse input signal, where `getsignal` and `getri` are given in Listings 3.22 and 3.23 respectively.

```
function data = genDataDeconv(conf)
% Signal
x       = getsignal(conf.data.N,1);
% Convolution
h       = getri(conf.data.Q);
y       = filter(h,1,x);
% Noise Variance calculation
SNR    = 10.^(conf.SNR/10);
signal = norm(y)^2/length(y);
noise  = sqrt(signal./SNR);
% Signal with noise
switch conf.data.typeNoise
    case 1 % Signal with Gaussian noise
        z = y + noise*randn(size(y));
    case 2 % Signal with uniform noise
        z = y + noise*rand(size(y));
    case 3 % Signal with laplacian noise
        z = y + laplace_noise(length(y),1,noise);
end
% Return data structure
data.z=z; data.y=y; data.h=h; data.x=x; data.u=[];
```

Listing 3.21 Data generation for sparse deconvolution (genDataDeconv.m).

```
function x  = getsignal(N,tipo)
x = zeros(N,1);
switch tipo
    case 1
        x([20 25 47 71 95]) = [8 6.845 -5.4 4 -3.6];
    case 2
        Q = 15;
        p = 0.05;
        while ~sum(x)
            aux = rand(N-Q,1);
            q   = (aux<=p);
            x(1:N-Q) = randn(N-Q,1).*q;
        end
end
```

Listing 3.22 Signal generation for sparse deconvolution (getsignal.m).

```
function h = getri(Q)
num = [1 -.6];
den = poly([.8*exp(1j*5*pi/12) .8*exp(-1j*5*pi/12)]);
h = impz(num,den,Q);
```

Listing 3.23 Impulsive response generation for sparse deconvolution (getri.m).

The free parameters for the ℓ_1-penalized and for the GM algorithms are λ and λ_{GM} respectively. Estimation quality was quantified by using

$$\mathrm{MSE} = \frac{1}{N+1}\sum_{n=0}^{N}(\hat{x}_n - x_n)^2, \tag{3.75}$$

$$F = \sum_{x_n \neq 0}(\hat{x}_n - x_n)^2, \tag{3.76}$$

$$F^{\mathrm{null}} = \sum_{x_n=0}(\hat{x}_n - x_n)^2 = \sum_{x_n=0}\hat{x}_n^2, \tag{3.77}$$

as proposed by O'Brien *et al.* (1994). These quality estimations are coded as seen in Listings 3.24, 3.25, and 3.26.

```
function value = MSE(ypred,y)
value = mean((ypred - y).^2);
```

Listing 3.24 MSE (MSE.m).

```
function f = F(ypred,y)
indx = find(y ~= 0);
f = norm(ypred(indx) - y(indx));
```

Listing 3.25 F measure for peak samples (F.m).

```
function fnull = Fnull(ypred,y)
fnull = norm(ypred(y == 0));
```

Listing 3.26 F^{null} measure for null samples (Fnull.m).

All the configuration parameters that define these experiments are fixed in the configuration file `config_deconv.m` on the book's webpage. See again the help and demos of the `simpleInterp` toolbox and Appendix 3.A.

Experimental Results

Records of estimated signals show a high variability. In Figure 3.22 we present some representative results of those corresponding to previously presented performance figures. Optimum values were adjusted for the free parameters in this example. The ℓ_1-penalized deconvolution detected almost all the peaks, but it also produced a number of small spurious peaks. In the example, the GM procedure reaches higher estimation performance. Table 3.4 summarizes the averaged performance in 100 realizations. The results can be readily visualized by running the specific code `display_results_deconv.m`.

3.4.7 Array Processing

In this example we simulate an array of $K = 5$ equally spaced elements and three signals modulated by a QPSK modulation, with unitary amplitude and illuminating the array

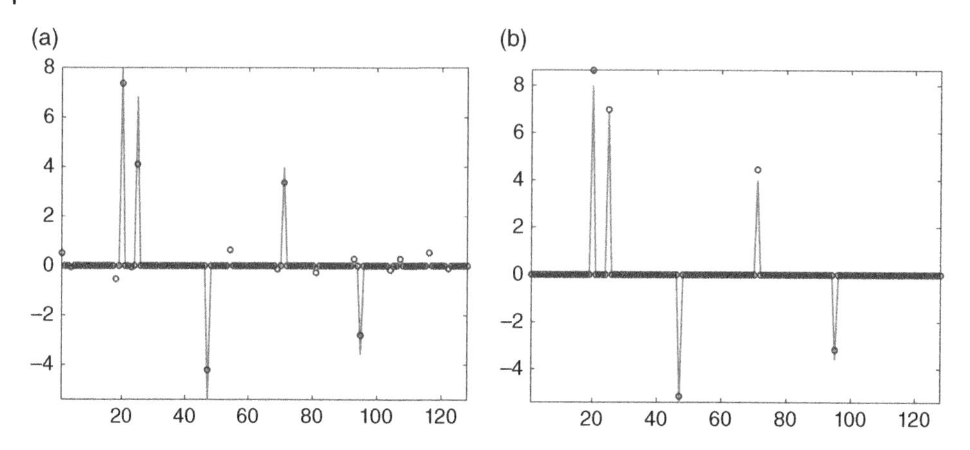

Figure 3.22 An example of sparse deconvolution results versus data with $SNR = 10$ dB for the ℓ_1-penalized algorithm (a) and the GM algorithm (b), with time in samples.

Table 3.4 S/E ratios (mean ± std) for Gaussian.

Method	MSE	F	F^{null}
ℓ_1-regularized	0.08 ± 0.04	2.72 ± 0.80	1.41 ± 0.42
GM	0.06 ± 0.06	2.53 ± 1.34	0.00 ± 0.00

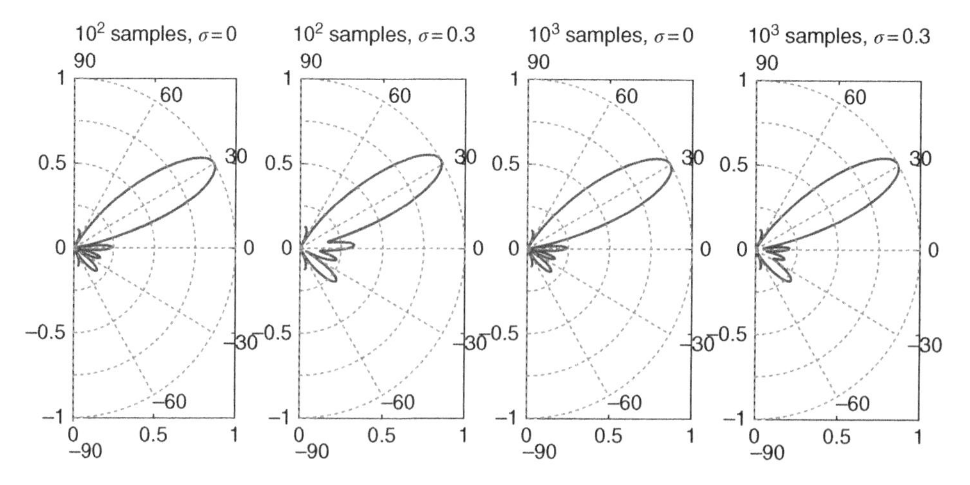

Figure 3.23 Experimental setup for the simulation and resulting array beams (training with 100 and 1000 samples).

from angles of arrival $\theta_1 = 30°$, $\theta_2 = 60°$, and $\theta_3 O = -60°$ with respect to the direction perpendicular to the array. Figure 3.23 shows the experimental setup of the simulation. The array processor output is $y_n = \boldsymbol{w}^{\mathrm{T}}\boldsymbol{x}_n = \sum_{k=0}^{K-1} w_k x_n^k$, where $\boldsymbol{x}_n$ represents the complex envelope of the current measured at each one of the elements of the array. Low-pass complex modulations are simulated for the three independent transmitters, and then a train of snapshots or antenna array measurements are further simulated by multiplying each modulation by its corresponding steering vector, as described in Section 3.3.5. The data are then corrupted with additive Gaussian noise, and the array output is linearly processed to detect the desired user signal arriving from $\theta_1 = 30°$. The parameters can be considered as a filter in the space, since all samples in snapshot $\boldsymbol{x}_n$ are taken at the same time at different directions of the arrival space.

In order to characterize the array, the FT of array parameter vector $\boldsymbol{w}$ has been computed and represented with respect to all spatial frequencies. Since the phase delay between elements corresponding to a DOA θ_l is $\phi_l = 2\pi(d/\lambda)\sin(\theta_l)$, we can make a representation with respect to the angle of arrival using the inverse of this function $\theta_l = \arcsin[(\lambda/2\pi d)\phi_l]$. The resulting graphs, represented in polar coordinates and normalized to 1, are the corresponding array beams. Listing 3.27 shows the code for this array processing example.

```
function array
% Transmitted signal
N = 100;
K = 5;
d = 0.51;
lambda = 1;
theta1 = 30*pi/180;   % Desired signal angle
theta2 = 60*pi/180;   % Interference 1
theta3 = -60*pi/180;  % Interference 2
% Visible angle
Theta = asin(-linspace(-pi,pi,1024)'*lambda/2/pi/d)*180/pi;
sigma = 0.00; % Noise variance
% Signal and interferences
[X1,b] = signal(N,K,lambda,d,theta1);
X2 = signal(N,K,lambda,d,theta2);
X3 = signal(N,K,lambda,d,theta3);
X = X1 + X2 + X3;
X = X + sigma*randn(size(X)); % Additive noise
% MMSE array optimization
w = pinv(X')*b;
p = abs(fftshift(fft(w,1024)));
p = p/max(p);
figure(1), polar(Theta*pi/180,p)

function [X,b] = signal(N,K,lambda,d,theta)
% QPSK symbol stream transmitted
b = sign(randn(N,1))+1j*sign(randn(N,1));
% Steering vector
a = exp(1j*2*pi*d/lambda*[0:K]*sin(theta));
% Complex envelope signal
X = a'*b';
```

Listing 3.27 Array processing example (array.m).

Figure 3.23 shows the resulting array beams after a training with 100 and 1000 samples, with no noise and with noise of standard deviation equal to 0.3. Under noise-free conditions, both beams are similar and place two zeros at 60° and −30°

in order to cancel the interfering signals, while the main lobe is pointed toward the desired signal at 30°. In the presence of noise, the performance of the beamforming is degraded, but a significant attenuation at both interfering angles of arrival can still be observed.

3.5 Questions and Problems

Exercise 3.5.1 Write the equation of a signal model different from the ones expressed in this chapter. Compare it with the general signal model.

Exercise 3.5.2 Write a simple program for generating the noise distributions described in this chapter.

Exercise 3.5.3 Write the code for generating the examples of the "Noise in Communication Systems" section.

Exercise 3.5.4 In the example for ECG analysis, change the filter settings to see the effect on the signal and on the noise distribution. Plot the residuals histograms and the residuals in time. What can you say?

Exercise 3.5.5 In the example of system identification, explore the results in terms of the estimated order of the system to be estimated. How does this affect the system estimation quality?

Exercise 3.5.6 In the examples of sinusoidal models, write down the equations for the MMSE solutions. For the synthetic example, analyze the effect of regularization, of the frequency grid, and of the presence of additive noise. For the HRV example, determine which error is taking place in the MMSE criterion, and propose an alternative aiming to correct it.

Exercise 3.5.7 Analyze the code of the free parameters tuning, provided with the toolbox. Think in a real application problem where the dataset can have dependencies, such as several measurements in the same individual. How would you change the free parameters tuning for ensuring the independence from the individual?

Exercise 3.5.8 In the example for sinc interpolation, plot the error in the time and frequency domains for each method. What can you say about the behavior of the different methods?

Exercise 3.5.9 In the deconvolution example, plot the effect of the regularization parameter on the estimated input sparse signal.

Exercise 3.5.10 Complete the intermediate steps for going from Equation 4.61 to the primal functional in Equation 4.65, to the Lagrangian functional in Equation 4.69, and to the dual functional in Equation 4.73.

Exercise 3.5.11 Prove the relationship between $\boldsymbol{w}$ and the residual cost function in Equation 4.61. Give a result for the MMSE cost and for the Vapnik cost.

Exercise 3.5.12 Prove Equation 5.33 for the robust expansion coefficients property.

Exercise 3.5.13 Propose two signal models for time series problems. Can the problems that you proposed be stated in the form of the general signal model in Equation 3.7? Why?

3.A MATLAB `simpleInterp` Toolbox Structure

This appendix describes the MATLAB implementation structure of the `simpleInterp` toolbox. The toolbox will be extensively used for interpolation and deconvolution problems, both using linear and kernel signal models. In particular, the toolbox will be used in Chapters 3, 5, and 7. In order to run experiments with the toolbox, we will need: (1) an experiment configuration file; (2) a data load or data generation function; (3) the algorithm function to be used; and (4) some code to visualize the results. In the toolbox you can find the script `simpleInterpDemo.m` that will set up everything needed (such as paths or experiment to run).

The main parts of the toolbox are:

1) The main experiments part, where each of the specific parts of a given experiment is invoked.
2) The shared functions for different experiments; namely, the calculation of measurement evaluation (such as MSE, SNRs, or bit error rate (BER)).
3) The necessary methods for tuning the free parameters in each algorithm, and including:
   - cross-validation (CV), which splits the training set into different partitions, possibly even for one partition per sample, which corresponds to the well-known *leave-one-out* (LOO) validation;
   - grid search;
   - sequential search.

The two last methods for free parameters tuning use a different set of observations for validating the performance. The choice of each search method can be achieved in terms of the needs and available resources from the reader. In general, with unlimited computational resources, the best option will be CV, or even LOO–CV. On the other hand, with time or computational burden limitations, the other two methods are recommendable. The grid search method can be recommendable from a strict point of view, as far as sequential search could suffer from local minima stacking in some problems, though in many problems it is enough with this sequential exploration in a valid range of parameter values. Even though the free parameters tuning has to be done with a dataset different from the test set (usually by splitting the original set), sometimes this will be not a simple and effective approach in signal processing tasks, where time correlation among samples is strong by their own nature.

We present in Listings 3.28 and 3.29 the shared parts of the code allowing the global and structured execution of the examples following this structure, and the main function.

```
% Configuration file is loaded to execute a particular example
config_file

% Double loop is allowed in order to evaluate different conditions
n1 = length(conf.vector_loop1);
n2 = length(conf.vector_loop2);
solution = cell(n1,n2); summary = solution;
for i = 1:n1
    eval(['conf.',conf.varname_loop1,'=conf.vector_loop1(i);
                                                    ']);
    for j = 1:n2
        eval(['conf.',conf.varname_loop2,'=conf.vector_loop2
                                                    (j);']);
        solution{i,j} = main(conf); % Common main structure is called
        summary{i,j} = conf.displayfunc(solution{i,j},conf); % ...
            Particular visualization
    end
end
```

Listing 3.28 Run execution (run.m).

```
function solution = main(conf)

for m = 1:conf.NREPETS
    conf.I = m; fprintf('Executing run %d...\n',m);
    % Specified dataset is generated or loaded
    [Xtrain,Ytrain,Xtest,Ytest] = load_data(conf);
    % Specified algorithms are used
    for ialg = 1:length(conf.machine.algs)
        conf.machine.function_name = conf.machine.algs{ialg};
        alg = func2str(conf.machine.function_name);
        % Free parameters are selected by a searching function
        conf.machine.value_range = ...
            conf.cv.value_range(conf.machine.ind_params{ialg});
        param_selec = conf.cv.searchfunc(Xtrain,Ytrain,conf,Xtest,Ytest);
        ind_params = conf.machine.ind_params{ialg};
        for ip = 1:length(ind_params)
            solution.(alg)(m).(conf.machine.params_name
            {ind_params(ip)}) = param_selec{ip};
        end
        % Adjusted algorithms are trained and applied to predict
        [Ypred,coefs] = conf.machine.function_name(Xtrain,Ytrain,Xtest,
                                            param_selec{:});
        solution.(alg)(m).coefs = coefs;
        solution.(alg)(m).Ytestpred = Ypred;
        solution.(alg)(m).Ytest = Ytest;
        solution.(alg)(m).Xtest = Xtest;
        solution.(alg)(m).Xtrain = Xtrain;
        solution.(alg)(m).Ytrain = Ytrain;
        for ie = 1:length(conf.evalfuncs)
            evalfunc = conf.evalfuncs{ie};
            evalname = func2str(evalfunc);
```

```
            solution.(alg)(m).(evalname) = evalfunc(Ypred,Ytest);
        end
    end
end
```

Listing 3.29 Main function (main.m).

The search functions can be seen in the code URL, and their names are `cross_validation.m`, `gridSearch.m`, and `sequentialSearch.m`. Though the functions calculating the evaluation measurements can be used in different examples, they will be presented in subsequent sections when their use is needed. Function `load_data.m` in Listing 3.30 addresses the unification of the data structures for two different kind of tasks; namely, convolutional problems and machine-learning problems.

```
function [Xtrain,Ytrain,Xtest,Ytest] = load_data(conf)

data = conf.data.loadDataFuncName(conf);
if isfield(data,'h') % Deconvolution problem
    Xtrain = data.z; Xtest = data.u;
    Ytrain = data.h; Ytest = data.x;
elseif isfield(data,'Xtrain') % Learning task
    Xtrain = data.Xtrain; Xtest = data.Xtest;
    Ytrain = data.Ytrain; Ytest = data.Ytest;
else
    error('ErrorTests:convertTest', 'Unexpected data structure ...
        generated...\nFor further information see: load_data.m');
end
```

Listing 3.30 Loading structured data (load_data.m).

The results obtained are always given back with the same structure, and the code for this is in Listing 3.31.

```
function solution_summary = results_summary(solution,conf)

algs = fields(solution);
addstr = ''; extra1 = '%.2f'; extra2 = extra1;
cname1 = conf.(conf.varname_loop1); cname2 = conf.(conf.varname_loop2);
if iscell(cname1); cname1 = cname1{1}; extra1 = '%s'; end
if iscell(cname2); cname2 = cname2{1}; extra2 = '%s'; end
cnf1 = sprintf(['%s = ',extra1],conf.varname_loop1,cname1);
cnf2 = sprintf(['%s = ',extra2],conf.varname_loop2,cname2);
cv1 = conf.varname_loop1; cv2 = conf.varname_loop2;
if ~strcmp(cv1,'idle1') && ~strcmp(cv2,'idle2');
    addstr = sprintf('(%s, %s)',cnf1,cnf2);
elseif ~strcmp(cv1,'idle1')
    addstr = sprintf('(%s)',cnf1);
elseif ~strcmp(cv2,'idle2')
    addstr = sprintf('(%s)',cnf2);
end
fprintf('\n-----------------------------------------------
                                            ----------\n');
fprintf(' Performance summary %s\n',addstr);
fprintf('-----------------------------------------------
                                            ----------\n');
for ie = 1:length(conf.evalfuncs)
```

```
    evalname = func2str(conf.evalfuncs{ie});
    fprintf('%s\n',evalname)
    for ia = 1:length(algs)
        err = cat(1,solution.(algs{ia}).(evalname));
        solution_summary.(algs{ia}).([evalname,'_mean']) = mean(err);
        solution_summary.(algs{ia}).([evalname,'_std']) = std(err);
        solution_summary.(algs{ia}).([evalname,'_vector']) = err;
        fprintf('  \t- %s \t %s: Mean = %.2e  Std = %.2e \n',...
            algs{ia}, evalname,...
            solution_summary.(algs{ia}).([evalname,'_mean']),...
            solution_summary.(algs{ia}).([evalname,'_std']));
    end
end
fprintf('-----------------------------------------------------
                                               ---------\n');
```

Listing 3.31 Displaying performance summary (results_summary.m).

Finally, and given that an SVM is going to be a very used algorithm throughout Chapters 3, 5, and 7 in terms of different kernel matrix used as inputs, we present in Listing 3.32 the shared part of each of the presented solutions using the SVM algorithm.

```
function [predict,model] = SVM(X,Y,Xtest,inparams)
% svm_train = @mysvmtrain;
% svm_predict = @mysvmpredict;
% Training SVM
model = mexsvmtrain(Y,X,inparams);
% Prediction
predict = zeros(size(Xtest,1),1);
if ~isempty(model) && ~isempty(model.SVs) && ~isempty(Xtest)
    predict = mexsvmpredict(zeros(size(Xtest,1),1),Xtest,model);
end
```

Listing 3.32 Common SVM code (SVM.m).

Note that the trained machine `model` is first obtained with a set of training data `X` and `Y`, and with some previously tuned free parameters `inparams`, and then the prediction `predict` is obtained for a given test set `Xtest`, as indicated.

Functions `mysvmtrain` and `mysvmpredict` are modified versions of LibSvm (Chang and Lin, 2002), and they allow us to include some additional parameters. The input parameters format is just the same as in LibSvm.

4

Kernel Functions and Reproducing Kernel Hilbert Spaces

Whereas Chapter 3 gave a (moderately deep) introduction to the signal processing concepts to be used in this book, in this chapter we put together fundamental and advanced relevant concepts on Mercer's kernels and reproducing kernel Hilbert spaces (RKHSs). The fundamental building block of the kernel learning theory is the *kernel function,* which provides an elegant framework to compare complex and nontrivial objects. After its introduction, we review the concept of an RKHS, and state the representer theorem. Then we study the main properties on kernel functions and their construction, as well as the basic ideas to work with complex objects and reproducing spaces. The support vector regression (SVR) algorithm is also introduced in detail, as it will be widely used and modified in Part II for building many of the DSP algorithms with kernel methods therein. We end up the chapter with some synthetic examples illustrating the concepts and tools presented.

4.1 Introduction

Kernel methods build upon the notion of *kernel functions* and RKHSs. Roughly speaking, a Mercer kernel in a Hilbert space is a function that computes the inner product between two vectors embedded in that space. These vectors are maps of vectors in an Euclidean space, where the *mapping function* can be nonlinear. This informal definition will be formalized through Mercer's theorem, which gives the analytical power to kernel methods.

We will see that kernel functions are very often dot products in inaccessible Hilbert spaces. This means that vectors in that space are not explicitly defined; hence, only the dot products are accessible, but not the mapped vector coordinates therein. In spite of this, the expression of the dot product is the only tool needed to operate in such space. The RKHSs that are summarized in this chapter allow us to implicitly map a finite-dimension vector from an Euclidean space (typically known as *input space* $\mathcal{X}$) into a higher dimension Hilbert space (referred as *feature space* $\mathcal{H}$) by expressing the dot product inside this space as a function of the input space vectors solely. This property has allowed *kernel methods* to become a preferred tool in machine learning and pattern recognition for years, being an attractive alternative to traditional methods in statistics and signal processing. In the past decade, methods based on kernels have gained popularity in almost all applications of machine learning because of several

Digital Signal Processing with Kernel Methods, First Edition. José Luis Rojo-Álvarez, Manel Martínez-Ramón, Jordi Muñoz-Marí, and Gustau Camps-Valls.

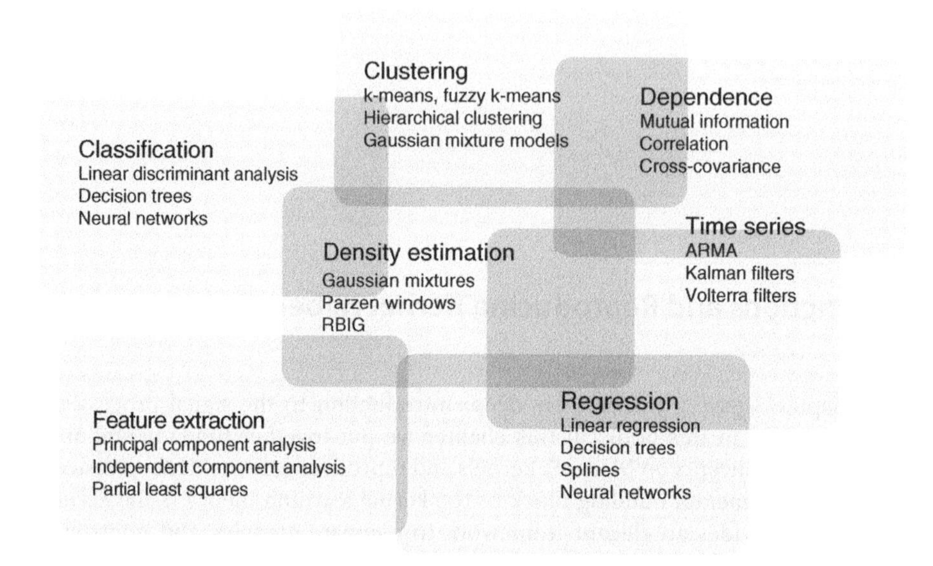

Figure 4.1 Different machine learning algorithms are available to tackle particular machine learning and signal-processing problems.

fundamental reasons. In particular, we can identify three main motivations for the use of kernel techniques in machine learning and signal processing.

The first one consists of the change of paradigm from classical nonlinear algorithms to kernel techniques that still rely on linear algebra. Linear methods used in statistics and machine learning typically resort to linear algebra operations, are very well established, and their properties and limits are mostly known. These facts give the user the tools needed to adapt linear algorithms to the application at hand and to obtain solutions very easily. Linear algorithms, however, are restricted to capturing linear relations between the features and dependent variables, and very often only capture second-order statistical relations. Such limitations call for extensions to nonlinear and higher order statistics algorithms. For every learning problem encountered in machine learning and signal processing, a full toolbox of nonlinear extensions is available (see Figure 4.1). Different approaches, however, include the nonlinearity in very different ways, as well as a particularly different set of parameters to tune and criteria to optimize their design. For example, the most well-known family of nonlinear algorithms is probably that of artificial NNs. In this case, a point-wise nonlinearity is included in every node/neuron in the network, and several parameters have to be tuned to control the model capacity (in this case, its architecture).

NNs dominated the field and applications in the 1980s and 1990s, and often produce improved performance with respect to linear algorithms. Nevertheless, a clear shortcoming was identified, which was that the knowledge about linear techniques cannot be easily translated into the design of the nonlinear ones. Actually, when using linear algorithms, a well-established theory and efficient methods are often available, but such advantages are lost when translating the linear neuron model into even a simple sigmoid-shaped neuron model. Kernel methods offer the opportunity to translate linear models into nonlinear models while still working with linear algebra. In this way, kernel methods will allow us to exploit all the intuitions and properties of linear algorithms. Essentially, kernel methods embed the data set S defined over the input or attribute feature space $\mathcal{X}$ ($S \subseteq \mathcal{X}$) into a higher (possibly infinite-dimensional) Hilbert space $\mathcal{H}$, also known as kernel *feature space*, and a linear algorithm is built on this last for yielding a nonlinear algorithm with respect to the input data space.

The second reason is related to the paradigm change introduced by kernel methods. They allow the user to take linear algorithms expressed in an Euclidean space and, by means of an often straightforward *kernelization* procedure, consisting of deriving a nonlinear counterpart of the algorithm, they provide nonlinear properties to the linear algorithms. The kernelization procedure consists of two basic steps. The first of them tries to find an expression of the linear algorithm as a function of dot products between data *only*. This representation is called the *dual* representation in contrast to the *primal* representation.[1] This is always possible for a linear algorithm by just taking into account that the parameters learned by the linear algorithm lie inside the subspace spanned by the data. A representation of these parameters as a linear combination of the mapped data straightforwardly leads to a dual representation, the combination parameters being the so-called *dual parameters*. The second step consists of the substitution of the Euclidean dot product by a dot product into an RKHS. From this point of view, kernel methods are methods that are linear in the parameters, or alternatively they are linear inside the RKHS. Notationally, the mapping function is denoted as $\boldsymbol{\phi} : \mathcal{X} \rightarrow \mathcal{H}$.

Whereas the nonlinear version of linear algorithms will exhibit increased flexibility, the computational load should increase, especially when we need to compute the sample new coordinates explicitly in that high-dimensional space. But this computation can be omitted by using the *kernel trick*; that is, when an algorithm can be expressed using dot products in the input space, then its (nonlinear) kernel version only needs the dot products between mapped samples. Kernel methods compute the similarity between examples (samples, objects, patterns, points) $S = \{\boldsymbol{x}_i\}_{i=1}^N$ using inner products between mapped examples, and thus the so-called kernel matrix $\boldsymbol{K}$, with entries $\boldsymbol{K}_{i,j} = K(\boldsymbol{x}_i, \boldsymbol{x}_j) = \langle \boldsymbol{\phi}(\boldsymbol{x}_i), \boldsymbol{\phi}(\boldsymbol{x}_j) \rangle$, contains all the necessary information to perform many classical linear algorithms in the feature space.

Figure 4.2 illustrates the concepts of kernel feature mapping, $\boldsymbol{\phi}$, and the exploitation of the kernel trick. The latter will allow us to measure similarities in the Hilbert space $\mathcal{H}$ using a reproducing kernel K that solely works with input examples from $\mathcal{X}$. In this example, the input space is not a proper representation space because the endowed Euclidean metric would wrongly tell that distant points along the manifold, $\boldsymbol{x}_i$ and $\boldsymbol{x}_j$, are close. This is remedied by mapping the input examples to a Hilbert spaces $\mathcal{H}$ such that unfolds the original banana-shaped data distribution such that the linear dot product is an appropriate one. Computing such a dot product in $\mathcal{H}$ would be impossible unless the feature mapping is explicit. Fortunately, as we will see later, neither knowing the mapping nor the coordinates of the mapped data are strictly necessary, as we can

1 The dual representation is known in *multivariate statistics* as the Q-mode as opposed to the primal R-mode. Operating in Q-mode typically endows the method with numerical stability and computational efficiency in high-dimensional problems.

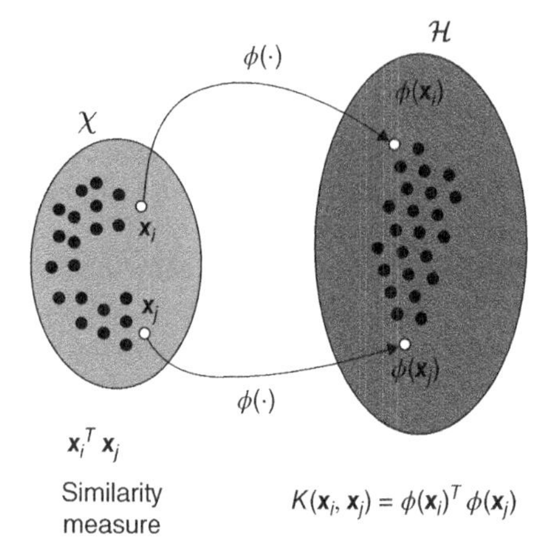

Figure 4.2 Illustration of the kernel feature mapping, ϕ, and the kernel trick by which we can measure similarities in $\mathcal{H}$ using a reproducing kernel that solely works with input examples from $\mathcal{X}$.

find a kernel function K that *reproduces* the similarity in $\mathcal{H}$ solely working with input examples from $\mathcal{X}$.

By comparison again with linear classical techniques, there is a third very important motivation for the use of kernel methods that is related to the *complexity of nonlinear machines*. If the structural complexity of a machine is too high, it is known that it may be unable to properly generalize. This is known as the problem of overfitting. In practice this problem boils down to models capable of adjusting the parameters so that the training data are fully explained, but then being unable to properly explain new, unseen, test data. Nonlinear learning machines typically introduce many parameters (on some occasions more than examples in the dataset) so they become extremely flexible and able to overfit the training set well. Limiting the capacity of the class of functions of interest is intimately related to the concept of *regularization*, already introduced in Chapter 3 for DSP signal models. We will see here how kernel methods include regularization in a very natural way, even working in inaccessible feature spaces. We will see also that kernel methods can generalize the concept of Tikhonov's regularization to infinite-dimensional feature spaces.

Finally, we should mention that kernel methods have been widely adopted in applied communities such as computer vision, time-series analysis and econometrics, physical sciences, as well as signal and image processing, because these methods typically lead to neat, simple, and intuitive algorithms in these fields. But, in addition, kernel methods allow the exploitation of a vast variety of kernel functions that can be designed and adapted to the application at hand. For example, combining and fusing heterogeneous signal sources (text, images, graphs) can be done easily via concepts of multiple kernel learning (MKL). Also, learning the kernel function directly from data is possible via generative models and exploitation of the wealth of information in the unlabeled data. Kernel design is an open research topic in machine learning and signal processing that does not stop giving theoretical results and designs for many different applications. In this chapter, we also review the main properties of kernel methods to construct new kernel functions.

4.2 Kernel Functions and Mappings

Kernel methods rely on the properties of kernel functions. As we will see next, they reduce to computing dot products of vectors mapped in Hilbert spaces through an implicit (not necessarily known) mapping function. A scalar product space (or pre-Hilbert space) is a space endowed with a scalar product. If the space is *complete* (i.e., every Cauchy sequence converges inside the space), then it is called a *Hilbert space.*

This section conveys a short introduction to kernel methods, functions approximation, and RKHSs. The tools reviewed here are of fundamental relevance for the next sections, and build upon solid branches of mathematics such as *linear algebra* (Golub and Van Loan, 1996) and *functional analysis* (Reed and Simon, 1981).

4.2.1 Measuring Similarity with Kernels

The dot (scalar or inner) product between two vectors is an algebraic operation that measures the similarity between them. Intuitively, a dot or scalar product measures how much of a vector is contained in the other, or alternatively how much the two vectors point in the same direction. The question raised here is whether one should naively rely on the dot product in the input space to measure the similarity of (possibly complex) objects or not. If the representation space of those objects is rich enough, a dot product in input space should suffice. Unfortunately, many times the input feature space is limited in resolution, and expressive power and a new richer representation is needed. Kernel methods rely on the notion of similarity between examples in a higher (possibly infinite-dimensional) Hilbert space. Consider the set of empirical data $(\boldsymbol{x}_1, y_1), \dots, (\boldsymbol{x}_n, y_n) \in \mathcal{X} \times \mathcal{Y}$, in which $\boldsymbol{x}_i$ are the *inputs* taken from $\mathcal{X}$ and $y_i \in \mathcal{Y}$ are called the *outputs*. The learning from samples process consists of using these data pairs to predict a new set of test examples $\boldsymbol{x} \in \mathcal{X}$. In order to develop learning machines capable of generalizing well, kernel methods often support a good approximation to the structure of the data and incorporate regularization in a natural way.

In some of those cases when $\mathcal{X}$ does not stand for a good support of similarity, examples can be mapped to a (dot product) space $\mathcal{H}$ by using a mapping $\boldsymbol{\phi} : \mathcal{X} \to \mathcal{H}, \boldsymbol{x} \mapsto \boldsymbol{\phi}(\boldsymbol{x})$. The mapping function can be defined explicitly (if some prior knowledge about the problem is available) or implicitly, as in the case of kernel methods. The similarity between the elements in $\mathcal{H}$ can now be measured using its associated dot product $\langle \cdot, \cdot \rangle_{\mathcal{H}}$. Here, we define a function that computes that similarity , $K : \mathcal{X} \times \mathcal{X} \to \mathbb{R}$, such that $(\boldsymbol{x}, \boldsymbol{x}') \mapsto K(\boldsymbol{x}, \boldsymbol{x}')$. This function, often called a *kernel*, is required to satisfy

$$K(\boldsymbol{x}, \boldsymbol{x}') = \langle \boldsymbol{\phi}(\boldsymbol{x}), \boldsymbol{\phi}(\boldsymbol{x}') \rangle_{\mathcal{H}}. \tag{4.1}$$

The mapping $\boldsymbol{\phi}$ is its *feature map*, the space $\mathcal{H}$ is the reproducing Hilbert *feature space*, and K is the reproducing kernel function since it *reproduces* dot products in $\mathcal{H}$ without even mapping data explicitly therein.

4.2.2 Positive-Definite Kernels

The types of kernels that can be written in the form of Equation 4.1 coincide with the class of *positive-definite kernels*. The property of positive definiteness is crucial in the field of kernel methods.

Definition 4.2.1 A function $K : \mathcal{X} \times \mathcal{X} \to \mathbb{R}$ is a positive-definite kernel *if and only if* there exists a Hilbert space $\mathcal{H}$ and a feature map $\boldsymbol{\phi} : \mathcal{X} \to \mathcal{H}$ such that for all $\boldsymbol{x}, \boldsymbol{x}' \in \mathcal{X}$ we have $K(\boldsymbol{x}, \boldsymbol{x}') = \langle \boldsymbol{\phi}(\boldsymbol{x}), \boldsymbol{\phi}(\boldsymbol{x}') \rangle_{\mathcal{H}}$ for some inner-product space $\mathcal{H}$ such that $\forall \boldsymbol{x} \in \mathcal{X}$ and $\boldsymbol{\phi} \in \mathcal{H}$.

Definition 4.2.2 A kernel matrix (or Gram matrix) $\boldsymbol{K}$ is the matrix that results from applying the kernel function K to all pairs of points in the set $\{\boldsymbol{x}_i\}_{i=1}^N$; that is:

$$\boldsymbol{K} = \begin{pmatrix} K(\boldsymbol{x}_1, \boldsymbol{x}_1) & K(\boldsymbol{x}_1, \boldsymbol{x}_2) & \cdots & K(\boldsymbol{x}_1, \boldsymbol{x}_N) \\ K(\boldsymbol{x}_2, \boldsymbol{x}_1) & K(\boldsymbol{x}_2, \boldsymbol{x}_2) & \cdots & K(\boldsymbol{x}_2, \boldsymbol{x}_N) \\ \vdots & \vdots & \ddots & \vdots \\ K(\boldsymbol{x}_N, \boldsymbol{x}_1) & K(\boldsymbol{x}_N, \boldsymbol{x}_2) & \cdots & K(\boldsymbol{x}_N, \boldsymbol{x}_N) \end{pmatrix}, \tag{4.2}$$

and whose entries are denoted as $\boldsymbol{K}_{i,j} = K(\boldsymbol{x}_i, \boldsymbol{x}_j)$.

Kernel functions must be symmetric since inner products are always symmetric. In order to show that K is a valid kernel, it is not always sufficient to show that a mapping $\boldsymbol{\phi}$ exists; rather, this is a nontrivial theoretical task. In practice, a real symmetric $N \times N$ matrix $\boldsymbol{K}$, whose entries are $K(\boldsymbol{x}_i, \boldsymbol{x}_j)$ or simply $\boldsymbol{K}_{i,j}$, is called *positive-definite* if, for all $\alpha_1, \ldots, \alpha_N \in \mathbb{R}$, $\sum_{i,j=1}^N \alpha_i \alpha_j \boldsymbol{K}_{i,j} \geq 0$. A positive-definite kernel produces a positive-definite Gram matrix in the RKHS.

Definition 4.2.3 Kernel matrices that are constructed from a kernel corresponding to a strict inner product space $\mathcal{H}$ are positive semidefinite.

This last property is easy to prove. Note that, by construction, we have $\boldsymbol{K}_{i,j} = K(\boldsymbol{x}_i, \boldsymbol{x}_j) = \langle \boldsymbol{\phi}(\boldsymbol{x}_i), \boldsymbol{\phi}(\boldsymbol{x}_j) \rangle_{\mathcal{H}}$; thus, for any vector $\boldsymbol{\alpha} \in \mathbb{R}^N$:

$$\begin{aligned} \boldsymbol{\alpha}^{\mathrm{T}} \boldsymbol{K} \boldsymbol{\alpha} &= \sum_{i=1}^{N} \sum_{j=1}^{N} \alpha_i \boldsymbol{K}_{i,j} \alpha_j = \sum_{i=1}^{N} \sum_{j=1}^{N} \alpha_i \langle \boldsymbol{\phi}(\boldsymbol{x}_i), \boldsymbol{\phi}(\boldsymbol{x}_j) \rangle_{\mathcal{H}} \alpha_j \\ &= \left\langle \sum_i \alpha_i \boldsymbol{\phi}(\boldsymbol{x}_i), \sum_j \alpha_j \boldsymbol{\phi}(\boldsymbol{x}_j) \right\rangle_{\mathcal{H}} = \left\| \sum_{i=1}^{N} \alpha_i \boldsymbol{\phi}(\boldsymbol{x}_i) \right\|_{\mathcal{H}}^2 \geq 0. \end{aligned} \tag{4.3}$$

Accordingly, any algorithm which operates on the data in such a way that it can be mathematically described in terms of dot products can be used with any positive-definite kernel by simply replacing $\langle \boldsymbol{\phi}(\boldsymbol{x}), \boldsymbol{\phi}(\boldsymbol{x}') \rangle_{\mathcal{H}}$ with kernel evaluations $K(\boldsymbol{x}, \boldsymbol{x}')$, a technique known as *kernelization* or a *kernel trick* (Schölkopf and Smola, 2002). Moreover, for a positive-definite kernel we do not need to know the explicit form of the feature map, but instead it is implicitly defined through the kernel calculation.

4.2.3 Reproducing Kernel in Hilbert Space and Reproducing Property

In this section we will define the notion of an RKHS through the reproducing property. Then, we will see that a kernel function is a positive-definite and symmetric function.

It is also relevant to observe that the kernel fully generates the space, and that for a given kernel there is a unique RKHS; conversely, every RKHS contains a single kernel.

Let us assume a Hilbert space $\mathcal{H}$ where its elements are functions, provided with a dot product $\langle \cdot, \cdot \rangle$. We will denote $f(\cdot)$ as one element of the space, and $f(\boldsymbol{x})$ as its value at a particular argument $\boldsymbol{x}$. We will assume that arguments belong to a real or complex Euclidean space; that is, $\boldsymbol{x} \in \mathbb{R}^N$ or $\boldsymbol{x} \in \mathbb{C}^N$ respectively.

Definition 4.2.4 RKHS (Aronszajn, 1950). A Hilbert space $\mathcal{H}$ is said to be an RKHS if: (1) the elements of $\mathcal{H}$ are complex- or real-valued functions $f(\cdot)$ defined on any set of elements $\boldsymbol{x}$; and (2) for every element $\boldsymbol{x}, f(\cdot)$ is bounded.

The name of these spaces comes from the so-called *reproducing property*. Indeed, in an RKHS $\mathcal{H}$, there exists a function $K(\cdot, \cdot)$ such that

$$f(\boldsymbol{x}) = \langle f(\cdot), K(\cdot, \boldsymbol{x}) \rangle, \quad f \in \mathcal{H} \tag{4.4}$$

by virtue of the Riesz representation theorem (Riesz and Nagy, 1955). This function is called a *kernel*.

Property 4 The kernel $K(\cdot, \cdot)$ is a positive-definite and symmetric function.

Assume that $f(\cdot) = K(\cdot, \boldsymbol{x})$ in Equation 4.4. Hence:

$$K(\boldsymbol{x}, \boldsymbol{x}') = \langle K(\cdot, \boldsymbol{x}), K(\cdot, \boldsymbol{x}') \rangle. \tag{4.5}$$

Applying the complex conjugate operator * in the kernel and the dot product leads to

$$K^*(\boldsymbol{x}, \boldsymbol{x}') = (\langle K(\cdot, \boldsymbol{x}), K(\cdot, \boldsymbol{x}') \rangle)^* = \langle K(\cdot, \boldsymbol{x}'), K(\cdot, \boldsymbol{x}) \rangle = K(\boldsymbol{x}', \boldsymbol{x}), \tag{4.6}$$

which proves that a kernel is symmetric. In addition, consider the series

$$\sum_{i=1}^{N} \alpha_i K(\cdot, \boldsymbol{x}_i), \tag{4.7}$$

where α_i is any finite set of complex numbers. The norm of this series can be computed as

$$\left\| \sum_{i=1}^{N} \alpha_i K(\cdot, \boldsymbol{x}_i) \right\|^2 = \left\langle \sum_{i=1}^{n} \alpha_i K(\cdot, \boldsymbol{x}_i), \sum_{j=1}^{N} \alpha_{j'} K(\cdot, \boldsymbol{x}_j) \right\rangle. \tag{4.8}$$

By virtue of the reproducing property, $K(\boldsymbol{x}_i, \boldsymbol{x}_j) = \langle K(\cdot, \boldsymbol{x}_i), K(\cdot, \boldsymbol{x}_j) \rangle$, and by the linearity of the dot product, Equation 4.8 can be written as

$$\left\| \sum_{i=1}^{N} \alpha_i K(\cdot, \boldsymbol{x}_i) \right\|^2 = \sum_{i=1}^{n} \sum_{j=1}^{n} \alpha_i \alpha_{j'} K(\boldsymbol{x}_i, \boldsymbol{x}_j) \geq 0, \tag{4.9}$$

which proves that the kernel function is positive definite.

Property 5 The RKHS $\mathcal{H}$ with a kernel K is generated by $K(\cdot, \boldsymbol{x})$, where $\boldsymbol{x}$ belongs to any set.

It follows from the reproducing property that if $\langle f(\boldsymbol{x}), K(\cdot, \boldsymbol{x})\rangle = 0$, then, necessarily $f(\boldsymbol{x}) = 0$; hence, all elements in $\mathcal{H}$ are generated by the kernel.

Property 6 An RKHS contains a single reproducing kernel. Conversely, a reproducing kernel uniquely defines an RKHS.

If we consider the linear space generated by K, we can define

$$\langle K(\cdot, \boldsymbol{x}), K(\cdot, \boldsymbol{x}')\rangle = K(\boldsymbol{x}, \boldsymbol{x}'). \tag{4.10}$$

Let us now consider two sets of scalars μ_i and λ_j, $1 \leq i, j \leq N$; then:

$$\left\langle \sum_{i=1}^{N} \mu_i K(\cdot, \boldsymbol{x}_i), \sum_{j=1}^{N} \lambda_j K(\cdot, \boldsymbol{x}_j) \right\rangle = \sum_{i=1}^{N} \sum_{j=1}^{N} \mu_i \lambda_j K(\boldsymbol{x}_i, \boldsymbol{x}_j). \tag{4.11}$$

This expression satisfies the reproducing property and the requirements of dot product as follows:

$$\begin{aligned}
K(\boldsymbol{x}, \boldsymbol{x}') &= K^*(\boldsymbol{x}', \boldsymbol{x}') \\
\langle K(\cdot, \boldsymbol{x}) + K(\cdot, \boldsymbol{x}'), K(\cdot, \boldsymbol{x}'')\rangle &= \langle K(\cdot, \boldsymbol{x}), K(\cdot, \boldsymbol{x}')\rangle + \langle K(\cdot, \boldsymbol{x}), K(\cdot, \boldsymbol{x}'')\rangle \\
\langle \lambda K(\cdot, \boldsymbol{x}), K(\cdot, \boldsymbol{x}')\rangle &= \lambda \langle K(\cdot, \boldsymbol{x}), K(\cdot, \boldsymbol{x}')\rangle \\
\langle K(\cdot, \boldsymbol{x}), K(\cdot, \boldsymbol{x})\rangle &= 0 \quad \text{if and only if} \quad K(\cdot, x) = 0.
\end{aligned}$$

The last requirement can be proven by using the Cauchy–Schwartz inequality:

$$\left\| \sum_{i=1}^{N} \lambda_i K(\cdot, \boldsymbol{x}_i), K(\cdot, \boldsymbol{x})\rangle \right\|^2 \leq \sum_{i=1}^{N} \lambda_i K(\cdot, \boldsymbol{x}_i), \sum_{j=1}^{N} \lambda_j K(\cdot, \boldsymbol{x}_j)\rangle \langle K(\cdot, \boldsymbol{x}), K(\cdot, \boldsymbol{x})\rangle. \tag{4.12}$$

Hence:

$$\left\langle \sum_{i=1}^{N} \lambda_i K(\cdot, \boldsymbol{x}_i), \sum_{j=1}^{N} \lambda_j K(\cdot, \boldsymbol{x}_j) \right\rangle = 0 \tag{4.13}$$

implies

$$\sum_{i=1} \lambda_i K(\boldsymbol{x}, \boldsymbol{x}_i) = 0 \tag{4.14}$$

for every $\boldsymbol{x}$. Therefore, K is unique in $\mathcal{H}$ and it generates a unique RKHS.

4.2.4 Mercer's Theorem

Mercer's theorem is one of the best known results of the mathematician James Mercer (January 15, 1883–February 21, 1932), and of fundamental importance in the context of kernel methods. Mercer's theorem is the key idea behind the so-called kernel trick, which allows one to solve a variety of nonlinear optimization problems through the construction of kernelized counterparts of linear algorithms. Mercer's theorem can be stated as follows.

Assume that $K(\cdot,\cdot)$ is a continuous kernel function satisfying the properties in Section 4.2.3. Assume further that the kernel belongs to the family of square integrable functions.

Theorem 4.2.5 (Mercer's theorem (Aizerman *et al.*, 1964)) Let $K(\boldsymbol{x},\boldsymbol{x}')$ be a bivariate function fulfilling the Mercer condition; that is, $\int_{\mathbb{R}^{N_r}\times\mathbb{R}^{N_r}} K(\boldsymbol{x},\boldsymbol{x}')f(\boldsymbol{x})f(\boldsymbol{x}')d\boldsymbol{x}d\boldsymbol{x}' \geq 0$ for any square integrable function $f(\boldsymbol{x})$. Then, there exists a RKHS $\mathcal{H}$ and a mapping $\boldsymbol{\phi}(\cdot)$ such that $K(\boldsymbol{x},\boldsymbol{x}') = \langle\boldsymbol{\phi}(\boldsymbol{x}),\boldsymbol{\phi}(\boldsymbol{x}')\rangle$.

We will now consider a set D embedded in a space; typically, we will use only $\mathbb{R}^d$ or $\mathbb{C}^d$. From Mercer's theorem, it follows that a mapping function $\boldsymbol{\phi} : D \rightarrow \mathcal{H}$ can be expressed as a (possibly infinite dimension) column vector:

$$\boldsymbol{\phi}(\boldsymbol{x}) = \{\sqrt{\lambda_i}\boldsymbol{\phi}_i(\boldsymbol{x})\}_{i=1}^{\infty}. \tag{4.15}$$

The dot product between two of these maps $\boldsymbol{\phi}(\boldsymbol{x}_i)$ and $\boldsymbol{\phi}(\boldsymbol{x}_2)$ is defined then as the kernel function of vectors $\boldsymbol{x}$ and $\boldsymbol{x}'$ as

$$\langle\boldsymbol{\phi}(\boldsymbol{x}),\boldsymbol{\phi}(\boldsymbol{x}')\rangle = \boldsymbol{\phi}(\boldsymbol{x})^{\mathrm{T}}\boldsymbol{\phi}(\boldsymbol{x}') = \sum_{i=1}^{\infty}\lambda_i\boldsymbol{\phi}_i(\boldsymbol{x})\boldsymbol{\phi}_i(\boldsymbol{x}') = K(\boldsymbol{x},\boldsymbol{x}'). \tag{4.16}$$

Mercer's theorem shows that a mapping function into an RKHS and a dot product $K(\cdot,\cdot)$ exist if and only if $K(\cdot,\cdot)$ is a positive-definite function. Hence, if a given function $\mathcal{D}\times\mathcal{D}\rightarrow\mathbb{R}$ (or $\mathbb{C}$) is proven to be positive definite, then it is the kernel of a given RKHS.

Example 4.2.6 Kernel function of band-limited signals. Consider the interval $\omega \in [-W, W]$, and the subspace of square integrable functions generated by complex exponentials as

$$\boldsymbol{\phi}(t) = \mathrm{e}^{\mathbb{i}\omega t}. \tag{4.17}$$

Since the dot product of square integrable functions is the integral of their product, the kernel function inside this RKHS is

$$\langle h(t_1), h(t_2)\rangle = \frac{1}{2W}\int_{-W}^{W}\mathrm{e}^{\mathbb{i}\omega t_1}\,\mathrm{e}^{-\mathbb{i}\omega t_2}\,\mathrm{d}\omega = \frac{\sin(W(t_1-t_2))}{W(t_1-t_2)}. \tag{4.18}$$

This is then the kernel of the RKHS of signals that are band-limited to interval W. Moreover, the scalar dot product between functions is then

$$\langle f(\cdot), g(\cdot)\rangle = \int_{-W}^{W} F(\omega)G^{*}(\omega)\,\mathrm{d}\omega, \tag{4.19}$$

where $F(\omega)$ and $G(\omega)$ are the FT of $f(t)$ and $g(t)$ respectively. Indeed, if we denote the FT operator as $\mathcal{F}()$ and use this dot product, the reproducing property

$$f(t) = \langle f(\cdot), K(\cdot, t)\rangle = \int_{-W}^{W} F(\omega)\mathcal{F}^{*}(K(\cdot, t))\,\mathrm{d}\omega \tag{4.20}$$

holds. From Equation 4.18, it follows that $\mathcal{F}(K(\cdot, t)) = \mathrm{e}^{-\mathbb{i}\omega t}$. Then:

$$f(t) = \langle f(\cdot), K(\cdot, t)\rangle = \int_{-W}^{W} F(\omega)\,\mathrm{e}^{\mathbb{i}\omega t}\,\mathrm{d}\omega. \tag{4.21}$$

Example 4.2.7 Second-order polynomial kernel. Kernel $K(\boldsymbol{x}_i, \boldsymbol{x}_j) = (1+\boldsymbol{x}_i^{\mathrm{T}}\boldsymbol{x}_j)^2$ is a particular case of nth-order polynomial kernels. Assuming that $\boldsymbol{x}_i \in \mathbb{R}^2$, it is straightforward to find the mapping function $\boldsymbol{\phi} : \mathbb{R}^2 \rightarrow \mathcal{H}$. If $\boldsymbol{x}_i = \{x_i^{(1)}, x_i^{(2)}\}$, the dot product is

$$K(\boldsymbol{x}_i, \boldsymbol{x}_j) = (1+\boldsymbol{x}_i^{\mathrm{T}}\boldsymbol{x}_j)^2 = \left(1 + x_i^{(1)}x_j^{(1)} + x_i^{(2)}x_j^{(2)}\right)^2, \tag{4.22}$$

which can be expanded as

$$\begin{aligned} K(\boldsymbol{x}_i, \boldsymbol{x}_j) = 1 &+ (x_i^{(1)}x_j^{(1)})^2 + (x_i^{(2)}x_j^{(2)})^2 + 2x_i^{(1)}x_j^{(1)} + 2x_i^{(2)}x_j^{(2)} \\ &+ 2x_i^{(1)}x_j^{(1)}x_i^{(2)}x_j^{(2)}. \end{aligned}$$

By visual inspection, the corresponding mapping can be found to be

$$\boldsymbol{\phi}(\boldsymbol{x}) = (1, (x^{(1)})^2, (x^{(2)})^2, \sqrt{2}x^{(1)}, \sqrt{2}x^{(2)}, \sqrt{2}x^{(1)}x^{(2)})^{\mathrm{T}}. \tag{4.23}$$

Each element of this vector is an eigenfunction of this space. The corresponding eigenvectors are respectively $\lambda_1 = \lambda_2 = \lambda_3 = 1$, and $\lambda_4 = \lambda_5 = \lambda_6 = \sqrt{2}$.

4.3 Kernel Properties

This section pays attention to important properties of kernel methods: the issue of regularization, the representer's theorem, and basic operations that can be implicitly done in RKHSs via kernels.

4.3.1 Tikhonov's Regularization

Regularization methods are those methods intended to turn an ill-posed problem into a well-posed one such that a stable solution exists. Well-posedness is a concept introduced by Hadamard in 1912 to determine the solvability of mathematical models of physical phenomena. He defined a problem as *well posed* if a solution of the problem exists and the solution is *unique*. Also, the solution must be *stable*; that is, when an initial condition slightly changes, the solution must not change abruptly. In particular, if a parameter estimation problem does not verify these three conditions, it is said to be *unstable*. A regularization method that assures well-posedness of problems is the Tikhonov minimization, which ensures stability of solutions in this sense (Tikhonov and Arsenin, 1977) (see Section 3.2.8).

Notationally, let us assume a set of N pairs of data examples $\{\boldsymbol{x}_i, y_i\}$; the problem consists of finding an estimation function f parameterized by a set of weights $\boldsymbol{a}$ such that we approximate observations as $y_i = f(\boldsymbol{x}_i, \boldsymbol{a}) + \epsilon_i$, where ϵ_i is the estimation error. The regularization procedure consists of constructing a functional (see Definition 3.2.11):

$$\mathcal{L} = \sum_{i=1}^{N} V(y_i, f(\boldsymbol{x}_i, \boldsymbol{a})) + \lambda\Omega(f), \tag{4.24}$$

where $V(\cdot)$ is a cost function over the empirical error or risk of the estimation procedure, and $\Omega(\cdot)$ plays the role of a regularizer over the parameters of the estimation function $f(\cdot)$. The idea behind the application of this functional is that, whereas the empirical risk is used in order to choose those parameters that produce the best prediction of y_i given $\boldsymbol{x}_i$, the regularizer is used to account for smoothness (flatness) of the solution. Let us exemplify the concept of smoothness through a simple example.

Example 4.3.1 Assume an arbitrary system expressed by

$$f(x) = -0.2x^3 - 0.5x + 0.5 \tag{4.25}$$

for $x \in \mathbb{R}$; this produces outputs y_i that are corrupted with zero mean white Gaussian noise of $\sigma_n = 0.03$. A 6th-order polynomial defined as

$$\hat{f}(x) = a_6x^6 + a_5x^5 + a_4x^4 + a_3x^3 + a_2x^2 + a_1x + a_0 \tag{4.26}$$

is proposed for approximating the system's output. This model clearly has a complexity much higher than needed, and seven parameters including a bias term need to be adjusted, which are collectively grouped in $\boldsymbol{a} = [a_1, \dots, a_M]$, $M = 7$. A simple LS adjustment can be used to approximate the data. The polynomial shows a small error over the training sample, as is seen by the dotted line in Figure 4.3. Nevertheless, since the output data contain a given amount of noise, the test results in poor estimations with respect to the real function (continuous line). A regularized functional of the form

$$\mathcal{L} = \sum_{i=1}^{N} \|y_i - f(x_i, \boldsymbol{a})\|^2 + \lambda \sum_{j=0}^{M} a_j^2, \tag{4.27}$$

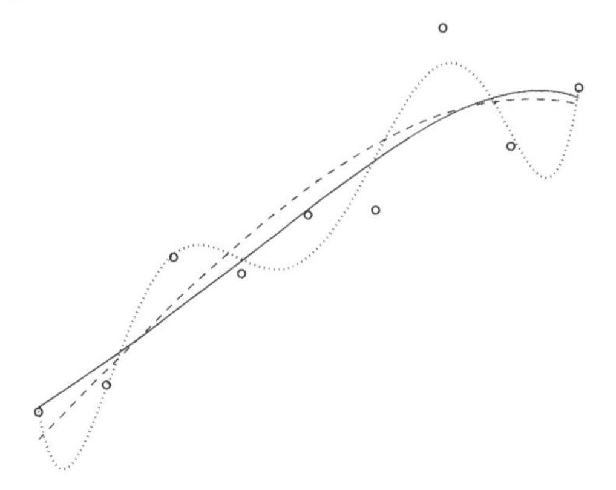

Figure 4.3 Nonlinear problem solved using two 6th-order polynomials. Circles represent points generated by the function (solid line) plus i.i.d. Gaussian noise. The dashed line corresponds to a regularized solution, while the dotted line corresponds to an unregularized solution ($\lambda = 0$).

which is known as ridge regression (Hoerl and Kennard, 1970) (see Exercise 4.9.3) can be applied in order to obtain a smoother function (dashed line) that better approximates the real function. The estimated parameters of the polynomial are as follows:

	a_0	a_1	a_2	a_3	a_4	a_5	a_6
LS	0.52	2.29	−22.6	91	−161.5	130.8	−39.73
RR	0.52	0.38	0.098	0.075	−0.117	−0.066	−0.035

which show how the range and variance of the obtained weights are constrained in the regularized solution, while the unregularized LS solution ($\lambda = 0$) yields big and uneven weights that produce unstable results (see Figure 4.3). Section 4.3.2 shows the extension of regularization to the field of RKHSs.

4.3.2 Representer Theorem and Regularization Properties

The representer theorem (Kimeldorf and Wahba, 1971) is a fundamental result in the field of kernel machines, since it establishes a generalization of the idea of regularization into an RKHS. The first motivation of the theorem is that, given a set of data, a linear expression of a solution based on a regularized functional exists in terms of a linear combination of the maps of the training data inside the RKHS. Moreover, recall that many kernels are expressed in infinite-dimension Hilbert spaces. This may completely preclude a good generalization, since the use of certain kernel functions may lead to a perfect fit of any data regardless of their statistical properties. Regularization becomes mandatory in all these cases. The second motivation is related to the fact that the implementation of smoothness is attached to the particular kernel chosen to solve the estimation problem. The theorem statement is as follows.

Theorem 4.3.2 (Representer theorem) (Kimeldorf and Wahba, 1971) Let $\Omega : [0, \infty) \rightarrow \mathbb{R}$ be a strictly monotonic increasing function; let $V : (\mathcal{X} \times \mathbb{R}^2)^N \rightarrow \mathbb{R} \cup \{\infty\}$ be an arbitrary loss function; and let $\mathcal{H}$ be an RKHS with reproducing kernel K. Then:

$$f^* = \min_{f \in \mathcal{H}} \{ V \left((f(\boldsymbol{x}_1), \boldsymbol{x}_1, y_1), \ldots, (f(\boldsymbol{x}_N), \boldsymbol{x}_N, y_N) \right) + \Omega(\|f\|_2^2) \} \tag{4.28}$$

admits a representation

$$f^*(\cdot) = \sum_{i=1}^{N} \alpha_i K(\cdot, \boldsymbol{x}_i), \quad \alpha_i \in \mathbb{R}, \quad \boldsymbol{\alpha} \in \mathbb{R}^{n \times 1} \tag{4.29}$$

That is, the function that minimizes the (regularized) optimization functional in Equation 4.28 is a linear function of dot products between data mapped into RKHS $\mathcal{H}$.

Since the mapped training data $\boldsymbol{\phi}(\boldsymbol{x}_i)$ span a subspace $\mathcal{H}_1$ inside the RKHS, a solution f may be expressed as a linear combination of these data plus a function g belonging to $\mathcal{H}_1 \perp \mathcal{H}_2$:

$$f = \sum_{i=1}^{N} \alpha_i K(\cdot, \boldsymbol{x}_i) + g(\cdot) \tag{4.30}$$

for which, obviously,

$$\langle \boldsymbol{\phi}(\boldsymbol{x}_i), g \rangle = 0. \tag{4.31}$$

Then, on the one hand, by virtue of the reproducing property, for any training data point $\boldsymbol{x}_j$ the approximation function is

$$f(\boldsymbol{x}_j) = \sum_{i=1}^{N} \alpha_i K(\boldsymbol{x}_j, \boldsymbol{x}_i) + \langle \boldsymbol{x}_j, g(\cdot) \rangle \tag{4.32}$$

and, on the other hand, the regularization function Ω satisfies

$$\begin{aligned} \Omega\left(\left\| \sum_{i=1}^{N} \alpha_i K(\cdot, \boldsymbol{x}_i) + g(\cdot) \right\| \right) &= \Omega\left(\sqrt{\left\| \sum_{i=1}^{N} \alpha_i K(\cdots, \boldsymbol{x}_i) \right\|^2} + \sqrt{\|g(\cdot)\|^2} \right) \\ &\geq \Omega\left(\| \sum_{i=1}^{N} \alpha_i K(\cdot, \boldsymbol{x}_i) \| \right), \end{aligned} \tag{4.33}$$

which proves the theorem. We should stress that the representer theorem (Kimeldorf and Wahba, 1971) states that the solutions of a large class of optimization problems can be expressed by only a finite number of kernel functions. Hence its importance.

Example 4.3.3 Let us consider a linear prediction model $f(\boldsymbol{x}) = \langle f, K(\cdot, \boldsymbol{x}) \rangle$. A common regularizer in machine learning and statistics is the ℓ_2-norm $\|f\|_2^2$.

By virtue of the representer theorem, we can express the learning function f as a linear combination of kernel functions $f = \sum_{i=1}^{N} \alpha_i K(\cdot, \boldsymbol{x}_i)$; then:

$$
\begin{aligned}
\|f\|^2 = \langle f, f\rangle &= \left\langle \sum_{i=1}^{N} \alpha_i K(\cdot, \boldsymbol{x}_i), \sum_{j=1}^{N} \alpha_j K(\cdot, \boldsymbol{x}_j) \right\rangle \\
&= \sum_{i=1}^{N}\sum_{j=1}^{N} \alpha_i \alpha_j \left\langle K(\cdot, \boldsymbol{x}_i) K(\cdot, \boldsymbol{x}_j) \right\rangle \\
&\sum_{i=1}^{N}\sum_{j=1}^{N} \alpha_i \alpha_j K(\boldsymbol{x}_i, \boldsymbol{x}_j) = \boldsymbol{\alpha}^{\mathrm{T}} \boldsymbol{K} \boldsymbol{\alpha}.
\end{aligned}
\tag{4.34}
$$

Note that if we express $f(\boldsymbol{x}) = \langle \boldsymbol{w}, K(\cdot, \boldsymbol{x})\rangle = \boldsymbol{w}^{\mathrm{T}} \boldsymbol{\phi}(\boldsymbol{x})$ then $\boldsymbol{w} = \sum_{i=1}^{N} \alpha_i \boldsymbol{\phi}(\boldsymbol{x}_i)$ and $\boldsymbol{w}^{\mathrm{T}} \boldsymbol{w} = \boldsymbol{\alpha}^{\mathrm{T}} \boldsymbol{K} \boldsymbol{\alpha}$.

4.3.3 Basic Operations with Kernels

We now review some basic properties with kernels. The interest here is to introduce some basic operations that allow us to operate with objects embedded in high-dimensional, often inaccessible, feature spaces. Essentially, we will show that we can compute distances, norms, and angles, as well as perform projections and normalizations, in the feature space $\mathcal{H}$ via kernels in an implicit way.

Translation. A translation in feature space can be expressed as the modified feature map $\tilde{\boldsymbol{\phi}}(\boldsymbol{x}) = \boldsymbol{\phi}(\boldsymbol{x}) + \Gamma$, $\Gamma \in \mathcal{H}$. Then, the translated dot product for $\langle \tilde{\boldsymbol{\phi}}(\boldsymbol{x}), \tilde{\boldsymbol{\phi}}(\boldsymbol{x}')\rangle_{\mathcal{H}}$ can be computed if we restrict Γ to lie in the span of the functions $\{\boldsymbol{\phi}(\boldsymbol{x}_1), \ldots, \boldsymbol{\phi}(\boldsymbol{x}_N)\} \in \mathcal{H}$.

Centering. The previous translation allows us to center data $\{\boldsymbol{x}_i\}_{i=1}^{N} \in \mathcal{X}$ in the *feature space*. The mean of the data in $\mathcal{H}$ is $\boldsymbol{\phi}_\mu = c(1/N) \sum_{i=1}^{N} \boldsymbol{\phi}(\boldsymbol{x}_i)$, which is a linear combination of the span of functions, and hence it fulfills the requirement for Γ. One can center data in $\mathcal{H}$ by computing $\tilde{\boldsymbol{K}} = \boldsymbol{HKH}$, where entries of $\boldsymbol{H}$ are $\boldsymbol{H}_{ij} = \delta_{ij} - (1/N)$, and the Kronecker symbol is $\delta_{i,j} = 1$ if $i = j$ and zero otherwise.

Subspace projections. Given two points $\boldsymbol{\psi}$ and $\boldsymbol{\gamma}$ in the feature space, the projection of $\boldsymbol{\psi}$ onto the subspace spanned by $\boldsymbol{\gamma}$ is

$$
\boldsymbol{\psi}' = \frac{\langle \boldsymbol{\gamma}, \boldsymbol{\psi}\rangle_{\mathcal{H}}}{\|\boldsymbol{\gamma}\|_{\mathcal{H}}^2} \boldsymbol{\gamma}.
$$

Therefore, one can compute the projection $\boldsymbol{\psi}'$ expressed solely in terms of kernel evaluations. If one has access to vectors $\boldsymbol{\psi}$ and $\boldsymbol{\gamma}$, the projection can be computed explicitly. Otherwise, one can compute the dot products of the pre-image vectors using reproducing kernels. For example, assuming pre-images $\boldsymbol{\psi} = \boldsymbol{\phi}(\boldsymbol{x})$ and $\boldsymbol{\gamma} = \boldsymbol{\phi}(\boldsymbol{z})$, the norm of a vector in Hilbert spaces can be computed using a reproducing kernel; that is, $\|\boldsymbol{\gamma}\|_{\mathcal{H}}^2 = \langle \boldsymbol{\phi}(\boldsymbol{z}), \boldsymbol{\phi}(\boldsymbol{z})\rangle_{\mathcal{H}} = K_\Gamma(\boldsymbol{z}, \boldsymbol{z})$, and $\langle \boldsymbol{\gamma}, \boldsymbol{\psi}\rangle_{\mathcal{H}} = \langle \boldsymbol{\phi}(\boldsymbol{z}), \boldsymbol{\phi}(\boldsymbol{x})\rangle_{\mathcal{H}} = K_\Psi(\boldsymbol{z}, \boldsymbol{x})$.

Computing distances. Given that a kernel corresponds to a dot product in a Hilbert space $\mathcal{H}$, we can compute distances between mapped samples entirely in terms of kernel evaluations:

$$\|\boldsymbol{\phi}(\boldsymbol{x}) - \boldsymbol{\phi}(\boldsymbol{x}')\|_{\mathcal{H}} = \sqrt{K(\boldsymbol{x},\boldsymbol{x}) + K(\boldsymbol{x}',\boldsymbol{x}') - 2K(\boldsymbol{x},\boldsymbol{x}')}. \tag{4.35}$$

Normalization. Exploiting the previous property, one can also normalize data in feature spaces implicitly:

$$K'(\boldsymbol{x},\boldsymbol{x}') = \left\langle \frac{\boldsymbol{\phi}(\boldsymbol{x})}{\|\boldsymbol{\phi}(\boldsymbol{x})\|}, \frac{\boldsymbol{\phi}(\boldsymbol{x}')}{\|\boldsymbol{\phi}(\boldsymbol{x}')\|} \right\rangle_{\mathcal{H}} = \frac{K(\boldsymbol{x},\boldsymbol{x}')}{\sqrt{K(\boldsymbol{x},\boldsymbol{x})K(\boldsymbol{x}',\boldsymbol{x}')}}, \tag{4.36}$$

where the result is a new kernel function K'. Note that the operation is useless for some kernel functions. For example, the RBF kernel is given by $K(\boldsymbol{x}_i,\boldsymbol{x}_j) = \exp(-\|\boldsymbol{x}_i - \boldsymbol{x}_j\|^2/(2\sigma^2))$, so sample self-similarities are $K(\boldsymbol{x}_i,\boldsymbol{x}_i) = 1$ and normalization does not affect the resulting kernel. Note that the RBF kernel function implicitly maps examples to a hypersphere with unit norm, $\|\boldsymbol{\phi}(\boldsymbol{x})\| = 1$.

4.4 Constructing Kernel Functions

4.4.1 Standard Kernels

Any kernel method has its foundations in the definition of a kernel mapping function $\boldsymbol{\phi}$ that accurately measures the similarity among samples in some sense. But not all the kernel similarity functions can be used; instead, valid kernels are only those fulfilling Mercer's theorem. Roughly speaking, kernel functions yield positive-definite similarity matrices for a set of data measurements, and the most common ones in this setting are the linear $K(\boldsymbol{x},\boldsymbol{z}) = \langle \boldsymbol{x},\boldsymbol{z}\rangle$, the polynomial $K(\boldsymbol{x},\boldsymbol{z}) = (\langle \boldsymbol{x},\boldsymbol{z}\rangle + 1)^d$, $d \in \mathbb{Z}^+$, and the RBF $K(\boldsymbol{x},\boldsymbol{z}) = \exp(-\|\boldsymbol{x}-\boldsymbol{z}\|^2/2\sigma^2)$, $\sigma \in \mathbb{R}^+$. Note that, by Taylor series expansion, the RBF kernel is a polynomial kernel with infinite degree. Thus, the corresponding Hilbert space is infinite dimensional, which corresponds to a mapping into the space of functions C^∞. Table 4.1 summarizes the most useful kernels and their main characteristics.

The relation between *similarity* and *distance* has produced a high number of kernel functions that rely on standard distance measures, such as Mahalanobis kernels (Hofmann *et al.*, 2008). Most of the methods rely on the observation that the Gaussian kernel is of the form $K(\boldsymbol{x},\boldsymbol{x}') = \exp(-a\, d(\boldsymbol{x},\boldsymbol{x}'))$, where $d : \mathcal{X} \times \mathcal{X} \rightarrow \mathbb{R}_+$ represents a distance function and $a > 0$. Unfortunately, this instantiation in general is not true (Cortes *et al.*, 2003). If $\boldsymbol{K}$ is positive definite then $-\boldsymbol{K}$ is negative definite, but negative definite, but the converse is not true in general. For example $K(\boldsymbol{x},\boldsymbol{x}') = (\boldsymbol{x}-\boldsymbol{x}')^b$ is negative definite for $b \in [0,2)$. The RBF kernel is also of practical convenience (stability and only one parameter to be tuned), and it is the preferred kernel function in standard applications. Nevertheless, specific applications need particular kernel functions. Reviewing all the possible kernels is beyond the scope of this chapter. For a more general overview, including examples for other data structures such as graphs, trees, strings, and others, we refer the reader to Hofmann *et al.* (2008), Schölkopf and Smola (2002), Bakır *et al.* (2007), and Shawe-Taylor and Cristianini (2004).

Table 4.1 Main kernel functions used in the literature.

Kernel function	Expression
Linear kernel	$K(\boldsymbol{x},\boldsymbol{y}) = \boldsymbol{x}^{\mathrm{T}}\boldsymbol{y} + c$
Polynomial kernel	$K(\boldsymbol{x},\boldsymbol{y}) = (\alpha\boldsymbol{x}^{\mathrm{T}}\boldsymbol{y} + c)^d$
RBF kernel	$K(\boldsymbol{x},\boldsymbol{y}) = \exp\left(-\frac{\Vert\boldsymbol{x}-\boldsymbol{y}\Vert^2}{2\sigma^2}\right)$
Exponential kernel	$K(\boldsymbol{x},\boldsymbol{y}) = \exp\left(-\frac{\Vert\boldsymbol{x}-\boldsymbol{y}\Vert}{2\sigma^2}\right)$
Laplacian kernel	$K(\boldsymbol{x},\boldsymbol{y}) = \exp\left(-\frac{\Vert\boldsymbol{x}-\boldsymbol{y}\Vert}{\sigma}\right)$
Analysis of variance kernel	$K(\boldsymbol{x},\boldsymbol{y}) = \sum_{k=1}^{d}\exp[-\sigma(\boldsymbol{x}^{(k)} - \boldsymbol{y}^{(k)})^2]^d$
Hyperbolic tangent (sigmoid) kernel	$K(\boldsymbol{x},\boldsymbol{y}) = \tanh(\alpha\boldsymbol{x}^{\mathrm{T}}\boldsymbol{y} + c)$
Rational quadratic kernel	$K(\boldsymbol{x},\boldsymbol{y}) = 1 - \frac{\Vert\boldsymbol{x}-\boldsymbol{y}\Vert^2}{\Vert\boldsymbol{x}-\boldsymbol{y}\Vert^2 + c}$
Multiquadric kernel	$K(\boldsymbol{x},\boldsymbol{y}) = \sqrt{\Vert\boldsymbol{x}-\boldsymbol{y}\Vert^2 + c^2}$
Inverse multiquadric kernel	$K(\boldsymbol{x},\boldsymbol{y}) = \frac{1}{\sqrt{\Vert\boldsymbol{x}-\boldsymbol{y}\Vert^2 + c^2}}$
Power kernel	$K(\boldsymbol{x},\boldsymbol{y}) = -\Vert\boldsymbol{x}-\boldsymbol{y}\Vert^d$
Log kernel	$K(\boldsymbol{x},\boldsymbol{y}) = -\log(\Vert\boldsymbol{x}-\boldsymbol{y}\Vert^d + 1)$
Cauchy kernel	$K(\boldsymbol{x},\boldsymbol{y}) = \frac{1}{1 + (\Vert\boldsymbol{x}-\boldsymbol{y}\Vert^2/\sigma^2)}$
Chi-square kernel	$K(\boldsymbol{x},\boldsymbol{y}) = 1 - \sum_{k=1}^{d}\frac{(\boldsymbol{x}^{(k)} - \boldsymbol{y}^{(k)})^2}{\frac{1}{2}(\boldsymbol{x}^{(k)} + \boldsymbol{y}^{(k)})}$
Histogram (or min) intersection kernel	$K(\boldsymbol{x},\boldsymbol{y}) = \sum_{k=1}^{d}\min(\boldsymbol{x}^{(k)}, \boldsymbol{y}^{(k)})$
Generalized histogram intersection kernel	$K(\boldsymbol{x},\boldsymbol{y}) = \sum_{k=1}^{m}\min(\vert\boldsymbol{x}^{(k)}\vert^{\alpha}, \vert\boldsymbol{y}^{(k)}\vert^{\beta})$
Generalized T-Student kernel	$K(\boldsymbol{x},\boldsymbol{y}) = \frac{1}{1 + \Vert\boldsymbol{x}-\boldsymbol{y}\Vert^d}$

4.4.2 Properties of Kernels

Taking advantage of some algebra and functional analysis properties (Golub and Van Loan 1996; Reed and Simon 1981), very useful properties of kernels can be derived. Let K_1 and K_2 be two positive-definite kernels on $\mathcal{X} \times \mathcal{X}$, $\mathcal{A}$ be a symmetric positive semidefinite matrix, $d(\cdot,\cdot)$ be a metric fulfilling distance properties, and $\mu > 0$. Then, the following kernels (Schölkopf and Smola, 2002) are valid:

$$K(\boldsymbol{x},\boldsymbol{x}') = K_1(\boldsymbol{x},\boldsymbol{x}') + K_2(\boldsymbol{x},\boldsymbol{x}') \tag{4.37}$$

$$K(\boldsymbol{x},\boldsymbol{x}') = \mu K_1(\boldsymbol{x},\boldsymbol{x}') \tag{4.38}$$

$$K(\boldsymbol{x},\boldsymbol{x}') = K_1(\boldsymbol{x},\boldsymbol{x}') \times K_2(\boldsymbol{x},\boldsymbol{x}') \tag{4.39}$$

$$K(\boldsymbol{x},\boldsymbol{x}') = \boldsymbol{x}^{\mathrm{T}}\mathcal{A}\boldsymbol{x}' \tag{4.40}$$

$$K(\boldsymbol{x},\boldsymbol{x}') = K(f(\boldsymbol{x}), f(\boldsymbol{x}')) \tag{4.41}$$

These basic properties make it easy to construct refined similarity measures better fitted to the data characteristics. One can sum dedicated kernels to different portions of the feature space, to different data representations, or even to different temporal or spatial scales through Equation 4.37. A scaling factor for each kernel can also be used (see Equation 4.38). Recent advances for kernel development also involve the following:

Convex combinations. By exploiting Equations 4.37 and 4.38, one can build new kernels by linear combinations of kernels:

$$K(\boldsymbol{x}, \boldsymbol{x}') = \sum_{m=1}^{M} d_m K_m(\boldsymbol{x}, \boldsymbol{x}'), \tag{4.42}$$

where each kernel K_m could, for instance, work with particular feature subsets with eventually a different kernel function. This field of research is known as MKL, and many algorithms have been proposed to optimize jointly the weights and the kernel parameters (Rakotomamonjy *et al.*, 2008). Note that this composite kernel offers some insight into the problem as well, since relevant features receive higher values of d_m, and the corresponding kernel parameters θ_m yield information about similarity scales.

Deforming kernels. The field of semi-supervised kernel learning deals with techniques to modify the values of the training kernel, including the information from the whole data distribution. In this setting, the kernel K is either deformed with a graph distance matrix built with both labeled and unlabeled samples, or using kernels built from clustering solutions (Belkin *et al.* 2006; Sindhwani *et al.* 2005).

Generative kernels. Exploiting Equation 4.41, one can construct kernels from probability distributions by defining $K(\boldsymbol{x}, \boldsymbol{x}') = K(\boldsymbol{p}, \boldsymbol{p}')$, where $\boldsymbol{p}$, $\boldsymbol{p}'$ are defined on the space $\mathcal{X}$ (Jaakkola and Haussler, 1999). This family of kernels is known as *probability product kernels between distributions* and defined as

$$K(\boldsymbol{p}, \boldsymbol{p}') = \langle \boldsymbol{p}, \boldsymbol{p}' \rangle = \int_{\mathcal{X}} \boldsymbol{p}(\boldsymbol{x})\boldsymbol{p}'(\boldsymbol{x})\, \mathrm{d}\boldsymbol{x}. \tag{4.43}$$

Joint input–output mappings. Kernels are typically built on a set of input samples. During recent years, the framework of *structured output learning* has dealt with the definition of joint input–output kernels, $K((\boldsymbol{x}, y), (\boldsymbol{x}', y'))$ (Bakır *et al.* 2007; Weston *et al.* 2003).

4.4.3 Engineering Signal Processing Kernels

The properties of kernel methods presented before have been widely used to develop new kernel functions suitable for tackling the peculiarities of the signals under analysis in a given DSP application. Let us review now some of the most relevant families of engineered kernels of special interest in signal processing.

Translation- or Shift-Invariant Kernels

A particular class of kernels is *translation-invariant kernels*, also known as shift-invariant kernels, which fulfill

$$K(\boldsymbol{u}, \boldsymbol{v}) = K(\boldsymbol{u} - \boldsymbol{v}). \tag{4.44}$$

A necessary and sufficient condition for a translation-invariant kernel to be Mercer's kernel (Zhang *et al.*, 2004) is that its FT must be real and nonnegative; that is:

$$\frac{1}{2\pi}\int_{\boldsymbol{v}=-\infty}^{+\infty} K(\boldsymbol{v})\,\mathrm{e}^{-\mathrm{i}2\pi\langle \boldsymbol{f},\boldsymbol{v}\rangle} d\boldsymbol{v} \geq 0 \quad \forall \boldsymbol{f} \in \mathbb{R}^d \tag{4.45}$$

Bochner's theorem (Reed and Simon, 1981) states that a continuous shift-invariant kernel $K(\boldsymbol{x}, \boldsymbol{x}') = K(\boldsymbol{x} - \boldsymbol{x}')$ on $\mathbb{R}^d$ is positive definite if and only if the FT of K is nonnegative. If a shift-invariant kernel K is properly scaled, its FT $p(\boldsymbol{\omega})$ is a proper probability distribution.

As we will see in Chapter 10, this property has been recently used to approximate kernel functions and matrices with linear projections on a number of D randomly generated features as follows:

$$K(\boldsymbol{x}, \boldsymbol{x}') = \int_{\mathbb{R}^d} p(\boldsymbol{\omega})\,\mathrm{e}^{-\mathrm{i}\boldsymbol{\omega}^{\mathrm{T}}(\boldsymbol{x}-\boldsymbol{x}')}\,\mathrm{d}\boldsymbol{\omega} \approx \sum\nolimits_{i=1}^{D} \frac{1}{D}\,\mathrm{e}^{-\mathrm{i}\boldsymbol{\omega}_i^{\mathrm{T}}\boldsymbol{x}}\,\mathrm{e}^{\mathrm{i}\boldsymbol{\omega}_i^{\mathrm{T}}\boldsymbol{x}'}, \tag{4.46}$$

where $p(\boldsymbol{\omega})$ is set to be the inverse FT of K, and $\boldsymbol{\omega}_i \in \mathbb{R}^d$ is randomly sampled from a data-independent distribution $p(\boldsymbol{\omega})$ (Rahimi and Recht, 2009). In this case, we define a D-dimensional feature map $\boldsymbol{z}(\boldsymbol{x}) \ : \ \mathbb{R}^d \mathrm{T} \mathbb{R}^D$, which can be *explicitly* constructed as $\boldsymbol{z}(\boldsymbol{x}) \ := \ [\mathrm{e}^{(\mathrm{i}\boldsymbol{\omega}_1^{\mathrm{T}}\boldsymbol{x})}, \ldots, \exp(\mathrm{i}\boldsymbol{\omega}_D^{\mathrm{T}}\boldsymbol{x})]^{\mathrm{T}}$. In matrix notation, given N data points, the kernel matrix $\mathbf{K} \in \mathbb{R}^{N\times N}$ can be approximated with the explicitly mapped data, $\boldsymbol{Z} = [\boldsymbol{z}_1 \cdots \boldsymbol{z}_n]^{\mathrm{T}} \in \mathbb{R}^{N\times D}$, and will be denoted as $\hat{\boldsymbol{K}} \approx \boldsymbol{Z}\boldsymbol{Z}^{\mathrm{T}}$. This property can be used to approximate any shift-invariant kernel. For instance, the familiar squared exponential (SE) Gaussian kernel $K(\boldsymbol{x}, \boldsymbol{x}') = \exp(-\|\boldsymbol{x} - \boldsymbol{x}'\|^2/(2\sigma^2))$ can be approximated by using $\boldsymbol{\omega}_i \sim \mathcal{N}(0, \sigma^{-2}\boldsymbol{I}), 1 \leq i \leq D$.

Autocorrelation Kernels

A kernel related to the shift-invariant kernels family is the *autocorrelation-induced kernel*. Notationally, let $\{h_n\}$ be an $(N+1)$-samples limited-duration discrete-time real signal (i.e., $h_n = 0, \forall n \notin (0, N)$), and let $R_n^h = h_n * h_{-n}$ be its autocorrelation function. Then, the following shift-invariant kernel can be built:

$$K^h(n, m) = R_n^h(n - m), \tag{4.47}$$

which is called an *autocorrelation-induced kernel*, or simply an autocorrelation kernel. As R_n^h is an even signal, its spectrum is real and nonnegative, as expressed in Equation 4.45; hence, an autocorrelation kernel is always a Mercer kernel.

The previous two types of kernels are clear examples of the importance of including classic concepts of DSP in kernel-based algorithms. We will come back to this kernel in chapters in Part II.

Convolution Kernels

Convolution kernels, sometimes called *part kernels* or *bag of words kernels*, constitute an interesting development. The underlying idea for them is that the signal components (objects) are structured in any sense; thus, rather than measuring similarities between complete objects, one aggregates similarities of individual components (or feature subsets). A simple definition of a convolution kernel between $\boldsymbol{x}$ and $\boldsymbol{x}'$ is

$$K(\boldsymbol{x}, \boldsymbol{x}') = \sum_{\bar{\boldsymbol{x}}_i} \sum_{\bar{\boldsymbol{x}}'_i} \prod_{p=1}^{P} K_p(\boldsymbol{x}_p, \boldsymbol{x}'_p), \tag{4.48}$$

where $\bar{\boldsymbol{x}}$ and $\bar{\boldsymbol{x}}'$ are the sets of parts ($p = 1, \ldots, P$) of examples $\boldsymbol{x}$ and $\boldsymbol{x}'$ respectively, and $K_p(\boldsymbol{x}_p, \boldsymbol{x}'_p)$ is a kernel between the pth part of $\boldsymbol{x}$ and $\boldsymbol{x}'$.

It is easy to show that the RBF kernel is a convolution kernel by simply letting each of the P dimensions of $\boldsymbol{x}$ be a part, and using Gaussian kernels $K_p(\boldsymbol{x}_p, \boldsymbol{x}'_p) = \exp(-\|\boldsymbol{x}_p - \boldsymbol{x}'_p\|^2/(2\sigma^2))$. However, note also that the linear kernel $K(\boldsymbol{x}, \boldsymbol{x}') = \sum_{p=1}^{P} \boldsymbol{x}_p \boldsymbol{x}'_p$ is not a convolution kernel, since we would need to sum products of more than one term.

Spatial Kernels

The latter convolution kernels have been extensively used in many fields of signal processing. In particular, the fields of image processing and computer vision have contributed with many kernel functions, especially designed to deal with the peculiarities of image features. Let us review two interesting spatial kernels to deal with *structured domains*, such as images.

A spatial kernel for image processing is inspired in the well-known concepts of *locality* and *receptive fields* of NNs. A kernel value is here computed by using not all features of an image, but only those which fall into a particular region inside a window (Schölkopf, 1997). Extensions of that kernel to deal with hierarchical organizations lead to the related *spatial pyramid match kernel* (Lazebnik *et al.*, 2006) and the *pyramid match kernel* (Grauman and Darrell, 2005). A standard approach in these methods is to divide the image into a grid of 1×1, 2×2, ... equally spaced windows, resembling a pyramid, whose depth is referred as *level*. Then, for each level l, a histogram $\boldsymbol{h}_l$ is computed by concatenating the histograms of all subwindows within the level. Two images $\boldsymbol{x}, \boldsymbol{x}'$ are compared by combining the similarity of the individual levels:

$$K(\boldsymbol{x}, \boldsymbol{x}') = \sum_{l=0}^{L-1} d_l K_l(\boldsymbol{h}_l, \boldsymbol{h}'_l), \tag{4.49}$$

where $d_l \in \mathbb{R}_+$ is an extra weighting parameter for each level, which must be optimized. As can be noted, the resulting kernel is just the result of applying the property of sum of kernels.

A second interesting kernel engineering was presented by Laparra *et al.* (2010) to deal with image denoising in the wavelet domain. Here, an SVR algorithm used a combination of kernels adapted to different scales, orientations, and frequencies in the *wavelet domain*. The specific signal relations were encoded in an anisotropic kernel obtained from mutual information measures computed on a representative image database. In particular, a set of Laplacian kernels was used to consider the intraband oriented relations within each wavelet subband:

$$K_\alpha(\boldsymbol{p}_i, \boldsymbol{p}_j) = \exp(-((\boldsymbol{p}_i - \boldsymbol{p}_j)^{\mathrm{T}} \boldsymbol{G}_\alpha^{\mathrm{T}} \boldsymbol{\Sigma}^{-1} \boldsymbol{G}_\alpha (\boldsymbol{p}_i - \boldsymbol{p}_j))^{1/2}), \tag{4.50}$$

where $\boldsymbol{\Sigma} = \mathrm{diag}(\sigma_1, \sigma_2)$, σ_1, and σ_2 are the widths of the kernels, $\boldsymbol{p}_i \in \mathbb{R}^2$ denotes the spatial position of wavelet coefficient y_i within a subband, and $\boldsymbol{G}_\alpha$ is the 2D rotation matrix with rotation angle α, corresponding to the orientation of each subband.

Time–Frequency and Wavelet Kernels

Time–frequency signal decompositions in general, and wavelet analysis in particular, are at the core of signal and image processing. These signal representations have been widely studied and employed in practice. Inspired by wavelet theory, particular wavelet kernels have been proposed (Zhang *et al.*, 2004), which can be roughly approximated as

$$K(\boldsymbol{x}, \boldsymbol{x}') = \prod_{i=1}^{N} h\left(\frac{x_i - c}{a}\right) h\left(\frac{x'_i - c}{a}\right), \tag{4.51}$$

where a and c represent the wavelet dilation and translation coefficients respectively. A translation-invariant version of this kernel can be given by

$$K(\boldsymbol{x}, \boldsymbol{x}') = \prod_{i=1}^{N} h\left(\frac{x_i - x'_i}{a}\right), \tag{4.52}$$

where in both kernels the function $h(x)$ denotes a *mother wavelet function*, which is typically chosen to be $h(x) = \cos(1.75x)\exp(-x^2/2)$, as it yields to a valid admissible kernel function (Zhang *et al.*, 2004). Note that, in this case, the wavelet kernel is the result of applying the properties of direct sum and direct product of kernels.

4.5 Complex Reproducing Kernel in Hilbert Spaces

The complex representation of a magnitude is of important interest in signal processing, particularly in communications, since it provides a natural and compact expression that makes signal manipulation easier. The justification for the use of complex numbers in communications arises from the fact that any band-pass *real* signal centered around a given frequency admits a representation in terms of *in-phase* and *quadrature* components. Indeed, assume a real-valued signal $x(t)$ whose spectrum lies between two limits ω_{min} and ω_{max}. This function can the be expressed as

$$x(t) = A_{\mathrm{I}}(t)\cos\omega_0 t - A_{\mathrm{Q}}(t)\sin\omega_0 t, \tag{4.53}$$

with $A_{\mathrm{I}}(t)$ and $A_{\mathrm{Q}}(t)$ being the in-phase and quadrature components. This expression can be rewritten as

$$x(t) = \mathrm{Re}\{(A_{\mathrm{I}}(t) + \mathbb{i}A_{\mathrm{Q}}(t))\, e^{\omega_0 t}\} = \mathrm{Re}\{A(t)\, \mathrm{e}^{\omega_0 t}\} \tag{4.54}$$

for some arbitrary frequency ω_0 called central frequency or carrier, $A(t)$ being any arbitrary function called a complex envelope of $x(t)$. Since in-phase and quadrature components are orthogonal with respect to the dot product of L^2 functions, it is straightforward that $A(t)$ can be modulated as in Equation 4.53 without loss of information. The central frequency is usually known and removed during the signal processing, thus obtaining the complex envelope. This signal can either be processed as a pair of real-valued signals or as a complex signal, thus obtaining a more compact notation.

Though the concept of a complex-valued Mercer kernel is classic (Aronszajn, 1950), it was proposed by Martínez-Ramón *et al.* (2005); Martínez-Ramón *et al.* (2007) for its use in antenna array processing, and with a more formal treatment and rigorous justification by Bouboulis and coworkers (Bouboulis and Theodoridis, 2010, 2011; Bouboulis *et al.*, 2012) and Ogunfunmi and Paul (2011) for its use with the kernel LMS (KLMS) (Liu *et al.*, 2008, 2009) with the objective to extend this algorithm to a complex domain, as given in

$$\boldsymbol{\phi}(\boldsymbol{x}) = \boldsymbol{\phi}(\boldsymbol{x}) + \mathbb{i}\boldsymbol{\phi}(\boldsymbol{x}) = K((\boldsymbol{x}_{\mathbb{R}}, \boldsymbol{x}_{\mathbb{I}}), \cdot) + \mathbb{i}K((\boldsymbol{x}_{\mathbb{R}}, \boldsymbol{x}_{\mathbb{I}}), \cdot), \tag{4.55}$$

which is a transformation of the data $\boldsymbol{x} = \boldsymbol{x}_{\mathbb{R}} + \mathbb{i}\boldsymbol{x}_{\mathbb{I}} \in \mathbb{C}^d$ into a complex RKHS. This is known as the *complexification trick*, where the kernel is defined over real numbers.

Since the complex LMS algorithm involves the use of a complex gradient, the Wirtinger calculus must be introduced in the notation. The reason is that the cost function of the algorithm is real valued, and it is defined over a complex domain. Hence, it is non-holomorphic and complex derivatives cannot be used. A convenient way to compute such derivatives is to use Wirtinger derivatives. Assume a variable $x = x_{\mathbb{R}} + \mathbb{i}x_{\mathbb{I}} \in \mathbb{C}$ and a non-holomorfic function $f(x) = f_{\mathbb{R}}(x) + f_{\mathbb{I}}(x)$, then its Wirtinger derivatives with respect to x and x^* are

$$\begin{aligned}\frac{\partial f}{\partial x} &= \frac{1}{2}\left(\frac{\partial f_{\mathbb{R}}}{\partial x_{\mathbb{R}}} + \frac{\partial f_{\mathbb{I}}}{\partial x_{\mathbb{I}}}\right) + \frac{\mathbb{i}}{2}\left(\frac{\partial f_{\mathbb{I}}}{\partial x_{\mathbb{R}}} - \frac{\partial f_{\mathbb{R}}}{\partial x_{\mathbb{I}}}\right)\\ \frac{\partial f}{\partial x^*} &= \frac{1}{2}\left(\frac{\partial f_{\mathbb{R}}}{\partial x_{\mathbb{R}}} - \frac{\partial f_{\mathbb{I}}}{\partial x_{\mathbb{I}}}\right) + \frac{\mathbb{i}}{2}\left(\frac{\partial f_{\mathbb{I}}}{\partial x_{\mathbb{R}}} + \frac{\partial f_{\mathbb{R}}}{\partial x_{\mathbb{I}}}\right).\end{aligned} \tag{4.56}$$

This concept is restricted to complex-valued functions defined in $\mathbb{C}$. Authors generalize this concept to functions defined in n RKHS through the definition of Fréchet differentiability. The kernel function in this RKHS is defined, from Equation 4.55, as

$$\hat{K}(\boldsymbol{x}, \boldsymbol{x}') = \boldsymbol{\phi}^{\mathrm{H}}(\boldsymbol{x})\boldsymbol{\phi}(\boldsymbol{x}') \quad = (\phi(\boldsymbol{x}) - \mathbb{i}\phi(\boldsymbol{x}))(\phi^{\mathrm{T}}(\boldsymbol{x}) + \mathbb{i}\phi^{\mathrm{t}}(\boldsymbol{x})) = 2K(\boldsymbol{x}, \boldsymbol{x}'). \tag{4.57}$$

The representer theorem (see Theorem 4.3.2) can be rewritten here as follows:

$$f^*(\cdot) = \sum_{i=1}^{N} \alpha_i K(\cdot, \boldsymbol{x}_i) + \mathbb{i}\beta_i K(\cdot, \boldsymbol{x}_i), \tag{4.58}$$

where $\alpha_i, \beta_i \in \mathbb{R}$. Note, however, that Theorem 4.3.2 is already defined over complex numbers. Bouboulis *et al.* (2012) use pure complex kernels; that is, kernels defined over $\mathbb{C}$. In this case, the representer theorem can be simply written as

$$f^*(\cdot) = \sum_{i=1}^{N} \left(\alpha_i + \mathbb{i}\beta_i\right) K^*(\cdot, \boldsymbol{x}_i), \tag{4.59}$$

which resembles the standard expansion when working with real data except for the complex nature of the weights and the conjugate operation on the kernel.

4.6 Support Vector Machine Elements for Regression and Estimation

This section introduces the instantiation of a kernel method for regression and function approximation; namely, the SVR (Schölkopf and Smola 2002; Smola and Schölkopf 2004; Vapnik 1995). This method will accompany us in the following chapters, and it conveys all the key elements to work with a variety of estimation problems, including convexity of the optimization problem, regularized solution, sparsity, flexibility for nonlinear modeling, and adaptation to different sources of noise. Let us review two important definitions; namely, the SVR data model and the loss function used in the minimization problem. We will come back to these definitions later in order to define alternative signal models and cost functions, as well as to study model characteristics. Departing from the *primal problem*, we derive the SVR equations, and then we summarize the main properties of the SVR signal model.

4.6.1 Support Vector Regression Signal Model and Cost Function

Definition 4.6.1 (Nonlinear SVR signal model) Let a labeled training i.i.d. data set $\{(\boldsymbol{v}_i, y_i), i = 1, \dots, N\}$, where $\boldsymbol{v}_i \in \mathbb{R}^d$ and $y_i \in \mathbb{R}$. The SVR signal model first maps the observed explanatory vectors to a higher dimensional kernel feature space using a nonlinear mapping $\boldsymbol{\phi} : \mathbb{R}^N \rightarrow \mathcal{H}$, and then it calculates a linear regression model inside that feature space; that is:

$$\hat{y}_i = \langle \boldsymbol{w}, \boldsymbol{\phi}(\boldsymbol{v}_i)\rangle + b, \tag{4.60}$$

where $\boldsymbol{w}$ is a weight vector in $\mathcal{H}$ and b is the regression bias term. Model residuals are given by $e_i = y_i - \hat{y}_i$.

In order to obtain the model coefficients, the SVR minimizes a cost function of the residuals, which is often regularized with the ℓ_2 norm of $\boldsymbol{w}$. That is, we minimize with regard $\boldsymbol{w}$ the following primal functional:

$$\frac{1}{2}\|\boldsymbol{w}\|^2 + \sum_{i=1}^{N} \mathcal{L}(e_i). \tag{4.61}$$

In the standard SVR formulation, the Vapnik ε-insensitive cost is often used (Vapnik, 1998).

Definition 4.6.2 (Vapnik's ε-insensitive cost) Given a set or residual errors e_i in an estimation problem, the ε-insensitive cost is given by

$$\mathcal{L}_{\varepsilon}(e_i) = C\max(|e_i| - \varepsilon, 0), \tag{4.62}$$

where C represents a trade-off between regularization and losses. Those residuals lower than ε are not penalized, whereas those larger ones have linear cost.

Here, we want to highlight an important issue regarding the loss. The Vapnik ε-insensitive cost function results in a suboptimal estimator in many applications when combined with a regularization term (Schölkopf and Smola 2002; Smola and Schölkopf 2004; Vapnik 1995). This is because a linear cost is not the most suitable one to deal with Gaussian noise, which will be a usual situation in a number of time-series analysis applications. This fact has been previously taken into account in the formulation of LS-SVM (Suykens *et al.*, 2002), also known as kernel ridge regression (KRR) (Shawe-Taylor and Cristianini, 2004), where a quadratic cost is used, but in this case, the property of sparsity is lost.

The ε-Huber cost was proposed by Rojo-Álvarez *et al.* (2004), combining both the quadratic and the ε-insensitive zones, and it was shown to be a more appropriate residual cost not only for time-series problems, but also for function approximation problems in general where the data can be fairly considered to be i.i.d. (Camps-Valls. *et al.*, 2007).

Definition 4.6.3 (ε-Huber cost function) The ε-Huber cost is given by

$$\mathcal{L}_{\varepsilon\mathrm{H}}(e_i) = \begin{cases} 0, & |e_i| \le \varepsilon \\ \frac{1}{2\delta}(|e_i| - \varepsilon)^2, & \varepsilon \le |e_i| \le e_C \\ C(|e_i| - \varepsilon) - \frac{1}{2}\delta C^2, & |e_i| \ge e_C \end{cases} \tag{4.63}$$

where $e_C = \varepsilon + \delta C$; ε is the insensitive parameter, and δ and C control the trade-off between the regularization and the losses.

The three different regions in ε-Huber cost can work with different kinds of noise. First, the ε-insensitive zone neglects absolute residuals lower than ε. Second, the quadratic cost zone is appropriate for Gaussian noise. Third, the linear cost zone is efficient for limiting the impact of possibly present outliers on the model coefficients estimation. Note that Equation 4.63 represents the Vapnik ε-insensitive cost function when δ is small enough, the MMSE criterion for $\delta C \to \infty$ and $\varepsilon = 0$, and Huber cost function when $\varepsilon = 0$ (see Figure 4.4).

4.6.2 Minimizing Functional

In order to adjust the model parameters, an optimality criterion must be chosen. Typically in machine learning, one selects a risk function based on the *pdf* of data, which is unfortunately unknown. Hence, an induction principle is needed to best fit the real *pdf* based on the available data. A common choice is to minimize the so-called

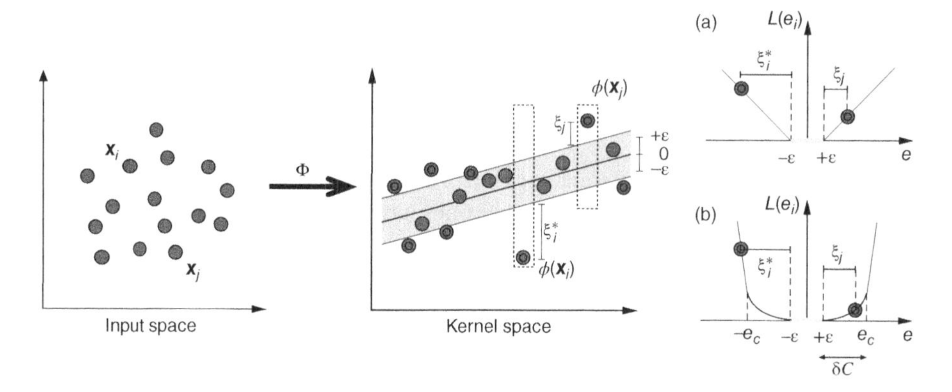

Figure 4.4 SVR signal model and cost functions. Samples in the input space are mapped onto an RKHS, and a linear regression is performed therein. All samples outside a fixed tube of size ε are penalized, and they are the support vectors (double circles). Penalization is given by (a) the Vapnik ε-insensitive or (b) the ε-Huber cost functions.

empirical risk; that is, the error in the training data set. However, to alleviate the problem of overfitting and control model complexity, regularization is usually adopted (Tikhonov and Arsenin, 1977), which is carried out in practice by minimizing the norm of the model parameters. This is intuitively equivalent to find the estimator which uses the minimum possible energy of the data to estimate the output. The resulting functional should take into account both this complexity term and an empirical error measurement term. The latter is defined according to an a-priori determined cost function of the committed errors.

Hence, the problem of estimating the coefficients can be stated as the minimization of the following functional:

$$F(\boldsymbol{w}, e_i) = \frac{1}{2}\|\boldsymbol{w}\|^2 + \sum_{i=1}^{N} \mathcal{L}_{\varepsilon\mathrm{H}}(e_i). \tag{4.64}$$

Introducing the previous loss in Equation 4.63 into Equation 4.64, we obtain the following functional:

$$\frac{1}{2}\|\boldsymbol{w}\|^2 + \frac{1}{2\delta}\sum_{i\in I_1}(\xi_i^2 + \xi_i^{*2}) + C\sum_{i\in I_2}(\xi_i + \xi_i^*) - \sum_{i\in I_2}\frac{\delta C^2}{2} \tag{4.65}$$

to be minimized with respect to $\boldsymbol{w}$ and $\{\xi_i^{(*)}\}$, and constrained to

$$y_i - \boldsymbol{w}^{\mathrm{T}}\boldsymbol{\phi}(\boldsymbol{v}_i) - b \le \varepsilon + \xi_i \tag{4.66}$$

$$-y_i + \boldsymbol{w}^{\mathrm{T}}\boldsymbol{\phi}(\boldsymbol{v}_i) + b \le \varepsilon + \xi_i^* \tag{4.67}$$

$$\xi_i, \xi_i^* \ge 0 \tag{4.68}$$

for $i = 1, \cdots, N$. For notation simplicity, $\{\xi_i^{(*)}\}$ will denote both $\{\xi_i\}$ and $\{\xi_i^*\}$ hereafter. Here, $\{\xi_i^{(*)}\}$ are *slack variables* or *losses*, which are introduced to handle the residuals according to the robust cost function. I_1 and I_2 are the sets of samples for which losses are required to have a quadratic or a linear cost respectively, and these sets are not necessarily static during the optimization procedure.

The optimization problem involves two operations: a minimization and a set of constraints that cannot be violated. This a standard problem typically solved using Lagrange multipliers.[2] Essentially, one has to include linear constraints in Equations 4.66–4.68 into Equation 4.65, which gives us the dual form of the problem. Then, the primal–dual functional (sometimes referred to as the Lagrange functional) is given by

$$L_{\mathrm{PD}} = \frac{1}{2}\|\boldsymbol{w}\|^2 + \frac{1}{2\delta}\sum_{i\in I_1}(\xi_i^2 + \xi_i^{*2}) + C\sum_{i\in I_2}(\xi_i + \xi_i^*) - \sum_{i\in I_2}\frac{\delta C^2}{2}$$
$$- \sum_i(\beta_i\xi_i + \beta_i^*\xi_i^*) + \sum_i(\alpha_i - \alpha_i^*)\left(y_i - \boldsymbol{w}^{\mathrm{T}}\boldsymbol{\phi}(\boldsymbol{v}_i) - b - \varepsilon - \xi_i\right) \tag{4.69}$$

constrained to $\alpha_i^{(*)}, \beta_i^{(*)}, \xi_i^{(*)} \geq 0$. By making zero the gradient of L_{PD} with respect to the primal variables (Rojo-Álvarez *et al.*, 2004), the following conditions are obtained:

$$\alpha_i^{(*)} = \frac{1}{\delta}\xi_i^{(*)} \quad (i \in I_1) \tag{4.70}$$

$$\alpha_i^{(*)} = C - \beta_i^{(*)} \quad (i \in I_2), \tag{4.71}$$

as well as the following expression relating the primal and dual model weights:

$$\boldsymbol{w} = \sum_n(\alpha_i - \alpha_i^*)\boldsymbol{\phi}(\boldsymbol{v}_i). \tag{4.72}$$

Constraints in Equations 4.70, 4.71, and 4.72 are then included into the Lagrange functional in Equation 4.69 in order to remove the primal variables. Note that the dual problem can be obtained and expressed in matrix form, and it corresponds to the maximization of

$$-\frac{1}{2}(\boldsymbol{\alpha} - \boldsymbol{\alpha}^*)^{\mathrm{T}}[\boldsymbol{K} + \delta\boldsymbol{I}](\boldsymbol{\alpha} - \boldsymbol{\alpha}^*) + (\boldsymbol{\alpha} - \boldsymbol{\alpha}^*)^{\mathrm{T}}\boldsymbol{y} - \varepsilon\mathbf{1}^{\mathrm{T}}(\boldsymbol{\alpha} + \boldsymbol{\alpha}^*) \tag{4.73}$$

constrained to

$$C \geq \alpha_i^{(*)} \geq 0, \tag{4.74}$$

---

2 Giuseppe Lodovico (Luigi) Lagrangia (1736–1813) was the father of a famous methodology for optimization of functions of several variables subject to equality and inequality constraints. The method of Lagrange multipliers essentially solves $\max_{x,y} f(x,y)$ s.t. $g(x,y) = c$. He proposed to introduce a new variable (called the "Lagrange multiplier" α) and optimize the new function: $\max_{x,y,\lambda}\{\Lambda(x,y,\alpha)\}$ s.t. $f(x,y) - \alpha(g(x,y) - c)$, which is equivalent to solving the dual problem $\max_{x,y,\lambda}\{f(x,y) - \alpha(g(x,y) - c)\}$ s.t. $\alpha \geq 0$. Interestingly enough, the method is applicable to any number of variables and constraints, $\min\{f_0(x)\}$ s.t. $f_i(x) \leq 0$ and $h_j(x) = 0$. Then, the dual problem reduces to minimizing $\Lambda(x,\alpha,\mu) = f_0(x) - \sum_i \alpha_i f_i(x) - \sum_j \mu_j h_i(x)$. The proposed procedure to solve these types of problems is very simple: first, one has to derive this primal–dual problem and equate to zero, $\nabla_\lambda \Lambda(x,y,\alpha) = 0$, then the constraints obtained are stationary points of the solution, which are eventually included in the problem again. These give rise to a linear or quadratic problem, for which there are very efficient solvers nowadays.

where $\boldsymbol{\alpha}^{(*)} = [\alpha_1^{(*)}, \cdots, \alpha_N^{(*)}]^{\mathrm{T}}$. Then, after obtaining Lagrange multipliers $\boldsymbol{\alpha}^{(*)}$, the time-series model for a new sample at time instant m can be readily expressed as

$$y_j = f(\boldsymbol{v}_j) = \sum_{i=1}^{N} (\alpha_i - \alpha_i^*) \boldsymbol{\phi}(\boldsymbol{v}_i)^{\mathrm{T}} \boldsymbol{\phi}(\boldsymbol{v}_j) = \sum_{i=1}^{N} (\alpha_i - \alpha_i^*) K(\boldsymbol{v}_i, \boldsymbol{v}_j), \tag{4.75}$$

where the dot product $\boldsymbol{\phi}(\boldsymbol{v}_i)^{\mathrm{T}} \boldsymbol{\phi}(\boldsymbol{v}_j)$ has been finally replaced by the kernel function working solely on input samples $\boldsymbol{v}_i$ and $\boldsymbol{v}_j$; that is: $K(\boldsymbol{v}_i, \boldsymbol{v}_j) = \boldsymbol{\phi}(\boldsymbol{v}_i)^{\mathrm{T}} \boldsymbol{\phi}(\boldsymbol{v}_j)$, which is a function of weights in the input space associated with nonzero Lagrange multipliers. More details on the derivation of these equations can be found in the literature for SVR (Smola *et al.*, 1998) and for linear SVM–ARMA (Rojo-Álvarez *et al.*, 2004).

By including the ε-Huber residual cost into Equation 4.64, the SVR coefficients can be estimated by solving a quadratic programming (QP) problem (Rojo-Álvarez *et al.* 2004; Smola and Schölkopf 2004). Several relevant properties can be highlighted, which can be shown by stating the Lagrange functional, setting the Karush–Khun–Tucker (KKT) conditions, and then obtaining the dual functional. These properties are summarized next.

Property 7 (SVR sparse solution and support vectors) The weight vector in $\mathcal{H}$ can be expanded in a linear combination of the transformed input data:

$$\boldsymbol{w} = \sum_{i=1}^{N} \eta_i \boldsymbol{\phi}(\boldsymbol{v}_i), \tag{4.76}$$

where $\eta_i = (\alpha_i - \alpha_i^*)$ are the model weights, and $\alpha_i^{(*)}$ are the Lagrange multipliers corresponding to the positive and negative residuals in the nth observation. Observations with nonzero associated coefficients are called *s*upport vectors, and the solution is expressed as a function of them solely.

This property describes the sparse nature of the SVM solution for estimation problems.

Property 8 (Robust expansion coefficients) A nonlinear relationship between the residuals and the model coefficients for the ε-Huber cost is given by

$$\eta_i = \left.\frac{\partial L_{\varepsilon\mathrm{H}}(e)}{\partial e}\right|_{e=e_i} = \begin{cases} 0, & |e_i| \le \varepsilon \\ \frac{1}{\delta} \cdot \mathrm{sgn}(e_i)(|e_i| - \varepsilon), & \varepsilon < |e_i| \le \varepsilon + \gamma C \\ C \cdot \mathrm{sgn}(e_i), & |e_i| > \varepsilon + \gamma C. \end{cases} \tag{4.77}$$

Therefore, the impact of a large residual e_i on the coefficients is limited by the value of C in the cost function, which yields estimates of the model coefficients that are robust in the presence of outliers.

The kernel trick in SVM consists of stating a data-processing algorithm in terms of dot products in the RKHS, and then substituting those products by Mercer kernels. The kernel expression is actually used in any kernel machine, but neither the mapping function $\boldsymbol{\phi}(\cdot)$ nor the RKHS need to be known explicitly. The Lagrangian of Equation 4.61 is used to obtain the dual problem, which in turn yields the Lagrange multipliers used as model coefficients.

Property 9 (Regularization in the dual) The dual problem of Equation 4.64 for the ε-Huber cost corresponds to the maximization of

$$-\frac{1}{2}(\boldsymbol{\alpha} - \boldsymbol{\alpha}^*)^{\mathrm{T}}(\boldsymbol{K} + \delta\boldsymbol{I})(\boldsymbol{\alpha} - \boldsymbol{\alpha}^*) + (\boldsymbol{\alpha} - \boldsymbol{\alpha}^*)^{\mathrm{T}}\boldsymbol{y} - \varepsilon\mathbf{1}^{\mathrm{T}}(\boldsymbol{\alpha} + \boldsymbol{\alpha}^*) \tag{4.78}$$

constrained to $0 \leq \alpha_i^{(*)} \leq C$. Here, $\boldsymbol{\alpha}^{(*)} = [\alpha_1^{(*)}, \cdots, \alpha_n^{(*)}]^{\mathrm{T}}$, $\boldsymbol{y} = [y_1, \cdots, y_{N_r}]^{\mathrm{T}}$, $\boldsymbol{K}$ represents the kernel matrix, given by $\boldsymbol{K}_{i,j} = K(\boldsymbol{v}_i, \boldsymbol{v}_j) = \langle\boldsymbol{\phi}(\boldsymbol{v}_i), \boldsymbol{\phi}(\boldsymbol{v}_j)\rangle$, $\mathbf{1}$ is an all-ones column vector, and $\boldsymbol{I}$ is the identity matrix.

The use of the quadratic zone in the ε-Huber cost function gives rise to a numerical regularization. The effect of δ in the solution was analyzed by Rojo-Álvarez *et al.* (2004).

Property 10 (Estimator as an expansion of kernels) The estimator is given by a linear regression in the RKHS, and it can be expressed only in terms of the Lagrange multipliers and Mercer kernels as

$$\hat{y}(\boldsymbol{v}) = \langle\boldsymbol{w}, \boldsymbol{\phi}(\boldsymbol{v})\rangle + b = \sum_{i=1}^{N} \eta_i K(\boldsymbol{v}_i, \boldsymbol{v}) + b, \tag{4.79}$$

where only the support vectors (i.e., training examples whose corresponding Lagrange multipliers are nonzero) contribute to the solution.

4.7 Tutorials and Application Examples

In this section, we first present some simple examples of the concepts developed with kernels so far. Then, we present a set of real examples with several data bases focusing on SVR performance and the impact of the cost function.

4.7.1 Kernel Calculations and Kernel Matrices

In these first examples we will see how to compute some of the kernel matrices in Table 4.1. We will assume we have a matrix of vector samples for training, $\boldsymbol{X} \in \mathbb{R}^{n \times d}$, where n is the number of vectors and d its dimensionality (or number of features), and a matrix of vector samples for testing, $\boldsymbol{X}_* \in \mathbb{R}^{m \times d}$, where m is the number of samples in the test set. The first two kernels in the table can be straightforwardly computed in Listing 4.1.

```
% Assuming we have Xt (training) and Xv (test)
Kt = Xt * Xt'; % Linear kernel for training
Kv = Xv * Xt'; % Linear kernel for validation/test
% Polynomial kernel
```

```
Kt = (Xt * Xt' + c).^d; % c: bias, d: polynomial degree
Kv = (Xv * Xt' + c).^d;
```

Listing 4.1 Linear and polynomial kernels (kernels1.m).

Kernel matrices must be positive definite to be valid. In MATLAB, there are several ways to test if a matrix is positive definite. In a positive-definite matrix all eigenvalues are positive; therefore, we can use `eig` to obtain all eigenvalues and check if all of them are positive (see Listing 4.2).

```
% Eigenvectors of Kt
e = eig(Kt);
if any(e < 0),
    error('Matrix is not P.D.')
end
```

Listing 4.2 Checking positive definiteness via eigenvalues (kernels2.m).

However, the preferred way to test for positive definiteness is using `chol`, as in Listing 4.3.

```
% Test is matris is P.D. using chol
[R,p] = chol(Kt);
if p > 0,
    error('Matrix is not P.D.')
end
```

Listing 4.3 Checking positive definiteness using `chol` (kernels3.m).

Note that sometimes this test can fail due to numerical precision. In those cases, the eigenvalues may have a very small negative value, but different from zero; therefore, the test is not passed. If you have the theoretical guarantee that your kernel function is positive definite, a way to work around numerical issues is to add a small value to the diagonal; for instance, like `Kt + 2*eps*eye(size(Kt))`.

In many cases it is very useful to inspect the kernel matrix to visualize structures or groups. Kernels measure the similarity between vectors in the feature space, and a visual representation of the kernel should reveal some structure. In the following example, in Listing 4.4, we will generate three random 2D sets, compute a basic linear kernel, and see its structure. Figure 4.5 shows the result of the example. It is clear that the kernel reveals the relationship between different groups.

```
np   = 200;
dist = 2.5;
m1   = [-dist ; dist];
m2   = [0 ; 0];
m3   = [2*dist ; 2*dist];
% Generate samples
X1 = randn(2,np) + repmat(m1,1,np);
X2 = randn(2,np) + repmat(m2,1,np);
X3 = randn(2,np) + repmat(m3,1,np);
% All samples matrix
X = [X1 X2 X3]';
% Normalization
Xn = zscore(X);
% Look at the generated examples
```

```
figure(1)
plot(Xn(1:np,1), Xn(1:np,2), 'xb', ...
     Xn((1:np)+1*np,1), Xn((1:np)+1*np,2), 'xr', ...
     Xn((1:np)+2*np,1), Xn((1:np)+2*np,2), 'xk')
% Linear kernel
K = Xn * Xn';
% Inspecting the kernel
figure(2), imagesc(K), axis square off
```

Listing 4.4 Generating random sets and inspecting the linear kernel (kernels4.m).

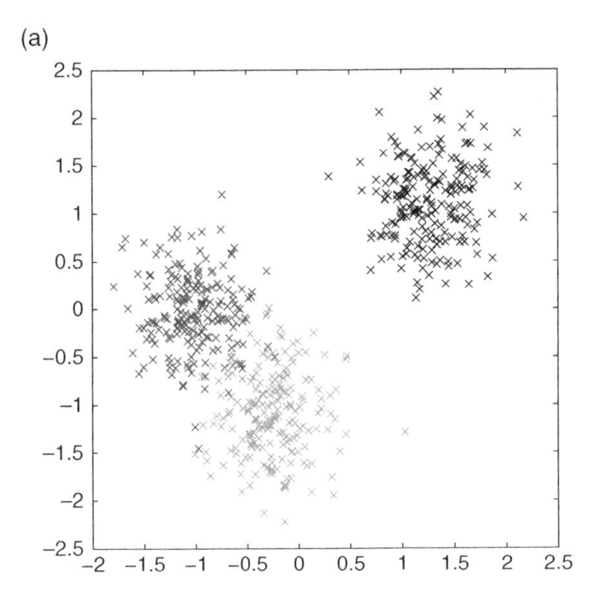

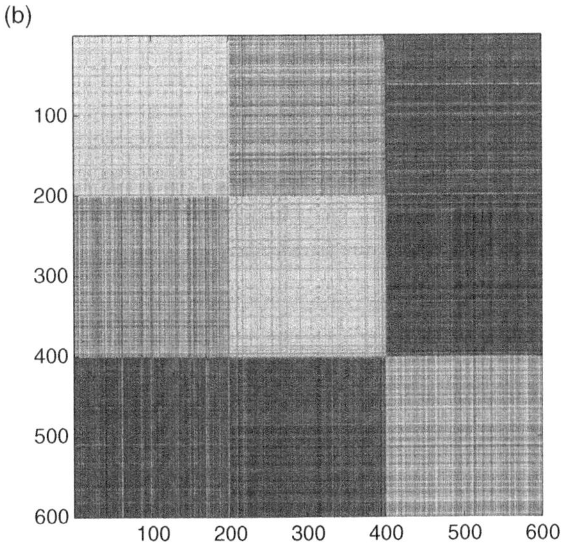

Figure 4.5 Example of kernel matrix: (a) the three sets; (b) generated linear kernel.

One the of most used kernels is the RBF or Gaussian kernel, also known as the squared exponential kernel, due to its simplicity (it only has one free parameter to adjust), and because it is a universal characteristic kernel that includes other kernels as particular cases. By exploiting the decomposition of the Euclidean distance we can efficiently compute this kernel, as seen in Listing 4.5.

```
% For the train kernel we have
nt = size(Xt,2);
ntsq = sum(Xt.^2,1);
Dt = 2 * (ntsq * ones(nt,1) - Xt' * Xt);
Kt = exp(-Dt / (2*sigma^2)); % sigma is the width of the RBF
% The test kernel can be computed as
nv = size(Xv,2);
nvsq = sum(Xv.^2,1);
Dv = ntsq' * ones(1,nv) + ones(nt,1) * nvsq - 2 * Xt' * Xv;
Kv = exp(-Dv / (2*sigma^2));
```

Listing 4.5 Computation of the RBF kernel (kernels5.m).

The exponential and Laplacian kernels can be obtained as in the RBF kernel just by taking the root of the Dt or Dv matrices in Listing 4.5. Similarly, it is easy to obtain the rational, multi-quadratic, power, log, Cauchy, and generalized T-Student kernels from these matrices. On the other hand, the chi-square and histogram intersection kernels are usually employed in computer vision, and they are particularly suitable for data in the form of normalized histograms. In general, given two vectors $\boldsymbol{x}$ and $\boldsymbol{z}$ of dimension d, we need to compute d kernels, one kernel per feature, and sum them all. For the chi-square kernel (other kernels are obtained similarly) a naive MATLAB implementation would be as given in Listing 4.6.

```
% Chi-Square kernel
d = size(X1,2); n1 = size(X1,1); n2 = size(X2,1);
K = 0;
for i = 1:d
    num = 2 * X1(:,i) * X2(:,i)';
    den = X1(:,i) * ones(1,n2) + ones(n1,1) * X2(:,i)';
    K = K + num./den;
end
```

Listing 4.6 Chi-square kernel (kernels6.m).

In the example in Listing 4.6, as the reader may notice, we have not computed the chi-square kernel as defined in Table 4.1. The reason, as pointed out by Vedaldi and Zisserman (2010), is that this kernel is only conditionally positive definite. It can be obtained as

$$K(\boldsymbol{x}, \boldsymbol{z}) = \sum_{i=1}^{d} \frac{2(x_i - z_i)^2}{(x_i + z_i)},$$

which is the kernel we calculate with Listing 4.6.

4.7.2 Basic Operations with Kernels

Let us start this subsection with a fundamental property in many kernel machines, which is the data centering in Hilbert space. This will need the property of translation of vectors. The example in Listing 4.7 shows how to center data, although it can easily modified to move the kernel to any other place in the Hilbert space.

```
% Assuming a pre-computed kernel matrix in Kt, centering is done as:
[Ni,Nj] = size(Kt);
Kt = Kt - ( mean(Kt,2)*ones(1,Nj) - ones(Ni,1)*mean(Kt,1) + mean(Kt(:)) );
% Sum columns and divide by the number of rows
S = sum(Kt) / Ni;
% Centered test kernel w.r.t. a train kernel
Kv = Kv - ( S' * ones(1,Nj) - ones(Ni,1) / Ni * sum(Kv) + sum(S) / Ni );
```

Listing 4.7 Centering kernels (kernels7.m).

Another important property of kernel methods is that one can estimate linear projections onto subspaces of the Hilbert feature space $\mathcal{H}$ explicitly. We are given a data matrix $\boldsymbol{X} \in \mathbb{R}^{N \times d} \subset \mathcal{X}$ which is mapped to an RKHS, $\boldsymbol{\phi} \in \mathbb{R}^{N \times d_{\mathcal{H}}} \subset \mathcal{H}$. Now we want to project the data onto a D-dimensional subspace of $\mathcal{H}$, where $D < d_{\mathcal{H}}$ given by a projection matrix $\boldsymbol{V} \in \mathbb{R}^{N \times N}$.

In order to map new input data $\boldsymbol{X}' \in \mathbb{R}^{m \times d}$ into a subspace of dimensionality D in $\mathcal{H}$, one has to first map the data to the Hilbert space $\boldsymbol{\Phi}'$, and then express the projection matrix $\boldsymbol{V}$ as a linear combination of the n examples therein, $\boldsymbol{V} = \boldsymbol{\Phi}^{\mathrm{T}} \mathcal{A}$:

$$\mathcal{P}_D(\boldsymbol{X}') = \boldsymbol{\Phi} \boldsymbol{V} = \boldsymbol{\Phi}' \boldsymbol{\Phi}^{\mathrm{T}} \mathcal{A} = \boldsymbol{K} \mathcal{A}, \tag{4.80}$$

where $\mathcal{A}, \boldsymbol{K} \in \mathbb{R}^{N \times N}$, $\boldsymbol{K}_{ij} := \langle \boldsymbol{\phi}(\boldsymbol{x}_i'), \boldsymbol{\phi}(\boldsymbol{x}_j) \rangle$, and hence projecting in a finite dimension D can be easily done by retaining (truncating) a number D of columns of $\mathcal{A}$.

This can be illustrated in Listing 4.8. Let us for a moment assume that the eigenvectors of the kernel matrix are the ones spanning the subspace we are interested in. The operation essentially reduces to an eigendecomposition of an $N \times N$ matrix, truncation of the eigenvectors matrix, and linear projection. This property and the latter application are widely used in kernel feature extraction methods, such as the KPCA, which will be revised thoroughly in Chapter 12.

```
K  = kernelmatrix('rbf',X,X,sigma);  % compute kernel matrix n x n
K2 = kernelmatrix('rbf',X2,X,sigma); % compute kernel matrix n x m
n = size(K,2);          % n: number of samples used in the kernel matrix
D = 10;                 % subspace dimensionality (D<n)
[A L] = eigs(K,n);      % extract the top D eigenvectors of the kernel matrix
P_X2 = K2*A;            % projection of X2 onto the subspace of size m x n
P_X2 = K2*A(:,1:D);     % projection of X2 onto the subspace of size m x D
```

Listing 4.8 Projections with kernels (kernels8.m).

Very often we aim to estimate distances in the Hilbert space explicitly. Given two data matrices $\mathbf{X}$ and $\mathbf{Y}$, one can compute the distances between the data points implicitly via kernels. Essentially, one is interested in mapping the data to a Hilbert space $\mathcal{H}$ which yields $\boldsymbol{\Phi}_x$ and $\boldsymbol{\Phi}_y$, and estimating the (squared) Euclidean distance therein:

$$d_{xy}^2 := \|\boldsymbol{\Phi}_x - \boldsymbol{\Phi}_y\|_{\mathcal{H}}^2 = \boldsymbol{K}_x + \boldsymbol{K}_y - 2\boldsymbol{K}_{xy}, \tag{4.81}$$

which can be done in MATLAB with the following code: `D2_H = Kx+Ky-2*Kxy;`. It is worth noting that if we use an RBF kernel function, self-similarities become one, $K(\boldsymbol{x}_i, \boldsymbol{x}_i) = 1$, and then all points are mapped to a hypersphere and hypersphere and computing distances reduces to simply `D2_H = 2*(1-Kxy)`.

Another useful operation when working with kernels is about computing distances to the empirical center of mass. As an extension of the previous exercise, the reader may consider computing the distance from the given data points to its (empirical) center of mass in feature spaces. This would require computing the empirical mean in Hilbert space, but we can do this implicitly with kernels, as in Listing 4.9. Related to this example, one may think of normalizing the energy of the data points in Hilbert spaces. This operation of normalization is also possible via kernel functions only. Given a kernel matrix $\boldsymbol{K}$, it is trivial to compute its normalized version as in Listing 4.10.

```
n = size(K,1);
D = sum(K) / ell;
E = sum(D) / ell;
D2 = diag(K) - 2 * D' + E * ones(n,1);
```

Listing 4.9 Computing distances to the empirical center of mass (kernels9.m).

```
D = diag(1./sqrt(diag(K)));
Kn = D * K * D;
```

Listing 4.10 Kernel normalization (kernels10.m).

Note that in all these operations, only linear algebra operations, such as matrix multiplications or SVDs, are typically involved.

A quite useful operation with kernels is to measure how well *aligned* two kernel matrices are, either to discard redundant kernel parameters or to optimize the (in)dependence among them. Imagine that you are given a supervised problem like the "two moons" shown in Figure 4.6 in which you need to classify samples belonging to one of the two classes colored in blue or red. Building a proper kernel matrix requires

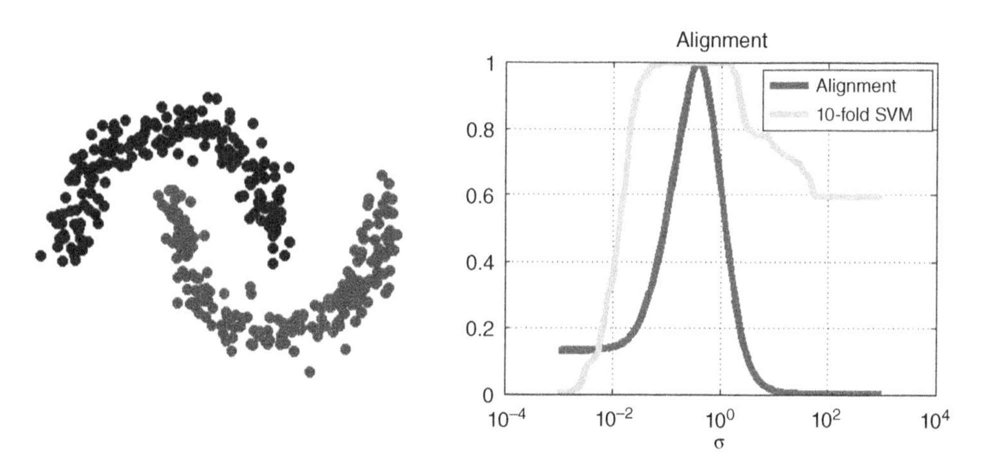

Figure 4.6 Illustration of the operation of alignment for the selection of hyperparameters.

typically tuning parameters, which is commonly done via cross-validation. A first guess of the proper parameter can be estimated via *kernel alignment* in a very cheap way without training any classifier, just by estimating the kernel hyperparameter that aligns best the mapped samples with the kernel matrix $\boldsymbol{K}_x(\boldsymbol{\theta})$ and the labels $\boldsymbol{y}$ (through the so-called *ideal kernel* $\boldsymbol{K}_{\text{ideal}} = \boldsymbol{y}\boldsymbol{y}^{\text{T}}$) in feature space. The problem then boils down to estimating the alignment measure

$$\boldsymbol{\theta}^* = \arg\min_{\boldsymbol{\theta}} \|\boldsymbol{K} - \boldsymbol{y}\boldsymbol{y}^{\text{T}}\|_{\text{F}}^2, \tag{4.82}$$

where $\|\cdot\|_{\text{F}}$ is the Frobenius norm. The (normalized) alignment measure for a particular hyperparameter can be easily computed as in Listing 4.11. Figure 4.6 shows the results obtained when using an RBF kernel parametrized with the length-scale σ parameter, comparing the results with an expensive training of an SVM by grid search of the hyperparameters.

```
temp = F_norm(K,Y);
A = temp/sqrt(F_norm(K,K)*F_norm(Y,Y));

function F = F_norm(A,B)
F = trace(A'*B);
```

Listing 4.11 Normalized alignment measure (kernels11.m).

4.7.3 Constructing Kernels

In this section we illustrate two examples of kernel construction, as a follow-up of Section 4.4.1. In particular, let us discuss the standard convex kernel combinations. Commonly used kernels families include the SE, periodic (Per), linear (Lin), and rational quadratic (RQ); see Table 4.1. We illustrate these basic kernels and some functions drawings in Figure 8.9. These basic kernels can be actually combined by following simple operations, such as summation, multiplication, or convolution. This way, one may build sophisticated kernels from simpler ones. The code in Listing 4.12 illustrates the construction of such kernels and several convex combinations.

```
%% Data
X_o = [-1.5 -1 -0.75 -0.4 -0.3 0]';
Y_o = [-1.6 -1.3 -0.5 0 0.3 0.6]';
%% Define a SE kernel
nu = 1.15; % scale of the kernel
s  = 0.30; % lengthscale sigma parameter of the kernel
kf = @(x,x2) nu^2 * exp( (x-x2).^2  /(-1*s^2) );
sn = 0.01; % known noise on observed data
ef = @(x,x2) sn^2 * (x==x2); % iid noise
% Let's sum the signal and noise kernel functions:
k = @(x,x2) kf(x,x2) + ef(x,x2);
% Let's plot the kernel function of domain x
np = 100; x = linspace(-1,1,np)';
y = kf(zeros(length(x),1),x);
figure(1), plot(x,y,'k','LineWidth',8), axis([-1 1 0 1.5]), axis off
% Now let's draw two random functions out of the kernel
K = zeros(length(x));
```

```
for i = 1:length(x)
    K(i,:) = kf(x(i) * ones(length(x),1), x);
end
% Due to numerical precision, we need to add a small factor to the ...
    diagonal
% to ensure positive definiteness
sphi = chol(K + 1e-12 * eye(size(K)))';
rf = sphi * randn(length(x),2);
figure(2), plot(x,rf(:,1)+1,x,rf(:,2)-1,'LineWidth',8), axis off

%% Done!

% As a proposed exercise, try to do the same with the
% rational-quadratic kernel, the periodic kernel and combinations.
% We give you some hint material:

% 1) Rational-quadratic kernel,
c = 0.05;
krq = @(x1,x2) sigma_f^2 * (1 - (x1-x2).^2 ./ ((x1-x2).^2 + c)) + ...
    ef(x1,x2);
% 2) Periodic kernel
p = 0.4; % periode
s = 1;   % SE kernel lengthscale
kper = @(x1,x2) sigma_f^2 * exp(-sin(pi*(x1-x2)/p).^2/l^2) + ef(x1,x2);
% 3) Linear kernel
offset = 0;
kl = @(x1,x2) sigma_f^2 * (x1 - offset) .* (x2 - offset) + 1e-5*ef(x1,x2);
```

Listing 4.12 Examples of kernel construction (kernels12.m).

In Figure 8.9 from Chapter 8, all the base kernels are 1D. Nevertheless, kernels over multidimensional inputs can actually be constructed by adding and multiplying kernels over individual dimensions; namely, (a) linear, (b) SEl (or RBF), (c) rational quadratic, and (d) periodic. See Table 4.1 for the explicit functional form of each kernel. Some simple kernel combinations are represented in the two last panels of Figure 8.9. For instance, a linear plus periodic covariances may capture structures that are periodic with trend (e), while a linear plus SE covariances can accommodate structures with increasing variation (f). By summing kernels, we can model the data as a superposition of independent functions, possibly representing different structures in the data. For example, in multitemporal image analysis, one could, for instance, dedicate a kernel for the time domain (perhaps trying to capture trends and seasonal effects) and another kernel function for the spatial domain (equivalently capturing spatial patterns and autocorrelations).

In time-series models, sums of kernels can express superposition of different processes, possibly operating at different scales. Very often, changes in geophysical variables through time occur at different temporal resolutions (hours, days, etc.), and this can be incorporated in the prior covariance with those simple operations. In multiple dimensions, summing kernels gives additive structure over different dimensions, similar to generalized additive models (Hastie *et al.*, 2009). Alternatively, multiplying kernels allows us to account for interactions between different input dimensions or different notions of similarity. In the following subsections, we also show how to design kernels that incorporate particular time resolutions, trends, and periodicities.

4.7.4 Complex Kernels

This example shows a nonlinear complex channel equalization with a linear filter and a kernel filter. A QPSK data burst $d_n = d_{r,n} + \mathbb{i}d_{i,n}$, where $d_{r,n}, d_{i,n} \in \{-1, 1\}$ is sent over a channel with a linear dispersive section of impulse response $h_n = \delta_n + (0.5 - 0.5\mathbb{i})\delta_{n-1} + (0.1 + 0.1\mathbb{i})\delta_{n-2}$, and a nonlinear section at its output with the function $x = f(u) = (1/3)u + u^{-1/3}$, where u is the linear channel output. The nonlinear channel output has a complex additive white Gaussian noise of standard deviation $\sigma_n = 0.05$. The linear transversal equalizer has the expression

$$y = \boldsymbol{w}\boldsymbol{x}_n^{\mathrm{H}}, \tag{4.83}$$

where $\boldsymbol{x}_n = (x_n \cdots x_{n-L+1})$, H is the complex transpose (Hermitian) operator, and L is the filter length.

The filter weights are adjusted using a ridge regression procedure, where regularization parameter λ is adjusted to the channel noise (i.e., $\lambda = \sigma_n^2$), and thus the filter is intended to minimize the cost function

$$L = \mathbb{E}[\|d_n - \boldsymbol{w}\boldsymbol{x}_n^{\mathrm{H}}\|] + \sigma_n^2\|\boldsymbol{w}\|^2. \tag{4.84}$$

The optimization is performed simply by nulling the gradient of the cost function with respect to the parameter vector, using the Wirtinger derivative. The expression of the optimal filter is simply

$$\boldsymbol{w} = \boldsymbol{d}\boldsymbol{X}(\boldsymbol{X}^{\mathrm{H}}\boldsymbol{X} + \sigma_n^2\boldsymbol{I})^{-1}, \tag{4.85}$$

where $\boldsymbol{X}$ is the matrix containing all training vectors $\boldsymbol{x}_n$, and $\boldsymbol{d}$ is a row vector containing the transmitted signals, assumed to be known by the receiver for training purposes.

The channel output data can be transformed using a nonlinear transformation into an RKHS, and then the expression for the weight vector is

$$\boldsymbol{w} = \boldsymbol{d}\boldsymbol{\Phi}(\boldsymbol{\Phi}^{\mathrm{H}}\boldsymbol{\Phi} + \sigma_n^2 I)^{-1}, \tag{4.86}$$

where $\boldsymbol{\Phi}$ is a matrix containing all row vectors $\boldsymbol{\phi}(\boldsymbol{x}_n)$, and $\boldsymbol{\phi}(\cdot)$ is a mapping into a given RKHS.

A kernelized version of this filter can be constructed using the complex version of the representer theorem (Equation 4.58), given by

$$\boldsymbol{w} = \sum_{i=1}^{N}(\alpha_i + \mathbb{i}\beta_i)K(\cdot, \boldsymbol{x}_i) = \sum_{i=1}^{N}(\alpha_i + \mathbb{i}\beta_i)\boldsymbol{\phi}(\boldsymbol{x}_i) = (\boldsymbol{\alpha} + \mathbb{i}\boldsymbol{\beta})\boldsymbol{\Phi}, \tag{4.87}$$

where $\boldsymbol{\alpha} + \mathbb{i}\boldsymbol{\beta}$ is a row vector containing all complex coefficients $\alpha_i + \mathbb{i}\beta_i$.

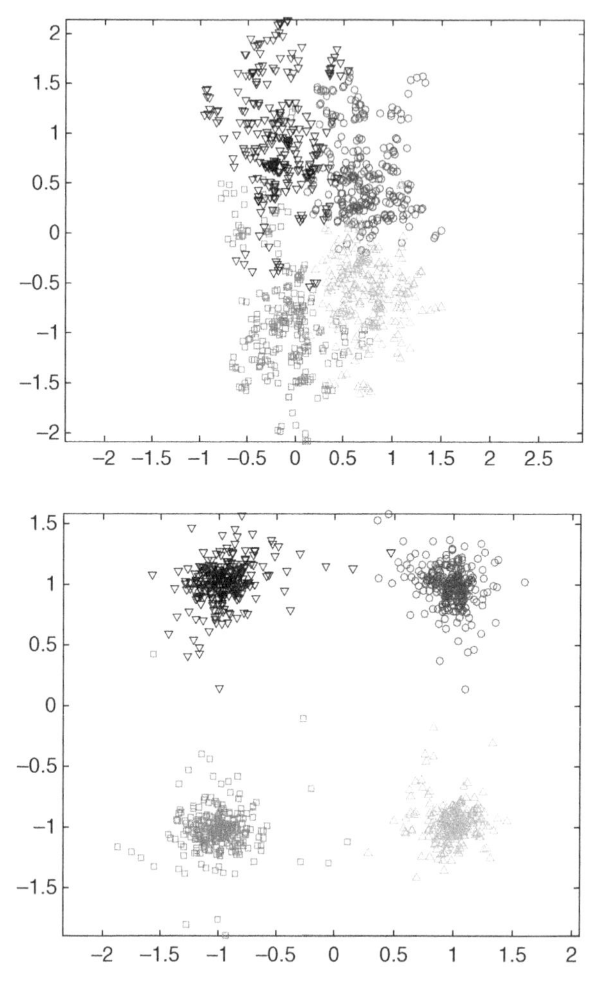

Figure 4.7 Estimated symbols from a burst of 1000 training data. Upper panel: the linear equalizer estimation; lower panel: the complex kernel equalizer estimation.

The combination of Equations 4.86 and 4.87 gives the result

$$\boldsymbol{\alpha} + \mathbb{i}\boldsymbol{\beta} = \boldsymbol{d}(\boldsymbol{\Phi}\boldsymbol{\Phi}^H + \sigma_n^2 \boldsymbol{I})^{-1}. \tag{4.88}$$

The complex kernel estimator is then expressed as

$$y_n = \sum_{i=1}^{N} (\alpha_i + \mathbb{i}\beta_i) K(\boldsymbol{x}_i, \boldsymbol{x}_n). \tag{4.89}$$

In our example, a Gaussian kernel is used. The upper panel of Figure 4.7 shows the representation of the estimated complex symbols of a test burst after training a linear equalizer with Equation 4.85, and the lower panel the symbols estimated with the equalizer in Equation 4.89 using the optimization in Equation 4.88. As can be shown, the linear equalizer produces a poor estimation due to the nonlinear behavior of the channel. Nevertheless, the complex kernel estimator can model and invert this nonlinear behavior. Listing 4.13 has been used to obtain these results.

```
function kernels13
sn = 0.05;        % Noise std
L = 4;            % Filter length
p = sqrt(2)*2;  % Kernel parameter
N = 1000;
h = [1 0.5-1j*0.5 .1+1j*.1]; % Linear channel impulse response

% Data
d = sign(randn(1,N)) + 1j*sign(randn(1,N));
d_test = sign(randn(1,N)) + 1j*sign(randn(1,N));

% Channel
x = channel(d,h,sn);
x_test = channel(d_test,h,sn);

% Data matrix
X = buffer(x,L,L-1)';
X_test = buffer(x_test,L,L-1)';

% Equalization with Linear MMSE
R = X' * X;
R = R + sn^2*eye(size(R));
w = d * X / R;
y = w * X_test';
figure(1), plotresults(y,d_test), title('Linear MMSE eq.')

% Equalization with Kernel MMSE
K = kernel_matrix(X,X,p);
alpha = d * pinv(K + sn^2*eye(size(K)));
K_test = kernel_matrix(X_test,X,p);
y = alpha * K_test;
figure(2), plotresults(y,d_test), title('Kernel MMSE eq.')

function x = channel(d,h,s)
% Channel: Linear section
xl = filter(h,1,d);
% Channel: Nonlinear section
xnl = xl.^(1/3) + xl / 3;
% Noise
x = xnl + s*(randn(size(xl)) + 1j*randn(size(xl)));

function K = kernel_matrix(X1,X2,p)
N1 = size(X1,1);
N2 = size(X2,1);
aux = kron(X1,ones(N2,1)) - kron(ones(1,N1),X2')';
D = buffer(sum(aux.*conj(aux),2),N2,0);
K = exp(-D/(2*p));

function plotresults(y,d)
plot(real(y(d == 1+1j)), imag(y(d == 1+1j)), 'bo'), hold on
plot(real(y(d == -1-1j)), imag(y(d == -1-1j)), 'rs')
```

```
plot(real(y(d == 1-1j)),  imag(y(d == 1-1j)),  'm^')
plot(real(y(d == -1+1j)), imag(y(d == -1+1j)), 'kv'), hold off
axis equal, grid on
```

Listing 4.13 Complex kernel (kernels13.m).

4.7.5 Application Example for Support Vector Regression Elements

In this subsection, three types of loss-based SVRs are benchmarked: (1) the standard ε-insensitive SVR (ε-SVR); (2) the squared-loss SVR (also known as kernel ridge regression or LS-SVM); and (3) the ε-Huber-SVR. We have chosen a real regression example that involves different noise sources for the sake of illustration purposes. Results are obtained with NNs and with classical empirical models. These models are compared in terms of accuracy and bias of the estimations, and also their robustness is scrutinized when a low number of training samples are available in two different data sets related to the estimation of oceanic chlorophyll concentration from satellite data. More details can be found in Camps-Valls *et al.* (2006c).

Data Description

We use here two different datasets for illustration purposes. The first dataset simulates the data acquired by the medium-resolution imaging spectrometer (MERIS) on-board at the Envisat satellite (MERIS dataset), and in particular the spectral behavior of chlorophyll concentration C in the subsurface waters (Cipollini *et al.*, 2001). The data were generated according to a fixed and noise-free model, and the total number of samples (pairs of in-situ concentrations and acquired radiances) available for our experiments was 5000. These samples were randomly divided into three sets: a training set (500 samples, model building), a validation set (500 samples, free parameters tuning), and a test set (4000 samples).

The second dataset (SeaBAM dataset, (O'Reilly and Maritorena, 1997; O'Reilly *et al.*, 1998)) consists of a compilation of 919 in-situ measurements of chlorophyll concentration around the USA and Europe, all of them related to five different multispectral remote-sensing reflectance corresponding to the SeaWiFS wavelengths. The available data were randomly split into three sets: 230 samples for training, 230 samples for validation, and the remaining 459 samples for testing the performance. In both cases, we transformed the concentration data logarithmically, $y = \log(C)$, according to Cipollini *et al.* (2001). Hereafter, units of all accuracy and bias measurements are referred to y [$\log(\text{mg/m}^3)$] instead of C [mg/m^3].

Model Accuracy and Comparison

Table 4.2 presents results in the test set for all the SVR models and datasets. Results are compared with a feedforward NN trained with backpropagation (NN-BP) for both datasets. In the case of the SeaBAM dataset, we also include results with the models Morel-1, Morel-3, and CalCOFI 2-band (cubic and linear), as they performed best among a set of 15 *empirical* estimation models of the chlorophyll-a concentration in O'Reilly *et al.* (1998), and results from the ε-SVR obtained in Zhan *et al.* (2003). Several prediction error measures are considered, including mean error (ME), root mean-square error (RMSE), and mean absolute error (ABSE), as well as the correlation

Table 4.2 Results in the test set in the two datasets.

	ME	RMSE	ABSE	r
MERIS dataset				
NN-BP, 4 hidden nodes	-7.68×10^{-4}	0.024	0.016	0.998
ε-SVR	-2.36×10^{-4}	0.015	0.061	0.998
L_2-loss SVR	-9.96×10^{-4}	0.031	0.018	0.998
Proposed ε-Huber-SVR	-3.26×10^{-6}	0.011	0.004	0.999
SeaBAM dataset				
NN-BP, 4 hidden nodes	−0.046	0.143	0.111	0.971
ε-SVR in Zhan *et al.* (2003)	—	0.138	—	0.973
ε-SVR	−0.070	0.139	0.105	0.971
L_2-loss SVR	−0.034	0.140	0.107	0.971
ε-Huber-SVR	−0.020	0.137	0.104	0.972

coefficient between the desired output and the output offered by the models as a measure of fit. Source code for running the experiments can be found in this book's repository.

Table 4.2 shows that the kernel technique is more accurate (RMSE, ABSE) and unbiased (ME). For the MERIS data, significant numerical (see Table 4.2) and statistical differences between ε-Huber-SVR and the rest of the models were observed for both bias ($F = 2.46, p < 0.05$) and accuracy ($F = 767.8, p < 0.001$). For the SeaBAM dataset, the kernel method improved the performance, but no statistical differences were observed in bias ($F = 0.08, p = 0.92$) or accuracy ($F = 0.04, p = 9.86$). These results are similar to those previously reported in Zhan *et al.* (2003), but we were able to use half of the training samples, a particularly interesting situation when working with few in-situ measures usually available in biophysical problems.

Model Flexibility and Uncertainty

In order to model the accuracy demonstrated in the previous section, the ε-Huber cost function provides some relevant information about the reliability of the available data, which is extremely important when dealing with uncertainty. Figure 4.8 shows the histograms for residuals obtained with the ε-Huber-SVR model. If no uncertainty is present in data (MERIS data, synthetic data generated according to a physical model), we can see that the δC parameter covers a wide range of residuals and mainly penalizes them with a squared loss function (assuming Gaussian noise), and that the residual tails are linearly penalized (Cipollini *et al.*, 2001).

Model Robustness to Reduced Datasets

The number of the available in-situ measurements available is often scarce. In this experiment, we test the capabilities of the models to deal with low-sized datasets.

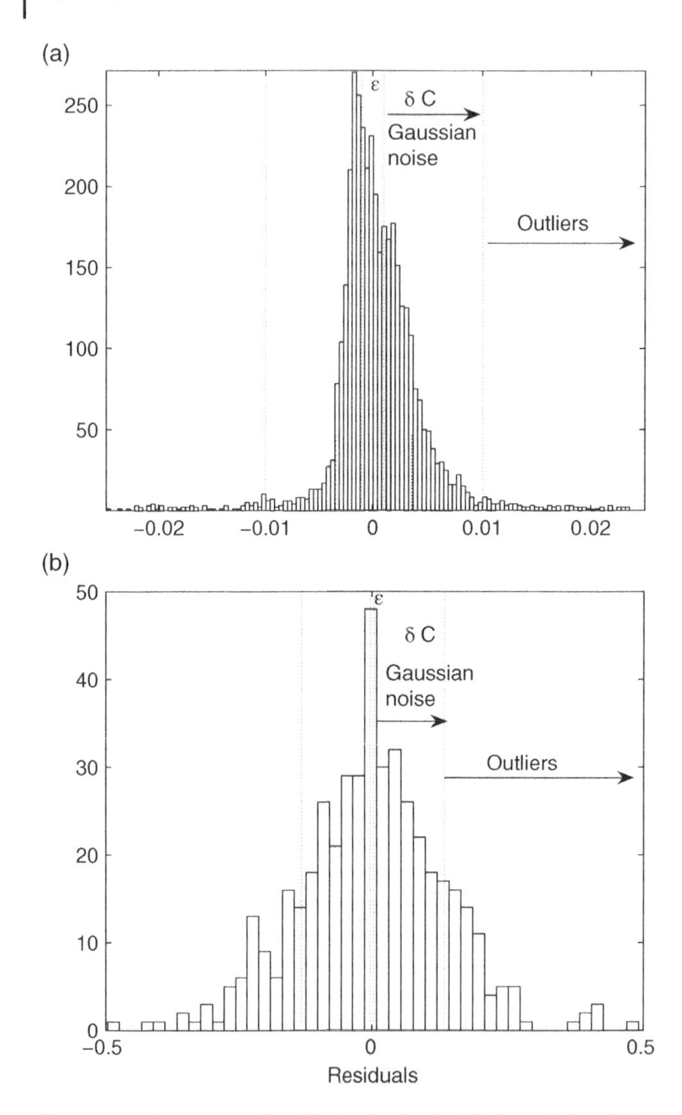

Figure 4.8 Histograms of residuals obtained with the ε-Huber-SVR model for (a) MERIS and (b) SeaBAM datasets.

Figure 4.9 shows the behavior of the RMSE versus the number of training samples in the SeaBAM data. We show the average value for 100 random different trials. The gain of the ε-SVR algorithm is noticeable, obtaining an average improvement of 8.5–13% in RMSE with respect to the other models. This effect is especially significant when working with very reduced training sets, as seen in the average gain of 16.67% in RMSE compared with ε-SVR when using only 10 training samples. We can conclude that, when the number of available samples is limited (low to moderate) and samples are affected by uncertainty factors, the versatility to accommodate different noise models to the available data makes ε-Huber-SVR an efficient and robust model.

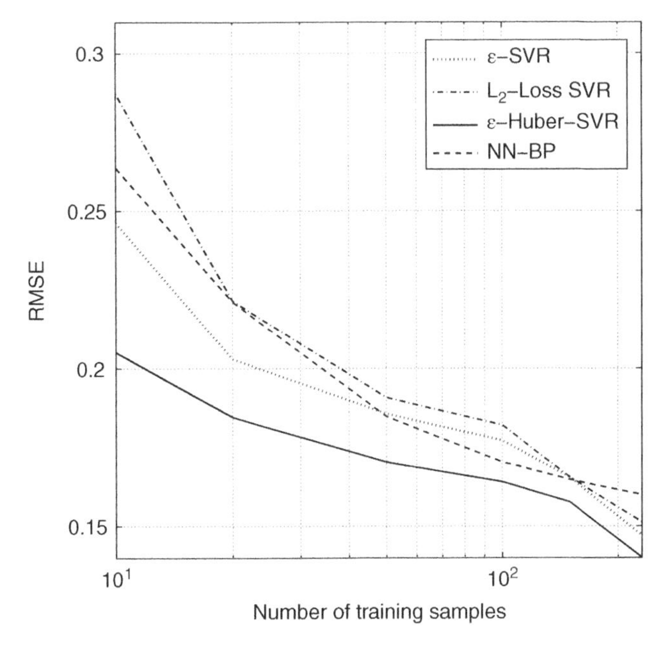

Figure 4.9 RMSE in the test SeaBAM dataset as a function of the number of training samples.

4.8 Concluding Remarks

In this chapter we introduced the most basic concepts of Mercer's kernels. The concepts of kernel function, feature mapping, positive definiteness and reproducing kernel Hilbert spaces were introduced as the core concepts that will accompany us throughout the remainder of the book.

The main idea in kernel methods is that positive definite kernels provide a measure of similarity between possibly complex objects. With the regularized risk framework one can implement the search over rich classes of functions and still obtain functions that can be expressed in finitely many terms of kernel evaluations. Another benefit is that this search can be made convex and thus yield problems which can be solved efficiently. Kernel methods generalize linear algorithms and allows us to develop nonlinear counterparts easily and still requiring only linear algebra. This is obviously of great practical convenience since linear algebra is perhaps one of the most solid and well-established part of mathematics.

The use of kernel methods in machine learning has expanded in the last decades, and they now constitute a preferred toolbox for practitioners in many disciplines, and in signal processing in particular. In the following chapters we will build upon the concepts provided here, and will subsequently introduce more advanced concepts to develop new kernel methods that tackle specific signal processing problems.

4.9 Questions and Problems

Exercise 4.9.1 Reproduce Example 4.2.6 for the case of a discrete finite FT.

Exercise 4.9.2 Consider a nonnegative function $F(\omega)$. Prove that the inverse FT of this function defines a shift-invariant kernel in a given RKHS of the form $K(\delta t)$ with $\delta t = t - t'$.

Exercise 4.9.3 The technique that optimizes parameters $\boldsymbol{w}$ of a linear estimator $\hat{y} = f(\boldsymbol{x}) = \boldsymbol{x}^{\mathrm{T}}\boldsymbol{a}$ by minimizing the regularized functional

$$\mathcal{L} = \sum_{i=1}^{N} \|y_i - f(\boldsymbol{x}_i)\|^2 + \lambda \sum_{j=0}^{M} a_j^2 \tag{4.90}$$

is known as ridge regression (Hoerl and Kennard, 1970). Derive the particular ridge regression algorithm for the example in Figure 4.3. In order to derive the algorithm, its parameters must be identified as the coefficients of the model; that is, $\boldsymbol{x} := [1, x, x^2, \dots, x^6]^{\mathrm{T}}$.

Exercise 4.9.4 Define $\mathcal{A}^2$ as the set of all complex-valued functions that are analytic inside the unit disk $D = \{\boldsymbol{x} : \|z\| < 1, \boldsymbol{x} \in \mathbb{C}\}$ of the complex plane, and square integrable. If an orthonormal basis of $\mathcal{A}^2$ is

$$\boldsymbol{\phi}(\boldsymbol{x}) = \sqrt{\frac{n+1}{\pi}}\boldsymbol{x}^n,$$

then prove that the kernel of $\mathcal{A}^2$ (Bergman kernel) (Halmos, 1974) is

$$K(\boldsymbol{x}_1, \boldsymbol{x}_2) = \frac{1}{\pi}\frac{1}{(1 - \boldsymbol{x}_1\boldsymbol{x}_2^*)}.$$

Exercise 4.9.5 Express the polynomial model in the example in Figure 4.3 as a function of Mercer kernels. Identify the kernel and its corresponding nonlinear mapping according to Mercer's theorem. Rewrite the ridge regression algorithm in Exercise 4.9.3 in terms of kernel dot products.

Exercise 4.9.6 Show that the following is a kernel function:

$$K(\boldsymbol{x}, \boldsymbol{y}) = \frac{1}{2}(\|\boldsymbol{x} + \boldsymbol{y}\|_K^2 + \|\boldsymbol{x} - \boldsymbol{y}\|_K^2),$$

for the norm defined by any kernel function K.

Exercise 4.9.7 Let us compute the difference in feature space between two mapped vectors $\boldsymbol{x}$ and $\boldsymbol{y}$; that is, $\boldsymbol{\phi} = \boldsymbol{\phi}(\boldsymbol{x}) - \boldsymbol{\phi}(\boldsymbol{y})$. Compute the reproducing kernel function that acts solely on examples in input space. Show that this is a valid kernel function.

Exercise 4.9.8 Let us compute the pointwise division in feature space between two mapped vectors $\boldsymbol{x}$ and $\boldsymbol{y}$; that is, $\boldsymbol{\phi} = \boldsymbol{\phi}(\boldsymbol{x})/\boldsymbol{\phi}(\boldsymbol{y})$. Compute the reproducing kernel function that acts solely on examples in input space. Show that this is a valid kernel function.

Exercise 4.9.9 Demonstrate the properties in Section 4.4.2.

Exercise 4.9.10 Demonstrate that a continuous kernel $K(\boldsymbol{x}, \boldsymbol{y}) = K(\boldsymbol{x} - \boldsymbol{y})$ on $\mathbb{R}^d$ is positive definite if and only if $K(\Delta)$ is the FT of a nonnegative measure. Using this theorem, also demonstrate that if a shift-invariant kernel $K(\Delta)$ is properly scaled, its FT $p(\boldsymbol{\omega})$ is a proper probability distribution.

Exercise 4.9.11 Build a simple script for performing SVR on a sinc function in terms of an RBF kernel for a nonregularly sampled time grid. Search for the optimal width of the RBF kernel. What can you conclude by observing the spectra of the observation signal, of the interpolated signal, and of the tuned RBF kernel? Plot the residuals in terms of the coefficients and discuss the related properties.

Part II

Function Approximation and Adaptive Filtering

5

A Support Vector Machine Signal Estimation Framework

5.1 Introduction

SVMs were originally conceived as efficient methods for pattern recognition and classification (Vapnik, 1995), and the SVR was subsequently proposed as the SVM implementation for regression and function approximation (Shawe-Taylor and Cristianini, 2004; Smola and Schölkopf, 2004). Nowadays, the SVR and other kernel-based regression methods have become a mature and recognized tool in DSP. This is not incidental, as the widespread adoption of SVM by researchers and practitioners in DSP is a direct consequence of their good performance in terms of accuracy, sparsity, and flexibility.

Early studies of time series with supervised SVM algorithms paid attention mainly to two DSP signal models; namely, nonlinear system identification and time series prediction (Drezet and Harrison 1998; Goethals *et al.* 2005b; Gretton *et al.* 2001b; Mattera 2005; Pérez-Cruz and Bousquet 2004; Suykens 2001; Suykens *et al.* 2001b). However, the algorithm used in both of them was the conventional SVR, just working on time-lagged samples of the available time series (i.e., essentially an ad hoc time embedding). Although good results have been reported with this approach, several concerns can be raised from a conceptual viewpoint of estimation theory:

1) The basic assumption for the regression problem statement, in an MMSE sense, is i.i.d. observations. This assumption is not at all fulfilled in time-series analysis, and algorithms neglecting temporal dependencies such as autocorrelations or cross-correlations can be missing crucial a priori information.
2) The use of Vapnik's ε-insensitive cost function might not be the most appropriate loss function in the case of Gaussian noise in the data.
3) The *kernel trick* (Aizerman *et al.*, 1964) used to develop nonlinear versions from well-established linear techniques is not the only advantage of the SVM methodology that would be desirable in DSP.

In this part, we introduce a framework coping with all these issues, which was previously proposed by Rojo-Álvarez *et al.* (2005, 2014). Starting from the last point, note that, for instance, SVMs are intrinsically regularized algorithms which, unlike MMSE methods, are quite resilient to overfitting and they are robust in low number of available training samples and high-dimensional datasets. SVMs also produce sparse solutions induced by the cost function used, which in turn is advantageous for the

Digital Signal Processing with Kernel Methods, First Edition. José Luis Rojo-Álvarez, Manel Martínez-Ramón, Jordi Muñoz-Marí, and Gustau Camps-Valls.

model interpretability and computational efficiency. SVMs also involve few model parameters to be tuned and lead to convex optimization problems. SVM algorithms are rooted on a solid mathematical background, and hence bounds of performance, uniqueness, stability, generalization, and optimality conditions can be established, and they can benefit from the theory of reproducing kernel functions to treat heterogeneous information in a unified way, as we will see in the following chapters.

In recent years, several SVM algorithms for DSP applications have been proposed aiming to overcome the aforementioned limitations. A first approach to nonparametric spectral analysis, using the robust SVM optimization criterion, instead of MMSE, was introduced by Rojo-Álvarez *et al.* (2003). The robustness properties of the SVM were further exploited by proposing linear approaches for γ filtering (Camps-Valls *et al.*, 2004), ARMA modeling (Rojo-Álvarez *et al.*, 2004), array beamforming (Martínez-Ramón *et al.* 2005; Pérez-Cruz and Bousquet 2004), and subspace-based spectrum estimation (Gonnouni *et al.*, 2012). The nonlinear generalization of ARMA filters with kernels (Martínez-Ramón *et al.*, 2006) and temporal and spatial reference antenna beamforming using kernels and SVM (Martínez-Ramón *et al.*, 2007) have also been proposed. The use of convolutional signal mixtures has been proposed for interpolation and sparse deconvolution problems (Rojo-Álvarez *et al.*, 2007, 2008), thanks to the autocorrelation kernel concept, a straightforward property which has opened the field for a number of unidimensional and multidimensional extensions of communications problems (Figuera *et al.*, 2012, 2014).

This chapter starts to pave the way to treat all these problems within the field of kernel machines, and it presents the fundamentals for a simple, yet unified, framework for tackling estimation problems in DSP using SVM, following the ideas presented by (Rojo-Álvarez *et al.*, 2014). The second part of this chapter and Chapters 6 and 7 are devoted to the particular models and approximations defined within the framework. Even though models in these chapters are formulated for SVM algorithms, they can be easily included in other kernel methods, as we will see later. Chapter 8 reviews some advances in kernel-based methods for regression and function approximation beyond SVMs. We end this part with an important topic in DSP: learning in online nonstationary scenarios with kernel-based adaptive filtering methods. The main approximations are reviewed and exemplified in Chapter 9.

Hence, in this chapter, we formalize the field in a simple, bottom-up framework for constructing kernel algorithms for DSP. In Section 5.2 we propose a three-block framework to treat SVM estimation in DSP problems. We think that such a framework is essential to develop new methods in a consistent and systematic way, but also to characterize recent methods in the kernel methods community. Then, we review the elements of the (nonlinear) SVR algorithm, which all have the fundamental properties required for general estimation problems.

A class of linear SVM-DSP algorithms come from the so-called primal signal model (PSM) (Rojo-Álvarez *et al.*, 2005). In this framework, rather than the prediction of the observed signal, the goal of the SVM is the estimation of a set of model coefficients containing the relevant information of that signal. This chapter presents the DSP algorithms using SVM for linear estimation, known as PSM algorithms. In this class of algorithms, the signal model is readily stated in the primal model of the SVM, which can be casted as a simple substitution of the linear regression data model by other particular time-series data models. A generic notation for this set of algorithms is given in terms of the expression of a signal using a given signal space expansion. The kernel trick is not used at this stage, but rather the use of different well-known signal models instead of regression or classification can be readily analyzed.

We conclude this chapter with some examples of use of SVR in these problems, and on PSM problem statements. Particular signal space expansions are given for nonparametric spectral analysis, for system identification and time series prediction, for digital communications, for convolutional models, and for array processing (temporal reference). Synthetic examples are also used to consolidate the main concepts, and application examples are further provided. In the following sections, taking into account the general signal model in previous chapters, a general PSM is stated which allows us to highlight the common and the problem-specific steps in the design of an SVM algorithm in this setting. Examples of the use of this general signal model for stating new SVM-DSP linear algorithms are given by making SVM algorithms for well-known signal processing estimation problems; namely, sinusoidal decomposition, ARX and γ-filter system identification, digital communications, sinc interpolation, sparse deconvolution, and array processing with temporal reference. Several application examples show the performance and scope of the PSM in SVM-DSP algorithms.

5.2 A Framework for Support Vector Machine Signal Estimation

This section formalizes a unified framework for developing SVM algorithms for *supervised estimation* in DSP (Rojo-Álvarez *et al.*, 2014). The framework focuses on time-series analysis, in which the time structure of the data is highly informative, but often discarded or at least mistreated. Figure 5.1 shows the main model building strategies in the SVM estimation framework, and examples of dedicated problem formulations that can be developed within it.

The treatment of other structure domains (e.g., images or video) can be also formalized under the same principles. Specific detailed model instantiations and empirical

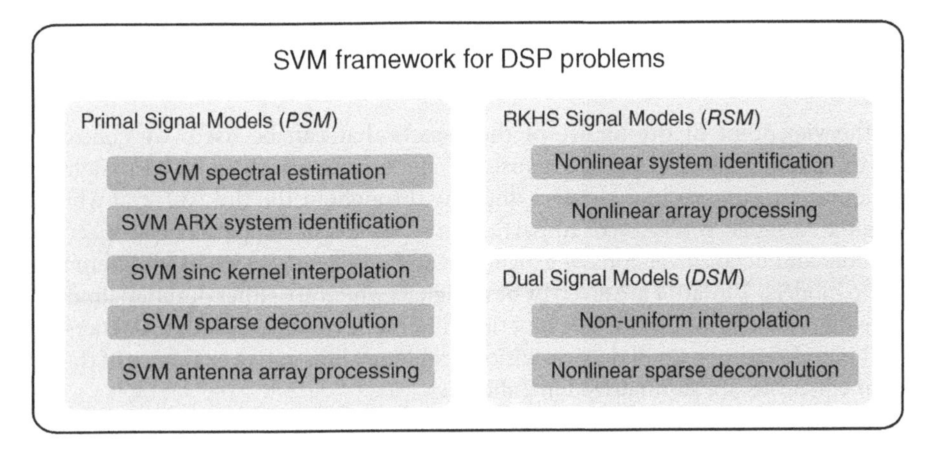

Figure 5.1 A schematic overview of the SVM framework for signal estimation.

evidence of performance will be given in this and in the following chapters. This framework can be summarized in the following types of signal models:

- **PSMs.** When a specific time-signal structure is assumed to model the underlying system that generated the data, the linear signal model (the so-called PSM) can actually be used. We will treat this formulation later in this chapter in detail. The statement of linear signal model equations in the primal, or SVM PSM for short, allows us to obtain robust estimators of the model coefficients (Rojo-Álvarez *et al.*, 2005) and to take advantage of almost all the characteristics of the SVM methodology in classical DSP problems, such as ARX time series modeling, spectral analysis (Camps-Valls *et al.* 2004; Rojo-Álvarez *et al.* 2003, 2004), and antenna array signal processing (Martínez-Ramón *et al.*, 2007). Then, nonlinear versions of the signal structure can be readily developed by the following two further approaches.
- **RKHS signal models (RSMs).** This is the first choice for nonlinear model building. The signal model equation is written in an RKHS by using the well-known RKHS model formulation and the kernel trick, in such a way that Mercer kernels can be readily used to replace dot products (Martínez-Ramón *et al.* 2006; Martínez-Ramón *et al.* 2007; Vapnik 1995). We will treat this formulation in Chapter 6 in detail. This constitutes the standard kernelization approach to signal processing, which has provided many kernel algorithms beyond SVM, as will be reviewed in Chapter 8.
- **Dual signal models (DSMs).** This second (and not so much usual) nonlinear alternative considers a signal expansion made by an auxiliary signal model equation, which in turn is given by a nonlinear regression of each time instant in the observed time series. We will treat this formulation in Chapter 7 with detail. DSMs have been previously proposed in an implicit way, and were based on the nonlinear regression of the time instants with appropriate Mercer kernels (Rojo-Álvarez *et al.*, 2007, 2008). While RSM allows us to scrutinize the statistical properties in the RKHS, the DSM can give us an interesting and straightforward interpretation of the SVM algorithm under study, in connection with classical LTI system theory.

This framework is summarized in Tables 5.1 and 5.2, where the key formulas are shown for better understanding the signal model equations. The tables give a principled framework for building efficient SVM linear and nonlinear algorithms in DSP applications.

Note that there is an almost endless variety of signal model problems and equations in DSP. Among them, we choose the following motivating pathways, following Rojo-Álvarez *et al.* (2014):

- From the viewpoint of the nature of the signals that can be used, we consider global-time and local-time signal expansions. The former are given by basis signals whose duration expands in a nondecaying way throughout the time interval where the estimated signal is observed. In particular, sinusoids in nonparametric spectral estimation, and delayed versions of exogenous and endogenous signals in difference equation models. The latter are given by basis signals which are either duration limited or decaying, which is the case of sinc functions in time-series interpolation, or energy-defined impulse responses in deconvolution problems. Illustrative examples of these kinds of equations are summarized in Table 5.1.
- A different, yet related approach comes when different applications can be stated according to different unknown terms of the same specific signal model equation. An

Table 5.1 Scheme of the DSP-SVM framework (I): equations of the time-series models for signal estimation.

		Global time, $\hat{y}_n$		Local time, $\hat{y}_n$	
	Regression, $\hat{y}$	Spectral	ARx	Sinc interpolation	Deconvolution
PSM	$\langle \boldsymbol{w}, \boldsymbol{x} \rangle + b$	$\sum_{k=0}^{K} a_k \cos(k\omega_0 t_n + \phi_k)$	$\sum_{p=1}^{Q} D_p y_{n-p} + \sum_{q=0}^{Q} E_q x_{n-q+1}$	$\sum_{k=0}^{N} a_k \text{sinc}(t - t_k)$	$x_n * h_n$
RSM	$\langle \boldsymbol{w}, \phi(\boldsymbol{x}) \rangle + b$	—	$\langle \boldsymbol{w}_d, \phi_d(\boldsymbol{y}_n) \rangle + \langle \boldsymbol{w}_e, \phi_e(\boldsymbol{x}_n) \rangle$	—	—
DSM	$\sum_{i=1}^{N} \alpha_i K(\boldsymbol{x}_i, \boldsymbol{x}) + b$	—	—	$K(t) * \sum_k \eta_k \delta(t - t_k)$	$\eta_n * R_n^h$

Table 5.2 Scheme of the DSP-SVM framework (II): equations for signal models in antenna array processing.

	Temporal reference	Spatial reference
PSM	$\langle \boldsymbol{a}, \boldsymbol{x}_n \rangle$	—
RSM	$\langle \boldsymbol{w}, \boldsymbol{\phi}(\boldsymbol{x}_n) \rangle$	$\langle \boldsymbol{w}, \boldsymbol{\phi}(b_i \boldsymbol{a}_0) \rangle - b$
DSM	—	—

excellent example in this setting is antenna array processing for beamforming, where the same signal model equation supports temporal reference signal detection, and spatial reference estimation problems. As illustrated in Table 5.2, PSM and RSM have been proposed for temporal reference problems, in a similar way to DSP problems in Table 5.1. And interestingly, a slightly different signal model is used in spatial reference, expressed in both cases in terms of possibly nonlinear mapping. This aims to illustrate that, in this case, we switch to an eigenproblem statement, which is a better representation for the data model, both for the linear and the nonlinear cases.

The equations in Tables 5.1 and 5.2 will be explained in detail in this and the following chapters, so the reader is encouraged to come back to these tables after their first reading. Note also that, in these tables, many problems have not been addressed yet in the literature (as indicated by "—"), showing that our intention is to motivate the interested reader to complete and expand this table according to their own DSP application needs.

5.3 Primal Signal Models for Support Vector Machine Signal Processing

We start by defining the transverse vector of a set of explanatory signals in the general signal model in Definition 3.2.10, and then formulating the SVM algorithm from that expansion. This will be called the PSM for building time-series DSP algorithms.

Definition 5.3.1 (Time-transversal vector of a signal expansion) Let $\{y_n\}$ be a discrete time series in a Hilbert space, and given the general signal model in Definition 3.2.10, then the nth time-transversal vector of the signals in the generating expansion set $\{s_n^{(k)}\}$ is defined as

$$\boldsymbol{s}_n = [s_n^{(0)}, s_n^{(1)}, \dots, s_n^{(K)}]^{\mathrm{T}}. \tag{5.1}$$

Hence, it is given by the nth samples of each of the signals generating the signal subspace where the signal approximation is made.

Figure 5.2 depicts an example of the time-transversal vector for a sinusoidal model. Note that, for determining the time-transversal vectors in a given signal model, we start

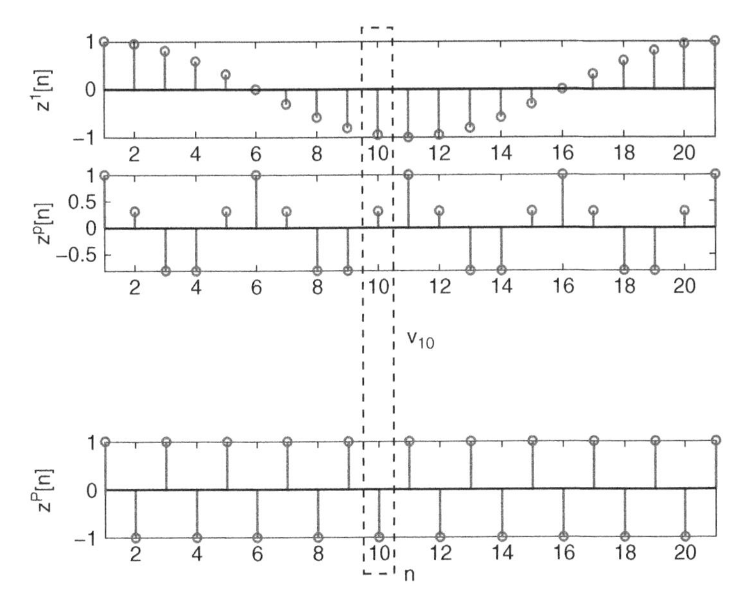

Figure 5.2 Time-transversal vectors used for PSM in the statement of linear SVM-DSP algorithms.

by identifying the signals considered in the expansions and then we work on a time basis for identifying the elements of each of those signals for a given discrete time. The SVM problem statement can be stated as follows.

Theorem 5.3.2 (PSM problem statement) Let $\{y_n\}$ be a discrete-time series in a Hilbert space, and given Definition 3.2.10, then the optimization of

$$\frac{1}{2}\|\boldsymbol{a}\|^2 + \sum_{n=0}^{N} \mathcal{L}_{\varepsilon\mathrm{H}}(e_n), \tag{5.2}$$

with $\boldsymbol{a} = [a_0, a_1, \dots, a_K]^{\mathrm{T}}$, gives an expansion solution whose signal model is

$$\hat{y}_n = \sum_{k=1}^{K} a_k s_n^{(k)} = \langle \boldsymbol{a}, \boldsymbol{s}_n \rangle \tag{5.3}$$

$$a_k = \sum_{n=0}^{N} \eta_n s_n^k \Rightarrow a = \sum_{n=0}^{N} \eta_n \boldsymbol{s}_n, \tag{5.4}$$

where η_n are the SVM Lagrange multipliers, and the solution at instant m is

$$\hat{y}_m = \sum_{n=0}^{N} \eta_n \langle \boldsymbol{s}_n, \boldsymbol{s}_m \rangle. \tag{5.5}$$

Only instants n with $\eta_n \neq 0$ are part of the solution (support time instants).

Therefore, each expansion coefficient a_k can be expressed as a linear combination of input space vectors. Note at this point that sparseness can be obtained in coefficients a_k, but not in coefficients η_n, and also that robustness is ensured for coefficients a_k when using the ε-Huber cost. The Lagrange multipliers are obtained from the dual problem, which is built in terms of a kernel matrix depending on the signal correlation.

Definition 5.3.3 (Correlation matrix from time-transverse vectors) Given a set of time-transverse vectors, the correlation matrix of the PSM is defined as

$$\boldsymbol{R}^s_{m,n} \equiv \langle \boldsymbol{s}_m, \boldsymbol{s}_n \rangle. \tag{5.6}$$

Property 11 (Correlation matrix and dual problem) Given the general PSM in Equation 5.3 and the correlation matrix from the time-transverse vectors in Equation 5.6, the dual problem yielding the Lagrange multipliers consists of maximizing

$$-\frac{1}{2}(\boldsymbol{\alpha} - \boldsymbol{\alpha}^*)^{\mathrm{T}}(\boldsymbol{R}^s + \delta \boldsymbol{I})(\boldsymbol{\alpha} - \boldsymbol{\alpha}^*) + (\boldsymbol{\alpha} - \boldsymbol{\alpha}^*)^{\mathrm{T}}\boldsymbol{y} - \varepsilon \boldsymbol{1}^{\mathrm{T}}(\boldsymbol{\alpha} + \boldsymbol{\alpha}^*) \tag{5.7}$$

constrained to $0 \le \alpha_n, \alpha_n^* \le C$.

This property can be readily shown from considerations on the Lagrange functional and the associated KKT conditions (Rojo-Álvarez *et al.*, 2005). Therefore, by taking into account the PSM for a given DSP problem, one can determine the signals $s_n^{(k)}$ generating the Hilbert subspace where the observations are projected to, and then the remaining elements and steps of the SVM methodology (such as the input space, the input space correlation matrix, the dual QP problem, and the solution) can be straightforwardly obtained.

5.3.1 Nonparametric Spectrum and System Identification

The first SVM-DSP algorithms that were proposed using the PSM framework were the sinusoidal decomposition (Rojo-Álvarez *et al.*, 2003), the ARX system identification (Rojo-Álvarez *et al.*, 2004), and the γ-filter structure (Camps-Valls *et al.*, 2004). We next point out the relevant elements that can be identified in these algorithms.

Property 12 (PSM coefficients for nonparametric spectral analysis) Given the signal model for nonparametric spectral analysis in Definition 3.3.1, the estimated coefficients using the PSM are

$$c_k = \sum_{n=0}^{N} \eta_n \cos(k\omega_0 t_n); \qquad d_k = \sum_{n=0}^{N} \eta_n \sin(k\omega_0 t_n). \tag{5.8}$$

Property 13 (PSM correlation and dual problem for nonparametric spectral analysis) Given the signal model hypothesis in Definition 3.3.1, the correlation matrix is given by the sum of two terms:

$$R^{\cos}_{m,n} = \sum_{k=0}^{K} \cos(k\omega_0 t_m)\cos(k\omega_0 t_n) \tag{5.9}$$

$$R^{\sin}_{m,n} = \sum_{k=0}^{K} \sin(k\omega_0 t_m)\sin(k\omega_0 t_n) \tag{5.10}$$

and the dual functional is given by Equation 5.7 using $\boldsymbol{R}^s = \boldsymbol{R}^{\cos} + \boldsymbol{R}^{\sin}$.

The identification of the corresponding time-transverse vectors is straightforward. The derivation of this algorithm in Rojo-Álvarez *et al.* (2003) is obtained by using these two properties in the PSM. Similar considerations can be drawn for the ARX system identification algorithm in Rojo-Álvarez *et al.* (2004) using the next two properties.

Property 14 (PSM coefficients for ARX system identification) Given the signal model for ARX system identification in Definition 3.3.2, the estimated PSM coefficients are given by

$$a_k = \sum_{n=0}^{N} \eta_n y_{n-k}; \qquad b_k = \sum_{n=0}^{N} \eta_n x_{n-k+1} \tag{5.11}$$

Property 15 (PSM Correlation and dual problem for ARX system identification) Given the signal model hypothesis for ARX system identification in Definition 3.3.2, the correlation matrix is given by the sum of two terms:

$$R^{y}_{m,n} = \sum_{k=1}^{P} y_{m-k} y_{n-k} \tag{5.12}$$

$$R^{x}_{m,n} = \sum_{k=0}^{Q} x_{m-k+1} x_{n-k+1} \tag{5.13}$$

These equations represent the time-local Pth- and Qth-order sample estimators of the values of the (non-Toeplitz) autocorrelation functions of the input and the output discrete time processes respectively. The dual functional to be maximized is given by Equation 5.7 using $\boldsymbol{R}^s = \boldsymbol{R}^y + \boldsymbol{R}^x$.

As described in previous chapters, the γ-filter in Camps-Valls *et al.* (2004) is defined by the following expressions:

$$y_n = \sum_{i=1}^{P} w_i x^i_n \tag{5.14}$$

$$x^i_n = \begin{cases} x_n, & i = 1 \\ (1-\mu)x^i_{n-1} + \mu x^{i-1}_{n-1}, & i = 2, \ldots, P \end{cases} \tag{5.15}$$

where y_n is the filter output signal, x_n is the filter input signal, x^i_n is the signal present at the input of the ith gamma tap, n is the time index, and μ is a free parameter. The SVM γ-filter algorithm in the PSM formulation is derived by means of the next property.

Property 16 (Time-transverse vector in γ-filter system identification) The time-transverse vector in the γ-filter for system identification in a PSM is given by

$$s^q = [x^q_0, x^q_1, \dots, x^q_N]^\mathrm{T}, \qquad q = 0, \dots, Q. \tag{5.16}$$

The signals generating the projection subspace are the input vector signals after each γ unit loop. For a previously fixed value of μ, the generating vectors of the Hilbert projection space are straightforwardly determined. By using this definition, the obtention of the signal model, the coefficients, and the dual problem can be readily found by just obtaining the autocorrelation matrix from Equation 5.6 and solving the dual problem in Equation 5.7.

5.3.2 Orthogonal Frequency Division Multiplexing Digital Communications

Fernández-Getino *et al.* (2006) proposed an SVM robust algorithm for channel estimation that is specifically adapted to the orthogonal frequency division multiplexing (OFDM) data structure. There were two main novelties in this proposal. First, the use of complex regression in SVM formulation provided with a simpler scheme than describing OFDM signal with either multilevel or nested binary SVM classification algorithms, as addressed by some previous approaches to OFDM with SVM. Second, the adequacy of free parameters in the ε-Huber robust cost function (Rojo-Álvarez *et al.*, 2004) was investigated in the presence of impulse noise. Although robustness of some digital communication receivers against impulse noise had been examined by using M-estimates (Bai *et al.* 2003; Ghosh 1996), there was no previous work about the performance of SVM algorithms in digital communications under this condition. We summarize here the main concepts and results in that work.

For simplicity, a linear dispersive channel with non-Gaussian noise is analyzed here. The discrete-time received OFDM signal for a system with N subcarriers is

$$r_n = \sum_{k=0}^{N-1} S_k H_k \, \mathrm{e}^{\mathbb{i}(2\pi/N)kn} + w_n + b_n g_n, \tag{5.17}$$

where r_n $(n = 0, \dots, N-1)$ are time-domain samples before DFT transformation, H_k is the channel's frequency response at kth frequency, S_k is the complex symbol transmitted at kth subcarrier, and w_n is the complex white Gaussian noise process $N(0, \sigma^2_w)$. The impulse noise is modeled as a BG process; that is, the product of a real Bernoulli process b_n with $Pr(b_n = 1) = p$ and a complex Gaussian process $g_n \sim N(0, \sigma^2_i)$ (Ghosh, 1996). Then, the residual noise at the receiver side is given by the sum of both terms $z_n = w_n + b_n g_n$.

In coherent OFDM systems, pilot symbols are usually inserted for channel estimation purposes. Then, the channel frequency response can be first estimated over a subset $\mathcal{K}_p$

of subcarriers, with cardinality $N_p = |\mathcal{K}_p|$, and then interpolated over the remaining subcarriers $(N - N_p)$ by using, for example, DFT-based techniques with zero-padding in the time domain (Edfors *et al.*, 1996). Now, the OFDM system can be expressed as

$$r_n = \sum_{k\in\{\mathcal{K}_p\}} P_k H_k \, e^{\mathbb{i}(2\pi/N)kn} + \sum_{k\notin\{\mathcal{K}_p\}} X_k H_k \, e^{\mathbb{i}(2\pi/N)kn} + z_n, \tag{5.18}$$

where P_k and X_k are respectively the complex pilot and data symbol transmitted at the kth subcarrier. It is well known that if the channel impulse response has a maximum of L resolvable paths (and hence of degrees of freedom), then N_p must be at least equal to L (Fernández-Getino García *et al.*, 2001). The signal model for OFDM-SVM is as follows:

$$r_n = \sum_{k\in\{\mathcal{K}_p\}} P_k H_k \, e^{\mathbb{i}(2\pi/N)kn} + e_n, \tag{5.19}$$

where $e_n = \sum_{k\notin\{\mathcal{K}_p\}} X_k H_k \, e^{\mathbb{i}(2\pi/N)kn} + z_n$ contains the residual noise plus the term due to data symbols. Here, these unknown symbols carrying information will be considered as noise during the training phases. Channel estimation via an MMSE cost function is no longer the ML criterion when dealing with this sort of noise (Papoulis, 1991). In order to improve the performance of the estimation algorithm, a robust cost function must be introduced.

Here, and for complex e_n, we define $\mathcal{L}_{\varepsilon H}(e_n) = \mathcal{L}_{\varepsilon H}(\mathrm{Re}\{e_n\}) + \mathcal{L}_{\varepsilon H}(\mathrm{Im}\{e_n\})$, where $\mathrm{Re}\{\cdot\}$ and $\mathrm{Im}\{\cdot\}$ denote real and imaginary parts respectively. The primal problem can be stated as minimizing

$$\begin{aligned}
&\frac{1}{2}\sum_{k\in\{\mathcal{K}_p\}} |H_k|^2 + \frac{1}{2\delta}\sum_{n\in I_1}(\xi_n + \xi_n^+)^2 + C\sum_{n\in I_2}(\xi_n + \xi_n^+) \\
&\quad + \frac{1}{2\delta}\sum_{n\in I_3}(\zeta_n + \zeta_n^+)^2 + C\sum_{n\in I_4}(\zeta_n + \zeta_n^+) - \sum_{n\in I_2, I_4} \delta C^2
\end{aligned} \tag{5.20}$$

constrained to

$$\mathrm{Re}\{r_n\} - \sum_{k\in\{\mathcal{K}_p\}} \mathrm{Re}\{P_k H_k \, e^{\mathbb{i}(2\pi/N)kn}\} \le \varepsilon + \xi_n \tag{5.21}$$

$$\mathrm{Im}\{r_n\} - \sum_{k\in\{\mathcal{K}_p\}} \mathrm{Im}\{P_k H_k \, e^{\mathbb{i}(2\pi/N)kn}\} \le \varepsilon + \zeta_n \tag{5.22}$$

$$-\mathrm{Re}\{r_n\} + \sum_{k\in\{\mathcal{K}_p\}} \mathrm{Re}\{P_k H_k \, e^{\mathbb{i}(2\pi/N)kn}\} \le \varepsilon + \xi_n^+ \tag{5.23}$$

$$-\mathrm{Im}\{r_n\} + \sum_{k\in\{\mathcal{K}_p\}} \mathrm{Im}\{P_k H_k \, e^{\mathbb{i}(2\pi/N)kn}\} \le \varepsilon + \zeta_n^+ \tag{5.24}$$

$$\xi_n^{(+)}, \zeta_n^{(+)} \ge 0 \tag{5.25}$$

for $n = 0, \dots, N-1$, where pairs of slack variables are introduced for both real ($\xi_n^{(+)}$) and imaginary ($\zeta_n^{(+)}$) residuals; superscript plus signs and no superscript stand for positive and negative components of residuals respectively; and I_1–I_2 (I_3–I_4) are the set of samples for which real (imaginary) parts of the residuals are in the quadratic–linear cost zone.

For giving support material on complex formulation, we next present the complete derivation of this algorithm. In brief, the primal–dual functional is obtained by introducing the constraints into the primal functional by means of Lagrange multipliers $\{\alpha_{\mathrm{R},n}\}$, $\{\alpha^{+}_{\mathrm{R},n}\}$, $\{\alpha_{\mathrm{I},n}\}$, and $\{\alpha^{+}_{\mathrm{I},n}\}$ for the real (subscript R) and imaginary (subscript I) parts of the residuals. By making zero the primal–dual functional gradient with respect to H_k, we have the following expression for channel estimated values at pilot positions:

$$\hat{H}_k = \sum_{n=0}^{N-1} \psi_n P_k, \tag{5.26}$$

where $\psi_n = (\alpha_{\mathrm{R},n} - \alpha^{+}_{\mathrm{R},n}) + \mathbb{i}(\alpha_{\mathrm{I},n} - \alpha^{+}_{\mathrm{I},n})$. For notation, we define the following column vector components:

$$\boldsymbol{v}_n(k) = [P_k \, \mathrm{e}^{\mathbb{i}(2\pi/N)kn}], \qquad k \in \{\mathcal{K}_p\}, \tag{5.27}$$

and the following Gram matrix as $\boldsymbol{R}_{n,m} = \boldsymbol{v}_n^H \boldsymbol{v}_m$. Now, by placing the optimal solution from Equation 5.26 into the primal–dual functional and grouping terms, a compact form of the functional problem can be stated in vector form, that consists of maximizing

$$-\frac{1}{2}\psi^{\mathrm{H}}(\boldsymbol{R} + \delta\boldsymbol{I})\psi + \mathrm{Re}(\psi^{\mathrm{H}}\boldsymbol{r}) - (\boldsymbol{\alpha}_{\mathrm{R}} + \boldsymbol{\alpha}^{+}_{\mathrm{R}} + \boldsymbol{\alpha}_{\mathrm{I}} + \boldsymbol{\alpha}^{+}_{\mathrm{I}})\mathbf{1}\varepsilon \tag{5.28}$$

constrained to $0 \leq \{\alpha_{\mathrm{R},n}\}, \{\alpha^{+}_{\mathrm{R},n}\}, \{\alpha_{\mathrm{I},n}\}, \{\alpha^{+}_{\mathrm{I},n}\} \leq C$, where $\boldsymbol{\psi} = [\psi_0, \dots, \psi_{N-1}]$; $\boldsymbol{I}$, $\mathbf{1}$ are the identity matrix and the all-ones column vector respectively; α_{R} is the vector containing the corresponding Lagrange multipliers, the other subsets being similarly represented; and $\boldsymbol{r} = [r_0, \dots, r_{N-1}]^{\mathrm{T}}$.

The channel values at pilot positions in Equation 5.26 can be obtained by optimizing Equation 5.28 with respect to $\{\alpha_{R,n}\}$, $\{\alpha^{+}_{R,n}\}$, $\{\alpha_{I,n}\}$ and $\{\alpha^{+}_{I,n}\}$ and then substituting into Equation 5.26.

5.3.3 Convolutional Signal Models

Convolutional signal models are simply those models that contain a convolutive mixture in their formulation. Some of the most representative ones are the nonuniform interpolation (using sinc kernels, RBF kernels, or others) and the sparse deconvolution, presented in Rojo-Álvarez *et al.* (2007, 2008). Their analysis also gives us the foundations of the DSM to be subsequently used in a variety of signal problem statements.

We next focus on summarizing the properties that are relevant for giving a signal processing block structure, which will be used for their analysis, following the framework proposed by Rojo-Álvarez *et al.* (2014).

Property 17 (PSM coefficients for sinc interpolation) Given the signal model in Definition 3.3.4 for sinc kernel interpolation, the PSM coefficients are

$$a_k = \sum_{n=0}^{N} \eta_n \mathrm{sinc}(\sigma_0(t_k - t_n)). \tag{5.29}$$

Property 18 (PSM correlation and dual problem for sinc interpolation) Given the signal model hypothesis in Definition 3.3.4, the correlation matrix is given by

$$\boldsymbol{R}^{\mathrm{sinc}}_{m,n} = \sum_{k=0}^{N} \mathrm{sinc}(\sigma_0(t_m - t_k))\mathrm{sinc}(\sigma_0(t_n - t_k)). \tag{5.30}$$

The dual functional to be maximized is given by Equation 5.7 using $\boldsymbol{R}^s = \boldsymbol{R}^{\mathrm{sinc}}$.

Coefficients in Equation 5.29 are proportional to the cross-correlation of coefficients η_n and a set of sinc functions, each centered at time instants t_n (Rojo-Álvarez *et al.*, 2007). Similar considerations can be made about the sparse deconvolution signal model, as follows:

Property 19 (PSM coefficients for sparse deconvolution) The estimated PSM coefficients of the signal model in Definition 3.3.5 are

$$\hat{x}_n = \sum_{i=0}^{N} \eta_i h_{i-n}. \tag{5.31}$$

Property 20 (PSM correlation and dual problem for sparse deconvolution) The dual problem corresponding to Definition 3.3.4 is found by using the time-transversal vector

$$\boldsymbol{s}^p = [h_n, h_{n-1}, h_{n-2}, \ldots, h_{n-p+1}]^{\mathrm{T}}. \tag{5.32}$$

The correlation matrix $\boldsymbol{R}^h$ is given by Equation 5.6, and in this case it represents the correlation matrix of h_n. The dual functional to be maximized is given by Equation 5.7.

In order to express the SVM-DSP algorithm for sparse deconvolution in terms of signal processing blocks, we can use a well-known relationship between model residuals and Lagrange multipliers, valid for general SVM algorithms.

Property 21 (Nonlinear relationship between Lagrange multipliers and residuals) Given the model residuals e_n and the corresponding model coefficient η_n in an SVM algorithm for estimation, the following relationship between them holds:

$$\eta_n = \left.\frac{\partial L_{\varepsilon\mathrm{H}}(e)}{\partial e}\right|_{e=e_n} = L'_{\varepsilon\mathrm{H}}(e_n), \tag{5.33}$$

which is a nonlinear relationship depending on the free parameters of the ε-Huber cost function.

Therefore, the Lagrange multipliers (or, equivalently, the model coefficients) are mapped from the model residuals by using a static nonlinear map which is given by the first derivative of the cost function (in our case, the ε-Huber cost). This property can be easily shown by using the appropriate set of KKT conditions (Rojo-Álvarez *et al.*, 2005), and it indicates that the model coefficients are a piecewise linear function of the model residuals.

According to this expression of the Lagrange multipliers as a time series, the sparse deconvolution model can be further analyzed as follows.

Property 22 (PSM block diagram for sparse deconvolution) Let a discrete time signal be defined by model coefficients η_n for $n = 0, \dots, N$, and being zero otherwise. Then, we can express the relationship between the model coefficients and the estimated signal as follows:

$$\hat{x}_n = \eta_n * h_{-n} * \delta_{n+M} \tag{5.34}$$

where δ_n is the Kronecker delta sequence (1 for $n = 0$ and 0 elsewhere), and $*$ denotes the discrete-time convolution operator.

This is an interesting property from the signal processing point of view, and it can be easily obtained by examining the KKT in the PSM of the sparse deconvolution problem (Rojo-Álvarez *et al.*, 2008). Hence, we can consider a joint equivalent closed-loop system, given in Figure 5.3a, which contains all the elements of the SVM algorithm expressed as signals or systems. Specifically, one is a nonlinear system, given by Property 21, and the remaining ones are linear, time-invariant systems. According to the preceding property, estimated signal $\hat{x}_n$ will not be sparse in general, because it is the sparseness of η_n that can be controlled with the ε parameter, but there is a convolutional relationship between $\hat{x}_n$ and η_n that will depend on the impulse response, which in general does not have to be sparse.

In addition, note that there is no Mercer kernel appearing explicitly in this problem statement of PSM for sparse deconvolution, as should be expected in an SVM approach. However, the block diagram highlights that there is an implicitly present autocorrelation kernel, given by

$$R_n^h = h_n * h_{-n}, \tag{5.35}$$

in the case we associate the two systems containing the original system impulse response and its reversed version. From linear system theory, the order of the blocks could be changed without modifying the total system. However, the solution signal is embedded between these two blocks, which precludes the explicit use of this autocorrelation kernel in this PSM formulation. Finally, the role of delay system δ_{n+M} can be interpreted as just an index compensation that makes the total system causal.

In summary, the PSM yields a regularized solution, in which an autocorrelation kernel is implicitly used, but it does not allow one to control the sparseness of the estimated signal. These properties will be used later in the DSM for proposing high-performance sparse deconvolution algorithms.

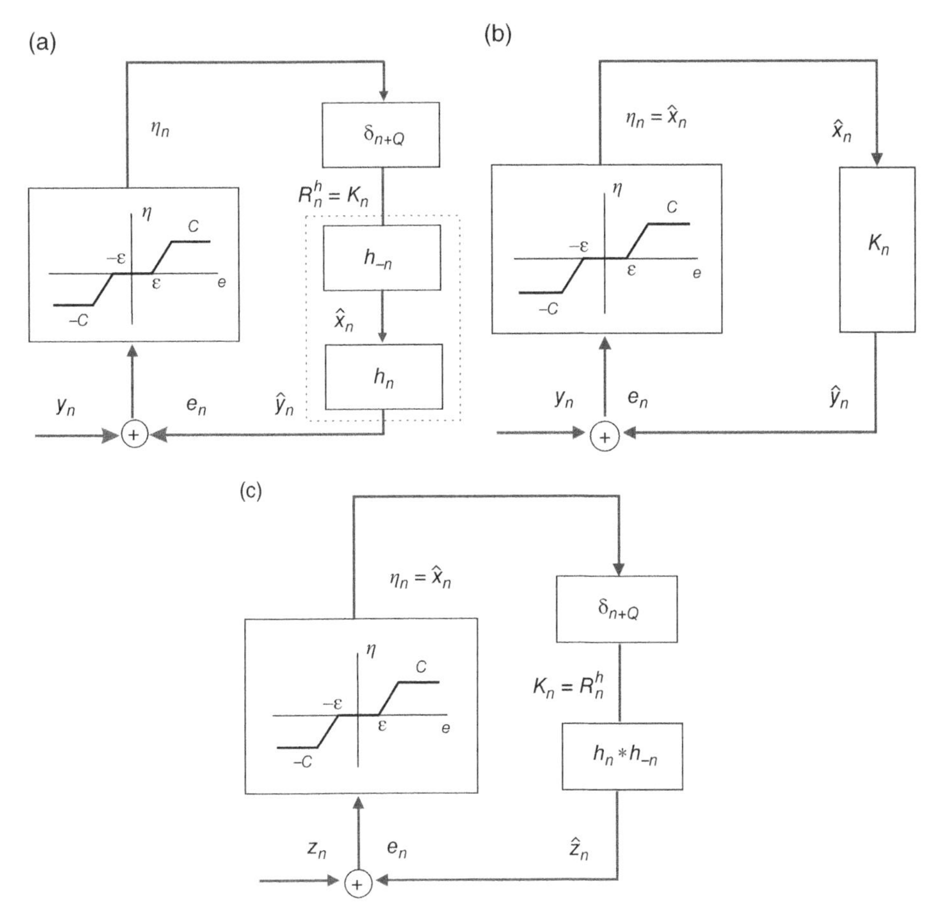

Figure 5.3 Signal model for SVM-DSP sparse deconvolution. (a) Block diagram, elements, and signals in the PSM. (b) Convolutional model for sinc interpolation using the DSM problem statement, where the impulse response of the convolutional model is a valid Mercer kernel. (c) Convolutional model for sparse deconvolution, for an arbitrary impulse response, where the Mercer kernel is the autocorrelation of the impulse response signal.

5.3.4 Array Processing

We present here the algorithm formulation from SVM-DSP framework both for temporal reference and for spatial reference in array processing. The array processing algorithm needs a complex-valued formulation. The complex Lagrange coefficients can be expressed as $\psi_n = \eta_n + \mathbb{i}\nu_n$, and $\eta_n = \alpha_n - \alpha_n^*$ and $\nu_n = \beta_n - \beta_n^*$ are the Lagrange coefficients generated by the real and imaginary parts of the error.

Property 23 (PSM for temporal reference array processing) Given the array processing signal defined in Definition 3.3.6, the PSM coefficients for this problem are given by

$$a_k = \sum_{n=0}^{N} \psi_n x_n^k. \tag{5.36}$$

Property 24 (Dual problem for temporal reference array processing) Let y_n, with $0 \leq n \leq N$, be a set of desired signals available for training purposes, then the problem is known as temporal reference array processing. The incoming signal kernel matrix is defined as

$$\boldsymbol{K}_{l,m} = \sum_{i=0}^{K} \overline{x}_l^k x_m^k. \tag{5.37}$$

The dual functional to be maximized is a complex-valued extension of Equation 5.7; that is:

$$\boldsymbol{\psi}^{\mathrm{T}} \boldsymbol{K} \boldsymbol{\psi} + \mathrm{Re}(\boldsymbol{\psi}^{\mathrm{T}} \overline{\boldsymbol{y}}) - \varepsilon \mathbf{1}^{\mathrm{T}} (\boldsymbol{\alpha} + \boldsymbol{\alpha}^* + \boldsymbol{\beta} + \boldsymbol{\beta}^*), \tag{5.38}$$

where $\overline{\boldsymbol{y}}$ stands for the complex conjugate of $\boldsymbol{y}$.

While the prior knowledge that is exploited in temporal reference algorithms is a set of training observations for which the corresponding transmitted symbols are y_n, this information is not available in spatial reference. Instead, the prior knowledge is the DOA of the signal of interest. Thus, the spatial filter must be optimized for minimum variance of its output, subject to the constraint of producing the right response in the presence of a signal coming from the given DOA.

The incoming signal is then filtered by a steering vector in the DOA (see Equation 3.49). The training data of the SVM must be a set of arbitrary signals containing this DOA (Martínez-Ramón *et al.*, 2007). Thus, we must satisfy the following minimization:

$$\min_{\boldsymbol{w}} \boldsymbol{w}^{\mathrm{H}} \boldsymbol{R} \boldsymbol{w}. \tag{5.39}$$

If prior to the minimization the data are whitened with the transformation

$$\tilde{x}_n = \boldsymbol{\Lambda}^{1/2} \boldsymbol{Q}^{\mathrm{H}} \boldsymbol{x}, \tag{5.40}$$

then the expression equivalent to the previous minimization is simply

$$\min_{\boldsymbol{w}} \boldsymbol{w}^{\mathrm{H}} \boldsymbol{w}, \tag{5.41}$$

where we used the eigendecomposition of autocorrelation matrix $\boldsymbol{R}$ as $\boldsymbol{R} = \boldsymbol{Q} \boldsymbol{\Lambda} \boldsymbol{Q}^{\mathrm{H}}$. Then, the minimization can be expressed as one of a regular SVM. The training vectors have the form of steering vectors $\boldsymbol{a}_0$ scaled with arbitrary amplitudes d_n. Thus, after whitening these vectors, the weight vector can be expressed as

$$\boldsymbol{w} = \boldsymbol{\Lambda}^{-1/2} \boldsymbol{Q} \boldsymbol{a}_0 \sum_{n=0}^{N} \psi_n d_n. \tag{5.42}$$

Property 25 (PSM spatial reference array processing) Let $\tilde{\boldsymbol{a}}_0 = \boldsymbol{\Lambda}^{-1/2} \boldsymbol{Q} \boldsymbol{a}_0$ be a whitened version of the steering vector $\boldsymbol{a}_0$ as expressed in Equation 3.49, coming from DOA θ_0. The PSM coefficients for this problem are given by

$$a_k = \sum_{n=0}^{N} \psi_n d_n \tilde{a}_0^k, \tag{5.43}$$

where d_n is an arbitrary complex amplitude.

Property 26 (Dual problem for spatial reference array processing) Let d_n, $0 \leq n \leq N$, be a set of arbitrary complex amplitudes. The incoming signal kernel matrix is defined as

$$\boldsymbol{K}_{l,m} = \sum_{i=0}^{K} \overline{d_l} \overline{\tilde{a}_0^k} \tilde{a}_0^k d_m, \tag{5.44}$$

where the last expression follows from the fact that the steering vector is normalized.

5.4 Tutorials and Application Examples

The cost function in terms of robustness to outlying samples, the effect of the δ parameter, and the sparsity property are analyzed with the SVM algorithm for nonparametric spectral analysis (Rojo-Álvarez *et al.*, 2003). We also illustrate here the performance of some of the preceding PSM algorithms; namely, linear system identification with the SVM γ-filter structure (Camps-Valls *et al.*, 2004; Rojo-Álvarez *et al.*, 2004), SVM-ARMA formulation for parametric spectral estimation, OFDM communications (Fernández-Getino *et al.*, 2006), and PSM algorithms performance for antenna processing.

5.4.1 Nonparametric Spectral Analysis with Primal Signal Models

A synthetic data example shows the usefulness of PSM for dealing with outliers in the data, where SVM-NSA has been considered (see MATLAB code in Listing 5.1 for more details).

```
function [predict,coefs] = SVM_NSA(~,Ytrain,~,gamma,epsilon,C)

% Initialization
ts=1; fs=1;
N = length(Ytrain);
t = ts*(0:N-1); % ts is the sampling period
w0 = 2*pi*fs/N; % fs is the sampling frequency, w0 is the angular ...
    frequency resolution

% Construct kernel matrix
KC = cos((1:N)'*w0*t);
KS = sin((1:N)'*w0*t);
H = KC'*KC + KS'*KS;

% Train SVM and predict
inparams=sprintf('-s 3 -t 4 -g %f -c %f -p %f -j 1', gamma, C, epsilon);
[predict,model]=SVM(H,Ytrain,H,inparams);
coefs=getSVs(Ytrain,model);
```

Listing 5.1 SVM-NSA algorithm (SVM_NSA.m).

Data Generation

Let $y_n = \sin(2\pi fn) + e_n^v + e_n^j$, where $f = 0.3$, e_n^v is a white, Gaussian noise sequence (zero mean, variance 0.1), and e_n^j is an impulsive noise process, generated as a sparse sequence for which 30% of the randomly placed samples have high amplitude values given by $\pm 10 + \mathcal{U}(-0.5, 0.5)$, where $\mathcal{U}()$ denotes the uniform distribution in the given interval, and the remaining are null samples. The length is 128 samples, and we set $N_\omega = 128$. We fixed $\varepsilon = 0$ and $\delta = 10$. The MATLAB code in Listing 5.2 can be used to create this synthetic data example.

```
function data = genDataSpectral(conf)

rng(500)
cf = conf.data;
% Signal
y = sin(2*pi*cf.f*(1:cf.n))';
% Gaussian noise
y = y + sqrt(cf.var_noise)*randn(size(y));
% Impulsive noise
nin = round(cf.p*cf.n);
affected_samples = randperm(cf.n,nin);
imp_noise = cf.A*sign(rand(1,nin)-.5)+(rand(1,nin)-0.5);
y(affected_samples)=imp_noise;
% Create data structure
data.Xtrain=cf.f; data.Ytrain=y; data.Xtest=cf.f; data.Ytest=y;
```

Listing 5.2 Generating spectral analysis data (genDataSpectral.m).

Experiment Settings and Results

Parameter C can be chosen according to Equation 5.33. Results can be easily generated using the code `config_spectral.m` and `display_results_spectral.m` on this book's web page. Results show that, for C low enough, large residual amplitudes can be present without distorting the solution.

5.4.2 System Identification with Primal Signal Model γ-filter

In this example, we focused on the main advantages of both the γ-filter structure (stability and memory depth) and the SVM with PSM (regularization). We compared the memory parameter μ in the LS, the SVM, and the regularized γ-filter algorithms. We identified the third-order elliptic low-pass filter with coefficients

$$\begin{aligned} y_n = 0.0563x_n - 0.0009x_{n-1} - 0.0009x_{n-2} + 0.0563x_{n-3} \\ + 2.1291y_{n-1} - 1.7834y_{n-2} + 0.5435y_{n-3} \end{aligned} \tag{5.45}$$

$$H(z) = \frac{0.0563 - 0.0009z^{-1} - 0.0009z^{-2} + 0.0563z^{-3}}{1 - 2.1291z^{-1} + 1.7834z^{-2} - 0.5435z^{-3}}, \tag{5.46}$$

which was proposed in Principe *et al.* (1993) because of its long impulse response.

A 100-sample input discrete process $\{x_n\}$ was a white, Gaussian noise sequence with zero mean and unit variance. The output signal $\{y_n\}$ was corrupted by additive, small variance ($\sigma_e^2 = 0.1$) noise. An independent set of 100 samples was used for testing, and

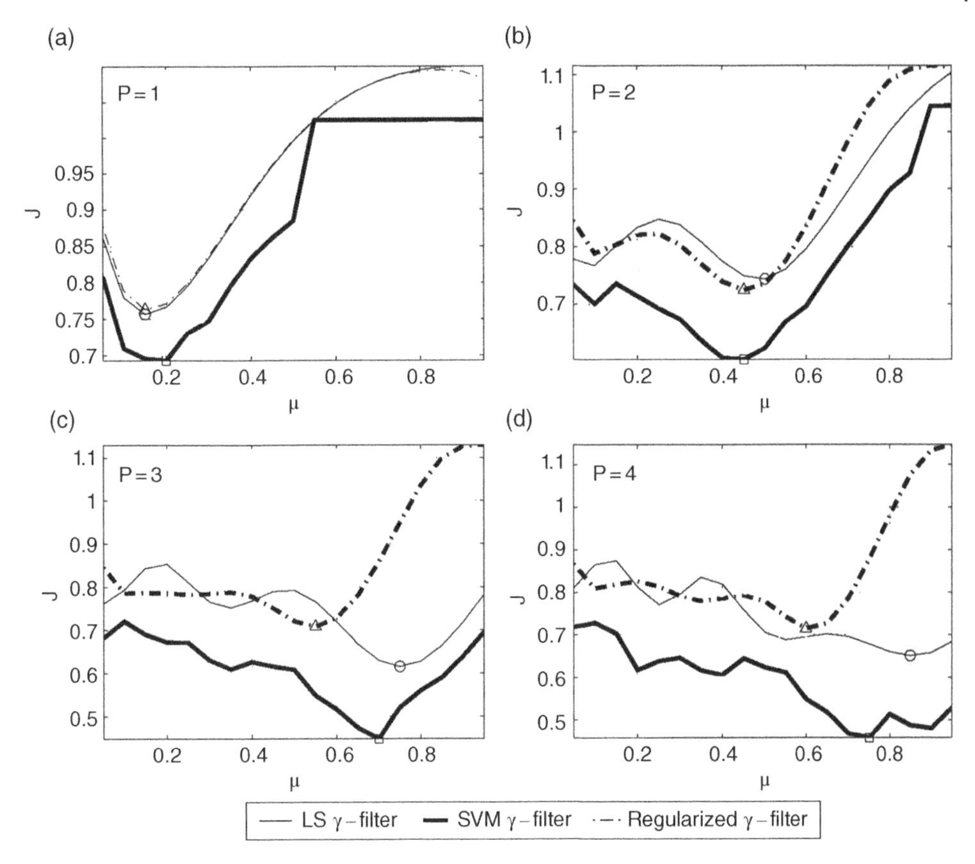

Figure 5.4 Performance for the system identification by the LS γ-filter and the SVM γ-filter for different orders P. The optimal μ parameter is indicated (o, □, △).

the experiment was repeated 100 times. Figure 5.4 shows the performance criterion ($J_{\min} = var(e_n)/var(y_n)$) (Principe *et al.*, 1993) as a function of μ and Q in the test set. In all cases, the adaline structure ($\mu = 1$) performed worse than the γ structures, and the SVM γ-filter clearly improved the results of the LS and the regularized versions. For all methods, memory depth M for a fixed Q increased with decreasing μ, especially for the regularized γ-filter, but at the expense of poor performance, whereas the SVM γ-filter still presented a good trade-off between memory depth and performance. Some code snippets for running the γ-filter are included in Listing 5.3.

```
% System
A_true=[1 0.5 0.3]; B_true=[1 0.5];
% Data
x  = randn(100,1); Pn = 0.01;
y  = filter(B_true,[1 A_true],x)+sqrt(Pn)*randn(size(x));
% gamma-SVM parameters
epsilon = 0; gamma = 1e-1; C = 10; p = 5; mu = (.1:.1:1.9);
% Try different mu values
for i = 1:length(mu)
    disp([num2str(i) ' de ' num2str(length(mu))]);
```

```
    [a, b, ypred,er(i)] = svm_gamma_wls(x,y,p,mu(i),epsilon,C,gamma);
end
% Get best results
[m,k] = min(er); mimu = mu(k); [a,b,ypred,ee] = ...
    svm_gamma_wls(x,y,p,mimu,epsilon,C,gamma);
% Plot figure
figure(1), clf, subplot(211), plot(y), hold on, plot(ypred,'r'),hold ...
    off; subplot(212),plot(mu,er);

function [a,b,ypred,er] = svm_gamma_wls(x,y,p,mu,e,C,gam)
%
%  SVM_GAMMA_WLS Support Vector for ARMA modeling with WLS optimization
%
%  Usage: [a b ypred] = svm_gamma_wls(x,y,p,mu,e,C,gam)
%
%  Parameters: x        - Plant input
%              y        - Plant output
%              p        - Order of a
%              mu       - gamma memory
%              e        - Insensitivity on prediction error
%              C        - Upper bound and linear cost param
%              gam      - Quadratic cost param
%              a  b     - Model coeffs
%              ypred    - Predicted output

% Init
y = y(:); x = x(:); N = length(y);
% Construct X matrix and w vector
k0 = p+1; X = []; b = mu; a = [1 -(1-mu)];
xold = x';
for i = 1:p
    X = [X;xold]; xold = filter(b,a,xold);
end
X = X(:,k0:end); yold = y; y = y(k0:end);
%%% >>> Insert here your SVM solver: [W,mu] = SVMsolver(X,y)  <<<<
ypred = X'*W; ypred = [zeros(p,1) ; ypred];
er = norm(ypred(p+1:end) - yold(p+1:end));
```

Listing 5.3 γ-filter (gammafilterDemo.m).

5.4.3 Parametric Spectral Density Estimation with Primal Signal Models

Most of the available frequency-domain estimation methods can be grouped into four main categories: parametric, nonparametric (or periodogram-based), high-resolution (or subspaces), and time–frequency methods (Marple, 1987). In parametric methods, the observed signal is modeled as the output of a linear, time-invariant system that is driven by white noise, and the coefficients of the linear system are estimated by Yule–Walker, Burg, or covariance methods (Ljung, 1999). Parametric spectral methods tend to produce better PSD estimates than classical nonparametric methods when the signal length is short, the spectrum is spiky, or some a priori knowledge about the signal is available.

The most widely used linear system model for parametric spectral estimation is the all-pole structure (Marple, 1987). The output y_n of such a filter for a white noise input is an AR process of order P, AR(P), which can be expressed as $y_n = \sum_{p=1}^{P} a_p y_{n-p} + e_n$, where a^p are the AR parameters and e_n denotes the samples of the innovation process. Once the coefficients a^p of the AR process are calculated, the PSD estimation is

$$P_y(f) = \frac{1}{f_s} \frac{\sigma_n^2}{\left|1 - \sum_{p=1}^{P} a_p \, \mathrm{e}^{-\mathrm{i}2\pi pf/fs}\right|^2}, \tag{5.47}$$

where f_s is the sampling frequency and σ_n^2 is the variance of the residuals.

We used data generated as an ARMA process with e_n given by white Gaussian noise with zero mean and unit variance. Two systems, previously introduced (Söderström and Stoica, 1987), were analyzed; namely, an AR(3) process, given by

$$y_n = e_n - 0.9816y_{n-1} - 0.9400y_{n-2} - 0.7799y_{n-3}, \tag{5.48}$$

and a narrow-band ARMA(4, 4) process, given by

$$\begin{aligned} y_n = e_n &+ 0.4800e_{n-1} + 0.6876e_{n-2} + 0.4476e_{n-3} + 0.3538e_{n-4} \\ &+ 1.0200y_{n-1} - 2.0902y_{n-2} + 0.9808y_{n-3} - 0.9275y_{n-4}. \end{aligned} \tag{5.49}$$

The input discrete process was an $\mathcal{N}(0, 1)$ sequence with sample length $L = 128$. The output signal was corrupted by additive noise $\mathcal{N}(0, 0.1)$, and 20% of samples were affected by impulsive noise from a zero mean and unit variance uniform distribution and randomly placed. These L samples were used for training the model, and 1000 samples more were used for validation. For all simulations, parameters C and δ were searched in the range $[10^{-5}, 10^5]$, and we fixed $\varepsilon = 0$. The performance criterion used for the general estimate of $P_Y(f)$ was the integrated mean-square error (IMSE), given by $\text{IMSE} = \frac{1}{N_F} \sum_{f=1}^{N_F} |P_Y(f) - \hat{P}_Y(f)|^2$, where N_F is the number of estimated frequencies in the spectrum. The experiment was repeated 100 times and the best model was selected according to the estimated IMSE in the validation set.

Figure 5.5 illustrates the effect of different power of outliers P_o on the estimation accuracy. In both systems, the SVM-AR method outperformed the standard methods in all situations, with an average gain of 1.5–2 dB. Differences between the methods are lower with increasing noise, specially for the ARMA(4, 4).

5.4.4 Temporal Reference Array Processing with Primal Signal Models

A linear array of six elements spaced $d = 0.51\lambda$, was used to detect the signal from a desired user in the presence of different interferences in an environment with Gaussian noise (Martínez-Ramón *et al.*, 2005). The desired signal was assumed to experiment small multipath propagation coming from DOAs -0.1π and -0.25π, with amplitudes 1 and 0.3 respectively and different phases. Two interferences signals came from -0.05π, 0.1π, and 0.3π, all of them with unit amplitude and phase 0. The desired signal was organized in bursts of 50 training samples plus 10^5 test samples.

The SVM was compared with the standard LS algorithm for array processing. Since the noise was assumed to be thermal, then its variance could be assumed to be approximately known. In communications, parameter validation is usually not affordable due

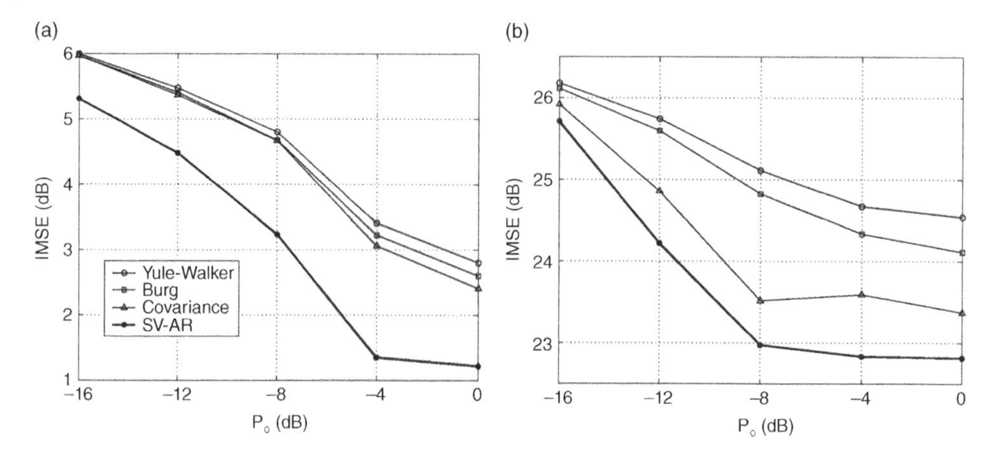

Figure 5.5 Evolution of IMSE for different power of outliers for (a) an AR(3) process and (b) an ARMA(4,4) process.

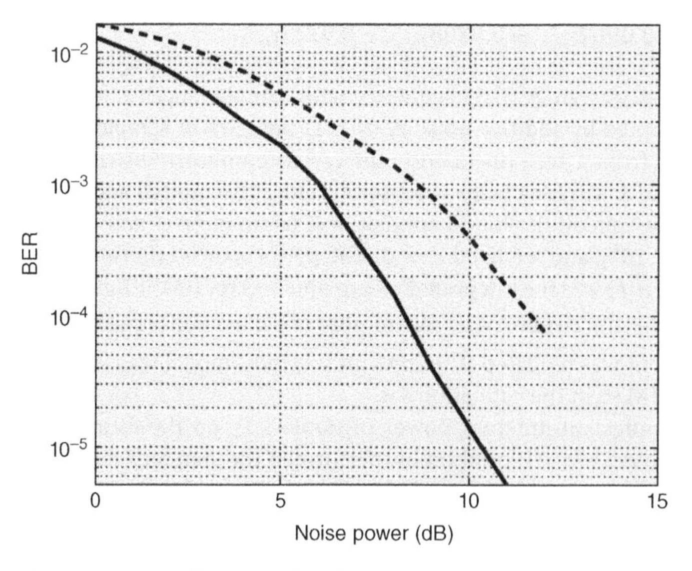

Figure 5.6 BER performance of the SVM (continuous) and regular LS (dashed) beamformers.

the small amount of available data and the low computational power of systems. Therefore, parameter δ of the SVM cost function was set to 10^{-6}, and then δC was adjusted to the thermal noise standard deviation. Hence, residuals for samples corrupted mainly by thermal noise were likely in the quadratic cost, and residuals for samples with high error (when interfering signals added in phase) were likely in the linear cost.

Results are shown in Figure 5.6. Each BER was evaluated by averaging 100 independent trials. The LS criterion is highly biased by the non-Gaussian nature of the data produced by the multipath environment plus the interfering signals. SVM is closer to the linear optimal and offers a processing gain of several decibels with respect to the LS.

5.4.5 Sinc Interpolation with Primal Signal Models

As commented in Section 3.4.5, standard interpolation algorithms can exhibit some limitations, such as poor performance in low signal-to-noise scenarios or in the presence of non-Gaussian noise. In addition, these methods result in non-sparse solutions. These limitations can be alleviated by accommodating the SVM formulation with the nonuniform sampling problem. In order to benchmark the PSM for sinc interpolation with methods Y1, Y2, S1, S2, and S3, which were described in Section 3.4.5, the PSM with sinc kernel is considered, and its MATLAB code is in Listing 5.4.

```
function [predict,coefs] = ...
    SVM_Primal(Xtrain,Ytrain,Xtest,T0,gamma,epsilon,C)

% Initials
tk   = Xtrain;
N = length(tk);
Ntest = length(Xtest);
% Construct kernel matrix
S = sinc((repmat(tk,1,N) - repmat(tk',N,1))/T0);
Stest = sinc((repmat(Xtest,1,N) - repmat(tk',Ntest,1))/T0);
H = zeros(N,N);
Htest = zeros(Ntest,N);
for m = 1:N
    H(:,m) = sum(S.*repmat(sinc((tk(m)-tk)/T0)',N,1),2);
    Htest(:,m) = sum(Stest.*repmat(sinc((tk(m)-tk)/T0)',Ntest,1),2);
end
% Train SVM and predict
inparams = ['-s 3 -t 4 -g ',num2str(gamma),' -c ',num2str(C),' -p ...
    ',num2str(epsilon)];
[predict,model] = SVM(H,Ytrain,Htest,inparams);
coefs = getSvmWeights(model);
```

Listing 5.4 SVM-P algorithm (SVM_Primal.m).

In this code, the `SVM` function is presented in the MATLAB code in Listing 3.32. At this moment, we postpone to look at the results of this algorithmic implementation until the following chapters, so that we can have a complete landscape for the comparison.

5.4.6 Orthogonal Frequency Division Multiplexing with Primal Signal Models

In order to test the performance of the OFDM-SVM scheme, detailed in Fernández-Getino *et al.* (2006), a scenario for IEEE 802.16 fixed Broadband Wireless Access Standard has been considered (Erceg *et al.*, 2003). In second-generation systems for these types of applications, non-line-of-sight conditions are present. To simulate this environment, we use the modified Stanford University interim SUI-3 channel model for omnidirectional antennas, with $L = 3$ taps, a maximum delay spread of $\tau_{max} = 1$ µs, and maximum Doppler frequency $f_m = 0.4$ Hz (Erceg *et al.*, 2003). The main parameters of this channel are summarized in Table 5.3. It can be observed that the channel exhibits an RMS delay spread of $\tau_{rms} = 0.305$ µs. Also, it must be noted that K-factors are given in linear values, and not in decibel values; values shown in Table 5.3 mean that 90% of the cell locations have K-factors greater than or equal to the K-factor specified. Finally, the specified Doppler is the maximum frequency parameter f_m of the round-shaped spectrum. Additionally, the SUI-3 channel models specify a normalization factor equal

Table 5.3 SUI-3 channel model parameters for multipath fading.

	Tap 1	Tap 2	Tap 3	Units
Delay	0	0.5	1	μs
Power	0	−5	−10	dB
K factor (90%)	1	0	0	Linear
Doppler	0.4	0.4	0.4	Hz

to −1.5113 dB, which must be added to each tap power to get 0 dB as the total mean power.

Subsequent distortion as impulse noise is modeled with a BG process ($p = 0.05$). This OFDM system consists of $N = 64$ subcarriers conveying QPSK symbols. We consider a packet-based transmission, where each packet consists of a header at the beginning of the packet with a known training sequence or preamble to carry out channel estimation, followed by a certain number $L_x = 20$ of OFDM data symbols. At the preamble, there are two OFDM symbols with $N_p = 16$ pilot subcarriers with randomly generated symbols. Each OFDM symbol is appended a cyclic prefix to overcome the delay spread τ_{max} of the channel. For transmission, we have chosen a channel bandwidth of W = 2 MHz. Since we sample at the Nyquist rate, this yields a sampling interval $T_s = 1/f_s$ = 0.5μs; this means a length for the cyclic prefix L_{CP} of two samples. The total length of each OFDM symbol becomes $(N + L_{\text{CP}})$ samples. Detection is carried out with a hard-decision slicer over the equalized data. The code used to generate this data is in Listing 5.5.

```
function data=genDataOFDM(conf)

cf=conf.data;

% Training sequence (preamble)
% (orthogonality is imposed with frequency-domain symbols)
preamble=zeros(cf.Npil,cf.Lp);
for kl=1:cf.Lp
    preamble(:,kl)=exp(1i*2*pi*rand(cf.Npil,1)); % Phase reference ...
        random (PN-seq)
end

% TRANSMITTER
s_bits=round(rand(1,cf.Nb_frm));  % generate random 0/1's
s_trx=modcohe(s_bits,cf.N,cf.PC,cf.M,preamble,cf.Nsymb_frm);
                                              %modulation

% CHANNEL
[~,~,paths]=chan_SUI3(conf.NREPETS,conf.I); % generate channel taps for ...
    SUI/3 model (independent channels)
xch=conv(s_trx,paths(:,conf.I)); % convolution with the channel
r_rx_chan=xch(1:length(s_trx));  % fix same length
if cf.sirFlag        % 'normal','only_preamble','only_data'
```

```
    r_rx=channelAWGN_IMPULSIVE(r_rx_chan,...
        cf.SNR,conf.SIR,cf.p,'only_preamble',cf.Npil);
        % IMPULSIVE + AWGN noise
else
    r_rx=channelAWGN_IMPULSIVE(r_rx_chan,conf.SNR(rr),conf.
    SIR, cf.p); % IMPULSIVE + AWGN noise
end
% Preamble reception
Lpil=cf.Npil+cf.PC;
for l=1:cf.Lp % take Lp preamble OFDM-symbols and eliminate PC
    preamble_rx(:,l) = ...
        r_rx((cf.inicSync+cf.PC:cf.inicSync+cf.PC+cf.Npil-1)+
                                        ((l-1)*Lpil)).';
end
% For Equalization: ZF
inic_dat=cf.inicSync+(Lpil*cf.Lp); %init of data samples at the frame
for l=1:cf.Nsymb_frm,
    s_datos_fd(:,l) = ...
        fft(r_rx((inic_dat+cf.PC:inic_dat+cf.PC+cf.N-1)+((l-1)
                                        *cf.L)).');
end

data.Xtrain=preamble; data.Ytrain=preamble_rx;
data.Xtest=s_datos_fd; data.Ytest.input=s_bits; ...
    data.Ytest.path=paths(:,conf.I);
```

Listing 5.5 OFDM data generation (genDataOFDM.m).

Channel Estimation Algorithms

Channel estimation for the coherent demodulation of this OFDM system is performed with the SVM algorithm (see Listing 5.6). For comparison purposes, LS channel estimates in the frequency domain (after DFT demodulation) are simultaneously obtained in all cases (see Listing 5.7). In both algorithms, a DFT-based technique with zero-padding in the time domain is used to interpolate the channel's frequency response for data subcarrier positions.

```
function [predict,coefs] = SVM_OFDM(X,Y,Xtest,N,M,epsilon,gamma,C)

% Initials
[Np,Lp]=size(X);
n = (0:Np-1)';
Y = Np*Y; % Denormalizing

coefs = zeros(Np,Lp);
for i=1:Lp
    % Construct kernel matrix
    H = zeros(Np);
    for j=1:Np
        H(:,j)=X(j,i)*exp(1i*2*pi/Np*(j-1)*n);
    end
    % Train SVM and obtain coefficients
    inparams=sprintf('-s 3 -t 4 -g %f -c %f -p %f -j 1', gamma, C,...
        epsilon);
    [~,model]=SVM(H,Y(:,i),H,inparams);
    coefs(:,i) = getSVs(Y(:,i),model);
```

```
end

% Channel estimation
ch_est_fd=mean(coefs,2); % mean over Lp OFDM-symbols (only Np subcarriers)
ch_est_td=ifft(ch_est_fd); % per sub-carrier (for N subcarriers)
ch_est_td_N=[ch_est_td(1:3); zeros(N-3,1)];
predict.ch_est_fd=fft(ch_est_td_N);

% Equalization (ZF) and detection
Nsymb_frm=size(Xtest,2);
predict.output=zeros(Nsymb_frm*N*M,1);
for l=1:Nsymb_frm
    bits_simbofdm=degray(Xtest(:,l)./predict.ch_est_fd,M);
    predict.output((1:N*M)+(l-1)*N*M)=bits_simbofdm;
end
```

Listing 5.6 SVM-OFDM algorithm (SVM_OFDM.m).

```
function [predict,coefs] = LS_OFDM(X,Y,Xtest,N,M)

coefs=zeros(size(Y));
for l=1:size(Y,2)
    coefs(:,l)=fft(Y(:,l));
end

%channel estimation
ch_est_fd=mean(coefs./X,2); % mean over Lp OFDM-symbols (only Npil ...
    subcarriers)
ch_est_td=ifft(ch_est_fd); % per sub-carrier (for N subcarriers)
ch_est_td_N=[ch_est_td(1:3); zeros(N-3,1)];
predict.ch_est_fd=fft(ch_est_td_N);

% Equalization (ZF) and detection
Nsymb_frm=size(Xtest,2);
predict.output=zeros(Nsymb_frm*N*M,1);
for l=1:Nsymb_frm
    bits_symbofdm=degray(Xtest(:,l)./predict.ch_est_fd,M);
    predict.output((1:N*M)+(l-1)*N*M)=bits_symbofdm;
end
```

Listing 5.7 LS-OFDM algorithm (LS_OFDM.m).

Experiment Settings and Results

The signal-to-impulse ratio (SIR) is defined as $SIR = \mathbb{E}[r_n - z_n]/\mathbb{E}[g_n]$, and it ranged from around −21 to +21 dB. The BER and the MSE improvements that can be attained with SVM are represented in Figure 5.7. For high SIR values, a performance similar to LS can be obtained, whereas for low SIR values the SVM outperforms LS by properly choosing C, δC, and ε.

The difference in BER between SVM and LS was analyzed, given the definition of $\Delta \log(\text{BER}) = \log(\text{BER}_{\text{SVM}}) - \log(\text{BER}_{\text{LS}})$, in order to easily detect working zones where SVM works better ($\Delta \log(\text{BER}) > 0$), similar ($\Delta \log(\text{BER}) = 0$) or worse ($\Delta \log(\text{BER}) < 0$) than LS. Figure 5.8 shows that the BER performance of the SVM algorithm can be superior to LS for properly chosen values of the free parameters, and

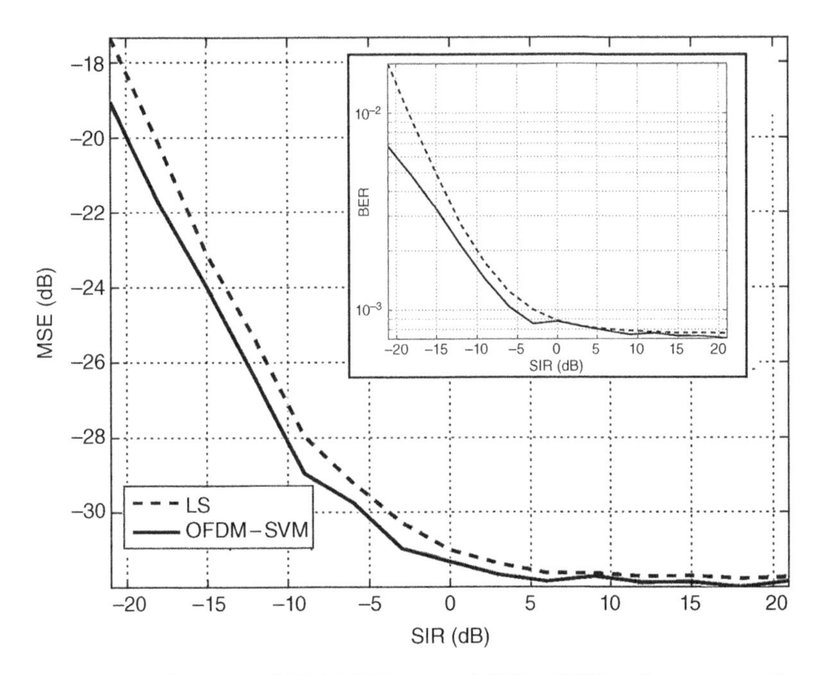

Figure 5.7 Performance of OFDM-SVM versus LS. BER and MSE performance as a function of SIR.

that there are a wide range of values for it. All the these configuration parameters are defined in the MATLAB file config_OFDM.m available in the simpleInterp toolbox. In code config_OFDM.m, BER and MSE evaluation measures are calculated as seen in Listing 5.8.

```
function values = BER_and_MSE(ypred,y)

% BER
[M,Nb_frm]=size(y.input);
values.ber=zeros(M,1);
for kt=1:M
    values.ber(kt,1)=sum(y.input(kt,:)~=ypred.output(kt,:))/
                            Nb_frm;
end

% MSE
N=length(ypred.ch_est_fd);
L=length(y.path);
% frequency domain
values.mse_fd=mean(abs(ypred.ch_est_fd-fft(y.path,N)).^2);
% temporal domain
ch_est_td=ifft(ypred.ch_est_fd);
values.mse_td=mean(abs(ch_est_td(1:L)-y.path).^2);
```

Listing 5.8 BER and MSE evaluation functions (BER_and _MSE.m).

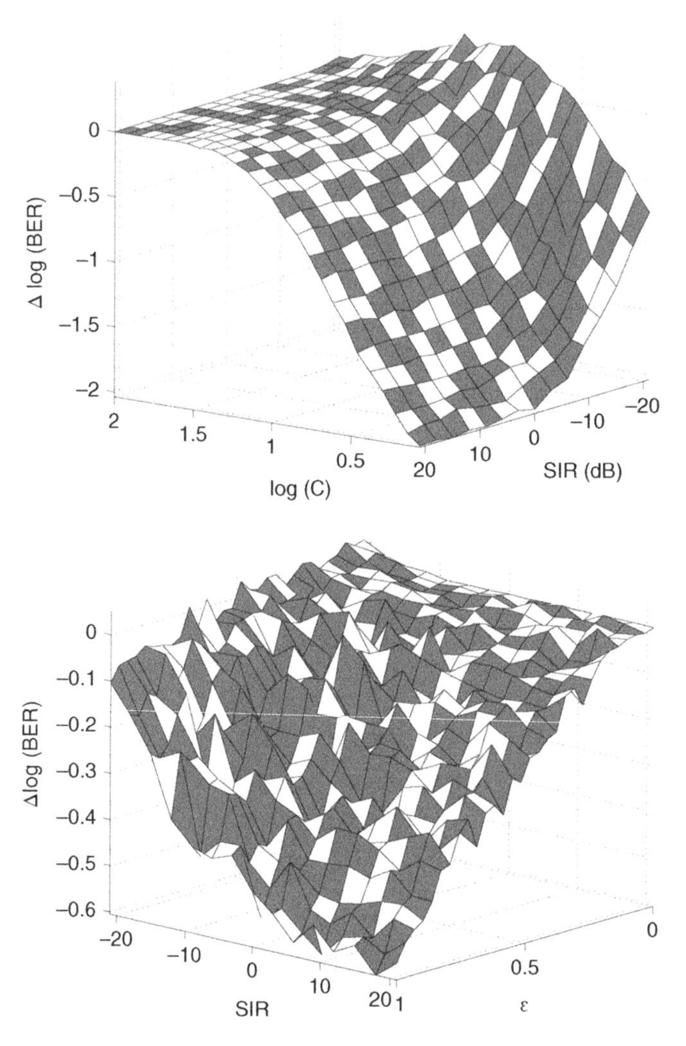

Figure 5.8 Performance of OFDM-SVM with free parameters, compared with LS in terms of Δ log(BER), for SIR versus C (up), and for SIR versus ε (down).

5.5 Questions and Problems

Exercise 5.5.1 Write the time-transverse vectors for nonparametric spectral analysis, for ARX system identification, for OFDM robust estimation, and for sinc interpolation.

Exercise 5.5.2 Show the form of the autocorrelation matrix in the dual problem for the OFDM complex formulation.

Exercise 5.5.3 Write the signal model, transverse functions, autocorrelation, and solution for a time-series expansion given by a triangular-shaped kernel.

Exercise 5.5.4 Make a spectral analysis with sunspot numbers and nonparametric spectral analysis with the PSM. Represent the residuals, the multipliers, and the correlation matrix for that problem.

Exercise 5.5.5 Make the spectral analysis now with AR decomposition of the signals. Represent the Lagrange multipliers and the coefficients. Represent graphically the matrix and study its main properties.

Exercise 5.5.6 Make an interpolation of sunspot numbers with the sinc kernel. Make the spectral representation. Represent the Lagrange multipliers, the coefficients, the matrix, and the error.

Exercise 5.5.7 Reproduce the example of Section 5.4.4 but using spatial reference. Construct a set of example signals coming from the two different desired directions of arrival, and synthesize a set of incoming signals including the interferences. Compare the results as a function of the number of example signals. Represent the spectrum of the resulting primal weight vectors.

Exercise 5.5.8 Work the code for showing the main results on the γ-filter application example, on the array with spatial reference example, and on the array with temporal reference example.

6

Reproducing Kernel Hilbert Space Models for Signal Processing

6.1 Introduction

In this chapter we introduce a set of signal models properly defined in an RKHS. The set of processing algorithms presented here are collectively termed an RSM because they share a distinct feature; namely, all of them intrinsically implement a particular signal model (like those previously described in Chapter 2) whose equations are written in the RKHS generated with kernels. For doing that, we exploit the well-known kernel trick (Schölkopf and Smola, 2002), and this is the most usual approach to kernel-based signal processing problems in the literature. Quite often one is interested in describing the signal characteristics by using a specific signal model whose performance is hampered by the (commonly strong) assumption of linearity. As we already know, the use of the theory of reproducing kernels can circumvent this problem by defining nonlinear algorithms by simply replacing dot products in the feature space by an appropriate Mercer kernel function (Figure 6.1). The most famous example of this kind of approaches is the support vector classification (SVC) algorithm, which has yielded a vast number of applications in the field of signal processing, from speech recognition to image classification.

We should note first that nonlinear SVM for DSP algorithms can be obtained from nonlinear versions of linear algorithms presented in Chapter 5. However, nonlinear SVM for DSP algorithms can be developed from two different general approaches, which are presented in this chapter and in Chapter 7. In this chapter, we will focus on a class of SVM for DSP algorithms that consists of stating the signal model of the time-series structure in the RKHS, and hence they are called RSM algorithms. Several examples of SVM-DSP in this setting are nonlinear system identification and antenna array processing. These and some other examples are summarized in this chapter, both in terms of theoretical foundations and of practical applications.

In particular, the chapter pays attention to the fundamental elements of the RSM approach. We concentrate on particular signal structures which have been previously studied. After their definition, nonlinear versions of the signal models will be developed. Specifically, the chapter is devoted to the study of:

- nonlinear ARX system identification techniques;
- nonlinear FIR and γ-filter structures;
- array processing, both with temporal and spatial reference;
- semiparametric regression (SR).

Digital Signal Processing with Kernel Methods, First Edition. José Luis Rojo-Álvarez, Manel Martínez-Ramón, Jordi Muñoz-Marí, and Gustau Camps-Valls.

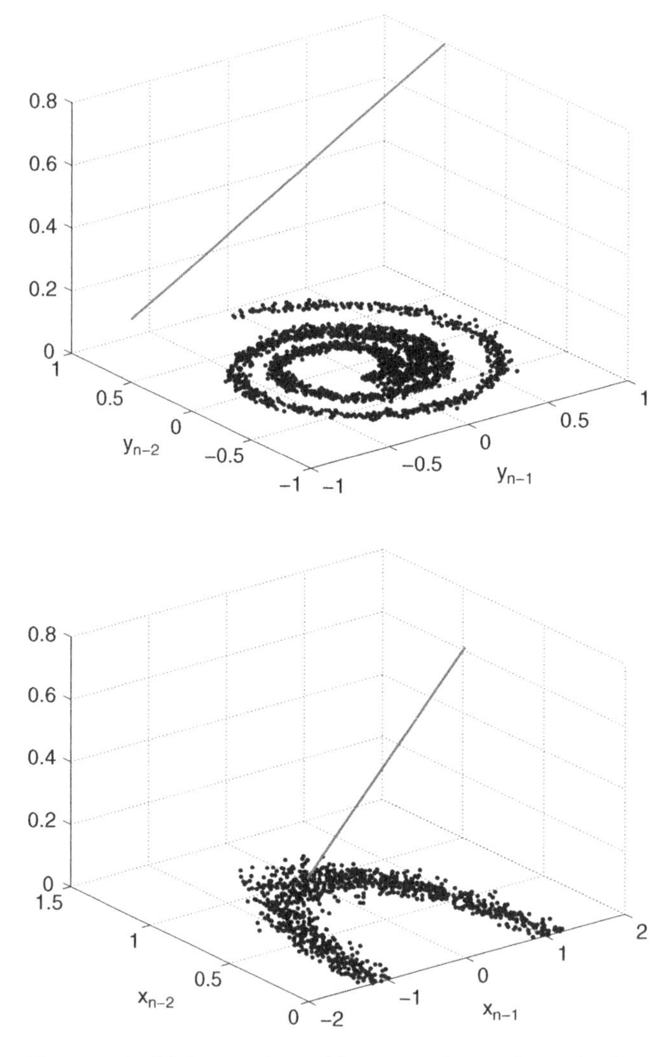

Figure 6.1 In SVM estimation problems, a nonlinear relationship between data points in the input space is transformed into a linear relationship between mapped data into the (higher dimensional) RKHS. Different signal model equations can be used in the RKHS, as far as the problem statement is expressed in terms of dot products.

Examples are used to consolidate the main concepts (nonlinear system identification), and application examples are further provided (electric network modeling and array structures from communication systems).

6.2 Reproducing Kernel Hilbert Space Signal Models

The background for stating signal models in the RKHS is well established, and it has been widely used in the kernel literature. In this section, we limit ourselves to stating a general signal model for estimation with discrete-time series notation, which will be useful for immediately summarizing the relevant elements of three SVM for DSP algorithms in this setting, following the framework in Rojo-Álvarez *et al.* (2014). On the one hand, nonlinear ARX system identification and γ-filter system identification use a signal model equation in the RKHS relating an exogenous time series and an observed data time-series. On the other hand, antenna array processing with spatial reference uses an energy expression in the RKHS, together with complex-valued algebra.

Theorem 6.2.1 (RSM problem statement) Let $\{y_n\}$ be a discrete-time series, whose signal model to be estimated can be expressed in the form of a dot product of a weight vector $\boldsymbol{a}$ and a vector observation for each time instant $\boldsymbol{v}_n$; that is:

$$\hat{y}_n = \langle \boldsymbol{a}, \boldsymbol{v}_n \rangle + b, \tag{6.1}$$

where b is a bias term that often will be convenient, given that an unbiased estimator in the input space will not necessarily correspond to an unbiased estimator when formulated in the RKHS. Then, a nonlinear signal model can be stated by transforming the weight vector and the input vectors at time instant n to an RKHS:

$$\hat{y}_n = \langle \boldsymbol{w}, \boldsymbol{\varphi}(\boldsymbol{v}_n) \rangle + b, \tag{6.2}$$

where the same signal model is now used with weight vector $\boldsymbol{w} = \sum_{n=0}^{N} \eta_n \boldsymbol{\varphi}(\boldsymbol{v}_n)$, and the solution can be expressed as

$$\hat{y}_m = \sum_{n=0}^{N} \eta_n \langle \boldsymbol{\varphi}(\boldsymbol{v}_n), \boldsymbol{\varphi}(\boldsymbol{v}_m) \rangle = \sum_{n=0}^{N} \eta_n K(\boldsymbol{v}_n, \boldsymbol{v}_m). \tag{6.3}$$

The proof is straightforward (and similar to the demonstration for deriving the PSM), by using the conventional Lagrangian functional and dual problem. This is the mostly used approach to tackle nonlinear problems with kernels in general and to tackle signal processing problems with SVM in particular. This theorem is next used to obtain the nonlinear equations for several DSP problems. Before doing that, we summarize two relevant properties that will be useful for further developments.

A concept that has been largely used in RKHS in SVM for DSP is the composite kernel, which will give flexibility for defining relationships between two observed signals (exogenous and system output), by means of an RKHS system identification model.

Property 27 (Composite summation kernel) A simple composite kernel comes from the concatenation of nonlinear transformations of $\boldsymbol{c} \in \mathbb{R}^c$ and $\boldsymbol{d} \in \mathbb{R}^d$. If we construct the transformation

$$\boldsymbol{\phi}(\boldsymbol{c}, \boldsymbol{d}) = [\boldsymbol{\phi}_1^{\mathrm{T}}(\boldsymbol{c}), \boldsymbol{\phi}_2^{\mathrm{T}}(\boldsymbol{d})]^{\mathrm{T}} \tag{6.4}$$

as the concatenation of (column) vectors $\boldsymbol{\phi}_1(\boldsymbol{c})$ and $\boldsymbol{\phi}_2(\boldsymbol{d})$ defined in $\mathcal{H}_1$ and $\mathcal{H}_2$, respectively, then the corresponding dot product between vectors is simply

$$K(\boldsymbol{c}_1, \boldsymbol{d}_1; \boldsymbol{c}_2, \boldsymbol{d}_2) = \langle \boldsymbol{\phi}(\boldsymbol{c}_1, \boldsymbol{d}_1), \boldsymbol{\phi}(\boldsymbol{c}_2, \boldsymbol{d}_2) \rangle = K_1(\boldsymbol{c}_1, \boldsymbol{c}_2) + K_2(\boldsymbol{d}_1, \boldsymbol{d}_1), \tag{6.5}$$

which is known as summation kernel.

Note that this measure of similarity between different data observations is different to the traditional *stacked approach* in the input space. The common approach when dealing with different data entities, variables, or observations, here $\boldsymbol{c}$ and $\boldsymbol{d}$, is to concatenate them in the input space and then to define a mapping function $\boldsymbol{\phi}$ and its corresponding kernel function to work with them. This approach has been widely used in time series analysis, as we will see in Section 6.2.1. By doing so, however, one loses the signal model structure.

An important second difference is that stacking vectors in feature spaces leads to considering dedicated kernels to each variable separately, whose similarity functions are then summed up. The model is limited in the sense that variable cross-relations are not considered. The summation kernel expression can be readily modified to account for the cross-information among the different variables, in our case between an exogenous and an output observed data time series.

Property 28 (Composite kernels for cross-information) Assume a nonlinear mapping $\boldsymbol{\varphi}(\cdot)$ into a Hilbert space $\mathcal{H}$ and three linear transformations $\boldsymbol{A}_i$ from $\mathcal{H}$ to $\mathcal{H}_i$, for $i = 1, 2, 3$. We construct the following composite vector:

$$\boldsymbol{\phi}(\boldsymbol{c}, \boldsymbol{d}) = [\boldsymbol{A}_1\boldsymbol{\varphi}(\boldsymbol{c}), \boldsymbol{A}_2\boldsymbol{\varphi}(\boldsymbol{d}), \boldsymbol{A}_3(\boldsymbol{\varphi}(\boldsymbol{c}) + \boldsymbol{\varphi}(\boldsymbol{d}))]^{\mathrm{T}}. \tag{6.6}$$

If the dot product is computed, we obtain

$$\begin{aligned} K(\boldsymbol{c}_1, \boldsymbol{c}_2; \boldsymbol{d}_1, \boldsymbol{d}_2) &= \boldsymbol{\varphi}^{\mathrm{T}}(\boldsymbol{c}_1)\boldsymbol{R}_1\boldsymbol{\varphi}(\boldsymbol{c}_2) + \boldsymbol{\varphi}^{\mathrm{T}}(\boldsymbol{d}_1)\boldsymbol{R}_2\boldsymbol{\varphi}(\boldsymbol{d}_2) + \boldsymbol{\varphi}^{\mathrm{T}}(\boldsymbol{c}_1)\boldsymbol{R}_3\boldsymbol{\varphi}(\boldsymbol{d}_2) \\ &+ \boldsymbol{\varphi}^{\mathrm{T}}(\boldsymbol{d}_1)\boldsymbol{R}_3\boldsymbol{\varphi}(\boldsymbol{c}_2) = K_1(\boldsymbol{c}_1, \boldsymbol{c}_2) + K_2(\boldsymbol{d}_1, \boldsymbol{d}_2) + K_3(\boldsymbol{c}_1, \boldsymbol{d}_2) + K_3(\boldsymbol{d}_1, \boldsymbol{c}_2), \end{aligned} \tag{6.7}$$

where $\boldsymbol{R}_1 = \boldsymbol{A}_1^{\mathrm{T}}\boldsymbol{A}_1 + \boldsymbol{A}_3^{\mathrm{T}}\boldsymbol{A}_3$, $\boldsymbol{R}_2 = \boldsymbol{A}_2^{\mathrm{T}}\boldsymbol{A}_2 + \boldsymbol{A}_3^{\mathrm{T}}\boldsymbol{A}_3$, and $\boldsymbol{R}_3 = \boldsymbol{A}_3^{\mathrm{T}}\boldsymbol{A}_3$ are three independent definite-positive matrices.

Note that $\boldsymbol{c}$ and $\boldsymbol{d}$ must have the same dimension for the formulation to be valid, otherwise K_3 cannot be computed.

6.2.1 Kernel Autoregressive Exogenous Identification

As we have seen in the introductory chapters, a common problem in DSP is to model a functional relationship between two simultaneously recorded discrete-time processes (Ljung, 1999). When this relationship is linear and time invariant it can be addressed by using an ARMA difference equation, and when a simultaneously observed set of signal samples $\{x_n\}$ is available, called an exogenous signal, the parametric model is called an ARX signal model for system identification.

Many approaches have been considered to tackle this important problem of ARX system identification. General nonlinear models, such as artificial neural networks, wavelets, and fuzzy models, are common and effective choices (Ljung, 1999; Nelles, 2000), though their temporal structure cannot be easily analyzed, because it remains inside a black-box model. In the last decade, enormous interest has been paid to kernel methods in general and SVM in particular in this setting. The first approaches considered the SVM version for regression, the so-called SVR method. In particular, Drezet and Harrison (1998), Goethals *et al.* (2005b), and Espinoza *et al.* (2005) used the SVR algorithm for nonlinear system identification. However, in all these studies the time series structure of the data was not scrutinized and the approach essentially consisted of stacking the signals that were then fed to the SVR. Alternatively, Rojo-Álvarez *et al.* (2004) explicitly formulated SVM for modeling linear time-invariant ARMA systems (linear SVM-ARMA), and this kind of formulation has been recently extended to a general framework for linear signal processing problems (Rojo-Álvarez *et al.*, 2005). However, if linearity cannot be assumed, then nonlinear system identification techniques are required.

We next summarize several SVM for DSP procedures for nonlinear system identification. The material has been presented in detail by Martínez-Ramón *et al.* (2006), Camps-Valls *et al.* (2009b), and Camps-Valls *et al.* (2007a), so we summarize the most relevant details here. First, the stacked SVR algorithm for nonlinear system identification is briefly examined in order to check that, though efficient, this approach does not correspond explicitly to an ARX model in the RKHS.

Let $\{x_n\}$ and $\{y_n\}$ be two discrete-time signals, which are the input and the output respectively of a nonlinear system. Let $\boldsymbol{y}_n = [y_{n-1}, y_{n-2}, \ldots, y_{n-M}]^{\mathrm{T}}$ and $\boldsymbol{x}_n = [x_n, x_{n-1}, \ldots, x_{n-Q+1}]^{\mathrm{T}}$ denote the states of input and output at time instant n. The stacked-kernel system identification algorithm (Gretton *et al.*, 2001a; Suykens *et al.*, 2001a) can be described as follows.

Property 29 (Stacked-kernel signal model in SVM for nonlinear system identification) Assuming a nonlinear transformation $\boldsymbol{\phi}([\boldsymbol{y}_n^{\mathrm{T}}, \boldsymbol{x}_n^{\mathrm{T}}])$ for the concatenation of the input and output discrete-time processes to a B-dimensional feature space, $\phi : \mathbb{R}^{M+Q} \rightarrow \mathcal{H}$, a linear regression model can be built in $\mathcal{H}$, its corresponding equation being

$$y_n = \langle \boldsymbol{w}, \boldsymbol{\phi}([\boldsymbol{y}_n^{\mathrm{T}}, \boldsymbol{x}_n^{\mathrm{T}}]) \rangle + e_n, \tag{6.8}$$

where $\boldsymbol{w}$ is a vector of coefficients in the RKHS, given by

$$\boldsymbol{w} = \sum_{n=0}^{N} \eta_n \boldsymbol{\phi}([\boldsymbol{y}_n^{\mathrm{T}}, \boldsymbol{x}_n^{\mathrm{T}}]), \tag{6.9}$$

and the following Gram matrix containing the dot products can be identified:

$$\boldsymbol{G}(m, n) = \langle \boldsymbol{\phi}([\boldsymbol{y}_m^{\mathrm{T}}, \boldsymbol{x}_m^{\mathrm{T}}]), \boldsymbol{\phi}([\boldsymbol{y}_n^{\mathrm{T}}, \boldsymbol{x}_n^{\mathrm{T}}]) = K([\boldsymbol{y}_m^{\mathrm{T}}, \boldsymbol{x}_m^{\mathrm{T}}], [\boldsymbol{y}_n^{\mathrm{T}}, \boldsymbol{x}_n^{\mathrm{T}}]), \tag{6.10}$$

where the nonlinear mappings do not need to be explicitly computed, but instead the dot product in the RKHS can be replaced by Mercer kernels. The predicted output for newly observed $[\boldsymbol{y}_m^{\mathrm{T}}, \boldsymbol{x}_m^{\mathrm{T}}]$ is given by

$$\hat{y}_m = \sum_{n=0}^{N} \eta_n K([\boldsymbol{y}_m^{\mathrm{T}}, \boldsymbol{x}_m^{\mathrm{T}}], [\boldsymbol{y}_n^{\mathrm{T}}, \boldsymbol{x}_n^{\mathrm{T}}]). \tag{6.11}$$

Note that this is the expression for a general nonlinear system identification, but it does not correspond to an ARX structure in the RKHS. Moreover, though the reported performance of the algorithm is high when compared with other approaches, this formulation does not allow us to scrutinize the statistical properties of the time series that are being modeled in terms of autocorrelation and/or cross-correlation between the input and the output time series.

Composite kernels can be introduced at this point, allowing us to next introduce a nonlinear version of the linear SVM-ARX algorithm by using actually an ARX scheme on the RSM. After noting that the cross-information between the input and the output is lost with the stacked-kernel signal model, the use of composite kernels is proposed for taking into account this information and for improving the model versatility.

Property 30 (SVM-DSP in RKHS for ARX nonlinear system identification) If we separately map the state vectors of both the input and the output discrete-time signals to $\mathcal{H}$, using a nonlinear mapping given by $\boldsymbol{\phi}_e(\boldsymbol{x}_n) : \mathbb{R}^M \to \mathcal{H}_e$ and $\boldsymbol{\phi}_d(\boldsymbol{y}_n) : \mathbb{R}^Q \to \mathcal{H}_d$, then a linear ARX model can be stated in $\mathcal{H}$, the corresponding difference equation being given by

$$y_n = \langle \boldsymbol{w}_d, \boldsymbol{\phi}_d(\boldsymbol{y}_n)\rangle + \langle \boldsymbol{w}_e, \boldsymbol{\phi}_e(\boldsymbol{x}_n)\rangle + e_n, \tag{6.12}$$

where $\boldsymbol{w}_d$ and $\boldsymbol{w}_e$ are respectively the vectors determining the AR and the MA coefficients of the system in (possibly different) RKHSs. The vector coefficients are given by

$$\boldsymbol{w}_d = \sum_{n=0}^{N} \eta_n \boldsymbol{\phi}_d(\boldsymbol{y}_n), \qquad \boldsymbol{w}_e = \sum_{n=0}^{N} \eta_n \boldsymbol{\phi}_e(\boldsymbol{x}_n). \tag{6.13}$$

Two different kernel functions can be further identified:

$$\boldsymbol{R}^y_{m,n} = \langle \boldsymbol{\phi}_d(\boldsymbol{y}_m), \boldsymbol{\phi}_d(\boldsymbol{y}_n)\rangle = K_d(\boldsymbol{y}_m, \boldsymbol{y}_n) \tag{6.14}$$

$$\boldsymbol{R}^x_{m,n} = \langle \boldsymbol{\phi}_e(\boldsymbol{x}_m), \boldsymbol{\phi}_e(\boldsymbol{x}_k)\rangle = K_e(\boldsymbol{x}_m, \boldsymbol{x}_n), \tag{6.15}$$

which account for the sample estimators of input and output time-series autocorrelation functions (Papoulis, 1991) respectively in the RKHS. Specifically, they are proportional to the non-Toeplitz estimator of each time series autocorrelation matrix.

The dual problem consists of maximizing the PSM dual problem (see Equation 5.7) with $\boldsymbol{R}^s = \boldsymbol{R}^x + \boldsymbol{R}^y$, and the output for a new observation vector is obtained as

$$\hat{y}_m = \sum_{n=n_0}^{N} \eta_n (K_d(\boldsymbol{y}_n, \boldsymbol{y}_m) + K_e(\boldsymbol{x}_n, \boldsymbol{x}_m)). \tag{6.16}$$

The kernels in the preceding equation correspond to correlation matrices computed into the direct summation of kernel spaces $\mathcal{H}_1$ and $\mathcal{H}_2$. Hence, the autocorrelation matrices' components given by $\boldsymbol{x}_n$ and $\boldsymbol{y}_n$ are expressed in their corresponding RKHS and the cross-correlation component is computed in the direct summation space. A third space can be used to compute the cross-correlation component, which introduces generality to the model.

Property 31 (Composite kernels for general cross-information in system identification) Assuming a nonlinear mapping $\boldsymbol{\phi}(\cdot)$ into a Hilbert space $\mathcal{H}$ and three linear transformations $\boldsymbol{A}_i$ from $\mathcal{H}$ to $\mathcal{H}_i$, for $i = 1, 2, 3$, we can construct the following composite vector:

$$\boldsymbol{\phi}(\boldsymbol{y}, \boldsymbol{x}) = \begin{pmatrix} \boldsymbol{A}_1\boldsymbol{\phi}(\boldsymbol{x}) \\ \boldsymbol{A}_2\boldsymbol{\phi}(\boldsymbol{y}) \\ \boldsymbol{A}_3(\boldsymbol{\phi}(\boldsymbol{x}) + \boldsymbol{\phi}(\boldsymbol{y})) \end{pmatrix}$$

According to Property 28, we have

$$\begin{aligned} K(\boldsymbol{y}_m, \boldsymbol{y}_n; \boldsymbol{x}_m, \boldsymbol{x}_n) &= \boldsymbol{\phi}^{\mathrm{T}}(\boldsymbol{y}_m)\boldsymbol{R}_1\boldsymbol{\phi}(\boldsymbol{y}_n) + \boldsymbol{\phi}^{\mathrm{T}}(\boldsymbol{y}_m)\boldsymbol{R}_2\boldsymbol{\phi}(\boldsymbol{x}_n) \\ &\quad + \boldsymbol{\phi}^{\mathrm{T}}(\boldsymbol{x}_m)\boldsymbol{R}_3\boldsymbol{\phi}(\boldsymbol{y}_n) + \boldsymbol{\phi}^{\mathrm{T}}(\boldsymbol{x}_m)\boldsymbol{R}_3\boldsymbol{\phi}(\boldsymbol{y}_n) \\ &= K_1(\boldsymbol{y}_m, \boldsymbol{y}_n) + K_2(\boldsymbol{x}_m, \boldsymbol{x}_n) + K_3(\boldsymbol{y}_m, \boldsymbol{x}_n) + K_3(\boldsymbol{x}_m, \boldsymbol{y}_n), \end{aligned}$$

where it is straightforward to identify $K_1 = K_d$ and $K_2 = K_e$.

Note that, in this case, $\boldsymbol{x}_n$ and $\boldsymbol{y}_n$ need to have the same dimension, which can be naively accomplished by zero completion of the embeddings.

Property 32 (General composite kernels) A general composite kernel, which can be obtained as a combination of the previous ones, is given by

$$\begin{aligned} K(\boldsymbol{x}_m, \boldsymbol{y}_m; \boldsymbol{y}_n, \boldsymbol{x}_n) &= K_1(\boldsymbol{y}_m, \boldsymbol{y}_n) + K_2(\boldsymbol{x}_m, \boldsymbol{x}_n) \\ &\quad + K_3(\boldsymbol{y}_m, \boldsymbol{x}_n) + K_3(\boldsymbol{x}_m, \boldsymbol{y}_n) + K_4(\boldsymbol{z}_m, \boldsymbol{z}_n). \end{aligned} \tag{6.17}$$

Therefore, despite the fact that SVM-ARX and SVR nonlinear system identifications are different problem statements, both models can be easily combined.

6.2.2 Kernel Finite Impulse Response and the γ-filter

As described in previous chapters, many NN structures with a linear memory stage followed by a non-linear memoryless stage are commonly used in signal processing, such as the time delay NN and the focused γ-network. These networks offer good performance at the expense of increasing the dimensionality of the state vector of the linear memory stage, and thus training the memoryless stage involves both high computational burden and risk of overfitting. In Camps-Valls *et al.* (2009b), a set of kernel methods are introduced in order to develop nonlinear γ-filters in a straightforward yet principled way. These RKHS algorithms for nonlinear system identification with γ-filters are summarized next.

Property 33 (Nonlinear γ-filter with the kernel trick) Let us express the SVM γ-filter PSM as

$$y_m = \sum_{n=1}^{N} \eta_n \sum_{q=0}^{Q} \langle s_n^q, s_m^q \rangle. \tag{6.18}$$

Then, the delayed line outputs $\boldsymbol{s}_n^q$ can now be transformed to an RKHS using a nonlinear mapping $\boldsymbol{\phi}$. The autocorrelation function is $\boldsymbol{R}^s_{n,m} := \sum_{q=0}^{Q} \langle \boldsymbol{\phi}(\boldsymbol{s}_n^q), \boldsymbol{\phi}(\boldsymbol{s}_m^q) \rangle$, and the solution becomes inherently nonlinear:

$$\hat{y}_m = \sum_{n=1}^{N} \eta_n \sum_{q=0}^{Q} \langle \boldsymbol{\phi}(\boldsymbol{s}_n^q), \boldsymbol{\phi}(\boldsymbol{s}_m^q) \rangle = \sum_{n=1}^{N} \eta_n \sum_{q=0}^{Q} K(\boldsymbol{s}_m^q, \boldsymbol{s}_n^q). \tag{6.19}$$

This property can be readily used to derive the nonlinear γ-filters with composite kernels and with tensor-product kernels, as detailed and benchmarked in Camps-Valls *et al.* (2009b).

6.2.3 Kernel Array Processing with Spatial Reference

An antenna array is a group of (usually identical) electromagnetic radiator elements placed in different positions of the space. This way, an electromagnetic flat wave illuminating the array produces currents that have different amplitudes and phases depending on the DOA of the wave and of the position of each radiating element. The discrete-time signals collected from the array elements can be seen as a time and space discrete process.

The fundamental property of the array is that it is able to detect the DOA of one or several incoming signals or it is able to discriminate one among various incoming signals provided they have different DOAs (Van Trees, 2002). Applications of antenna array processing range from radar systems (which minimize the mechanical components by electronic positioning of the array radiation beam), to communication systems (in which the system capacity is increased by the use of spatial diversity), and to radioastronomy imaging systems, among many others.

The array processing problem stated in Equation 3.51 can be solved when there are no training symbols available, but just a set of incoming data and information about the angle of arrival of the desired user. In this case, the algorithm to be applied consists of a processor that detects without distortion (distortionless property) the signal from the desired DOA while minimizing the total output energy. The signal can be easily mapped to an RKHS, and the algorithm must optimize

$$\min_{\boldsymbol{w}} \{\boldsymbol{w}^{\mathrm{H}} \boldsymbol{\phi}(\boldsymbol{x}_n) \boldsymbol{\phi}(\boldsymbol{x}_n)^{\mathrm{H}} \boldsymbol{w}\} = \min_{\boldsymbol{w}} \{\boldsymbol{w}^{\mathrm{H}} \boldsymbol{R} \boldsymbol{w}\} \approx \min_{\boldsymbol{w}} \{\boldsymbol{w}^{\mathrm{H}} \boldsymbol{\Phi}\boldsymbol{\Phi}^{\mathrm{H}} \boldsymbol{w}\} \tag{6.20}$$

for a given set of previously collected snapshots, and $\boldsymbol{\Phi}$ is a matrix containing all mapped snapshots $\boldsymbol{\phi}(\boldsymbol{x}_n)$.

Property 34 (Spatial reference signal model in RKHS) In order to introduce the aforementioned distortionless property, constraints must be applied to a set of canonical signals whose steering vector in Equation 3.49 contains the desired direction of arrival θ_0, carrying a set of symbols b_i. These are the spatial reference signals. The reference signal model is then

$$b_i = \boldsymbol{w}^{\mathrm{H}} \boldsymbol{\phi}(b_i \boldsymbol{a}_0) - b, \tag{6.21}$$

where $\boldsymbol{a}_0$ is the steering vector corresponding to the desired signal. Then, a primal functional must contain the following constraints:

$$\begin{aligned} \mathrm{Re}\{b_i - \boldsymbol{w}^{\mathrm{H}}\boldsymbol{\phi}(b_i\boldsymbol{a}_0) - b\} &\leq \varepsilon + \xi_i \\ -\mathrm{Re}\{b_i - \boldsymbol{w}^{\mathrm{H}}\boldsymbol{\phi}(b_i\boldsymbol{a}_0) - b\} &\leq \varepsilon + \xi_i' \\ \mathrm{Im}\{b_i - \boldsymbol{w}^{\mathrm{H}}\boldsymbol{\phi}(b_i\boldsymbol{a}_0) - b\} &\leq \varepsilon + \zeta_i \\ -\mathrm{Im}\{b_i - \boldsymbol{w}^{\mathrm{H}}\boldsymbol{\phi}(b_i\boldsymbol{a}_0) - b\} &\leq \varepsilon + \zeta_i' \end{aligned} \quad (6.22)$$

with s_i being all possible transmitted symbols in a given amplitude range, and ξ_i, ζ_i, ξ_i', and ζ_i' being the slack variables corresponding to the real and imaginary constraints.

Property 35 (Spatial reference primal coefficients) An SVM procedure applied to this constrained optimization problem must include the minimization of Equation 6.21, and it gives the solution

$$\boldsymbol{w} = \sum_i \boldsymbol{R}^{-1}\boldsymbol{\phi}(b_i\boldsymbol{a}_0)\psi_i. \quad (6.23)$$

Property 36 (Spatial reference kernel) The application of Equation 6.23 in Equation 6.21 implicitly gives the kernels

$$K(b_i\boldsymbol{a}_0, b_j\boldsymbol{a}_0) = \boldsymbol{\phi}(b_i\boldsymbol{a}_0)^{\mathrm{T}}\boldsymbol{R}^{-1}\boldsymbol{\phi}(b_j\boldsymbol{a}_0). \quad (6.24)$$

These kernels cannot be directly used because an expression for $\boldsymbol{R}$ is not available in an infinite-dimension RKHS. A kernel eigenanalysis introduced by Schölkopf *et al.* (1998) leads to the kernel expression

$$K(b_i\boldsymbol{a}_0, b_j\boldsymbol{a}_0) = N\boldsymbol{\phi}(b_i\boldsymbol{a}_0)^{\mathrm{T}}\boldsymbol{\Phi}\boldsymbol{K}^{-1}\boldsymbol{K}_0\boldsymbol{\Phi}^{\mathrm{T}}\boldsymbol{\phi}(b_j\boldsymbol{a}_0), \quad (6.25)$$

where $\boldsymbol{\Phi}$ is a matrix containing all the incoming data used to compute the autocorrelation matrix $\boldsymbol{R}$, and $\boldsymbol{K}_0$ is a kernel matrix containing all dot products $\boldsymbol{\phi}(s_n^0\boldsymbol{a}_0)^{\mathrm{T}}\boldsymbol{\phi}(s_m^0\boldsymbol{a}_0)$. These kernels can be used to solve a dual problem equal to the one of Property 24. The primal coefficients can be expressed as

$$\boldsymbol{w} = N\boldsymbol{\Phi}\boldsymbol{K}^{-1}\boldsymbol{K}_0\boldsymbol{\psi}, \quad (6.26)$$

where $\psi_n = \eta_n + \mathbb{i}\nu_n$ are complex-valued dual coefficients.

6.2.4 Kernel Semiparametric Regression

SR has been a widely studied topic in conventional statistics. It supports the idea that some phenomena under analysis can be represented with parametric models, especially using linear regression, and by nonparametric models, when the explicit relationship between the input and the output turns to be more complicated to know. Given that the knowledge path can be thought of as usually going from nonparametric to parametric models, SR combines both approaches. The most widely known method for SR is the

Nadayara–Watson (NW) estimator. In this section, we start defining the fundamentals of this classical estimator, and then we show how it can be readily expressed in terms of a composite kernel in the RSM framework.

In addition, we introduce in this chapter the use of bootstrap resampling techniques (BRTs) in SVM for RSM models. We focus on the role that it has played for model diagnosis and for analyzing the statistics of the model parameters, specially for SR. This section paves the way toward the use of these popular nonparametric methods for working with confidence intervals and for establishing statistical tests in this setting.

Nadayara–Watson Estimator for Semiparametric Regression

Let y_i be the ith observation of a response variable and $\boldsymbol{x}_i = [1, x_i^1, \dots, x_i^K]^{\mathrm{T}}$ be the K-dimensional ith vector containing the observed predictor variables ($i = 1, \dots, N$), where T denotes the transposed vector. The constant unit level is introduced to take into account the interception component.

The simplest parametric regression model is the general linear model:

$$\mathbb{E}[y_i|\boldsymbol{x}_i] = \sum_{k=0}^{K} \beta^k x_i^k = \langle \boldsymbol{\beta}, \boldsymbol{x}_i \rangle, \tag{6.27}$$

where $\mathbb{E}[\cdot|\cdot]$ denotes conditional statistical expectation, and $\langle \cdot, \cdot \rangle$ is the dot product. Model coefficients $\beta^1, \dots, \beta^K$, and interception β_0, are estimated by ordinary LS.

Nonparametric kernel regression (Ruppert *et al.*, 2003) computes a local weighted average of the criterion variable given the values of the predictors; that is:

$$\mathbb{E}[y_i|\boldsymbol{x}_i] = m(\boldsymbol{x}_i), \tag{6.28}$$

where $m(\cdot)$ is a nonparametric function. For instance, the NW estimator consists of a constant-kernel approximation given by

$$m(\boldsymbol{x}_i) = \sum_{t=1}^{N} w_t(\boldsymbol{x}_i) y_i, \tag{6.29}$$

with

$$w_t(\boldsymbol{x}_i) = \frac{K\left(\frac{\boldsymbol{x}_i - \boldsymbol{x}_t}{\sigma}\right)}{\sum_{s=1}^{N} K\left(\frac{\boldsymbol{x}_i - \boldsymbol{x}_s}{\sigma}\right)} \tag{6.30}$$

and

$$K(\boldsymbol{x}) = (2\pi)^{K/2} \prod_{i=0}^{K} \exp\left(\frac{-x_i^2}{2}\right). \tag{6.31}$$

Parameter σ is called the *bandwidth*, and it represents the neighborhood of influence for each observation. A smaller bandwidth will give a lower bias but increased variance in the estimator, whereas a larger bandwidth will produce higher bias (model mismatch) despite a reduced variance estimator.

The SR model can be very useful when mixed nature predictor variables are available (for instance, metric and dichotomic), because different components can be modeling the contribution of each subset of variables. Also, sometimes we have some a priori knowledge about the model that we can assume for a subset of variables, but the remaining subset has a completely unknown nature. Without loss of generality, let us assume observation vectors that are composed of D dichotomic variables and M metric variables, $\boldsymbol{x}_i = [\boldsymbol{x}_i^{m\mathrm{T}}, \boldsymbol{x}_i^{d\mathrm{T}}]^{\mathrm{T}}$. Let us use a parametric, linear component for the dichotomic variables, and a nonparametric, nonlinear component for the metric variables. The corresponding SR model is

$$\mathbb{E}[y_i|\boldsymbol{x}_i] = m(\boldsymbol{x}_i^m) + \langle \boldsymbol{\beta}, \boldsymbol{x}_i^d \rangle. \tag{6.32}$$

The estimation method is described in Heerde *et al.* (2001) and Lee and Nelder (1996), and it is briefly presented here. The model can be written down as

$$y_i = m(\boldsymbol{x}_i^m) + \langle \boldsymbol{\beta}, \boldsymbol{x}_i^d \rangle + e_i, \tag{6.33}$$

where e_i denotes the ith residual. By taking the conditional average with respect to the parametric variables, we obtain

$$\mathbb{E}[y_i|\boldsymbol{x}^m] = m(\boldsymbol{x}_i^m) + \langle \boldsymbol{\beta}, \mathbb{E}[\boldsymbol{x}_i^d|\boldsymbol{x}^m] \rangle. \tag{6.34}$$

Then, by subtracting Equations 6.33 and 6.34, we have

$$y_i = \mathbb{E}[y_i|\boldsymbol{x}^m] + \langle \boldsymbol{\beta}, (\boldsymbol{x}_i^d - \mathbb{E}[\boldsymbol{x}_i^d|\boldsymbol{x}^m]) \rangle + e_i. \tag{6.35}$$

Therefore, a three-step procedure can be stated. First, averages conditional to the nonparametric variables are estimated for the response variable and for the parametric variables, given by

$$\hat{\mathbb{E}}[y_i|\boldsymbol{x}^m] = \frac{\sum_{j\neq i} K[(\boldsymbol{x}_j^m - \boldsymbol{x}_i^m)/\sigma_1] y_j}{\sum_{j\neq i} K((\boldsymbol{x}_j^m - \boldsymbol{x}_i^m)/\sigma_1)}, \tag{6.36}$$

$$\hat{\mathbb{E}}[\boldsymbol{x}_i^d|\boldsymbol{x}^m] = \frac{\sum_{j\neq i} K((\boldsymbol{x}_j^m - \boldsymbol{x}_i^m)/\sigma_2) \boldsymbol{x}_j^d}{\sum_{j\neq i} K[(\boldsymbol{x}_j^m - \boldsymbol{x}_i^m)/\sigma_2]}. \tag{6.37}$$

Note that bandwidths σ_1 and σ_2 must be properly chosen for the statistical estimation. Second, the new response variable $\tilde{y} = y - \hat{\mathbb{E}}[y|\boldsymbol{x}^m]$ and the new predictor variables $\tilde{\boldsymbol{x}}^d = \boldsymbol{x}^d - \hat{\mathbb{E}}[\boldsymbol{x}^d|\boldsymbol{x}^m]$ are used to find $\tilde{\boldsymbol{\beta}}$ by solving

$$\tilde{y}_i = \langle \tilde{\boldsymbol{\beta}}, \tilde{\boldsymbol{x}}_i^d \rangle + e_i \tag{6.38}$$

by means of ordinary LS. Note that the intercept term disappears at this step because of the mean subtraction, so that it remains inside the nonparametric component. Third, the nonparametric component is obtained by nonparametric regression on $\tilde{y}$; that is:

$$\hat{m}(\boldsymbol{x}_i^m) = \frac{\sum_{j=1}^N K((\boldsymbol{x}_i^m - \boldsymbol{x}_j^m)/\sigma_3)\tilde{y}_j}{\sum_{j=1}^N K[(\boldsymbol{x}_i^m - \boldsymbol{x}_j^m)/\sigma_3]}, \tag{6.39}$$

where bandwidth σ_3 must be also properly fixed. The choice of three different bandwidths must be addressed, with two of them coming from a nonparametric estimation of the *pdf* and the third one coming from the nonparametric regression process. Several bandwidth selection techniques are available, among which are Lee's rule of thumb, cross-validation, or subjective approach (Heerde *et al.* 2001; Lee and Nelder 1996).

Some limitations of the procedure could be present according to this formulation:

- Automatic selection procedures can deteriorate when low-sized data sets are analyzed.
- The denominators in Equations 6.36, 6.37, and 6.39 can become very small for a newly tested point that is far enough from the observations, and this produces an unbounded predicted output. Again, this situation can be more present when reduced data sets are under analysis. A solution for this drawback can be the introduction of a small threshold parameter in the exponential exponent, which is, in fact, a regularization procedure. This threshold should be also found as a free parameter.
- Finally, all the available observations are used for building the solution in Equation 6.39, independently of their adequacy or noise level. This solution will not be operational for studies with high number of observations.

These limitations can be alleviated by the SVM formulation for SR, which is presented next.

Semiparametric Regression Approach using the Support Vector Machine

An alternative formulation of SR can be stated by using the RSM framework. Let us assume that the model can be expressed with two additive contributions: one from the parametric and another from the nonparametric predictor variables. Let us assume also that there exists a possibly nonlinear transformation of the metric predictor variables into a higher (possibly infinite) dimensionality space, $\boldsymbol{\varphi}(\boldsymbol{x}^m) : \mathbb{R}^M \to \mathcal{F}_m$, where $\mathcal{F}_m$ is known as the *feature space* for the metric model component. The main property of this nonlinear transformation is that a linear regression operator can be found in the feature space, given by vector $\boldsymbol{w}^m \in \mathcal{F}_m$. Let us finally assume that, for the dichotomic model component, there exists another transformation of the dichotomic vectors $\boldsymbol{\phi}(\boldsymbol{x}^d) : \mathbb{R}^D \to \mathcal{F}_d$ to a different feature space $\mathcal{F}_d$ where another linear regression operator $\boldsymbol{v}^d \in \mathcal{F}_d$ can be properly adjusted.

In these conditions, the joint regression model is given by

$$y_i = \langle \boldsymbol{w}^m, \boldsymbol{\varphi}(\boldsymbol{x}_i^m)\rangle + \langle \boldsymbol{v}^d, \boldsymbol{\phi}(\boldsymbol{x}_i^d)\rangle + b_{\mathrm{r}} + e_i. \tag{6.40}$$

Here, the reference interception term b_{r} can be previously fixed, in order to make easy the comparison to a given level. For instance, b_{r} could be the average sales level constrained to nonpromotional periods, so that the resulting model will provide us with information about the predictors either increasing or decreasing that level.

The SVM-SR algorithm is stated as the minimization of the ϵ-Huber cost plus a regularization term given by the ℓ_2 norm of the weights in the feature spaces; that is:

$$\frac{1}{2}(\|\boldsymbol{w}^m\|^2 + \|\boldsymbol{v}^d\|^2) + \frac{1}{2\gamma}\sum_{i\in I_1}(\xi_i^2 + \xi_i^{*2}) + C\sum_{i\in I_2}(\xi_i + \xi_i^*) - \sum_{i\in I_2}\frac{\gamma C^2}{2} \tag{6.41}$$

constrained to

$$y_i - \langle \boldsymbol{w}^m, \boldsymbol{\varphi}(\boldsymbol{x}_i^m)\rangle - \langle \boldsymbol{v}^d, \boldsymbol{\phi}(\boldsymbol{x}_i^d)\rangle - b_r \le \varepsilon + \xi_i, \tag{6.42}$$

$$-y_i + \langle \boldsymbol{w}^m, \boldsymbol{\varphi}(\boldsymbol{x}_i^m)\rangle + \langle \boldsymbol{v}^d, \boldsymbol{\phi}(\boldsymbol{x}_i^d)\rangle + b_r \le \varepsilon + \xi_i^*, \tag{6.43}$$

$$\xi_i, \xi_i^* \ge 0, \tag{6.44}$$

where ξ_i and ξ_i^* (in the following, denoted jointly by $\xi_i^{(*)}$) are the slack variables used to account for the residuals in the model; I_1 is the set of samples for which $\varepsilon \le \xi_i^{(*)} \le e_C$, and I_2 is the set of samples for which $\xi_i^{(*)} > e_C$. Following the usual SVM formulation methodology, Lagrangian functional $\mathcal{L}$ can be written down, and by making zero its gradient with respect to the primal variables we obtain

$$\nabla_{\boldsymbol{w}^m}\mathcal{L} = 0 \Rightarrow \boldsymbol{w}^m = \sum_{i=1}^{N}(\alpha_i - \alpha_i^*)\boldsymbol{\varphi}(\boldsymbol{x}_i^m), \tag{6.45}$$

$$\nabla_{\boldsymbol{v}^d}\mathcal{L} = 0 \Rightarrow \boldsymbol{v}^d = \sum_{i=1}^{N}(\alpha_i - \alpha_i^*)\boldsymbol{\phi}(\boldsymbol{x}_i^d), \tag{6.46}$$

$$\frac{\partial \mathcal{L}}{\partial \xi_i^{(*)}} = 0 \Rightarrow 0 \le \alpha_i^{(*)} \le C, \tag{6.47}$$

with α_i and α_i^* denoting the Lagrange multipliers that correspond to Equation 6.45 and Equation 6.46 respectively. Matrix notation is introduced as follows:

$$\boldsymbol{y} = [y_1, \dots, y_N]^{\mathrm{T}}, \tag{6.48}$$

$$\boldsymbol{\alpha}^{(*)} = [\alpha_1^{(*)}, \dots, \alpha_N^{(*)}]^{\mathrm{T}}, \tag{6.49}$$

$$M_{ij} = \langle \boldsymbol{\varphi}(\boldsymbol{x}_i^m), \boldsymbol{\varphi}(\boldsymbol{x}_j^m)\rangle, \tag{6.50}$$

$$D_{ij} = \langle \boldsymbol{\phi}(\boldsymbol{x}_i^d), \boldsymbol{\phi}(\boldsymbol{x}_j^d)\rangle. \tag{6.51}$$

Finally, the dual problem can be stated (Rojo-Álvarez *et al.*, 2004) as the maximization of

$$\begin{aligned} &-\frac{1}{2}(\boldsymbol{\alpha} - \boldsymbol{\alpha}^*)^{\mathrm{T}}(\boldsymbol{M} + \boldsymbol{D} + \gamma \boldsymbol{I})(\boldsymbol{\alpha} - \boldsymbol{\alpha}^*) \\ &+(\boldsymbol{\alpha} - \boldsymbol{\alpha}^*)^{\mathrm{T}}(\boldsymbol{y} - b_r) - \varepsilon(\boldsymbol{\alpha} + \boldsymbol{\alpha}^*)^{\mathrm{T}}\mathbf{1} \end{aligned} \tag{6.52}$$

constrained to Equation 6.47, with respect to dual variables $\alpha_i^{(*)}$. After this quadratic programming problem is solved, and according to Equations 7.29, 6.45, and 6.46, the final expression of the solution can be easily shown to be given by the following expression:

$$\hat{y} = \sum_{i=1}^{N} \eta_i (\langle \boldsymbol{\varphi}(\boldsymbol{x}_i^m), \boldsymbol{\varphi}(\boldsymbol{x}^m) \rangle + \langle \boldsymbol{\phi}(\boldsymbol{x}_i^d), \boldsymbol{\phi}(\boldsymbol{x}^d) \rangle) + b_r, \tag{6.53}$$

where $\eta_i = (\alpha_i - \alpha_i^*)$. Note that the estimated response variable $\hat{y}$ is calculated from a weighted expansion of dot products in the feature spaces. With this expression for the solution, the explicit calculation of $\boldsymbol{w}^m$ and $\boldsymbol{v}^d$ is not required.

Here, we are mainly concerned about two properties of Mercer kernels: (1) the sum of two Mercer kernels is a Mercer kernel; and (2) the product of a Mercer kernel times a positive constant is a Mercer kernel. These simple properties allows us to propose the use of a scaled linear kernel that generates the parametric component:

$$\langle \boldsymbol{\phi}(\boldsymbol{x}_i^d), \boldsymbol{\phi}(\boldsymbol{x}_j^d) \rangle = K_l(\boldsymbol{x}_i^m, \boldsymbol{x}_j^d) = \delta \langle \boldsymbol{x}_i^d, \boldsymbol{x}_j^d \rangle, \tag{6.54}$$

with $\delta \in \mathbb{R}^+$, plus a nonlinear kernel that generates the nonparametric component:

$$\langle \boldsymbol{\varphi}(\boldsymbol{x}_i^m), \boldsymbol{\varphi}(\boldsymbol{x}_j^m) \rangle = K_{nl}(\boldsymbol{x}_i^m, \boldsymbol{x}_j^m). \tag{6.55}$$

Constant δ can be chosen for giving a balance between the parametric and nonparametric components. Thus, the final solution of SVM-SR can be readily expressed as

$$\hat{y} = \sum_{i=1}^{N} \eta_i K_{nl}(\boldsymbol{x}_i^m, \boldsymbol{x}^m) + \delta \langle \boldsymbol{\beta}, \boldsymbol{x}^d \rangle + b_r. \tag{6.56}$$

Taking Equations 6.53 and 6.56 into account, coefficients η_i determine completely both the parametric and the nonparametric components.

Bootstrap Resampling for Model Diagnosis

One of the main limitations of current SR methods is the difficulty in establishing clear cut-off tests for the nonparametric variables of the model, and much effort is being done in this framework (Ruppert *et al.*, 2003). Also, this aspect has not yet been completely solved in the SVM literature, and systematic procedures for establishing feature selection, significance levels, and confidence intervals (CIs) for model diagnosis have been developed (Lal *et al.*, 2004). An interesting approach to the model diagnosis and feature selection issues in SVM for SR can be given by BRTs, which were first proposed as possibly nonparametric procedures for estimating the *pdf* of an estimator from a limited, but informative enough, set of observations (Efron and Tibshirani, 1998). BRTs have been successfully used before for fixing the free parameters of SVM classifiers (Rojo-Álvarez *et al.*, 2002) and as a feature selection strategy using the robust SVM linear maximum margin classifier (Soguero-Ruiz *et al.*, 2014a,b). We propose here to extend their use to model diagnosis and free parameter selection for SVM problems with the SR algorithm.

For a given set of N observations $\boldsymbol{v}$, the dependence between the predictor variables and the response variable can be fully described by using their joint distribution:

$$p_{y,\boldsymbol{x}}(\boldsymbol{x}^m, \boldsymbol{x}^d, y) \mapsto \boldsymbol{v} = \{(\boldsymbol{x}_i^m, \boldsymbol{x}_i^d, y_i) | i = 1, \ldots, N\}. \tag{6.57}$$

In order to obtain the SVM-SR model, Equation 7.33 is maximized. This estimation process is denoted by operator $s(\cdot)$, and it depends on observations $\boldsymbol{v}$, and on the free parameters of the model that have been fixed a priori. Those free parameters can be grouped in a vector $\boldsymbol{\theta}$, that consists of the ε-Huber cost parameters and of the kernel-related parameters; that is, for the Gaussian kernel, $\boldsymbol{\theta} = \{\varepsilon, C, \gamma, \delta, \sigma\}$. The SVM-SR Lagrange multipliers obtained by using the observations and a given a priori fixed $\boldsymbol{\theta}$ are

$$\boldsymbol{\eta} = s(\boldsymbol{v}, \boldsymbol{\theta}). \tag{6.58}$$

The model performance can be measured with the *empirical risk*, which can be defined as a merit figure of the model that is evaluated at the observations used for building the model. It can be expressed as

$$\hat{R}_{\text{emp}} = t(\boldsymbol{\eta}, \boldsymbol{v}), \tag{6.59}$$

where $t(\cdot)$ represents an operator that stands for the empirical risk calculation. Two usual merit figures for the model are the coefficient of determination R^2 and the RMSE:

$$R^2 = \frac{\sum_{i=1}^{N} (y_i - m_y)(\hat{y}_i - m_{\hat{y}})}{\sqrt{\sum_{i=1}^{N} (y_i - m_y)^2 \sum_{i=1}^{N} (y_i - m_{\hat{y}})^2}}, \tag{6.60}$$

$$\text{RMSE} = \frac{1}{N} \sqrt{\sum_{i=1}^{N} (y_i - \hat{y}_i)^2}, \tag{6.61}$$

where m_y and $m_{\hat{y}}$ are the averages of the observations and of the model predicted response respectively. Note that ρ should be as small (close to zero) as possible, whereas R^2 should be as high (close to +1) as possible. As merit figures are random variables that depend on the observations, they are more accurately described in terms of CIs, which can be denoted by

$$P_{R^2}(R_\text{l}^2 \leq R^2 \leq R_\text{u}^2 | y, \boldsymbol{x}) \geq q, \tag{6.62}$$

$$P_\rho(\rho_\text{l} \leq \rho^2 \leq \rho_\text{u}^2 | y, \boldsymbol{x}) \geq q, \tag{6.63}$$

where $P_{R^2}(\cdot)$ and $P_\rho(\cdot)$ are the *pdfs* of each merit figure; $q \in (0, 1)$ is the confidence level; and subscripts "l" and "u" are the lower and upper limits respectively of the CI.

Given that the SVM-SR model does not rely on any a priori distribution of the data, it is not easy to know the functional form of the *pdf* of the merit figures. Moreover, the sample merit figures' estimators can be optimistically biased, especially for some degenerate choices of the free parameters; for instance, when too much emphasis is put on the cost of the residuals or when a too small bandwidth is used. Therefore, the empirical risk criterion is not a good criterion for fixing $\boldsymbol{\theta}$. Cross-validation can be useful in some cases, but it requires one to reduce the training set, which can lead to important information loss in the model when a low number of observations are available. A method for estimating the joint *pdf* of the observations, necessary for a statistical characterization of the SVM-SR model, is given by a BRT, and it can compensate the optimistic bias in the merit figures' estimators (Rojo-Álvarez *et al.*, 2002).

A *bootstrap resample* is a data subset that is drawn from the observation set according to their empirical distribution $\hat{p}_{y,\boldsymbol{x}}(\boldsymbol{x}^m, \boldsymbol{x}^d, y)$. Hence, the true *pdf* is approximated by the empirical *pdf* of the observations, and the bootstrap resample can be seen as a sampling with replacement process of the observed data; that is:

$$\hat{p}_{y,\boldsymbol{x}}(\boldsymbol{x}^m, \boldsymbol{x}^d, y) \mapsto \boldsymbol{v}^* = \{(\boldsymbol{x}_i^{m*}, \boldsymbol{x}_i^{d*}, y_i^*)|i = 1, \dots, N\}, \tag{6.64}$$

where superscript $*$ represents, in general, any observation, functional, or estimator that arises from the bootstrap resampling process. Therefore, resample $\boldsymbol{v}^*$ contains elements of $\boldsymbol{v}$ appearing none, one, or several times. The resampling process is repeated for $b = 1, \dots, B$ times.

A partition of $\boldsymbol{v}$ in terms of resample $\boldsymbol{v}^*(b)$ is given by

$$\boldsymbol{v} = \{\boldsymbol{v}_{\text{in}}^*(b), \boldsymbol{v}_{\text{out}}^*(b)\}, \tag{6.65}$$

where $\boldsymbol{v}_{\text{in}}^*(b)$ is the subset of observations that are included in resample b, and $\boldsymbol{v}_{\text{out}}^*(b)$ is the subset of nonincluded observations. The SVM-SR coefficients for resample (b) are obtained by

$$\boldsymbol{\eta}^*(b) = s(\boldsymbol{v}_{in}^*(b), \boldsymbol{\theta}). \tag{6.66}$$

A *bootstrap replication* of an estimator is given by its calculation constrained to the observations included in the bootstrap resample. The bootstrap replication of the empirical risk estimator is

$$\hat{R}_{\text{emp}}^*(b) = t(\boldsymbol{\eta}^*(b), \boldsymbol{v}_{in}^*(b)). \tag{6.67}$$

The scaled histogram obtained from B resamples is an approximation to the pfd of the empirical risk. However, further advantage can be obtained by calculating the bootstrap replication of the risk estimator on the nonincluded observations. By doing so, rather than estimating the empirical risk, we are in fact obtaining the replication of the *actual risk*; that is:

$$\hat{R}_{\text{act}}^*(b) = t(\boldsymbol{\eta}^*(b), \boldsymbol{v}_{\text{out}}^*(b)). \tag{6.68}$$

The bootstrap replication of the averaged actual risk can be obtained by just taking the average of $\hat{R}_{\text{act}}^*(b)$ for $b = 1, \dots, B$. Simple search strategies can be used for finding the free parameters that minimize the averaged actual risk (Rojo-Álvarez *et al.*, 2003; Soguero-Ruiz *et al.*, 2014b). Moreover, the replications of the *pdf* of the model merit figures can provide CIs for the performance, by fulfilling

$$P^*_{R^2}(R^{2*}_{\mathrm{l}} \le R^{2*} \le R^{2*}_{\mathrm{u}} | y^*, \boldsymbol{x}^*) \ge q, \tag{6.69}$$

$$P^*_{\rho}(\rho^*_{\mathrm{l}} \le \rho^* \le \rho^*_{\mathrm{u}} | y^*, \boldsymbol{x}^*) \ge q. \tag{6.70}$$

A typical range for B in practical applications can be from 50 to 500 bootstrap resamples. Other useful statistical characterizations can be readily achieved, for instance, CIs for the response variable, CIs for the parametric coefficients, and significance levels for the inclusion of nonparametric variables.

Confidence Intervals for the Response Variable

Frequently, it is not enough to report the predicted response variable, but it is also convenient to characterize the uncertainty on this prediction. This can be done by reporting the CI for the average of each prediction, if no strong statistical dependence is present in the response, or by reporting confidence bands, if output samples are strongly dependent. For the first case, the CI for each average output level can be readily obtained by calculating the *pdf* of the replications for each response variable in Equation 6.56 given by model in Equation 6.66 as

$$P^*_{y_i}(y^*_{i,l} \le y^*_i \le y^*_{i,u} | y, \boldsymbol{x}) \ge q, \tag{6.71}$$

where P_{y_i} denotes the *pdf* of the ith observation of the response, $i = 1, \ldots, N$. When statistical independence of the response variable can no longer be assumed, confidence bands should be used instead of CIs (Politis *et al.*, 1992). Prediction intervals for the output level can also be calculated.

Inference estimators can be obtained for each of the kth parametric coefficients of Equation 6.56 by obtaining the bootstrap replications of the parametric coefficients in each model in Equation 6.66 as follows:

$$P^*_{\beta^k}(\beta^{k*}_{l} \le \beta^{k*} \le \beta^{k*}_{u} | y, \boldsymbol{x}) \ge q, \tag{6.72}$$

where P_{β^k} denotes the bootstrap estimated *pdf* of the kth parametric coefficient, $k = 1, \ldots, K$. For a cut-off test, a CI overlapping zero level corresponds to a nonsignificant parametric variable, and it can be excluded from the model.

Significance Level for Nonparametric Features

It is not possible, in general, to obtain CIs for coefficients related to the nonparametric variables, as these variables remain in a nonlinear, difficult to observe, equation. However, the performance of the complete model (all the nonparametric variables included) can be statistically compared with the performance of a reduced model (only a subset of them included). This can be made by comparing the CI of the merit figure of the reduced model with the bias-corrected merit figure of the complete model. For instance, let $\bar{R}^{2^*}$ denote the bootstrap bias-corrected correlation coefficient for the complete model, and let S^2 be the correlation coefficient for the reduced model, with CI estimated by fulfilling

$$P^*_{S}(S^{2*}_{\mathrm{l}} \le S^{2*} \le S^{2*}_{\mathrm{u}} | y, \boldsymbol{x}) \ge q. \tag{6.73}$$

A possible test is the following:

$$\begin{cases} H_0: & R^2 = S^2, \quad \text{or} \quad \bar{R^2}^* \in (S_l^{2*}, S_u^{2*}) \\ H_1: & R^2 > S^2, \quad \text{or} \quad \bar{R^2}^* < S_u^{2*}. \end{cases} \tag{6.74}$$

That is, the null hypothesis H_0 states that *the reduced model is sufficient*, whereas the alternative hypothesis H_1 indicates that there is an important loss of fitness when considering only the reduced model.

6.3 Tutorials and Application Examples

This section illustrates three RKHS signal model formulations; namely, SVM kernel ARX framework for system identification, the family of γ-filters with kernels for time series prediction and system identification, the SVM for SR in two real examples (telecontrol network modeling and promotional impact prediction), and the spatial reference for antenna array processing.

6.3.1 Nonlinear System Identification with Support Vector Machine–Autoregressive and Moving Average

The performance of RSM for nonlinear system identification was benchmarked by our group in Martínez-Ramón *et al.* (2006). We used different kernel combinations; namely, separated kernels for input and output processes (called SVM-ARMA$_{2K}$), accounting for the input–output cross-information (SVM-ARMA$_{4K}$), and different combinations of nonlinear SVR and SVM-ARMA models (see Listing 6.1). We used the RBF kernel in all of them.

```
function [K,K2] = BuildKernels(X,Y,X2,Y2,ker,params,method)

switch lower(method)

case {'svr'}
    K = kernelmatrix(ker,[X;Y],[X;Y],params.z);
    K2 = kernelmatrix(ker,[X;Y],[X2;Y2],params.z);
case {'2k'} % SVM-ARMA_{2K}
    K = kernelmatrix(ker,X,X,params.x)+kernelmatrix(ker,Y,Y,
                                                     params.y);
    K2 = kernelmatrix(ker,X,X2,params.x)+kernelmatrix(ker,Y,Y2,
                                                    params.y);
case {'svr+2k'} % SVR-ARMA_{2K}
    K = kernelmatrix(ker,X,X,params.x)+kernelmatrix(ker,Y,Y,
        params.y) +kernelmatrix(ker,[X;Y],[X;Y],params.z);
    K2 = kernelmatrix(ker,X,X2,params.x)+kernelmatrix(ker,Y,Y2,
         params.y) + kernelmatrix(ker,[X;Y],[X2;Y2],params.z);
case {'4k'}        % SVM-ARMA_{4K}
    K = kernelmatrix(ker,X,X,params.x)+kernelmatrix(ker,Y,Y,
                                                   params.y)
    + kernelmatrix(ker,X,Y,params.xy) +kernelmatrix(ker,Y,X,params.xy);
    K2 = kernelmatrix(ker,X,X2,params.x)+kernelmatrix(ker,Y,Y2,
                                                   params.y)
```

```
    + kernelmatrix(ker,X,Y2,params.xy)+kernelmatrix(ker,Y,X2,
                                                   params.xy);
case {'svr+4k'} % SVR-ARMA_{4K}
    K = kernelmatrix(ker,X,X,params.x)+kernelmatrix(ker,Y,Y,
                                                 params.y)
    + kernelmatrix(ker,X,Y,params.xy) +kernelmatrix(ker,Y,X,params.xy)
    + kernelmatrix(ker,[X;Y],[X;Y],params.z);
    K2 = kernelmatrix(ker,X,X2,params.x)+kernelmatrix(ker,Y,Y2,
                                                       params.y)
    kernelmatrix(ker,X,Y2,params.xy) +kernelmatrix(ker,Y,X2,params.xy)
    kernelmatrix(ker,[X;Y],[X2;Y2],params.z);
end
```

Listing 6.1 Kernel combinations for SVR, SVM–ARMA$_2K$, SVR–ARMA$_2K$, SVM–ARMA$_4K$, and SVR–ARMA$_4K$ (BuildKernels.m).

The system that generated the data is illustrated in Figure 6.2. The input discrete-time signal to the system was generated by sampling the Lorenz system, given by differential equations $\mathrm{d}x/\mathrm{d}t = -\rho x + \rho y$, $\mathrm{d}y/\mathrm{d}t = -xz + rx - y$, and $dz/dt = xy - bz$, with $\rho = 10$, $r = 28$, and $b = 8/3$ for yielding a chaotic time series. Only the x component was used as input signal to the system. This signal was then passed through an eighth-order low-pass filter $H(z)$ with cutoff frequency $\Omega_n = 0.5$ and normalized gain of -6 dB at

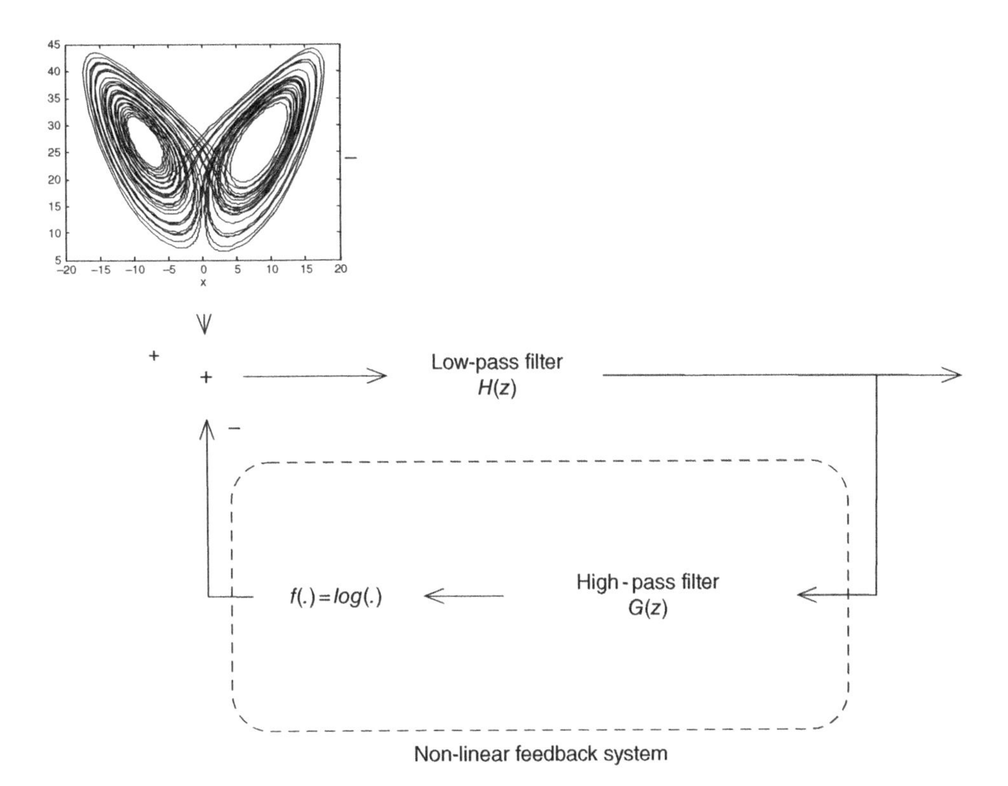

Figure 6.2 System that generates the input–output signals to be modeled in the SVM-DSP nonlinear system identification example.

Table 6.1 ME, MSE, MAE, and RMSE ρ of models in the test set.

	ME	MSE	MAE	ρ
SVR	0.05	30.37	4.63	0.76
SVM-ARMA$_{2K}$	−0.21	39.77	5.11	0.94
SVM-ARMA$_{4K}$	2.95	20.64	2.99	0.96
SVR+SVM-ARMA$_{2K}$	−0.00	0.01	0.07	0.99
SVR+SVM-ARMA$_{4K}$	0.03	0.02	0.11	0.99

Ω_n. The output signal was then passed through a feedback loop consisting of a high-pass minimum-phase channel, given by $o_n = g_n - 2.01o_{n-1} - 1.46o_{n-2} - 0.39o_{n-3}$, where o_n and g_n denote the output and the input signals to the channel. Output o_n was nonlinearly distorted with $f(\cdot) = \log(\cdot)$. We generated 1000 input–output sets of observations, and these were split into a cross-validation dataset (free parameter selection, 100 samples) and a test set (model performance, following 500 samples). The experiment was repeated 100 times with randomly selected starting points, and the free parameters were adjusted with a cross-validation method in all the experiments.

Table 6.1 shows the averaged results. The best models were obtained when combining the SVR and SVM-ARMA models, though no numerical differences were observed between SVR+SVM-ARMA$_{2K}$ and SVR+SVM-ARMA$_{4K}$. In this example, all models considering cross-terms in the kernel formulation significantly improved the results from the standard SVR, indicating the gain given by the inclusion of input–output cross-information in the model.

6.3.2 Nonlinear System Identification with the γ-filter

In this section we study the performance of the kernel γ-filter for nonlinear system identification and time series prediction. We first define the following composite kernels for their use in this section:

- summation composite kernel (SK)

$$K(\boldsymbol{x}_{n-1}, \boldsymbol{x}_{k-1}) = K_1(\boldsymbol{x}^i_{n-1}, \boldsymbol{x}^i_{k-1}) + K_2(\boldsymbol{x}^{i-1}_{n-1}, \boldsymbol{x}^{i-1}_{k-1}); \tag{6.75}$$

- tensor product composite kernel (TP)

$$K(\boldsymbol{x}_{n-1}, \boldsymbol{x}_{k-1} = K_1(\boldsymbol{x}^i_{n-1}, \boldsymbol{x}^i_{k-1}) \cdot K_2(\boldsymbol{x}^{i-1}_{n-1}, \boldsymbol{x}^{i-1}_{k-1}); \tag{6.76}$$

- cross-information composite kernel (CT)

$$\begin{aligned} K(\boldsymbol{x}_{n-1}, \boldsymbol{x}_{k-1}) &= K_1(\boldsymbol{x}^i_{n-1}, \boldsymbol{x}^i_{k-1}) + K_2(\boldsymbol{x}^{i-1}_{n-1}, \boldsymbol{x}^{i-1}_{k-1}) \\ &+ K_3(\boldsymbol{x}^i_{n-1}, \boldsymbol{x}^i_{k-1}) + K_3(\boldsymbol{x}^{i-1}_{n-1}, \boldsymbol{x}^{i-1}_{k-1}); \end{aligned} \tag{6.77}$$

- and extended composite kernels obtaining by using three transformations (K+S)

$$K(\boldsymbol{x}_{n-1}, \boldsymbol{x}_{k-1}) = K_{zz}(z_{n-1}, z_{k-1}) + K_1(x^i_{n-1}, x^i_{k-1}) + K_2(x^{i-1}_{n-1}, x^{i-1}_{k-1}) \tag{6.78}$$

  or by defining a mapping that leads to the summation of the cross-terms composite kernel and the KT matrix (K+T)

$$\begin{aligned} K(\boldsymbol{x}_{n-1}, \boldsymbol{x}_{k-1}) = {} & K_1(x^i_{n-1}, x^i_{k-1}) + K_2(x^{i-1}_{n-1}, x^{i-1}_{k-1}) \\ & + K_3(x^i_{n-1}, x^{i-1}_{k-1}) + K_3(x^{i-1}_{n-1}, x^i_{k-1}) + K_{zz}(z_{n-1}, z_{k-1}). \end{aligned} \tag{6.79}$$

Model Development

Model building requires tuning different free parameters depending on the SVM formulation (σ_{ker}, C, ε) and filter parameters (μ, P). A nonexhaustive iterative search strategy (T iterations) was used, and values of $T = 3$ and $M = 20$ exhibited good performance in our simulations in terms of the averaged normalized MSE:

$$\text{nMSE} = \log_{10}\left[\frac{1}{N\hat{\sigma}^2}\sum_{i=1}^{N}(y_i - \hat{y}_i)^2\right], \tag{6.80}$$

where the $\hat{\sigma}^2$ is the estimated variance of the data. Most of MATLAB source code for the experiments is available and linked in the book's web page.

Nonlinear Feedback System Identification

We now consider the system previously described and illustrated in Figure. 6.2. This system was used to generate 10 000 input–output sample pairs $(x_n, f(g(x_n)))$, that were split into a training set (50) and a test set (following 500 samples). The experiment was repeated 100 times with randomly selected starting points. Table 6.2 shows the average results for all composite kernels. The best results are obtained with the summation kernel, followed by the kernel trick.

The Mackey–Glass Time Series

Our next experiment deals with the standard Mackey–Glass time series prediction problem, which is generated by the delay differential equation $\mathrm{d}x/\mathrm{d}t = -0.1x_n + 0.2x_{n-\Delta}/(1 + x^{10}_{n-\Delta})$, with delays $\Delta = 17$ and $\Delta = 30$, thus yielding the time series MG17 and MG30 respectively. We considered 500 training samples and used the next 1000 for free parameter selection (validation set), following the same approach as Mukherjee *et al.* (1997). The results are shown in Table 6.2 for both time series, suggesting that a more complex model is necessary to obtain good results in the prediction of this time series, which exhibits more complex dynamics.

Electroencephalogram Prediction

This additional and real-life experiment deals with the EEG signal prediction four samples ahead. This is a very challenging nonlinear problem with high levels of noise and uncertainty. We used file "SLP01A" from the MIT-BIH Polysomnographic Database.[1]

---

1 Data available at http://www.physionet.org/physiobank/database/slpdb/slpdb.shtml.

Table 6.2 The nMSE for the kernel γ-filters in nonlinear feedback system identification (NLSYS), Mackey–Glass time series prediction with $\Delta = 17$ and $\Delta = 30$, and EEG prediction. Bold and italics respectively indicate the best and second best results for each problem. The left side of the table includes the results of Casdagli and Eubank (1992) for comparison.

						KT	SK	TP	CT	K+S	K+T
Method	**Poly**	**Rat**	**Loc1**	**Loc2**	**MLP**	**Eq. 6.19**	**Eq. 6.75**	**Eq. 6.76**	**Eq. 6.77**	**Eq. 6.78**	**Eq. 6.79**
NLSYS	−0.04	−0.11	−0.12	−0.71	−0.78	*−1.23*	**−1.26**	−1.08	−1.005	−0.72	−1.06
MG17	−1.95	−1.14	−1.48	−1.89	−2.00	−2.33	**−2.35**	−2.33	−2.34	*−2.35*	−2.35
MG30	−1.40	−1.33	−1.24	−1.42	−1.50	−1.64	**−1.75**	−1.68	*−1.75*	−1.72	−1.69
EEG	−0.05	−0.13	−0.33	−0.32	−0.46	−0.49	−0.66	−0.68	*−0.73*	*−0.73*	**−0.77**

The file contains 10 000 samples; hence, we used 100 as training samples, the next 1000 samples for free parameter selection (validation set), and the remainder for testing.

Average test results are shown in Table 6.2, showing that the kernel trick combined with the summation or cross-terms kernel performs best, and suggesting that the high complexity of the underlying signal model has been retrieved by the data model.

On Model Complexity and Nonlinear Time Scales

Attending to the numerical results (nMSE) in Table 6.2, one could identify EEG and NLSYS as high-complexity problems, and MG17 and MG30 as moderate-complexity problems. However, different kernel structures may accommodate the problem difficulty better than others. Certainly, complexity and versatility is an important aspect for time-series analysis. In this sense, Figure 6.3 reports the results for the four nonlinear time-series problems in terms of machine complexity (SVs (%)), needed tap delays P, memory requirements μ, and its attendant memory depth $M = P/\mu$, which quantifies the past information retained and it has units of time samples (Principe *et al.*, 1993).

The memory depth M serves to uncover the efficiency in modeling the (nonlinear) time scales. On the one hand, it is worth noting that in complex problems (NLSYS and EEG) the kernel trick (KT) yields slightly higher values of M at the expense of poor numerical results (see Table 6.2).

The code used for the last three examples (MG17, MG30 and EEG) is shown in Listing 6.2. First, we select the problem and the method, and then we load the data and generate the input–output data matrices. Finally, we build the kernel using the code previously described and train the regression algorithm.

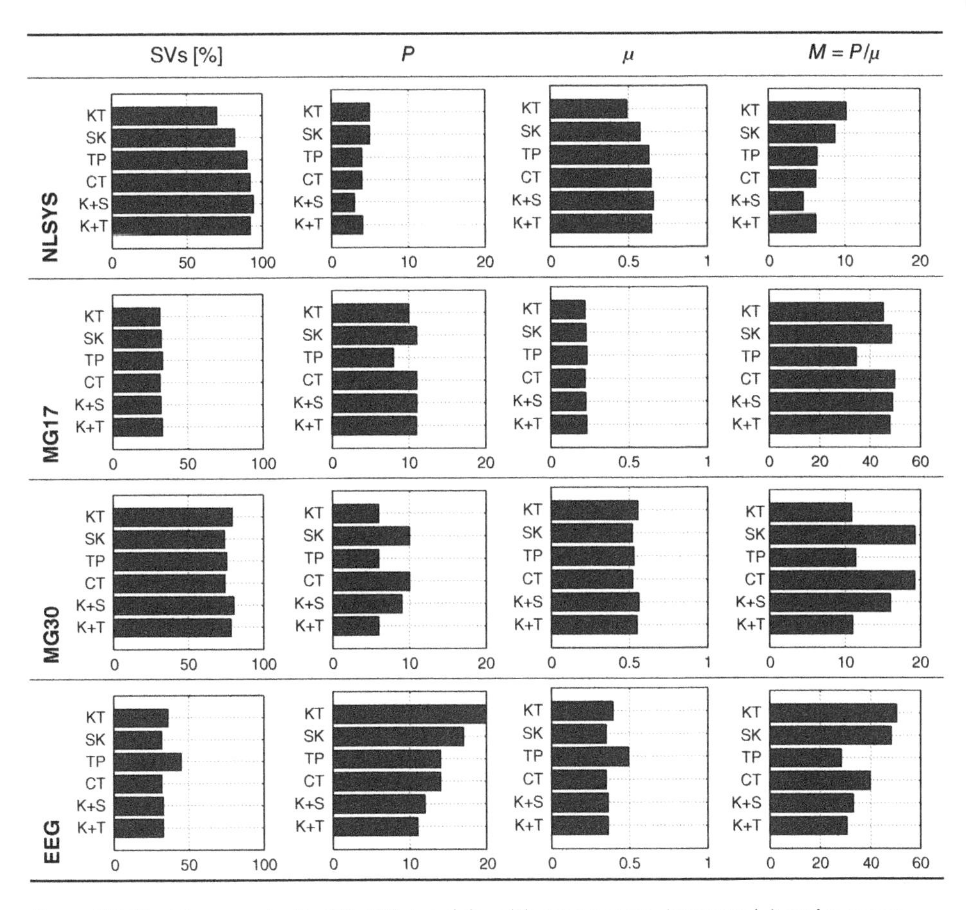

Figure 6.3 Machine complexity (SVs (%)), tap delays (P), memory requirements (μ), and memory depth ($M = P/\mu$) for all kernel methods and problems.

```
% -----------------------------------------------------------
% Problem and method
% problems={'mg17','mg30', 'EEG_SLP01A'};
% methods = {'svr' '2k' 'svr+2k' '4k' 'svr+4k'};
method  = '4k'; problem = 'mg17';
% -----------------------------------------------------------
% Free Parameters (cost function and data)
D = 0; % no delay of the signal
p = 2; % number of taps
e = 1e-5; % epsilon insensitivity zone
C = 1e2; % Penalization parameter
gam = 1e-4; % Regularization factor of the kernel matrix (gam*C
% is the amount of Gaussian noise region in the loss)
% ---------------------------------------------
                                                   %-----------

% Free Parameters (kernel)
ker       = 'rbf';
kparams.x     = 2;     % Gaussian kernel width for the input
kparams.y     = 2;    % Gaussian kernel width for the output
```

```
kparams.z    = 2;   % Gaussian kernel width for the stacked
kparams.xy = 2; % Gaussian kernel width for the cross-information
% -----------------------------------------------------------
                                                        %-----------

% Load data
load mg17.dat
N = 500; M = 1000;
X = mg17(1:N-1); Y = mg17(2:N); X2 = mg17(N+1:N+M-1); Y2 = mg17(N+2:N+M);
% -----------------------------------------------------------
                                                        %---------

% Generate I/O data matrices with a given signal delay D and tap order p
Hx  = buffer(X(D+2:end),p,p-1,'nodelay');             % input, train
Hy  = buffer(Y(1:end-1-D),p,p-1,'nodelay');           % output, train
Hx2 = buffer(X2(D+2:end),p,p-1,'nodelay');  % input, test
Hy2 = buffer(Y2(1:end-1-D),p,p-1,'nodelay');     % output, test
% -----------------------------------------------------------
                                                   %---------------

% Build kernel matrices from these data matrices:
[K,K2] = BuildKernels(Hx,Hy,Hx2,Hy2,ker,kparams,method);
% -----------------------------------------------------------
                                                   %---------------

% Train a regression algorithm with the previous kernel matrices
[Y2hat,results,model] = TrainKernel(K,K2,Y,Y2,D,p,gam,e,C);
```

Listing 6.2 Nonlinear system identification with γ filter (NonlinSysIdGammaFilter.m).

6.3.3 Electric Network Modeling with Semiparametric Regression

High reliability communication networks (HRCNs), as is the case with electric networks, are characterized by very high performance in terms of availability periods, and very low failure rates, as the classical problems of network optimization. In these networks, an extremely low number of events can be observed each year. Then, special models have to be built. It is well known that SVMs have been demonstrated to be especially efficient in scenarios where a low number of samples are available, such as signal analysis or image processing, among others (Camps-Valls *et al.*, 2006c; Soguero-Ruiz *et al.*, 2014b). An SVM approach was followed in this application encouraged by its previous performance and robustness.

Network System and Network Model

In the Spanish electrical grid, power flows continuously from the power generators to the consumption centers (Feijoo *et al.*, 2010). To achieve reliability, one of the main issues of this electrical grid consists of designing the telecontrol service with two redundant but physically different paths. Figure 6.4 shows an example of path calculation in which a simple description of the link availability based on an exponential failure probability with the distance link has been used to determine a reliable double path from a given origin node to the destination node assigned by the telecontrol system using Bayesian networks. The reliability of the telecommunication system can be estimated from data obtained and depends critically on the accuracy of the link and node availability estimates. In what follows, a method based on composite kernel and multiresolution is introduced and studied.

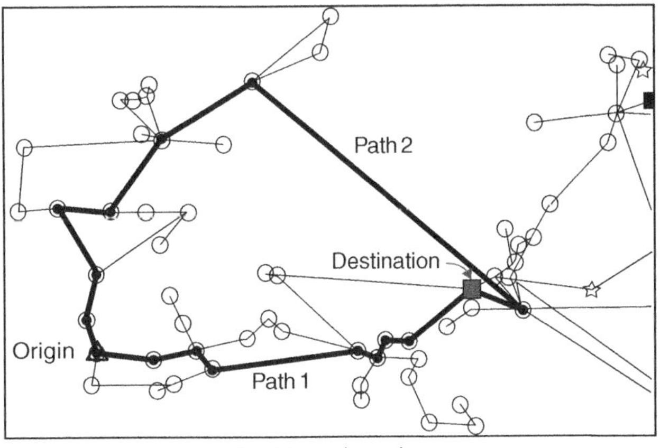

Figure 6.4 Example of double path calculation in the telecontrol network model with a previously developed Bayesian network.

Composite Kernels and Multiresolution

In this application example, we used SVR for grouping and dealing with subsets of features of a different nature. This is usually addressed by using a Gaussian kernel, and in that case a free parameter σ is used for nonlinearly mapping input vectors. Input feature vectors $\boldsymbol{x}_i$ can be redefined into L disjoint feature groups: $\boldsymbol{x}_i = [\boldsymbol{x}_i^{1\mathrm{T}}, \dots, \boldsymbol{x}_i^{L\mathrm{T}}]^{\mathrm{T}}$, with cardinality D_l, such that $D = \sum_{l=1}^{L} D_l$. In that case, we can start from a multiple or composite (multiple kernels) regression model, with a vector accounting for each subset of input variables:

$$y_i = \sum_{l=1}^{L} \langle \boldsymbol{w}^l, \boldsymbol{\phi}^l(\boldsymbol{x}_i^l) \rangle + b + e_i. \tag{6.81}$$

Following Equation 6.11, a solution can be obtained for a new input vector:

$$\hat{y} = \sum_{i=1}^{N} \eta_i \sum_{l=1}^{L} \langle \boldsymbol{\phi}^l(\boldsymbol{x}^l), \boldsymbol{\phi}^l(\boldsymbol{x}_i^l) \rangle + b. \tag{6.82}$$

A scaled kernel for each term in the sum can be used (see Camps-Valls *et al.* (2006c) and Soguero-Ruiz *et al.* (2016b)), according to

$$\langle \boldsymbol{\phi}^l(\boldsymbol{x}_i^l), \boldsymbol{\phi}^l(\boldsymbol{x}_j^l) \rangle = \lambda_l K_l(\boldsymbol{x}_i^l, \boldsymbol{x}_j^l), \tag{6.83}$$

with $\lambda_l \geq 0$. Thus, the final solution of the SVM can be readily expressed as

$$\hat{y} = \sum_{i=1}^{N} \eta_i \left(\sum_{l=1}^{L} \lambda_l K_l(\boldsymbol{x}^l, \boldsymbol{x}_i^l) \right) + b, \tag{6.84}$$

where λ_l represents the contribution of the lth group of input variables to the model. We can get mutiresolution by adjusting the widths σ_l for each kernel, so that it can adapt to different sources of different nature. Note also that weight vectors in feature spaces $\boldsymbol{w}^l$ have no physical meaning, but instead they are a mathematical tool for giving support to the kernel trick.

The methodology is evaluated in both a simulated and a real network. In the simulated network, the statistical distribution for the link failure – the rate consisting of geographical effects and link length effects – is known and allows us to generate failures on the links. Instead of using the simulated network to have a realistic link failure model, its final goal is to benchmark the performance of link failure estimators based on SVM and NN methods in a known scenario. Furthermore, it can be useful to obtain conclusions about the selection of the free model parameters, which were addressed in the HRCN power system.

Surrogate Data Model

A surrogate model was built for the underlying law of failure rate according to the real links and nodes of the network; that is, using the coordinates. Three different effects were considered:

- *Smooth spatial variation*, giving higher link failure probability P_1 for northern than for southern links (the relative difference when comparing northern and southern links is five times larger than when comparing eastern and western links (see Figure 6.5, left), following a linear trend (see Listing 6.3), as follows:

$$P_1(\text{lat}, \text{lon}) = 5 \cdot \text{lat} + \text{long} + c_1, \tag{6.85}$$

where c_1 is an offset constant.

```
%% Smooth Spatial Variation
syms xlon ylat
v= -5*ylat + xlon;
```

Listing 6.3 Smooth spatial variation (SmoothSpatialVariation.m).

- *Fast spatial variation*, yielding higher failure probabilities in network regions with higher grid density. This variation was given by a smoothed version (Gaussian kernel filtering) of the bidimensional histogram of the number of links in a region (see Matlab code in Listing 6.4). We denoted it as $P_2(\text{lat}, \text{lon})$ (see Figure 6.5, middle).

```
%% Fast Spatial Variation
% Create a grid
minx = min(X(:,1)); miny = min(X(:,2));
maxx = max(X(:,1)); maxy = max(X(:,2));
nlinks = length(X(:,1));
npoints = 100;
lineaLat = linspace(miny,maxy,npoints); % Latitud
lineaLon = linspace(minx,maxx,npoints); % Longitud
[Llon,Llat] = meshgrid(lineaLon,lineaLat);
aux = zeros(size(Llon));
for i=1:nlinks
    [kk,indx] = min(abs( X(i,1) - lineaLon ));
    [kk,indy] = min(abs( X(i,2) - lineaLat ));
```

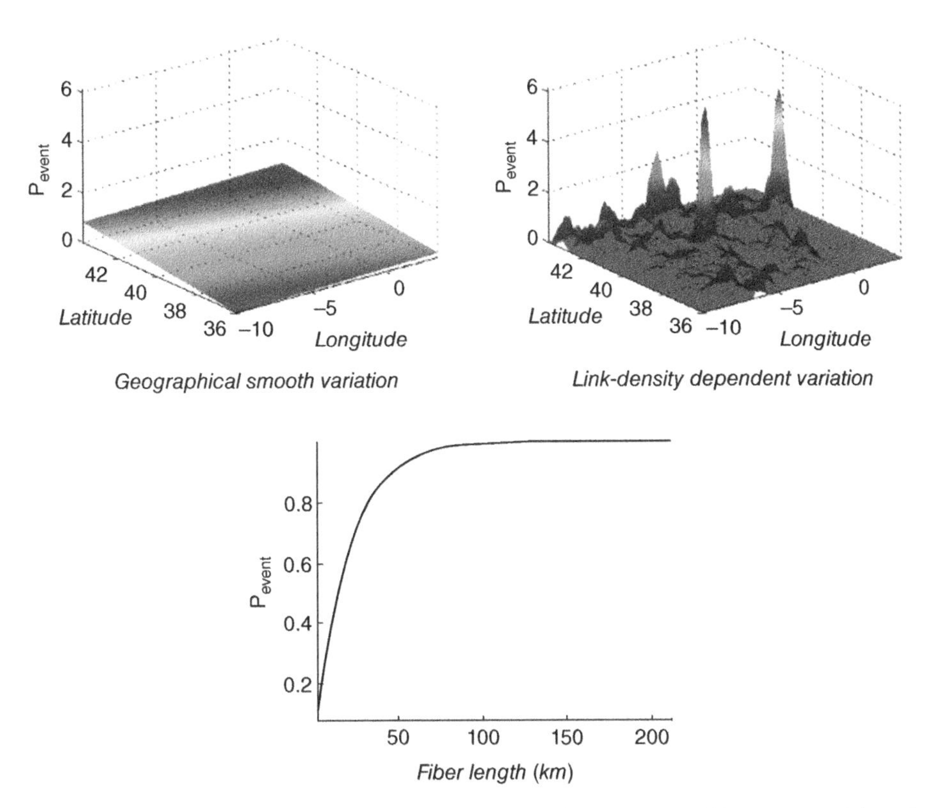

Figure 6.5 Surrogate model for link failure, and conditional components of the event model used.

```
    aux(indy,indx) = aux(indy,indx) + 1;
end
```

Listing 6.4 Fast spatial variation (FastSpatialVariation.m).

- *Link failure probability*, related to the link length. The fiber length increases exponentially to unity, following

$$P_3(L) = 1 - \exp(-L/20), \tag{6.86}$$

and it is depicted in Figure 6.5 (right) (see Listing 6.5).

```
% Link failure probability
fiberLength = X(:,3);
failFiber = 1 - exp(-5e-2*sort(fiberLength));
```

Listing 6.5 Link failure probability (LinkFailure.m).

The final probability of a link failure happening in a given link was given by an additive mixture of P_1 and P_2 spatial effects, and a multiplicative mixture of the spatial and the fiber length, as follows:

$$P_{\text{event}}(L, \text{lat}, \text{lon}) = c \cdot P_3(L)(P_1(\text{lat}, \text{lon}) + P_2(\text{lat}, \text{lon})), \tag{6.87}$$

where c is a numerical normalization constant. We use the *pdfs* for defining the link events in terms of its geometric enter and its length (see their smooth variation in Figure 6.5) as follows:

$$f(\text{lat}, \text{lon}, L) = P_{\text{event}}(\text{lat}, \text{lon}, L) \times \sum_{k=1}^{K} \delta(\text{lat} - \text{lat}_k, \text{lon} - \text{lon}_k, L - L_k). \tag{6.88}$$

Annual and Asymptotic Data Generation

We generated an annual event series given by a list of events that have been observed $s[n]$ during the nth year, with N_y the number of annual events, given by $\boldsymbol{s}_n$, with $n = 1, \dots, N_y$, where for each event $\boldsymbol{s}_n$ consists of a three-tuple including the coordinates and length of each link. The annual event can be accumulated as follows:

$$\boldsymbol{S}_n = \bigcup_{m=1}^{n} \boldsymbol{s}_m, \tag{6.89}$$

allowing us to use a frequentist estimation of the event rates at each link:

$$g_n^k = \frac{\text{Number of events in link } k \text{ in } S_n}{nN_y}, \tag{6.90}$$

and it can be trivially shown to converge asymptotically to the actual event rate of the kth link; that is, $\lim_{n\to\infty} g_n^k = P_{\text{event}}(\text{lat}_k, \text{lon}_k, L_k)$. These asymptotic event probabilities were used as a benchmark for comparison of the estimated event probability given by learning from algorithms.

Simulations with Surrogate Data

We used SVM and NN algorithms to estimate g_n^k obtaining $\hat{g}_n^k$; that is, the frequentist estimation of the events for each link up to year n, given the inputs variables latitude, longitude, and fiber length, for each link $(\text{lat}_k, \text{lon}_k, L_k)$. Three different approaches were built using SVM algorithms with a Gaussian kernel in all of them: (1) *SVM-1K* uses just a single kernel for a vector containing the three input variables; (2) *SVM-2K* has two kernels, one for coordinates and one for length; and (3) *SVM-3K* using three kernels, one for each input variable. The results obtained were benchmarked with a generalized regression NN (GRNN) for function approximation (Demuth and Beale, 1993), which is also a nonlinear method and only requires a free parameter (Gaussian width) to be previously fixed.

In the SVM models, a cross-validation strategy (50% for training and 50% for validation) was applied to tune the free parameters; namely, the ones from the cost function (C, γ, ϵ), the width (σ_i) for each kernel and the relevance parameter λ_i. For each model, and using given free parameters, the MAE was calculated. We want to tune the set of free parameters that minimize the MAE. To that end, we started with fixed initial values for $C, \varepsilon, \sigma_i, \lambda_i$ $(i = 1, 2, \text{ or } 3)$, we obtained the variation of MAE_n with γ. We then fixed parameter γ to the value minimizing $\text{MAE}[n]$, and we obtained the variation of MAE_n

as a function of C (while keeping the rest of the parameters to their fixed values). We continued this process until we explored all parameters.

The purpose of this surrogate data model consists of evaluating the methodology in a theoretic way, and comparing the estimated and the asymptotic results:

$$\mathrm{MAE}_n^{\infty} = \frac{1}{K/2} \sum_{k=K/2+1}^{K} |gk^{k,\infty} - \hat{g}_n^k|, \tag{6.91}$$

where $\hat{g}_n^k$ is the estimated output after tuning the free parameters of the SVM with the training and validation sets, and $g^{k,\infty}$ is the asymptotic corresponding value ($n = 10\,000$ for Equation 6.90, as previously described).

A similar procedure was followed for the GRNN, in which only the width parameter had to be searched in the rank $\sigma \in (10^{-2}, 10)$. The same considerations for the training and test set were followed. MAE_n, and MAE_n^{∞} were also calculated for comparison purposes.

Nonobserved Events

In this application, we studied two possibilities for dealing with null events in the links of the networks: (1) including the set of samples for building the model (both training and testing) by giving them a null numerical value; and (2) excluding them from the training and validation process.

Results on Observed and Asymptotic Data

In this section we show results in terms of observed error and asymptotic error, which are given by MAE_n, and $\mathrm{MAE}^{\infty,n}$ respectively, when evaluating SVM-1K, SVM-2K, and SVM-3K, and an NN scheme using the GRNN. We used 10 independent realizations. Listing 6.6 shows how to obtain both the asymptotic model and when considering 2–20 years.

```
function SurrogateDataModel
% First, we create the asymptotic model
[X,Ysim] = synteticModel(50000,0);
% For different years
nyears = logspace(0,4,20);
err = [];
for n=round(nyears)
    [X,Y] = synteticModel(n,0);
    err = [err,mean(abs(Y-Ysim))];
end
% Compute MAE among Y real and Ysim
err = [err,mean(abs(Y-Ysim))];
% Plot the evolution of the error
figure(4), semilogx(nyears,err,'.-.');
xlabel('# years'), ylabel('MAE');
% Plot the asymptotic probability
figure(5), plot(Ysim), xlabel('# node'), ylabel('P_{event}'), axis tight
```

Listing 6.6 Surrogate data model for asymptotic data and considering different number of years (SurrogateDataModel.m).

Figure 6.6 shows the box plots for all the methods in the cases of including and not including the null events in the learning procedure. Panels (a) and (c) show the MAE_n

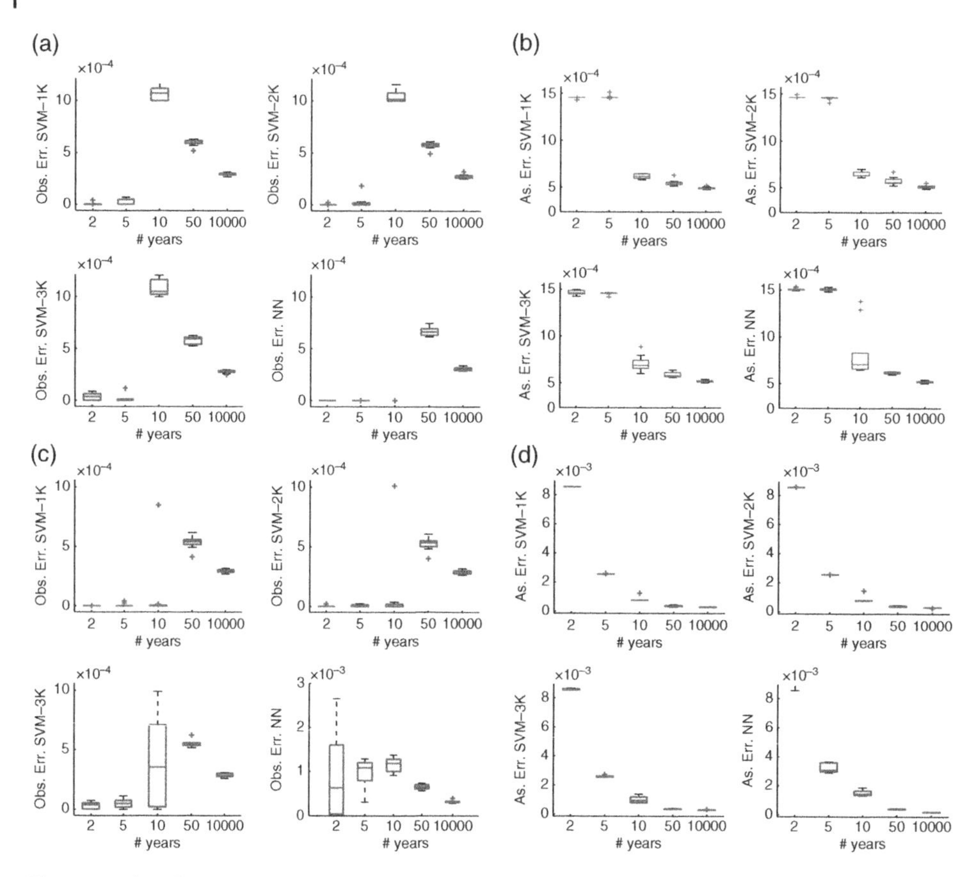

Figure 6.6 Simulations with surrogate data including null events into the training and test set (a, b) and without including them (c, d), after 10 independent realizations. (a, c) Box plot of MAE_n (observed error in n years). (b, d) Box plot of $MAE^{\infty,n}$ (asymptotic error, computed using Equation 6.91).

that should be observed with the events available up to year n, and it can be compared with the gold standard given by MAE_n^{∞} in panels (b) and (d) in terms of the box plots. Box plots use a box and whisker plot for the merit figures from each number of available years n in the conventional way, and have lines at the lower quartile, median, and upper quartile values. The plus sign represents the outliers.

The observed error (i.e., the MAE_n in n years) showed a too-optimistic bias with very few years of events available, in particular, up to 5 years and 10 years respectively when considering and when not considering the null events in the training procedure. It also can be concluded that the asymptotic error (i.e., the MAE_n^{∞}) clearly stabilized after an initial period of 10 years when considering the null events and of 5 years when not considering them. Finally, and overall, including the links with no events in the training procedure yielded lower asymptotic error in this case, which can be seen for instance in Figure 6.6b and d, with lower asymptotic error (MAE_n^{∞}) being obtained for 2–20 years for this approach. The synthetic model created for the surrogate data model is shown in Listing 6.7.

```
function [Xsim,Ysim,Llon,Llat,failGeo,fiberLength,failFiber] =
    synteticModel(numbYears,plotflag)

% Load estimated coordinates (lon,lat)
load network
[X,Y]=loadcoordinatesReal(red);
% Number of year
if nargin==0, numbYears  = 1; end
% Smooth Spatial Variation
SmoothSpatialVariation
% Fast Spatial Variation
FastSpatialVariation
% Normalized variations
B = ones(5); aux2 = filter2(B,aux,'same');
failSmooth = zeros(npoints);
for i=1:npoints;
    for j=1:npoints;
        xlon = lineaLon(i); ylat = lineaLat(j);
        failSmooth(j,i) = eval(v);
    end
end
failSmooth = failSmooth - min(failSmooth(:));
failSmooth = failSmooth/max(failSmooth(:));
failDens = aux2/10; failGeo = failSmooth + failDens;
%% Link failure probability
LinkFailure
% Generate events with uniforme probability
list = zeros(nlinks,1);
for i=1:nlinks;
    [kk,indx] = min(abs( X(i,1) - lineaLon ));
    [kk,indy] = min(abs( X(i,2) - lineaLat ));
    list(i) = failGeo(indy,indx);
end
list = list.*failFiber; list = cumsum(list); list = list/list(end);
```

Listing 6.7 Synthetic model (SyntheticModel.m).

It can be concluded that there are smooth partial relationships among variables, such as spatial or link fiber dependence. The methodology allows the study of complex interaction in HRCNs, increasing the accuracy with the number of availability events. We analyzed the effect of including or excluding the links with unobserved events in the training set. Finally, we benchmarked that better performance is obtained with SVM models, especially with few years of available data.

A Case Study

We also analyzed the historical data consisting of the events and the failures in the optical links from a real HRCN during 2 years. We used the same three input variables (longitude, latitude, and fiber length) to characterize each link. This real HRCN had a low number of observed events, 45 and 50 for $\boldsymbol{S}_1$ and $\boldsymbol{S}_2$ respectively. The same SVM algorithms were tested by using one, two, and three Gaussian kernels. Free parameters were tuned by using a cross-validation technique, by randomly splitting the available data into training and validation subsets. We used MAE_2 on the validation to assess the performance of the model. To give a statistical description of the accuracy, the estimations were repeated 200 times with different randomization.

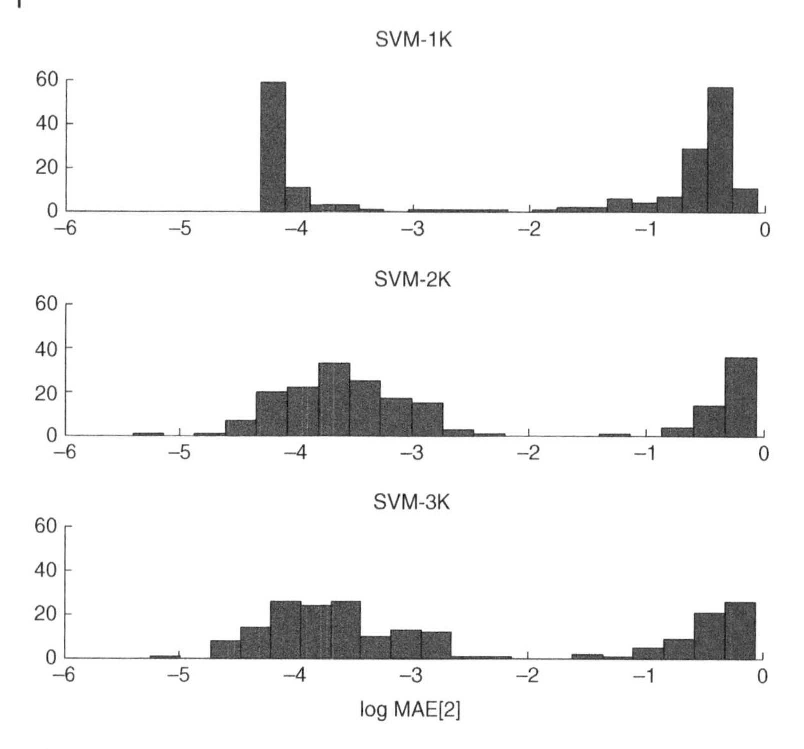

Figure 6.7 Histograms of the log(MAE_2) in the validation sets for each SVM algorithm in the real case study.

Figure 6.7 shows the histograms of MAE_2, in logarithmic units, for the three SVM algorithms. It is important to emphasize the bimodality in the error histograms, showing that there are two cluster of training subsets: one provides suboptimal results, whereas the other yields good performance. The MAE_2 values for SVM-1K, SVM-2K, and SVM-3K algorithms were −0.84, −3.41, and −3.42 respectively. The conclusion is that using two kernels provides a significant improvement, whereas the inclusion of a third kernel did not further enhance performance.

6.3.4 Promotional Data

In this subsection, we first analyze from simulations the effect on the number of input features in different SR models. Then, we use SVM-SR for an application example based on price promotions effects. In both cases, the use of bootstrap resampling for different model diagnoses is presented.

Effect of the Number of Input Features

One of the main limitations of SR is that, due to the nonparametric component of the model, its performance sensibly deteriorates with the number of input features (curse of dimensionality). For classification and regression problems, the SVM has been shown to be robust when working with high-dimensional input spaces, in part due to the sparsity enforced on the solution. We present a simple simulation example that compares the

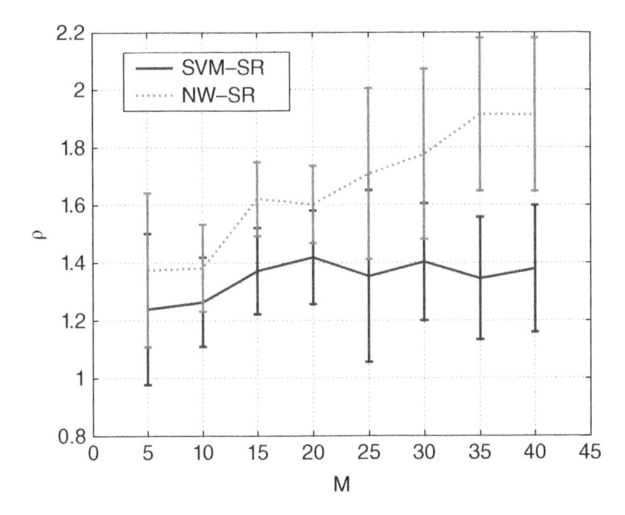

Figure 6.8 Quality score ρ for SVM-SR and for NW-SR in the simulation example as a function of the number of input features M.

behavior of SVM-SR with the SR using the NW estimator (NW-SR), in terms of the dimension of the input space.

The following model was used to generate the observations:

$$y^m = \arctan(\mathbf{Q} \cdot \mathbf{x}^m), \tag{6.92}$$

$$y^d = \langle \boldsymbol{\beta}, \mathbf{x}^d \rangle, \tag{6.93}$$

$$y = \frac{y^m}{\sigma_m} + \frac{y^d}{\sigma_d} + e_n, \tag{6.94}$$

where σ_m and σ_d are the standard deviations of the y_n^m and y_n^d processes respectively; $\mathbf{Q}$ is an $M \times M$ constant matrix, whose elements are i.i.d. and drawn from a rectified Gaussian *pdf*, $N(0,1)$; β is a $D \times 1$ vector, whose elements are i.i.d. and drawn from a uniform *pdf* $U(-1,1)$; $\mathbf{x}^m$ is an $M \times 1$ random vector, drawn from an $N(0,2)$; $\mathbf{x}^d$ is a $D \times 1$ random vector, drawn from a Bernoulli *pdf* with 1-probability of 0.5; and e_n is an $N(0,0.2)$ perturbation. Performance is evaluated for SVM-SR and for NW-SR, over a rank $M \in (5,40)$, and $M = D$. The training and the validation sets consist of 50 and 500 observations respectively. The experiment was repeated for 10 runs.

Figure 6.8 shows the runs-averaged ρ for both methods. Whereas the performance of NW-SR gets worse with the number of input variables, the error in SVM-SR remains almost at the same level. For $M > 30$, performance is significantly different for both methods.

Deal Effect Curve Estimation in Marketing

As a real application example, we describe next an approach to the analysis of the deal effect curve shape used in promotional analysis, by using the SVM-SR in an available database (Soguero-Ruiz *et al.*, 2012). Our data set is constructed from store-level scanner data of a Spanish supermarket for the period 2 January until 31 December 1999.

Table 6.3 Qualitative and quantitative prices (in pesetas) of coffee brands considered in the study.

Brand	#1	#2	#3	#4	#5	#6
Price	Low	High	High	High	High	Low
Min–max	159–225	189–259	185–240	189–249	187–235	157–195

The number of days in the data set is 304. To account for the effects of price discounts with promotional periods, data were represented on a daily basis. Ground coffee category is considered, as it is a storable product with a daily rate of sales. Brands occupying the major positions in the market (more than 80% of sales) and being sold on promotion were selected, leading to the selection of six brands: two national low-priced brands (#1, #6), and four national high-priced brands (#2 to #5), as seen in Table 6.3.

The predicted variable is the sales units y_i^k sold, at day i, $i = 1, \dots, 304$, for a certain brand k, $k = 1, \dots, 5$, in the category. Brand #6 was neither modeled nor considered in the model for the other brands, because it had no promotional discounts.

To capture the influence of the day of the week on the sales obtained on each day of the promotional period, we introduce two groups of dichotomic variables: for brand k and day i, variables $x_{i,1}^{k,d}, \dots, x_{i,6}^{k,d}$ are the indicators of the day of week (Monday (1) to Saturday (6)) during promotional periods, whereas $x_{i,7}^{k,d}, \dots, x_{i,12}^{k,d}$ are the indicators of the day of week (Monday (7) to Saturday (12)) during nonpromotional periods in brand k. By distinguishing between both groups of variables, we can observe the gap in sales between promotional and nonpromotional periods due to the seasonal component.

One of the characteristics of price discounts that researchers have commonly analyzed is the influence of promotional prices ending in the digit 9 in the sales obtained by the retailer (Blattberg and Neslin, 1993). To capture the influence of 9-ending promotional prices in the sales obtained in the category, we introduced an indicator variable $x_{i,13}^{k,d}$ that takes the unit value when the promotional price of brand k is 9-ending.

We also considered in our model the influence of the promotional price. In order to remove the effect of the price, the amplitude of the promotional discount was considered instead of the actual price, as proposed in Heerde *et al.* (2001). The price index, or ratio between the discounted price and the regular price, was introduced using a metric variable for each brand, $x_{i,1}^{k,m}, \dots, x_{i,6}^{k,m}$, with $m = 1, \dots, 6$. Although we know that the retailer had used some kind of feature advertising and displays during the period considered, we do not have any information referring to their usage, so these important effects could not be included in the model.

For each SVM-SR model, the free parameters were found using $B = 20$ bootstrap resamples for the R^2 estimator. The observations were split into training (75%) and testing. For the best set of free parameters, the CI for merit figures R^2, ρ, and the CI for parametric variables were obtained (95% content). As collinearity between metric variables for price indexes and dichotomic variables for promotional day of the week was suspected, a test for the exclusion of the metric variables was performed. Interaction effects between pairs of price indexes were obtained for each model.

Table 6.4 Significance for the inclusion of metric variables.

	R^2	CI S^2	ρ	CI ϱ
#1	0.78†	(0.27,0.65)	2.61†	(2.84,4.40)
#2	0.85†	(0.61,0.83)	5.54†	(5.66,8.51)
#3	0.84†	(0.55,0.70)	6.37†	(6.57,9.76)
#4	0.82†	(0.39,0.76)	3.00†	(3.02,4.95)
#5	0.69	(0.43,0.76)	1.85	(1.77,2.34)

† indicates significant at the 95% level.

Models for all the brands reached a significant R^2 (see Table 6.4). Two examples of model fitting are shown in Figure 6.9a and d for a high-quality and a low-quality brand respectively. Their corresponding CIs for the parametric variables are depicted in Figure 6.9b and e. It can be observed that the weekly pattern is significantly different in both of them, with the amplitude of the amplitude of the promotional oscillations being greater than the nonpromotional. The highest rates of sale correspond to promotional weekend periods. This behavior is present in all the other models (not shown). With respect to the 9-ending effect, it significantly increased the sales in brand #1, but not in brand #2. The self-effects of the in brand #2. The self-effects of the promotion are shown in Figure 6.9c and f. The increment of sales in brand #2 (high quality).

Table 6.4 shows the results for the test of significance of the price indexes in the model. In all but one (brand #5), nonparametric variables were relevant to explain the sales jointly with the weekly oscillations. This can be seen as two different effects due to the promotion: an increase in the average level of sales (function of the price index), and a fixed increase in the weekly oscillations amplitude.

Once the relevance of the price indexes has been established, it is worth exploring the complex, nonlinear relationship among them. We only describe here two examples. Figure 6.9g shows the cross-effects of brand #1 on brand #3 model. For the simultaneous promotion situation, brand in a stronger competence than brand #3, as sales of the later fall. However, Figure 6.9h illustrates a weaker competence, as promotion in brand #4 increases the sales even despite the simultaneous promotion in brand #5.

6.3.5 Spatial and Temporal Antenna Array Kernel Processing

We benchmarked in Martínez-Ramón *et al.* (2006) the kernel temporal reference (SVM-TR) and the spatial reference (SVM-SRef) array processors, together with their kernel LS counterparts (kernel-TR and kernel-SRef) and to the linear with temporal reference (MMSE) and with spatial reference (MVDM). We used a Gaussian kernel in all cases. The scenario consisted of a multiuser environment with four users, one being the desired user and the rest being the interferences. The modulated signals were independent QPSK. The noise was assumed to be thermal, simulated by additive white Gaussian noise. The desired signal was structured in bursts containing 100 training symbols, followed by 1000 test symbols. Free parameters were chosen in the first experiment and fixed after checking the stability of the algorithms with respect to them.

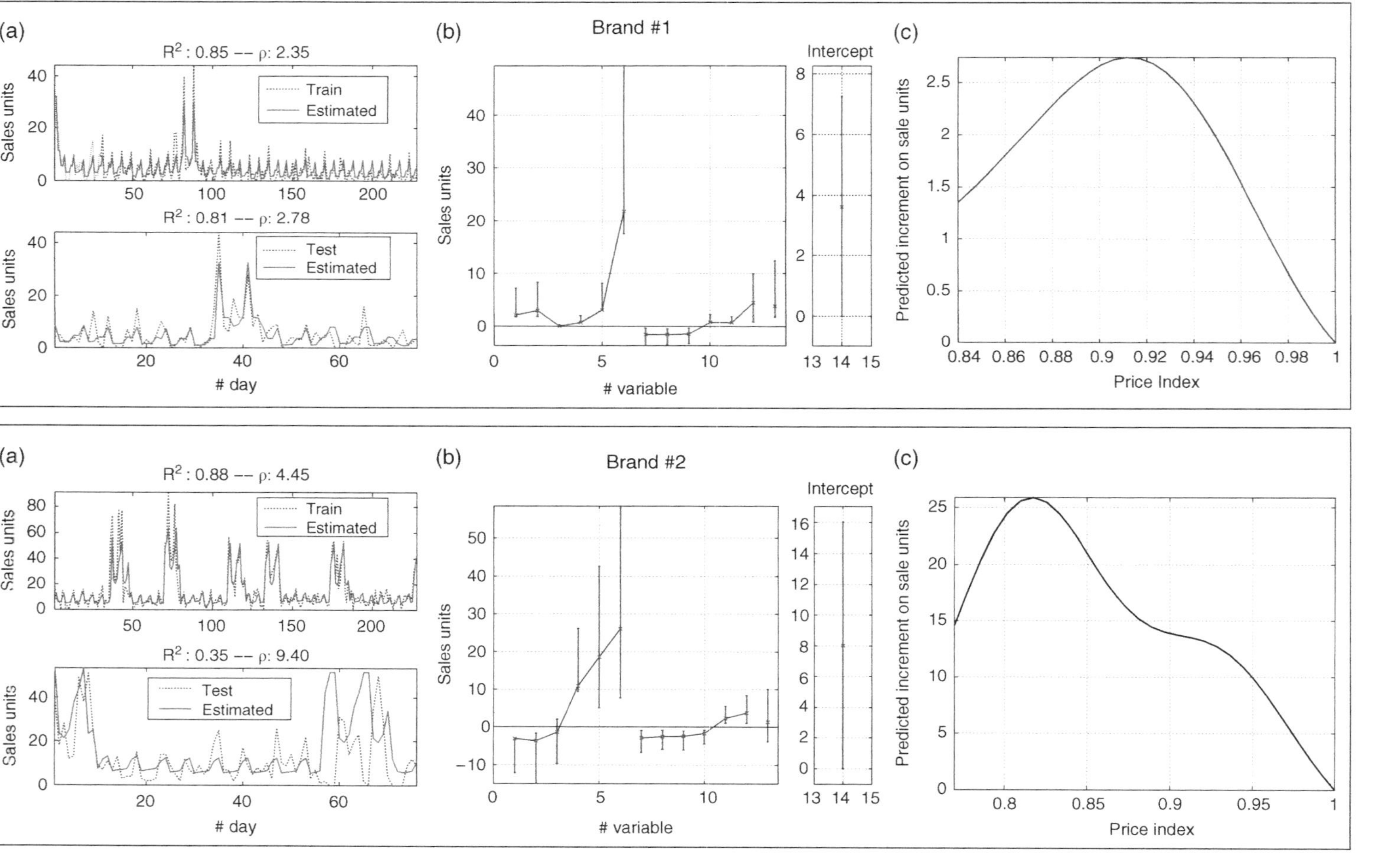

Figure 6.9 Examples of results for deal effect curve analysis, showing model fitting (a), CI for parametric variables and intercept (b), and self-effect of discounts (c) for brands #1 (upper row) and #2 (middle row). Cross-item effect are shown for brand #1 on brand #3 model (a) and for brand #5 on brand #4 model. Confidence bands for averaged output are shown in brand #2 (detail).

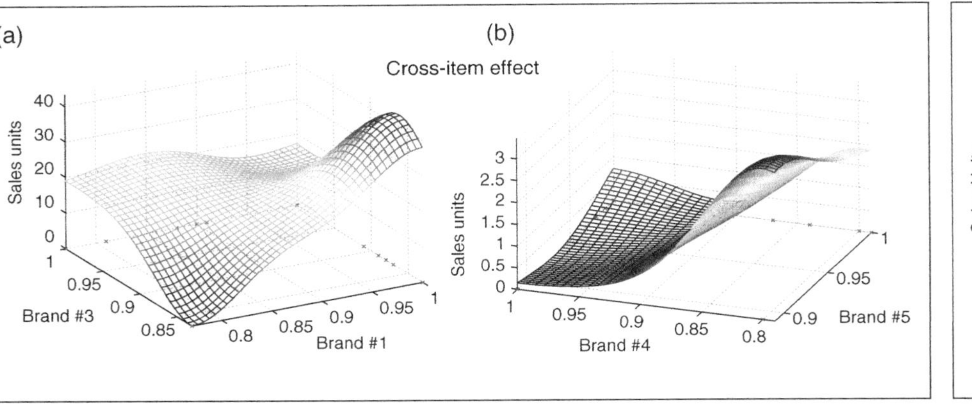

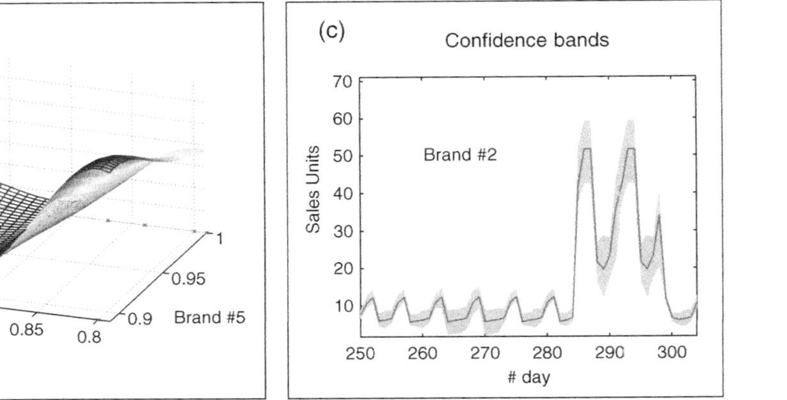

Figure 6.9 (Continued)

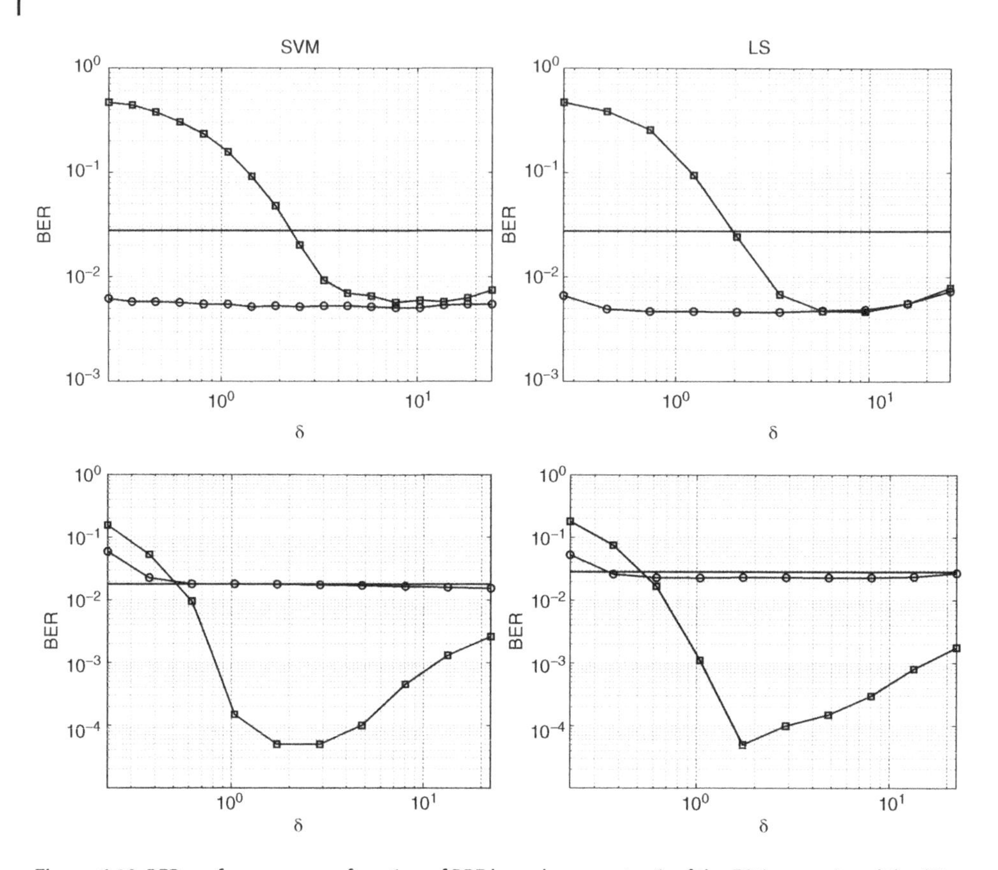

Figure 6.10 BER performance as a function of RBF kernel parameter δ, of the TR (squares) and the SR (circles) in an array of seven (top) and five (bottom) elements and with three interferent signals. Continuous line corresponds to the performance of the linear algorithms.

The BER was measured as a function of kernel parameter δ for arrays of five and seven elements, in an environment of three interferences from angles of arrival of 10°, 20° and −10°, and unitary amplitudes, while the desired signal came from an angle of arrival of 0° with the same amplitude as the interferences. Results in Figure 6.10 show the BER as a function of the RBF kernel width for the temporal and spatial reference SVM algorithms.

These results were compared with the temporal and spatial kernel LS algorithms (i.e., kernel-TR and -SR), and for seven and five array-elements. The noise power was of −1 dB for seven elements and −6 dB for five elements. The results of the second experiment are shown in Figure 6.11. The experiment measured the BER of the four nonlinear processors as a function of the thermal noise power in an environment with three interfering signals from angles of arrival of −10°, 10°, and 20°. The desired signal direction of arrival was 0°.

Performance was compared with the linear MVDR and MMSE algorithms. All nonlinear approaches showed similar performance, and an improvement of several decibels with respect to the linear algorithms. SVM approaches showed similar or slightly better

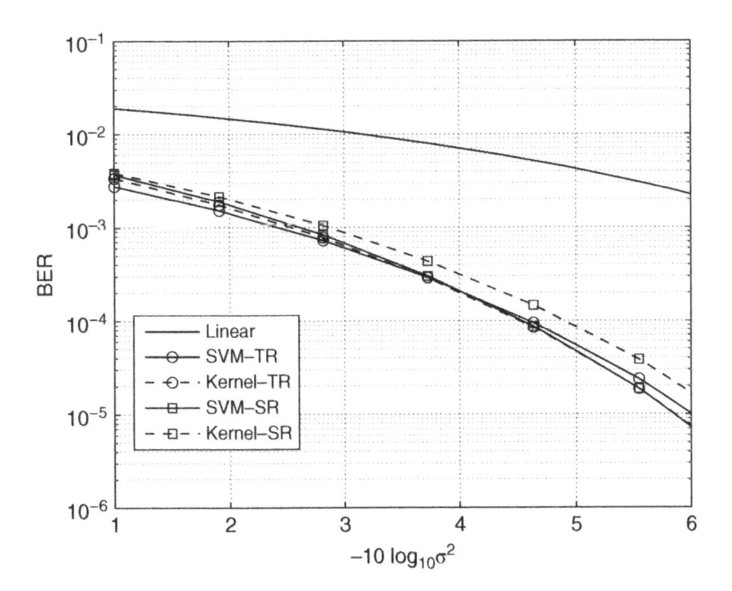

Figure 6.11 BER performance as a function of thermal noise power for linear algorithms, SVM SVM-TR, SVM-SR, Kernel-TR, and Kernel-SR.

performance than nonlinear LS algorithms, and with lower test computational burden due to their sparseness properties.

6.4 Questions and Problems

Exercise 6.4.1 Propose a data model in which composite kernels can provide an RKHS signal model. You can search in classic statistics, in DSP literature, or in machine learning literature.

Exercise 6.4.2 Propose an example of signal or image analysis where you need to combine different sources of data, and hence where the use of a composite kernel can be advantageous.

Exercise 6.4.3 Program the simple simulation example used to compare NW with SVM for SR in terms of the number of input features.

Exercise 6.4.4 With the previous example, adjust the NW estimator and the SVM estimator. Are your results consistent with the ones shown in the chapter?

Exercise 6.4.5 Provide CIs for the statistical estimates described in the bootstrap resampling section for this same synthetic problem.

7

Dual Signal Models for Signal Processing

7.1 Introduction

In this chapter, a third set of signal processing algorithms with kernels is introduced: the so-called DSMs. The shared property among them is that the primal model consists of a nonlinear mapping from sampled or discrete time instants of a unidimensional time series to an RKHS in a conventional nonlinear regression signal model, and then one uses the kernel trick, and a time series expansion is stated in terms of a kernel comparing each two different time instants of the series. The use of *autocorrelation kernels* at this point, with well-known properties in time and spectral domains, supports the generation of algorithms with underlying convolutional signal models, such as for classical sinc interpolation and nonblind deconvolution. Advanced topics on autocorrelation kernels are also given in terms of the fundamentals for spectrally adapted Mercer kernels for signal interpolation. In the second part of the chapter we give empirical evidence of the performance of these dual-form signal processing models via dedicated tutorials and real-life examples: HRV estimation, Doppler ultrasound processing for fault detection, Doppler cardiac images in M-mode for heart-filling monitoring, indoor location based on power measurements from mobile devices, and interpolation of cardiac meshes representations from cardiac navigation systems.

7.2 Dual Signal Model Elements

As explained in Chapter 4, and following the framework introduced in Rojo-Álvarez *et al.* (2014), a particular class of kernels are the shift-invariant kernels, which are those fulfilling $K(u,v) = K(u-v)$. A sufficient condition for shift-invariant kernels to be Mercer's kernels (Zhang *et al.*, 2004) is their FT being nonnegative.

Let s_n be a discrete-time and real signal, with $s_n = 0 \; \forall n \notin [0, S-1]$, and let $R^s_n = s_n * s_{-n}$ be its autocorrelation. Then, the following kernel can be built:

$$K^s_{m,n} = R^s_{m-n}, \tag{7.1}$$

which is called the autocorrelation kernel induced by signal s_n. The spectrum of an autocorrelation is nonnegative, and given that Equation 7.1 is also a shift-invariant kernel, this is always a valid Mercer kernel.

Digital Signal Processing with Kernel Methods, First Edition. José Luis Rojo-Álvarez, Manel Martínez-Ramón, Jordi Muñoz-Marí, and Gustau Camps-Valls.

Now, an additional class of nonlinear SVM for DSP algorithms can be obtained by considering the nonlinear regression on the time lags or the time instants of the observed signals and using an appropriate choice of Mercer kernel. These methods are termed *DSM-based SVM algorithms*. Here, we summarize this approach and pay attention to the interesting and simple interpretation of these SVM algorithms under study in connection with LTI system theory. The interested reader can see Rojo-Álvarez *et al.* (2008) and Figuera *et al.* (2014) for further details.

We use the SVR problem statement as support for the algorithm, by making a nonlinear mapping of each *time instant* to an RKHS; however, the signal model of the DSP to be implemented will be the resulting kernel-based solution by using autocorrelation kernels suitable with the problem at hand. We summarize these ideas in the following theorem. Although here we use nonlinear kernels in a nonlinear regression problem statement as a starting point, the resulting DSP models are linear (and time invariant) in terms of the model equation obtained.

Theorem 7.2.1 (DSM problem statement) Let $\{y_n\}$ be a finite set of samples for a discrete-time signal in a Hilbert space, which is to be approximated in terms of the SVR model in Definition 4.6.1, and let the explanatory signals be just the (possibly nonuniformly sampled) time instants t_n that are mapped to an RKHS. Then, the signal model is given by

$$y_n = y(t)_{|t=t_n} = \langle \boldsymbol{w}, \boldsymbol{\phi}(t_n) \rangle \tag{7.2}$$

and the expansion solution has the following form:

$$\hat{y}_{|t=t_m}(t) = \sum_{n=0}^{N} \eta_n K^h(t_n, t_m) = \sum_{n=0}^{N} \eta_n R^h(t_n - t_m), \tag{7.3}$$

where $K^h(\cdot)$ is an autocorrelation kernel originated from signal $h(t)$. Model coefficients η_n can be obtained from the optimization of Equation 4.61 (nonlinear SVR signal model hypothesis in Property 4.6.1), with the kernel matrix given by

$$\boldsymbol{K}^h_{n,m} = \langle \boldsymbol{\phi}(t_n), \boldsymbol{\phi}(t_m) \rangle = R^h(t_n - t_m). \tag{7.4}$$

Hence, the problem is equivalent to nonlinearly transforming time instants t_n and t_m and making the dot product in the RKHS. For discrete-time DSP models, it is straightforward to use discrete time n for nth sampled time instant $t_n = nt_s$, where t_s is the sampling period in seconds.

This theorem is used later to obtain the nonlinear equations for several DSP problems. In particular, the statement of the classical sinc interpolation SVM algorithm can be addressed from a DSM (Figuera *et al.*, 2014), and its interpretation in terms of linear system theory allows one to propose a DSM algorithm also for sparse deconvolution, even in some cases when the impulse response is not an autocorrelation (Rojo-Álvarez *et al.*, 2008). In addition, the use of autocorrelation kernels provides an interesting interpretation of the SVM with DSM which can be connected to the well-known Wiener filter in estimation theory. All these cases are studied in what follows.

7.3 Dual Signal Model Instantiations

7.3.1 Dual Signal Model for Nonuniform Signal Interpolation

The sinc function in the sinc interpolator has a nonnegative FT, and hence it can be used as a Mercer kernel in SVM algorithms. Note that for a signal to be an autocorrelation kernel R_n^h, we do not need to know explicitly signal h_n that is its origin.

Property 37 (DSM for sinc interpolation) Given the sinc interpolation signal model in Definition 3.3.4, and given that the sinc function is an autocorrelation kernel, a DSM algorithm can be obtained by using an expansion solution as follows:

$$\hat{y}_m = \sum_{n=0}^{N} \eta_n K(t_n, t_m) = \sum_{n=0}^{n} \eta_n \text{sinc}(\sigma_0(t_n - t_m)), \tag{7.5}$$

where Lagrange multipliers η_n can be obtained accordingly.

This equation can be compared with the sinc interpolation signal model in Definition 3.3.4 in terms of model coefficients and explanatory signals. For uniform sampling conditions, Equation 7.5 can be seen as the continuous-time reconstruction of the samples given by an LTI filter with impulse response given by the sinc function, and the input signal given by the sequence of the Lagrange multipliers corresponding to each time instant.

A Dirac delta train can be used for highlighting this continuous-time equivalent signal model. For uniform sampling, if we assume that η_n are observations from a discrete-time process and that $K^h(t_n)$ is the continuous-time version of the sinc kernel given by K_n^h, then solution x_n can be written down as a convolutional model; that is:

$$x_n = \eta_n * K_n. \tag{7.6}$$

Figure 5.3b illustrates this situation. However, a noncausal impulse response h_n is used here, which for some applications is not an acceptable assumption, so that this scheme cannot be proposed for all the convolutional problems. Nevertheless, by allowing ε to be nonzero, only a subset of the Lagrange multipliers will be nonzero, thus providing a sparse solution, a highly desirable property in a number of convolutional problems.

In order to qualitatively compare the sinc kernel PSMs and DSMs for nonuniform interpolation, the following expansion of the solution for the PSM approach given in Equation 3.28 can be written as

$$\begin{aligned} \hat{y}_m &= \sum_{n=0}^{N} a_n \text{sinc}(\sigma_0(t_n - t_m)) \\ &= \sum_{n=0}^{N} \left(\sum_{r=0}^{N} \eta_r \text{sinc}(\sigma_0(t_m - t_r) \right) \text{sinc}(\sigma_0(t_n - t_m)). \end{aligned} \tag{7.7}$$

A comparison between Equation 7.7 and Equation 7.5 reveals that these are quite different approaches using SVM for solving a similar signal processing problem. For the PSM formulation, limiting the value of C prevents these coefficients from growing without control; that is, exhibiting low robustness to outliers and heavy tailed noise. For the DSM formulation, the SRM principle, which is implicit in the SVM formalism (Vapnik, 1995), will lead to a reduced number of nonzero coefficients.

7.3.2 Dual Signal Model for Sparse Signal Deconvolution

Given the observations of two discrete-time sequences y_n and h_n, we recall that deconvolution consists of finding the discrete-time sequence $\hat{x}_n$ fulfilling the following convolutional signal model:

$$y_n = \hat{x}_n * h_n + e_n. \tag{7.8}$$

In many practical situations, x_n is a sparse signal, so that solving this problem using an SVM algorithm can provide with its sparsity property in the dual coefficients. If impulse response h_n is itself an autocorrelation signal, then the deconvolution problem can be stated in a similar way to the sinc interpolation problem in Section 7.3.1, using h_n signal instead of the sinc signal as a Mercer kernel. If an autocorrelation signal is used as a kernel (as we did in the Section 7.3.1 for the sinc interpolation), then h_n is necessarily a noncausal LTI system.

For a causal system, the impulse response cannot be an autocorrelation, and in this case a first approach is the statement of the PSM in Section 5.3. The solution can be expressed as

$$\hat{x}_n = \sum_{i=0}^{N} \eta_i h_{i-n}. \tag{7.9}$$

Hence, an implicit signal model can be written down, which is given by

$$\hat{x}_n = \sum_{i=M}^{N} \eta_i h_{i-n} = \eta_n * h_{-n+M} = \eta_n * h_{-n} * \delta_{n+M}. \tag{7.10}$$

This means that the estimated signal is built as the convolution of the Lagrange multipliers with the time-reversed impulse response and with an M-lagged time-offset delta function δ_n.

Figure 5.3b and c shows the schemes of both SVM algorithms. According to the KKT conditions, the residuals between the observations and model's output are used to control the Lagrange multipliers. In the DSM-based SVM algorithms, the Lagrange multipliers are the input to an LTI noncausal system whose impulse response is the Mercer kernel (Figure 5.3b and c). Interestingly, for the PSM-based SVM algorithms, the Lagrange multipliers can be seen as the input to a single LTI system, whose global input response is $h_n^{\text{eq}} = h_n * h_{-n} * \delta_{n-M}$ (see Figure 5.3a). It is easy to show that h_n^{eq} is the expression for a Mercer kernel, which emerges naturally from the PSM formulation. This provides us with a new direction to explore the properties of the DSM algorithms in connection with classical LTI system theory, which is described in the following.

Property 38 (DSM for sparse deconvolution problem statement) Given the sparse deconvolution signal model in Definition 3.3.5, and given a set of observations $\{y_n\}$, these observations can be transformed into

$$z_n = y_n * h_{-n} * \delta_{n-M}, \tag{7.11}$$

and hence a DSM-SVM algorithm can be obtained by using an expansion solution with the following form:

$$\hat{y}_m = \sum_{n=0}^{n} \eta_n K(n, m) = \eta_n * h_n * h_{-n} = \eta_n * R_n^h, \tag{7.12}$$

where R_n^h is the autocorrelation of h_n, and Lagrange multipliers η_n can be readily obtained according to the DSM theorem.

Figure 5.3c depicts this new situation, which now can be easily solved. This simple transformation of the observations allows one to address the sparse deconvolution problem for a number of impulse responses h_n to be considered in practical applications, especially for FIRs.

7.3.3 Spectrally Adapted Mercer Kernels

Shannon's work on uniform sampling (García 2000; Shannon 1998) states that a noise-free, band-limited, uniformly sampled continuous-time signal can be perfectly recovered whenever the sampling rate is larger than or equal to twice the signal bandwidth. These results from yesteryear have been extended both in theoretical studies (Jerri 1977; Meigering 2002; Unser 2000; Vaidyanathan 2001) and in practical applications (Choi and Munson 1998a; Jackson *et al.* 1991; Strohmer 2000). However, the interpolation problem when these assumptions are not met becomes a very hard one from an estimation theory point of view, and many approaches have been proposed by extending Shannon's original idea.

A seminal work in this setting was Yen's algorithm (Yen, 1956), where an expression for the uniquely defined interpolator of a nonuniformly sampled band-limited signal is computed by minimizing the energy of the reconstructed signal. The solution is given as the weighted sum of sinc kernels with the same bandwidth as the signal. This algorithm suffers from ill-posing due to the degrees of freedom of the solution (Choi and Munson, 1998a), and this limitation can be alleviated with the inclusion of a regularization term (Diethron and Munson, 1991). Other interpolation algorithms using the sinc kernel have been proposed (Choi and Munson, 1998a,b), in which the sinc weights are obtained according to the minimization of the maximum error on the observed data. These algorithms, which use the sinc kernel as their basic interpolation function, implicitly assume a band-limited signal to be interpolated. For non-band-limited signals, other algorithms have considered a non-band-limited kernel, such as the Gaussian kernel (Vaidyanathan, 2001). Additionally, very efficient methods have been developed to reduce the computational complexity of the interpolator; for example, by using filter banks (Tertinek and Vogel, 2008), or a modified weighted version of the Lagrange interpolator (Selva, 2009) (see also references therein). It is interesting to note that the well-known Wiener filter has received less attention than the use of Gaussian or sinc kernel expansions for nonuniform-sampled signal interpolation problems (Kay, 1993), probably due to the difficulty of formulation in nonuniform signals problems.

As explained before, SVM algorithms have been proposed for nonuniform-sampled signal interpolation using sinc and Gaussian kernels, showing good performance for low-pass signals in terms of robustness, sparseness, and regularization capabilities. The suitability of using the autocorrelation of the observed process as the SVM kernel for interpolation problems was addressed in Figuera *et al.* (2014), where an analysis was made on the SVM kernel choice better taking into account the spectral adaptation between the observed signal, the kernel itself, and the Lagrange multipliers yielded by the model. In this section, we summarize the theoretical contributions in that work, and we introduce several Mercer kernels that are spectrally adapted to the signal to be interpolated. We first analyze the relationship between the Wiener filter and the SVM algorithm for this problem, by using the previously presented spectral interpretation of both algorithms. Then, according to this analysis, we examine different SVM interpolation kernels accounting for different degrees of spectral adaptation and performance; namely, band-pass kernels, estimated signal autocorrelation kernels, and actual signal autocorrelation kernels.

Comparison with Support Vector Machine Interpolation

We compare DSM for signal interpolation with a Wiener filter (Kay, 1993) and Yen interpolator (Yen, 1956) for two reasons. First, they are representative cases of optimal algorithms (with a different optimality concept for each case). And second, they have a straightforward spectral interpretation, which allows an interesting comparison with new and standard algorithms.

As we have seen before, in the DSM formulation we can assume a nonuniform interpolator of the form

$$\hat{z}(t'_k) = \langle \boldsymbol{a}, \boldsymbol{\varphi}(t'_k) \rangle, \tag{7.13}$$

where $\boldsymbol{a}$ is a weight vector which defines the solution and $\boldsymbol{\varphi}(t'_k)$ is a nonlinear transformation of the time instants to a Hilbert space $\mathcal{H}$, provided with a dot product

$$\langle \boldsymbol{\varphi}(t_1), \boldsymbol{\varphi}(t_2) \rangle = K(t_1, t_2), \tag{7.14}$$

with $K(\cdot,\cdot)$ being a kernel that satisfies the Mercer theorem. The solution of the SVM is stated in terms of dots products of the transformed input samples. Hence, Equation 7.14 indicates that the nonlinear transformation in Equation 7.13 will be done implicitly by means of a kernel function.

In order to construct the interpolator, vector $\boldsymbol{a}$ must be found. The primal functional has to be optimized in order to obtain $\boldsymbol{a}$, and we can use the ε-Huber cost. The solution is

$$\boldsymbol{a} = \sum_{n=1}^{N} (\alpha_n - \alpha_n^*) \boldsymbol{\varphi}(t_n) = \sum_{n=1}^{N} \beta_n \boldsymbol{\varphi}(t_n), \tag{7.15}$$

where $\beta_n = \alpha_n - \alpha_n^*$ are the Lagrange multipliers for constraints in the residuals. Finally, by combining and expanding the scalar product into a summation, the interpolated signal is given by

$$\hat{z}(t'_k) = \sum_{n=1}^{N} \beta_n \langle \boldsymbol{\varphi}(t_n), \boldsymbol{\varphi}(t'_k) \rangle = \sum_{n=1}^{N} \beta_n K(t_n, t'_k) = \sum_{n=1}^{N} \beta_n K(t_n - t'_k), \tag{7.16}$$

where for the last equality we have assumed that the kernel fulfills the condition $K(u, v) = K(u - v)$. Note again that, in that case, the kernel can be thought of as an LTI system, and this provides us with a convolutional model for the solution.

It has been seen that both Yen and Wiener filter algorithms use the LS (regularized for Yen) criterion. However, the Wiener algorithm is linear with the observations, and it does not assume any a priori decomposition of the signal in terms of building functions. Instead, it relies on the knowledge of the autocorrelation function, which can be hard to estimate in a number of applications. Alternatively, the Yen algorithm is nonlinear with respect to the observations and assumes an a priori model based on sinc kernels. Hence, the knowledge of the signal autocorrelation is not always needed. The SVM interpolation uses a different optimization criterion, which is the SRM, and its solution is nonlinear with respect to the observations since it assumes a signal decomposition in terms of a given Mercer kernel.

Continuous-Time Equivalent System for Interpolation

Here, we present a continuous-time equivalent system for nonuniform interpolation (CESNI), which represents the solution of the interpolation problem based on the SVM approach. The objective of presenting a continuous-time equivalent system is to establish a frequency-domain description of the interpolation SVM algorithm. Based on the analysis of the CESNI model, several effective Mercer kernels are proposed for SVM-based nonuniform sampling interpolation. These kernels account for different degrees of spectral adaptation to the observed data.

Definition 7.3.1 (CESNI) Given the DSM algorithm described in Equation 7.5, we define its continuous-time equivalent system as

$$\hat{z}(t) = \mathcal{T}\{x(t)\}, \tag{7.17}$$

with $x(t) = z(t) + w(t)$, $\hat{z}(t)$ the estimation of $z(t)$, and $\mathcal{T}\{\cdot\}$ a continuous-time nonlinear feedback system. If $\mathcal{T}\{\cdot\}$ is evaluated in a set of N time instants $\{t_n, n = 1, \dots, N\}$, taken from a uniform random distribution, the system defined by Equation 7.5 is obtained.

In order to define $\mathcal{T}\{\cdot\}$, recall that Lagrange coefficients are related to the observed data by the derivative of the cost function; that is, $\beta_n = \mathcal{L}'_{\varepsilon\text{H}}(e_n) \equiv \text{d}\mathcal{L}_{\varepsilon\text{H}}(e_n)/de$ and that $e_n = x(t_n) - \hat{z}(t_n)$. Using these results, it can be seen that the solution defined in Equation 7.16 can be modeled as a feedback system, and we will define $\mathcal{T}\{\cdot\}$ as its continuous-time version, which is represented in Figure 7.1. CESNI elements are next scrutinized.

Property 39 (Residual continuous-time signal) Given the CESNI of SVM algorithm for unidimensional signal interpolation, the residual continuous-time signal is given by

$$e(t) = x(t) - \hat{z}(t), \tag{7.18}$$

and it corresponds to the continuous-time signal from which the residuals are sampled.

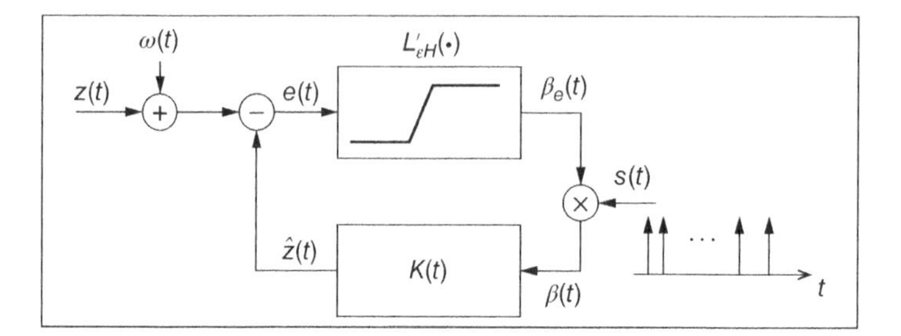

Figure 7.1 CESNI for the SVM interpolation algorithm. The interpolated signal $\hat{z}(t)$ is built by filtering the continuous-time sampled version of the Lagrange coefficients $\beta(t)$ with the SVM kernel $K(t)$.

Property 40 (Model coefficient continuous-time signal) In the CESNI of SVM algorithm for unidimensional signal interpolation, the model coefficient continuous-time signal is given by the following set of equations:

$$\beta_e(t) = \mathcal{L}'_{\varepsilon\mathrm{H}}(e(t)) \tag{7.19}$$

$$s(t) = \sum_{n=1}^{N} \delta(t_n) \tag{7.20}$$

$$\beta(t) = \beta_e(t)s(t) = \sum_{n=1}^{N} \beta_n \delta(t_n), \tag{7.21}$$

where $\beta_e(t)$ is the equivalent continuous signal for the Lagrange coefficient sequence, $\beta(t)$ is its sampled version, and $\delta(t)$ represents the Dirac delta function. Hence, Equation 7.21 represents the discrete set of the model coefficients given by the SVM algorithm as obtained by random sampling of a continuous-time signal $\beta_e(t)$.

Property 41 (Recovered continuous-time signal) In the CESNI of SVM algorithm for unidimensional signal interpolation, the recovered continuous-time signal is given by

$$\hat{z}(t) = K(t) * \beta(t), \tag{7.22}$$

which shows that the kernel works as a linear, time-invariant filter and that the Lagrange coefficients are the inputs to that filter.

Consequently, by denoting the PSD of $\hat{z}(t)$, $K(t)$, and $\beta(t)$ as $P_{\hat{Z}}(f)$, $P_{\mathcal{K}}(f)$, and $P_{\mathcal{B}}(f)$, respectively, the recovered signal PSD is given by $P_{\hat{Z}}(f) = P_{\mathcal{K}}(f)P_{\mathcal{B}}(f)$. Hence, we can conclude that the kernel is shaping the output in the frequency domain.

On the one hand, an appropriate adaptation of the kernel spectrum to that of the original signal will improve the interpolation performance. On the other hand, if the signal and kernel spectra are not in the same band, the performance will be really poor. This is the case of the sinc or the Gaussian kernels when used for band-pass signal interpolation. This suggests that Mercer kernels represent the transfer function which should emphasize the recovered signal in those bands with higher SNR. Looking at the Wiener filter transfer function in Equation 3.36, we can see that the signal autocorrelation could be used for this purpose, since its FT is the PSD of the signal.

Nevertheless, despite these being well-known principles of signal processing, little attention has been paid to the possibility of using spectrally adapted Mercer kernels in SVM-based interpolation algorithms. According to these considerations, several Mercer kernels were proposed by Figuera *et al.* (2014) with different degrees of spectral adaptation; namely, modulated and autocorrelation kernels.

Property 42 (Modulated kernels) If $z(t)$ is a band-pass signal centered at f_0, modulated versions of RBF and *sinc* kernels given by

$$K(t_n, t'_k) = \text{sinc}[\sigma_0(t_n - t'_k)]\sin[2\pi f_0(t_n - t'_k)] \tag{7.23}$$

$$K(t_n, t'_k) = \exp\left[-\frac{(t_n - t'_k)^2}{2\sigma_0^2}\right]\sin[2\pi f_0(t_n - t'_k)] \tag{7.24}$$

are suitable Mercer kernels. Moreover, their spectra are adapted to the signal spectrum. Note that, in this case, an additional free parameter ω_0 has to be settled for the kernel.

Property 43 (Autocorrelation kernels) Similar to the Wiener filter, the autocorrelation of the signal to be interpolated ($z(t)$) or its noisy observations ($x(t_n)$) can be used to define the following kernels:

$$K_{\text{ideal}}(t_n, t'_k) = r_{zz}(t_n - t'_k) \tag{7.25}$$

$$K_{\text{est}}(t_n, t'_k) = r_{xx}(t_n - t'_k) \tag{7.26}$$

which are respectively the ideal (actual) autocorrelation function computed from the underlying process and autocorrelation function estimated from the observations.

If the second-order statistics of the process are known, the kernel defined in Equation 7.25 can be used. When the autocorrelation of the process is not known, an estimation procedure must be considered. Note that this problem is not exclusive of the SVM interpolator, but is also present in the Wiener filter problems in general. However, owing to the robustness of the SVM algorithm, simple procedures for estimating the autocorrelation functions can often be used.

Figure 7.2 illustrates the effect of using different kernels. The signal to be interpolated is band pass, so its interpolation with low-pass kernels, either RBF (Figure 7.2a) or sinc (Figure 7.2b) can be a loose spectral adaptation which indeed emphasizes the noise in the low-pass band. In Figure 7.2c, the use of a modulated band-pass sinc kernel allows us to enhance the transfer function spectral adaptation to the signal spectral profile, which is further refined in Figure 7.2d when using the estimated autocorrelation as interpolation kernel.

7.4 Tutorials and Application Examples

This section highlights the differences between PSM and DSM algorithms for DSP, focusing on nonuniform interpolation and sparse deconvolution. Whereas the theoretical sections in this chapter dealt with unidimensional signals, application examples

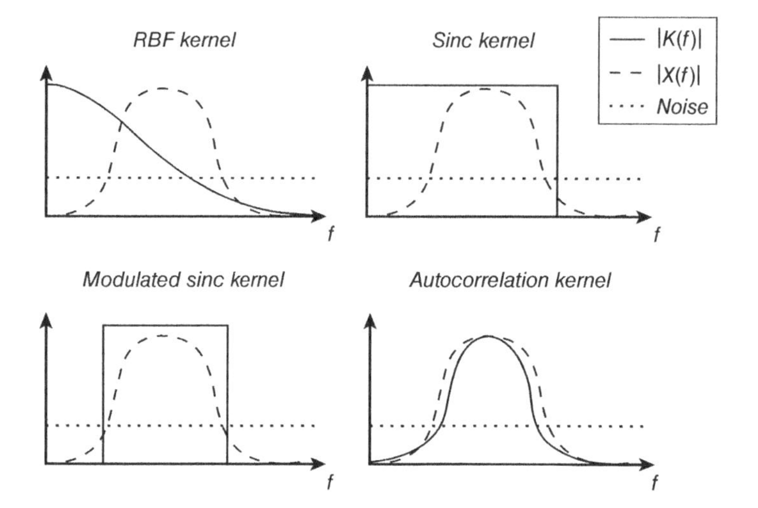

Figure 7.2 Illustration of the spectral adaptation of the kernels to the observations for a band-pass signal.

are extended to higher dimensional cases. The following set of application examples are included:

- First, we benchmark the interpolation algorithms scrutinized in the previous chapters with the DSM interpolation algorithm, and we analyze the implications when using a sinc and an RBF interpolation kernel.
- Next, we show the performance of the DSP sparse deconvolution algorithm, also compared with implementations in previous chapters, both from classical and from SVM problem statements and for synthetic data.
- A real data application example is provided for using the DSM algorithm for sparse deconvolution, from the fault detection in materials through exploration with ultrasound probes.
- Another set of examples analyzes the different spectrally adapted Mercer kernels in synthetic and known-solution signals.
- A specific example in real data is devoted to reconstruct the cardiac signals obtained from HRV measurement in long-term monitoring recordings, in terms of the estimated autocorrelation of a stochastic process.
- A real data application shows the use of autocorrelation kernels in a 2D problem, given by the denoising of Doppler ultrasound medical images.
- Afterwards, another real data application shows the use of autocorrelation kernels in a different problem, which can be approached by 2D and 3D formulations, consisting of the indoor location of mobile devices.
- And finally, a real data application is used to show the 3D approach to reconstruction of meshes of the heart in cardiac navigation systems for arrhythmia ablation.

7.4.1 Nonuniform Interpolation with the Dual Signal Model

In this subsection we evaluate the performance of the three SVM-based signal interpolators. Methods are compared with the standard interpolation techniques defined in Section 3.4.5. As described therein, conventional approaches to interpolation problems can exhibit some limitations, such as loose performance in low signal-to-noise scenarios, or in the presence of non-Gaussian noise. In addition, these methods usually result in nonsparse solutions. These limitations can be alleviated by accommodating the SVM formulation with the nonuniform sampling problem in the DSM problem statements.

The following additional signal interpolators are considered here for benchmarking SVM-DSP interpolation with methods Y1, Y2, S1, S2, and S3 described in Section 3.4.5, as well as with the SVM-PSM algorithm in Section 5.4.5:

1) SVM-DSM with *sinc* kernel (SVM-D), as seen in Listing 7.1.

```
function [predict,coefs] = ...
    SVM_Dual(Xtrain,Ytrain,Xtest,T0,gamma,epsilon,C)

% Initials
tk = Xtrain;
N = length(tk);
Ntest = length(Xtest);
% Construct kernel matrix
H = sinc((repmat(tk,1,N)-repmat(tk',N,1))/T0);
Htest = sinc((repmat(Xtest,1,N)-repmat(tk',Ntest,1))/T0);
% Train SVM and predict
inparams = sprintf('-s 3 -t 4 -g %f -c %f -p %f -j 1', gamma, C, ...
    epsilon);
[predict,model] = SVM(H,Ytrain,Htest,inparams);
coefs = getSVs(Ytrain,model);
```

Listing 7.1 SVM-D algorithm (SVM_Dual.m).

2) SVM-DSM with RBF kernel (SVM-R), as seen in Listing 7.2.

```
function [predict,coefs] = ...
    SVM_RBF(Xtrain,Ytrain,Xtest,T0,gamma,epsilon,C)

% Initials
tk = Xtrain;
N = length(tk);
Ntest = length(Xtest);
% Construct kernel matrix
H = exp(-(repmat(tk,1,N)-repmat(tk',N,1)).^2/(2*T0^2));
Htest = exp(-(repmat(Xtest,1,N)-repmat(tk',Ntest,1)).^2/
                                                  (2*T0^2));
% Train SVM and predict
inparams=sprintf('-s 3 -t 4 -g %f -c %f -p %f -j 1', gamma, C, epsilon);
[predict,model] = SVM(H,Ytrain,Htest,inparams);
coefs = getSVs(Ytrain,model);
```

Listing 7.2 SVM-R algorithm (SVM_RBF.m).

Free Parameters and Settings

Four free parameters have to be tuned in the SVM algorithms, which are cost function parameters (ε, C, γ) and the parameter σ_0 (or, equivalently, the time duration T_0) for the RBF Gaussian kernel. These free parameters need to be a priori fixed, either by theoretical considerations or by cross-validation search with an additional validation dataset, whenever this is available. Note that, in the case of working with signals, the available training set is the set of available samples. Note that, in general, σ_0 and γ can be optimized by using the same methodology as in Kwok (2000), but such analysis is beyond the scope of the present example. For each interpolator here, the optimal free parameters were searched according to the reconstruction on the validation set. For SVM interpolators, cost function parameters and kernel parameter were optimally adjusted. For the other algorithms, the best kernel width was obtained, and for Y2 the best regularization parameter was determined by using a specified validation set.

For comparison purposes with algorithms in Section 3.4.5 we use here the same signal generation and experimental configuration settings therein (see again Listings 3.15 and 3.16). In addition, the required code lines were included in the configuration for validating the free parameters and then executing the new scrutinized algorithms, as indicated in Listing 7.3.

```
% Parameters defining algorithms and theirs free parameters
conf.machine.algs = {@Y1, @Y2, @S1, @S2, @S3, @SVM_Primal, @SVM_Dual, ...
    @SVM_RBF};
conf.machine.params_name = {'T0', 'gamma', 'epsilon', 'C'};
conf.machine.ind_params = {1, 1:2, 1, 1, 1, 1:4, 1:4, 1:4};
conf.cv.value_range{1} = linspace(conf.data.T/2, 2*conf.data.T, 10); % T0
conf.cv.value_range{2} = logspace(-9,1,10);    % gamma
conf.cv.value_range{3} = logspace(-6,1,5);     % epsilon
conf.cv.value_range{4} = logspace(-1,3,5);     % C
```

Listing 7.3 Aditional configuration parameters for SVM-based interpolators (config_interp_SVM.m).

Results

Table 7.1 shows the performance of the algorithms in the presence of additive, Gaussian noise as a function of SNR. Recall that Table 3.3 in Section 3.4.5 compared the set of algorithms Y1, Y2, S1, S2, and S3 and allowed us to observe that Y2 shows the best performance for all noise levels in these experiments, according to its theoretical optimality (from an ML and regularization point of view) for Gaussian noise, and the last one is only included in this table for comparison with SVM methods. Overall, all the SVM approaches remain close to this optimum for high SNR, and for worsening SNR the SVM algorithms tend to improve. Interestingly, SVM-R provides with the best performance in these conditions.

7.4.2 Sparse Deconvolution with the Dual Signal Model

In this section the SVM-DSP algorithm for sparse deconvolution is benchmarked with the methods presented in Section 3.4.6. We denote here this algorithm as AKSM, as seen in Listing 7.4.

Table 7.1 *S/E* ratios (mean plus/minus std) for Gaussian and sinc kernels.

Method	No noise	40 dB	30 dB	20 dB	10 dB
Y2	**53.4 ± 1.9**	39.5±1.4	29.8±1.3	20.1±1.2	10.6±1.2
SVM-P	49.1±4.3	39.1±1.3	29.9±1.2	20.5±1.1	12.3±1.6
SVM-D	50.2±3.2	39.5±1.3	29.9±1.2	20.4±1.1	12.4±1.6
SVM-R	49.8±1.4	**39.6 ± 1.1**	**30.0 ± 1.2**	**21.4 ± 1.3**	**13.5 ± 2.0**

```
function [predict,coefs] = deconv_AKSM(z,h,u,gamma,epsilon,C)

% Initials
ht     = h(end:-1:1);      % Anticausal
z1     = filter(ht,1,z); % Anticausal filtering
y      = [z1(length(h):end)' zeros(1,length(h)-1)]';
N      = length(y);
hauto  = xcorr(h);         % Autocorr. del filtro
L      = (length(hauto)+1)/2;
aux    = hauto(L:end);
Kh     = [aux' zeros(1,N-L)];
% Construct kernel matrix
H      = toeplitz(Kh);
% Deconvolution
inparams = ['-s 3 -t 4 -g ',num2str(gamma),' -c ',num2str(C),' -p ...
    ',num2str(epsilon) ' -j 1']; %,' -q'];
[~,model] = SVM(H,y,[],inparams);
predict = getSVs(y,model);
coefs = filter(h,1,predict);
if ~isempty(u); aux=predict; predict=coefs; coefs=aux; end
```

Listing 7.4 AKSM deconvolution algorithm (deconv_AKSM.m).

Free Parameters and Settings

The merit figures were obtained by averaging the results of the algorithms in 100 realizations. The same set of 100 realizations was used for each SNR and for each tested trade-off parameter in order to analyze their impact on the solution. Figure 7.3 shows the smooth averaged trend in the free parameters for these kinds of problems.

Aiming to benchmark SVM-DSP algorithms for sparse deconvolution with conventional algorithms in Section 3.4.6, we used the same signal generation process and the same configuration parameters in the experiments (see Listings 3.21, 3.22, and 3.23).In addition, Listing 7.5 shows the configuration file for validating the free parameters and running the new algorithm presented in this section.

```
% Parameters defining algorithms and theirs free parameters
conf.machine.algs={@deconv_L1, @deconv_GM, @deconv_AKSM};
conf.machine.params_name={'lambda','gamma','alfa','gamma',
                                            'epsilon','C'};
conf.machine.ind_params={1, 2:3, 4:6};
conf.cv.value_range{1} = 2.0;  % linspace(0,8,10);  % lambda
conf.cv.value_range{2} = 0.8;  % logspace(-9,1,10); % gamma (GM)
conf.cv.value_range{3} = 8.0;  % linspace(0,8,10);  % alfa
conf.cv.value_range{4} = 1e-3; % logspace(-6,1,10); % gamma (AKSM)
conf.cv.value_range{5} = 2.4;  % logspace(-9,1,10); % epsilon
conf.cv.value_range{6} = 100;  % logspace(-1,3,5);  % C

% Parameters defining visualization function
conf.displayfunc = @display_results_deconv;

% Vector with different values of SNR (Signal Noise Rate)
conf.varname_loop1 = 'SNR';
conf.vector_loop1 = [4,10,16,20]; %9
```

Listing 7.5 Aditional configuration parameters for AKSM deconvolution (config_interp_AKSM.m).

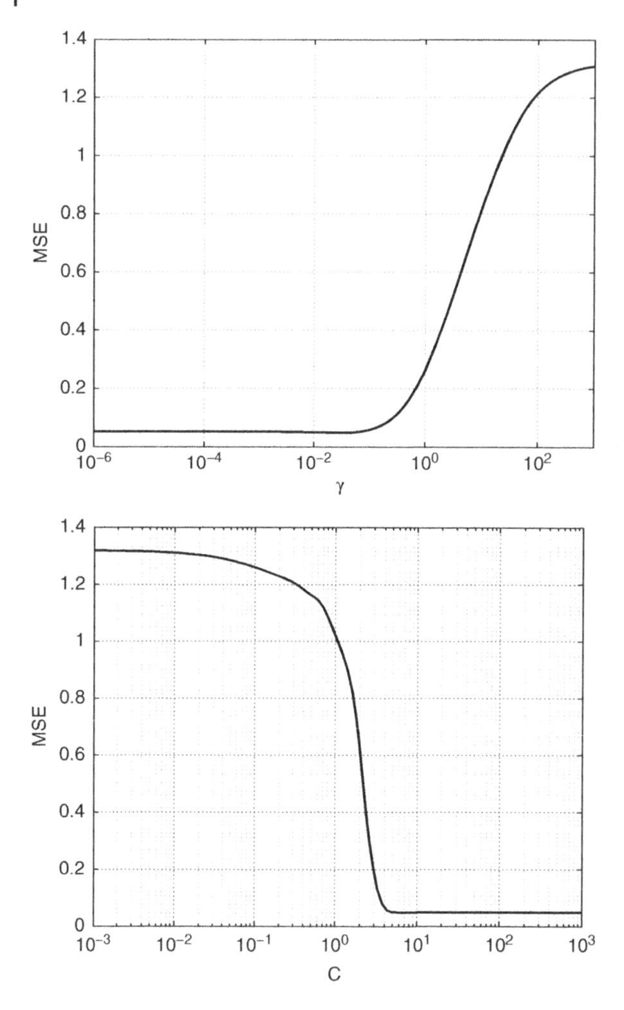

Figure 7.3 MSE of the AKSM algorithm as a function of free parameters γ and C in the ε-Huber cost function, with SNR = 4 dB.

Results

In order to show the advantages of the AKSM algorithm when other algorithms fail due to low SNR, a summary of performances for four different SNR values (4, 10, 16, and 20 dB) is shown in Table 7.2. As seen, AKSM outperforms other methods for low SNR values. This advantage can be highlighted in Figure 7.4, when SNR $=$ 9 dB (only 1 dB less than in Figure 3.22). In this case, AKSM reaches a high estimation performance and detects all the peaks. However, the GM algorithm does not detect a large peak and ℓ_1 penalized deconvolution produces a number of low-amplitude spurious peaks.

7.4.3 Doppler Ultrasound Processing for Fault Detection

Rojo-Álvarez *et al.* (2008) benchmarked sparse deconvolution with SVM for analyzing a B-scan given by an ultrasonic transducer array from a layered composite material.

Table 7.2 *S/E* ratios (mean plus/minus std) for Gaussian.

Method	4 dB	10 dB	16 dB	20 dB
L_1-regularized	0.29 ± 0.13	0.08 ± 0.05	0.02 ± 0.01	0.01 ± 0.00
GM	0.67 ± 0.90	0.05 ± 0.05	0.04 ± 0.03	0.03 ± 0.01
AKSM	$\mathbf{0.12 \pm 0.06}$	0.06 ± 0.02	0.05 ± 0.01	0.05 ± 0.01

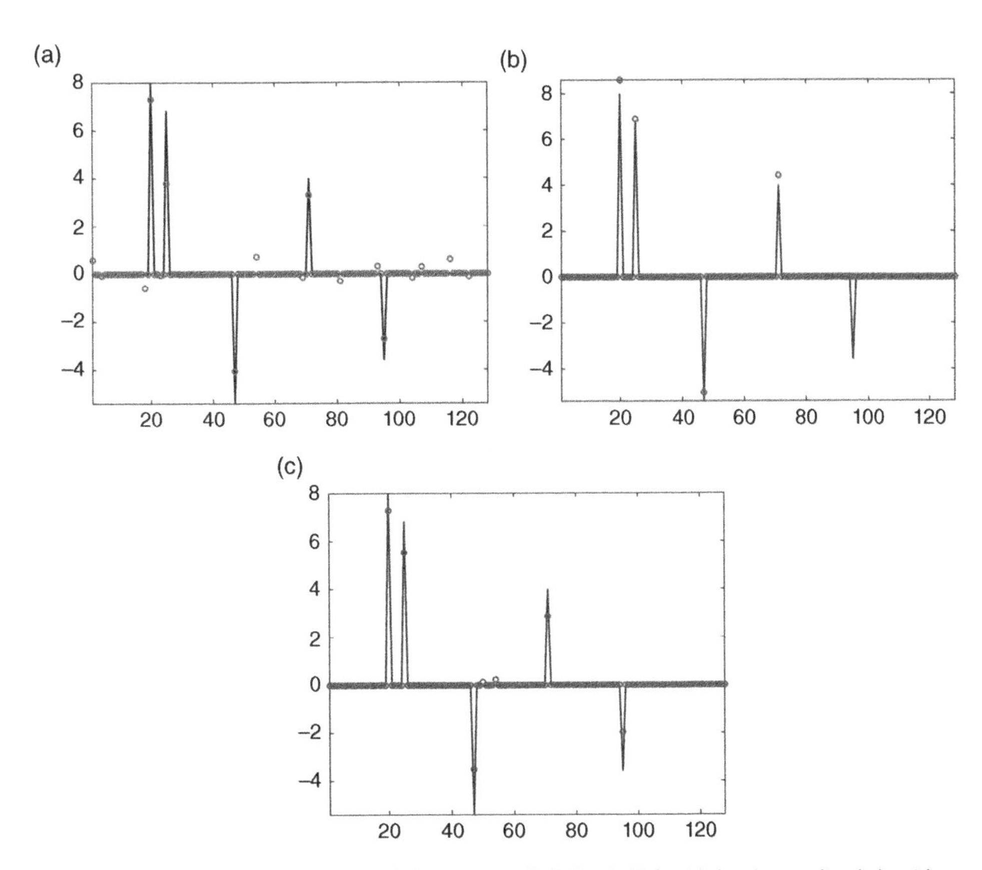

Figure 7.4 An example of spurious peak detection with SNR = 9 dB for (a) the ℓ_1-penalized algorithm, (b) the GM algorithm, and (c) the AKSM algorithm.

Complete details on this application can be found in Olofsson (2004). In order to compare the deconvolution algorithms described in Sections 3.4.6 and 7.4.2 (see Listings 3.19, 3.20, and 7.4 for details) in this real-world application, it is only necessary to load the data structure (see Listing 7.6) and to define the experiment settings in `config_ultrasound.m` in the book supplementary material.

```
function data = genDataUltrasound(conf)

% Each column is a scan (time signal) from a location
load CompPlate4
% Normalization
Bscan4 = Bscan4/max(abs(Bscan4(:)));
% Return structured data
data.h = Bscan4(225:end-60,5);
data.z = Bscan4(:,conf.I);
data.u = 1;
data.x = data.z;
```

Listing 7.6 Loading and preprocessing ultrasound data (genDataUltrasound.m).

Figure 7.5a shows a signal example (A-scan) of the sparse signal estimated by interpolation methods. The same panel also shows the reconstructed observed signal. The ℓ_1 deconvolution yielded a good quality solution with a noticeably number of spurious peaks, the DSM algorithm yielded a good quality solution with less spurious peaks, and the GM algorithm often failed at detecting the low-amplitude peaks. Figure 7.5b shows the reconstruction of the complete B-scan data obtained by each algorithm. These results can be displayed by using the visualization code in Listing 7.7.

```
function solution_summary = display_results_ultrasound(solution,conf)

algs=fields(solution);
y = solution.(algs{1})(50).Ytest;
n = length(y); nalgs=length(algs);
figure
for ia=1:nalgs
    xpred=solution.(algs{ia})(50).coefs;
    ypred=solution.(algs{ia})(50).Ytestpred;
    subplot(nalgs,2,(ia-1)*2+1); plot(1:n,xpred); ...
        title(algs{ia}(8:end)), axis tight; xlim([1 n]); ylim([-.1 .5])
    subplot(nalgs,2,(ia-1)*2+2); plot(1:n,y,':g',1:n,ypred(1:n),'r'); ...
        title(algs{ia}(8:end)), axis tight; xlim([1 n]); ylim([-.6 .6])
end
figure
Bscan4 = cat(2,solution.deconv_AKSM.Ytest);
subplot(2,2,1), surf(Bscan4), title('B-scan')
axis tight, shading interp, alpha(.5)
for ia=1:nalgs
    alg=func2str(conf.machine.algs{ia});
    Xpred = cat(2,solution.(algs{ia}).coefs);
    subplot(2,2,ia+1), surf(Xpred), title(algs{ia}(8:end))
    axis tight, shading interp, alpha(.5)
end
% Performance summary
solution_summary = results_summary(solution,conf);
```

Listing 7.7 Displaying results for ultrasound example (display_results_ultrasound.m).

7.4.4 Spectrally Adapted Mercer Kernels

In this section the DSM algorithms for spectrally adapted Mercer kernels are benchmarked in detail. We first analyze their performance when interpolating a band-pass signal. Then, we evaluate the interpolation of two signals with very different spectral

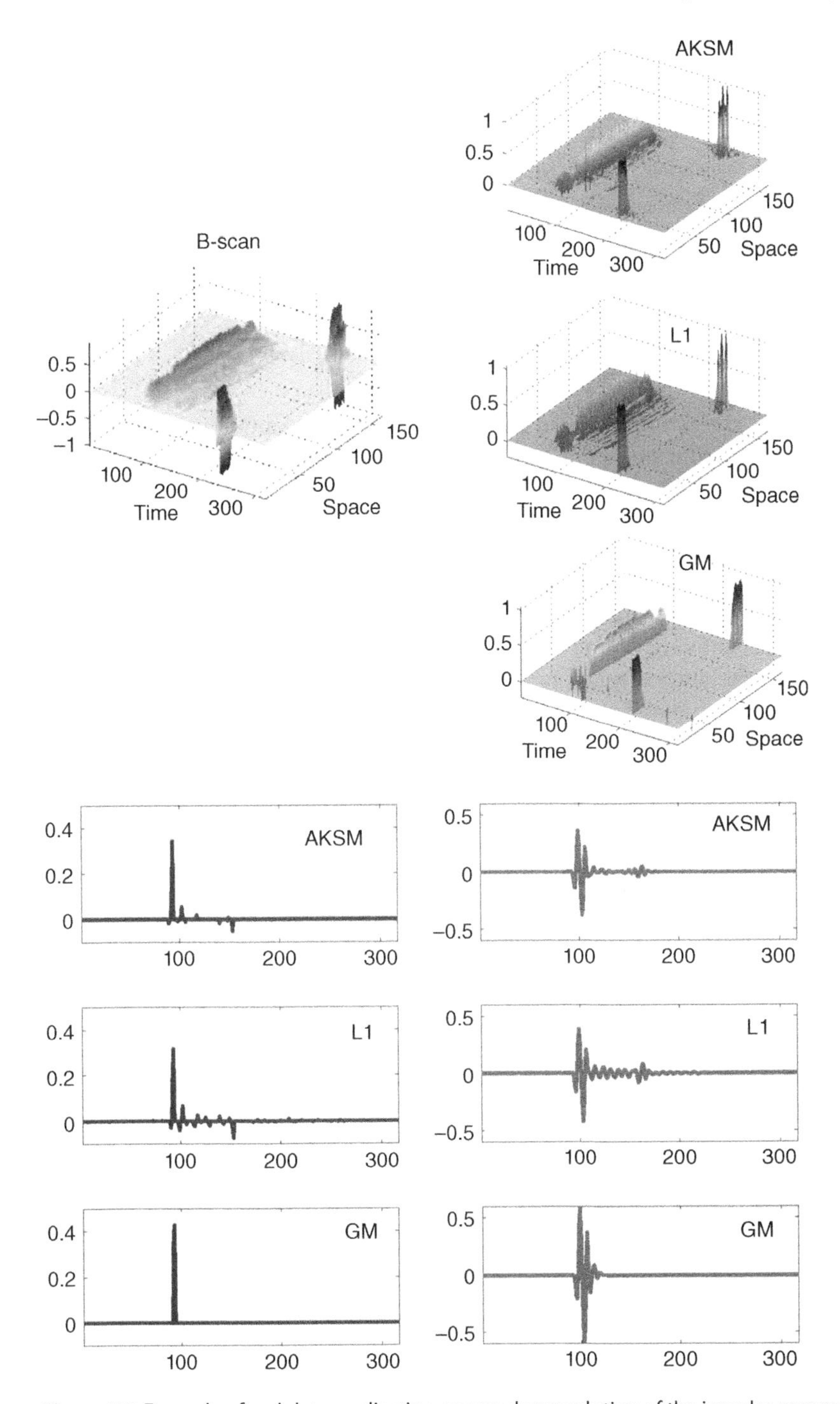

Figure 7.5 Example of real data application: sparse deconvolution of the impulse response in an ultrasound transducer for the analysis of a layered composite material.

Table 7.3 List of algorithms benchmarked in the experiments.

Algorithm description	Label
Yen algorithm with regularization	Yen
Wiener filter with estimated autocorrelation	Wiener
SVM with low-pass RBF kernel	SVM-RBF
SVM with low-pass sinc kernel	SVM-Dual
SVM with estimated autocorrelation kernel	SVM-Corr

profiles in order to assess the effect of the kernel spectral adaptation. We also test different levels of nonuniformity in the sampling process, different number of training samples, and non-Gaussian noise. In a second set of experiments, we test the algorithms for several 1D functions with different and representative spectral profiles.

Interpolation Algorithms for Benchmarking

We benchmarked the interpolation algorithms in Table 7.3. For the Wiener and SVM-Corr algorithms, the autocorrelation function had to be estimated from the observed samples. Note that the autocorrelation had to be computed for every time shift $\tau = t_n - t'_k$, so it had to be estimated over a grid with a resolution much higher than that of the observed samples. Hence, two steps can be carried out:

1) Estimating the autocorrelation from the observed samples.
2) Interpolating it for every time shift $\tau = t_n - t'_k$.

Although many methods exist for these purposes, let us analyze here a simple procedure based on frequency-domain interpolation. The main reason for this choice is that the overall procedure is simple and well established. Specifically, the method consist of:

1) Using a Lomb periodogram to estimate the PSD of the signal (Laguna *et al.*, 1998).
2) Using a zero-padding technique in the frequency domain to carry out the interpolation step.
3) Finally, using inverse FT of the zero-padded PSD for computing the autocorrelation function.

Listings 7.8 and 7.9 show the implementation of these two new algorithms.

```
function [predict,coefs] = Wiener(Xtrain,Ytrain,Xtest,path_acorr)

% Initials
if nargin == 4 && ischar(path_acorr)
    load(path_acorr) % Precomputed autocorrelation of all available data
else
    % Computing autocorrelation with training data
    if nargin < 4; paso = 1; else paso = path_acorr; end
    [acorr,acs] = aCorr(Xtrain,Ytrain,paso);
end
% Contruct kernel matrix
resolution = mean(diff(acs));
% Train
[x,y] = meshgrid(Xtrain,Xtrain);
```

```
tau = abs(x-y);
idx = round(tau / resolution) + 1;
H = acorr(idx);
% Test
[x,y] = meshgrid(Xtrain,Xtest);
tau = abs(x-y);
idx = round(tau / resolution) + 1;
Htest = acorr(idx);
% Train and predict
coefs = H \ Htest';
predict = coefs'*Ytrain;
coefs = mean(coefs,2);
```

Listing 7.8 Wiener filter with estimated autocorrelation – Wiener algorithm (Wiener.m).

```
function [predict,coefs] = ...
    SVM_aCorr(Xtrain,Ytrain,Xtest,gamma,epsilon,C,path_acorr)

% Initialization
if nargin == 7 && ischar(path_acorr)
    load(path_acorr) % Precomputed autocorrelation of all available data
else
    % Computing autocorrelation with training data
    if nargin < 7; paso = 1; else paso = path_acorr; end
    [acorr,acs] = aCorr(Xtrain,Ytrain,paso);
end
% Contruct kernel matrix
resolution = mean(diff(acs));
% Train
[x,y] = meshgrid(Xtrain,Xtrain);
tau = abs(x-y);
idx = round(tau / resolution) + 1;
H = acorr(idx);
% Test
[x,y] = meshgrid(Xtrain,Xtest);
tau = abs(x-y);
idx = round(tau / resolution) + 1;
Htest = acorr(idx);
% Train SVM and predict
inparams=sprintf('-s 3 -t 4 -g %f -c %f -p %f -j 1', gamma, C, epsilon);
[predict,model] = SVM(H,Ytrain,Htest,inparams);
coefs = getSVs(Ytrain,model);
```

Listing 7.9 SVM with estimated autocorrelation kernel – SVM-Corr algorithm (SVM_aCorr.m).

Note that function `aCorr` is used in both algorithms (see Listing 7.10), and it estimates the autocorrelation required to construct the kernel matrix by means a Lomb periodogram.

```
function [rxx,t_rxx] = aCorr(Xtrain,Ytrain,paso)

% Initials
K = 4;
N = length(Xtrain);
Nc_ent = K*N;
fs_corr = 1/paso;
fs = mean(diff(Xtrain));
T = paso * round(1/(fs*paso));
```

```
fs = 1/T;
F = fs_corr/fs;
Nc_test = K*N*F;

% Autocorrelation
deltaf = fs / Nc_ent;
f = linspace(-fs/2,fs/2-deltaf,Nc_ent);
Yf = sqrt(lombperiod(Ytrain,Xtrain,f));
Yf_flip = Yf(end:-1:1);
Yf_flip = [0; Yf_flip(1:end-1)];
Yf = fftshift(Yf + Yf_flip);
Yfcent = fftshift(Yf); Yfcent = Yfcent(:);
N_der = ceil((Nc_test-Nc_ent)/2); N_izq = N_der;
Yf_padded=fftshift([zeros(N_izq,1); Yfcent; zeros(N_der,1)]);
acorr = fftshift(real(ifft(abs(Yf_padded).^2)));
rxx = (acorr(ceil(Nc_test/2)+1:end))./max(acorr);
t_rxx = (0:1/fs_corr:(length(rxx)-1)*(1/fs_corr));
```

Listing 7.10 Estimating the autocorrelation with Lomb periodogram (aCorr.m).

Signals Generation

For synthetic experiments, a 1D signal with spectral information contained in $[-B/2, B/2]$ was interpolated. This signal was sampled in a set of L uneven time instants, different for each realization, with an average sampling interval T, such that $BT = 1$. The interpolation instants lie on a uniform grid with step $T_{\text{int}} = T/F$, with F the interpolation factor. The nonuniform sampling time instants were simulated by adding a random quantity taken from a uniform distribution in the range $[-u, u]$ to the equally spaced time instants $t_k = kT, k = 1, 2, \dots, L$.

In order to simplify the computation of the kernels, each time instant was rounded to be a multiple of T_{int}. The performance of each algorithm was measured by using the S/E indicator; that is, the ratio between the power of the signal and the power of the error in decibels. Each experiment was repeated 50 times.

Two new signals are generated in order to show the advantages of these new algorithms:

1) A modulated squared sinc function (MSSF), defined by

$$f(t) = \text{sinc}^2\left(\frac{\pi}{T_0}t\right)\cos(2\pi f_1 t), \tag{7.27}$$

where T_0 and f_1 are chosen in order that the signal bandwidth fulfills $BT = 1$. The spectrum of this signal is a triangle centered at f_1. The experiment was carried out with $L = 32$ samples, $T = 0.5$ s, a nonuniformity parameter $u = T/10$, and for Gaussian noise with different values of SNR.

2) A signal consisting of two Gaussian functions modulated at different frequencies and added together – called a double modulated Gaussian function (DMGF) – given by

$$f(t) = \frac{1}{\sigma\sqrt{2\pi}}\exp\left(-\frac{|t-\mu|}{2\sigma^2}\right)[\cos(2\pi f_1 t) + \cos(2\pi f_2 t)], \tag{7.28}$$

with $\sigma = 3, f_1 = 0.75$ Hz, and $f_2 = 0.25$ Hz.

Experiment Settings and Results

Three different results are displayed in order to provide insight on the advantages of the new algorithms. To do this, the additional experimental settings for interpolation are in Listing 7.11.

```
% Parameters defining the data generation (in @genDataInterp)
conf.data.FNAME = 'MSSF';
conf.data.f = 0.8;
conf.vector_loop1 = conf.data.T/10; % (u)
conf.vector_loop2 = [40 30 20 10]; % dB (SNR)
% Parameters defining algorithms and theirs free parameters
conf.machine.algs={@Y2, @SVM_Dual, @SVM_RBF, @SVM_aCorr, @Wiener};
conf.machine.params_name={'T0','gamma','epsilon','C','paso'};
conf.machine.ind_params={1:2, 1:4, 1:4, 2:5, 5};
conf.cv.value_range{1} = linspace(conf.data.T/2, 2*conf.data.T, 10); % T0
conf.cv.value_range{2} = logspace(-9,1,10);     % gamma
conf.cv.value_range{3} = logspace(-6,1,5);      % epsilon
conf.cv.value_range{4} = logspace(-1,3,5);      % C
conf.cv.value_range{5} = [1e-3 1e-2 1e-1];      % paso
```

Listing 7.11 Experimental settings for spectrally adapted Mercer kernels (config_interp_Mercer.m).

Interpolation of Band-Pass Signals

To get a first insight of the impact of the kernel spectral adaptation on the algorithm performance, we compared them when interpolating a test signal generated with the MSSF function. Figure 7.6 shows the spectra of the original and reconstructed signals and the error of reconstruction of the Yen and the SVM algorithms. The error at low frequencies (where there is no significant signal power) is high for the SVM with low-pass kernels, since in this band the noise is enhanced by the kernel spectrum. On the contrary, it can be seen that the error produced by the autocorrelation kernel is lower, since it is adapted to the signal spectrum.

Table 7.4 represents the performance of all the algorithms for different SNRs. It can be observed that the SVM with estimated autocorrelation kernels has a good performance. Note that it clearly outperforms the version of the Wiener filter. Finally, SVM with low-pass kernels and Yen algorithms provide a performance lower than that of the others.

Spectral Adaptation of Support Vector Machine Kernel

In order to investigate the importance of the spectral adaptation of the kernel, we analyzed the performance of two SVM-based interpolators on the two test signals MSSF and DMGF. Following the same setup as in the previous experiments, we interpolated these two functions by using two SVM algorithms, one with the ideal autocorrelation kernel and the other with the modulated sinc band-pass kernel (SVM-Dual) given by Equation 7.23, where σ_0 and w_0 have been chosen to obtain the same spectrum as the one of the MSSF function.

In Figure 7.7a, the spectra of both functions and the modulated sinc kernel are shown. It can be seen that the modulated sinc kernel is spectrally adapted to the MSSF, but not to the DMGF. Figure 7.7b shows the S/E performance for both algorithms when interpolating both functions. The modulated sinc kernel performs well for the MSSF, since the spectrum is similar, but the performance degrades for the DMGF. Notably, the autocorrelation kernel is able to adapt its spectrum to both signals, and therefore it performs well with both of them.

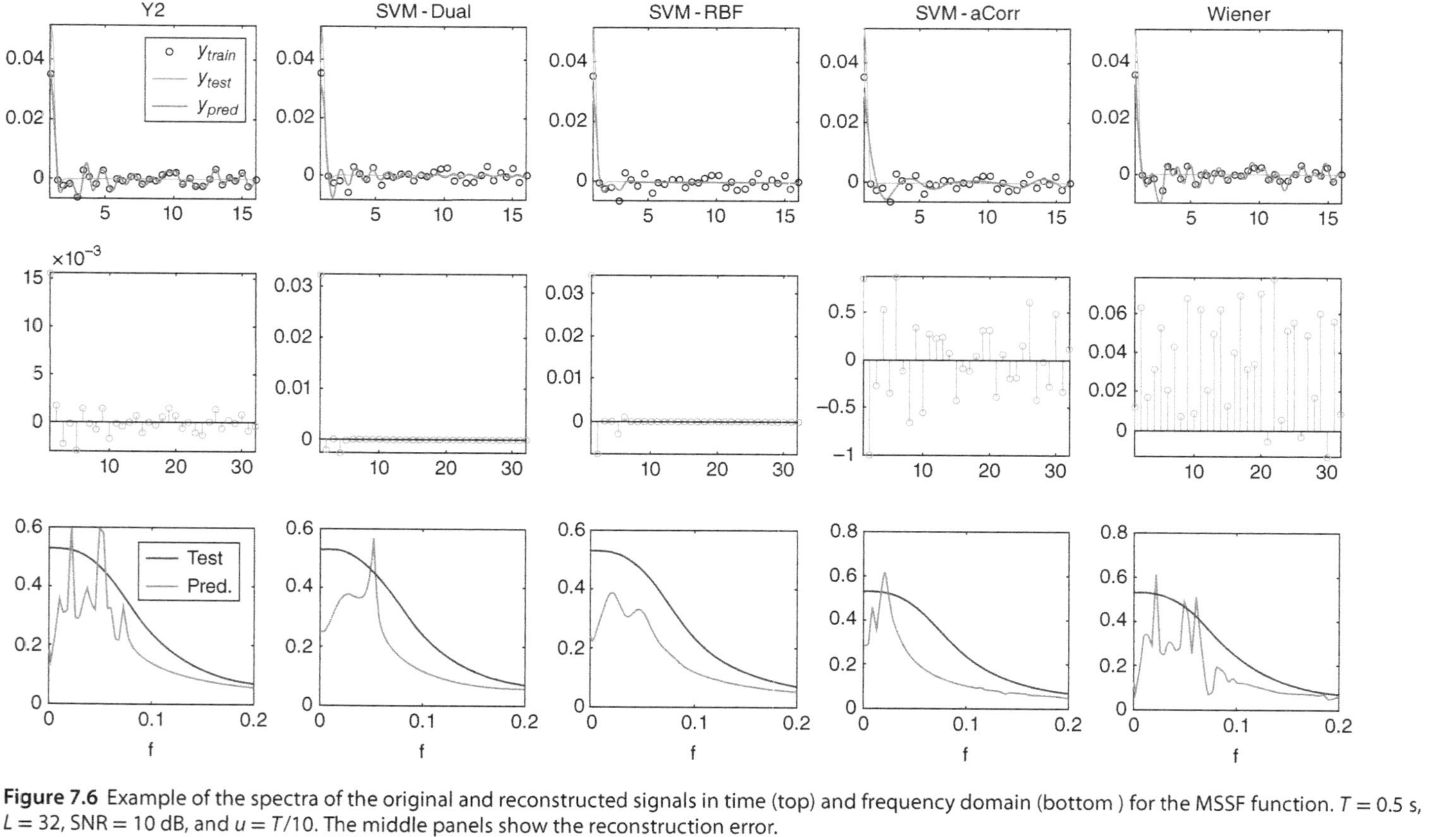

Figure 7.6 Example of the spectra of the original and reconstructed signals in time (top) and frequency domain (bottom) for the MSSF function. $T = 0.5$ s, $L = 32$, SNR = 10 dB, and $u = T/10$. The middle panels show the reconstruction error.

Table 7.4 Mean *S/E* and std (in parentheses) with SNR for a band-pass signal interpolation, $T = 0.5$ s, $L = 32$, and $u = T/10$.

Algorithm	40 dB	30 dB	20 dB	10 dB
Yen	35.1 (0.5)	29.3 (0.9)	20.3 (1.1)	10.4 (1.3)
Wiener	39.8 (1.2)	30.0 (1.0)	20.2 (1.1)	10.1 (1.3)
SVM-Corr	39.5 (1.9)	30.9 (1.4)	22.0 (1.3)	12.6 (1.6)
SVM-RBF	27.4 (1.0)	26.1 (0.7)	19.2 (0.7)	10.8 (1.2)
SVM-Dual	34.1 (0.7)	28.9 (1.1)	20.2 (1.1)	10.9 (1.2)

(a)

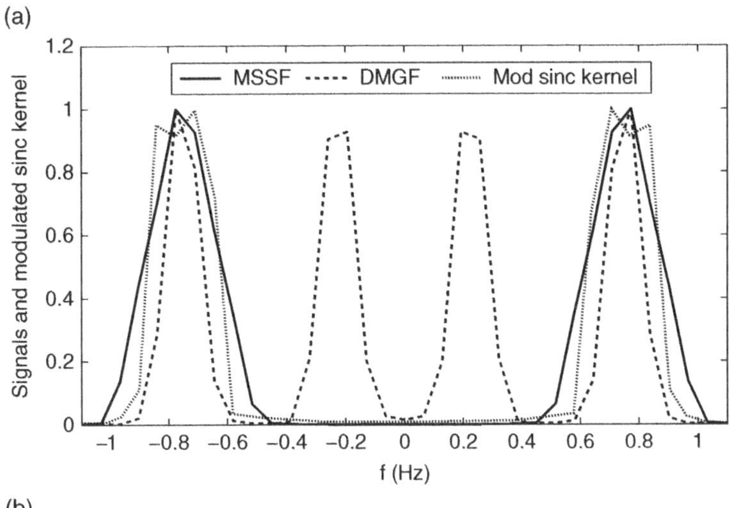

(b)

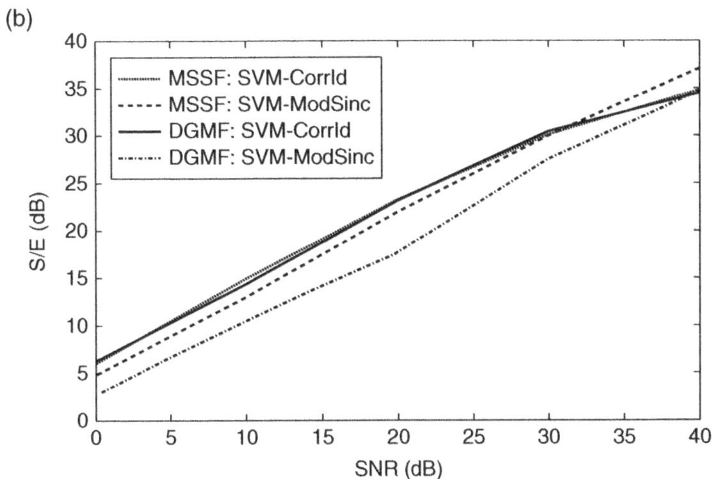

Figure 7.7 Spectrally adapted Mercer kernels. (a) Spectrum for the MSSF, DMGF, and modulated sinc kernel. (b) *S/E* ratio for SVM-Corrld (solid lines) and SVM-ModSinc (dashed lines) algorithms when interpolating MSSF (circles) and DMGF (triangles). $T = 0.5$ s, $L = 32$, $u = T/10$.

Table 7.5 Mean *S/E* and std (in parentheses) with the number of samples *L*, for a band-pass signal interpolation, and $u = T/10$.

Algorithm	$L = 32$	$L = 64$	$L = 128$
Yen	19.6 (0.9)	19.6 (0.8)	15.7 (2.1)
Wiener	19.2 (1.0)	19.6 (0.9)	19.9 (0.5)
SVM-Corr	21.9 (1.2)	21.9 (0.9)	21.8 (0.8)
SVM-RBF	19.2 (1.0)	19.0 (0.8)	18.9 (0.5)
SVM-Dual	19.5 (1.0)	19.8 (0.7)	19.5 (0.5)

Effect of the Sampling Process

In this experiment we further investigated the performance of all the algorithms for different numbers of samples, using the same function and parameters as in the previous sections. Note that if the function bandwidth is not changed, the comparison would be unfair. Hence, we increased it accordingly with the number of samples. Table 7.5 shows the results for different lengths of the training set. Most algorithms perform similarly for all the values of L. However, the Yen algorithm has a poorer behavior, since the numerical problems associated with the matrix inversion become more patent. To obtain these results, the following two code lines have been modified in the experimental settings:

```
conf.varname_loop1 = 'data.L';
conf.vector_loop1 = [32 64 128];
```

7.4.5 Interpolation of Heart Rate Variability Signals

HRV is a relevant marker of the autonomic nervous system control on the heart. This marker has been proposed for risk stratification of lethal arrhythmias after acute myocardial infarction, as well as for prognosis of sudden death events (Task Force of the European Society of Cardiology and the North American Society of Pacing and Electrophysiology, 1996). When analyzing the HRV time series, the sequence of time intervals between two consecutive beats (called the RR-interval time series) is often used, which is by nature sampled at unevenly spaced time instants.

Advanced methods for spectral analysis have shown that the HRV signal contains well-defined oscillations that account for different physiological information. The spectrum of the HRV could be divided into three bands: very low frequency (VLF) band, between 0 and 0.03 Hz; low-frequency (LF) band, between 0.03 and 0.15 Hz; and high-frequency (HF) band, between 0.15 and 0.4 Hz. LF and HF bands have been shown to convey information about the autonomic nervous system control on the heart rhythm, representing the balance between the sympathetic and parasympathetic contributions. Spectral-based methods, such as FT or AR modeling, require the RR-interval time series to be resampled into a uniform sampling grid.

The analysis of the HRV is often performed on 24 h Holter recordings, and a common procedure is to divide the RR-intervals time series into 5 min segments, in order to study the evolution of the spectral components along time. Classic techniques for computing the spectrum of HRV signals aim to obtain a good estimate of LF and HF components, but due to the nonuniform sampling and the noisy nature of the measurements, estimating the HRV spectrum is a very hard problem, especially in the LF and HF ranges. In this example, we applied the SVM algorithms for interpolating two HRV signals with this purpose.

Methodology

A 24 h Holter recording from a patient with congestive heart failure (labeled with a "D") and another one from a healthy patient (labeled with an "H") have been used for this experiment. These recordings were divided into 5 min segments, and two preprocessing steps were done:

1) Discarding the segments with more than a 10% of invalid measurements, usually due to low signal amplitude or ectopic origin of the beat.
2) Applying a detrending algorithm to subtract the mean value and the constant trend of each segment, for reducing their distortion in the VLF region.

Two algorithms were compared: SVM-Corr and SVM-RBF. Both algorithms were used to interpolate each segment, by using the RR-intervals in each segment as the training samples and interpolating the signal over a uniform grid of 500 ms. The SVM-RBF and SVM-Corr solutions were obtained by using Listings 7.2 and 7.12 respectively.

```
function [predict,coefs] = ...
    SVM_Corr(Xtrain,Ytrain,Xtest,gamma,epsilon,C,path_acorr)

% Initials
load(path_acorr)
resolution = mean(diff(t_rxx));

% Contruct kernel matrix
% Train
[x,y] = meshgrid(Xtrain,Xtrain);
tau = abs(x-y);
idx = round(tau / resolution) + 1;
H = rxx(idx); %H.H_ent = interp1(t_rxx,rxx,tau,'linear');
% Test
[x,y] = meshgrid(Xtrain,Xtest);
tau = abs(x-y);
idx = round(tau / resolution) + 1;
Htest = rxx(idx); %interp1(t_rxx,rxx,tau,'linear');

% Train SVM and predict
inparams=sprintf('-s 3 -t 4 -g %f -c %f -p %f -j 1', gamma, C, epsilon);
[predict,model]=SVM(H,Ytrain,Htest,inparams);
coefs = getSvmWeights(model);
```

Listing 7.12 SVM-Corr algorithm (SVM_Corr.m).

Here, MATLAB code `rxx` and `t_rxx` are the data autocorrelation and its temporal indices precomputed in Listing 7.13.

```
function data = genDataHRV(conf)

load('patient_HRV_preprocessed.mat')

if  strcmp(conf.recordType, 'NN') % NN intervals
    auxX = patient.sNNx2;
    auxY = patient.sNNy2;
else                                % RR intervals
    auxX = patient.sRRx;
    auxY = patient.sRRy;
end

% Segments (with less than 10% of NaNs) to train and test
X = auxX(patient.idx_segmentos_utiles);
Y = auxY(patient.idx_segmentos_utiles);

data.Xtrain = X{conf.I};
data.Ytrain = Y{conf.I} - mean(Y{conf.I});
data.Xtest = data.Xtrain(1):conf.Tinterp:data.Xtrain(end);
data.Ytest = zeros(length(data.Xtest),1);

% Segments used to obtain autocorrelation vector
idx_without_nan = patient.idx_segmentos_sin_nan; % segments without NaNs
if ~conf.load_acorr && conf.I==1 % rxx is calculated only the first time
    X_withoutNaNs = auxX(idx_without_nan(selected_idx_for_rxx));
    Y_withoutNaNs = auxY(idx_without_nan(selected_idx_for_rxx));
    [rxx,t_rxx] = calcularRxx(X_withoutNaNs, Y_withoutNaNs, conf);
    save(conf.path_acorr,'rxx','t_rxx')
end
```

Listing 7.13 Preprocessing HRV data procedure (genDataHRV.m).

The autocorrelation kernel for each patient was estimated as follows:

1) A set of segments (around 20) with low noise and high power in the LF and HF regions were previously selected.
2) An estimate of the autocorrelation of each of these segments was computed by using the method described in Section 7.3.1, over a fine grid with a step of 5 ms.
3) Mean estimated autocorrelation was calculated from this set of estimates, in order to reduce the noise level.

A subjective evaluation based on the spectrograms (see Listing 7.14 for more details) and some examples were used to compare them.

```
function [axisF, axisT, EG] = getSpectrogram(X, Y, conf)

% Getting spectrogram
EG = zeros(length(Y),conf.Lfft/2);
for n = 1:length(Y)
    y = detrend(Y{n});
    aux = abs(fft(y,conf.Lfft));
    aux = aux(1:length(aux)/2);
    EG(n,:) = aux;
end
EG = EG ./ max(max(EG));
% Getting frequency and time axis
```

```
f = X;
if iscell(X),
    T = 0; lX = length(X);
    for n = 1:lX
        T = T + mean(diff(X{n}))/lX;
    end
    f = linspace(0, (1000/T)/2, conf.Lfft/2);
end
t = 1:length(Y);
[axisF, axisT] = meshgrid(f, t);
```

Listing 7.14 Getting spectrogram of a signal (getSpectrogram.m).

All of these experiment settings are defined in the configuration file in Listing 7.15.

```
% Paramaters of data generation
% conf.path_acorr='data/Section_8.7.4/acorr_HRV.mat';
conf.path_acorr='acorr_HRV.mat';
conf.NREPETS=279;          % It is used to load different segments
conf.load_acorr=1;         % 0: calculate a new rxx, 1: load a precomputed one
conf.recordType = 0;       % 0: in order to use RR intervals, 1: NN intervals
conf.Finterp  = 2;         % interpolation average factor
conf.Lfft = 2048;
conf.Tinterp = 500;        %ms % DELTA T is the time vector to interpolate
conf.T = 810;              %ms Delta T approximated average, used for the ...
    first spectral estimation in autocorr kernel
conf.resolution = 5;       %ms
conf.fs = 1/(conf.Tinterp/1000);
conf.f = linspace(-conf.fs/2,conf.fs/2,conf.Lfft);
%conf.filterBy_p_vlf = 1;
conf.correct_vlf = 0;      % 0: do not correct the power excess in very low ...
    freqs (vlf, f<0.04Hz).
                           % 1: the ls is corrected (fft(rxx)) to remove ...
                                content of vlf freqs
                           % 2: the vlf is corrected with a ramp
                           % 3: the vlf is corrected with a parabola


% Parameters defining algorithms
conf.evalfuncs={@MSE};
conf.cv.searchfunc=@gridSearch;
conf.cv.evalfunc = @MSE;
conf.machine.params_name={'T0','gamma_RBF','epsilon_RBF',
'C_RBF', 'gamma_Corr','epsilon_Corr','C_corr','path_acorr'};
conf.machine.algs={@SVM_RBF, @SVM_Corr};
conf.machine.ind_params={1:4, 5:8};

conf.cv.value_range{1} = 600;                % T0      (of SVM_RBF)
conf.cv.value_range{2} = 1e-3;               % gamma   (of SVM_RBF)
conf.cv.value_range{3} = 1e-3;               % epsilon (of SVM_RBF)
conf.cv.value_range{4} = 100;                % C       (of SVM_RBF)
conf.cv.value_range{5} = 0.1;                % gamma   (of SVM_Corr)
conf.cv.value_range{6} = 0.1;                % epsilon (of SVM_Corr)
conf.cv.value_range{7} = 50;                 % C       (of SVM_Corr)
conf.cv.value_range{8} = conf.path_acorr;    % acorr   (of SVM_Corr)

conf.displayfunc = @display_results_HRV;
```

Listing 7.15 Experiment settings for HRV example (config_HRV.m).

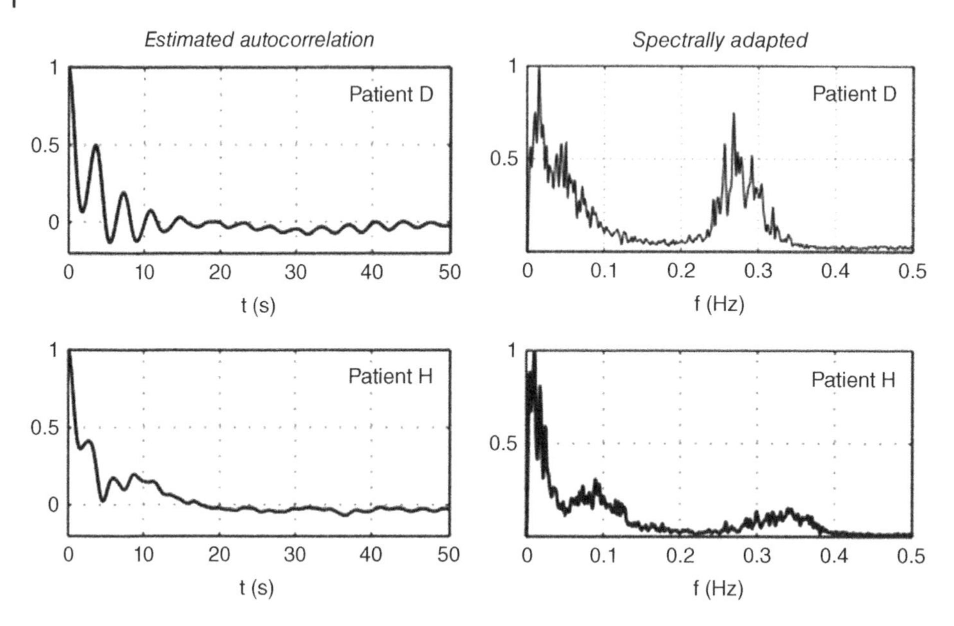

Figure 7.8 Autocorrelation kernels in time and frequency for the HRV segments of patients D and H.

Results

The autocorrelation kernels and their spectra for both patients D and H are shown in Figure 7.8. Although they are still noisy, note that in patient D only one peak is present (probably due to the disease) and both peaks LF and HF are present in patient H. Using the SVM-RBF and the SVM-Corr with these kernels, we interpolated each segment. The main effect was that SVM-Corr was able to filter the noise in order to highlight the LF and HF peaks better than the RBF algorithm, especially where the density of noise was very high in frequency bands out of the regions of interest, as can be seen in the examples for both patients shown in Figure 7.9a for patient D and Figure 7.9b for patient H. For the latter, two details for a region of interest and for a region with noise are shown in the lower plots. In the three examples it can be checked that the SVM-Corr was able to reduce the noise level better than the RBF algorithm. Note that the SVM-Corr algorithm is able to filter these misleading measurements much better than SVM-RBF. These results can be displayed by using Listing 7.16 for visualization.

```
function solution_summary = display_results_HRV(solution,conf)

% Init. params.
fmin = 0.04;
% Loading data
load('patient_HRV_preprocessed.mat')
if  strcmp(conf.recordType, 'NN') % NN intervals
    auxX = patient.sNNx2;
    auxY = patient.sNNy2;
else                              % RR intervals
    auxX = patient.sRRx;
    auxY = patient.sRRy;
end
```

```
X = auxX(patient.idx_segmentos_utiles);
Y = auxY(patient.idx_segmentos_utiles);
% Figures
fs = 1000 / conf.Tinterp;
f_interp = linspace(0, fs/2, conf.Lfft/2);
algs=['Original';fields(solution)];
nalgs=length(algs);
for ia=1:nalgs
    alg=algs{ia};
    if ia==1
        [F.(alg), T.(alg), EG.(alg)] = getSpectrogram(X, Y, conf);
    else
        for i=1:length(solution.(alg)); ...
            Ypred{i}=solution.(alg)(i).Ytestpred; end
        [F.(alg), T.(alg), EG.(alg)] = getSpectrogram(f_interp, Ypred, ...
            conf);
    end
    fmin2 = find(F.(alg)(1,:) > fmin, 1); fmax2 = find(F.(alg)(1,:) > ...
        0.5, 1);
    subplot(3,1,ia);
    mesh(F.(alg)(:,fmin2:fmax2), T.(alg)(:,fmin2:fmax2), ...
        EG.(alg)(:,fmin2:fmax2)); view(10, 30)
    title([strrep(alg,'_','-'),' Interpolated HRV spectrogram']);
    axis([0 0.5 1 length(X) 0 0.2]);
    xlabel('f (Hz)'); ylabel('Segment index');
end
figure;
extra={' segment',' interpolated',' interpolated'};
for ia=1:nalgs
    alg=algs{ia};
    subplot(3,1,ia); hold on;
    for n = 1:size(EG.(alg),1),
        plot(F.(alg)(1,:),EG.(alg)(n,:),'k');
    end
    axis([0 0.5 0 1]);
    title([strrep(alg,'_','-'),extra{ia},' spectra (accumulated plot)']);
end
xlabel('Hz');
%% Performance summary
solution_summary = results_summary(solution,conf);
```

Listing 7.16 Displaying results for HRV example (display_results_HRV.m).

7.4.6 Denoising in Cardiac Motion-Mode Doppler Ultrasound Images

Doppler echocardiography has become a widespread noninvasive technique to assess cardiovascular function in clinical practice. It has been really useful to measure the blood flow within the heart (Weyman, 1994). A frequency shift occurs when the ultrasound waves interact with objects in motion, such as red cells within blood. This detectable frequency shift is used to determine the velocity of the blood flow. Not only the blood velocity, but also the intracardiac pressure differences can be obtained noninvasively by using ultrasound images under certain conditions (Bermejo *et al.*, 2002). Motion mode (M-mode) echocardiography is used to evaluate the morphology of structures, movement, and velocity of cardiac values and timing of

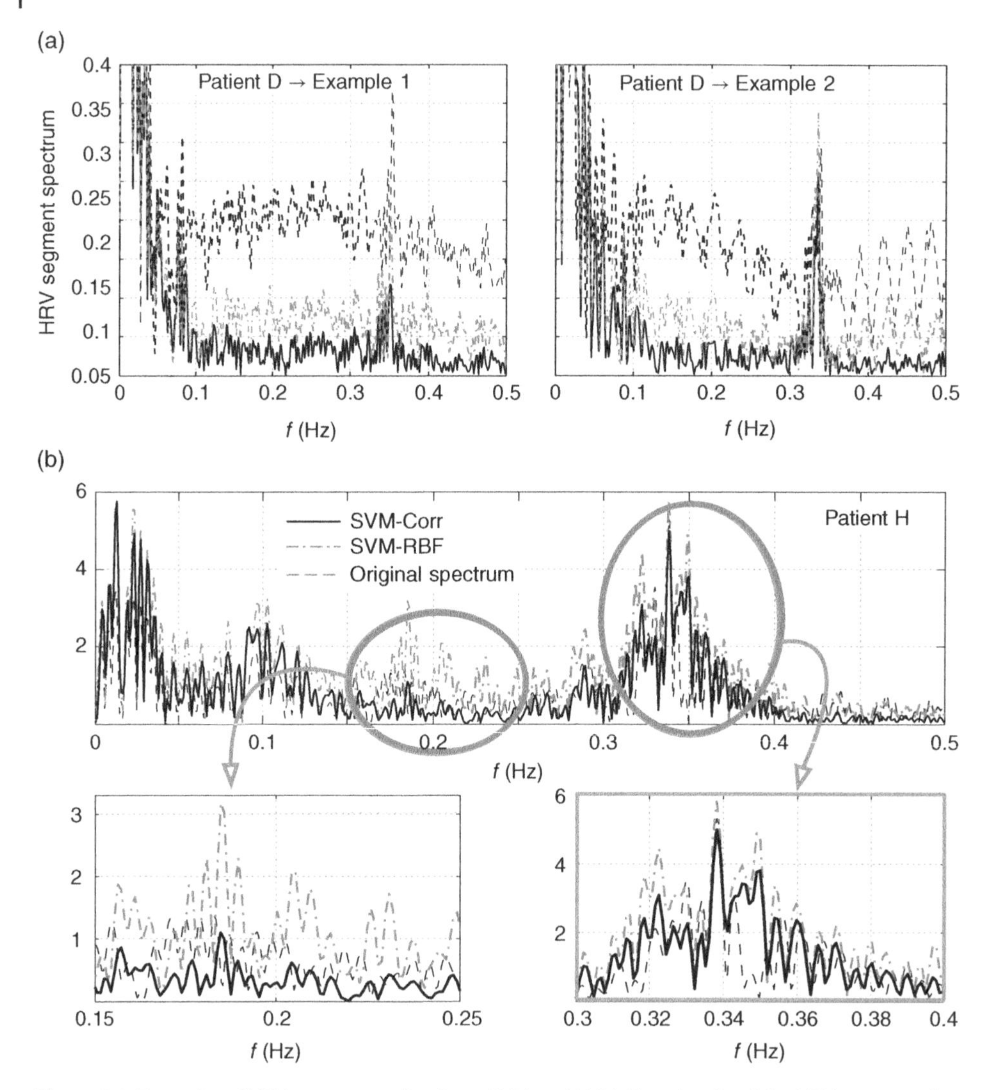

Figure 7.9 Examples of HRV segments of patients D (a) and H (b). Two details of the HRV spectrum for patient H are shown in the lower plots.

cardiac events. Whereas more advanced modes of ultrasound in echocardiography have been developed, M-mode is still a fundamental part of the routine echocardiographic exams.

In order to denoise the blood velocity from color Doppler M-mode (CDMM) images, we study here an SVM algorithm with a kernel able to get a better adaptation to these types of images, such as the kernel based on the image autocorrelation function described so far in this chapter. Conde-Pardo *et al.* (2006) earlier proposed an SVM algorithm based on a multiple-width RBF kernel for this goal. Extending those results, Soguero-Ruiz *et al.* (2016a) used an autocorrelation kernel to characterize the smoothness of these types of images. We include this application here to show how to generalize the autocorrelation concepts from 1D to higher dimensional domains. The next applications will also scrutinize the usefulness of the autocorrelation kernel in multidimensional signals and problems.

Dual Signal Model for Color Doppler Motion-Mode Image Denoising

A nonlinear regression model for CDMM images denoising using SVM can be described as follows. We denote the velocity field and the acquired image by $v_b(s,t)$ and $\{V_{i,j} = v(i\delta s, j\delta t),\ i = 1,\dots,N_s,\ j = 1,\dots,N_t\}$ respectively. In this case, $[i,j]$ denotes the image coordinates of each pixel $V_{i,j}$, and I denotes the set of coordinates for all the image pixels. This set can be divided into a training I^{tr} and a test I^{test} subsets. Then, the acquired image can be expressed as follows:

$$V_{ij} = \langle \boldsymbol{w}, \boldsymbol{\phi}([i,j])\rangle + b + e_{i,j}, \tag{7.29}$$

where $e_{i,j}$ is the model residual for each pixel, $\boldsymbol{\phi}([i,j])$ is a nonlinear application of coordinate vector $[i,j]$ to a high-dimensional feature space $\mathcal{F}$, and b is a bias term. A linear regression for the pixel value is given by the dot product of nonlinearly transformed pixel coordinates and $\boldsymbol{w} \in \mathcal{F}$. Given this image model, and using as usual the ε-Huber cost, the primal model can be stated as minimizing

$$\frac{1}{2}\|\boldsymbol{w}\|^2 + \frac{1}{2\delta}\sum_{[i,j]\in I_1^{\text{tr}}}(\xi_{i,j}^2 + \xi_{i,j}^{\star 2}) + C\sum_{[i,j]\in I_2^{\text{tr}}}(\xi_{i,j} + \xi_{i,j}^{\star}) - \sum_{[i,j]\in I_2^{\text{tr}}}\frac{\delta C^2}{2} \tag{7.30}$$

with respect to w^p, $\{\xi_{i,j}\}$, $\{\xi_{i,j}^{\star}\}$, and b, and constrained to

$$V_{i,j} - \langle \boldsymbol{w}, \boldsymbol{\phi}([i,j])\rangle - b \le \varepsilon + \xi_{i,j} \tag{7.31}$$

$$-V_{i,j} + \langle \boldsymbol{w}, \boldsymbol{\phi}([i,j])\rangle + b \le \varepsilon + \xi_{i,j}^{\star}, \tag{7.32}$$

and to $\xi_{i,j}, \xi_{i,j}^{\star} \ge 0$, for $[i,j] \in I^{\text{tr}}$; $\{\xi_{i,j}\}$ and $\{\xi_{i,j}^{\star}\}$ are *slack variables* or *losses*, and I_1^{tr} and I_2^{tr} are the subsets of pixels for which losses are in the quadratic or in the linear cost zone respectively.

The dual problem can be expressed in terms of the correlation matrix of input space pixel pairs, $\boldsymbol{R}([i,j],[k,l]) \equiv \langle \boldsymbol{\phi}([i,j]), \boldsymbol{\phi}([k,l])\rangle$, maximizing

$$-\frac{1}{2}(\boldsymbol{\alpha} - \boldsymbol{\alpha}^{\star})^{\text{T}}[\boldsymbol{R} + \delta \boldsymbol{I}](\boldsymbol{\alpha} - \boldsymbol{\alpha}^{\star}) + (\boldsymbol{\alpha} - \boldsymbol{\alpha}^{\star})^{\text{T}}\boldsymbol{V} - \varepsilon \mathbf{1}^{\text{T}}(\boldsymbol{\alpha} + \boldsymbol{\alpha}^{\star}) \tag{7.33}$$

constrained to $C \ge \alpha_{i,j}^{(\star)} \ge 0$, where $\alpha_{i,j}$ and $\alpha_{i,j}^{*}$ are the Lagrange multipliers corresponding to Equations 7.31 and 7.32; and $\boldsymbol{\alpha}^{(\star)} = [\alpha_{i,j}^{(\star)}]$ and $\boldsymbol{V} = [V_{i,j}]$, for $[i,j] \in I^{\text{tr}}$, are column vectors. After obtaining $\boldsymbol{\alpha}^{(\star)}$, the velocity for a pixel at $[k,l]$ is

$$\hat{V}_{k,l} = \sum_{[i,j]\in I^{\text{tr}}} \beta_{i,j}\langle \boldsymbol{\phi}([i,j]), \boldsymbol{\phi}([k,l])\rangle + b, \tag{7.34}$$

with $\beta_{i,j} = \alpha_{i,j} - \alpha^{\star}_{i,j}$, which is a weighted function of the nonlinearly observed times in the feature space.

By using Mercer kernels, we can substitute $K([i,j],[k,l]) = \langle \phi([i,j]), \phi([k,l]) \rangle$. We will use here the *Gaussian kernel*, given by

$$K_{\rm G}([i,j],[k,l]) = \exp\left(\frac{-\|[i,j]-[k,l]\|^2}{2\sigma^2}\right),$$

where σ is the width parameter. In Conde-Pardo *et al.* (2006), a double-width RBF kernel was proposed, given by

$$K_{\rm D}([i,j],[k,l]) = \exp\left(\frac{-|i-k|^2}{2\sigma_s^2}\right)\exp\left(\frac{-|j-l|^2}{2\sigma_t^2}\right),$$

in order to separate the contribution of the two different dimensions (space and time).

Taking into account that CDMMs are characterized by their smoothness, we study the following autocorrelation kernel, adapted from results in this chapter so far:

$$K([i,j],[k,l]) = \rho_v(k-i, l-j), \tag{7.35}$$

with ρ_v denoting the image autocorrelation. Thus, the CDMM image model can finally be expressed as

$$\hat{V}_{k,l} = \sum_{[i,j]\in I^{\rm tr}} \beta_{i,j} K([i,j],[k,l]) + b. \tag{7.36}$$

We used a cross-validation strategy to tune the free parameters of the SVM cost function (ε, δ, C), as seen in Listing 7.17.

```
function [predict,coefs] = ...
    SVM_AutoCorr(Xtrain,Ytrain,Xtest,gamma,nu,C,path_acorr)

% Initialize params
if nargin==7 && ischar(path_acorr)
    load(path_acorr) % Precomputed autocorrelation of all available data
else
    % Computing autocorrelation with training data
    if nargin<7; paso=path_acorr; else paso=1; end
    x = (min(Xtrain(:)):paso:max(Xtrain(:))+paso);
    axs=cell(1,size(Xtrain,2)); axs(:)={x(:)};
    [ACs{1:size(Xtrain,2)}]=ndgrid(axs{:});
    interpAxis = griddatan(Xtrain,Ytrain,cat(2,ACs{:}));
    interpAxis(isnan(interpAxis)) = 0;
    [acorr,acs] = autocorr(reshape(interpAxis,size(ACs{1})),ACs{:});
end
% Construct kernel matrix
H = getAutoCorrKernel(Xtrain,Xtrain,acorr,acs{:});
% Train SVM and predict
nchunks=16; ntest=length(Xtest); % Chunks partition for speeding up
chunks=floor(ntest/nchunks); % If Xtest is an image: 4x4 chunks
chunksTest = mat2cell(Xtest,...
```

```
    [repmat(chunks,1,nchunks-1),ntest-(nchunks-1)*chunks],2);
%inparams=sprintf('-s 3 -t 4 -g %f -c %f -p %f -j 1', gamma, C, epsilon);
inparams=sprintf('-s 4 -t 4 -g %f -c %f -n %f -j 1', gamma, C, nu);
chunksPredict=cell(1,nchunks);
for i=1:nchunks
    Htest = getAutoCorrKernel(Xtrain,chunksTest{i},acorr,acs{:});
    [chunksPredict{i},model]=SVM(H,Ytrain,Htest,inparams);
end
predict=cell2mat(chunksPredict');
coefs = getSvmWeights(model);
```

Listing 7.17 Autocorrelation calculation for CDMM image denoising (SVM_Autocorr.m).

A simple synthetic model of diastolic transmitral flow in CDMM was created by addition of three bivariate Gaussian components, given by

$$v_b(s,t) = \sum_{i=1}^{3} a_i \exp\left\{-\frac{1}{2}[s_i,t_i]\Sigma_i^{-1}[s_i,t_i]^{\mathrm{T}}\right\}, \tag{7.37}$$

where $[s_i, t_i]$ denotes a bidimensional row vector, Σ_i is the covariance matrix of each component, and s_i, t_i, and Σ_i are given in Table 7.6. This table shows the parameters that were adjusted to match physiological values of both waves. Time zero was defined at the QRS onset. Listing 7.18 provides the code implementation, which is represented in Figure 7.10.

```
function velocityModel
% Simulated velocity model of diastole, with three waves (E1, E2, A).

syms f v s t ro Sigma

% Parameters
s1=8;    s2=4;                    %  limits of the spatial integration
Sigma = [1 -0.5; -0.5 1];
at= 0.05;    as= 1.5;

v0 = 100*exp(-(1/2) *[s,t] * inv(Sigma)*[s;t]);          % E (symbolic)
v1 = subs(v0,{s,t},{(s-6)/as, ((t+0.25)/at)});           % Substitution for  E1
v2 = 0.5 * subs(v0,{s,t},{(s-7)/as,(t+0.08)/at});        % Substitution for E2
Sigma2 = [.8 -.05; -.05 .1];
v4 = 100*exp(-(1/2) *[s,t] * inv(Sigma2)*[s;t]);         % A (symbolic)
v3 = .75*subs(v4,{s,t},{(s-6)/as, ((t+0.3)/at)});        % Substitution for A

% To avoid values higher than 100 cm/s
v1 = .9*v1; v2 = .9*v2; v3 = .65*v3;
v = v1+v2+v3;
```

Listing 7.18 Velocity model image for CDMM image (velocityModel.m).

Sample Selection Criteria

As described before, the dataset was divided into two subsets: one for training and one for testing. Note that this is a special application of machine learning principles, in which we actually have available all the existing samples (i.e., the pixels in an image for denoising). Therefore, different strategies for training samples selection are expected

Table 7.6 Parameter values of color-Doppler transmitral flow model (a_i cm/s, t_i s).

i	1	2	3
t_i	$\frac{t+0.25}{0.05}$	$\frac{t+0.08}{0.05}$	$\frac{t+0.25}{0.05}$
a_i	90	45	48.75
s_i	$\frac{s-6}{1.5}$	$\frac{s-7}{1.5}$	$\frac{s-6}{1.5}$
Σ_i	$\begin{pmatrix} 1 & -0.5 \\ -0.5 & 1 \end{pmatrix}$	$\begin{pmatrix} 1 & -0.5 \\ -0.5 & 1 \end{pmatrix}$	$\begin{pmatrix} 0.8 & -0.5 \\ -0.5 & 0.1 \end{pmatrix}$

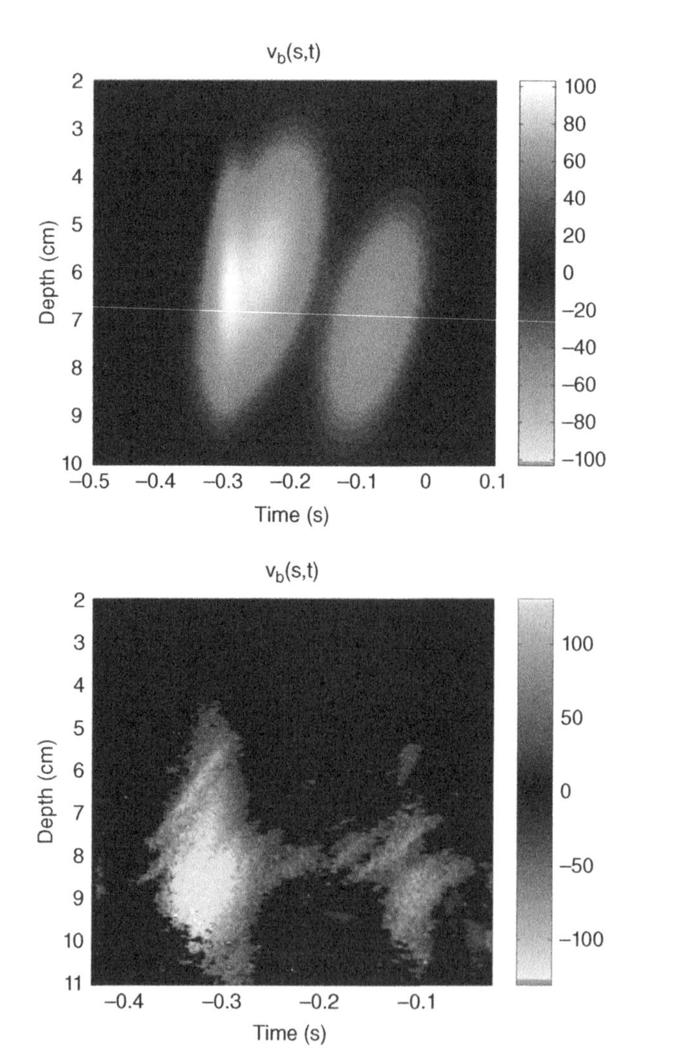

Figure 7.10 Synthetic (up) and real (down) CDMM images used in this example.

to have an impact on the final performance. Therefore, we scrutinized several training set selection criteria:

1) *Random.* Samples for training and test subsets are selected randomly.
2) *Based on amplitudes.* All pixels are ranked in a descending way based on their amplitude values. Greater weights are given to those samples (pixels) with greater amplitudes following a Gaussian function, and the training and test sets are finally constituted.
3) *Based on gradients.* The image pixels are ranked in order of their gradient amplitude from greatest to least. The amplitude gradient is obtained by merging (in module) the amplitude changes between adjacent pixels in both image directions. Then, by using a Gaussian function, greater weights are given to those samples (pixels) corresponding to greater gradients values. Finally, the training and test sets are constituted.
4) *Based on second derivatives.* This criterion is based on the gradient one. The steps to follow are the same as those in the gradient-based criterion by considering the second derivatives as the changes in the amplitude gradient between adjacent pixels in both image directions.
5) *Based on edges.* By using a 2D edge filter (Sobel filter (Farid and Simoncelli, 2004)), only those samples from image regions with sharp amplitude changes are considered to build the machine. A threshold can be considered to obtain a training set with a sufficient numbers of samples. Finally, the training and test sets are constituted.

See `randomCriterion.m`, `amplitudeCrierion.m`, `gradientCriterion.m`, `SeconderivativeCriterion.m`, and `EdgeCriterion.m` for the implementation of each of these criteria.

Experiments and Results

A DSM algorithm with three different kernels was used to denoise the CDMM image; specifically, RBF, D-RBF, and autocorrelation kernels, considering different sample selection criteria and 250 and 1000 training samples in each approach. Table 7.7 shows the results obtained in terms of MAE, concluding that better approximations are obtained when an autocorrelation kernel is considered and when the number of training samples increases. Then, we evaluate different sample selection criteria (random, edges, amplitude, gradient, second derivative) considering a higher number of training samples (100, 500, 1000, and 2000) (see Table 7.8). In general, lower MAE values were obtained as the number of training samples increased. Hence, a sample selection criterion based on edges provides good approximations.

Table 7.7 Synthetic image. MSE in CDMMI approximation considering different kernels.

No. training pixels	RBF	D-RBF	Corr
250	7.93	1.52	1.23
1000	7.01	0.75	0.72

Table 7.8 Synthetic image. MSE in CDMMI approximation considering different sample selection criteria.

No. training pixels	Random	Edges	Amplitude	Gradient	Second derivative
100	2.15	1.96	6.31	4.03	5.85
500	0.94	0.96	3.45	1.48	3.22
1000	0.77	0.79	2.14	0.96	2.35
2000	0.69	0.66	1.79	0.71	0.92

Table 7.9 Real image. MSE in CDMMI approximation considering different sample selection criteria.

No. training pixels	Random	Edges	Amplitude	Gradient	Second derivative
100	4.97	4.50	17.1	8.24	6.09
500	2.99	3.00	11.6	5.25	5.08
1000	2.52	2.67	7.86	5.03	4.14
2000	2.32	2.20	5.35	4.24	3.50

After obtaining results based on synthetic data, we analyzed results based on a real CDMM image (126×171) from a healthy volunteer. In this case, an SVM algorithm with an autocorrelation kernel was considered for the image approximation, as well as the same sample selection criterion as the one evaluated with the synthetic image, for 100, 500, 1000, and 2000 training pixels, respectively (see Table 7.9). As expected, the higher the number of training samples, the better the approximation obtained. Figure 7.11 shows the flow velocity recording for a healthy volunteer (a) and the residual obtained when using different sample selection criteria: (b) random; (c) based on edges; (d) based on amplitudes, (e) based on gradients; and (f) based on second derivative considering 1000 training pixels. In general, the CDMM image residuals are lower with an edges sample selection criteria.

7.4.7 Indoor Location from Mobile Devices Measurements

In recent years, a great number of location-dependent computer services have been proposed using wireless mobile devices. For these services to work properly, a key issue is the knowledge of the position of the user, which is a complex task, especially in the context of indoor scenarios. For cost-effective systems, one of the most common approaches to provide indoor location is to use the received signal strength (RSS) measured by 802.11 devices to estimate the user location (Bahl and Padmanabhan, 2000). Fingerprinting techniques are used for this purpose, which consist of finding the relationship between the RSS and the position from a set of measurements in known locations (which can be dealt with as a training set of samples) and then using the learned relationship to estimate the position of the user according to its received power. The k-nearest neighbors (k-NN) scheme is a very simple statistical approach quite often used for location purposes (Bahl and Padmanabhan 2000; Figuera *et al.* 2009, 2011).

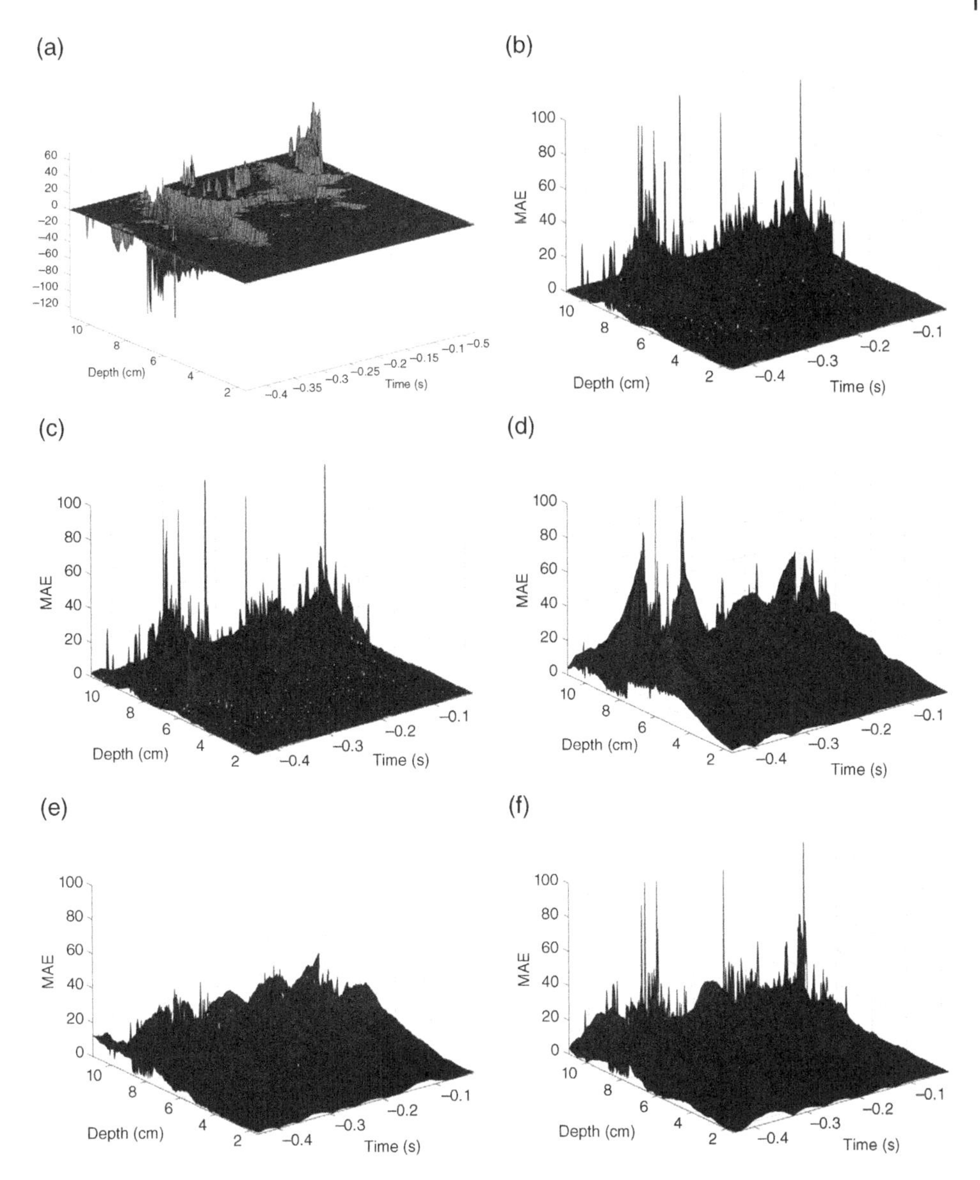

Figure 7.11 Color Doppler M-mode trasmitral flow velocity recording for a healthy volunteer (a) and residuals using different sample selection criteria: (b) random; (c) based on edges; (d) based on amplitudes, (e) based on gradients; and (f) based on second derivative.

With this method, the measured RSS is compared with those in the training set, finding the k most similar entries and then averaging them to estimate the location. Several probabilistic algorithms and other technically advanced approaches have also been proposed to perform signal fingerprinting.

Though statistical learning approaches can estimate the unknown mapping from a set of examples, the result can be significantly improved by incorporating a priori knowledge about the problem, in the same way that better inference is provided when a priori knowledge is considered in a Bayesian framework. To exploit the SVM potential

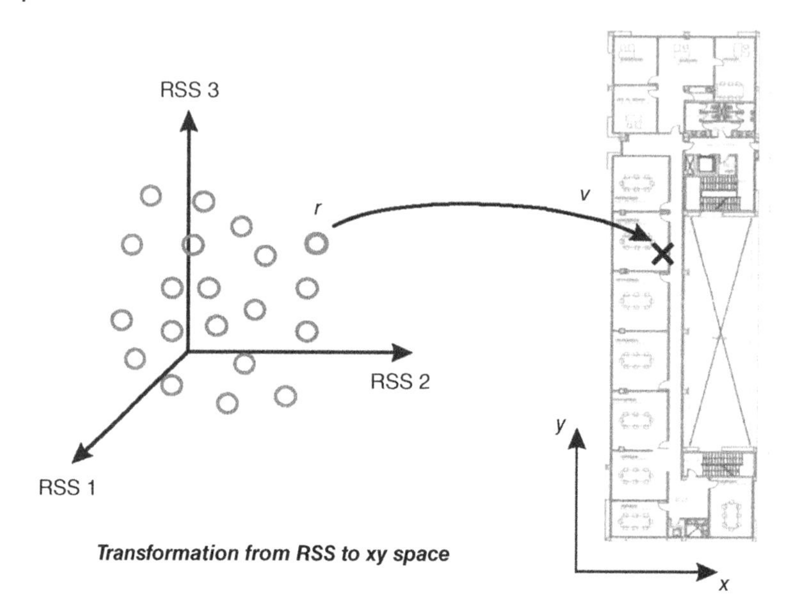

Figure 7.12 Conceptual representation of function $\mathbf{v} = \boldsymbol{f}(\boldsymbol{r})$ for three APs.

in the location context from regression algorithms, two ways were proposed by Figuera *et al.* (2012) for incorporating a priori knowledge in the SVM design:

1) Choosing a suitable kernel, adapted to the specific task, and consisting of the multidimensional autocorrelation of RSS space from multiple WiFi nodes.
2) Taking advantage of the potential relationship between the two spatial coordinates (in bidimensional location) and designing a single SVM scheme to provide the two coordinates simultaneously by means of complex algebra. This takes advantage of the fact that the two coordinates are in fact strongly correlated.

Regarding the first method, the standard kernel is replaced by an alternative one that incorporates the spatial structure of the training set. This adapted kernel is obtained as the autocorrelation function relating RSS to spatial position. The second proposal considers the possible dependence between the two spatial coordinates by applying a complex SVM formulation (Martínez-Ramón *et al.*, 2007), so that real and imaginary parts of the complex output correspond to the x and y coordinates respectively. A third proposal was developed therein by simply combining the autocorrelation kernel and the complex output. In this section, we present some relevant results from the work in Figuera *et al.* (2012). The equations of the SVM models can be found detailed in that reference.

Signal Model for Received Signal Strength Indoor Location

The 2D location problem based on fingerprinting consists of estimating a function which provides the user position $\nu = [\nu^x, \nu^y]^{\mathrm{T}}$ from the measured RSS (see Figure 7.12) as follows:

$$\nu = (\mathbf{r}) + \mathbf{e}(\mathbf{r}), \tag{7.38}$$

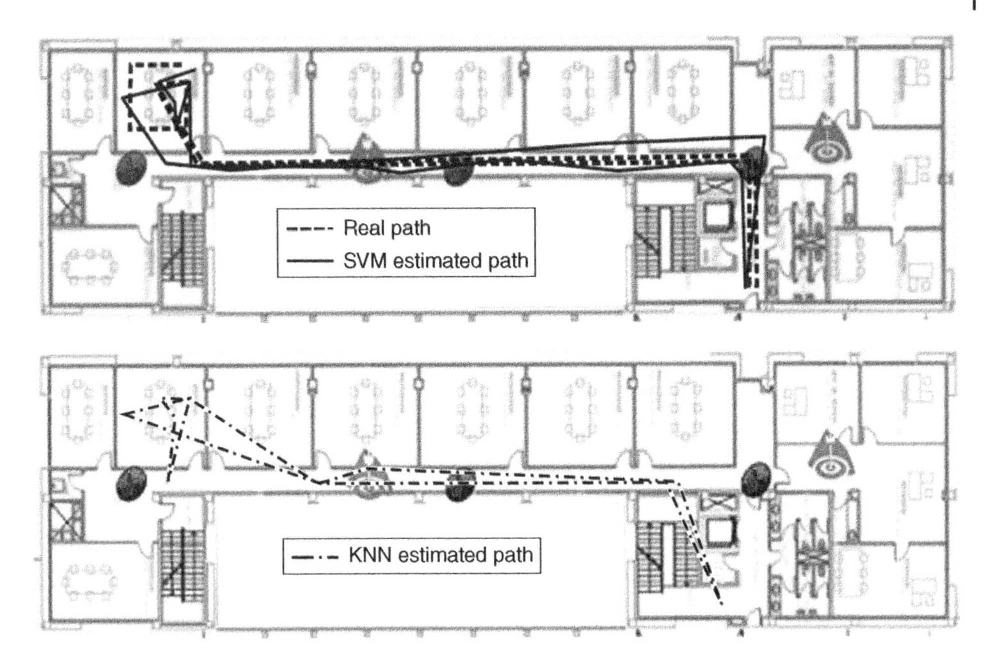

Figure 7.13 Estimation of the location of a user walking from the office to the elevator and back obtained with the SVM-CC algorithm and with the k-NN algorithm.

where $\boldsymbol{r}$ is the $\mathbb{R}^Q$ vector containing the RSS from the Q access points (APs) used for the location services, $\boldsymbol{f}$ is the mapping $\mathbb{R}^Q \longrightarrow \mathbb{R}^2$, and $\boldsymbol{e}$ accounts for the noise and the nondeterministic part of the mapping between the RSS and position.

The general procedure of a fingerprinting technique based on SVM methodology is as follows. First, a set of measurements is registered at L known locations. This set of measurements is then preprocessed to build the input space for the algorithm. The output of this process is the training set $\mathcal{A} \equiv \{A_1, \dots, A_L\}$ with $A_l = (\boldsymbol{r}_l, \boldsymbol{v}_l)$, where $\boldsymbol{r}_l = [r_l^1, \dots, r_l^Q]^{\mathrm{T}}$ are the RSS measurements from the available Q APs at location $\boldsymbol{v}_l$. Using this set $\mathcal{A}$ the algorithm can then "learn" the dependence between the received power and the position. For this purpose, several parameters have to be adjusted. The first one is the kernel of the SVM, which in this section can be either the RBF kernel or the autocorrelation kernel. The rest of the parameters are related to the cost function to optimize (here, the ϵ-Huber cost), and they are regularization parameter δ and trade-off parameter C.

Usually, the APs used in the location system are placed when the WiFi network is deployed only for communications purposes. In these cases, when the location system is implemented, the position of the APs is not a design parameter, but a fixed constraint. In our setup, the only APs which could be used for location tasks were those represented in Figure 7.13. We selected AP1, AP2, and AP3 for our experiments since they had a full coverage over the experimental area.

A Complex Support Vector Machine Algorithm

With this general procedure in mind, a complex SVR with autocorrelation kernel is used to solve the indoor location problem, which is given in Listing 7.19.

```
function [predict,coefs] = ...
    SVM_CC(Xtrain,Ytrain,Xtest,gamma,epsilon,C,path_acorr)

% Initialize params
if nargin==7 && ischar(path_acorr)
    load(path_acorr) % Precomputed autocorrelation of all available data
else
    % Computing autocorrelation with training data
    Ytrain = Ytrain(:,1) + 1i * Ytrain(:,2);
    if nargin<7; stepA=path_acorr; else stepA=1; end
    x = (min(Xtrain(:)):stepA:max(Xtrain(:))+stepA);
    axs=cell(1,size(Xtrain,2)); axs(:)={x(:)};
    [ACs{1:size(Xtrain,2)}]=ndgrid(axs{:});
    interpAxis = griddatan(Xtrain,Ytrain,reshape(cat(2,ACs{:}),numel
                                   (ACs{1}),length(ACs)));
    interpAxis(isnan(interpAxis)) = 0;
    [acorr,acs] = autocorr(reshape(interpAxis,size(ACs{1})),ACs{:});
    Ytrain = [real(Ytrain); imag(Ytrain)];
end
% Construct kernel matrix
H = real(getAutoCorrKernel(Xtrain,Xtrain,acorr,acs{:}));
Htest = real(getAutoCorrKernel(Xtrain,Xtest,acorr,acs{:}));
H = blkdiag(H,H);
Htest = blkdiag(Htest,Htest);

% Train SVM and predict
%params.ggamma=1e-1; params.epsilon=1e-6; params.C=100;
inparams=sprintf('-s 3 -t 4 -g %f -c %f -p %f -j 1', gamma, C, epsilon);
[ypred,model]=SVM(H,Ytrain,Htest,inparams);
Ntest=size(Xtest,1);
Ytestpred = ypred(1:Ntest) + 1i*ypred(Ntest+1:end);
predict = [real(Ytestpred), imag(Ytestpred)];
coefs = getSvmWeights(model);
```

Listing 7.19 A complex SVR with autocorrelation kernel algorithm (SVM_CC.m).

Note that autocorr function is used to compute the autocorrelation required to construct the kernel matrix. This autocorrelation kernel matrix is then obtained by the code in Listings 7.20 and 7.21.

```
function [ac,Taus] = autocorr(f,varargin)

[sizes{1:nargin-1}]=size(varargin{1});
t=cell(1,nargin-1);
for i=1:nargin-1
    ind=prod(cell2mat(sizes(1:i-1)));
    Ts = varargin{i}(ind+1)-varargin{i}(1);
    Rng = max(varargin{i}(:))-min(varargin{i}(:));
    t{i} = -Rng:Ts:Rng;
end
if nargin==2      % 1D
    ac = xcorr(f);
elseif nargin==3  % 2D
    ac = xcorr2(f);
else              % 3D
    ac = convn(f,f(end:-1:1,end:-1:1,end:-1:1));
end
```

```
ac = ac/max(ac(:));
[Taus{1:length(varargin)}] = ndgrid(t{:});
```

Listing 7.20 Estimating the autocorrelation (autocorr.m).

```
function H = getAutoCorrKernel(Xtrain,Xtest,acorr,acx,acy,acz)

[str1,str2]=size(Xtrain);
stst1=size(Xtest,1);

deltaR = zeros(stst1,str1,str2);
for nx=1:stst1
    deltaR(nx,:,:) = bsxfun(@minus,Xtest(nx,:),Xtrain);
end
deltaRv = reshape(deltaR,[stst1*str1,str2]);
dRv = num2cell(deltaRv,1);
NN=1e6; % Maximum size for batch execution
NTEST=size(deltaRv,1);
if NTEST<NN; % Batch execution
    if str2==2
        Hv=interp2(acx',acy',acorr',dRv{:},'*linear');
    elseif str2==3
        acx=permute(acx,[2 1 3]); acy=permute(acy,[2 1 3]);
        acz=permute(acz,[2 1 3]); acorr=permute(acorr,[2 1 3]);
        Hv=interp3(acx,acy,acz,acorr,dRv{:},'*linear');
    end
else % Obtaining kernel matrix by chunks
    Hv=zeros(NTEST,1);
    L=ceil(NTEST/NN);
    for k=1:L
        range=(k-1)*NN+1:min(k*NN,NTEST);
        dRv = num2cell(deltaRv(range,:),1);
        if str2==2
            Hv(range)=interp2(acx',acy',acorr',dRv{:}, '*linear');
        else
            acx=permute(acx,[2 1 3]); acy=permute(acy,[2 1 3]);
            acz=permute(acz,[2 1 3]); acorr=permute(acorr,[2 1 3]);
            Hv(range)=interp3(acx,acy,acz,acorr,dRv{:}, '*linear');
        end
    end
end
H = reshape(Hv,stst1,str1);
```

Listing 7.21 Getting the autocorrelation kernel matrix (getAutoCorrKernel.m).

The reader is encouraged to check the code presented with the original equations in Figuera *et al.* (2012).

Experiment Settings and Results

In order to test the solution, a 2 min walk was recorded in real time. Figure 7.13 displays the estimation of this walk with the SVM-CC algorithm against the real user walking, as well as the estimation obtained with the classical k-NN method, which is used as a reference algorithm. The code for setting the experiment and for visualizing the estimations is given in supplementary code `config_indoorLocation.m` and `display_results_indoorLocation.m`.

The estimation of the position was made during a walk from an office to the elevator and back. The experiment was performed during working hours, so several people performing normal tasks were present during the experiment. Also, no attempt was made to control the environmental conditions. For example, orientation of the mobile device, or whether the doors remained open or closed. Moreover, training and free parameter tuning were based on the usual measurement settings. These measurements, taken during day 1, were obtained under a completely different set of conditions compared with training data. The RSSs of AP1, AP2, and AP3 are used, and the SVM-CC is used for the location estimation. The whole path was completed in 1 min, and the estimation of each location lasted less than 100 ms with a MATLAB program running on a laptop with a 2.5 GHz microprocessor. As can be observed in the figures, small acceptable errors were observed along the path.

In summary, including a priori information provided by the autocorrelation of the training set and using the complex output leads to improvements in the performance of the learning algorithm, which in this case is an SVM-based technique. Among the proposed algorithms, the SVM-CC uses the autocorrelation kernel and the complex output, and provides the best performance in terms of all the quality indicators tested. Also, these results show that the combined complex SVM with autocorrelation kernel is a powerful tool for dealing with multidimensional estimation problems.

7.4.8 Electroanatomical Maps in Cardiac Navigation Systems

In the cardiac arrhythmia treatment, the knowledge of the arrhythmia mechanisms is decisive for the application of a successful therapy. Among the current treatments, the cardiac ablation in electrophysiology (EP) studies is one of the most effective for some kind of arrhythmia, such as ventricular or supraventricular tachycardia. During ablation procedures, several catheters are introduced inside the heart with the purpose of searing the tissue that is producing the arrhythmia using radiofrequency or intense cold (Morady, 1999). Until the late 1990s, 2D X-ray images were used to guide and locate the catheters inside the heart during EP studies. The projection of bones and the heart shape were used as reference.

However, this technique did not provide electric information about the arrhythmia, and the exposure to radiation was a hazard for both the patient and the medical team. For that reason, cardiac navigation systems (CNSs) were created as an alternative to the X-ray technique. These systems are able to build a 3D electroanatomic map (EAM) using catheters inside the heart, which register and store both their 3D anatomical position and the electrical signal associated with each position (Cross *et al.*, 2015).

EAMs represent the anatomical cardiac surface by means of a triangular mesh (vertices corresponding to the spatial position joined by triangular faces) and an electrical feature, such as activation time or voltage amplitude, projected onto the surface and associated with each vertex. The electrical feature is obtained by processing the registered electrical signal, a so-called electrogram, measured with the catheters. These EAMs support the electrophysiologist's decisions and determining the ablation targets (Juneja 2009; Shpun *et al.* 1997).

CNSs also show a representation of the catheters inside the EAM with high accuracy during the EP procedure and allow including relevant anatomical information of cardiac

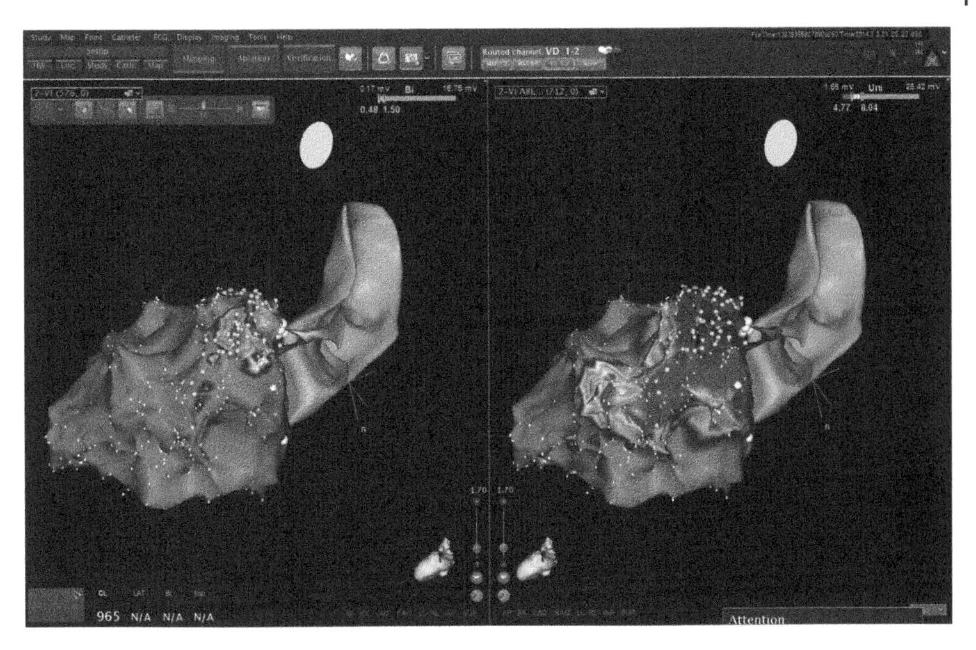

Figure 7.14 Bipolar (left) and unipolar (right) voltage EAM of a left ventricle generated by the Carto 3® system (Biosense Webster®).

geometry obtained from the segmentation of computed tomography or magnetic resonance images (Cross *et al.*, 2015). Figure 7.14 shows the bipolar and unipolar voltage EAM of a left ventricle during the EP study of a ventricular tachycardia.

Although current CNSs use multi-electrode catheters for the fast acquisition of a high number of anatomical vertices (the 3D position), the associated electrical feature requires several seconds to be registered, and hence not all vertices have an associated electrical feature. For that reason, an interpolation method is used to estimate the electrical feature in the vertices where the electrogram was not registered.

The accuracy of the EAM is determined by the number of registered vertices with associated electrical information and the interpolation method; hence, the aim was to evaluate visually and quantitatively the effect of the interpolation method in the electrical feature comparing the EAM provided by the CNS with the EAM obtained with three interpolation methods; namely, (1) linear interpolation (LI), (2) thin-plate spline (TPS), and (3) the ν-SVR with RBF kernel.

Methods

The LI method creates first a Delaunay triangulation formed by a set of adjacent, continuous, and nonoverlapping triangles, which represent a convex surface of the set of vertices $\boldsymbol{v}_i$. A triangulation is a Delaunay triangulation when the circumcircle about each triangle contains no other vertices in its interior. It has several advantageous properties;, for instance, it maximizes the minimum angle in the plane or it is the convex hull of the vertices. Readers are referred to de Berg *et al.* (2008) for a detailed explanation of the Delaunay triangulation properties. The value of the new interpolated vertex is

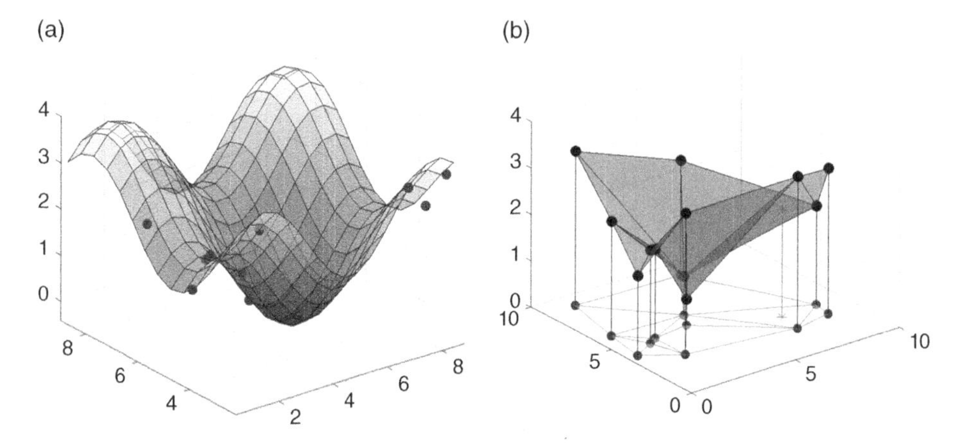

Figure 7.15 (a) Example of function $f(\mathbf{x},\mathbf{y}) = \sin(\mathbf{x}) + \sin(\mathbf{y}) + 2$ used to evaluate the analyzed interpolation methods (i.e., LI, TPS, and ν-SVR). The vertices (samples) used to interpolate are marked with blue points. (b) LI of the function in (a) using a Delaunay triangulation and a set of 12 vertices (blue points). First, the triangulation is computed from the vertices, and then the value of new vertex (red asterisk) is obtained by using the triangle which encloses this new vertex.

obtained by finding the triangle which encloses it and computing the weighted sum of the corresponding three vertices associated with the enclosed triangle. Figure 7.15b shows an example of the Delaunay triangulation for a set of 12 selected vertices (marked with blue points) from the function $f(\boldsymbol{x},\boldsymbol{y}) = \sin(\boldsymbol{x}) + \sin(\boldsymbol{y}) + 2$ (see Figure 7.15a), and the LI of a new vertex (marked with a red asterisk) by using this triangulation.

The TPS method estimates a thin, smooth surface that passes through all given vertices. This surface is constructed by selecting a function f minimizing the bending energy of a surface (Bookstein 1989; Friedman *et al.* 2001; Powell 1994). Thus, the name *TPS* refers to the analogy of bending a thin sheet of metal. For a set of N vertices in $\mathbb{R}^2$, where $\boldsymbol{v}_i = [\boldsymbol{x}_i^{\mathrm{T}}, \boldsymbol{y}_i^{\mathrm{T}}]^{\mathrm{T}}$ and $\boldsymbol{z}_i = f(\boldsymbol{v}_i) = f(\boldsymbol{x}_i, \boldsymbol{y}_i)$, the functional to minimize is defined as

$$\sum_{i=1}^{N} (h_i - f(\boldsymbol{v}_i))^2 + \lambda I(f), \tag{7.39}$$

where $I(f)$ is a penalty functional, which takes the following form in $\mathbb{R}^2$:

$$I(f) = \int\int_{\mathbb{R}^2} \left(\frac{\mathrm{d}^2 f}{\mathrm{d}x^2} + 2\frac{\mathrm{d}^2 f}{\mathrm{d}x\, \mathrm{d}y} + \frac{\mathrm{d}^2 f}{\mathrm{d}y^2} \right) \mathrm{d}x\, \mathrm{d}y. \tag{7.40}$$

λ is a positive constant that controls the smoothing in order to relax the interpolation in the presence of noise. $\lambda = 0$ corresponds to the interpolation function without regularization (no smoothing), whereas $\lambda \to \infty$ shows a least-squares plane solution.

Figure 7.16 shows the interpolation of the aforementioned set of 12 vertices using the TPS method and different λ values (i.e., 0, 1, 10, 100, 1000, and 10 000). Meanwhile, the surface passes through all the vertices (no regularization) for $\lambda = 0$; the surface is a plane for $\lambda = 10\,000$.

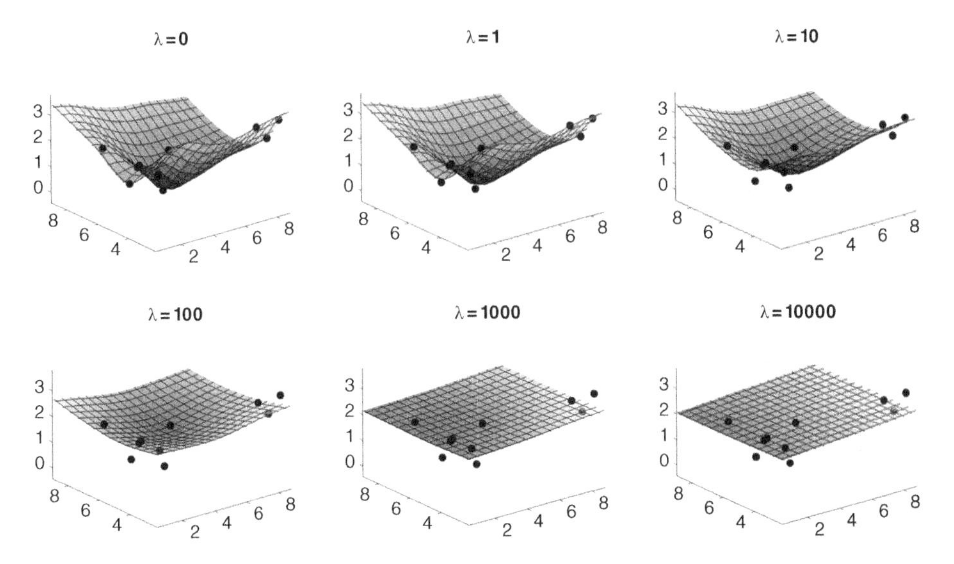

Figure 7.16 TPS interpolation of the function f defined in Figure 7.15a using $\lambda = 0, 1, 10, 100, 1000,$ and 10 000.

The solution to this minimization problem is as follows:

$$f(\mathbf{v}_j) = \beta_0 + \boldsymbol{\beta}^{\mathrm{T}}\mathbf{v}_j + \sum_{i=1}^{N} \alpha_i s(r), \tag{7.41}$$

where $s(r) = r^2 \log(r)$, $r = \|\mathbf{v}_j - \mathbf{v}_i\|$ is an RBF expression, β_0 is a constant, $\boldsymbol{\beta}$ represent a linear regression (i.e., $\beta_x \boldsymbol{x}_i + \beta_y \boldsymbol{y}_i$), and the coefficients $\boldsymbol{\alpha}_i$ have a set of constraints:

$$\sum_{i=1}^{N} \boldsymbol{\alpha}_i = 0; \quad \sum_{i=1}^{N} \boldsymbol{\alpha}_i \boldsymbol{x}_i = 0; \quad \sum_{i=1}^{N} \boldsymbol{\alpha}_i \boldsymbol{y}_i = 0. \tag{7.42}$$

The solution can also be obtained by following a matrix notation:

$$\begin{bmatrix} \boldsymbol{K} + \lambda \boldsymbol{I} & \boldsymbol{P}^{\mathrm{T}} \\ \boldsymbol{P} & \mathbf{0} \end{bmatrix} \begin{bmatrix} \boldsymbol{\alpha} \\ \boldsymbol{\beta} \end{bmatrix} = \begin{bmatrix} \boldsymbol{z} \\ \boldsymbol{o} \end{bmatrix}, \tag{7.43}$$

where $\boldsymbol{K}$ is an $N \times N$ matrix, the elements of $\boldsymbol{K}$ are defined as $K_{ij} = s(r_{ij}) = s((\|\mathbf{v}_j - \mathbf{v}_i\|)^2 \log(\|\mathbf{v}_j - \mathbf{v}_i\|)$. $\boldsymbol{P}$ is a $3 \times N$ matrix where the ith row of $\boldsymbol{P}$ is $(1, x_i, y_i)$. $\mathbf{0}$ is a 3×3 matrix of zeros, $\boldsymbol{z}$ is a column vector of target values, and $\boldsymbol{o}$ is a column vector of three zeros. The column vectors $\boldsymbol{\alpha}$ and $\boldsymbol{\beta}$ correspond to the coefficients of Equation 7.41 and they can be computed by solving the following matrix equation:

$$\begin{bmatrix} \boldsymbol{\alpha} \\ \boldsymbol{\beta} \end{bmatrix} = \begin{bmatrix} \boldsymbol{K} + \lambda \boldsymbol{I} & \boldsymbol{P}^{\mathrm{T}} \\ \boldsymbol{P} & 0 \end{bmatrix}^{-1} \begin{bmatrix} \boldsymbol{z} \\ \boldsymbol{o} \end{bmatrix}. \tag{7.44}$$

Although TPS was initially defined for a 2D interpolation, the formulation can be extended to higher dimensions (Goshtasby, 2005).

The ν-SVR is a learning algorithm based on the SRM principle (Chang and Lin 2002; Schölkopf *et al.* 2000). This method estimates a regression hyperplane in a high-dimensional space using the nonlinear mapping $\boldsymbol{\phi}(\boldsymbol{v})$ of the vertex coordinates (as input vectors). The regression function is defined as

$$f(\boldsymbol{v}_i) = \langle \boldsymbol{w}, \boldsymbol{\phi}(\boldsymbol{v}_i)\rangle + b + e, \tag{7.45}$$

where $\boldsymbol{w}$ is a linear weight vector, b is the bias term of the nonlinear regression data model, and e is the error term. In order to obtain both $\boldsymbol{w}$ and b in the previous model, the functional to minimize in the ν-SVR is as follows:

$$\begin{aligned} \min_{\boldsymbol{w},\epsilon,\xi^{(*)},b} \quad & \frac{1}{2}\|\boldsymbol{w}\|^2 + C\nu\epsilon + \frac{1}{N}\sum_{i=1}^{N}(\xi_i + \xi_i^*) \\ \text{subject to} \quad & (\langle \boldsymbol{w}, \boldsymbol{\phi}(\boldsymbol{v}_i)\rangle + b) - z_i \le \epsilon + \xi_i, \\ & z_i - (\langle \boldsymbol{w}, \boldsymbol{\phi}(\boldsymbol{v}_i)\rangle + b) \le \epsilon - \xi_i^*, \\ & \xi_i^{(*)} \ge 0, \epsilon \ge 0, \end{aligned} \tag{7.46}$$

where $i = 1, \dots, N$, $C \in (0, \infty)$ is the regularization parameter, and $\nu \in (0, 1)$ parameter is an upper bound on the fraction of margin errors and it is also used to control the amount of support vectors in the resulting model; that is, it is a lower bound on the fraction of support vectors.

The ϵ-insensitive loss function is used and ϵ is automatically minimized (Schölkopf *et al.*, 2000), and ξ is a positive slack variable. Including the constraints in Equation 7.46, the dual form of the problem is obtained and the Lagrangian is defined as

$$\begin{aligned} L(\boldsymbol{w}, \boldsymbol{\xi}^{(*)}, b, \epsilon, \boldsymbol{\alpha}^{(*)}, \boldsymbol{\beta}, \eta^{(*)}) = & \frac{1}{2}\|\boldsymbol{w}\|^2 + C\nu\epsilon + \frac{C}{N}\sum_{i=1}^{N}\eta_i\xi_i + \eta_i^*\xi_i^* \\ & - \sum_{i=1}^{N}\alpha_i(\xi_i + z_i - (\boldsymbol{w}^{\mathrm{T}}\boldsymbol{\phi}(\boldsymbol{v}_i)) - b + \epsilon) \\ & - \sum_{i=1}^{N}\alpha_i^*(\xi_i^* + (\boldsymbol{w}^{\mathrm{T}}\boldsymbol{\phi}(\boldsymbol{v}_i)) + b - z_i + \epsilon), \end{aligned}$$

where (β_i, $\alpha_i^{(*)}$, and $\eta^{(*)}$) ≥ 0 are the Lagrange multipliers. This function has to be maximized with respect to the dual variables β_i, $\alpha_i^{(*)}$, and $\eta^{(*)}$. The KKT conditions are obtained by deriving with the respect to the primal variables and equating to zero. A new problem of optimization, known as the Wolfe dual problem, is achieved by substituting the previous conditions into L.

Rewriting the constraints and substituting a kernel K for the dot product – that is, $K(\boldsymbol{v}_i, \boldsymbol{v}_j) = \phi(\boldsymbol{v}_i) \cdot \phi(\boldsymbol{v}_j)$ – the ν-SVR optimization problem is defined as

$$\begin{aligned}
\underset{\alpha_i^*, \alpha_i}{\text{maximize}} \quad & W(\boldsymbol{\alpha}^{(*)}) = \sum_{i=1}^{N} (\alpha_i^* - \alpha_i) y_i - \frac{1}{2} \sum_{i,j=1}^{N} (\alpha_i^* - \alpha_i)(\alpha_j^* - \alpha_j) K(\boldsymbol{v}_i, \boldsymbol{v}_j) \\
\text{subject to} \quad & \sum_{i=1}^{N} (\alpha_i^* - \alpha_i) = 0, \\
& \alpha_i^{(*)} \in \left[0, \frac{C}{N}\right], \\
& \sum_{i=1}^{N} (\alpha_i^* + \alpha_i) \leq C \cdot \nu.
\end{aligned}$$

The regularization parameter C, the percentage of support vectors ν, and the Gaussian width σ have to be searched and tuned during the training process. Finally, Equation 7.45 takes the form

$$f(\boldsymbol{v}) = \sum_{i=1}^{N} (\alpha_i^* - \alpha_i) K(\boldsymbol{v}_i, \boldsymbol{v}) + b, \tag{7.47}$$

where b can be computed by using the constraints of Equation 7.46 and $\xi_i^{(*)} = 0$. Here, we use an RBF kernel; that is, $K(\boldsymbol{v}_i, \boldsymbol{v}_j) = \exp(-\|\boldsymbol{v}_i - \boldsymbol{v}_j\|^2/(2\sigma^2))$, where $\sigma \in \mathbb{R}^+$.

Figure 7.17 shows an example of different values of C, ν, and σ when ν-SVR estimates the function shown in the Figure 7.15a. Overfitting occurs for low values of σ (i.e., $\sigma = 0.1$ or 0.5) because a very narrow Gaussian is used in the model and the regression model is being adjusted exclusively to the training data. When σ values are increased, the estimated interpolation function approximates better to the real one until the regression model starts to soften the interpolation function. Low values of the C parameter also soften the interpolation function. For values of $C = 100$ or 1000, the interpolated function is better approximated to the original one. For the ν parameter, the interpolation function is obtained when $\nu \geq 0.3$.

Results and Conclusions

Visual and quantitative comparisons of the EAM obtained with the three interpolation methods LI, TPS, and ν-SVR are presented and compared with the EAM of a CNS.

The CNS provides N irregularly spaced vertices $\boldsymbol{V} = \{\boldsymbol{v}_1, \dots, \boldsymbol{v}_N\}$, where $\boldsymbol{v}_i := [x_i, y_i, z_i]$ and x_i, y_i and z_i are the 3D coordinates of the vertex i. These vertices have associated a measured and known electrical feature h_i (activation time or voltage amplitude); that is, $h_i = f(\boldsymbol{v}_i)$. This set of vertices allows us to define the function f by using the LI, TPS, and ν-SVR interpolation methods.

Also, the CNS provides a triangular mesh (surface) with a new set of m irregularly spaced vertices $\boldsymbol{U} = \{\boldsymbol{u}_1, \boldsymbol{u}_2, \dots, \boldsymbol{u}_m\}$, which describes the anatomy of the cardiac chamber. In this new set of vertices, the spatial position of the vertices is known but the electrical feature is not registered; however, the CNS interpolates the feature with

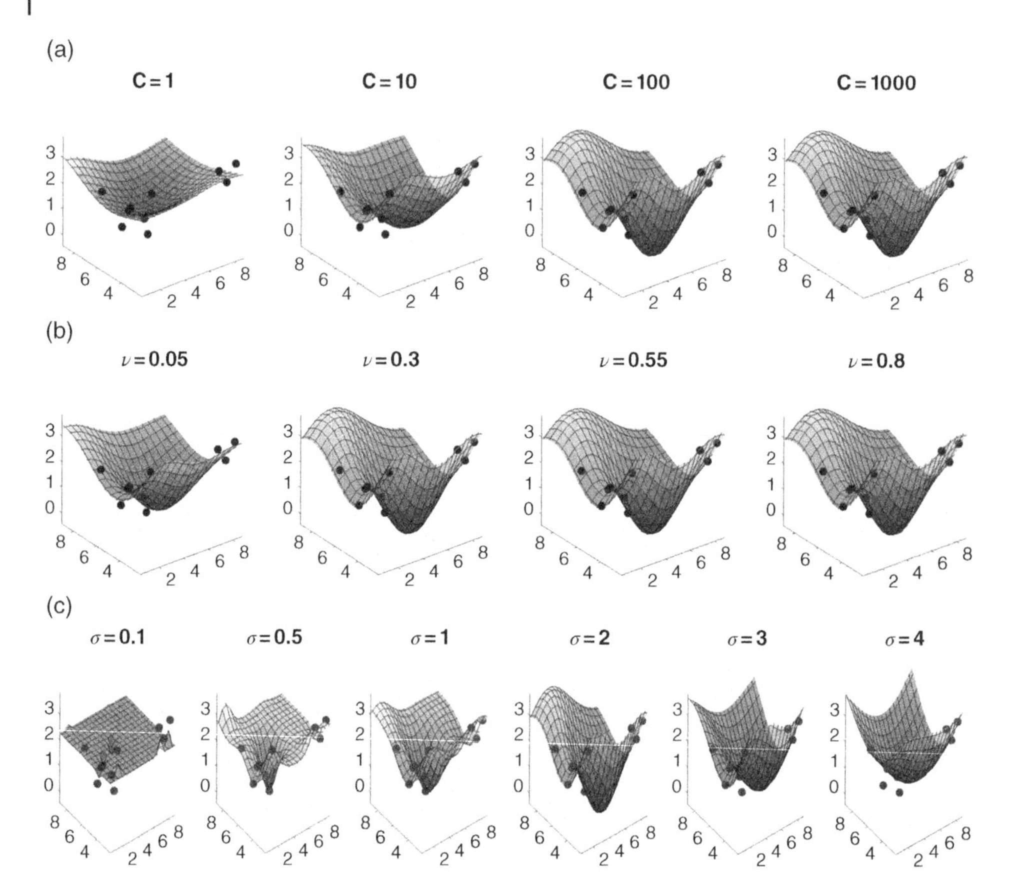

Figure 7.17 ν-SVR interpolation of the function f defined in Figure 7.15a using different values of C, ν, and σ: (a) $C = [1, 10, 100, 1000]$, $\nu = 0.5$, and $\sigma = 2$; (b) $C = 100$, $\nu = [0.05, 0.3, 0.55, 0.8]$, and $\sigma = 2$; (c) $C = 100$, $\nu = 0.5$, and $\sigma = [0.1, 0.5, 1, 2, 3, 4]$.

its own interpolation method and the set of vertices $\boldsymbol{v}$ in order to show a representation of spatial anatomical distribution of the electrical feature in this mesh.

As was mentioned earlier, a CNS is able to register the spatial position of the catheters inside the heart instantaneously, but the electrical feature is registered by capturing the electrical signal during several seconds; hence, the number of vertices in the set $\boldsymbol{V}$, which includes spatial position and measured feature, is lower than the number of vertices in $\boldsymbol{U}$; that is, $N \ll m$. Note that both sets of vertices $\boldsymbol{V}$ and $\boldsymbol{U}$ have to be different and they have not to lie on a straight line, and this means $N \geq 3$ and $m \geq 3$ to avoid incorrect estimations. In this application, the set of vertices $\boldsymbol{V}$ is used as training data and the set of vertices $\boldsymbol{U}$ as test data.

Ten bipolar voltage and 10 activation time EAMs from CNSs were used for the evaluation of these three methods. The CNS EAMs have an average and standard deviation of 189.4 ± 110.4 vertices for the set $\boldsymbol{V}$ and 1264.4 ± 360.4 vertices for the set $\boldsymbol{U}$ in the bipolar voltage EAMs, and 330.4 ± 321.3 vertices for the set $\boldsymbol{V}$ and 2567.7 ± 3991.6

vertices for the set $\boldsymbol{U}$ in the activation time EAMs. The MSE between the interpolated feature and the feature provided by the CNS for the set of vertices $\boldsymbol{U}$ is used as a merit figure. Note that the feature provided by the CNSs is not a gold standard because these systems use their own interpolation algorithm to build the EAM; however, it is the EAM used by the electrophysiologist to apply the treatment and it is a quality reference.

As defined in Chapter 3, the script to start the execution is `run.m`, while the additional code needed is included in the supplementary material: data preprocessing and training–test split in `genDataEAM.m`, while `config_file.m` and `config_EAM.m` contain configuration files. In particular, the function `genDataEAM.m` reads the `.mat` files, which contain the set of vertices $\boldsymbol{V}$ and $\boldsymbol{U}$. Note that the set of vertices $\boldsymbol{U}$ has several vertices with the feature value equal to −10 000, which corresponds to nonactive vertices. Although these vertices are not considered in the evaluation of the methods, they are not deleted because they are included in the triangular mesh. The function `config_EAM.m` defines the path of data and the CNS color map, the interpolation method functions with their associated parameters, the range of the parameters for these methods, the strategy of search the optimum parameter and the function to display the results. In this application, the tuning of the λ parameter for the TPS, and C, ν, and σ parameters for ν-SVR is done by searching the most appropriate value following a grid-search strategy defined in `gridSearch.m`.

The LI and TPS interpolation methods are defined in Listings 7.22 and 7.23 respectively. Listing 7.26 shows the ν-SVR interpolation method

```
function [predict, coefs] = LI(Xtrain,Ytrain,Xtest,~)

F=scatteredInterpolant(Xtrain(:,1),Xtrain(:,2),Xtrain(:,3),
                                                Ytrain,'linear');
predict=F(Xtest(:,1),Xtest(:,2),Xtest(:,3));
coefs=[];
```

Listing 7.22 LI method (LI.m).

```
function [predict,coefs] = TPS(Xtrain,Ytrain,Xtest,lambda)

[N,D]=size(Xtrain);
Ntest=size(Xtest,1);
% Contruct kernel matrix
distances = sqrt(dist2(Xtrain,Xtrain));
K = K_matrix(distances);
P = [ones(N,1) Xtrain];
% Train
K_reg = K + lambda*eye(size(K,1));
L = [K_reg P; P' zeros(D+1,D+1)];
Q = [Ytrain; zeros(D+1,1)];
coefs = L\Q;
% Predict
distances = sqrt(dist2(Xtest,Xtrain));
matrix_eta= K_matrix(distances);
predict = [ones(Ntest,1) Xtest]*coefs(N+1:end)+matrix_eta*coefs(1:N);
```

Listing 7.23 TPS method (TPS.m).

The functions `dist2.m` and `K_matrix.m` are defined in Listings 7.24 and 7.25.

```
function distance = dist2(x1,x2)

n1 = size(x1,1);
n2 = size(x2,1);
distance = (ones(n2,1) * sum((x1.^2)', 1))' + ...
   ones(n1,1) * sum((x2.^2)',1) - ...
   2.*(x1*(x2'));
distance(distance<0) = 0;
```

Listing 7.24 Distance ℓ_2 computation (dist2.m).

```
function K = K_matrix(r)

[controlP,newP] = size(r);
new_r = r(:);
K = zeros(size(r));
repVert = find(new_r>0);

K(repVert) = single(log(r(repVert).^2).*r(repVert).^2);
K = reshape(K,controlP,newP);
```

Listing 7.25 K_matrix (K_matrix.m).

```
function [predict,coefs] = SVR(Xtrain,Ytrain,Xtest,sigma,nu,C)

% Train SVM and predict
gamma = 1/(2*sigma)^2;
inparams=sprintf('-s 4 -t 2 -g %f -c %f -n %f -j 1',gamma,C,nu);
[predict,model]=SVM(Xtrain,Ytrain,Xtest,inparams);
coefs = getSvmWeights(model);
```

Listing 7.26 Code for training the ν-SVR (SVR.m).

Finally, the function to store the results and display the MSE results and the visual interpolation is `display_results_EAM.m`, which is given in the supplementary material.

Table 7.10 shows the average and the standard deviation of the MSE between the interpolated feature and the feature provided by the CNS for the set of vertices $\boldsymbol{U}$ and for each interpolation method. LI and TPS yielded better performance than ν-SVR for both bipolar voltage and activation time EAMs. Activation time EAMs also yielded lower MSE than bipolar voltage EAMs. Note that this comparison is made by using the EAM provided by the CNS and used by electrophysiologist, but it is not the gold standard. Figures 7.18 and 7.19 show the original and the interpolated bipolar voltage and activation time EAMs using LI, TPS, and ν-SVR.

Table 7.10 Average and standard deviation MSE between the electrical feature provided by the CNS and the interpolated feature with LI, TPS, and ν-SVR methods using 10 bipolar voltage EAMs and 10 activation time EAMs.

EAM	LI	TPS	ν-SVR
Bipolar voltage (mV)	0.166 ± 0.101	0.144 ± 0.069	0.248 ± 0.156
Activation time (ms)	132.8 ± 65.0	98.1 ± 59.1	162.0 ± 110.9

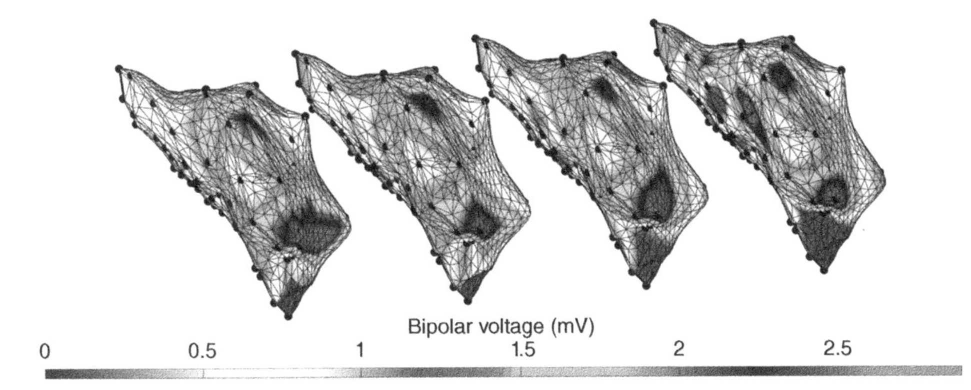

Figure 7.18 Original and interpolated bipolar voltage (mV) EAMs using LI, TPS, and ν-SVR (from left to right). The black points show the position of the known feature positions.

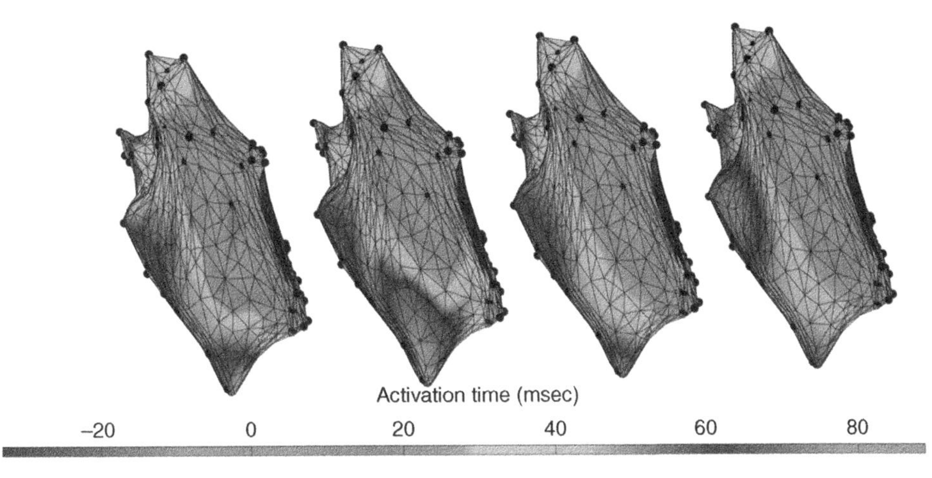

Figure 7.19 Original and interpolated activation time (ms) EAMs using LI, TPS, and ν-SVR (from left to right). The black points show the position of the known feature positions.

Slight differences between EAMs are presented mainly in the places where there are less vertices with a measured feature. In general, LI and TPS yielded better performance than ν-SVR in terms of similarity to the EAM provided by the CNS, but the EAM accuracy is dependent on the EAM type, on the cardiac chamber, and on the number of vertices with an associated measured feature (size of set $\boldsymbol{\nu}$).

7.5 Questions and Problems

Exercise 7.5.1 Show that h_n^{eq} fulfills the properties to be a Mercer kernel.

Exercise 7.5.2 Show that 2D and 3D autocorrelation functions are valid Mercer kernels.

Exercise 7.5.3 Show that the multidimensional Lomb periodogram provides a suitable expression for its use as Mercer kernel.

Exercise 7.5.4 Can you propose and implement two more ways for estimating multidimensional autocorrelation functions? Hint: think on multidimensional spline interpolation, as well as on nearest neighbors interpolation.

Exercise 7.5.5 Starting from the sinc nonuniform interpolation code provided, make the necessary modifications to benchmark the impact of using the following three methods for estimating the single-dimensional autocorrelation kernel from data: (a) Lomb periodogram; (b) splines; (c) k-NN. What is the impact on the interpolation quality of the method used?

Exercise 7.5.6 For the indoor location example, check the impact on the performance of each of the methods presented for estimating the multidimensional autocorrelation. Explain the observed differences with the single-dimensional case.

Exercise 7.5.7 For the HRV code, benchmark the impact of using a single signal for the estimation of the autocorrelation kernel (instead of an averaged one). Explain the results obtained.

8

Advances in Kernel Regression and Function Approximation

8.1 Introduction

Kernel methods constitute a proper framework to tackle regression problems that encompass fitting and regularization. In this chapter we will pay attention to two particularly interesting ways of treating the regression problem: based on *discriminative* kernel regression, and based on *generative* Bayesian nonparametric regression. Both families have found wide application in signal processing in the last decade (Pérez-Cruz *et al.*, 2013; Rojo-Álvarez *et al.*, 2014).

The chapter is divided into two main sections. In Section 8.2, we will depart from the standard SVR method (Smola and Schölkopf, 2004) extensively used in the previous chapters. The method allows many alternative formulations to deal with the specificities of the DSP problems; for example, dealing with multiple outputs, unlabeled data, and signal-dependent noise sources and heteroscedastic models. Recent approaches to treat such problems will be introduced and experimentally evaluated. In all of them, the important role of both the loss and the regularizer will appear, as well as the design of the kernel function. In Section 8.3, as an alternative to this treatment based on the SVR algorithm, we will review two instantiations of the field of Bayesian nonparametrics: relevance vector machines (RVMs) (Tipping, 2000) and Gaussian processes (Rasmussen and Williams, 2006). We study the important issue of model selection and hyperparameters search procedures in both Sections 8.2 and 8.3. Then in Section 8.4 we illustrate the algorithms presented in both synthetic and real application examples.

8.2 Kernel-Based Regression Methods

This section treats the problem of regression and function approximation with kernels. The basic approach here is to cast the problem in a supervised way. As we have seen before, a large class of regression problems in particular are defined as the joint minimization of a loss function accounting for errors of the function $f \in \mathcal{H}$ to be learned, and a regularization term $\Omega(f)$ that controls its smoothness and, as a consequence, its excess of flexibility. More specifically, in this chapter we focus on the problem of inferring the function $f(\boldsymbol{x}) : \mathcal{X} \subseteq \mathbb{R}^d \to \mathbb{R}$, given a set of observed data pairs $\{(\boldsymbol{x}_i, y_i)\}_{i=1}^N$, where $\boldsymbol{x} := [x_1, \dots, x_d]^{\mathrm{T}} \in \mathbb{R}^d$, $y \in \mathbb{R}$, and we assume an additive noise

Digital Signal Processing with Kernel Methods, First Edition. José Luis Rojo-Álvarez, Manel Martínez-Ramón, Jordi Muñoz-Marí, and Gustau Camps-Valls.

observation model, $y_i = f(\boldsymbol{x}_i) + e_i$, $i = 1, \ldots, N$, where e_i represents the perturbation errors. As we have see in previous chapters, the problem of inferring the function f reduces in general to selecting a class of models and to optimizing a functional that considers both a fitting cost or loss function and a regularizer term.

We also saw that there exist several possible choices of both the loss and the regularizer, each one leading to different kernel regression algorithms; see Chapter 4.

8.2.1 Advances in Support Vector Regression

The SVR is the SVM implementation for regression and function approximation (Schölkopf and Smola, 2002; Smola and Schölkopf, 2004); see Chapter 5. The standard SVR formulation uses Vapnik's ε-insensitive cost function

$$\mathcal{L}_\varepsilon(e) = C\max(0, |e| - \varepsilon), \qquad C > 0, \tag{8.1}$$

in which an error e up to ε is not penalized, otherwise a linear penalization will be incurred. The regression estimation problem is regarded as finding the mapping between an incoming vector $\boldsymbol{x} \in \mathbb{R}^d$ and an observable (unidimensional) output $y \in \mathbb{R}$, from a given training set of N data pairs, $\{(\boldsymbol{x}_i, y_i)\}_{i=1}^N$. The standard SVR (Smola and Schölkopf 2004; Smola *et al.* 1998; Vapnik 1998) solves this problem by finding the weights of a linear regressor $f(\boldsymbol{x}) = \boldsymbol{w}^\mathrm{T}\boldsymbol{\phi}(\boldsymbol{x}) + b$ – that is, $\boldsymbol{w}$ and b – where as usual $\boldsymbol{\phi}(\cdot)$ is a feature mapping to a higher (possibly infinite-dimensional) Hilbert space. Briefly, SVR estimates weights $\boldsymbol{w}$ by minimizing the following regularized functional:

$$\min_{\boldsymbol{w},b}\left\{C\sum_{i=1}^{N}\mathcal{L}_\varepsilon(y_i - \boldsymbol{w}^\mathrm{T}\boldsymbol{\phi}(\boldsymbol{x}_i) - b) + \frac{1}{2}\|\boldsymbol{w}\|^2\right\}, \tag{8.2}$$

where C is the penalization parameter applied to the errors and acts as a trade-off parameter between the minimization of errors and smoothness of the solution (flatness of the weights, $\|\boldsymbol{w}\|$). Solving the problem with arbitrary losses is, in general, done by introducing the Lagrange theory of multipliers. In the case of the ε-insensitive cost, we define in the following ξ_i and $\xi_i^{(*)}$ as positive slack variables, to deal with training samples with a prediction error larger than ε. Then, the problem in Equation 8.2 can be shown to be equivalent to the constrained optimization problem with objective function

$$\min_{\boldsymbol{w},b,\xi_i,\xi_i^*}\left\{C\sum_{i=1}^{N}(\xi_i + \xi_i^*) + \frac{1}{2}\|\boldsymbol{w}\|^2\right\} \tag{8.3}$$

constrained to

$$y_i - \langle \boldsymbol{w}, \boldsymbol{\phi}(\boldsymbol{x}_i)\rangle - b \le \varepsilon + \xi_i \qquad \forall i = 1, \ldots, N \tag{8.4}$$

$$\langle \boldsymbol{w}, \boldsymbol{\phi}(\boldsymbol{x}_i)\rangle + b - y_i \le \varepsilon + \xi_i^* \qquad \forall i = 1, \ldots, N \tag{8.5}$$

$$\xi_i, \xi_i^* \ge 0 \qquad \forall i = 1, \ldots, N. \tag{8.6}$$

This is a QP problem. The usual procedure for solving SVRs introduces the linear restrictions in Equations 8.4–8.6 into Equation 8.2 using Lagrange multipliers α_i and α_i^* (one per inequality constraint in Equation 8.4 and 8.5 respectively), computes the

KKT conditions, and solves the dual problem using QP procedures (Fletcher, 1987), which yields

$$\boldsymbol{w} = \sum_{i=1}^{N}(\alpha_i - \alpha_i^*)\boldsymbol{\phi}(\boldsymbol{x}_i), \tag{8.7}$$

which expresses the solution as a linear combination of the mapped examples. Note that the primal problem, whose solution is essentially expressed as a function of $\boldsymbol{w}$, was transformed into a dual problem expressed as a function of the dual weights $\boldsymbol{\alpha} = [\alpha_1 - \alpha_1^*, \dots, \alpha_N - \alpha_N^*]^{\mathrm{T}}$. These weights are directly obtained from solving the corresponding problem; see Chapter 5. Note that one can easily obtain the prediction for an input test $\boldsymbol{x}_j$ as

$$\hat{y}_j = f(\boldsymbol{x}_j) = \langle \boldsymbol{w}, \boldsymbol{\phi}(\boldsymbol{x}_j)\rangle = \sum_{i=1}^{N}(\alpha_i - \alpha_i^*)\langle\boldsymbol{\phi}(\boldsymbol{x}_i), \boldsymbol{\phi}(\boldsymbol{x}_j)\rangle = \sum_{i=1}^{N}(\alpha_i - \alpha_i^*)K(\boldsymbol{x}_i, \boldsymbol{x}_j). \tag{8.8}$$

As in the SVM classifier, here only the examples with nonzero multipliers are relevant for regression, and are called *support vectors*. Sparsity in the SVR is a direct consequence of the loss function adopted; as the value of ε increases, the number of support vectors is reduced. The SVR methodology allows us to develop new, powerful SVR methods by modifying the functional. Let us now see two simple modifications.

The ε-Huber Support Vector Regression

In the previous chapters we extensively focused on the use of the ε-Huber loss $\mathcal{L}_{\varepsilon\mathrm{H}}(e)$ instead of the Vapnik's $\mathcal{L}_{\varepsilon}(e)$. The ε-Huber cost is expressed as

$$\mathcal{L}_{\varepsilon\mathrm{H}}(e) = \begin{cases} 0, & |e| \le \varepsilon \\ \frac{1}{2\delta}(|e| - \varepsilon)^2, & \varepsilon \le |e| \le e_C \\ C(|e| - \varepsilon) - \frac{1}{2}\delta C^2, & |e| \ge e_C, \end{cases} \tag{8.9}$$

where $e_C = \varepsilon + \delta C$; ε is the insensitive parameter, and δ and C control the trade-off between the regularization and the losses. The cost considers a particular combination of three losses: insensitive for tolerated errors, quadratic to deal with Gaussian noise, and linear cost to cope with outliers. Once optimized, the corresponding optimal parameters ε, δ, and C yield some insight on the signal and noise characteristics. Interestingly, introducing this cost function does not alter much the QP problem to be solved; see Section 4.6.2. This particular SVR was successfully used in several signal processing models in Chapters 4–7.

The ν-Support Vector Regression

Another alternative functional was proposed in Schölkopf *et al.* (1999). Here, the tube size ε is traded off against model complexity and slack variables via an additional parameter $\nu \ge 0$:

$$\min_{\boldsymbol{w}, b, \xi_i, \xi_i^*} \left\{ C\sum_{i=1}^{N}(\xi_i + \xi_i^*) + CN\nu\varepsilon + \frac{1}{2}\|\boldsymbol{w}\|^2 \right\} \tag{8.10}$$

subject to

$$\begin{aligned} y_i - \langle \boldsymbol{w}, \boldsymbol{\phi}(\boldsymbol{x}_i)\rangle - b &\leq \varepsilon + \xi_i \\ \langle \boldsymbol{w}, \boldsymbol{\phi}(\boldsymbol{x}_i)\rangle + b - y_i &\leq \varepsilon + \xi_i^* \\ \xi_i, \xi_i^* &\geq 0 \\ \varepsilon &\geq 0, \end{aligned} \tag{8.11}$$

where ν acts as an upper bound on the fraction of errors and a lower bound on the fraction of support vectors (Schölkopf *et al.*, 1999). The so-called ν-SVR formulation reduces to solving again a QP problem. Here, we show an alternative procedure known as the iterated reweighted LS (IRWLS), which has already been demonstrated to be more efficient in both time and memory requirements (Pérez-Cruz and Artés-Rodríguez, 2001). This procedure is based on the fact that a Lagrangian can be constructed with the form

$$L_P = \frac{1}{2}\|\boldsymbol{w}\|^2 + \frac{1}{2}\sum_{i=1}^{N}(a_i e_i^2 + a_i^*(e_i^*)^2) + \varepsilon(CN\nu) \tag{8.12}$$

with the definitions

$$\begin{aligned} e_i &= y_i - \langle \boldsymbol{w}, \boldsymbol{\phi}(\boldsymbol{x}_i)\rangle - b - \varepsilon \\ e_i^* &= \langle \boldsymbol{w}, \boldsymbol{\phi}(\boldsymbol{x}_i)\rangle + b - y_i - \varepsilon \\ a_i &= \frac{2\alpha_i}{y_i - \langle \boldsymbol{w}, \boldsymbol{\phi}(\boldsymbol{x}_i)\rangle - b - \varepsilon} = \frac{2\alpha_i}{e_i} \\ a_i^* &= \frac{2\alpha_i^*}{\langle \boldsymbol{w}, \boldsymbol{\phi}(\boldsymbol{x}_i)\rangle + b - y_i - \varepsilon} = \frac{2\alpha_i^*}{e_i^*} \end{aligned} \tag{8.13}$$

(see Pérez-Cruz and Artés-Rodríguez (2001) for the derivation details). The optimization consists in minimizing Equation 8.12 with respect to $\boldsymbol{w}$, b, and ε, while keeping a_i and a_i^* constant, and then recalculating a_i and a_i^* with

$$a_i^{(*)} = \begin{cases} 0, & e_i^{(*)} < 0 \\ \dfrac{2C}{e_i^{(*)}}, & e_i^{(*)} \geq 0, \end{cases} \tag{8.14}$$

where $e_i = y_i - \langle \boldsymbol{w}, \boldsymbol{\phi}(\boldsymbol{x}_i)\rangle - b - \varepsilon$ and $e_i^* = \langle \boldsymbol{w}, \boldsymbol{\phi}(\boldsymbol{x}_i)\rangle + b - y_i - \varepsilon$. The process is repeated until convergence.

The first step, consisting of computing the gradient of Equation 8.12 with respect to $\boldsymbol{w}, b$ and ε, can be solved in block. By virtue of the representer's theorem (see Section 4.3.2), $\boldsymbol{w}$ is a linear combination of a subset of training samples $\boldsymbol{w} = \sum_{i=1}^{N} \beta_i \boldsymbol{\phi}(\boldsymbol{x}_i)$. By making this substitution in the following, we can write the following algorithm:

1) Solve the linear system:

$$\begin{bmatrix} \boldsymbol{K}+\boldsymbol{D}^{-1}_{(\boldsymbol{a}+\boldsymbol{a}^*)} & \mathbf{1} & \boldsymbol{E} \\ \mathbf{1}^{\mathrm{T}} & 0 & 0 \\ \boldsymbol{E}^{\mathrm{T}} & 0 & 0 \end{bmatrix} \begin{bmatrix} \boldsymbol{\beta} \\ b \\ \varepsilon \end{bmatrix} = \begin{bmatrix} \boldsymbol{y} \\ 0 \\ CN^*\nu \end{bmatrix}, \tag{8.15}$$

where we have defined

$$K_{ij} = \langle \boldsymbol{\phi}(\boldsymbol{x}_i)\boldsymbol{\phi}(\boldsymbol{x}_j)\rangle = K(\boldsymbol{x}_i,\boldsymbol{x}_j)$$

$$(\boldsymbol{D}^{-1}_{\boldsymbol{a}+\boldsymbol{a}^*})_{ij} = \frac{\delta(i-j)}{a_i + a_i^*}$$

$$\boldsymbol{E} = \left[\frac{a_1 - a_1^*}{a_1 + a_1^*}, \ldots, \frac{a_N - a_N^*}{a_N + a_N^*}\right]^{\mathrm{T}}.$$

2) Recompute a_i and a_i^* with Equation 8.14.
3) Repeat until convergence.

The column vectors $\boldsymbol{y}$, $\boldsymbol{a}$, $\boldsymbol{a}^*$, $\boldsymbol{\beta}$, and $\mathbf{1}$ present the obvious expressions, and $\boldsymbol{K}$ is known as the *kernel* matrix, since it is only formed by inner products of the training samples in the feature space. Consequently, neither the minimizing procedure nor the use of the regressor needs to know the explicit form of the nonlinear mapping $\boldsymbol{\phi}(\cdot)$, but only its kernel representation $K(\cdot,\cdot)$. The transformations needed to obtain the IRWLS procedure from the minimization of Equation 8.10 are detailed in Pérez-Cruz and Artés-Rodríguez (2001).

The Profile-Dependent Support Vector Regression

The profile-dependent SVR (PD-SVR) exploits the previous ν-SVR to tailor specific penalization C and tolerated error ε for every example (Camps-Valls *et al.*, 2001). The inspiration comes from the classification setting. In classification problems with unbalanced classes one typically sets different penalization factors for each class (Lin *et al.*, 2000; Orr and Müller, 1998). This way, the SVM learns to classify patterns independently from the class they belong to, which is not possible when using an overall constraint C. In the field of time-series prediction, we can extrapolate this fact and consider that the most recent samples contain, in principle, more information and consequently should receive more penalization. Therefore, problems with nonstationary processes can be alleviated using a different penalization factor for each training sample i according to a certain *confidence function* c_i on the samples. This idea can be extended by using a different margin (insensitivity zone) for each sample. This allows the SVR machine to follow the *pdf* variations over time.

The PD-SVR functional implies a simple modification of the original SVR:

$$\min_{\boldsymbol{w},\xi_i,\xi_i^*} \left\{ C\sum_i c_i(\xi_i + \xi_i^*) + \frac{1}{2}\|\boldsymbol{w}\|^2 \right\} \tag{8.16}$$

where we introduce a penalization per example c_i, and now restrictions over slack variables become sample-dependent too:

$$y_i - \boldsymbol{\phi}^{\mathrm{T}}(\boldsymbol{x}_i)\boldsymbol{w} - b \leq \frac{\varepsilon}{c_i} + \xi_i \qquad \forall i = 1, \dots, N \tag{8.17}$$

$$\boldsymbol{\phi}^{\mathrm{T}}(\boldsymbol{x}_i)\boldsymbol{w} + b - y_i \leq \frac{\varepsilon}{c_i} + \xi_i^* \qquad \forall i = 1, \dots, N \tag{8.18}$$

$$\xi_i, \xi_i^* \geq 0 \qquad \forall i = 1, \dots, N. \tag{8.19}$$

With regard to the ν-SVR (Schölkopf *et al.*, 1999), the primal function becomes

$$\min_{\boldsymbol{w},\xi_i,\xi_i^*} \left\{ C \sum_{i=1}^{N} c_i(\xi_i + \xi_i^*) + CN^* \nu\varepsilon + \frac{1}{2}\|\boldsymbol{w}\|^2 \right\}, \tag{8.20}$$

where $N^* = \sum_{i=1}^{N} c_i$ in order to restrict ν to be in the range $[0, 1]$. By including linear restrictions in Equations 8.17–8.19 in the corresponding functional, we can rewrite Equation 8.16 or 8.20 to its dual description, which once again constitutes a QP problem, which can be solved with the efficient IRWLS procedure as well.

New Support Vector Regression Algorithms

The previous SVR formulation has allowed developing powerful algorithms by modifying the loss function or the regularizer. An alternative pathway to develop improved SVR methods is to focus on introducing new kernel functions that respect the signal structure and characteristics. The ease of algorithmic design and the convex solutions provided yielded a plethora of exciting developments. Interestingly, the SVR methodology allows us to go beyond, and to study key concepts in DSP, such as the signal and noise characteristics and relations, the heteroscedasticity of the noise, or to exploit unlabeled data to improve regression and to accommodate multi-output problems. We will present techniques to study precisely these issues in the next sections: the kernel signal-to-noise regression (KSNR) to cope with signal and noise dependencies as an efficient form of regularization, the semi-supervised SVR (SS-SVR) to incorporate the manifold structure through the use of unlabeled data, and the multi-output SVR (MSVR) to deal with several, eventually correlated output variables.

8.2.2 Multi-output Support Vector Regression

Let us now treat a more challenging problem than single-output regression. The so-called MSVR was introduced in Pérez-Cruz *et al.* (2002) and Mati04 for DSP and MIMO channel equalization, and later applied to remote sensing and geoscience problems of multiple biophysical parameter estimation in Tuia *et al.* (2011a). Notationally, imagine that we aim to predict $Q > 1$ several variables simultaneously. Of course, a trivial approach is to develop Q different SVR models to approximate each one of the variables. This, however, has two main problems: the computational cost of both training and testing and, perhaps more importantly, the fact that the output variable relations are neglected. In the case the observable output is a vector with Q variables to be predicted, $\boldsymbol{y} \in \mathbb{R}^Q$, a more appropriate strategy is to solve a multidimensional regression estimation problem, in which we have to find a regressor $\boldsymbol{w} = [w_1, \dots, w_M]^{\mathrm{T}} \in \mathbb{R}^M$

per output variable; that is, a weight matrix $\boldsymbol{W} = [\boldsymbol{w}_1|\cdots|\boldsymbol{w}_Q] \in \mathbb{R}^{M\times Q}$. Then one can directly generalize the 1D SVR to solve the multidimensional case, whose problem can be stated as the minimization of the following cost function:

$$\min_{\boldsymbol{W}}\left\{C\sum_{i=1}^{N}L(u_i)+\frac{1}{2}\|\boldsymbol{W}\|^2\right\}=\min_{\boldsymbol{w}_k}\left\{C\sum_{i=1}^{N}L(u_i)+\frac{1}{2}\sum_{k=1}^{Q}\|\boldsymbol{w}_k\|^2\right\}, \tag{8.21}$$

where

$$L(u)=\begin{cases}0, & u<\varepsilon\\ u^2-2u\varepsilon+\varepsilon^2, & u\geq\varepsilon\end{cases}, \tag{8.22}$$

$u_i = \|\boldsymbol{e}_i\| = \sqrt{\boldsymbol{e}_i^{\mathrm{T}}\boldsymbol{e}_i}$, and the residuals are computed as $\boldsymbol{e}_i = \boldsymbol{y}_i - \boldsymbol{W}^{\mathrm{T}}\boldsymbol{\phi}(\boldsymbol{x}_i)$. Note that we here use the cost $L(u)$. This is due to the fact that extending the Vapnik ε cost function to multiple dimensions makes the problem complexity grow linearly with the number of dimensions (besides, note that the inequality constraints should be coupled for all pairs of data points). If, instead of this piecewise ℓ_1-norm, we use a ℓ_2-based norm as the one in $L(u)$, all dimensions can be considered into a unique restriction yielding a single support vector for all dimensions.

For $\varepsilon = 0$ this problem reduces to an independent regularized kernel LS regression for each component, but for a nonzero ε the solution takes into account all outputs to construct each individual regressor. This way, a structured-output model is built and the cross-output relations are exploited, thus leading to obtain in principle more robust predictions. The price to pay is that the resolution of the proposed problem cannot be done straightforwardly and we have to rely on an iterative procedure to obtain the desired solution. A quasi-Newton approach has been devised, in which each iteration has at most the same computational complexity as an LS procedure for each component. In particular, the MSVR is solved using the IRWLS explained before, which boils down to a weighted LS problem, and the number of iterations needed to obtain the final result is typically small, making the procedure only slightly more computationally demanding than LS regression for each component.

8.2.3 Kernel Ridge Regression

An alternative formulation to SVR consists of replacing the ℓ_1 with an ℓ_2-norm cost function. Such a kernel-based method is known as KRR (also known as the LS SVM), and can be seen as the kernel version of the regularized LS linear regression (Shawe-Taylor and Cristianini 2004; Suykens *et al.* 2002). Here, we introduce a simple *multi-output formulation*, similarly to the framework considered in the previous subsection.

Notationally, let $\boldsymbol{x}_i \in \mathbb{R}^d$ (for simplicity, here we consider $\mathcal{X} \subseteq \mathbb{R}^d$) and the multiple-output $\boldsymbol{y}_i \in \mathbb{R}^Q$, $i = 1,\ldots,N$. We define an output (target) matrix $\boldsymbol{Y} = [\boldsymbol{y}_1|\cdots|\boldsymbol{y}_N]^{\mathrm{T}} \in \mathbb{R}^{N\times Q}$. We want to perform a regularized linear LS regression where the samples have been transformed by the nonlinear mapping $\boldsymbol{\phi}(\boldsymbol{x}) \in \mathcal{H}$. Now, using all mapping functions for all examples $\boldsymbol{x}_i$, we define data matrix $\boldsymbol{\Phi} = [\boldsymbol{\phi}(\boldsymbol{x}_1)|\cdots|\boldsymbol{\phi}(\boldsymbol{x}_N)]^{\mathrm{T}}$ and weight matrix $\boldsymbol{W} = [\boldsymbol{w}_1|\cdots|\boldsymbol{w}_Q]$, where column vectors $\boldsymbol{w}_k \in \mathcal{H}$.

The assumed observation model in this case is

$$Y = \boldsymbol{\Phi} W + E, \tag{8.23}$$

where $\boldsymbol{E}$ is an i.i.d. Gaussian noise, $\boldsymbol{E}$ is an $N \times Q$ noise matrix with entries $e_{i,k}$, and $i = 1, \dots, N$, $k = 1, \dots, Q$, and $e_{i,k} \sim \mathcal{N}(0, \lambda)$, where λ denotes the variance of each component of the noise. Now, following the regularized LS linear regression setting, the KRR minimizes the following cost function:

$$\min_{\boldsymbol{W}} \{ \|\boldsymbol{Y} - \boldsymbol{\Phi W}\|_F^2 + \lambda \|\boldsymbol{W}\|_F^2 \}, \tag{8.24}$$

where the Frobenius norm is indicated as $\|\cdot\|_F$, and the parameter λ is a positive scalar value which should be tuned by the user. We have deliberately dropped the bias term for the sake of simplicity.[1] Solving the problem with respect to $\boldsymbol{W}$, one obtains an analytical solution:

$$\boldsymbol{W}^* = (\boldsymbol{\Phi}^T \boldsymbol{\Phi} + \lambda \boldsymbol{I})^{-1} \boldsymbol{\Phi}^T \boldsymbol{Y}. \tag{8.25}$$

The prediction vector at $\boldsymbol{x}$ can be expressed as

$$\hat{\boldsymbol{F}}(\boldsymbol{x}) = \boldsymbol{W}^{*T} \boldsymbol{\phi}(\boldsymbol{x}). \tag{8.26}$$

Applying the representer theorem, we can readily express the solution as a linear combination of mapped samples:

$$\boldsymbol{W} = \boldsymbol{\Phi}^T A, \tag{8.27}$$

where $\boldsymbol{A}$ is an $N \times Q$ matrix with components $\boldsymbol{A} = [\boldsymbol{a}_1 | \cdots | \boldsymbol{a}_Q]$ with $\boldsymbol{a}_k = [a_{k,1}, \dots, a_{k,N}]^T$ (i.e., we have one weight per sample, $i = 1, \dots, N$ and output observation variable, $k = 1, \dots, Q$). It is straightforward to show that the optimal dual weights are

$$A^* = (\boldsymbol{\Phi\Phi}^T + \lambda \boldsymbol{I})^{-1} Y = (\boldsymbol{K} + \lambda \boldsymbol{I})^{-1} Y, \tag{8.28}$$

where $\boldsymbol{K} = \boldsymbol{\Phi\Phi}^T$. Using Equation 8.27 and Equation 8.28, we can rewrite the solution as

$$\hat{\boldsymbol{F}}(\boldsymbol{x}) = \boldsymbol{W}^{*T} \boldsymbol{\phi}(\boldsymbol{x}) = (\boldsymbol{\phi}^T(\boldsymbol{x}) \boldsymbol{\Phi}^T (\boldsymbol{K} + \lambda \boldsymbol{I})^{-1} \boldsymbol{Y})^T. \tag{8.29}$$

Now let us define the kernel vector

$$\boldsymbol{k} = \boldsymbol{\Phi\phi}(\boldsymbol{x}) = [K(\boldsymbol{x}, \boldsymbol{x}_1), \dots, K(\boldsymbol{x}, \boldsymbol{x}_N)]^T \in \mathbb{R}^{N \times 1}, \tag{8.30}$$

which contains the kernel dot products between point $\boldsymbol{x}$ and all the N training data points. The final expression of the vector of predictions is

$$\hat{\boldsymbol{F}}(\boldsymbol{x}) = \boldsymbol{k}^T (\boldsymbol{K} + \lambda \boldsymbol{I})^{-1} \boldsymbol{Y}. \tag{8.31}$$

---

1 The bias term can be actually considered an additional constant weight for all input data points in $\boldsymbol{W}$.

Note also that by combining Equation 8.24 with Equation 8.27 we obtain the dual functional

$$\min_{\boldsymbol{A}}\{\|\boldsymbol{Y}-\boldsymbol{K}\boldsymbol{A}\|_{\mathrm{F}}^{2}+\lambda\boldsymbol{A}^{\mathrm{T}}\boldsymbol{K}\boldsymbol{A}\} \tag{8.32}$$

by simply using expression $\boldsymbol{K}=\boldsymbol{\Phi}\boldsymbol{\Phi}^{\mathrm{T}}$.

8.2.4 Kernel Signal-to-Noise Regression

DSP problems typically deal with observations that are corrupted by different types of noise sources. Most of the algorithms, however, assume that the noise is i.i.d. and additive. Also, signal and noise may exhibit relations that need to be disentangled. In this context, one typically looks for transformations of the observed signal such that the SNR is maximized, or alternatively for transformations that minimize the fraction of noise.[2] This is the case of the minimum noise fraction (MNF) transform (Green *et al.*, 1998), which extends PCA by maximizing the signal variance while minimizing the estimated noise variance. Although efficient in many applications, MNF cannot cope with problems when signal and noise are nonlinearly correlated.

The kernel SNR (KSNR) is the kernelization of the canonical MNF transformation. In the standard case, observations and the estimated noise in the original input space are transformed to a high-dimensional Hilbert feature space via suitable mappings endorsed with the reproducing kernel property and the SNR is maximized therein (Nielsen, 2011). This formulation has been extended in Gómez-Chova *et al.* (2017), where a KSNR scheme able to deal with signal and noise relations *explicitly* in Hilbert space has been presented. The KSNR is a modification of the KRR method (Shawe-Taylor and Cristianini, 2004) described in the previous subsection, by including the covariance matrix of the noise defined in the Hilbert space into the model. This matrix is incorporated into the regularizer of the primal cost function. Clearly, the realization of the noise and the corresponding covariance matrix is unknown, and the estimation of the covariance matrix is a key point of the KSNR technique. We will discuss this issue later.

Let us consider the same multi-output formulation as in the previous sections. In this case, we want to minimize the following cost function:

$$\min_{\boldsymbol{W}}\{\|\boldsymbol{Y}-\boldsymbol{\Phi}\boldsymbol{W}\|_{\mathrm{F}}^{2}+\lambda\boldsymbol{W}^{\mathrm{T}}\boldsymbol{\Sigma}_{e}\boldsymbol{W}\}, \tag{8.33}$$

where $\boldsymbol{\Sigma}_e=\boldsymbol{\Phi}_e^{\mathrm{T}}\boldsymbol{\Phi}_e$ represents the noise covariance in the Hilbert space, and

$$\boldsymbol{\Phi}_e=[\boldsymbol{\phi}(\boldsymbol{e}_1)|\cdots|\boldsymbol{\phi}(\boldsymbol{e}_N)]^{\mathrm{T}},$$

2 SNR is defined as the ratio of signal power to the noise power. This concept has been extraordinarily useful in signal processing for decades since it allows us to quantify the quality and robustness of systems and models. Noise reduction is an issue in signal processing, typically tackled by controlling the acquisition environment. Alternatively, one can filter the signal by looking at the noise characteristics. Actually, filter design and regularization are intimately related as maximizing the SNR can be casted as a way to seek for smooth solutions preserving signal characteristics and discarding features mostly influenced by the noise.

where $\boldsymbol{e}_i$, $i = 1, \dots, N$, are the *estimated* noise samples using some technique (see Section 8.2.4). Minimizing Equation 8.33, we obtain

$$\boldsymbol{W}^* = (\boldsymbol{\Phi}^{\mathrm{T}}\boldsymbol{\Phi} + \lambda \boldsymbol{\Sigma}_{\boldsymbol{e}})^{-1}\boldsymbol{\Phi}^{\mathrm{T}}\boldsymbol{Y}. \tag{8.34}$$

Alternatively, using the representer theorem, $\boldsymbol{W} = \boldsymbol{\Phi}^{\mathrm{T}}\boldsymbol{A}$, we obtain the dual cost function

$$\min_{\boldsymbol{A}}\{\|\boldsymbol{Y} - \boldsymbol{K}\boldsymbol{A}\|_F^2 + \lambda \boldsymbol{A}^{\mathrm{T}}\boldsymbol{K}_{\boldsymbol{xe}}\boldsymbol{K}_{\boldsymbol{ex}}\boldsymbol{A}\} \tag{8.35}$$

that is minimized with respect to $\boldsymbol{A}$ and where $\boldsymbol{K}_{\boldsymbol{ex}} = \boldsymbol{\Phi}_{\boldsymbol{e}}\boldsymbol{\Phi}^{\mathrm{T}}$ and $\boldsymbol{K}_{\boldsymbol{xe}} = \boldsymbol{\Phi}\boldsymbol{\Phi}_{\boldsymbol{e}}^{\mathrm{T}}$ are $N \times N$ matrices containing the kernel dot products between the observations and their estimated noise counterparts; that is, the i, jth element of $\boldsymbol{K}_{\boldsymbol{ex}}$ is a scalar product in the Hilbert space, $\langle \boldsymbol{\phi}(\boldsymbol{x}_i), \boldsymbol{\phi}(\boldsymbol{e}_j)\rangle_{\mathcal{H}}$. It is important to note that, in general, $\boldsymbol{K}_{\boldsymbol{ex}} \neq \boldsymbol{K}_{\boldsymbol{xe}}$, but the asymmetry is resorted by the appearance of the product $\boldsymbol{K}_{\boldsymbol{xe}}\boldsymbol{K}_{\boldsymbol{ex}}$ in the functional. The solution of the problem expressed as a function of the dual weights $\boldsymbol{A}$ is

$$\boldsymbol{A}^* = (\boldsymbol{K}^2 + \lambda \boldsymbol{K}_{\boldsymbol{xe}}\boldsymbol{K}_{\boldsymbol{ex}})^{-1}\boldsymbol{K}\boldsymbol{Y}, \tag{8.36}$$

which can be alternatively written as

$$\boldsymbol{A}^* = (\boldsymbol{K} + \lambda \boldsymbol{K}^{-1}\boldsymbol{K}_{\boldsymbol{xe}}\boldsymbol{K}_{\boldsymbol{ex}})^{-1}\boldsymbol{Y} = (\boldsymbol{K} + \lambda \boldsymbol{\Omega})^{-1}\boldsymbol{Y}. \tag{8.37}$$

This equation draws some intuition about the method: the regularization term $\boldsymbol{\Omega} = \boldsymbol{K}^{-1}\boldsymbol{K}_{\boldsymbol{xe}}\boldsymbol{K}_{\boldsymbol{ex}}$ is essentially discounting the impact in the solution of the noisy samples and reinforcing the noise-free ones. This essentially follows the line of discovering relevant directions in feature spaces mainly governed by signal and less affected by noise (Braun *et al.* 2008; Mika *et al.* 2003). As in the standard KRR, we finally need to show that we never actually need access to the mapped feature vector $\boldsymbol{\phi}(\boldsymbol{x})$, which could be of infinite dimension. Indeed, the prediction for $\boldsymbol{x}$ can be simply written as in Equation 8.31.

Note that the KSNR regression generalizes KRR to cases of non-i.i.d. noise. For i.i.d. noise in Hilbert space, $\boldsymbol{\Sigma}_{\boldsymbol{e}} = \sigma_n^2\boldsymbol{I}$, it is trivial to show that the solution (Equation 8.36) reduces to the standard KRR solution, and λ is related to the noise variance σ_n^2. Off-diagonal entries in $\boldsymbol{\Omega}$ stand out and account for signal-to-noise feature relations not accounted for when assuming signal and noise to be independent.

Noise Estimation

The important question of noise estimation remains unanswered for the KSNR method, which can be a difficult task. In audio and image processing and time-series analysis, the most common approach consists of assuming locally stationary signals that allow one to estimate the noise as a simple difference between observations, $\hat{\boldsymbol{n}}_i \approx \boldsymbol{x}_i - \boldsymbol{x}_{i-1}$. Other, more elaborated, approaches approximate the observed signal using autoregressive models to describe the local relations in structured domains. Good examples of these local relations are previous values in a time series or close pixels in an image, which allow one to estimate the noise as $\hat{\boldsymbol{n}}_i \approx \boldsymbol{x}_i - \sum_{j\in W_i} a_j\boldsymbol{x}_j$, denoting W_i the neighborhood for the sample $\boldsymbol{x}_i$. In problems in which there is not a clear structured domain, it is possible to calculate k-NN estimates of the noise $\hat{\boldsymbol{n}}_i \approx \boldsymbol{x}_i - 1/k \sum_{j \in C} \boldsymbol{x}_j$, denoting C the set of k neighbors of $\boldsymbol{x}_i$. This simple way of noise estimation follows the line of the *delta test*, which was proposed for time-series analysis by Pi and Peterson (1994), and intuitively seeks to estimate the residuals support. For convenience, we typically use simple differentiation for the noise estimation in structured domains, and k-NN approximation in unstructured domains. Other strategies allow estimating the noise explicitly in Hilbert spaces (Gómez-Chova *et al.*, 2017).

8.2.5 Semi-supervised Support Vector Regression

Kernel machines in DSP estimation problems typically exploit the labeled examples only. Both the unlabeled information (that accounts for the *pdf* characteristics) and the geometrical relationship between labeled and unlabeled samples is very often obviated. Including the unlabeled information in the regression method may improve the results, which is the focus of a field called semi-supervised learning (SSL) (Chapelle *et al.* 2006; Sindhwani *et al.* 2005).

The key issue in SSL is the assumption of consistency, which means that nearby points are likely to have the same label, and points on the same structure (typically referred to as "cluster" or "manifold") are likely to have the same label. Note that the first assumption is local, whereas the second one is global. Therefore, depending on the data complexity, either the cluster or manifold assumption may be not completely fulfilled, and thus no clear improvement may be eventually obtained with semi-supervised approaches (Chapelle *et al.* 2006; Sindhwani *et al.* 2005).

A simple yet effective way to estimate the marginal data distribution and then include this information into any kernel method consists of "deforming" the structure of the kernel matrix according to the unlabeled data structure. The deformation can be designed by clustering algorithms (Chapelle *et al.*, 2003), or by deforming a valid kernel with *graph-based methods* that account for the geometrical relations between labeled and unlabeled. In this section, we exploit this latter regularization framework proposed in Sindhwani *et al.* (2005) and Belkin and Niyogi (2004) to develop a semi-supervised version of the SVR method.

In SSL, we are given a set of labeled data $\{\boldsymbol{x}_i, y_i\}_{i=1}^{l}$ and some unlabeled data $\{\boldsymbol{x}_i\}_{i=l+1}^{l+u}$. Typically having access to large labeled datasets is not easy, and one faces problems where $u \gg l$. Let us define the evaluation map of the predictive function considering all available examples $S(\boldsymbol{f}) = [f(\boldsymbol{x}_1), \dots, f(\boldsymbol{x}_{l+u})]^{\mathrm{T}}$, and its semi-norm $\|S(\boldsymbol{f})\|^2 = \boldsymbol{f}^{\mathrm{T}}\boldsymbol{M}\boldsymbol{f}$ given by a symmetric semi-definite matrix $\boldsymbol{M}$. The explicit form of the corresponding reproducing kernel $\tilde{K}(\boldsymbol{x}_i, \boldsymbol{x}_j)$ can be defined (Sindhwani *et al.*, 2005) as

$$\tilde{K}(\boldsymbol{x}_i, \boldsymbol{x}_j) = K(\boldsymbol{x}_i, \boldsymbol{x}_j) - \boldsymbol{k}_i^{\mathrm{T}}(\boldsymbol{I} + \boldsymbol{M}\boldsymbol{K})^{-1}\boldsymbol{M}\boldsymbol{k}_j, \tag{8.38}$$

where $i, j \in \{1, \dots, l+u\}$; $\boldsymbol{K}$ is the (complete) kernel matrix, $\boldsymbol{I}$ is the identity matrix, and $\boldsymbol{k}_i = [K(\boldsymbol{x}_1, \boldsymbol{x}_i), \dots, K(\boldsymbol{x}_{l+u}, \boldsymbol{x}_i)]^{\mathrm{T}}$. Intuitively, the similarity between examples is modified by discounting the unexplained information conveyed by the labeled samples only, and that it is present when accounting for all, labeled and unlabeled, samples.

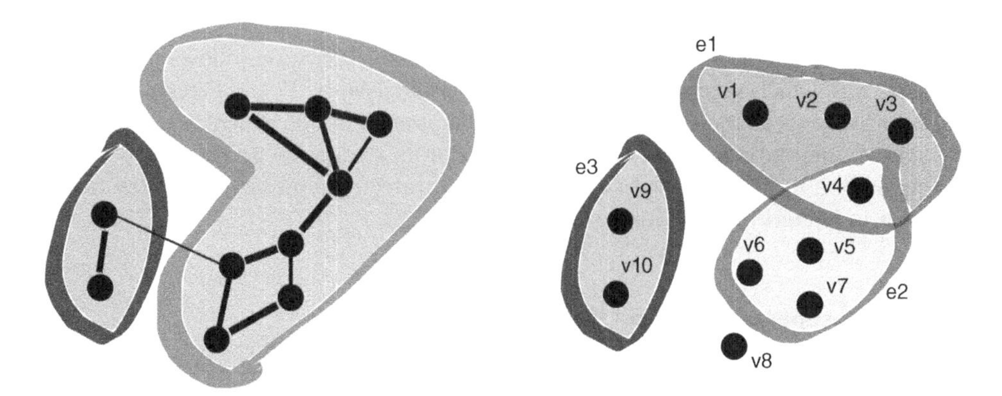

Figure 8.1 In the toy graph (left), edges thickness represents the similarity between samples. Graph methods group the unlabeled samples according to the weighted distance, not just to the shortest path lengths. The two clusters are intuitively correct, even being connected by a thin weak edge. In the toy hypergraph (right), the same vertices and three sets contain different potentially interconnected subgraphs.

The geometry of the data (both labeled and unlabeled) is included through a proper definition of the matrix $\boldsymbol{M}$, which is usually defined in terms of a *graph* or *hypergraph* containing both labeled and unlabeled points:

- *Graph Laplacians.* A graph $G(V, E)$ is defined with a set of nodes V connected by a set of edges E. The edge connecting nodes i and j has an associated weight W_{ij} (Chapelle *et al.*, 2006). The nodes are the samples, and the edges represent the similarity among samples in the dataset (see Figure 8.1 (left)). A proper definition of the graph is the key to accurately introducing data structure in the regression machine. For computing W_{ij}, the k-NN algorithm is typically used.

  The common choice takes $\boldsymbol{M} = \gamma \boldsymbol{L}$, where $\gamma \in [0, \infty)$ is a free parameter that controls the deformation of the kernel, and $\boldsymbol{L}$ is the graph Laplacian, which measures the (smooth) variation of the function along the graph (Chapelle *et al.*, 2006). The graph Laplacian is given by $\boldsymbol{L} = \boldsymbol{D} - \boldsymbol{W}$, where $\boldsymbol{D}$ is the diagonal degree matrix of $\mathbf{W}$; that is, $D_{ii} = \sum_{j=1}^{l+u} W_{ij}$.
- *Hypergraph Laplacian.* A hypergraph is a generalization of a graph, where edges can connect any number of vertices (Zhou *et al.*, 2007). While graph edges are pairs of nodes, hyperedges are arbitrary sets of nodes, and can therefore contain an arbitrary number of nodes (see Figure 8.1 (right)). A hypergraph is also called a *set system* or a *family of sets* drawn from the universal set. Hypergraphs can be viewed as incidence structures and vice versa.

  Notationally, a hypergraph can be represented by a matrix $H(V, E)$ with entries $h(v, e) = 1$ if $v \in e$ and 0 otherwise, called the *incidence matrix*. The hypergraph Laplacian is given by $\boldsymbol{L} = \frac{1}{2}(\boldsymbol{I} - \boldsymbol{D}_{\mathrm{v}}^{-1/2}\boldsymbol{H}\boldsymbol{W}\boldsymbol{D}_{\mathrm{e}}^{-1}\boldsymbol{H}^{\mathrm{T}}\boldsymbol{D}_{v}^{-1/2})$, where $\boldsymbol{D}_{\mathrm{v}}$ and $\boldsymbol{D}_{\mathrm{e}}$ denote the diagonal matrices containing the vertex and hyperedge degrees respectively. One can build the hypergraph by first performing k-means clustering, and then taking the centroids as hyperedges. The weights for all hyperedges are then simply set to 1. Other more sophisticated forms of defining $\boldsymbol{H}$ exist. Since hypergraph links may have any cardinality, there are multiple notions of the concept of a subgraph: subhypergraphs, partial hypergraphs, and section hypergraphs.

Independently of using either the graph or hypergraph Laplacian, note that by fixing $\gamma = 0$ the original (undeformed) kernel is obtained. Therefore, a proper selection of this free parameter should lead to improved performance over supervised regression. The use of the SS-SVR in practice reduces to simply first computing the deformed kernel using Equation 8.38 and then plugging it into a standard kernel solver, such as the SVR. It is worth noting that the very same methodology could be equally applied to any kernel-based method, either for regression, classification, or feature extraction.

8.2.6 Model Selection in Kernel Regression Methods

All kernel methods seen in previous sections have a series of hyperparameters that should be correctly tuned to obtain good results. The parameters of the kernel function are common to all methods, and some algorithms use several kernel functions and corresponding parameters (e.g., KSNR). The polynomial kernel and the Gaussian kernel are perhaps the most widely used kernel functions as they include just one parameter, the polynomial degree d and the length-scale – spread or width of the Gaussian function – parameter σ.

Besides kernel parameters, each method introduces an additional regularization hyperparameter, which trades off penalizing errors to finding smooth solutions in order to combat overfitting. For instance, the KRR method introduces λ (Equation 8.28), which for the case of the SVR, sometimes referred to as ε-SVR, we used $C = 2/\lambda$. In SVR, we also have an additional parameter to control the complexity of the solution: recall that ε essentially controls the sparsity (see Figure 4.4). The ε-Huber cost function introduces an extra hyperparameter, δ, which together with C controls penalization to account for Gaussian noise.

In order to find a set of optimal hyperparameters, a typical approach is to partition the training set into subsets and conduct a grid search where several parameter combinations are tested, until the best combination is found (Schölkopf *et al.*, 1999). A standard two-thirds–one-third partition is commonly used. However, most practitioners prefer to use more exhaustive (and often more reliable) cross-validation techniques, such as the so-called v-fold (Duda *et al.*, 1998). In v-fold, the training set is divided into v subsets, and each combination of hyperparameters is tried on each one, using $v - 1$ subsets to train a model and the vth subset to test it. The hyperparameters that on average work best for all v subsets are selected.

A different perspective to the issue of parameter tuning is that offered by bootstrapping (Efron and Tibshirani, 1998). There are two main approaches to using bootstrapping to select hyperparameters, and both of them require a previous partition of the training set, for instance a two-thirds training–one-third test partition:

1) For each combination of hyperparameters, the training set is sampled with repetition to obtain many training sets, which are then used to train models that are evaluated in the separated test partition. As in v-fold cross-validation, the combination that obtains the best performance on average is selected.
2) In the second approach, for each combination of hyperparameters a single model is obtained using the training partition and evaluated on the test subset. After this step,

the residuals are sampled with repetition to obtain a better and reliable estimation of the mean error. In this setup, the hyperparameters with the lowest error (which in the end is the same as obtaining the best performance) are selected.

The main problem of bootstrapping approaches is that they are computationally very demanding (especially the first one), as they require the repetition of the experiment in the order of hundreds of times to obtain a valid estimation (Efron and Tibshirani, 1998).

Although the techniques described are the most widely used, it is possible (and advisable) to take into account the properties of the signals: looking at the dynamic range of the signal and the histograms of the features and observations may help define a proper range of variation. In the following subsections, we provide some suggestions to select proper values or ranges for the most used hyperparameters in kernel regression models.

Gaussian Kernel Length-scale Parameter

One of the most used kernel functions is the RBF Gaussian kernel. The are several reasons to explain this selection: it covers other kernel functions as a particular case (for instance, the polynomial or the linear kernel (Schölkopf and Smola, 2002)), it (usually) obtains similar or better performances than other kernel functions, and, last but not least, it only has one parameter to tune, the width of the Gaussian function, σ. Let us remember the form of the RBF Gaussian kernel:

$$K(\boldsymbol{x},\boldsymbol{x}') = \exp\left(\frac{-\|\boldsymbol{x}-\boldsymbol{x}'\|^2}{2\sigma^2}\right).$$

Intuitively, one can see that large values of σ will cover large "spaces," thus making the difference of vectors $\boldsymbol{x}$ and $\boldsymbol{x}'$ less evident. On the other hand, small values of σ will cause that even close vectors (in the Euclidean distance sense) obtain low values of the kernel function, meaning these two vectors will be considered as "different." Given this, a way to have an idea of a proper value for σ is to take into account some statistics about the distances between the vectors in the training set; for example, the mean distance, the median, or the mode. Indeed, given a dataset, it does not make much sense to use values of σ that are much further than, for instance, $m_\mathrm{d} \pm 2\sigma_\mathrm{d}$, where m_d and σ_d would be the mean and standard deviation of all Euclidean distances in the dataset; σ values much lower than this mean that all samples in the training set would be treated as "different" (orthogonal), whereas large σ values would treat all samples as similar to each other. Intuitively, one wants values of σ that are able to distinguish between groups of samples, not to treat all of them as different, or as equal. Therefore, a rule of thumb to choose a good value for σ would be to set it as m_d. For a more extensive tuning, reasonable values of σ should varied in the interval $m_\mathrm{d} \pm 2\sigma_\mathrm{d}$.

Some other heuristics can be found in the literature. The code in Listing 8.1 gives the code for the implementation of some of the most widely used.

```
% Assume X data matrix: N samples, d features: N x d
% 1: 'mean': Average distance between all samples
D = pdist(X); sigma.mean = mean(D(D>0));
% 2: 'median': Median of the distances between all samples
D = pdist(X); sigma.median = median(D(D>0));
```

```
% 3: 'quantiles': 10 values in the range of 0.05-0.95 distance ...
    percentiles
D = pdist(X); sigma.quantiles = quantile(D(D>0),linspace(0.05,0.95,10));
% 4: 'histo': Sigma proportional to the dimensionality and feature ...
    variance
sigma.sampling = sqrt(d) * median(var(X));
% 5: 'range': 10 values to try in the range of 0.2-0.5 of the feature ...
    range
mm = minmax(X'); sigma.range = ...
    median(mm(:,2)-mm(:,1))*linspace(0.2,0.5,10);
% 6: 'silverman': Median of the Silverman's rule per feature
sigma.silverman = median( ((4/(d+2))^(1/(d+4))) * n^(-1/(d+4)).*std(X,1) ...
    );
% 7: 'scott': Median of the Scott's rule per feature
sigma.scott = median( diag( n^(-1/(d+4))*cov(X).^(1/2)) );
```

Listing 8.1 Estimation of the with in the RBF kernel function.

The previous rule of thumb criterion can be extremely useful in the case of unlabeled datasets. When some labeled samples are available, *kernel alignment* may help; see Chapter 4. Assume a data matrix $\boldsymbol{X}$ and the corresponding label vector $\boldsymbol{y}$. In kernel alignment one essentially constructs an ideal kernel using the training labels, $\boldsymbol{K}_{\text{ideal}} = \boldsymbol{y}\boldsymbol{y}^{\text{T}}$, and selects the RBF length-scale σ that minimizes the Frobenius norm of the generated kernel matrix and the ideal kernel, $\sigma^* = \min_\sigma\{\|\boldsymbol{K}_\sigma - \boldsymbol{K}_{\text{ideal}}\|_{\text{F}}^2\}$. Note that this heuristic exploits the labeled information, and it does not require costly cross-validation strategies as it only involves constructing kernel matrices and involves simple norm calculations.

Regularization and Sparsity Parameters

The second parameter in all kernel-based algorithms has to do with the selected cost function. When an ℓ_2-norm is selected, as in KRR, this is the only one parameter to be selected besides the kernel parameters. For the SVR, nevertheless, two parameters in the loss need to be chosen: the ε error insensitivity (to control the sparseness of the solution) and the regularization parameter C (to control the capacity of the class of functions used). Let us review the main prescriptions and intuitions behind tuning these.

Regularization Parameter

This parameter controls the trade-off between penalizing errors and obtaining smooth (therefore, not overfitted) solutions. Large values of C will penalize errors more aggressively, making the model function to tightly fit the training samples. In principle, this is not bad per se, but it may lead to lack of generalization on unseen, testing samples. In this case, we say that the model is *overfitted* to the training set. On the other hand, small values of C allow for errors when developing the model, giving more importance to obtain a smooth solution. Having a smooth solution is rather desirable (most of the natural signals are smooth) because it will probably work better on unseen (test) samples. Nevertheless, if C is too loose (very small values of C), the committed errors will be so large that the model just will not describe the problem adequately.

Many prescriptions for selecting reasonable values of C are available in the literature. For example, Cherkassky and Ma (2004) propose an analytic parameter value selection from the training data. It can be shown that $|\hat{y}(\boldsymbol{x})| \leq C \cdot N_{SV}$, where N_{SV} is the number of support vectors. The problem is that one does not know this number until all hyperparameters have been tuned (specifically, ε is the parameter related to the number of support vectors (Schölkopf *et al.*, 2000)). Therefore, an initial good guess for the value of C is $|\hat{y}(\boldsymbol{x})|$; that is, the range of output values. However, this value is very sensitive to the presence of outliers. To make the selection robust to outliers, Cherkassky and Ma (2004) proposed to select C as $\max(|\bar{y} + 3\sigma_y|, |\bar{y} - 3\sigma_y|)$, where $\bar{y}$ and σ_y are the mean and the standard deviation of the training samples y.

Insensitivity Error

This parameter defines the insensitive zone of the SVR (see Figure 4.4), and it is related to the noise present in the input signal. Several authors (Cherkassky and Mulier 1998; Kwok 2001; Smola *et al.* 1998) have shown that there is a linear relationship between the noise and the value of ε. If the level of noise σ_n is known, or can be estimated, one can start with a value of $\varepsilon = \sigma_n$. However, this value does not take into account the size of the training set and the eventual noise correlations either. With the same level of noise, ε should intuitively be smaller for a large number of training samples than for a small one. For linear regression, the variance of observations is proportional to σ_n^2/N, where σ_n^2 is the noise variance and N the number of training samples. Therefore, a better selection for epsilon is $\varepsilon \propto \sigma_n/N$. However, Cherkassky and Ma (2004) empirically found that large values of N yield values of ε that are too small, and proposed setting $\varepsilon = 3\sigma_n(\ln N/N)^{1/2}$.

Gaussian Noise Penalization

This parameter has to be tuned when the ε-Huber cost function is used. It is related to the amount of Gaussian noise present in the observation, and together with C it defines the area in the cost function dealing with this kind of noise. Therefore, it is usual to tune δC together instead or C and δ separately, although of course the latter is possible. Interestingly, the use of the ε-Huber cost function is beneficial to obtain some knowledge about the underlying noise process, since the optimal value of δC will return a good estimate of the Gaussian noise variance in the observed signal. And vice versa, if we know the amount of Gaussian noise present in the signal, given a value for C, δ can be deduced so their product matches the amount of Gaussian noise in the input.

8.3 Bayesian Nonparametric Kernel Regression Models

This section deals with nonparametric models following Bayesian principles. We will depart from a standard linear basis function model in Hilbert spaces, but here we will give a Bayesian treatment. The primal and dual forms will readily become apparent, giving rise to useful model instantiations: the Gaussian process regression (GPR) model and the RVM. The former algorithm has captured the attention of researchers in signal processing. GPR has shown great utility to solve real-life signal processing problems. Actually, in the last decade, many variants of GPs have been published to deal with specificities of the signal structure (through kernel design) and the noise characteristics (via heteroscedastic GPs). We will review some of these recent developments for DSP problem solving as well.

8.3.1 Gaussian Process Regression

The linear basis function model is a a simple extension of the standard linear Bayesian regression model (Bishop, 2006), and it is the basis of the well-known GPs (O'Hagan and Kingman, 1978; Rasmussen and Williams, 2006). GPs are Bayesian state-of-the-art tools for discriminative machine learning; that is, regression (Williams and Rasmussen, 1996), classification (Kuss and Rasmussen, 2005), and dimensionality reduction (Lawrence, 2005). GPs were first proposed in statistics by Tony O'Hagan (O'Hagan and Kingman, 1978) and they are well known to the geostatistics community as kriging. However, owing to their high computational complexity they did not become widely applied tools in machine learning until the early 21st century (Rasmussen and Williams, 2006). GPs can be interpreted as a family of kernel methods with the additional advantage of providing a full conditional statistical description for the predicted variable, which can be primarily used to establish credible intervals and to set hyperparameters. In a nutshell, GPs assume that a GP prior governs the set of possible latent functions (which are unobserved), and the likelihood (of the latent function) and observations shape this prior to produce posterior probabilistic estimates. Consequently, the joint distribution of training and test data is a multidimensional Gaussian and the predicted distribution is estimated by conditioning on the training data. As we will see later, GPs for regression can be cast as a natural nonlinear extension to optimal Wiener filtering.

The linear basis function model is based on a parametric approach that it is strictly related to other the nonparametric methods described later. More specifically, this approach employs nonlinear basis functions, but the model is still linear with respect to the parameters. Indeed, the observation model is again given by

$$y_i = \boldsymbol{w}^{\mathrm{T}}\boldsymbol{\phi}(\boldsymbol{x}_i) + e_i, \qquad i = 1, \dots, N \tag{8.39}$$

with $\boldsymbol{x}_i \in \mathbb{R}^d$, $\boldsymbol{\phi}(\boldsymbol{x}_i) \in \mathbb{R}^M$, $\boldsymbol{w} \in \mathbb{R}^M$, $e_i, y_i \in \mathbb{R}$ or, in a more compact matrix form:

$$\boldsymbol{y} = \boldsymbol{\Phi}\boldsymbol{w} + \mathbf{e}, \tag{8.40}$$

where $\boldsymbol{\Phi} = [\boldsymbol{\phi}(\boldsymbol{x}_1)| \cdots |\boldsymbol{\phi}(\boldsymbol{x}_N)]^{\mathrm{T}} \in \mathbb{R}^{N\times M}$ is a matrix formed by input vectors in rows, $\boldsymbol{y} = [y_1, \dots, y_N]^{\mathrm{T}} \in \mathbb{R}^{N\times 1}$ is the output column vector, $\boldsymbol{w} \in \mathbb{R}^{M\times 1}$ is a vector of weights which we desire to infer, and $\boldsymbol{e} \in \mathbb{R}^{N\times 1}$ is an additive noise vector in which each component is Gaussian distributed with zero mean and variance σ_{n}^2 (i.e., $e_i \sim \mathcal{N}(0, \sigma_{\mathrm{n}}^2)$ for all $i = 1, \dots, N$). Therefore, the *likelihood function* associated with the model in Equation 8.39 is

$$\begin{aligned} p(\boldsymbol{y}|\boldsymbol{\Phi}, \boldsymbol{w}) &= \prod_{i=1}^{N} p(y_i|\boldsymbol{\phi}(\boldsymbol{x}_i), \boldsymbol{w}) = \prod_{i=1}^{N} \frac{1}{\sqrt{2\pi\sigma_{\mathrm{n}}^2}} \exp\left[-\frac{(y_i - \boldsymbol{w}^{\mathrm{T}}\boldsymbol{\phi}(\boldsymbol{x}_i))^2}{2\sigma_{\mathrm{n}}^2}\right] \\ &= \frac{1}{(2\pi\sigma_{\mathrm{n}}^2)^{N/2}} \exp\left(-\frac{\|\boldsymbol{y} - \boldsymbol{\Phi}\boldsymbol{w}\|^2}{2\sigma_{\mathrm{n}}^2}\right), \end{aligned} \tag{8.41}$$

which represents the *pdf* of a normal distribution $\mathcal{N}(\boldsymbol{\Phi}\mathbf{w}, \sigma_\mathrm{n}^2\mathbf{I})$. Note that we assume that the observations y_i are i.i.d. so we can factorize the joint *pdf* as a product of N marginal *pdfs*. Furthermore, since we consider a Bayesian approach, we assume a *prior pdf* $p(\mathbf{w})$ over the vector of weights $\mathbf{w}$. More specifically, we consider

$$p(\mathbf{w}) = \frac{1}{\sqrt{(2\pi)^N|\boldsymbol{\Sigma}_p|}} \exp\left(-\frac{1}{2}\mathbf{w}^\mathrm{T}\boldsymbol{\Sigma}_p^{-1}\mathbf{w}\right), \tag{8.42}$$

which is Gaussian with zero mean and covariance matrix $\boldsymbol{\Sigma}_p$ of size $M \times M$; that is, $\mathbf{w} \sim \mathcal{N}(\mathbf{0}, \boldsymbol{\Sigma}_p)$ (note that $|\boldsymbol{\Sigma}_p|$ is determinant of $\boldsymbol{\Sigma}_p$). Using Bayes' rule, we obtain the expression of the *posterior pdf* of the parameter vector $\mathbf{w}$ given the observations $\mathbf{y}$ and the matrix $\boldsymbol{\Phi}$:

$$p(\mathbf{w}|\mathbf{y}, \boldsymbol{\Phi}) = \frac{p(\mathbf{y}|\boldsymbol{\Phi}, \mathbf{w})p(\mathbf{w})}{p(\mathbf{y}|\boldsymbol{\Phi})}. \tag{8.43}$$

The *marginal likelihood* $p(\mathbf{y}|\boldsymbol{\Phi})$ acts as a normalizing factor in Equation 8.43, and it is obtained by integration:

$$p(\mathbf{y}|\boldsymbol{\Phi}) = \int p(\mathbf{y}|\boldsymbol{\Phi}, \mathbf{w})p(\mathbf{w})\,\mathrm{d}\mathbf{w}, \tag{8.44}$$

which does not depend on $\mathbf{w}$. Hence, plugging the likelihood and prior (Equations 8.41 and 8.42) into Equation 8.43 leads to

$$p(\mathbf{w}|\mathbf{y}, \boldsymbol{\Phi}) \propto \exp\left[-\frac{(\mathbf{y}-\boldsymbol{\Phi}\mathbf{w})^\mathrm{T}(\mathbf{y}-\boldsymbol{\Phi}\mathbf{w})}{2\sigma_\mathrm{n}^2}\right]\exp\left(-\frac{\mathbf{w}^\mathrm{T}\boldsymbol{\Sigma}_p^{-1}\mathbf{w}}{2}\right). \tag{8.45}$$

We consider the computation of the MMSE estimator, which consists of computing the expected value of the posterior *pdf*:

$$\bar{\mathbf{w}} = \mathbb{E}[\mathbf{w}|\mathbf{y}, \boldsymbol{\Phi}] = \int \mathbf{w}\, p(\mathbf{w}|\mathbf{y}, \boldsymbol{\Phi})\,\mathrm{d}\mathbf{w}. \tag{8.46}$$

Given the assumed model, it is possible to compute $\mathbb{E}[\mathbf{w}|\mathbf{y}, \boldsymbol{\Phi}]$ analytically and to obtain

$$\bar{\mathbf{w}} = \frac{1}{\sigma_\mathrm{n}^2}\left(\frac{1}{\sigma_\mathrm{n}^2}\boldsymbol{\Phi}^\mathrm{T}\boldsymbol{\Phi} + \boldsymbol{\Sigma}_p^{-1}\right)^{-1}\boldsymbol{\Phi}^\mathrm{T}\mathbf{y} = (\boldsymbol{\Phi}^\mathrm{T}\boldsymbol{\Phi} + \sigma_\mathrm{n}^2\boldsymbol{\Sigma}_p^{-1})^{-1}\boldsymbol{\Phi}^\mathrm{T}\mathbf{y}. \tag{8.47}$$

It is important to remark on the close relationship between this estimator and the (primal) solutions of the KRR and KSNR methods provided in the previous sections. This MMSE estimator can also be expressed as

$$\bar{\mathbf{w}} = \frac{1}{\sigma_\mathrm{n}^2}\boldsymbol{\Lambda}^{-1}\boldsymbol{\Phi}^\mathrm{T}\mathbf{y}, \tag{8.48}$$

where

$$\boldsymbol{\Lambda} = \frac{1}{\sigma_{\mathrm{n}}^2}\boldsymbol{\Phi}^{\mathrm{T}}\boldsymbol{\Phi} + \boldsymbol{\Sigma}_p^{-1} \tag{8.49}$$

also represents the precision of the posterior; that is, the inverse of its covariance matrix. Hence, we can write

$$\begin{aligned} p(\boldsymbol{w}|\boldsymbol{y},\boldsymbol{\Phi}) &\propto \exp\left[-\frac{1}{2}(\boldsymbol{w}-\bar{\boldsymbol{w}})^{\mathrm{T}}\boldsymbol{\Lambda}(\boldsymbol{w}-\bar{\boldsymbol{w}})\right], \\ &= \mathcal{N}(\boldsymbol{w}|\bar{\boldsymbol{w}},\boldsymbol{\Lambda}^{-1}) = \mathcal{N}\left(\boldsymbol{w}\left|\frac{1}{\sigma_{\mathrm{n}}^2}\boldsymbol{\Lambda}^{-1}\boldsymbol{\Phi}^{\mathrm{T}}\boldsymbol{y},\boldsymbol{\Lambda}^{-1}\right.\right), \end{aligned}$$

where $\mathcal{N}(\boldsymbol{w}|\bar{\boldsymbol{w}},\boldsymbol{\Lambda}^{-1})$ denotes a Gaussian *pdf* with mean $\bar{\boldsymbol{w}}$ and covariance matrix $\boldsymbol{\Lambda}^{-1}$. Therefore, considering the assumed parametric form of the predictor $\boldsymbol{w}^{\mathrm{T}}\boldsymbol{\phi}(\boldsymbol{x})$, we can consider the prediction vector $\widehat{f}(\boldsymbol{x})$ at some test input $\boldsymbol{x}$ as

$$\widehat{f}(\boldsymbol{x}) = \bar{\boldsymbol{w}}^{\mathrm{T}}\boldsymbol{\phi}(\boldsymbol{x}) = \boldsymbol{\phi}(\boldsymbol{x})^{\mathrm{T}}\bar{\boldsymbol{w}} = \frac{1}{\sigma_{\mathrm{n}}^2}\boldsymbol{\phi}(\boldsymbol{x})^{\mathrm{T}}\boldsymbol{\Lambda}^{-1}\boldsymbol{\Phi}^{\mathrm{T}}\boldsymbol{y}. \tag{8.50}$$

However, in the Bayesian framework, the predictions are not just pointwise estimates, but are the result of summarizing (e.g., with the average estimator) the predictive density. A complete characterization of the posterior *pdf* of the hidden function $f(\cdot)$ evaluated at some test input $\boldsymbol{x}$ can be provided. The predictive mean is obtained by averaging over all possible parameter values weighted by their posterior probability. In this case, for instance, we can compute analytically the following predictive *pdf*:

$$\begin{aligned} p(f(\boldsymbol{x})|\boldsymbol{x},X,\boldsymbol{y}) &= \int p(f(\boldsymbol{x})|\boldsymbol{x},\boldsymbol{w})p(\boldsymbol{w}|\boldsymbol{y},\boldsymbol{\Phi})\,\mathrm{d}\boldsymbol{w} \\ &= \mathcal{N}\left(f(\boldsymbol{x})\left|\frac{1}{\sigma_{\mathrm{n}}^2}\boldsymbol{\phi}(\boldsymbol{x})^{\mathrm{T}}\boldsymbol{\Lambda}^{-1}\boldsymbol{\Phi}^{\mathrm{T}}\boldsymbol{y}\boldsymbol{\phi}(\boldsymbol{x})^{\mathrm{T}}\boldsymbol{\Lambda}^{-1}\boldsymbol{\phi}(\boldsymbol{x})\right.\right) \\ &= \mathcal{N}(f(\boldsymbol{x})|\widehat{f}(\boldsymbol{x}),\widehat{\sigma}(\boldsymbol{x})), \end{aligned} \tag{8.51}$$

where the predictive mean is

$$\widehat{f}(\boldsymbol{x}) = \frac{1}{\sigma_{\mathrm{n}}^2}\boldsymbol{\phi}(\boldsymbol{x})^{\mathrm{T}}\boldsymbol{\Lambda}^{-1}\boldsymbol{\Phi}^{\mathrm{T}}\boldsymbol{y} = \boldsymbol{\phi}(\boldsymbol{x})^{\mathrm{T}}(\boldsymbol{\Phi}^{\mathrm{T}}\boldsymbol{\Phi} + \sigma_{\mathrm{n}}^2\boldsymbol{\Sigma}_p^{-1})^{-1}\boldsymbol{\Phi}^{\mathrm{T}}\boldsymbol{y}, \tag{8.52}$$

which coincides with Equation 8.50, and the predictive variance is

$$\widehat{\sigma}(\boldsymbol{x}) = \boldsymbol{\phi}(\boldsymbol{x})^{\mathrm{T}}\boldsymbol{\Lambda}^{-1}\boldsymbol{\phi}(\boldsymbol{x}) = \boldsymbol{\phi}(\boldsymbol{x})^{\mathrm{T}}\left(\frac{1}{\sigma_{\mathrm{n}}^2}\boldsymbol{\Phi}^{\mathrm{T}}\boldsymbol{\Phi} + \boldsymbol{\Sigma}_p^{-1}\right)^{-1}\boldsymbol{\phi}(\boldsymbol{x}). \tag{8.53}$$

Remark. In the linear basis function model, we can easily draw random functions according to the posterior distribution of the weights given the observed data following the procedure

(a) draw a weight $\boldsymbol{w}_s$ according to the posterior $p(\boldsymbol{w}|\boldsymbol{y}, \boldsymbol{\Phi})$ and then
(b) compute the sample function $f_S(\boldsymbol{x}) = \boldsymbol{w}_s^{\mathrm{T}}\boldsymbol{\phi}(\boldsymbol{x})$ for all the test points $\boldsymbol{x}$.

See an experimental example in Figure 8.7.

The possibility to obtain a predictive variance is one of the benefits of a Bayesian probabilistic treatment with respect to the strategies considered in Section 8.2. Furthermore considering the observation model and an input test $\boldsymbol{x}_*$, we have $y_* = f(\boldsymbol{x}^*) + e^*$, and then we can also find the predictive distribution of the noisy observations; that is:

$$p(y_*|\boldsymbol{x}, \boldsymbol{X}, \boldsymbol{y}) = \mathcal{N}(y|\hat{f}(\boldsymbol{x}), \hat{\sigma}(\boldsymbol{x}) + \sigma_{\mathrm{n}}^2). \tag{8.54}$$

In the following subsections, we derive some interesting alternative formulations for the predictive mean and variance.

Alternative Formulation for the Predictive Mean

Now we work on the expression $\boldsymbol{\Lambda}^{-1}\boldsymbol{\Sigma}_p\boldsymbol{\Phi}^{\mathrm{T}}$ in order to obtain an interesting alternative formulation of the predictive mean. Noting that

$$\begin{aligned}\boldsymbol{\Lambda}\boldsymbol{\Sigma}_p\boldsymbol{\Phi}^{\mathrm{T}} &= \left(\frac{1}{\sigma_{\mathrm{n}}^2}\boldsymbol{\Phi}^{\mathrm{T}}\boldsymbol{\Phi} + \Sigma_p^{-1}\right)\Sigma_p\boldsymbol{\Phi}^{\mathrm{T}} = \frac{1}{\sigma_{\mathrm{n}}^2}\boldsymbol{\Phi}^{\mathrm{T}}\boldsymbol{\Phi}\Sigma_p\boldsymbol{\Phi}^{\mathrm{T}} + \Sigma_p^{-1}\Sigma_p\boldsymbol{\Phi} \\ &= \frac{1}{\sigma_{\mathrm{n}}^2}\boldsymbol{\Phi}^{\mathrm{T}}\boldsymbol{\Phi}\Sigma_p\boldsymbol{\Phi}^{\mathrm{T}} + \boldsymbol{\Phi}^{\mathrm{T}}\end{aligned} \tag{8.55}$$

we finally obtain

$$\boldsymbol{\Lambda}\Sigma_p\boldsymbol{\Phi}^{\mathrm{T}} = \frac{1}{\sigma_{\mathrm{n}}^2}\boldsymbol{\Phi}^{\mathrm{T}}(\boldsymbol{\Phi}\Sigma_p\boldsymbol{\Phi}^{\mathrm{T}} + \sigma_{\mathrm{n}}^2\boldsymbol{I}) \tag{8.56}$$

Given this latter equality, multiplying both sides by $\boldsymbol{\Lambda}^{-1}$ from the left and by $(\boldsymbol{\Phi}\Sigma_p\boldsymbol{\Phi}^{\mathrm{T}} + \sigma_{\mathrm{n}}^2\boldsymbol{I})^{-1}$ from the right, we arrive at the expression

$$\Sigma_p\boldsymbol{\Phi}^{\mathrm{T}}(\boldsymbol{\Phi}\Sigma_p\boldsymbol{\Phi}^{\mathrm{T}} + \sigma_{\mathrm{n}}^2\boldsymbol{I})^{-1} = \frac{1}{\sigma_{\mathrm{n}}^2}\boldsymbol{\Lambda}^{-1}\boldsymbol{\Phi}^{\mathrm{T}}. \tag{8.57}$$

Hence, using that equality, we can rewrite $\hat{f}(\boldsymbol{x})$ as

$$\begin{aligned}\hat{f}(\boldsymbol{x}) &= \frac{1}{\sigma_{\mathrm{n}}^2}\boldsymbol{\phi}(\boldsymbol{x})^{\mathrm{T}}\boldsymbol{\Lambda}^{-1}\boldsymbol{\Phi}^{\mathrm{T}}\boldsymbol{y} = \boldsymbol{\phi}(\boldsymbol{x})^{\mathrm{T}}\left(\frac{1}{\sigma_{\mathrm{n}}^2}\boldsymbol{\Lambda}^{-1}\boldsymbol{\Phi}^{\mathrm{T}}\right)\boldsymbol{y} \\ &= \boldsymbol{\phi}(\boldsymbol{x})^{\mathrm{T}}\Sigma_p\boldsymbol{\Phi}^{\mathrm{T}}(\boldsymbol{\Phi}\Sigma_p\boldsymbol{\Phi}^{\mathrm{T}} + \sigma_{\mathrm{n}}^2\boldsymbol{I})^{-1}\boldsymbol{y}.\end{aligned} \tag{8.58}$$

Now let us define the kernel function

$$K(\boldsymbol{x}, \boldsymbol{x}') = \boldsymbol{\phi}(\boldsymbol{x})^{\mathrm{T}}\Sigma_p\boldsymbol{\phi}(\boldsymbol{x}') = \boldsymbol{\psi}(\boldsymbol{x})^{\mathrm{T}}\boldsymbol{\psi}(\boldsymbol{x}'), \tag{8.59}$$

where $\boldsymbol{\psi}(\boldsymbol{x}) := \Sigma_p^{1/2}\boldsymbol{\phi}(\boldsymbol{x})$ (recall that Σ_p is definite positive). Also, we can define

$$\boldsymbol{k} = \boldsymbol{\phi}(\boldsymbol{x})^{\mathrm{T}}\Sigma_p\boldsymbol{\Phi}^{\mathrm{T}} = [K(\boldsymbol{x}, \boldsymbol{x}_1), \dots, K(\boldsymbol{x}, \boldsymbol{x}_N)]^{\mathrm{T}} \in \mathbb{R}^{N\times N} \tag{8.60}$$

and

$$K = \Phi \Sigma_p \Phi^{\mathrm{T}}. \tag{8.61}$$

With these definitions we can write

$$\hat{f}(x) = k^{\mathrm{T}}(K + \sigma_{\mathrm{n}}^2 I)^{-1} y, \tag{8.62}$$

which plays the role of the dual solution and it is essentially the same predictive function in the previous KRR and KSNR methods.

Alternative Formulation for the Predictive Variance

Considering the following generic matrices Z of size $M \times M$, U of size $N \times M$, L of size $N \times N$, and V of size $M \times N$, the following *matrix inversion lemma* (see Press *et al.* (1992) and Appendix A in Rasmussen and Williams (2006)) is satisfied:

$$(Z + ULV^{\mathrm{T}})^{-1} = Z^{-1} - Z^{-1}U(L^{-1} + V^{\mathrm{T}}Z^{-1}U)^{-1}V^{\mathrm{T}}Z^{-1}. \tag{8.63}$$

Using this matrix inversion lemma with $Z^{-1} = \Sigma_p$, $L^{-1} = \sigma_{\mathrm{n}}^2 I$, and $U = V = \Phi^{\mathrm{T}}$, we can develop the following term appearing in Equation 8.53:

$$\left(\Sigma_p^{-1} + \frac{1}{\sigma_{\mathrm{n}}^2} \Phi^{\mathrm{T}} \Phi \right)^{-1} = \Sigma_p - \Sigma_p \Phi^{\mathrm{T}} (\Phi \Sigma_p \Phi^{\mathrm{T}} + \sigma_{\mathrm{n}}^2 I)^{-1} \Phi \Sigma_p, \tag{8.64}$$

which is plugged into Equation 8.53 to obtain

$$\begin{aligned} \hat{\sigma}(x) &= \phi(x)^{\mathrm{T}} [\Sigma_p - \Sigma_p \Phi^{\mathrm{T}} (\Phi \Sigma_p \Phi^{\mathrm{T}} + \sigma_{\mathrm{n}}^2 I)^{-1} \Phi \Sigma_p] \phi(x), \\ &= \phi(x)^{\mathrm{T}} \Sigma_p \phi(x) - \phi(x)^{\mathrm{T}} \Sigma_p \Phi^{\mathrm{T}} (\Phi \Sigma_p \Phi^{\mathrm{T}} + \sigma_{\mathrm{n}}^2 I)^{-1} \Phi \Sigma_p \phi(x). \end{aligned} \tag{8.65}$$

Using the expressions in Equations 8.59 and 8.60 we obtain

$$\hat{\sigma}(x) = k - k^{\mathrm{T}} (K + \sigma_{\mathrm{n}}^2 I)^{-1} k, \tag{8.66}$$

where we have taken into account that Σ_p is symmetric; that is, $\Sigma_p = \Sigma_p^{\mathrm{T}}$.

Model Selection and Hyperparameter Inference

The choice of hyperparameters of the model (such as the variance of the noise σ_{n}^2 and any other parameters of the basis functions) often involves the study of the *marginal likelihood* $p(y|\Phi)$ in Equation 8.44, also known as *model evidence* or just *evidence*.

The corresponding hyperparameters θ are typically selected by type-II ML, maximizing the marginal likelihood of the observations, which is also analytical:

$$\begin{aligned} \log p(y|\theta) &= \log \mathcal{N}(y|\mathbf{0}, K + \sigma_{\mathrm{n}}^2 I) \\ &\propto -\log |(K + \sigma_{\mathrm{n}}^2 I)| - y^{\mathrm{T}} \left(K + \sigma_{\mathrm{n}}^2 I\right)^{-1} y. \end{aligned} \tag{8.67}$$

When the derivatives of Equation 8.67 are also analytical, which is often the case, conjugated gradient ascent is typically used for optimization; see related code in Listing 8.8.

It goes without saying that, when the number of hyperparameters is not high, one could apply standard cross-validation techniques to infer them. We intentionally omit all these technicalities in this book; the reader is addressed to Chapter 5 in Rasmussen and Williams (2006) for a more rigorous treatment.

There are other strategies to infer hyperparameters. For instance, an alternative is to consider some prior over the hyperparameters in order to make inference about them and provide a full Bayesian solution (Robert and Casella, 2004). Both strategies, optimization or sampling, can be performed by using Monte Carlo methods (Doucet and Wang 2005; Liu 2004; Robert and Casella 2004) such as importance sampling techniques (Bugallo *et al.* 2015; Martino *et al.* 2015a, 2017; Remondo *et al.* 2000; Yuan *et al.* 2013) and Markov chain–Monte Carlo (MCMC) (Andrieu *et al.* 2003; Larocque and Reilly 2002; Martino *et al.* 2015b, 2016) algorithms.

On the Covariance Function

The core of a kernel method like GPs is the appropriate definition of the covariance (or kernel) function. A standard, widely used covariance function is the squared exponential, $K(\boldsymbol{x}_i, \boldsymbol{x}_j) = \exp(-\|\boldsymbol{x}_i - \boldsymbol{x}_j\|^2/(2\sigma^2))$, which captures sample similarity well in most of the (unstructured) problems, and only one hyperparameter σ needs to be tuned.

In the context of GPs, kernels with more hyperparameters can be efficiently inferred. This is an opportunity to exploit asymmetries in the feature space by including a parameter per feature, as in the very common anisotropic SE kernel function:

$$K(\boldsymbol{x}_i, \boldsymbol{x}_j) = \nu \exp\left(-\sum_{k=1}^{d} \frac{(x_{i,k} - x_{j,k})^2}{2\sigma_k^2}\right), \tag{8.68}$$

where ν is a scaling factor, and we have one σ_k for each input feature, $k = 1, \dots, d$. This is a very flexible covariance function that typically suffices to tackle most of the problems. Furthermore, this kernel function is also used for the so-called *automatic relevance determination* (Bishop, 2006); that is, studying the influence of each input component to the output. Table 4.1 summarized the most common kernel functions in standard applications with kernel methods.

Time-Based Covariance for Gaussian Process Regression

Signals to be processed typically show particular characteristics, with time-dependent cycles and trends. One could include time t_i as an additional feature in the definition of the input samples. This *stacked approach* (Camps-Valls *et al.*, 2006a) essentially relies on a covariance function $K(z_i, z_j)$, where $z_i = [t_i, \boldsymbol{x}_i^{\mathrm{T}}]^{\mathrm{T}}$. The shortcoming is that the time relations are naively left to the nonlinear regression algorithm, and hence no explicit time structure model is assumed. To cope with this, one can use a linear combination (or composite) of different kernels, one dedicated to capture the different temporal characteristics and the other to the feature-based relations, to construct a composite time-based covariance for GPR (TGPR).

The issue here is how to design kernels capable of dealing with nonstationary processes. A possible approach is to use a *stationary* covariance operating on the variable of interest after being mapped with a nonlinear function engineered to discount such undesired variations. This approach was used by Sampson and Guttorp (1992) to model *spatial patterns* of solar radiation with GPR. It is also possible to adopt an SE as stationary covariance acting on the *time* variable mapped to a 2D *periodic space* $z(t) = [\cos(t), \sin(t)]^{\mathrm{T}}$, as explained in Rasmussen and Williams (2006):

$$K(t_i, t_j) = \exp\left(-\frac{\|z(t_i) - z(t_j)\|^2}{2\sigma_t^2} \right), \tag{8.69}$$

which gives rise to the following periodic covariance function:

$$K(t_i, t_j) = \exp\left(-\frac{2\sin^2[(t_i - t_j)/2]}{\sigma_t^2} \right), \tag{8.70}$$

where σ_t is a hyperparameter characterizing the periodic scale and needs to be inferred. It is not clear, though, that the seasonal trend is exactly periodic, so we modify this equation by taking the product with an SE component, to allow a decay away from exact periodicity:

$$K_2(t_i, t_j) = \gamma \exp\left(-\frac{2\sin^2[\pi(t_i - t_j)]}{\sigma_t^2} - \frac{(t_i - t_j)^2}{2\tau^2} \right), \tag{8.71}$$

where γ gives the magnitude, σ_t the smoothness of the periodic component, τ represents the *decay time* for the periodic component, and the period has been fixed to 1 year. Therefore, our final covariance is expressed as

$$K([\boldsymbol{x}_i, t_i], [\boldsymbol{x}_j, t_j]) = K_1(\boldsymbol{x}_i, \boldsymbol{x}_j) + K_2(t_i, t_j), \tag{8.72}$$

which is parameterized by only three more hyperparameters collected in $\boldsymbol{\theta} = \{\nu, \sigma_1, \dots, \sigma_d, \sigma_{\mathrm{n}}, \sigma_t, \tau, \gamma\}$. See Table 8.3 for a real-life application of this temporal covariance.

Variational Heteroscedastic Gaussian Process Regression

The standard GPR is essentially homoscedastic; that is, assumes constant noise power σ_{n}^2 for all observations. This assumption can be too restrictive for some problems. Heteroscedastic GPs, on the other hand, let noise power vary smoothly throughout input space, by changing the prior over $e_i \sim \mathcal{N}(0, \sigma_{\mathrm{n}}^2)$ to $e_i \sim \mathcal{N}(0, \exp(g(\boldsymbol{x}_i)))$, and placing a GP prior over $g(\boldsymbol{x}) \sim \mathcal{GP}(\boldsymbol{0}, K_g(\boldsymbol{x}, \boldsymbol{x}'))$. Note that the exponential is needed[3] in order to describe the nonnegative variance. The hyperparameters of the covariance functions of both GPs are collected in $\boldsymbol{\sigma}_f$ and $\boldsymbol{\sigma}_g$, accounting for the signal and the noise relations respectively.

Relaxing the homoscedasticity assumption into heteroscedasticity yields a richer, more flexible model that contains the standard GP as a particular case corresponding to a constant $g(\boldsymbol{x})$. Unfortunately, this also hampers analytical tractability, so approximate methods must be used to obtain posterior distributions for $f(\boldsymbol{x})$ and $g(\boldsymbol{x})$, which

3 Of course, other transformations are possible, but are just not as convenient.

are in turn required to compute the predictive distribution over y_*. Next, we summarize previous approaches to deal with the problem and the proposed variational alternative.

The heteroscedastic GP model was first described in Goldberg *et al.* (1998), where an MCMC procedure was used in order to implement full Bayesian inference (Andrieu *et al.*, 2003; Liu, 2004; Robert and Casella, 2004). A faster but more limited method is presented in Kersting *et al.* (2007) in order to perform MAP estimation. These approaches have certain limitations: MCMC is in general slower, but this provides much more statistical information (Andrieu *et al.*, 2003; Doucet and Wang, 2005; Martino *et al.*, 2015b), whereas MAP estimation does not integrate out all latent variables and is prone to overfitting. As an alternative to these costly previous approaches, variational techniques allow one to approximate intractable integrals arising in Bayesian inference and machine learning in general. They are typically used to: (1) provide analytical approximations to the posterior probability of the unobserved variables, and hence do statistical inference over these variables; and (2) derive a lower bound for the marginal likelihood (or "evidence") of the observed data, which allows model selection because higher marginal likelihoods relate to greater probabilities of a model generating the data.

In order to overcome the aforementioned problems, a sophisticated variational approximation called the *marginalized variational* approximation was introduced in Lázaro-Gredilla and Titsias (2011). The marginalized variational approximation renders (approximate) Bayesian inference in the heteroscedastic GP model both fast and accurate. In Lázaro-Gredilla and Titsias (2011), an analytical expression for the Kullback–Leibler (KL) divergence between a proposal distribution and the true posterior distribution of $f(\boldsymbol{x})$ and $g(\boldsymbol{x})$ (up to a constant) was provided. Minimizing this quantity with regard to both the proposal distribution and the hyperparameters yields an accurate estimation of the true posterior while simultaneously performing model selection. Furthermore, the expression of the approximate mean and variance of the posterior of y_* (i.e., predictions) can be computed in closed form. We will refer to this variational approximation for heteroscedastic GP regression as variational heteroscedastic GPR (VHGPR).

Warped Gaussian Process Regression

Even though GPs are very flexible priors for the latent function, they might not be suitable to model all types of data. It is often the case that applying a logarithmic transformation to the target variable of some regression task (e.g., those involving stock prices, measured sound power) can enhance the ability of GPs to model it. One can use a GP model that automatically learns the optimal transformation. The proposed method is called a *warped GP* (WGP) (Snelson *et al.*, 2004) or *Bayesian WGP* (Lázaro-Gredilla, 2012). The models assume a warping function $g(\cdot)$ and a new model:

$$y_i = g(f(\boldsymbol{x}_i)) + e_i,$$

where f is a possibly noisy latent function with d inputs, and g is an arbitrary warping function with scalar inputs. The Bayesian WGP essentially places another prior for $g(\boldsymbol{x}) \sim \mathcal{GP}(f, c(f, f'))$.

Snelson *et al.* (2004) showed that it is possible to include a nonlinear preprocessing of output data $h(y)$ (called a *warping function* in this context) as part of the modeling process and learn it. In more detail, a parametric form for $z = h(y)$ is selected, and then z (which depends on the parameters of $h(y)$) is regarded as a GP; and finally, the parameters of $h(y)$ are selected by maximizing the evidence of such GP (i.e., an ML approach). The authors suggested using

$$h(y) = \sum_{l=1}^{L} a_l \tanh(b_l(y + c_l))$$

as the parametric form of the warping function, which is parametrized as $\theta = [a_l, b_l, c_l]$ for $l = 1, \dots, L$, and where L is the number of piecewise functions needed to warp the signal that has to be selected by the user. Any alternative option resulting in a monotonic function is still valid too. A nonparametric version of warped GPs using a variational approximation has been proposed by (Lázaro-Gredilla, 2012). See the related example in Heteroscedastic and Warped Gaussian Process Regression in Section 8.4.7 for a real-life application of both the VHGPR and the WGP regression (WGPR) methods.

Ensembles of Gaussian Processes

An interesting possibility of GPR is about exploiting the credible intervals that can be derived. This is very often a valuable piece of information to understand the problem, discard outliers, or to design more appropriate sampling methodologies. Another possibility is to use them to give more credit to highly confident models, and downweight the unsure ones in committees of GPR experts. Given R trained GPR models, the ensemble is conducted here by exploiting the posterior variance. Essentially, one can weight the posterior means $\hat{f}_r$ with the posterior variances $\hat{\sigma}_r^2$, $r = 1, \dots, R$. This will give the overall posterior mean and variance estimates respectively:

$$\hat{f} = \frac{\sum_{r=1}^{R} \hat{f}_r / \sigma_r^2}{\sum_{r=1}^{R} 1/\sigma_r^2}, \qquad \hat{\sigma}^2 = \frac{1}{\sum_{r=1}^{R} 1/\hat{\sigma}_r^2},$$

where intuitively one downweights the contribution of the less-confident model predictions. Figure 8.10 and Listing 8.9 in Ensemble Gaussian Process Regression Models in Section 8.4.7 show an example of ensemble GPR.

Efficiency in Gaussian Process Regression

The reduction of the computation complexity of GPs is usually targeted by approximating the matrix inverse. The approximation methods can be broadly classified as in sparse methods, localized regression methods, and matrix multiplication methods.

Sparse methods, also known as *low-rank covariance matrix approximation methods*, are based on approximating the full posterior by expressions using matrices of lower rank $M \ll N$, where the M samples are typically selected to represent the dataset (e.g., from clustering or smart sampling). These global methods are well suited for modeling smooth-varying functions with long length scales. Methods in this family are based on substituting the joint prior with a reduced one using a set of m latent variables $\boldsymbol{u} = [u_1, \dots, u_M]^{\mathrm{T}}$ called *inducing variables* (Quiñonero-Candela and Rasmussen,

Table 8.1 Predictive distribution for the low-rank approximation methods described in Section 8.3.1, including the computational complexity for training, predictive mean, and predictive variance. N is the number of samples, M is the number of latent inducing variables, and $B = M/N$ is the number of blocks for methods that use them. We denote $\boldsymbol{Q}_{a,b} \equiv \boldsymbol{K}_{a,u}\boldsymbol{K}_{u,u}^{-1}\boldsymbol{K}_{u,b}$.

Method	Predictive mean, μ_*	Predictive variance, σ_*	Training	Test mean	Test variance
SoR	$\boldsymbol{Q}_{*f}(\boldsymbol{Q}_{ff} + \sigma^2\boldsymbol{I})^{-1}\boldsymbol{y}$	$\boldsymbol{Q}_{f*} - \boldsymbol{Q}_{*f}(\boldsymbol{Q}_{ff} + \sigma^2\boldsymbol{I})^{-1}\boldsymbol{Q}_{f*}$	$\mathcal{O}(NM^2)$	$\mathcal{O}(M)$	$\mathcal{O}(M^2)$
DTC	$\boldsymbol{Q}_{*f}(\boldsymbol{Q}_{ff} + \sigma^2\boldsymbol{I})^{-1}\boldsymbol{y}$	$\boldsymbol{K}_{f*} - \boldsymbol{Q}_{*f}(\boldsymbol{Q}_{ff} + \sigma^2\boldsymbol{I})^{-1}\boldsymbol{Q}_{f*}$	$\mathcal{O}(NM^2)$	$\mathcal{O}(M)$	$\mathcal{O}(M^2)$
FITC	$\boldsymbol{Q}_{*f}(\boldsymbol{Q}_{ff} + \boldsymbol{\Lambda})^{-1}\boldsymbol{y}$	$\boldsymbol{K}_{f*} - \boldsymbol{Q}_{*f}(\boldsymbol{Q}_{ff} + \boldsymbol{\Lambda})^{-1}\boldsymbol{Q}_{f*}$	$\mathcal{O}(NM^2)$	$\mathcal{O}(M)$	$\mathcal{O}(M^2)$
PITC	As FITC, but $\boldsymbol{\Lambda} \equiv \text{blkdiag}[\boldsymbol{K}_{ff} - \boldsymbol{Q}_{ff} + \sigma^2\boldsymbol{I}]$		$\mathcal{O}(NM^2)$	$\mathcal{O}(M)$	$\mathcal{O}(M^2)$
PIC	$\boldsymbol{K}_{*f}^{\text{PIC}}(\boldsymbol{Q}_{ff} + \boldsymbol{\Lambda})^{-1}\boldsymbol{y}$	$\boldsymbol{K}_{f*} - \boldsymbol{K}_{*f}^{\text{PIC}}(\boldsymbol{Q}_{ff} + \boldsymbol{\Lambda})^{-1}\boldsymbol{Q}_{f*}$	$\mathcal{O}(NM^2)$	$\mathcal{O}(M + B)$	$\mathcal{O}((M + B)^2)$

2005). These latent variables are values of the GP corresponding to a set of input locations $\boldsymbol{X}_{\boldsymbol{u}}$, called *inducing inputs*. By adopting a "subsets of data" approach, the computational complexity drastically reduces to $\mathcal{O}(M^3)$, being $M \ll N$. Examples of these approximation methods are the subsets of regressors (SoR), deterministic training conditional (DTC), fully independent training conditional (FICT), partially independent training conditional (PICT) (Quiñonero-Candela and Rasmussen, 2005), and partially independent conditional (PIC) (Snelson and Ghahramani, 2007). Table 8.1 summarizes their predictive distributions and their computational complexities.

Localized regression methods have also been proposed in this setting. When M is too small, then the representation of the whole set is poor and the performance of the associated GP is low. Local GPs are obtained by just dividing the region of interest and training a GP in each division, as far as dividing in B blocks such as $B = N/M$ reduces the computational complexity from $\mathcal{O}(N^3)$ to $\mathcal{O}(NM^2)$. Some disadvantages, such as discontinuities at the limits and loose performance when predicting in regions far from their locality, have been solved by recent new approximate methods, including PIC (Snelson and Ghahramani, 2007), FITC, and pure local GP (see Snelson and Ghahramani (2007) for details).

Matrix vector multiplication approximation methods are based on speeding up the solving of the linear system $(\boldsymbol{K} + \sigma^2\boldsymbol{I})\boldsymbol{\alpha} = \boldsymbol{y}$ using an iterative method, such as the conjugate gradient (CG). Each iteration of the CG method requires a matrix vector multiplication which takes $O(N^2)$. The CG method obtains the exact solution if iterated N times, but one can obtain an approximate solution if the method is stopped earlier.

In recent years, we have witnessed a huge improvement in GP runtime and memory demands. Inducing methods became popular but may lack expressive power of the kernel. A very useful approach is the sparse spectrum GPs (Lázaro-Gredilla *et al.*, 2010). On the other hand, there are methods that try to exploit structure in the kernel, either based on Kronecker or Toeplitz methods. The limitations of these methods to deal with data in a grid have been remedied recently with the KISS GP (Wilson and Nickisch, 2015), which generalizes inducing point methods for scalable GPs, and scales $\mathcal{O}(N)$ in time and storage for GP inference.

8.3.2 Relevance Vector Machines

The RVMs regression method – seeTipping (2001) and Chapter 7 in Bishop (2006) – is a Bayesian sparse kernel technique which shares many features with SVM. In particular, the RVM is a special case of linear basis function model where

a) $M = N$; that is, the number of basis functions is equal to the number of data points N.
b) The basis functions are localized functions centered around the data points, and satisfy all the properties required for the kernel functions. Namely, we have N localized basis functions $\boldsymbol{\phi}_i(\cdot)$, $i = 1, \dots, N$, which are kernel functions $K(\boldsymbol{x}, \cdot) = \boldsymbol{\phi}_i(\boldsymbol{x})$.
c) The prior *pdf* over $\boldsymbol{w}$ is

$$p(\boldsymbol{w}|\boldsymbol{\alpha}) = \prod_{i=1}^{N} \mathcal{N}(w_i|0, \alpha_i^{-1}) = \prod_{i=1}^{N} \sqrt{\frac{\alpha_i}{2\pi}} \exp(-\frac{1}{2}\alpha_i w_i^2) \\ \propto \exp(-\frac{1}{2}\boldsymbol{w}^{\mathrm{T}} \boldsymbol{\Sigma}_p^{-1} \boldsymbol{w}), \tag{8.73}$$

where $\boldsymbol{\alpha} = [\alpha_1, \dots, \alpha_N]^{\mathrm{T}}$ and $\boldsymbol{\Sigma}_p = \mathrm{diag}(\boldsymbol{\alpha})^{-1}$.

Namely, we have N basis functions and hence N weights, one per data point, as well as N independent hyperparameters $\boldsymbol{\alpha} = [\alpha_1, \dots, \alpha_N]^{\mathrm{T}}$, one per weight (or basis function) which moderate the scale of the prior *pdf*. Recall that an additional hyperparameter is the noise variance σ_{n}^2, then the total number of hyperparameters in RVM is $N + 1$.

Remark. The previous observation is a key to obtain sparsity. Tuning the hyperparameters (i.e., maximizing the evidence), a significant proportion of $\boldsymbol{\alpha} = [\alpha_1, \dots, \alpha_M]^{\mathrm{T}}$ diverges, and the corresponding posterior *pdf*s of the weights w_j are concentrated in zero. Namely, the posterior *pdf* of the weights w_j with $\alpha_j \to \infty$ tends to be a delta function centered around zero. Thus, the corresponding weight is virtually deleted from the model. The corresponding basis functions are effectively pruned out. The remaining examples with nonzero associated weights are called the relevance vectors (RVs), resembling the support vectors in the SVM framework (Bishop, 2006).

The predictive mean in an RVM is expressed as a linear combination of M basis functions, and can be expressed as

$$\widehat{f}(\boldsymbol{x}) = \sum_{i=1}^{M} \bar{w}_i \boldsymbol{\phi}_i(\boldsymbol{x}) = \sum_{i=1}^{N} \bar{w}_i K(\boldsymbol{x}, \boldsymbol{x}_i), \tag{8.74}$$

where in the second equality we used $K(\boldsymbol{x}, \boldsymbol{x}_i) = \boldsymbol{\phi}_i(\boldsymbol{x})$ because we have as many basis functions as points $N = M$. Observe that an RVM is something in the middle of the primal and dual approaches explained before. The mean weight vector $\bar{\boldsymbol{w}} = [\bar{w}_1, \dots, \bar{w}_N]^{\mathrm{T}}$ is provided in Equations 8.47 and 8.48; that is:

$$\bar{\boldsymbol{w}} = (\boldsymbol{\Phi}^{\mathrm{T}}\boldsymbol{\Phi} + \sigma_n^2 \boldsymbol{\Sigma}_p^{-1})^{-1} \boldsymbol{\Phi}^{\mathrm{T}} \boldsymbol{y}, \tag{8.75}$$

where, in this case, $\boldsymbol{\Phi}$ is an $N \times M$ or $N \times N$ matrix, $\boldsymbol{\Phi} = [\boldsymbol{\phi}(\boldsymbol{x}_1)| \cdots |\boldsymbol{\phi}(\boldsymbol{x}_M)]^{\mathrm{T}} = [\boldsymbol{\phi}(\boldsymbol{x}_1)| \cdots |\boldsymbol{\phi}(\boldsymbol{x}_N)]^{\mathrm{T}}$, and $\boldsymbol{\phi}(\boldsymbol{x}) = [\boldsymbol{\phi}_1(\boldsymbol{x}), \dots, \boldsymbol{\phi}_N(\boldsymbol{x})]^{\mathrm{T}} : \mathbb{R}^d \to \mathbb{R}^N$. Since we defined the kernel function as $K(\boldsymbol{x}, \boldsymbol{x}_i) = \boldsymbol{\phi}_i(\boldsymbol{x})$, in this case $\boldsymbol{\Phi} \equiv \boldsymbol{K}$, where $\boldsymbol{K}$ is the kernel matrix.

Now, given the aforementioned definitions of $\boldsymbol{\phi}_i(\boldsymbol{x})$ and $\boldsymbol{\Phi}$, the predictive mean and variance are respectively

$$\hat{f}(\boldsymbol{x}) = \boldsymbol{\phi}(\boldsymbol{x})^{\mathrm{T}}(\boldsymbol{\Phi}^{\mathrm{T}}\boldsymbol{\Phi} + \sigma_n^2\Sigma_p^{-1})^{-1}\boldsymbol{\Phi}^{\mathrm{T}}\boldsymbol{y} \tag{8.76}$$

and

$$\hat{\sigma}(\boldsymbol{x}) = \boldsymbol{\phi}(\boldsymbol{x})^{\mathrm{T}}\left(\frac{1}{\sigma_n^2}\boldsymbol{\Phi}^{\mathrm{T}}\boldsymbol{\Phi} + \Sigma_p^{-1}\right)^{-1}\boldsymbol{\phi}(\boldsymbol{x}), \tag{8.77}$$

as derived previously in this section. The tuning of the hyperparameters $\boldsymbol{\alpha}$ and σ_{n}^2 in the RVM is obtained by maximizing the marginal likelihood $p(\boldsymbol{y}|\boldsymbol{\Phi}, \boldsymbol{\alpha}, \sigma_{\mathrm{n}}^2)$.

Remark. When localized basis functions are employed, the predictive variance for linear regression models becomes small in the regions of input space where no basis functions are located. This is the case of when the basis functions are centered on data points (see Listing 8.10). This behavior is clearly counterintuitive and undesirable; see Rasmussen and Quiñonero-Candela (2005) and Chapter 7 in Bishop (2006). The posterior distribution in GPR does not suffer from this problem, and this is one of the main advantages of a GPR approach (see Figure 8.12c in Section 8.4.8).

8.4 Tutorials and Application Examples

This section gives empirical evidence on synthetic and real examples of the methods presented. In addition, some worked examples, tutorials, and pointers to useful toolboxes in MATLAB are provided for selected methods.

8.4.1 Comparing Support Vector Regression, Relevance Vector Machines, and Gaussian Process Regression

Let us start by illustrating the behavior of different models we considered in this chapter in a standard toy example function approximation problem of the sinc function. Specifically, we compare the SVR, RVM, and GPR methods. We generated a noisy version $\boldsymbol{y}$ from the sinc($\boldsymbol{s}$) function with additive Gaussian noise $\mathcal{N}(0, 0.01)$. We used 20% of the points for training the models and the rest for prediction. Training was done via standard fivefold cross-validation for SVR and by maximizing the log-likelihood for RVM and GPR. Figure 8.2 shows the clean signal $\boldsymbol{s}$, the observations (black circles), the selected support vectors/RVs (red circles), and the approximation function $\hat{\boldsymbol{y}}$. For the GPR model, we also give the CIs for the predictions. It is clear that all methods perform very well (high correlation coefficients with the signal) and offer different advantages: (1) the SVR gives the best overall performance in fit but is less sparse than expected (of course, this issue could be trimmed by properly tuning the ε parameter); (2) the RVM offers a rather sparse model but its performance is compromised; and (3) the GPR performs well in fit, but it is unfortunately not a sparse model, though advantageously we can derive CIs for the predictions. The code in Listing 8.2 generates Figure 8.2 and needs the `simpleR` MATLAB toolbox available on the book's webpage.

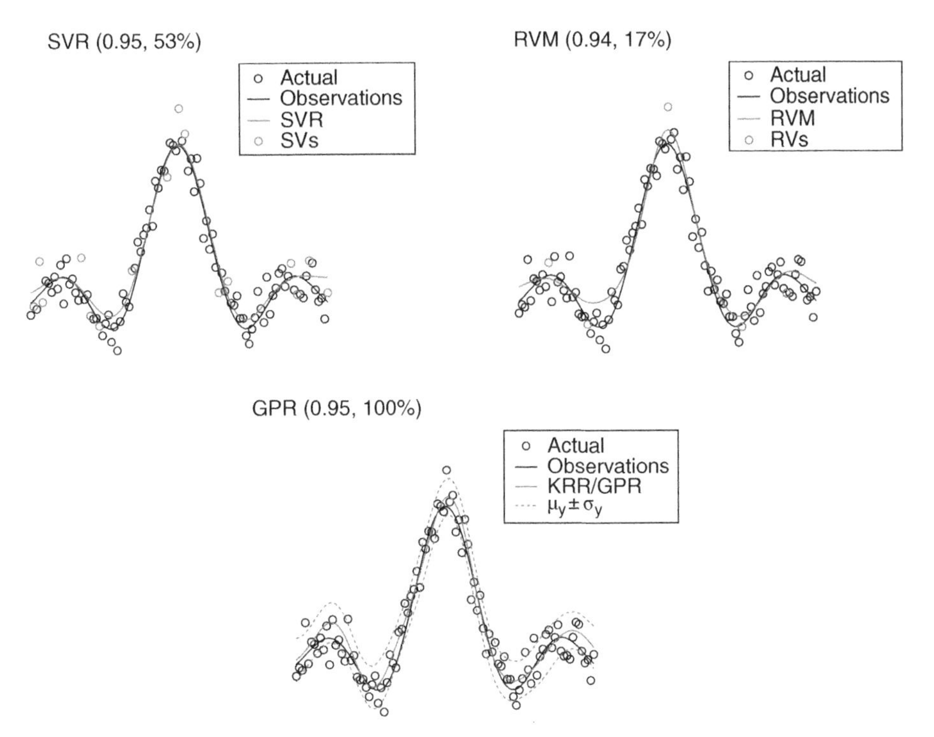

Figure 8.2 Solutions of the different regression models revised in this chapter. Pearson's correlation coefficient and sparsity level are shown in parentheses.

```
% Comparing SVR, KRR and GPR (simpleR MATLAB toolbox needed)
addpath(genpath('../simpleR/')), addpath('../libsvm/')
% Generating Data
n = 100; X = linspace(-pi,pi,n)'; s = sinc(X); y = s + 0.1*randn(n,1);
% Split training and testing sets
n = length(y); r = randperm(n); ntr = round(0.3*n);
idtr = r(1:ntr); idts = r(ntr+1:end);
Xtr = X(idtr,:); ytr = y(idtr,:); Xts = X(idts,:); yts = y(idts,:);
%% SVR
model1 = trainSVR(Xtr,ytr); ypred1 = testSVR(model1,X);
ypred1ts = testSVR(model1,Xts);
%% RVM
model2 = trainRVM(Xtr,ytr); ypred2 = testRVM(model2,X);
ypred2ts = testRVM(model2,Xts);
%% GPR
model3 = trainGPR(Xtr,ytr); [ypred3 s3] = testGPR(model3,X);
ypred3ts = testGPR(model3,Xts);
%% Results in the test set
r1 = assessment(yts,ypred1ts,'regress');
r2 = assessment(yts,ypred2ts,'regress');
r3 = assessment(yts,ypred3ts,'regress');
%% Plots
figure(1), clf, plot(X,y,'ko'), hold on, plot(X,s,'k')
plot(X,ypred1,'r'), plot(Xtr(model1.idx),ytr(model1.idx),'ro'),
legend('Actual','Observations','SVR','SVs'), axis off
```

```
figure(2), clf, plot(X,y,'ko'), hold on, plot(X,s,'k')
plot(X,ypred2,'r'), plot(Xtr(model2.used),ytr(model2.used),'ro')
legend('Actual','Observations','RVM','RVs'), axis off
figure(3), clf, plot(X,y,'ko'), hold on, plot(X,s,'k')
plot(X,ypred3,'r'), plot(X,ypred3+sqrt(s3),'r--'),
plot(X,ypred3-sqrt(s3),'r--'),
legend('Actual','Observations','KRR/GPR','\mu_y \pm \sigma_y'), axis off
```

Listing 8.2 Code related to Figure 8.2.

In contrast to the SVR-based algorithms in the previous sections, the solution obtained with GPR/KRR is not sparse, as the weights $\boldsymbol{A}$ in Equation 8.28 are in principle all different from zero. This is indeed a drawback compared with SVR, where the sparsity of the models drives to less operations and memory consumption in order to obtain predictions. In addition, dense solutions may give rise to strong overfitting because all examples seen in the training set are stored/memorized in the model's weights. On the other hand, the GPR/KRR solution reduces to minimizing a matrix inversion problem, for which there are very efficient solvers and approximate solutions.

8.4.2 Profile-Dependent Support Vector Regression

Let us now exemplify the performance of the PD-SVR, for which one has to design profiles for the variation of C and ε (Camps-Valls *et al.*, 2001, 2002). A standard approach in time-series prediction reduces to considering *exponential memory decays* in which one assumes that recent past samples contain relatively more information than distant samples. This leads to

$$c_i = \lambda^{t_N - t_i}, \quad \lambda \in (0,1], \tag{8.78}$$

where t_N is the actual time sample and t_i is the time instant for sample i. This profile reduces the penalization parameter and enlarges the ε-insensitive region of previous samples as new samples are obtained, as illustrated in Figure 8.3. The inclusion of a temporal confidence function in the SVR formulation offers some advantages. The overall number of support vectors remains constant through time, and better results are obtained when abrupt changes appear in the time series.

We test the performance of PD-SVR in a bioengineering signal processing problem. Specifically, we are concerned about the estimation of concentration of erythropoietin A for its dosage individualization. The high cost of this medication, its side effects, and the phenomenon of potential resistance which some individuals suffer all justify the need for a model that is capable of optimizing dosage individualization. A group of 110 patients and several patient factors were used to develop the model. To build the predictive models, we collected several factors for each patient such as age (years), weight (kilograms), plasmatic concentration of hemoglobin (grams per deciliter), hematocrit concentration (percent), ferritin (milligrams per liter), dosage of intravenous iron (milligrams per month), number of administrations, isoform and dosage of erythropoietin (IU per week). The variable to be predicted is the plasmatic concentration of hemoglobin (Hb). Patients were randomly assigned to two groups: one for training the models (77 patients, 495 data points) and another for their validation (33 patients, 174 data points).

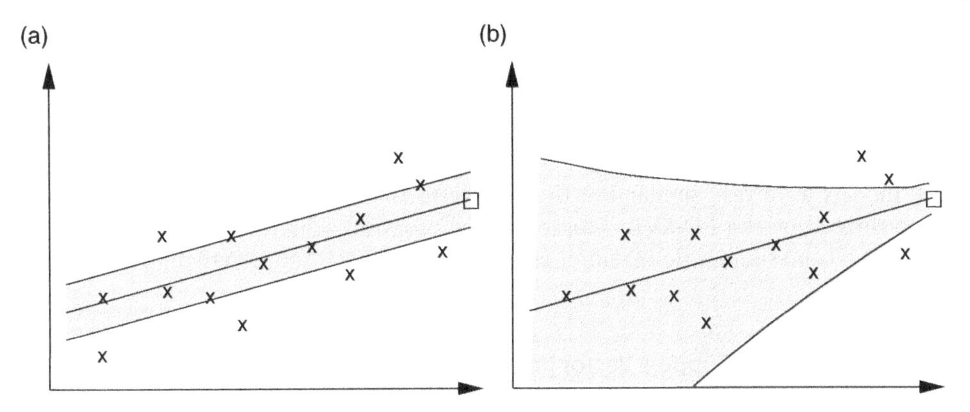

Figure 8.3 The ε-insensitive region for the standard SVR (a) and for the PD-SVR with an exponential memory decay profile (b). Only points outside the shaded regions contribute to the cost function. The square symbol indicates the actual sample to predict using the past samples. In the SVR, all samples outside a fixed tube of size ε are penalized, independently from their time sample. In the PD-SVR case, only the recent past samples are considered to be relevant to the solution. In the PD-SVR, only the points next to the actual sample are penalized and, consequently, become support vectors. Since recent samples contain, in principle, more information about the future, only these are penalized in a PD-SVR. This, in turn, yields solutions with a lower fraction of support vectors.

Table 8.2 ME, MSE, and correlation coefficient ρ of models in the validation set. Recognition rates (RRs) with admissible error of 0.5 and 0.25 g/dL are also given.

	ARCH	**MLP**	**RBF kernels**	
Model	**(1)**	**16 × 8 × 1**	**ν-SVR**	**PD-SVR**
ρ	0.7672	0.7944	0.6376	0.8011
ME (g/dL)	0.0167	−0.0118	−0.0131	−0.0111
RMSE (g/dL)	0.7726	0.2081	0.2840	0.1944
% RR (error < 0.5 g/dL)	59.20	98.28	93.68	98.88
% RR (error < 0.25 g/dL)	35.06	83.33	66.09	84.46

In Table 8.2 the results of the PD-SVR are benchmarked with an MLP trained using the back-propagation algorithm, with the standard ν-SVR, and an autoregressive conditional heteroscedasticity (ARCH) model which serves as a natural benchmark for the forecast performance of heteroscedastic models. The table also shows the recognition rates for (1) correct predictions according to the experts (when committed errors are less than 0.5 g/dL) and for (2) predictions with a high grade of precision (when committed errors are less than 0.25 g/dL).

The ARCH model achieved acceptable results, but its performance is inferior to the other models. In fact, excellent results are obtained in prediction using the MLP and the PD-SVR models. Good results were offered by the standard SVR model, but worse than those obtained with MLP and PD-SVR, possibly because of the high inter-subjects variability (coefficient of variation, CV = 48%), intra-subjects variability (CV = 28%), and the shortness of the time series (three to ten samples per patient). The simple exponential decay profile improved results. In general, the results of MLP and PD-SVR models were very similar, but the best performance in the validation group was demonstrated by the PD-SVR. Moreover, our proposal achieved a smoother (lower values of C) and simpler (lower number of support vectors) solution than the standard formulation.

8.4.3 Multi-output Support Vector Regression

Let us illustrate the MSVR algorithm for MIMO nonlinear channel estimation (Sánchez-Fernández *et al.*, 2004). Comparison is done for different channels, SNRs measured at the receiver inputs, and training sequence lengths. The goal in the channel estimation problem is to obtain a good approximation to the actual channel, modeling the dependence between transmitted and received signals. With the MMSE method, this relation is restricted to be linear, and the channel estimate can be explicitly given. This holds for the MSVR and SVR methods with linear kernel, but it is no longer possible when other (nonlinear) kernels are used. In pilot-aided channel estimation, it is necessary to use a training sequence known a priori by both the transmitter and the receiver. Once the channel has been modeled, the expected received vector $\boldsymbol{y}$ without noise, corresponding to each possible transmitted QPSK codeword $\boldsymbol{x}$, is calculated. During operation, each received signal is decoded using the nearest-neighbor criterion.

We here reproduce some simulation results for MIMO systems ($n_{\mathrm{T}} = 4$, $n_{\mathrm{R}} = 3$) originally presented in Sánchez-Fernández *et al.* (2004). For this purpose, two different channels are used. In the first channel model, at the receiver end, there is a lineal mixture of the transmitted symbols even though each of them is a nonlinear function of the information symbol; noise considered here is Gaussian and white. In the second channel model, a nonlinear mixture of all the transmitted symbols at the receiver and the noise can no longer be considered Gaussian. The number of test words has been chosen to assure that at most one erroneous bit occurs for each 100 bits received, and all results have been averaged over 20 trials.

The experiment shows results when the nonlinearities between input and output signals of the channel are introduced by the transmitter equipment due to, for example, amplifiers driven near their saturation zone. The channel is modeled with coefficients $\alpha_1 = 0.15$ and $\alpha_2 = 0$. MSVR is able to parameterize nonlinearities effectively, as is seen in Figure 8.4, and obtains lower BER than the RBF network for the variable SNR level. The improvement of the MSVR method is especially noticeable for short training lengths, although the difference is only slightly reduced for the larger training sets. The saturation point of the curves, for which the BER is no longer improved, increases as the SNR grows. In any case, this point is reached in first place by the MSVR. We have split the results into two plots, grouping in each one alternatives SNRs. Results of the SVR are comparable to those of the MSVR, but come at a higher computational cost. While MSVR requires just a few iterations of the IRWLS algorithm to converge, SVR needs

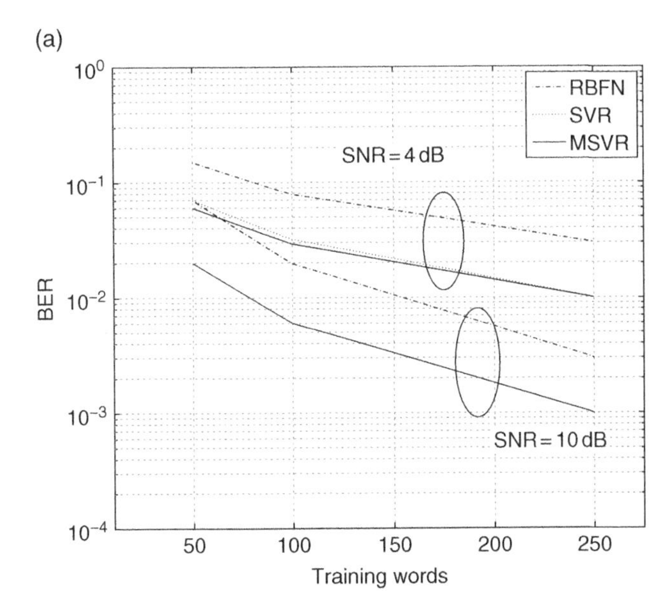

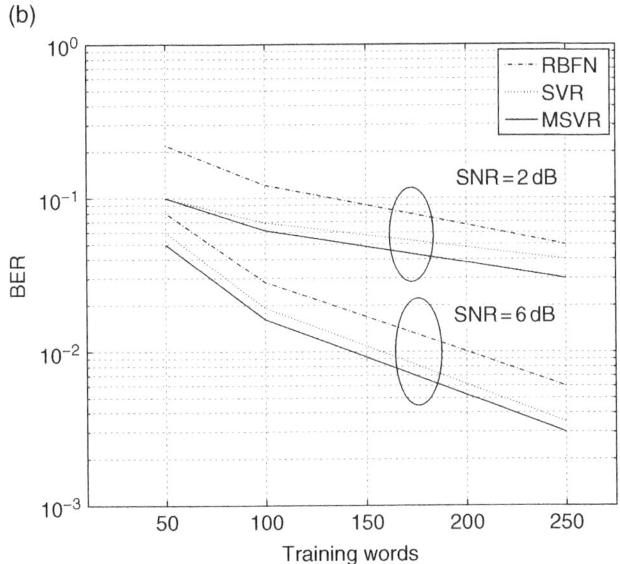

Figure 8.4 BER with MSVR, SVR, and RBF network (RBFN) as a function of the length of the training sequence used during the training phase, generated with SNR = 4 dB and 10 dB (a), and with SNR = 2 dB and 6 dB (b). The nonlinearity phenomenon occurs in the transmitter, then affecting solely the input signals. SVR-based methods improvement is evident.

approximately two orders of magnitude more iterations. Besides, the complexity of SVR increases both with n_R and the length of the training sequence, while that of MSVR does not depend on n_R. Listing 8.3 shows a simple example of MSVR with arbitrary data and hyperparameters.

```
%% Multioutput Support Vector Rregression (MSVR)
% - C       : penalization (regularization) parameter
% - epsilon: tolerated error, defines the insensitivity zone
% - sigma   : lengthscale of the RBF kernel
% - tol     : early stopping tolerance of the IRWLS algorithm
ker = 'rbf'; epsilon = 0.1; C = 2; sigma = 1; tol = 1e-10;
% Training with pairs Xtrain-Ytrain
[Beta,NSV,val] = msvr(Xtrain,Ytrain,ker,C,epsilon,sigma,tol);
% Prediction on new test data Xtest
Ktest = kernelmatrix(ker,Xtest',Xtrain',sigma);
Ntest = size(Xtest,1); testmeans = repmat(my,Ntest,1);
Ypred = Ktest * Beta + testmeans;
```

Listing 8.3 Example of MSVR (demoMSVR_simpler.m).

An operational demo for MSVR is available on the book web page.

8.4.4 Kernel Signal-to-Noise Ratio Regression

In this section we illustrate the main features of the KSNR in a simple toy example with non-Gaussian i.i.d noise. A total of 1000 samples were drawn from the sinc function $s_t = \sin(t)/t$ in the range $[-\pi, +\pi]$, and different noises were added, $y_t = s_t + e_t$: (1) Gaussian noise, $e_t \sim \mathcal{N}(0, \sigma_\text{n}^2)$; (2) uniform noise, $e_t \sim \mathcal{U}(0, 1)$; (3) Poisson noise, $e_t \sim \mathcal{P}(\lambda)$, $\lambda \in [0, 0.3]$; and (4) scale-dependent multiplicative noise, $e_t = m_t \times |s_t|$, where $m \sim \mathcal{N}(0, \sigma_\text{n}^2)$. We used 500 samples for cross-validation and the remaining examples for testing. Figure 8.5 shows the averaged test results in all situations as a function of the SNR for KRR and the KSNR methods. We also show the bound of performance in which KSNR works with the true (inaccessible) signal samples s_t. These results confirm some basic facts: low gain is obtained with uncorrelated Gaussian and uniform noise sources, but noticeable gains can be obtained in either correlated noise or strong non-Gaussian noise levels. Related MATLAB code can be found on the book's web page.

Now, let us test the KSNR model in a real example involving signal-to-noise relations. We test the performance of KSNR considering a real dataset. Specifically, the motorcycle dataset from Silverman (1985), consisting of 133 accelerometer readings through time following a simulated motorcycle crash during an experiment to determine the efficacy of crash-helmets. The code in Listing 8.4 compares KRR and KSNR in this scenario.

```
% Comparing KRR and KSNR
clear, clc
% Setup paths
addpath(genpath('../simpleR/'))
% Load Data
load data/motorcycle.mat; Y = y;
%% Split training-testing data
rate = 0.1; [n d] = size(X);           % samples x dimensions
r = randperm(n);                       % random index
ntrain = round(rate*n);                % #training samples
Xtrain = X(r(1:ntrain),:);             % training set
Ytrain = Y(r(1:ntrain),:);             % observed training variable
Xtest = X(r(ntrain+1:end),:);          % test set
Ytest = Y(r(ntrain+1:end),:);          % observed test variable
ntest = size(Ytest,1);
```

```
%% Remove the mean of Y for training only
my = mean(Ytrain); Ytrain = Ytrain - repmat(my,ntrain,1);
% KRR
modelKRR = trainKRR(Xtrain,Ytrain);
Yp_KRR = testKRR(modelKRR,Xtest) + repmat(my,ntest,1);
results_KRR = assessment(Yp_KRR,Ytest,'regress')
% KSNR
modelKSNR = trainKSNR(Xtrain,Ytrain);
Yp_KSNR = testKSNR(modelKSNR,Xtest) + repmat(my,ntest,1);
results_KSNR = assessment(Yp_KSNR,Ytest,'regress')
% Scatter plots and marginal distributions
figure(1), scatterhist(Ytest, Yp_KRR), title('KRR'); grid on
xlabel('Observed signal'); ylabel('Predicted signal')
figure(2), scatterhist(Ytest, Yp_KSNR), title('KSNR'); grid on
xlabel('Observed signal'); ylabel('Predicted signal')
```

Listing 8.4 Regression with KSNR considering the motorcycle dataset.

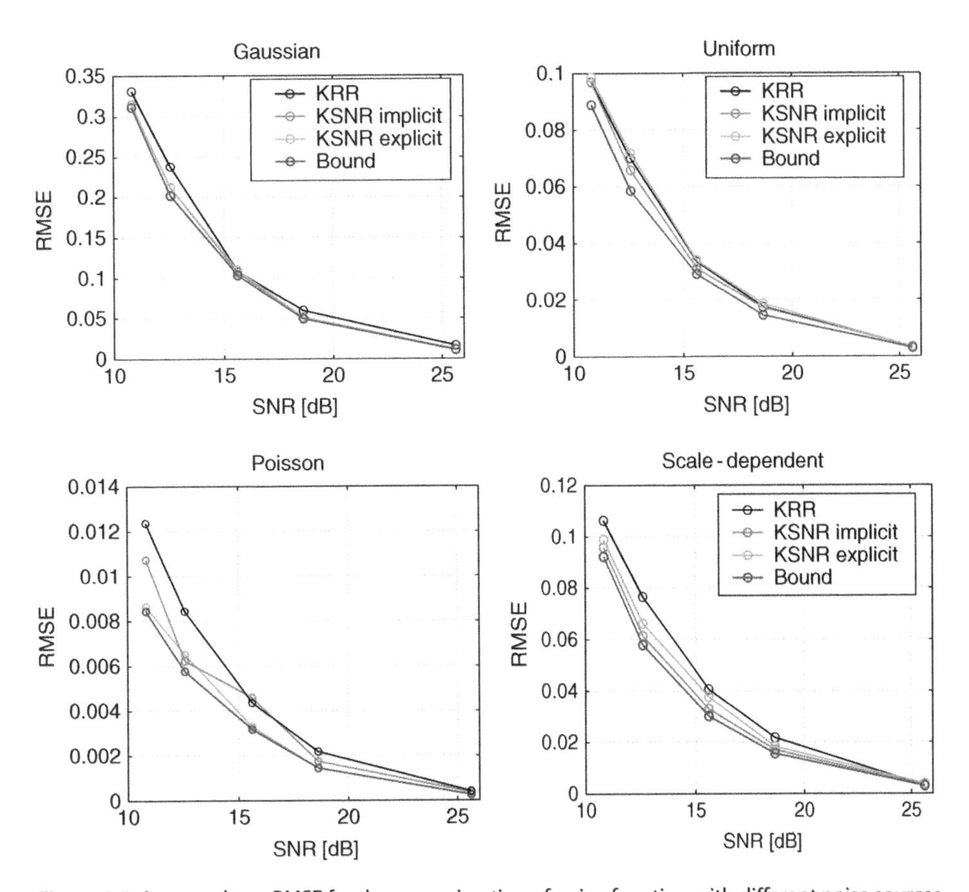

Figure 8.5 Averaged test RMSE for the approximation of a sinc function with different noise sources and levels.

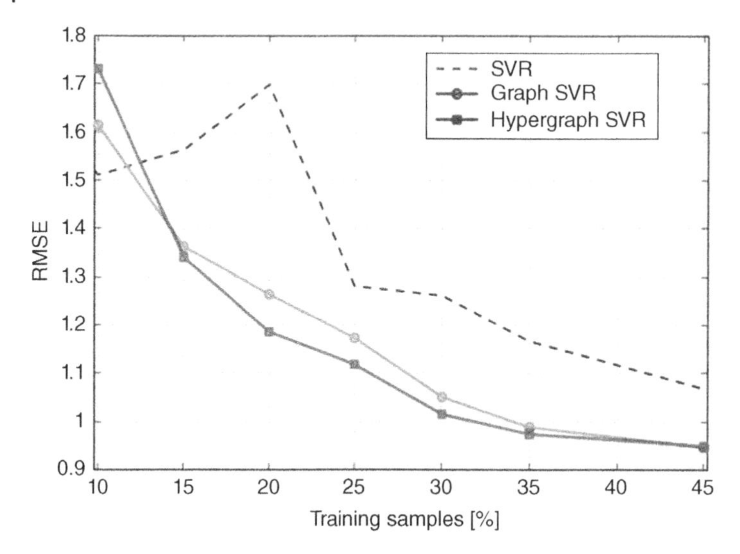

Figure 8.6 RMSE in the test set as a function of the rate of training samples for LAI estimation from CHRIS/PROBA data.

8.4.5 Semi-supervised Support Vector Regression

This experiment shows results in a complex real problem in geosciences: estimating physical parameters from satellite sensor images (Camps-Valls *et al.*, 2009a). In particular, here we are concerned with estimating the leaf-area index (LAI), which characterizes plant canopies, from hyperspectral satellite images. The problem is characterized by high uncertainty and ill-conditioned data.

We used data from the ESA Spectra Barrax Campaigns (SPARC). During the campaign, 17 multiangular hyperspectral CHRIS/PROBA images were acquired. Field nondestructive measurements of LAI were made by means of the digital analyzer LI-COR LAI-2000. The LAI parameter corresponds to the effective plant area index since no corrections of leaf clumping were done and the influence of additional plant components was not removed. The final database consists of only 139 LAI measurements and their associated surface reflectance spectra from the (very high-dimensional) 62 CHRIS reflectance bands (Vuolo *et al.*, 2006), which leads to a 62-dimensional input space.

Figure 8.6 compares the standard SVR, and the graph and hypergraph SVR methods. The proposed graph-based methods clearly improve the results of the supervised SVR (between a 10% and 50% improvement in RMSE) when a sufficient number of labeled training samples are used (>12%; i.e., 16 samples). In this application, the use of hypergraphs shows only a slight improvement over the graph. This could be due to the fact that data may be governed by simple one-to-one cluster relations, or to the low number of available data and the high uncertainty in the data acquisition. A simple code snippet showing the calculation of the deformed kernel using the graph Laplacian is given in Listing 8.5. A complete MATLAB demo can be obtained from the book web page.

```
% SS-SVR on LAI estimation
clear, clc
rng('default'); rng(0); addpath('../libsvm/')
% Supervised free parameters
sigma = 1; C = 1; epsilon = 0.1;
% Semi-supervised free parameters
gamma = 1e-2; nn = 5;
% Generate dummy data: labeled (X,Y) and unlabeled (Xu, Yu)
% X, Y, Xu, Yu (only for testing)
% Kernel matrix: (l+u) x (l+u)
K = kernelmatrix('rbf',[X;Xu]',[X;Xu]',sigma);
% Graph Laplacian
lap_options.NN = nn; % neighbors
L = laplacian([X;Xu],'nn',lap_options);
M = gamma*L; I = eye(size(L,1));
INVERSE = (I + M*K) \ M;
Khat = zeros(size(K));
for x = 1:length(K)
   Khat(x,:) = K(x,:) - K(:,x)' * INVERSE * K;
end
% Train and test with the deformed kernel
model = mexsvmtrain(Y,Khat,params);
Yp = mexsvmpredict(Y,Khat,model);
```

Listing 8.5 Related code of SS-SVR for LAI estimation (DemoSemiSVR.m).

8.4.6 Bayesian Nonparametric Model

Let us now illustrate an important feature provided by a Bayesian treatment of the regression problem. Since we have access to the full posterior distribution, we can calculate moments as the predictive mean and variance. Listing 8.6 generates Figure 8.7, which shows the predictive mean and variance in a Bayesian nonparametric model considering artificial data and nonlocalized polynomial basis functions.

```
% Generate data
sig_e = 0.5; % std of the noise
X = [-12 -10 -3 -2  2 8 12 14]';
N = length(X);
Y = sin(X) + sig_e * randn(N,1); % N x 1
% Basis functions
M = 4; % number of basis functions
phi = @(x) [1 x.^(1:M-1)];
% Basis function matrix
PHI = zeros(N,M);
for i = 1:N
    PHI(i,:) = phi(X(i)); % N x M
end
% Prior variance
sig_prior = 10;
Sigma_p = sig_prior^2 * eye(M); % Cov-Matrix of the prior over w
% Moments of the posterior of w
w_mean = ((PHI' * PHI) + sig_e^2 * inv(Sigma_p)) \ PHI' * Y; % M x 1
w_Covar_Matrix = ((1/sig_e^2) * (PHI'*PHI) + inv(Sigma_p)) \ eye(M); % M ...
    x M
```

```
% Test inputs
x_pr = -16:0.2:16;
% posterior mean and variance of f
f_mean = zeros(1,length(x_pr)); sigma_post = f_mean;
for i = 1:length(x_pr)
    f_mean(i) = phi(x_pr(i)) * w_mean;
    sigma_post(i) = phi(x_pr(i)) * w_Covar_Matrix * phi(x_pr(i))';
end
% Sampling functions from Posterior of w
Num_Funct = 3;
w_sample = mvnrnd(w_mean, w_Covar_Matrix, Num_Funct);
f_samples_from_post = zeros(length(x_pr), Num_Funct);
for i = 1:length(x_pr)
    f_samples_from_post(i,:) = phi(x_pr(i)) * w_sample';
end
% Plot results
V1 = f_mean - sqrt(sigma_post); V2 = f_mean + sqrt(sigma_post);
figure(1), clf
fill([x_pr, fliplr(x_pr)], [V1, fliplr(V2)], 'k', ...
    'FaceAlpha', 0.3, 'EdgeAlpha', 0.7); hold on
plot(x_pr,f_mean, 'k', 'LineWidth', 5)
plot(x_pr, f_samples_from_post, '--', 'LineWidth', 3);
plot(X, Y, 'ro', 'MarkerFaceColor', 'r', 'MarkerSize', 10)
axis([-16 16 -4.2 4.2]), set(gca, 'Fontsize', 20), box on
% Alternative procedure: Sampling functions from Posterior
PHI_pr = zeros(length(x_pr),M);
for i = 1:length(x_pr)
    PHI_pr(i,:) = phi(x_pr(i));
end
CovMatr_post = PHI_pr * w_Covar_Matrix * PHI_pr';
Num_Funct = 3;
f_samples_from_post = mvnrnd(f_mean, CovMatr_post, Num_Funct);
% Alternative formulation of f_mean
w_mean_alt = Sigma_p * PHI' / (PHI*Sigma_p*PHI' + sig_e^2*eye(N)) * Y;
w_mean - w_mean_alt  % check equality
w_Covar_Matrix_alt = Sigma_p - Sigma_p * PHI' / ...
                     (PHI*Sigma_p*PHI' + sig_e^2*eye(N)) * PHI*Sigma_p;
w_Covar_Matrix - w_Covar_Matrix_alt % check equality
f_mean_alt = zeros(1,length(x_pr));
for i = 1:length(x_pr)
    f_mean_alt(i) =  phi(x_pr(i)) * w_mean_alt;
end
figure(2), clf, plot(x_pr, sigma_post, 'k--', 'LineWidth', 3), hold on,
plot(X, zeros(1,N), 'ro', 'MarkerFaceColor', 'r', 'MarkerSize', 10)
axis([-22 22 0 1.5]), set(gca, 'Fontsize', 20), box on
```

Listing 8.6 Code of a linear basis function regression model related to Figure 8.7 using polynomial basis functions (Demo_LinBasisFun.m).

8.4.7 Gaussian Process Regression

In this section we will give intuitions and empirical evidence of the performance of the vanilla GPR and advanced versions in both synthetic and real examples, as well as the design of the covariance function.

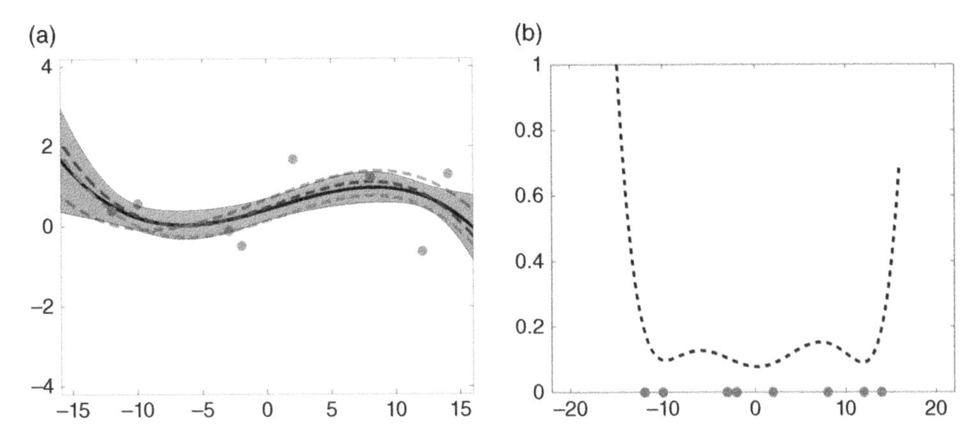

Figure 8.7 Example of linear basis regression model using nonlocalized polynomial basis functions (see Listing 8.6). (a) Posterior mean $\widehat{f}(\boldsymbol{x})$ (solid line) and $\widehat{f}(\boldsymbol{x}) \pm \sqrt{\widehat{\sigma}(\boldsymbol{x})}$ (colored area), three random functions drawn according to the posterior distribution (dashed line), as shown in Listing 8.6. (b) Posterior variance $\widehat{\sigma}(\boldsymbol{x})$ as function of $\boldsymbol{x}$. We can observe that, with nonlocalized basis, $\widehat{\sigma}(\boldsymbol{x})$ is smaller closer to the observed data and higher far away (it is the expected behavior).

Toy Examples

Let us start with a simple illustration of the GPR in a toy example. In Figure 8.8 we include an illustrative example with six training points in the range between −2 and +2. We first depict several random functions drawn from the GP prior and then we include functions drawn from the posterior. We have chosen an isotropic Gaussian kernel and $\sigma = 0.1$. We have plotted the mean function plus/minus two standard deviations (corresponding to a 95% CI). Typically, the hyperparameters are unknown, as well as the mean, covariance, and likelihood functions.

We assumed an SE covariance function and learned the optimal hyperparameters by minimizing the negative log marginal likelihood (NLML) with respect to the hyperparameters. We observe three different regions in the figure. Below $x = -1.5$, we do not have samples and the GPR provides the solution given by the prior (zero mean and ± 2). At the center, where most of the data points lie, we have a very accurate view of the latent function with small error bars (close to $\pm 2\sigma_{\rm n}$). For $x > 0$, we do not have training samples either, so we have the same behavior. GPs typically provide an accurate solution where the data lies and high error bars where we do not have available information; consequently, we presume that the prediction in that area is not accurate. This is why in regions of the input space without points the CIs are wide, resembling the prior distribution.

Structured, Nonstationary, and Multiscale in Gaussian Process Regression

Commonly used kernels families include the SE, periodic (Per), linear (Lin), and rational quadratic (RQ); see Table 4.1 and Figure 8.9. These base kernels can be actually combined following simple operations: summation, multiplication or convolution. This way one may build sophisticated covariances from simpler ones. Note that the same essential property of kernel methods apply here: a valid covariance function must be

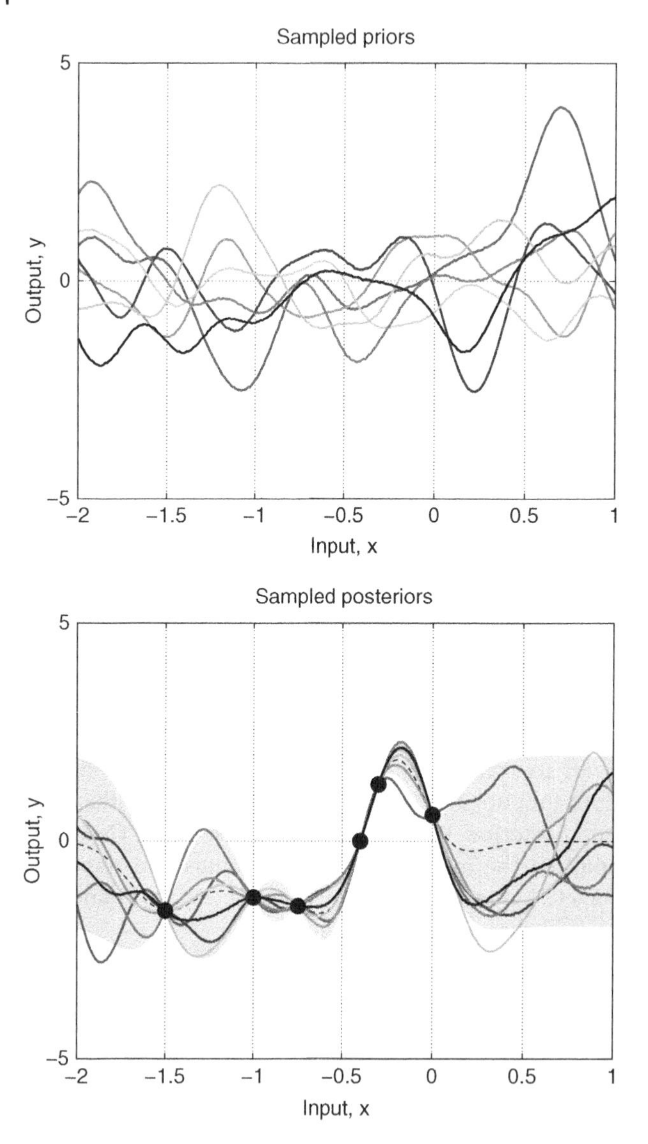

Figure 8.8 Example of a Gaussian process. *Up*: some functions drawn at random from the GP prior. *Down*: some random functions drawn from the posterior; that is, the prior conditioned on six noise-free observations indicated in big black dots. The shaded area represents the pointwise mean plus and minus two times the standard deviation for each input value (95% confidence region). Note the larger CIs for regions far from the observations.

positive semidefinite. In general, the design of the kernel should rely on the information that we have for each estimation problem and should be designed to get the most accurate solution with the least amount of samples. The related MATLAB code to generate random draws from such base kernels and GP priors is given in Listing 8.7.

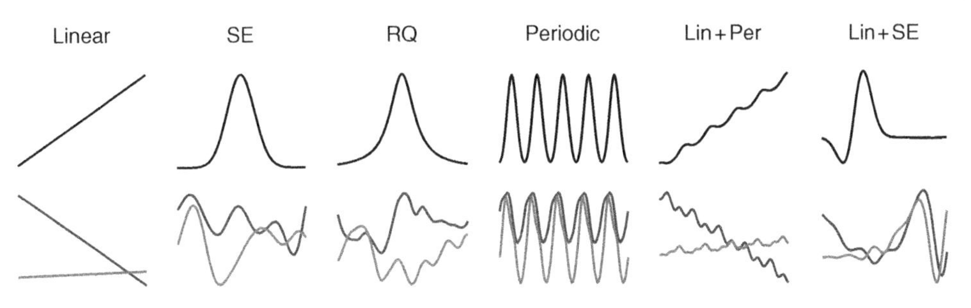

Figure 8.9 Base kernels (top) and two random draws from a GP with each respective kernel and combination of kernels (bottom).

```
% Drawing from a GP prior
kernelType = 'SE';  % squared exponential kernel
switch  kernelType
    case 'SE' % SE kernel
        sig = 0.2; k = @(x1,x2) exp(-(x1-x2).^2/sig^2);
    case 'RQ' % Rational-quadratic kernel
        c = 0.05; k = @(x1,x2) (1 - (x1-x2).^2 ./ ((x1-x2).^2 + c));
    case 'Per' % Periodic kernel
        p = 0.5; % period
        sig = 0.7; % SE kernel length
        k = @(x1,x2) exp(-sin(pi*(x1-x2)/p).^2/sig^2);
    case 'Lin' % Linear kernel
        offset = 0; k = @(x1,x2)  (x1 - offset) .* (x2 - offset);
    case 'Lin+Per' %% Linear+PER kernel
        offset = -1; p = 0.5; % period
        sig = 0.7; % SE kernel length
        k = @(x1,x2) (x1 - offset).*(x2 - offset) + ...
            exp(-sin(pi*(x1-x2)/p).^2/sig^2);
    case 'Lin+SE' % Linear+SE kernel
        offset = 0; sig=0.1;
        k = @(x1,x2)  (x1 - offset).*(x2 - offset) + ...
            exp(-(x1-x2).^2/sig^2);
end
% Generate plot of correspoding kernels
stepInp = 0.04; xforFig = 0; x = -1:0.02:1;
figure(1), clf, plot(x,k(x,xforFig),'k','LineWidth',8)
axis([-1 1 -1 max(k(x,xforFig))+0.5]), axis off
% Create kernel matrix
x = -1:stepInp:1; z = meshgrid(x); K = k(z,z');
% Proxy to phi transform using Cholesky decomp.
sphi = chol(K + 1e-6*eye(size(K)))';
Num_of_Funct = 5; % number of i.i.d. functions from GP Prior
% Genration of iid random functions - vectors
Functions_from_GP_Prior = sphi * randn(length(x), Num_of_Funct);
% Generate plot of the drawn functions
figure(2), clf, plot(x,Functions_from_GP_Prior,'LineWidth',8), axis off
```

Listing 8.7 Code generating GPs of Figure 8.9 (demo_GP_FunFromPrior.m).

In Listing 8.8 we show how to sample from the GP posterior and how to compute the marginal likelihood.

```
%% Sampling GP Posterior and Computation of Marginal Likelihood
clear, clc
% Generating data
sig_e = 0; %  std of the noise
X = [-10 -3 2 8 12]; N = length(X); Y = sin(X) + sig_e * randn(1,N);
% kernel function
delta = 3; k = @(z1,z2) exp(-(z1-z2).^2/(2*delta^2));
% kernel matrix
z = meshgrid(X); K = k(z,z');
Lambda = (K + sig_e^2 * eye(N));
% inputs for test
x_pr = -20:0.2:20;
% posterior mean and variance
Kxx = zeros(length(x_pr),N);
f_mean = zeros(1,length(x_pr)); sig_post = f_mean;
for i = 1:length(x_pr)
    kx = k(x_pr(i),X);
    Kxx(i,:) = kx';
    f_mean(i) = kx / Lambda * Y'; %  posterior mean
    sig_post(i) = k(x_pr(i),x_pr(i)) - kx / Lambda * kx'; %  posterior ...
        var.
end
% Sampling GP Posterior
z = meshgrid(x_pr); K_pr = k(z,z');
Sigma_pr = K_pr - Kxx / Lambda * Kxx';
f_samples_from_post = mvnrnd(f_mean, Sigma_pr, 2);
% Plot
figure(1), clf
V1 = f_mean - sqrt(sig_post); V2 = f_mean + sqrt(sig_post);
fill([x_pr, fliplr(x_pr)], [V1, fliplr(V2)], ...
    'k','FaceAlpha', 0.3, 'EdgeAlpha', 0.7);
hold on; plot(x_pr, f_samples_from_post, '--', 'LineWidth', 3)
    plot(x_pr, f_mean, 'k', 'LineWidth', 5)
    plot(X, Y, 'ro', 'MarkerFaceColor', 'r', 'MarkerSize', 15);
axis([-20 20 -3.5 3.5]), set(gca, 'Fontsize', 20), box on
% Computation of the Marginal Likelihood
LogMargLike = -1/2 * (Y / Lambda * Y' + log(det(Lambda)) + N * log(2*pi));
```

Listing 8.8 Code for sampling from the GP posterior and computing the marginal likelihood (Demo_GP_posterior.m).

Gaussian Process Regression with a Composite Time Covariance

We show the advantage of encoding such prior knowledge and structure in the relevant problem of solar irradiation prediction using GPR. Noting the cyclostationary behavior of the signal, we developed a particular time-based composite covariance to account for the relevant seasonal signal variations. Data from the AEMET, http://www.aemet.es/, radiometric observatory of Murcia (southern Spain, 38.0°N, 1.2°W) were used.

Table 8.3 reports the results obtained with GPR models and several statistical regression methods: regularized linear regression (RLR), SVR, RVM, and GPR. All methods were run with and without using two additional dummy time features containing the year and day-of-year. We will indicate the former case with a subscript (e.g., SVR_t). First, including time information improves all baseline models. Second, the best overall results are obtained by the GPR models, whether including time information or not.

Table 8.3 Results for the estimation of the daily solar irradiation of linear and nonlinear regression models. Subscript Method$_t$ indicates that the method includes time as input variable. Best results are highlighted in bold, and the second best in italics.

Method	ME	RMSE	MAE	*R*
RLR	0.27	4.42	3.51	0.76
RLR$_t$	0.25	4.33	3.42	0.78
SVR	0.54	4.40	3.35	0.77
SVR$_t$	0.42	4.23	3.12	0.79
RVM	0.19	4.06	3.25	0.80
RVM$_t$	0.14	3.71	3.11	0.81
GPR	0.14	3.22	2.47	*0.88*
GPR$_t$	0.13	*3.15*	*2.27*	*0.88*
TGPR	*0.11*	**3.14**	**2.19**	**0.90**

Third, in particular, the TGPR model outperforms the rest in accuracy (RMSE, MAE) and goodness-of-fit ρ, and closely follows the elastic net in bias (ME). TGPR performs better than GPR and GPR$_t$ in all quality measures. From the analysis of variance, statistical differences between TGPR and the rest of the models were observed both for bias ($F = 23.1, p < 0.01$) and accuracy ($F = 11.1, p < 0.01$).

Ensemble Gaussian Process Regression Models

We now give an example of an ensemble of GPR machines where the use of the posterior variance is used to determine the confidence of the models obtained. The code is provided in Listing 8.9, and the resulting results are given in Figure 8.10.

```
% Example of EGPR
clear, clc
% GENERATING DATA
sig_e = 0.6; % std of the noise
X = -15:0.6:15; N = length(X); Y = sin(X) + sig_e * randn(1,N);
% kernel function
delta = 2; k = @(z1,z2) exp(-(z1-z2).^2/(2*delta^2));
% Inputs test
x_pr = -20:0.2:20;
% Complete GPR
% Kernel matrix
z1 = meshgrid(X); K = k(z1,z1');
Lambda = (K + sig_e^2 * eye(N));
% Posterior mean and variance of the complete GP
```

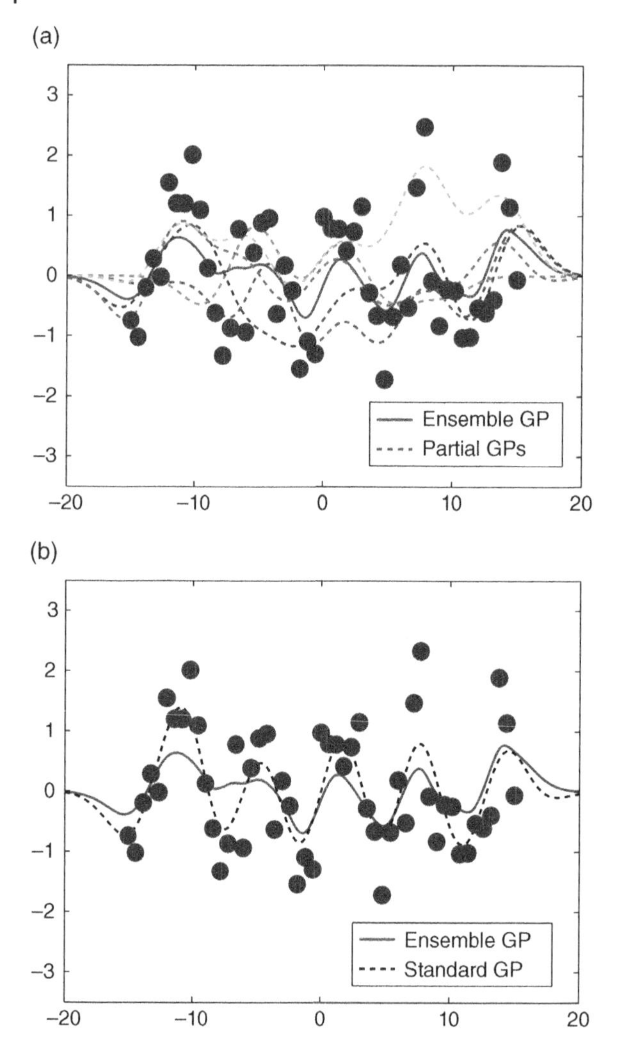

Figure 8.10 Example of ensemble GPR considering $R = 5$ partial GPs. (a) Ensemble solution $\widehat{f}$ and $R = 5$ partial solutions $\widehat{f}_r$. (b) Ensemble posterior mean $\widehat{f}$ and its corresponding posterior mean obtained by a standard complete GP considering jointly all the data.

```
f_mean = zeros(1,length(x_pr)); sig_post = f_mean;
for i = 1:length(x_pr)
    kx = k(x_pr(i),X);
    f_mean(i) = kx / Lambda * Y'; % posterior mean
    sig_post(i) = k(x_pr(i),x_pr(i)) - kx / Lambda * kx'; % posterior ...
        var.
end
% Partial GPs
R = 5; % number of partial GPs
Nr = fix(N/5); % number of data per GP (other choices are possible)
fr = zeros(R,length(x_pr)); sr = fr;
for r = 1:R
```

```
    pos = randperm(N,Nr); % random pick
    Xr = X(pos); Yr = Y(pos);
    zr = meshgrid(Xr); Kr = k(zr,zr');
    Lambda_r = (Kr + sig_e^2 * eye(Nr)); % for the partial GPs
    % posterior mean and variance for the partial GPs
    for i = 1:length(x_pr)
        kx_r = k(x_pr(i),Xr);
        fr(r,i) = kx_r / Lambda_r * Yr'; % partial posterior mean
        sr(r,i) = k(x_pr(i),x_pr(i)) - kx_r / Lambda_r * kx_r'; % p.p. ...
            var.
    end
end
% Ensemble GP solution
Prec = 1./sr; % precision
Den = sum(1./sr); Wn_prec = Prec./repmat(Den,R,1);
% normalized weights
f_EGP = sum(Wn_prec.*fr); sig_ESP = mean(Wn_prec);
% Represent
figure(1), clf, plot(x_pr,f_EGP,'b','LineWidth',5), hold on
    plot(x_pr,fr,'--','LineWidth',2)
    plot(X,Y,'ro','MarkerFaceColor','r','MarkerSize',12)
axis([-20 20 -3.5 3.5]), set(gca,'Fontsize',20)
legend('Ensemble GP','Partial GPs'); box on
figure(2), clf, plot(x_pr,f_EGP,'b','LineWidth',5), hold on
    plot(x_pr,f_mean,'k--','LineWidth',2)
    plot(X,Y,'ro','MarkerFaceColor','r','MarkerSize',12)
axis([-20 20 -3.5 3.5]), set(gca,'Fontsize',20)
legend('Ensemble GP','Standard GP'); box on
```

Listing 8.9 Simple example of ensemble GPR code.

Heteroscedastic and Warped Gaussian Process Regression

We focus now on a real problem in geosciences: the estimation of chlorophyll-a concentrations from remote sensing upwelling radiance just above the ocean surface. A variety of bio-optical algorithms have been developed to relate measurements of ocean radiance to *in situ* concentrations of phytoplankton pigments, and ultimately most of these algorithms demonstrate the potential of quantifying chlorophyll-a concentrations accurately from multispectral satellite ocean color data. In this context, robust and stable non-linear regression methods that provide inverse models are desirable. In addition, we should note that most of the bio-optical models (such as Morel, CalCOFI and OC2/OC4 models) often rely on empirically adjusted nonlinear transformation of the observed variable (which is traditionally a ratio between bands).

The experiment uses the SeaBAM dataset (Maritorena and O'Reilly, 2000; O'Reilly *et al.*, 1998), which gathers 919 *in situ* pigment measurements around the United States and Europe. The dataset contains coincident *in situ* chlorophyll concentration and remote sensing reflectance measurements (Rrs(λ), [sr^{-1}]) at some wavelengths (412, 443, 490, 510 and 555 nm) that are present in the SeaWiFS ocean color satellite sensor. The chlorophyll concentration values range from 0.019 to 32.79 mg/m^3 (revealing a clear exponential distribution). Actually, even though SeaBAM data originate from various researchers, the variability in the radiometric data is limited. In fact, at high Chl-a concentrations, Ca [mg/m^3], the dispersion of radiance ratios Rrs(490)/Rrs(555)

Table 8.4 Bias (ME), accuracy (RMSE, MAE) and fitness (Pearson's ρ) for different rates of training samples using both raw and empirically transformed observation variables.

	Raw				Empirically based			
	ME	RMSE	MAE	ρ	ME	RMSE	MAE	ρ
Rate = 10%								
GPR	0.38	2.54	0.51	0.52	0.26	2.12	0.42	0.71
VHGPR	0.37	2.49	0.49	0.56	0.26	2.12	0.42	0.71
WGPR	0.30	2.32	0.45	0.67	0.31	2.36	0.45	0.61
Rate = 40%								
GPR	0.02	1.74	0.33	0.82	0.15	1.69	0.29	0.86
VHGPR	0.29	2.51	0.46	0.65	0.15	1.70	0.29	0.85
WGPR	0.08	1.71	0.30	0.83	0.17	1.75	0.30	0.86

increases, mostly because of the presence of Case II waters. The shape of the scatterplots is approximately sigmoidal (in log-log space). At lowest concentrations the highest Rrs(490)/Rrs(555) ratios are slightly lower than the theoretical limit for clear natural waters. See analysis in (Camps-Valls *et al.*, 2006b). More information about the data can be obtained from http://seabass.gsfc.nasa.gov/seabam/seabam.html.

Table 8.4 shows different scores (ME, accuracy, Pearson correlation) between the observed and predicted variable when using the raw data (no ad hoc transform at all) and the empirically adjusted transform. Results are shown for three flavors of GPs: the standard GPR (Rasmussen and Williams, 2006), the VHGPR (Lázaro-Gredilla *et al.*, 2014), and the proposed WGPR (Lázaro-Gredilla, 2012; Snelson *et al.*, 2004) for different rates of training samples. Several conclusions can be obtained: (1) as expected, better results are obtained by all models when using more training samples to learn the hyperparameters; (2) empirically based warping slightly improves the results over working with raw data for the same number of training samples, but this requires prior knowledge about the problem, and time and effort to fit an appropriate function; (3) WGPR outperforms the rest of the GPs in all comparisons over standard GPR and VHGPR (≈1–10%); and finally (4) WGPR nicely compensates the lack of prior knowledge about the (possibly skewed) distribution of the observation variable.

An interesting advantage of WGPR is to observe the learned warping function. We plot these for different rates of training samples in Figure 8.11, which shows that: (a) as more samples are provided for learning, the warping function becomes more nonlinear for low concentration values; (b) the learned warping function actually looks linear (in log scale) for high observation values and strongly nonlinear for low values. The empirically based warping function typically used in most bio-optical models is a log function. Therefore, it seems that the WGPR accuracy comes from the better modeling of the nonlinearity for low chlorophyll values, which are the great majority in the database. The interested reader, may find the code of the VHGPR, along with other methods described in this chapter, such as SVR, KRR, RVM, and GPR, on the book web page.

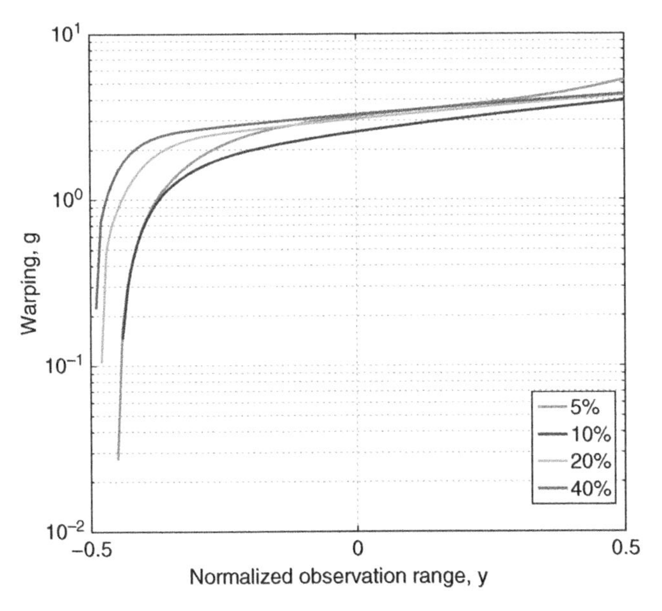

Figure 8.11 Learned warping with WGPR for different rates of training samples.

8.4.8 Relevance Vector Machines

This section exemplifies the RVM in simple 1D function approximation toy examples, and in complex time-series prediction in chaotic time series. We also give some remarks and analysis concerning the CIs obtained compared with those obtained with GPR.

Toy Examples

Listing 8.10 contains an example of an RVM model for regression and its comparison with the result of a GP for regression. Figure 8.12 shows the result.

```
% RVM compared with the GPR solution
clear, clc
% Generating data
sig_e = 0.5; % std of the noise
X = [-12 -10 -3 -2  2 8 12 14]';
N = length(X); % number of data
Y = sin(X) + sig_e * randn(N,1); % N x 1
% Basis functions
M = N; % number of basis functions  =  data size (RVM)
sig = 2; phi = @(x1,x2) exp(-(x1-x2).^2/sig^2);
% Basis function matrix
z = meshgrid(X); PHI = phi(z,z'); % N x M
% Prior variance
% For simplicity, we assume that we learn in training
% the same sigma_prior for each basis function/weight
% hence, we assume the same relevance for each basis function
sig_prior = 1;
Sigma_p = sig_prior^2 * eye(M); % Cov-Matrix of the prior over w
```

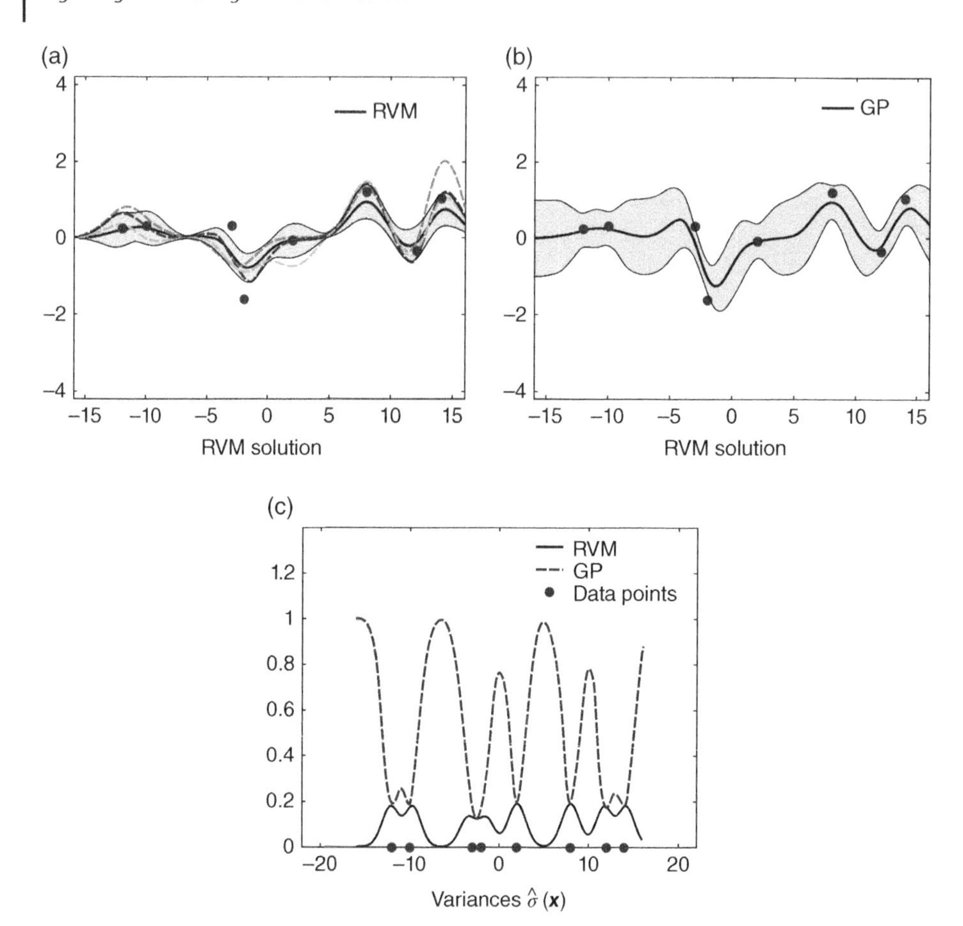

Figure 8.12 Example of an RVM model. (a) Using localized Gaussian basis functions with three random functions drawn according to the posterior distribution (dashed line); see Listing 8.10. (b) GPR solution given the same data points. (c) Comparison of the two predictive and posterior variances. Note that the variance of the RVM with localized basis functions is higher when closer to the data points, which is an undesirable effect. Big dots show the positions of the data ($\mathbf{x}_i$, $i = 1,\ldots,N$).

```
% moments of the posterior of w
w_mean = ((PHI'*PHI) + sig_e^2/sig_prior^2*eye(M)) \ PHI' * Y; % M x 1
w_Covar_Matrix = ((1/sig_e^2)*(PHI'*PHI)+inv(Sigma_p)) \ eye(M); % M x M
% Test inputs
x_pr = -16:0.2:16;
Lambda = (PHI + sig_e^2 * eye(N)); % FOR GP comparison
% Posterior mean and variance of f
f_mean = zeros(1,length(x_pr)); sigma_post = f_mean;
f_meanGP = f_mean; sigma_postGP = f_meanGP;
for i = 1:length(x_pr)
    Phix = phi(x_pr(i),X)';
    f_mean(i) =  Phix * w_mean; % RVM
    sigma_post(i) = Phix * w_Covar_Matrix * Phix'; % RVM
    f_meanGP(i) = Phix / Lambda * Y; % GP
    sigma_postGP(i) = phi(x_pr(i),x_pr(i)) - Phix / Lambda * Phix'; % GP
```

```
end
% Sampling functions from RVM Posterior
Num_Funct = 3;
w_sample = mvnrnd(w_mean, w_Covar_Matrix, Num_Funct);
f_samples_from_post = zeros(length(x_pr), Num_Funct);
for i = 1:length(x_pr)
    Phix = phi(x_pr(i),X)';
    f_samples_from_post(i,:) = Phix * w_sample';
end
% RVM
V1 = f_mean - sqrt(sigma_post); V2 = f_mean + sqrt(sigma_post);
figure(1), clf, plot(x_pr, f_mean, 'k', 'LineWidth', 5), hold on
fill([x_pr, fliplr(x_pr)], [V1, fliplr(V2)], 'k', ...
    'FaceAlpha', 0.3, 'EdgeAlpha', 0.7)
plot(x_pr, f_samples_from_post,'--', 'LineWidth', 3);
plot(X, Y, 'ro', 'MarkerFaceColor', 'r', 'MarkerSize', 10)
axis([-16 16 -4.2 4.2]), set(gca, 'Fontsize', 20)
h = legend('RVM'); set(h,'Box','off'), box on
% GP
V1 = f_meanGP - sqrt(sigma_postGP); V2 = f_meanGP + sqrt(sigma_postGP);
figure(2), clf, plot(x_pr, f_meanGP, 'k', 'LineWidth', 5), hold on
fill([x_pr, fliplr(x_pr)], [V1, fliplr(V2)], 'k', ...
    'FaceAlpha', 0.3, 'EdgeAlpha', 0.7)
plot(X, Y, 'ro', 'MarkerFaceColor', 'r', 'MarkerSize', 10)
axis([-16 16 -4.2 4.2]), set(gca, 'Fontsize', 20),
h = legend('GP'); set(h,'Box','off'), box on
% RVM & GP
figure(3); clf, plot(x_pr, sigma_post, 'k', 'LineWidth', 5),
hold on
plot(x_pr, sigma_postGP, 'r--', 'LineWidth', 5),
plot(X, zeros(1,N), 'ro', 'MarkerFaceColor', 'r', 'MarkerSize', 10)
axis([-22 22 0 max(sigma_postGP)+0.4]), set(gca, 'Fontsize', 20),
h = legend('RVM','GP','Data points'); set(h,'Box','off'), box on
```

Listing 8.10 Code comparing RVM and GP solutions.

Time Series Prediction

Despite these good characteristics, the use of RVM for nonlinear system identification and time-series prediction is limited to a few studies (Anguita and Gagliolo 2002; Bishop and Tipping 2000; D'Souza *et al.* 2004; Nikolaev and Tino 2005; Quiñonero-Candela 2004; Wipf *et al.* 2004), and in all cases the approach consisted of stacking the input and output discrete time processes into a training sample and then applying the traditional RVM formulation. This approach, though powerful, does not consider the input–output discrete time process relationships in the modeling, which can ultimately lead to suboptimal results. Here, we introduce a more general class of RVM-based system identification algorithms by means of *composite kernels* following the RKHS models introduced in Chapter 6, and show that the previous stacked approach is a particular case. Several algorithms for nonlinear system identification are presented, accounting for the input and output time processes either separately or jointly, allowing different levels of flexibility and sophistication for model development.

Table 8.5 Results for the MG time series prediction problem.

	RVM	RVM_{2K}	$RRVM_{2K}$	RVM_{4K}	$RRVM_{4K}$
MG17					
nMSE	−1.687	−2.055	−1.976	*−2.080*	**−2.088**
P	8	14	12	13	13
$\hat{\sigma}_n$	0.0010	0.0008	0.0007	0.0006	0.0007
%RVs	37.88	24.32	29.97	25.51	24.07
MG30					
nMSE	−1.257	−1.293	−1.296	*−1.298*	**−1.299**
P	6	6	6	6	6
$\hat{\sigma}_n$	0.0092	0.0075	0.0065	0.0071	0.0064
%RVs	10.34	13.38	20.28	15.01	20.28

We test the RKHS models introduced in Chapter 6 in the RVM for time-series prediction. More results and theoretical analysis of the methods can be found in Camps-Valls *et al.* (2007a). In particular, we compare the models performance in the standard Mackey–Glass time-series prediction problem, which is well known for its strong nonlinearity. This classical high-dimensional chaotic system is generated by the delay differential equation: $\mathrm{d}x/\mathrm{d}t = -0.1x_n + 0.2x_{n-\Delta}/(1 + x_{n-\Delta}^{10})$, with delays $\Delta = 17$ and $\Delta = 30$, thus yielding the time series MG17 and MG30 respectively. We considered 500 training samples and used the next 1000 for free parameter selection (validation set).

Results are shown in Table 8.5. The RVM yields very good results, compared with previous results in the literature (Mattera, 2005). The RKHS models for RVM nevertheless outperform the standard RVM, especially significant for the MG17 time series. In the case of MG30 the differences are not significant, but there is still a preference for RVM_{4K}-based models, suggesting that this is a more complicated system. Several interesting issues can be noticed. First, the fact that this dataset has virtually no output noise is better detected with all composite methods (lower $\hat{\sigma}_n$ than the standard RVM). Second, the expected zero noise variance along with the chaotic nature of the time series prevent sparsity from arising in a trivial way. We observe, however, that the number of RVs retained by the methods is smaller than the standard RVM in MG17, but as the dynamics complexity is increased (i.e., MG30), kernel methods need more RVs to attain competitive results. The MATLAB code to reproduce the results can be downloaded from the book web page.

8.5 Concluding Remarks

This chapter treated the problem of advanced regression and function approximation with kernels, under the two main existing approximations: methods based on the SVR and the alternative Bayesian nonparametric treatment of regression modeling.

We reviewed several techniques in each one of the approaches and illustrated their performance in real scenarios in signal and data processing problems. In particular, for the SVR-based regression approaches, we studied an SVR version for multi-output function approximation, a kernel-based regression to cope with signal and noise dependencies (KSNR), and the SS-SVR to incorporate the wealth of unlabeled samples in the regression function. On the other, we summarized some important achievements in the field of Bayesian nonparametrics and focused on the RVM and different developments of GPR algorithms. Both families have found wide application in signal processing in the last decade (Pérez-Cruz *et al.*, 2013; Rojo-Álvarez *et al.*, 2014). In both cases, we paid attention to the important problem of hyperparameter estimation, and provided some prescriptions and procedures for this core task. Methods performance was illustrated in both synthetic and real application examples, and tutorials and code snippets were provided.

The field of regression and function approximation is very vast, and there are plenty of new algorithmic developments in both families, and more algorithms are to come. In the near future, we foresee two main directions of research: (1) design methods that adapt to the signal characteristics, and (2) algorithms that self-tune or infer their hyperparameters from the data in an autonomous way. Chapter 9 will complete the topic by treating adaptive filtering and dynamical systems with kernels.

8.6 Questions and Problems

Exercise 8.6.1 Derive some of the iterative weighted recursive least squares (RLS) equations for ν-SVR. Use the primal Equation 8.10 and constraints in Equation 8.11 to prove that a Lagrangian of the form

$$\begin{aligned} L_P = \frac{1}{2}\|\boldsymbol{w}\|^2 + \sum_{i=1}^{N} \alpha_i(y_i - \langle \boldsymbol{w}, \boldsymbol{\phi}(\boldsymbol{x}_i)\rangle - b - \varepsilon) \\ + \sum_{i=1}^{N} \alpha_i^*(\langle \boldsymbol{w}, \boldsymbol{\phi}(\boldsymbol{x}_i)\rangle) + b - y_i - \varepsilon) + \varepsilon(CN\nu - \lambda) \end{aligned} \tag{8.79}$$

exists, and then compute its gradient from ξ_i, ξ_i^* to prove, in combination with the definitions in Equation 8.13, that the values of a_i, a_i^* are the ones in Equation 8.14.

Exercise 8.6.2 Derive some more of the iterative weighted RLS equations for ν-SVR. Using the definitions in Equation 8.13 derive Equation 8.79. Then, computing its gradient with respect variables $\boldsymbol{w}$, b, and ε and using the representer theorem, find Equation 8.15.

Exercise 8.6.3 Prove that the equivalent IRWLS updates for a_i and a_i^* in the PD-SVR are

$$a_i^{(*)} = \begin{cases} 0, & e_i^{(*)} < 0 \\ \dfrac{2c_i C}{e_i^{(*)}}, & e_i^{(*)} \geq 0. \end{cases}$$

Exercise 8.6.4 The solution of the KRR (and equivalently the GPR one) is closed-form; that is, $\boldsymbol{\alpha} = (\boldsymbol{K} + \lambda \boldsymbol{I})^{-1}\boldsymbol{y}$. Derive the equations for solving the problem using gradient descent.

Exercise 8.6.5 Kernel methods in general are gray machines. A possibility to understand the implemented function comes from the concept of sensitivity analysis, which measures how the approximating function varies with respect to the inputs. The sensitivity of feature x_{nj} in example $\boldsymbol{x}_n := [x_{n1}, \ldots, x_{nd}]^{\mathrm{T}}$ is defined as

$$s_j = \int_{\mathcal{X}} \left(\frac{\partial \phi(\boldsymbol{x})}{\partial \boldsymbol{x}_j} \right)^2 p(\boldsymbol{x})\, \mathrm{d}\boldsymbol{x},$$

where $p(\boldsymbol{x})$ is the *pdf* over the d-dimensional input vector $\boldsymbol{x}$ and $\phi(\boldsymbol{x})$ represents either the predictive mean, $\mu_{\mathrm{GP}*}$, or variance, $\sigma^2_{\mathrm{GP}*}$. The empirical estimate of the sensitivity for the jth feature can be written as

$$s_j = \frac{1}{N} \sum_{n=1}^{N} \left(\frac{\partial \phi(\boldsymbol{x}_n)}{\partial x_{nj}} \right)^2,$$

where N denotes the number of training samples. Derive the sensitivity maps for both the predictive mean and the variance assuming an SE kernel function.

Exercise 8.6.6 Assuming that the joint process $\{\boldsymbol{y}, \hat{f}(\boldsymbol{x})\}$, where $\hat{f}(\boldsymbol{x})$ is the predictive mean over $\boldsymbol{x}$ in a GP, has a Gaussian distribution with zero mean, prove that its covariance function is

$$\begin{bmatrix} \boldsymbol{K} + \sigma^2 \boldsymbol{I} & \boldsymbol{k} \\ \boldsymbol{k}^{\mathrm{T}} & K(\boldsymbol{x}, \boldsymbol{x}) \end{bmatrix}.$$

Using Bayes' rule and this joint distribution, derive the predictive variance of the GP shown in Equation 8.53 (Rasmussen and Williams, 2006).

Exercise 8.6.7 A crucial point in kernel regression methods is the inversion of a potentially big $N \times N$ matrix. An alternative is to perform a Cholesky decomposition such that $\boldsymbol{K} = \boldsymbol{R}\boldsymbol{R}^{\mathrm{T}}$. Write a MATLAB code to do the inversion of the (regularized) kernel matrix, and analyze the computational gain in terms of the rank of $\boldsymbol{K}$.

Exercise 8.6.8 Derive the equations of a randomized GPR in which the kernel function is approximated with random Fourier features (Rahimi and Recht, 2007). Given D random features, what is the order of memory and computational gains.

Exercise 8.6.9 Study and develop a MATLAB function to run the sparsification (reduced-rank) procedure for KRR presented in Cawley and Talbot (2002).

Exercise 8.6.10 Derive the leave-one-out estimates for GPs predictive mean and predictive variance.

Exercise 8.6.11 Discuss the issue of considering nonlocalized basis functions in the context of the RVM. Hint: use the remark in Section 8.3.2, and the study of Rasmussen and Quiñonero-Candela (2005).

9

Adaptive Kernel Learning for Signal Processing

9.1 Introduction

Adaptive filtering is a central topic in signal processing. An adaptive filter is a filter structure provided with an adaptive algorithm that tunes the transfer function, typically driven by an error signal. Adaptive filters are widely applied in nonstationary environments because they can adapt their transfer function to match the changing parameters of the system that generates the incoming data (Hayes 1996; Widrow *et al.* 1975). They have become ubiquitous in current DSP, mainly due to the increase in computational power and the need to process streamed data. Adaptive filters are now routinely used in all communication applications for channel equalization, array beamforming, or echo cancellation, to cite just a few, and in other areas of signal processing, such as image processing or medical equipment.

By applying linear adaptive filtering principles in the kernel feature space, powerful nonlinear adaptive filtering algorithms can be obtained. This chapter introduces the wide topic of adaptive signal processing, and it explores the emerging field of kernel adaptive filtering (KAF). Its orientation is different from the preceding ones, as adaptive processing can be used in a variety of scenarios. Attention is paid to kernel LMS and RLS algorithms, to previous taxonomies of adaptive kernel methods, and to emergent kernel methods for online and recursive KAF. MATLAB code snippets are included as well to illustrate the basic operations of the most common kernel adaptive filters. Tutorial examples are provided on applications, including chaotic time-series prediction, respiratory motion prediction, and nonlinear system identification.

9.2 Linear Adaptive Filtering

Let us first define some basic concepts of linear adaptive filtering theory. The goal of adaptive filtering is to model an unknown, possibly time-varying system by observing the inputs and outputs of this system over time. We will denote the input to the system on time instant n as $\boldsymbol{x}_n$, and its output as d_n. The input signal $\boldsymbol{x}_n$ is assumed to be zero-mean. We represent it as a vector, and it will often represent a time-delay vector of L taps of a signal x_n on time instant n as $\boldsymbol{x}_n = [x_n, x_{n-1}, \dots, x_{n-L+1}]^{\mathrm{T}}$.

Digital Signal Processing with Kernel Methods, First Edition. José Luis Rojo-Álvarez, Manel Martínez-Ramón, Jordi Muñoz-Marí, and Gustau Camps-Valls.

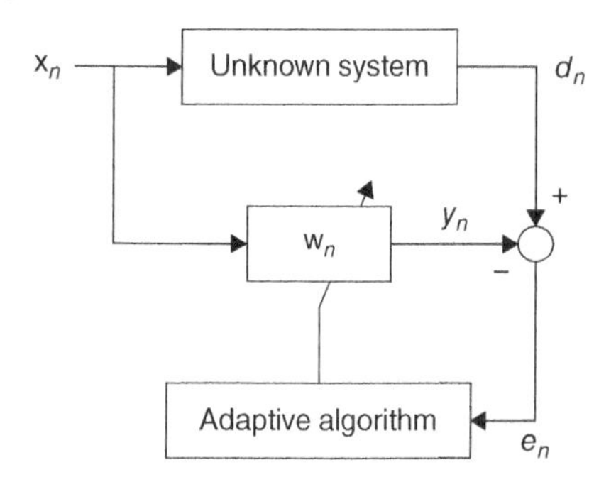

Figure 9.1 A linear adaptive filter for system identification.

A diagram for a linear adaptive filter is depicted in Figure 9.1. The input to the adaptive filter at time instant n is $\boldsymbol{x}_n$, and its response y_n is obtained through the following linear operation:

$$y_n = \boldsymbol{w}_n^{\mathrm{H}} \boldsymbol{x}_n, \tag{9.1}$$

where H is the Hermitian operator.

Linear adaptive filtering follows the *online learning* framework, which consists of two basic steps that are repeated at each time instant n. First, the online algorithm receives an observation $\boldsymbol{x}_n$ for which it calculates the estimated image y_n, based on its current estimate of $\boldsymbol{w}_n$. Next, the algorithm receives the desired output d_n (also known as a *symbol* in communications), which allows it to calculate the estimation error $e_n = d_n - y_n$ and update its estimate for $\boldsymbol{w}_n$. In some situations d_n is known a priori; that is, the received signal $\boldsymbol{x}_n$ is one of a set of *training* signals provided with known labels. The procedure is then called *supervised*. When d_n belongs to a finite set of quantized labels, and it can be assumed that the error will be likely much smaller than the quantization step or minimum Euclidean distance between labels, the desired label is estimated by quantizing y_n to the closest label. In these cases, the algorithm is called *decision directed*.

9.2.1 Least Mean Squares Algorithm

The classical adaptive optimization techniques have their roots in the theoretical approach called the steepest-descent algorithm. Assume that the expectation of the squared error signal, $J_n = \mathbb{E}[|e_n|^2]$ can be computed. Since this error is a function of the vector $\boldsymbol{w}_n$, the idea of the algorithm is to modify this vector toward the direction of the steepest descent of J_n. This direction is just opposite to its gradient $\nabla_{\boldsymbol{w}} J_n$. Indeed, assuming complex stationary signals, the error expectation is

$$\begin{aligned} \mathbb{E}[|e_n|^2] &= \mathbb{E}[|d_n - \boldsymbol{w}_n^{\mathrm{H}} \boldsymbol{x}_n|^2] \\ &= \mathbb{E}[|d_n|^2 + \boldsymbol{w}_n^{\mathrm{H}} \boldsymbol{x}_n \boldsymbol{x}_n^{\mathrm{H}} \boldsymbol{w}_n - 2\boldsymbol{w}_n^{\mathrm{H}} \boldsymbol{x}_n d_n^*] \\ &= \sigma_d^2 + \boldsymbol{w}_n^{\mathrm{H}} \boldsymbol{R}_{xx} \boldsymbol{w}_n - 2\boldsymbol{w}_n^{\mathrm{H}} \boldsymbol{p}_{xd}, \end{aligned} \tag{9.2}$$

where $\boldsymbol{R}_{xx}$ is the signal autocorrelation matrix, $\boldsymbol{p}_{xd}$ is the cross-correlation vector between the signal and the filter output, and σ_d^2 is the variance of the system output. Its gradient with respect to vector $\boldsymbol{w}_n$ is expressed as

$$\nabla_w J_n = 2R_{xx} w_n - 2p_{xd}. \tag{9.3}$$

The adaptation rule based on steepest descent thus becomes

$$w_{n+1} = w_n - \eta \nabla_w J_n, \tag{9.4}$$

where η is the *step size* or *learning rate* of the algorithm.

The LMS algorithm, introduced by Widrow in 1960 (Widrow *et al.*, 1975), is a very simple and elegant method of training a linear adaptive system to minimize the MSE that approximates the gradient $\nabla_w J_n$ using an instantaneous estimate of the gradient. From Equation 9.3, an approximation can be written as

$$\nabla_w J_n \approx 2x_n x_n^{\mathrm{H}} w_n - 2x_n d_n^*. \tag{9.5}$$

Using this approximation in Equation 9.4 leads to the well-known stochastic gradient descent update rule, which is the core of the LMS algorithm:

$$\begin{aligned} w_{n+1} &= w_n - \eta x_n (x_n^{\mathrm{H}} w_n - d_n^*) \\ &= w_n + \eta x_n e_n^*. \end{aligned} \tag{9.6}$$

This optimization procedure is also the basis for tuning nonlinear filter structures such as NNs (Haykin, 2001) and some of the kernel-based adaptive filters discussed later in this chapter. A detailed analysis including that of convergence and misadjustment is given in Haykin (2001). The MATLAB code for the LMS training step on a new data pair (x, d) is displayed in Listing 9.1.

```
y = x' * w; % evaluate filter output
err = d - y; % instantaneous error
w = w + mu * x * err'; % update filter coefficients
```

Listing 9.1 Training step of the LMS algorithm on a new datum (x, d).

Under the stationarity assumption, the LMS algorithm converges to the Wiener solution in mean, but the weight vector w_n shows a variance that converges to a value that is a function of η. Therefore, low variances are only achieved at low adaptation speed. A more sophisticated approach with faster convergence is found in the the RLS algorithm.

9.2.2 Recursive Least-Squares Algorithm

The RLS algorithm was first introduced in 1950 (Plackett, 1950). In a stationary scenario, it converges to the Wiener solution in mean and variance, improving also the slow rate of adaptation of the LMS algorithm. Nevertheless, this gain in convergence speed comes at the price of a higher complexity, as we will see later.

Recursive Update

The basis of the RLS algorithm consists of recursively updating the vector w that minimizes a regularized version of the cost function J_n:

$$J_n = \sum_{i=1}^{n} |d_i - x_i^{\mathrm{H}} w|^2 + \delta w^{\mathrm{H}} w, \tag{9.7}$$

where δ is a positive constant *regularization factor*. The regularization factor penalizes the squared norm of the solution vector so that this solution does not apply too much weight to any specific dimension.[1] The solution that minimizes the LS cost function (9.7) is well known and given by

$$w = (R_{xx} + \delta I)^{-1} p_{xd}. \tag{9.8}$$

The regularization δ guarantees that the inverse in Equation 9.8 exists. In the absence of regularization (i.e., for $\delta = 0$), the solution requires one to invert the matrix R_{xx}, which may be rank deficient. For a detailed derivation of the RLS algorithm we refer the reader to Haykin (2001) and Sayed (2003). In the following, we will provide its update equations and a short discussion of its properties compared with LMS.

We denote the autocorrelation matrix for the data x_1 till x_n as R_n:

$$R_n = \sum_{i=1}^{n} x_i^H x_i. \tag{9.9}$$

RLS requires the introduction of an *inverse autocorrelation matrix* P_n, defined as

$$P_n = (R_n + \delta I)^{-1}. \tag{9.10}$$

At step $n-1$ of the recursion, the algorithm has processed $n-1$ data, and its estimate w_{n-1} is the optimal solution for minimizing the squared cost function (Equation 9.7) at time step $n-1$. When a new datum x_n is obtained, the inverse autocorrelation matrix is updated as

$$P_n = P_{n-1} - g_n g_n^H P_{n-1}, \tag{9.11}$$

where g_n is the *gain vector* of the RLS algorithm:

$$g_n = \frac{P_{n-1}}{1 + x_n^H P_{n-1} x_n}. \tag{9.12}$$

The update of the solution itself reads

$$w_n = w_{n-1} + g_n e_n, \tag{9.13}$$

in which e_n represents the usual prediction error $e_n = d_n - x_n^H w_{n-1}$. The RLS algorithm starts by initializing its solution to $w_0 = 0$, and the estimate of the inverse autocorrelation matrix P_n to $P_0 = \delta^{-1} I$. Then, it recursively updates its solution by including one datum x_i at a time and performing the calculations from Equations 9.11–9.13. Owing to the matrix multiplications involved in the RLS updates, the computational complexity per time step for RLS is quadratic in terms of the data dimension, $\mathcal{O}(L^2)$, while the LMS algorithm has only linear complexity, $\mathcal{O}(L)$.

1 A slightly more general formulation involves a *regularization matrix* which can penalize the individual elements of the solution differently (Sayed, 2003).

Exponentially Weighted Recursive Least-Squares

The RLS algorithm takes into account all previous data when it updates its solution with a new datum. This kind of update yields a faster convergence than LMS, which guides its update based only on the performance on the newest datum. Nevertheless, by guaranteeing that its solution is valid for all previous data, the RLS algorithm is in essence looking for a stationary solution, and thus it cannot adapt to nonstationary scenarios, where a tracking algorithm is required. LMS, on the other hand, deals correctly with nonstationary scenarios, thanks to the instantaneous nature of its update, which forgets older data and only adapts to the newest datum.

A tracking version of RLS can be obtained by including a *forgetting factor* $\lambda \in (0, 1]$ in its cost function as follows:

$$J_n = \sum_{i=1}^{n} \lambda^{n-i} \|d_i - \boldsymbol{x}_i^{\mathrm{H}} \boldsymbol{w}\|^2 + \lambda^n \delta \boldsymbol{w}^{\mathrm{H}} \boldsymbol{w}. \tag{9.14}$$

The resulting algorithm is called exponentially weighted RLS (Haykin, 2001; Sayed, 2003). The inclusion of the forgetting factor assigns lower weights to older data, which allows the algorithm to adapt gradually to changes. The update for the inverse autocorrelation matrix becomes

$$\boldsymbol{P}_n = \lambda^{-1} \boldsymbol{P}_{n-1} - \lambda^{-1} \boldsymbol{g}_n \boldsymbol{x}_n^{\mathrm{H}} \boldsymbol{P}_{n-1}, \tag{9.15}$$

and the new gain vector becomes

$$\boldsymbol{g}_n = \frac{\lambda^{-1} \boldsymbol{P}_{n-1}}{1 + \lambda^{-1} \boldsymbol{x}_n^{\mathrm{H}} \boldsymbol{P}_{n-1} \boldsymbol{x}_n}. \tag{9.16}$$

The MATLAB code for the training step of the exponentially weighted RLS algorithm is displayed in Listing 9.2.

```
y = x' * w; % evaluate filter output
err = d - y; % instantaneous error
g = P * x / (lambda + x'*P*x); % gain vector
w = w + g * err; % update filter coefficients
P = lambda \ (P - g*x'*P); % update inverse autocorrelation matrix
```

Listing 9.2 Training step of the exponentially weighted RLS algorithm on a new datum $(\boldsymbol{x}, d)$.

Recursive estimation algorithms play a crucial role for many problems in adaptive control, adaptive signal processing, system identification, and general model building and monitoring problems (Ljung, 1999). In the signal-processing literature, great attention has been paid to their efficient implementation. Linear AR models require relatively few parameters and allow closed-form analysis, while ladder or lattice implementation of linear filters has long been studied in signal theory. However, when the system generating the data is driven by nonlinear dynamics, the model specification and parameter estimation problems increase their complexity, and hence nonlinear adaptive filtering becomes strictly necessary.

Note that, in the field of control theory, a range of sequential algorithms for nonlinear filtering have been proposed since the 1960s, notably the extended KF

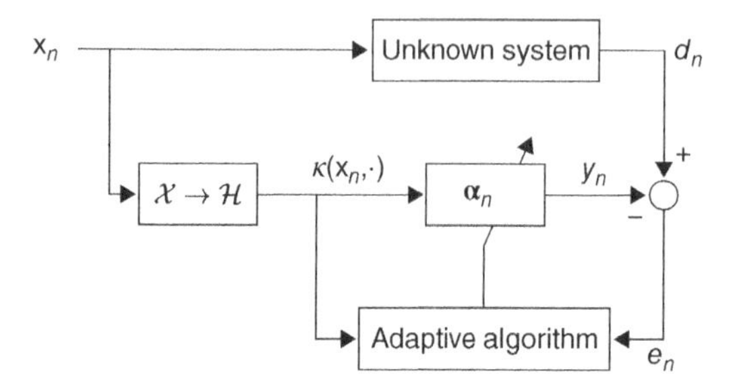

Figure 9.2 A kernel adaptive filter for nonlinear system identification.

(Lewis *et al.*, 2007) and the unscented KF (Julier and Uhlmann, 1996), which are both nonlinear extensions of the celebrated KF (Kalman, 1960), and particle filters (Del Moral, 1996). These methods generally require knowledge of a state-space model, and while some of them are related to adaptive filtering, we will not deal with them explicitly in this chapter.

9.3 Kernel Adaptive Filtering

The nonlinear filtering problem and the online adaptation of model weights were first addressed by NNs in the 1990s (Dorffner 1996; Narendra and Parthasarathy 1990). Throughout the last decade, a great interest has been devoted to developing nonlinear versions of linear adaptive filters by means of kernels (Liu *et al.*, 2010). The goal is to develop machines that learn over time in changing environments, and at the same time adopt the nice characteristics of convexity, convergence, and reasonable computational complexity, which was not successfully implemented in NNs.

Kernel adaptive filtering aims to formulate the classic linear adaptive filters in RKHS, such that a series of convex LS problems is solved. Several basic kernel adaptive filters can be obtained by applying a linear adaptive filter directly on the transformed data, as illustrated in Figure 9.2. This requires the reformulation of scalar-product-based operations in terms of *kernel evaluations*. The resulting algorithms typically consist of algebraically simple expressions, though they feature powerful nonlinear filtering capabilities. Nevertheless, the design of these online kernel methods requires dealing with some of the challenges that typically arise when dealing with kernels, such as overfitting and computational complexity issues.

In the following we will discuss two families of kernel adaptive filters in detail, namely KLMS and kernel RLS (KRLS) algorithms. Several related kernel adaptive filters will be reviewed briefly as well.

9.4 Kernel Least Mean Squares

The early approach to kernel adaptive filtering introduced a kernel version of the celebrated ADALINE in Frieß and Harrison (1999), though this method was not online. Kivinen *et al.* (2004) proposed an algorithm to perform stochastic gradient descent in

RKHS: the so-called naive online regularized risk minimization algorithm (NORMA) introduces a regularized risk that can be solved online and can be shown to be equivalent to a kernel version of leaky LMS, which itself is a regularized version of LMS.

9.4.1 Derivation of Kernel Least Mean Squares

As an illustrative guiding example of a kernel adaptive filter, we will take the kernelization of the standard LMS algorithm, known as KLMS (Liu *et al.*, 2008). The approach employs the traditional kernel trick. Essentially, a nonlinear function $\boldsymbol{\phi}(\cdot)$ maps the data $\boldsymbol{x}_n$ from the input space to $\boldsymbol{\phi}(\boldsymbol{x}_n)$ in the feature space. Let $\boldsymbol{w}_{\mathcal{H}}$ be the weight vector in this space such that the filter output is $y_n = \boldsymbol{w}_{\mathcal{H},n}\mathrm{T}\boldsymbol{x}_n$, where $\boldsymbol{w}_{\mathcal{H},n}$ is the estimate of $\boldsymbol{w}_{\mathcal{H}}$ at time instant n. Note that we will be taking scalar products of real-valued vectors from now on. Given a desired response d_n, we wish to minimize squared loss, $J_{\boldsymbol{w}_{\mathcal{H},n}}$, with respect to $\boldsymbol{w}_{\mathcal{H}}$. Similar to Equation 9.6, the stochastic gradient descent update rule obtained reads

$$\boldsymbol{w}_{\mathcal{H},n} = \boldsymbol{w}_{\mathcal{H},n-1} + \eta e_n \boldsymbol{\phi}(\boldsymbol{x}_n). \tag{9.17}$$

By initializing the solution as $\boldsymbol{w}_{\mathcal{H},0} = 0$ (and hence $e_0 = d_0 = 0$), the solution after n iterations can be expressed in closed form as

$$\boldsymbol{w}_{\mathcal{H},n} = \eta \sum_{i=1}^{n} e_i \boldsymbol{\phi}(\boldsymbol{x}_i). \tag{9.18}$$

By exploiting the kernel trick, one obtains the prediction function

$$y_* = \eta \sum_{i=1}^{n} e_i \langle \boldsymbol{\phi}(\boldsymbol{x}_i), \boldsymbol{\phi}(\boldsymbol{x}_*)\rangle = \eta \sum_{i=1}^{n} e_i K(\boldsymbol{x}_i, \boldsymbol{x}_*), \tag{9.19}$$

where $\boldsymbol{x}_*$ represents an arbitrary input datum and $K(\cdot,\cdot)$ is the kernel function; for instance, the commonly used Gaussian kernel $K(\boldsymbol{x}_i, \boldsymbol{x}_j) = \exp(-\|\boldsymbol{x}_i - \boldsymbol{x}_j\|^2/2\sigma)$ with *kernel width* σ. Note that the weights $\boldsymbol{w}_{\mathcal{H},n}$ of the nonlinear filter are not used explicitly in the KLMS algorithm. Also, since the present output is determined solely by previous inputs and all the previous errors, it can be readily computed in the input space. These error samples are similar to innovation terms in sequential state estimation (Haykin, 2001), since they add new information to improve the output estimate. Each new input sample results in an output, and hence a corresponding error, which is never modified further and incorporated in the estimate of the next output. This recursive computation makes KLMS especially useful for online (adaptive) nonlinear signal processing.

Liu *et al.* (2008) showed that the KLMS algorithm is well posed in RKHS without the need of an extra regularization term in the finite training data case, because the solution is always forced to lie in the subspace spanned by the input data. The lack of an explicit regularization term leads to two important advantages. First of all, it has a simpler implementation than NORMA, as the update equations are straightforward kernel versions of the original linear ones. Second, it can potentially provide better results because regularization biases the optimal solution. In particular, it was shown that a small enough step size can provide a sufficient "self-regularization" mechanism.

Moreover, since the space spanned by the mapped samples is possibly infinite dimensional, the projection error of the desired signal d_n could be very small, as is well known from Cover's theorem (Haykin, 1999). On the downside, the speed of convergence and the misadjustment also depend upon the step size. As a consequence, they conflict with the generalization ability.

9.4.2 Implementation Challenges and Dual Formulation

Another important drawback of the KLMS algorithm becomes apparent when analyzing its update, Equation 9.19. In order to make a prediction, the algorithm requires storing all previous errors e_i and all processed input data $\boldsymbol{x}_i$, for $i = 1, 2, \dots, n$. In online scenarios, where data are continuously being received, the size of the KLMS network will continuously grow, posing implementation challenges. This becomes even more evident if we recast the weight update from Equation 9.17 into a more standard filtering formulation, by relying on the representer theorem (Schölkopf *et al.*, 2001). This theorem states that the solution $\boldsymbol{w}_{\mathcal{H},n}$ can be expressed as a linear combination of the transformed input data:

$$\boldsymbol{w}_{\mathcal{H},n} = \sum_{i=1}^{n} \alpha_i \boldsymbol{\phi}(\boldsymbol{x}_i). \tag{9.20}$$

This allows the prediction function to be written as the familiar kernel expansion

$$y_* = \sum_{i=1}^{n} \alpha_i K(\boldsymbol{x}_i, \boldsymbol{x}_*). \tag{9.21}$$

The expansion coefficients α_i are called the *dual variables* and the reformulation of the filtering problem in terms of α_i is called the *dual formulation*. The update from Equation 9.17 now becomes

$$\sum_{i=1}^{n} \alpha_i \boldsymbol{\phi}(\boldsymbol{x}_i) = \sum_{i=1}^{n-1} \alpha_i \boldsymbol{\phi}(\boldsymbol{x}_i) + \eta e_n \boldsymbol{\phi}(\boldsymbol{x}_n), \tag{9.22}$$

and after multiplying both sides with the new datum $\boldsymbol{\phi}(\boldsymbol{x}_n)$ and adopting a vector notation, we obtain

$$\boldsymbol{\alpha}_n \mathrm{T} \boldsymbol{k}_n = \boldsymbol{\alpha}_{n-1} \mathrm{T} \boldsymbol{k}_{n-1} + \eta e_n k_{nn}, \tag{9.23}$$

where $\boldsymbol{\alpha}_n = [\alpha_1, \alpha_2, \dots, \alpha_n]\mathrm{T}$, the vector $\boldsymbol{k}_n$ contains the kernels of the n data and the newest point, $\boldsymbol{k}_n = [K(\boldsymbol{x}_1, \boldsymbol{x}_n), K(\boldsymbol{x}_2, \boldsymbol{x}_n), \dots, K(\boldsymbol{x}_n, \boldsymbol{x}_n)]$, and $k_{nn} = K(\boldsymbol{x}_n, \boldsymbol{x}_n)$. KLMS resolves this relationship by updating $\boldsymbol{\alpha}_n$ as

$$\boldsymbol{\alpha}_n = \begin{bmatrix} \boldsymbol{\alpha}_{n-1} \\ \eta e_n \end{bmatrix}. \tag{9.24}$$

The MATLAB code for the complete KLMS training step on a new data pair $(\boldsymbol{x}, d)$ is displayed in Listing 9.3.

```
k = kernel(dict,x,kerneltype,kernelpar); % kernels between dictionary ...
    and x
y = k' * alpha; % evaluate function output
err = d - y; % instantaneous error
kaf.dict = [kaf.dict; x]; % add base to dictionary
kaf.alpha = [kaf.alpha; kaf.eta*err]; % add new coefficient
```

Listing 9.3 Training step of the KLMS algorithm on a new datum (***x***, *d*).

The update in Equation 9.24 emphasizes the growing nature of the KLMS network, which precludes its direct implementation in practice. In order to design a practical KLMS algorithm, the number of terms in the kernel expansion in Equation 9.21 should stop growing over time. This can be achieved by implementing an *online sparsification technique*, whose aim is to identify terms in the kernel expansion that can be omitted without degrading the solution. We will discuss several different sparsification approaches in Section 9.7. Finally, observe that the computational complexity and memory complexity of the KLMS algorithm are both linear in terms of the number of data it stores, $\mathcal{O}(n)$. Recall that the complexity of the LMS algorithm is also linear, though not in terms of the number of data but in terms of the data *dimension*.

9.4.3 Example on Prediction of the Mackey–Glass Time Series

We demonstrate the online learning capabilities of the KLMS kernel adaptive filter by predicting the Mackey–Glass time series, which is a classic benchmark problem (Liu *et al.*, 2010). The Mackey–Glass time series is well known for its strong nonlinearity. It corresponds to a high-dimensional chaotic system, and its output is generated by the following time-delay differential equation:

$$\frac{\mathrm{d}x_n}{\mathrm{d}n} = -bx_n + \frac{ax_{n-\Delta}}{1+x_{n-\Delta}^{10}}. \tag{9.25}$$

We focus on the sequence with parameters $b = 0.1$, $a = 0.2$, and time delay $\Delta = 30$, better known as the MG30 time series. The prediction problem consists of predicting the nth sample, given all samples of the time series up till the $n-1$th sample.

Time-series prediction with kernel adaptive filters is typically performed by considering a time-delay vector $\boldsymbol{x}_n = [x_n, x_{n-1}, \ldots, x_{n-L+1}]\mathrm{T}$ as the input and the next sample of the time series as the desired output, $d_n = x_{n+1}$. This approach casts the prediction problem into the well-known filtering framework.[2] Prediction of several steps ahead can be obtained by choosing a prediction horizon $h > 1$, and $d_n = x_{n+h}$. For time series generated by a deterministic process, a principled tool to find the optimal embedding is Takens' theorem (Takens, 1981). In the case of the MG30 time series, Takens' theorem indicates that the optimal embedding is around $L = 7$ (Van Vaerenbergh *et al.*, 2012a).

We consider 500 samples for online training of the KLMS algorithm and use the next 100 data for testing. The step size of KLMS is fixed to 0.2, and we use the Gaussian kernel with $\sigma = 1$. Figure 9.3 displays the prediction results after 500 training steps. The left plot shows the learning curve of the algorithm, obtained as the MSE of the prediction on the test set, at each iteration of the online training process. As a reference,

2 Note that some particular time-series models have been exploited to define explicit recursivity in the RKHS (Li and Príncipe 2016; Tuia *et al.* 2014), as we will see later on in this chapter.

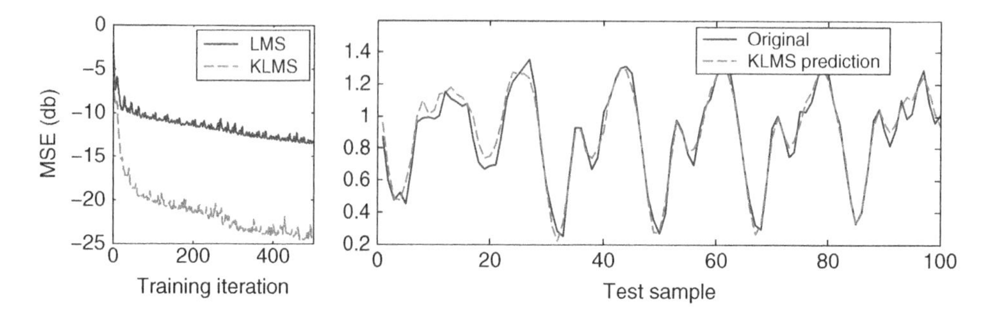

Figure 9.3 KLMS predictions on the Mackey–Glass time series. Left: learning curve over 500 training iterations. Right: test samples of the Mackey–Glass time-series and the predictions provided by KLMS.

we include the learning curve of the linear LMS algorithm with a suitable step size. The right plot shows the 100 test samples of the original time series, as the full line, and KLMS' prediction on these test samples after 500 training steps. These predictions are calculated by evaluating the prediction equation (Equation 9.19) on the test samples. The code for this experiment and all subsequent ones is included in the accompanying material in the book webpage.

9.4.4 Practical Kernel Least Mean Squares Algorithms

In the Mackey–Glass experiment described, the KLMS algorithm requires to store 500 coefficients α_i and the 500 corresponding data $\boldsymbol{x}_i$. The stored data $\boldsymbol{x}_i$ are referred to as its *dictionary*. If the online learning process were to continue indefinitely, the algorithm would require *ever-growing* memory and computation per time step. This issue was identified as a major roadblock early on in the research on kernel adaptive filters, and it has led to the development of several procedures to slow down the dictionary growth by *sparsifying* the dictionary.

A sparsification procedure based on Gaussian elimination steps on the Gram matrix was proposed in Pokharel *et al.* (2009). This method is successful in limiting the dictionary size in the nth training step to some $m < n$, but in order to do so it requires $\mathcal{O}(m^2)$ complexity, which defeats the purpose of using a KLMS algorithm.

Kernel Normalized Least Mean Squares and Coherence Criterion

Around the same time the KLMS algorithm was published, a kernelized version of the affine projection algorithm was proposed (Richard *et al.*, 2009). Affine projection algorithms hold the middle ground between LMS and RLS algorithms by calculating an estimate of the correlation matrix based on the p last data. For $p = 1$ the algorithm reduces to a kernel version of the normalized LMS algorithm (Haykin, 2001), called kernel normalized least-mean squares (KNLMS), and its update reads

$$\boldsymbol{\alpha}_n = \begin{bmatrix} \boldsymbol{\alpha}_{n-1} \\ 0 \end{bmatrix} + \frac{\eta}{\epsilon + \|\boldsymbol{k}_n\|} e_n \boldsymbol{k}_n. \tag{9.26}$$

Note that this algorithm updates all coefficients in each iteration. This is in contrast to KLMS, which updates just one coefficient.

The kernel affine projection and KNLMS algorithms introduced in Richard *et al.* (2009) also included an efficient dictionary sparsification procedure, called the *coherence criterion*. Coherence is a measure to characterize a dictionary in sparse approximation problems, defined in a kernel context as

$$\mu = \max_{i \neq j} |K(\boldsymbol{x}_i, \boldsymbol{x}_j)|. \tag{9.27}$$

The coherence of a dictionary will be large if it contains two bases $\boldsymbol{x}_i$ and $\boldsymbol{x}_j$ that are very similar, in terms of the kernel function. Owing to their similarity, such bases contribute almost identical information to the algorithm, and one of them may be considered redundant. The online dictionary sparsification procedure based on coherence operates by only including a new datum $\boldsymbol{x}_n$ into the current dictionary $\mathcal{D}_{n-1}$ if it maintains the dictionary coherence below a certain threshold:

$$\max_{j \in \mathcal{D}_{n-1}} |K(\boldsymbol{x}_n, \boldsymbol{x}_j)| < \mu_0. \tag{9.28}$$

If the new datum fulfills this criterion, it is included in the dictionary, and the KNLMS coefficients are updated through Equation 9.26. If the coherence criterion in Equation 9.28 is not fulfilled, the new datum is not included in the dictionary, and a *reduced* update of the KNLMS coefficients is performed:

$$\boldsymbol{\alpha}_n = \boldsymbol{\alpha}_{n-1} + \frac{\eta}{\epsilon + \|\boldsymbol{k}_n\|} e_n \boldsymbol{k}_n. \tag{9.29}$$

This update does not increase the number of coefficients, and therefore it maintains the algorithm's computational complexity fixed during that iteration. The MATLAB code for the complete KNLMS training step on a new data pair $(\boldsymbol{x}, d)$ is displayed in Listing 9.4.

```
k = kernel(dict,x,kerneltype,kernelpar); % kernels between dictionary ...
    and x
if (max(k) <= mu0), % coherence criterion
    dict = [dict; x]; % add base to dictionary
    alpha = [alpha; 0]; % reserve spot for new coefficient
end
k = kernel(dict,x,kerneltype,kernelpar); % kernels with new dictionary
y = k' * alpha; % evaluate function output
err = d - y; % instantaneous error
alpha = alpha + eta/(eps + k'*k)*err*k; % update coefficients
```

Listing 9.4 Training step of the KNLMS algorithm on a new datum ($\boldsymbol{x}$, d).

The coherence criterion is computationally efficient, in that it has a complexity that does not exceed that of the kernel adaptive filter itself, and it has been demonstrated to be successful in practical situations (Van Vaerenbergh and Santamaría, 2013).

Quantized Kernel Least Mean Squares

Recently, a KLMS algorithm was proposed that combines elements from the original KLMS algorithm and the coherence criterion, called quantized KLMS (QKLMS) (Chen *et al.*, 2012). In particular, when the sparsification criterion decides to include a datum into the dictionary, the algorithm updates its coefficients as follows:

$$\boldsymbol{\alpha}_n = \begin{bmatrix} \boldsymbol{\alpha}_{n-1} \\ \eta e_n \end{bmatrix}. \tag{9.30}$$

When the datum does not fulfill the coherence criterion, it is not included in the dictionary. Instead, the closest dictionary element is retrieved, and the corresponding coefficient is updated as follows

$$\alpha_{n,j} = \alpha_{n-1,j} + \eta e_n, \tag{9.31}$$

where j is the dictionary index of the element that is closest. Though conceptually very simple, this algorithm obtains state-of-the-art results in several applications when only a low computational budget is available. The MATLAB code for the complete QKLMS training step on a new data pair $(\boldsymbol{x}, d)$ is displayed in Listing 9.5.

```
k = kernel(dict,x,kerneltype,kernelpar); % kernels between dictionary ...
    and x
y = k' * alpha; % evaluate function output
err = d - y; % instantaneous error
[d2,j] = min(sum((dict - repmat(x,m,1)).^2,2)); % distance to dictionary
if d2 <= epsu^2,
    alpha(j) = alpha(j) + eta*err; % reduced coefficient update
else
    dict = [dict; x]; % add base to dictionary
    alpha = [alpha; eta*err]; % add new coefficient
end
```

Listing 9.5 Training step of the QKLMS algorithm on a new datum $(\boldsymbol{x}, d)$.

9.5 Kernel Recursive Least Squares

In linear adaptive filtering, the RLS algorithm represents an alternative to LMS, with faster convergence and typically lower bias, at the expense of a higher computational complexity. RLS is obtained by designing a recursive solution to the LS problem. Analogously, a recursive solution can be designed for the KRR problem, yielding KRLS algorithms.

9.5.1 Kernel Ridge Regression

Let us review the KRR algorithm seen in Chapter 8. This will be the key to obtain the kernel-based version of the regularized LS cost function in Equation 9.7. We essentially first transform the data into the kernel feature space:

$$\begin{aligned} J_n &= \sum_{i=1}^{n} |d_i - \boldsymbol{\phi}(\boldsymbol{x}_i)\mathrm{T}\boldsymbol{w}_{\mathcal{H}}|^2 + \lambda \boldsymbol{w}_{\mathcal{H}}\mathrm{T}\boldsymbol{w}_{\mathcal{H}}, \\ &= \|\boldsymbol{d} - \boldsymbol{K}\boldsymbol{\alpha}\|^2 + \lambda \boldsymbol{\alpha}\mathrm{T}\boldsymbol{K}\boldsymbol{\alpha}, \end{aligned} \tag{9.32}$$

where we have applied the kernel trick to obtain the second equality. Here, vector $\boldsymbol{d}$ contains the n desired values, $\boldsymbol{d} = [d_1, d_2, \ldots, d_n]\mathrm{T}$, and $\boldsymbol{K}$ is the kernel matrix with elements $K_{ij} = K(\boldsymbol{x}_i, \boldsymbol{x}_j)$. Equation 9.32 represents the *KRR* problem (Saunders *et al.*, 1998), and its solution is given by

$$\boldsymbol{\alpha} = (\boldsymbol{K} + \lambda \boldsymbol{I})^{-1}\boldsymbol{d}. \tag{9.33}$$

The prediction for a new datum $\boldsymbol{x}_*$ is obtained as

$$y_* = \boldsymbol{k}_*\mathrm{T}\boldsymbol{\alpha} = \boldsymbol{k}_*\mathrm{T}(\boldsymbol{K} + \lambda \boldsymbol{I})^{-1}\boldsymbol{d}. \tag{9.34}$$

9.5.2 Derivation of Kernel Recursive Least Squares

The KRLS algorithm (Engel *et al.*, 2004) formulates a recursive procedure to obtain the solution of the regression problem in Equation 9.32 in the absence of regularization, $\lambda = 0$. Without regularization, the solution (Equation 9.33) reads

$$\boldsymbol{\alpha} = \boldsymbol{K}^{-1}\boldsymbol{d}. \tag{9.35}$$

KRLS guarantees the invertibility of the kernel matrix $\boldsymbol{K}$ by excluding those data $\boldsymbol{x}_i$ from the dictionary that are linearly dependent on the already included data, in the feature space. As we will see, this is achieved by applying a specific online sparsification procedure, which guarantees both that $\boldsymbol{K}$ is invertible and that the algorithm's dictionary stays compact.

Assume the solution after processing $n-1$ data is available, given by

$$\boldsymbol{\alpha}_{n-1} = \boldsymbol{K}_{n-1}^{-1}\boldsymbol{d}_{n-1}, \tag{9.36}$$

In the next iteration n, a new data pair $(\boldsymbol{x}_n, d_n)$ is received and we wish to obtain the new solution $\boldsymbol{\alpha}_n$ by applying a low-complexity update on the previous solution (Equation 9.36). We first calculate the predicted output

$$y_n = \boldsymbol{k}_n\mathrm{T}\boldsymbol{\alpha}_{n-1}, \tag{9.37}$$

and we obtain the a-priori error for this datum, $e_n = d_n - y_n$. The updated kernel matrix can be written as

$$\boldsymbol{K}_n = \begin{bmatrix} \boldsymbol{K}_{n-1} & \boldsymbol{k}_n \\ \boldsymbol{k}_n\mathrm{T} & k_{nn} \end{bmatrix}. \tag{9.38}$$

By introducing the variables

$$\boldsymbol{a}_n = \boldsymbol{K}_{n-1}^{-1}\boldsymbol{k}_n \tag{9.39}$$

and

$$\gamma_n = k_{nn} - \boldsymbol{k}_n\mathrm{T}\boldsymbol{a}_n, \tag{9.40}$$

the update for the inverse kernel matrix can be written as

$$\boldsymbol{K}_n^{-1} = \frac{1}{\gamma_n} \begin{bmatrix} \gamma_n \boldsymbol{K}_{n-1}^{-1} + \boldsymbol{a}_n \boldsymbol{a}_n \mathrm{T} & -\boldsymbol{a}_n \\ -\boldsymbol{a}_n & 1 \end{bmatrix}. \tag{9.41}$$

Equation 9.41 is obtained by applying the Sherman–Morrison–Woodbury formula for matrix inversion; see, for instance, Golub and Van Loan (1996). Finally, the updated solution $\boldsymbol{\alpha}_n$ is obtained as

$$\boldsymbol{\alpha}_n = \begin{bmatrix} \boldsymbol{\alpha}_{n-1} \\ 0 \end{bmatrix} - e_n/\gamma_n \begin{bmatrix} \boldsymbol{a}_n \\ -1 \end{bmatrix}. \tag{9.42}$$

Equations 9.41 and 9.42 are efficient updates that allow one to obtain the new solution in $\mathcal{O}(n^2)$ time and memory, based on the previous solution. Directly applying Equation 9.35 at iteration n would require $\mathcal{O}(n^3)$ cost, so the recursive procedure is preferred in online scenarios. A detailed derivation of this result can be found in Engel *et al.* (2004) and Van Vaerenbergh *et al.* (2012b).

Online Sparsification by Approximate Linear Dependency

The KRLS algorithm from Engel *et al.* (2004) follows the described recursive solution. In order to slow down the dictionary growth, shown in Equations 9.41 and 9.42, it introduces a sparsification criterion based on *approximate linear dependency* (ALD). According to this criterion, a new datum $\boldsymbol{x}_n$ should only be included in the dictionary if $\boldsymbol{\phi}(\boldsymbol{x}_n)$ cannot be approximated sufficiently well in feature space by a *linear combination* of the already present data.

Given a dictionary $\mathcal{D}$ of data $\boldsymbol{x}_j$ and a new training point $\boldsymbol{x}_n$, we need to find a set of coefficients $\boldsymbol{a} = [a_1, a_2, \dots, a_m]^{\mathrm{T}}$ that satisfy the ALD condition

$$\min_{\boldsymbol{a}} \left\| \sum_{j=1}^{m} a_j \boldsymbol{\phi}(\boldsymbol{x}_j) - \boldsymbol{\phi}(\boldsymbol{x}_n) \right\|^2 \leq \nu, \tag{9.43}$$

where m is the cardinality of the dictionary. Interestingly, it can be shown that these coefficients are already calculated by the KRLS update itself, and they are available at each iteration n as $\boldsymbol{a}_n = \boldsymbol{K}_{n-1}^{-1} \boldsymbol{k}_n$, see Equation 9.39. The ALD condition can therefore be verified by simply comparing γ_n with the ALD threshold:

$$\gamma_n = k_{nn} - \boldsymbol{k}_n \mathrm{T} \boldsymbol{a}_n \leq \nu. \tag{9.44}$$

If $\gamma_n > \nu$, then we must add the newest datum $\boldsymbol{x}_n$ to the dictionary, $\mathcal{D}_n = \mathcal{D}_{n-1} \cup \{\boldsymbol{x}_n\}$, before updating the solution through Equation 9.42. If $\gamma_n \leq \nu$ then the datum $\boldsymbol{x}_n$ is already represented sufficiently well by the dictionary. In this case the dictionary is not expanded, $\mathcal{D}_n = \mathcal{D}_{n-1}$, and a *reduced* update of the solution is performed; see Engel *et al.* (2004) for details. The MATLAB code for the complete KRLS training step on a new data pair $(\boldsymbol{x}, d)$ is displayed in Listing 9.6.

```
k = kernel(dict,x,kerneltype,kernelpar); % kernels between dictionary ...
    and x
kxx = kernel(x,x,kaf.kerneltype,kaf.kernelpar); % kernel on x
a = Kinv * k; % coefficients of closest linear combination in feature ...
    space
gamma = kxx - k' * a; % residual of linear approximation in feature space
y = k' * alpha; % evaluate function output
err = d - y; % instantaneous error
if gamma>nu % new datum is not approximately linear dependent
    dict = [dict; x]; % add base to dictionary
    Kinv = 1/gamma*[gamma*Kinv+a*a',-a;-a',1]; % update inv. kernel matrix
    Z = zeros(size(P,1),1);
    P = [P Z; Z' 1]; % add linear combination coeff. to projection matrix
    ode = 1/gamma*err;
    alpha = [alpha - a*ode; ode]; % full update of coefficients
else % perform reduced update of alpha
    q = P * a / (1 + a'*P*a);
    P = P - q * (a' * P); % update projection matrix
    alpha = alpha + Kinv * q * err; % reduced update of coefficients
end
```

Listing 9.6 Training step of the KRLS algorithm on a new datum ($\mathbf{x}$, d).

9.5.3 Prediction of the Mackey–Glass Time Series with Kernel Recursive Least Squares

The update equations for KRLS require substantially more computation than KLMS. In particular, KRLS has quadratic complexity, $\mathcal{O}(m^2)$, in terms of its dictionary size, and KLMS has linear complexity, $\mathcal{O}(m)$. On the other hand, KRLS has faster convergence and lower bias. We illustrate these properties by applying KRLS on the Mackey–Glass prediction experiment from Section 9.4.3.

Figure 9.4 shows the results of training the KRLS algorithm on the Mackey–Glass time series. KRLS is applied with a Gaussian kernel with $\sigma = 1$, and its precision parameter was fixed to $\nu = 10^{-4}$. The left plot compares the learning curves of KLMS and KRLS, demonstrating a slightly faster initial convergence rate for KRLS, after which the algorithm converges to a much lower MSE than KLMS. The low bias is also visible in the right plot, which shows the prediction results on the test data.

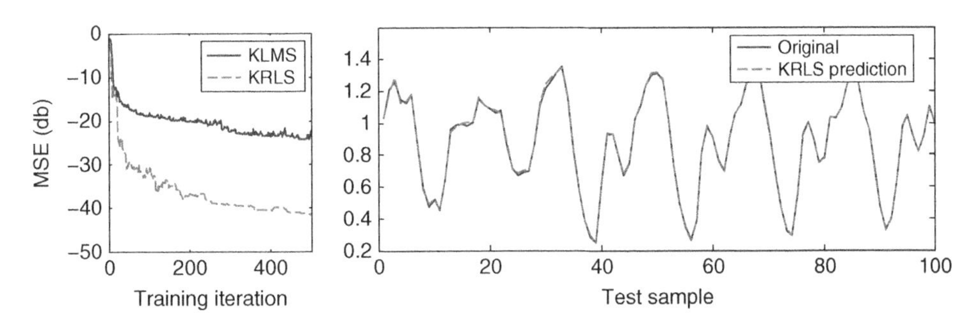

Figure 9.4 KRLS predictions on the Mackey–Glass time series. Left: learning curve over 500 training iterations, including comparison with KLMS. Right: test samples and KRLS predictions.

9.5.4 Beyond the Stationary Model

An important limitation of the KRLS algorithm is that it always assumes a stationary model, and therefore it cannot track changes in the true underlying data model. This is a somewhat odd property for an *adaptive* filter, though note that this is also the case for the original RLS algorithm; see Section 9.2.2. In order to enable tracking and make a truly adaptive KRLS algorithm, several modifications have been presented in the literature. An exponentially weighted KRLS algorithm was proposed by including a forgetting factor, and an extended KRLS algorithm was designed by assuming a simple state-space model, though both algorithms show numerical instabilities in practice (Liu *et al.*, 2010). In the following we briefly discuss two different approaches that successfully allow KRLS to adapt to changing environments.

Sliding-Window Kernel Recursive Least Squares

The KRLS algorithm summarizes past information into a compact formulation that does not allow easy manipulation. For instance, there does not exist a straightforward manner to include a forgetting factor to exclude the influence of older data.

Van Vaerenbergh *et al.* (2006) proposed sliding-window KRLS (SW-KRLS). This algorithm stores a window of the last m data as its dictionary, and once a datum is older than m time steps it is simply discarded. In each step the algorithm adds the new datum and discards the oldest datum, leading to a sliding-window approach. The algorithm stores the inverse regularized kernel matrix, $(\boldsymbol{K}_n + \lambda \boldsymbol{I})^{-1}$, calculated on its current dictionary, and a vector of the corresponding desired outputs, $\boldsymbol{d}_n$. By storing these variables it can calculate the solution vector by simply evaluating $\boldsymbol{\alpha}_n = (\boldsymbol{K}_n + \lambda \boldsymbol{I})^{-1}\boldsymbol{d}_n$; see Equations 9.33 and 9.34.

The inverse kernel matrix is updated in two steps. First, the new datum is added, which requires expanding the matrix with one row and one column. This is carried out by performing the operation from Equation 9.41, similar to in the KRLS algorithm. Second, the oldest datum is discarded, which requires removing one row and column from the inverse kernel matrix. This can be achieved by writing the kernel matrix and its inverse as follows:

$$\boldsymbol{K}_{n-1} = \begin{bmatrix} a & \boldsymbol{b}^{\mathrm{T}} \\ \boldsymbol{b} & \boldsymbol{D} \end{bmatrix}, \quad \boldsymbol{K}_{n-1}^{-1} = \begin{bmatrix} e & \boldsymbol{f}^{\mathrm{T}} \\ \boldsymbol{f} & \boldsymbol{G} \end{bmatrix}, \tag{9.45}$$

after which the inverse (regularized) kernel matrix is found as

$$\boldsymbol{D}^{-1} = \boldsymbol{G} - \boldsymbol{f}\boldsymbol{f}^{\mathrm{T}}/e. \tag{9.46}$$

Details can be found in Van Vaerenbergh *et al.* (2006). Figure 9.5 illustrates the kernel matrix updates when using a sliding window, compared with the classical growing-window approach of KRLS. The MATLAB code for the SW-KRLS training step on a new data pair $(\boldsymbol{x}, d)$ is displayed in Listing 9.7.

```
dict = [dict; x]; % add base to dictionary
dict_d = [dict_d; d];    % add d to output dictionary
k = kernel(dict,x,kerneltype,kernelpar); % kernels between dictionary ...
    and x
Kinv = grow_kernel_matrix(Kinv,k,c); % calculate new inverse kernel matrix
if (size(dict,1) > M) % prune
```

```
    dict(1,:) = []; % remove oldest base from dictionary
    dict_d(1) = []; % remove oldest d from output dictionary
    Kinv = prune_kernel_matrix(Kinv); % prune inverse kernel matrix
end
alpha = Kinv * dict_d; % obtain new filter coefficients
```

Listing 9.7 Training step of the SW-KRLS algorithm on a new datum ($\boldsymbol{x}, d$). The functions "grow_kernel_matrix" and "prune_kernel_matrix" implement the operations in Equations 9.41 and 9.46.

SW-KRLS is a conceptually very simple algorithm that obtains reasonable performance in a wide range of scenarios, most notably in nonstationary environments. Nevertheless, its performance is limited by the quality of the bases in its dictionary, over which it has little control. In particular, it has no means to avoid redundancy in its dictionary or to maintain older bases that are relevant to its kernel expansion. In order to improve this performance, a fixed-budget KRLS (FB-KRLS) algorithm was proposed in Van Vaerenbergh *et al.* (2010). Instead of discarding the oldest data point in each iteration, FB-KRLS discards the data point that causes the least error upon being discarded, using a least a-posteriori-error-based pruning criterion we will discuss in Section 9.7. In stationary scenarios, FB-KRLS obtains significantly better results.

Kernel Recursive Least Squares Tracker

The tracking limitations of previous KRLS algorithms were overcome by development of the KRLS tracker (KRLS-T) algorithm (Van Vaerenbergh *et al.*, 2012b), which has its roots in the probabilistic theory of GP regression. Similar to the FB-KRLS algorithm, this algorithm uses a fixed memory size and has a criterion on which data to discard in each iteration. But unlike FB-KRLS, the KRLS-T algorithm incorporates a forgetting mechanism to gradually downweight older data.

We will provide more details on this algorithm in the discussion on probabilistic kernel adaptive filtering of Section 9.8. As a reference, we list the MATLAB code for the KRLS-T training step on a new data pair ($\boldsymbol{x}, d$) in Listing 9.8. While this algorithm has similar complexity to other KRLS-type algorithms, its implementation is more complex owing its fully probabilistic treatment of the regression problem. Note that some additional checks to avoid numerical problems have been left out. The complete code

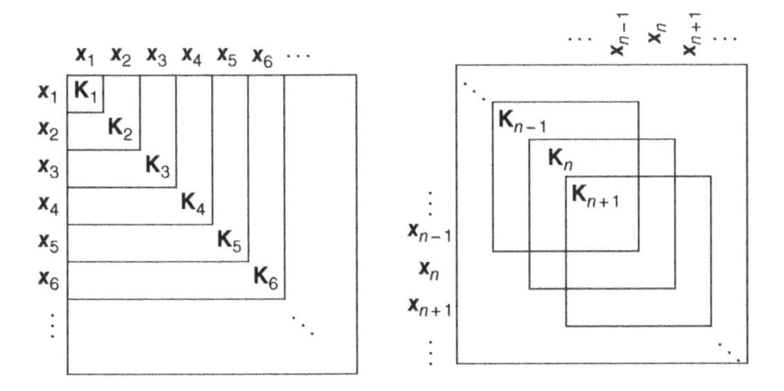

Figure 9.5 Different forms of updating the kernel matrix during online learning. In KRLS-type algorithms the update involves calculating the inverse of each kernel matrix, given the inverse of the previous matrix. Left: growing kernel matrix, as constructed in KRLS (omitting sparsification for simplicity). Right: sliding-window kernel matrix of a fixed size, as constructed in SW-KRLS.

can be found in the kernel adaptive filtering MATLAB toolbox (KAFBOX), discussed in Section 9.10, and available on the book web page.

```
% perform one forgetting step
Sigma = lambda * Sigma + (1-lambda) * K; % forgetting on covariance matrix
mu = sqrt(lambda) * mu; % forgetting on mean vector
% kernels
k = kernel(dict,x,kerneltype,kernelpar); % kernels between dictionary ...
    and x
kxx = kernel(x,x,kaf.kerneltype,kaf.kernelpar); % kernel on x

q = Q * k;
y_mean = q' * mu; % predictive mean of new datum
gamma2 = kxx - k' * q; % projection uncertainty
h = Sigma * q;
sf2 = gamma2 + q' * h; % noiseless prediction variance
sy2 = sn2 + sf2; % unscaled predictive variance of new datum
y_var = s02 * sy2; % predictive variance of new datum

% include new sample and add a basis
Qold = Q; % old inverse kernel matrix
p = [q; -1];
Q = [Q zeros(m,1);zeros(1,m) 0] + 1/gamma2*(p*p');
% updated inverse matrix

err = d - y_mean; % instantaneous error
p = [h; sf2];
mu = [mu; y_mean] + err / sy2 * p; % posterior mean
Sigma = [Sigma h; h' sf2] - 1 / sy2 * (p*p'); % posterior covariance
dict = [dict; x]; % add base to dictionary

% estimate scaling power s02 via ML
nums02ML = nums02ML + lambda * (y - y_mean)^2 / sy2;
dens02ML = dens02ML + lambda; s02 = nums02ML / dens02ML;

% delete a basis if necessary
m = size(dict,1);
if m > M
    % MSE pruning criterion
    errors = (Q*mu) ./ diag(Q);
    criterion = abs(errors);
    [~, r] = min(criterion); % remove element which incurs in the min. ...
        err.
    smaller = 1:m; smaller(r) = [];
    if r == m, % remove the element we just added (perform reduced update)
        Q = Qold;
    else
        Qs = Q(smaller, r);
        qs = Q(r,r); Q = Q(smaller, smaller);
        Q = Q - (Qs*Qs') / qs; % prune inverse kernel matrix
    end
    mu = mu(smaller); % prune posterior mean
    Sigma = Sigma(smaller, smaller); % prune posterior covariance
    dict = dict(smaller,:); % prune dictionary
end
```

Listing 9.8 Training step of the KRLS-T algorithm on a new datum ($\mathbf{x}$, d).

9.5.5 Example on Nonlinear Channel Identification and Reconvergence

In order to demonstrate the tracking capabilities of some of the reviewed kernel adaptive filters we perform an experiment similar to the setup described in Van Vaerenbergh *et al.* (2006) and Lázaro-Gredilla *et al.* (2011). Specifically, we consider the problem of online identification of a communication channel in which an abrupt change (switch) is triggered at some point.

A signal $x_n \in \mathcal{N}(0,1)$ is fed into a nonlinear channel that consists of a linear FIR channel followed by the nonlinearity $y = \tanh(z)$, where z is the output of the linear channel. During the first 500 iterations the impulse response of the linear channel is chosen as $\mathcal{H}_1 = [1, -0.3817, -0.1411, 0.5789, 0.191]$, and at iteration 501 it is switched to $\mathcal{H}_2 = [1, -0.0870, 0.9852, -0.2826, -0.1711]$. Finally, 20 dB of Gaussian white noise is added to the channel output.

We perform an online identification experiment with the algorithms LMS, QKLMS, SW-KRLS, and KRLS-T. Each algorithm performs online learning of the nonlinear channel, processing one input datum (with a time-embedding of five taps) and one output sample per iteration. At each step, the MSE performance is measured on a set of 100 data points that are generated with the current channel model. The results are averaged out over 10 simulations.

The kernel adaptive filters use a Gaussian kernel with $\sigma = 1$. LMS and QKLMS use a learning rate $\eta = 0.5$. The sparsification threshold of QKLMS is set to $\epsilon_{\mathbb{U}} = 0.3$, which leads to a final dictionary of size around $m = 300$ at the end of the experiment. The regularization of SW-KRLS and KRLS-T is set to match the true value of the noise-to-signal ratio, 0.01. Regarding memory, SW-KRLS and KRLS-T are given a maximum dictionary size of $m = 50$. Finally, KRLS-T uses a forgetting factor of $\lambda = 0.998$.

The results are shown in Figure 9.6. LMS performs worst, as it is not capable of modeling the nonlinearities in the system. QKLMS shows good results, given its low complexity, but a slow convergence. SW-KRLS and KRLS-T converge to a value that is mostly limited by its dictionary size ($m = 50$), and both show fast convergence rates.

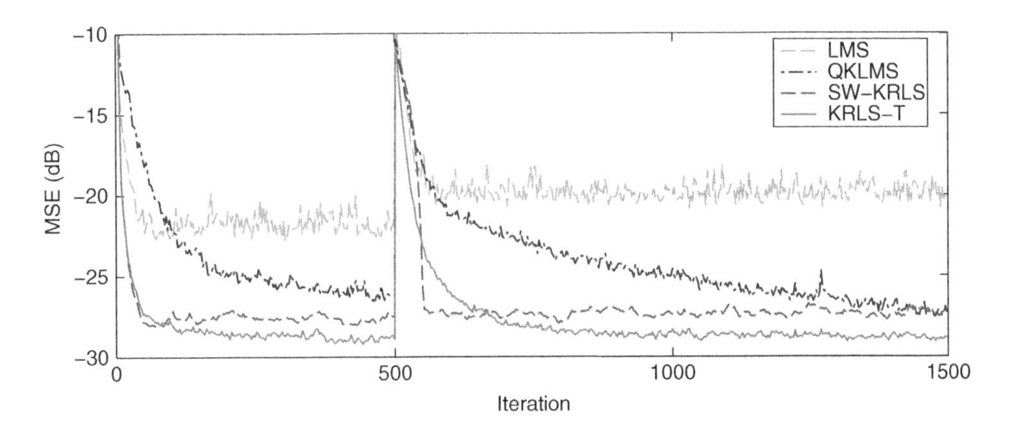

Figure 9.6 MSE learning curves of different kernel adaptive filters on a communications channel that shows an abrupt change at iteration 500.

All algorithms are capable of reconverging after the switch, though their convergence rate is typically slower at that point.

9.6 Explicit Recursivity for Adaptive Kernel Models

Recursivity plays a key role for many problems in adaptive estimation, including control, signal processing, system identification, and general model building and monitoring problems (Ljung, 1999). Much effort has been devoted to the efficient implementation of these kinds of algorithms in DSP problems, including linear AR models (which need few parameters and allow closed-form analysis) and ladder or lattice implementation of linear filters. For nonlinear dynamics present in the system generating the data, the model specification and parameter estimation problems become far more complex to handle, and methods such as NNs were proposed and intensely scrutinized in this scenario in the 1990s (Dorffner 1996; Narendra and Parthasarathy 1990).

Recursion and adaptivity concepts are strongly related to each other. As seen in this chapter so far, kernel-based adaptive filtering, in which model weights are updated online, has been visited starting from the sequential SVM in De Freitas *et al.* (1999), which used the Kalman recursions for updating the model weights. Other recent proposals are kernel KFs (Ralaivola and d'Alche Buc, 2005), KRLS (Engel *et al.*, 2004), KLMS, and specific kernels for dynamic modeling (Liu *et al.*, 2008). Mouattamid and Schaback (2009) focused on subspaces of native spaces induced by subsets, which ultimately led to a recursive space structure and its associated recursive kernels for interpolation.

Tuia *et al.* (2014) presented and benchmarked an elegant formalization of recursion in kernel spaces in online and adaptive settings. In a few words, the fundamentals of this formalization rise from defining the signal model recursion *explicitly in* the RKHS rather than mapping the samples and then just embedding them into a kernel. The methodology was defined in three steps, starting from defining the (recursive) model structure directly in a suitable RKHS, then exploiting a proper representer's theorem for the model weights, and finally applying standard recursive equations from classical DSP now in the RKHS. We next describe the recursive formalization therein, in order to provide the necessary equations and background for supporting with detail some application examples.

9.6.1 Recursivity in Hilbert Spaces

We start by considering a set of observed N time sample pairs $\{(x_n, y_n)\}$, and defining an arbitrary Markovian model representation of recursion between input–output time-series pairs:

$$\begin{aligned} y_n &= f(\langle w_i, x_n^i \rangle | \theta_f) + e_x \\ x_n^i &= g(\langle x_{n-k}^i, y_{n-l} \rangle | \theta_g) + e_y, \quad \forall k > 1, \ l \geq 0, \end{aligned} \tag{9.47}$$

where x_n^i is the signal at the input of the ith filter tap at time n, y_{n-l} is the previous output at time $n - l$, e_x is the state noise, and e_y is the measurement noise. Here, f and g are linear functions parametrized by model parameters w_i and hyperparameters θ_f and θ_g.

Now we can define a feature mapping $\boldsymbol{\phi}(\cdot)$ to an RKHS $\mathcal{H}$ with the reproducing property. One could replace samples by their mapped counterparts in Equation 9.47 with

$$y_n = f(\langle w_i, \boldsymbol{\phi}(g(\langle \boldsymbol{\phi}(x^i_{n-k}), y_{n-l}\rangle | \theta_g))\rangle | \theta_f) \tag{9.48}$$

and then try to express this in terms of dot products to replace them by kernel functions. However, this operation may turn into an extremely complex one, and even be unsolvable, depending on the parameterization of the recursive function g at hand. While standard linear filters can be easily "kernelized," some other filters introducing nonlinearities in g are difficult to handle with the standard kernel framework. Existing smart solutions still require pre-imaging or dictionary approximations, such as the kernel Kalman filtering and the KRLS (Engel *et al.*, 2004; Ralaivola and d'Alche Buc, 2004).

The alternative in the recursive formalization instead proposes to write down the recursive equations directly in the RKHS model. In this case, the linear model is defined *explicitly* in $\mathcal{H}$ as

$$\begin{aligned} y_n &= f(\langle \mathbf{w}_i, \boldsymbol{\phi}^i_n\rangle | \theta_{\mathcal{F}}), + n_x \\ \boldsymbol{\phi}^i_n &= g(\langle \boldsymbol{\phi}^i_{n-k}, y_{n-l}\rangle | \theta_{\mathcal{G}}) + n_y, \quad \forall k > 1, l \geq 0. \end{aligned} \tag{9.49}$$

Now, samples $\boldsymbol{\phi}^i_n$ do not necessarily have $\boldsymbol{\phi}(x^i_n)$ as a pre-image, and model parameters $\mathbf{w}_i$ are defined in the possibly infinite-dimensional feature space $\mathcal{H}$, while the hyperparameters have the same meaning (as they define recursion according to an explicit model structure regardless of the space where they are defined).

This problem statement may be solved by first defining a convenient and tractable reproducing property. For example, we define

$$\mathbf{w}_i = \sum_{m=1}^{N} \beta^i_m \boldsymbol{\phi}^i_m, \tag{9.50}$$

which provides us with a fully recursive solution, even when defined by assumption, as seen later. In order to formally define a reproducing property, we still need to demonstrate the associated representer theorem. We will alternatively show in the next pages the existence and uniqueness of the mapping for a particular instantiation of a digital filter in RKHS. In other words, the infinite-dimensional feature vectors are spanned by linear combinations of signals filtered in the feature space. Note that this is different from the traditional filter-and-map approach.

For now, let us focus on the reproducing property defined in Equation 9.50. Note that this is different from the traditional filter-and-map approach. Plugging Equation 9.50 into Equation 9.49 and applying the kernel trick yields

$$\hat{y}_n = f\left(\left\langle \sum_{m=1}^{N} \beta^i_m \boldsymbol{\phi}^i_m, \boldsymbol{\phi}^i_n \right\rangle | \theta_{\mathcal{F}}\right) = f\left(\sum_{m=1}^{N} \beta^i_m K^i(m,n) \,|\theta_{\mathcal{F}}\right). \tag{9.51}$$

As a direct consequence of using this reproducing property, the solution is solely expressed in terms of current and previous kernel values. When a number N of training samples is available, one can explicitly compute the kernel and eventually adapt their parameters. In the more interesting online case, the previous expression will also be useful because, as we will see next, the kernel function $K^i(m,n)$ for a particular filter instantiation can be expressed as a function only of previous kernel evaluations.

A model length can be assumed for the input samples and then mapping can be applied to vectors made up of delayed windows of the signals. Note that $K^i(m,n)$ is not necessarily equivalent to $K(x_m^i, x_n^i)$, and hence the problem is far different from those defined in previous kernel adaptive filters. To compute this kernel, one can actually resort to applying recursion formulas in Equation 9.49 and explicitly define $K^i(m,n)$ as a function of previously computed kernels. This methodology is next illustrated for the particular recursive model structure of the γ-filter. In this way, not only do we attain a recursive model in feature spaces, but we also avoid applying approximate methods and pre-images.

9.6.2 Recursive Filters in Reproducing Kernel Hilbert Spaces

Here, we illustrate the proposed methodology for building recursive filters in the RKHS. First we review the recursive AR and MA filters in kernel feature spaces, and underline their application shortcomings. Then, noting that inclusion of short-term memory in learning machines is essential for processing time-varying information, we analyze the particular case of the γ-filter (Principe *et al.*, 1993), which provides a remarkable compromise between stability and simplicity of adaptation.

Recursive Autoregressive and Moving-Average Filters in Reproducing Kernel Hilbert Spaces

The MA filter structure for time-series prediction is defined as

$$y_n = \sum_{i=0}^{P} b_i x_n^i + e_n, \tag{9.52}$$

where $x_n^i = x_{n-i}$. Accordingly, the corresponding structure can be described explicitly in RKHS:

$$y_n = \sum_{i=0}^{P} \boldsymbol{b}_i^{\mathrm{T}} \boldsymbol{\phi}_n^i + e_n, \tag{9.53}$$

and given that $\boldsymbol{\phi}_n^i = \boldsymbol{\varphi}(x_{n-i})$ and that we can approximately expand each vector $\boldsymbol{b}_i = \sum_{m=1}^{N} \beta_m^i \boldsymbol{\phi}_m^i$, we have

$$y_n = \sum_{i=0}^{P} \sum_{m=1}^{N} \beta_m^i \langle \boldsymbol{\varphi}(x_{n-i}), \boldsymbol{\varphi}(x_{m-i}) \rangle + e_n = \sum_{i=0}^{P} \sum_{m=1}^{N} \beta_m^i K(x_{n-i}, x_{m-i}) + e_n. \tag{9.54}$$

The AR filter structure is defined according to the following expression:

$$y_n = \sum_{i=1}^{P} a_i y_n^i + e_n, \tag{9.55}$$

where $y_n^i := y_{n-i}$. Accordingly, the corresponding nonlinear structure can be described by

$$y_n = \sum_{i=1}^{P} \boldsymbol{a}_i^{\mathrm{T}} \boldsymbol{\phi}_n^i + e_n. \tag{9.56}$$

Given that $\boldsymbol{\phi}_n^i := \boldsymbol{\varphi}(y_{n-i})$ and that we can approximately expand each vector $\boldsymbol{a}_i = \sum_{m=1}^{N} \beta_m^i \boldsymbol{\phi}_m^i$, we have

$$y_n = \sum_{i=1}^{P} \sum_{m=1}^{N} \beta_m^i \boldsymbol{\varphi}(y_{n-i})^{\mathrm{T}} \boldsymbol{\varphi}(y_{m-i}) + e_n = \sum_{i=1}^{P} \sum_{m=1}^{N} \beta_m^i K(y_{n-i}, y_{m-i}) + e_n. \tag{9.57}$$

The definition of recursive kernel ARMA filters poses challenging problems, mainly due the specification of proper parameters to ensure stability. This problem is even harder when working *explicitly* in RKHS, because we do not have direct access to the mapped samples. We circumvent these problems both in the linear and kernel versions by considering the γ-filter structure next.

The Recursive γ-Filter in Reproducing Kernel Hilbert Spaces

The standard γ-filter is defined by

$$y_n = \sum_{i=1}^{P} w_i x_n^i \tag{9.58}$$

$$x_n^i = \begin{cases} x_n, & i = 1 \\ (1-\mu)x_{n-1}^i + \mu x_{n-1}^{i-1}, & 2 \leq i \leq P, \end{cases} \tag{9.59}$$

where y_n is the filter output signal, x_n is the filter input signal, x_n^i is the signal present at the input of the ith gamma tap, n is the time index, and $\theta_f = P$, and $\theta_g = \mu$ are free parameters controlling stability and memory depth.

A formulation of the γ-filter is possible in the RKHS as follows. Essentially, the same recursion can be expressed in a Hilbert space:

$$y_n = \sum_{i=1}^{P} \boldsymbol{w}_i^{\mathrm{T}} \boldsymbol{\phi}_n^i \tag{9.60}$$

$$\boldsymbol{\phi}_n^i = \begin{cases} \boldsymbol{\varphi}(x_n), & i = 1 \\ (1-\mu)\boldsymbol{\phi}_{n-1}^i + \mu \boldsymbol{\phi}_{n-1}^{i-1}, & 2 \leq i \leq P, \end{cases} \tag{9.61}$$

where $\boldsymbol{\varphi}(\cdot)$ is a nonlinear transformation in an RKHS, and $\boldsymbol{\phi}_n^i$ is a vector in this RKHS that may not be an image of a vector of the input space; that is, $\boldsymbol{\phi}_n^i \neq \boldsymbol{\varphi}(x_n^i)$. As mentioned, this model assumes linear relations between samples in the RKHS. If this assumption is insufficient to solve the identification problem at hand, the scalar sample x_n can be easily changed by vector $\boldsymbol{z}_n = [x_n, x_{n-1} \cdots x_{n-P}]^{\mathrm{T}}$, where P is the selected time-embedding.

Nevertheless, the weight vectors $\boldsymbol{w}_i$ of Equation 9.60 are linearly spanned by the N training vectors, $\boldsymbol{w}_i = \sum_{m=1}^{N} \beta_m^i \boldsymbol{\phi}_m^i$. By including this expansion in Equation 9.60 and applying the kernel trick, $K^i(m,n) = \langle \boldsymbol{\phi}_m^i, \boldsymbol{\phi}_n^i \rangle$, we obtain

$$y_n = \sum_{i=1}^{P} \sum_{m=1}^{N} \beta_m^i (\boldsymbol{\phi}_m^i)^{\mathrm{T}} \boldsymbol{\phi}_n^i = \sum_{i=1}^{P} \sum_{m=1}^{N} \beta_m^i K^i(m,n). \tag{9.62}$$

The dot products can be computed by using the kernel trick and model recursion defined in Equation 9.61 as

$$K^i(m,n) = \begin{cases} K(x_m, x_n), & i = 1 \\ \langle (1-\mu)\boldsymbol{\phi}_{m-1}^i + \mu\boldsymbol{\phi}_{m-1}^{i-1}, (1-\mu)\boldsymbol{\phi}_{n-1}^i + \mu\boldsymbol{\phi}_{n-1}^{i-1} \rangle & 2 \leq i \leq P. \end{cases} \tag{9.63}$$

Arranging terms in the second case of Equation 9.63, we get

$$K^i(m,n) = \begin{cases} K(x_m, x_n), & i = 1 \\ (1-\mu)^2 K^i(m-1, n-1) & \\ \quad + \mu^2 K^{i-1}(m-1, n-1) & 2 \leq i \leq P \\ \quad + \mu(1-\mu)\langle \boldsymbol{\phi}_{m-1}^i, \boldsymbol{\phi}_{n-1}^{i-1} \rangle & \\ \quad + \mu(1-\mu)\langle \boldsymbol{\phi}_{m-1}^{i-1}, \boldsymbol{\phi}_{n-1}^i \rangle & \end{cases} \tag{9.64}$$

The second part still has two (interestingly nonsymmetric) dot products that are not straightforwardly computable. Nevertheless, applying recursion again in Equation 9.61, this can be rewritten as

$$\langle \boldsymbol{\phi}_{m-1}^i, \boldsymbol{\phi}_{n-1}^{i-1} \rangle = \begin{cases} 0, & i = 1 \\ \langle (1-\mu)\boldsymbol{\phi}_{m-2}^i + \mu\boldsymbol{\phi}_{m-2}^{i-1}, \boldsymbol{\phi}_{n-1}^{i-1} \rangle & 2 \leq i \leq P, \end{cases}$$

which, in turn, can be rearranged as

$$\langle \boldsymbol{\phi}_{m-1}^i, \boldsymbol{\phi}_{n-1}^{i-1} \rangle = \begin{cases} 0, & i = 1 \\ (1-\mu)\langle \boldsymbol{\phi}_{m-2}^i, \boldsymbol{\phi}_{n-1}^{i-1} \rangle + \mu K^{i-1}(m-2, n-1) & 2 \leq i \leq P. \end{cases}$$

Term $K^{i-1}(m-2, n-1)$ and the dot product in the second case can be recursively computed using Equations 9.64 and 9.63. Assuming that $\boldsymbol{\phi}_n^i = 0$ for $n < 0$, we obtain the recursion

$$\langle \boldsymbol{\phi}_{m-1}^i, \boldsymbol{\phi}_{n-1}^{i-1} \rangle = \begin{cases} 0, & i = 1 \\ \mu \sum_{j=2}^{m} (1-\mu)^{j-2} K^{i-1}(m-j, n-1) & i \leq P \end{cases}$$

and finally, the recursive kernel can be rewritten as

$$K^i(m,n) = \begin{cases} K(x_m, x_n), & i = 1 \\ (1-\mu)^2 K^i(m-1, n-1) & \\ \quad + \mu^2 K^{i-1}(m-1, n-1) & \\ \quad + \mu^2 \sum_{j=2}^{m-1} (1-\mu)^{j-1} [K^{i-1}(m-j, n-1) & \\ \quad + K^{i-1}(m-1, n-j)], & 2 \leq i \leq P. \end{cases} \tag{9.65}$$

9.7 Online Sparsification with Kernels

The idea behind sparsification methods is to construct a sparse dictionary of bases that represent the remaining data sufficiently well. As a general rule in learning theory, it is desirable to design a network with as few processing elements as possible. Sparsity reduces the complexity in terms of computation and memory, and it usually gives better generalization ability to unseen data (Platt, 1991; Vapnik, 1995). In the context of kernel methods, sparsification aims to identify the bases in the kernel expansion $y_* = \sum_{i=1}^{n} \alpha_i K(\boldsymbol{x}_i, \boldsymbol{x}_*)$ (see Equation 9.21) that can be discarded without incurring a significant performance loss.

Online sparsification is typically performed by starting with an empty dictionary, $\mathcal{D}_0 = \emptyset$, and, in each iteration, adding the input datum $\boldsymbol{x}_i$ if it fulfills a chosen sparsification criterion. We denote the dictionary at time instant $n-1$ as $\mathcal{D}_{n-1} = \{\boldsymbol{c}_i\}_{i=1}^{m_{n-1}}$, where $\boldsymbol{c}_i$ is the ith stored center, taken from the input data $\boldsymbol{x}$ received up till this instant, and m_{n-1} is the dictionary cardinality at this instant. When a new input–output pair $(\boldsymbol{x}_n, d_n)$ is received, a decision is made whether or not $\boldsymbol{x}_n$ should be added to the dictionary as a center. If the sparsification criterion is fulfilled, $\boldsymbol{x}_n$ is added to the dictionary, $\mathcal{D}_n = \mathcal{D}_{n-1} \cup \{\boldsymbol{x}_n\}$. If the criterion is not fulfilled, the dictionary is maintained, $\mathcal{D}_n = \mathcal{D}_{n-1}$, to preserve its sparsity.

Figure 9.7 illustrates the dictionary construction process for different sparsification approaches. Each horizontal line represents the presence of a center in the dictionary. At any given iteration, the elements in the dictionary are indicated by the horizontal lines that are present at that iteration. In the following we discuss each approach in detail.

9.7.1 Sparsity by Construction

We will first give a general overview of the different online sparsification methods in the literature, some of which we have already introduced in the context of the algorithms for which they were proposed. We distinguish three criteria that achieve sparsity by construction: novelty criterion, ALD criterion, and coherence criterion. If the dictionary is not allowed to grow beyond a specified maximum size, it may be necessary to discard bases at some point. This process is referred to as *pruning*, and we will review the most important pruning criteria later.

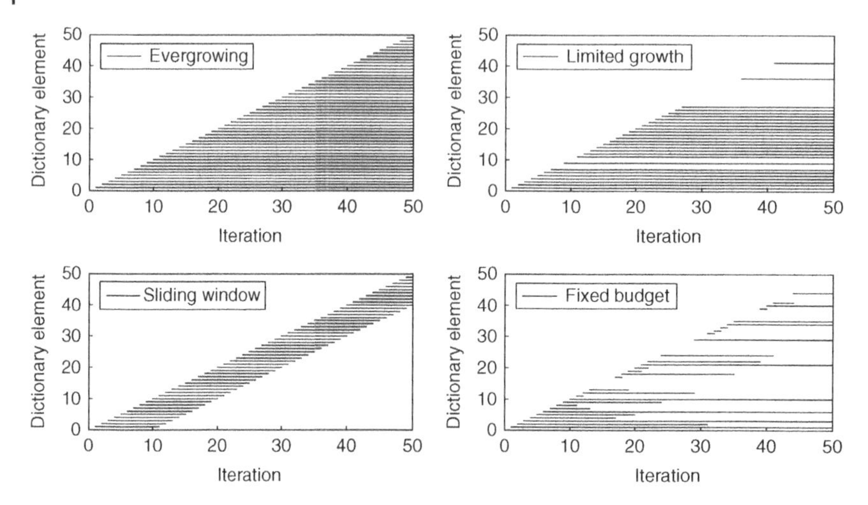

Figure 9.7 Dictionary construction processes for different sparsification approaches. Each horizontal line marks the presence of a center in the dictionary. Top left: the ever-growing dictionary construction, in which the dictionary contains *n* elements in iteration *n*. Top right: online sparsification by slowing down the dictionary growth, as obtained by the coherence and ALD criteria. Bottom left: sliding-window approach, displayed with 10 elements in the dictionary. Bottom right: fixed-budget approach, in which the pruning criterion discards one element per iteration, displayed with dictionary size 10.

Novelty criterion. The novelty criterion is a data selection method introduced by Platt (1991). It was used to construct resource-allocating networks, which are essentially growing RBF networks. When a new data point $\boldsymbol{x}_n$ is obtained by the network, the novelty criterion calculates the distance of this point to the current dictionary, $\min_{j\in\mathcal{D}_{n-1}} \|\boldsymbol{x}_n - \boldsymbol{c}_j\|$. If this distance is smaller than some preset threshold, $\boldsymbol{x}_n$ is added to the dictionary. Otherwise, it computes the prediction error, and only if this error e_n is larger than another preset threshold will the datum $\boldsymbol{x}_n$ be accepted as a new center.

ALD criterion. A more sophisticated dictionary growth criterion was introduced for the KRLS algorithm in Engel *et al.* (2004): each time a new datum $\boldsymbol{x}_n$ is observed, the ALD criterion measures how well the datum can be approximated in the feature space as a linear combination of the dictionary bases in that space. It does so by checking if the ALD condition holds; see Equation 9.43:

$$\min_{\boldsymbol{a}} \left\| \sum_{j=1}^{m} a_j \boldsymbol{\phi}(\boldsymbol{c}_j) - \boldsymbol{\phi}(\boldsymbol{x}_n) \right\|^2 \leq \nu.$$

Evaluating the ALD criterion requires quadratic complexity, $\mathcal{O}(m^2)$, and therefore it is not suitable for algorithms with linear complexity such as KLMS.

Coherence criterion. The coherence criterion is a straightforward criterion to check whether the newly arriving datum is sufficiently informative. It was introduced in the context of the KNLMS algorithm (Richard *et al.*, 2009). Given the dictionary D_{n-1} at iteration $n-1$ and the newly arriving datum $\boldsymbol{x}_n$, the coherence criterion to include the datum reads

$$\max_{j \in \mathcal{D}_{n-1}} |K(\boldsymbol{x}_n, \boldsymbol{c}_j)| < \mu_0. \tag{9.66}$$

In essence, the coherence criterion checks the similarity, as measured by the kernel function, between the new datum and the most similar dictionary center. Only if this similarity is below a certain threshold μ_0 is the datum inserted into the dictionary. The higher the threshold μ_0 is chosen, the more data will be accepted in the dictionary. It is an effective criterion that has linear computational complexity in each iteration: it only requires to calculate m kernel functions, making it suitable for KLMS-type algorithms. Chen *et al.* (2012) introduced a similar criterion, $\min_{j \in \mathcal{D}_{n-1}} \|\boldsymbol{x}_n - \boldsymbol{c}_j\| > \epsilon_u$, which is essentially equivalent to the coherence criterion with a Euclidean distance-based kernel.

9.7.2 Sparsity by Pruning

In practice, it is often necessary to specify a maximum dictionary size m, or *budget*, that may not be exceeded; for instance, due to limitations on hardware or execution time. In order to avoid exceeding this budget, one could simply stop including any data in the dictionary once the budget is reached, hence *locking* the dictionary. Nevertheless, it is very probable that at some point after locking the dictionary a new datum is received that is very informative. In this case, the quality of the algorithm's solution may improve by pruning the least relevant center of the dictionary and replacing it with the new, more informative datum.

The goal of a pruning criterion is to select a datum out of a given set, such that the algorithm's performance is least affected. This makes pruning criteria conceptually different from the previously discussed online sparsification criteria, whose goal is to decide whether or not to include a datum. Pruning techniques have been studied in the context of NN design (Hassibi *et al.* 1993; LeCun *et al.* 1989) and kernel methods (De Kruif and De Vries, 2003; Hoegaerts *et al.*, 2004). We briefly discuss the two most important pruning criteria that appear in kernel adaptive filtering: sliding-window criterion and error criterion.

Sliding window. In time-varying environments, it may be useful to discard the oldest bases, as these were observed when the underlying model was most different from the current model. This strategy is at the core of sliding-window algorithms such as NORMA (Kivinen *et al.*, 2004) and SW-KRLS (Van Vaerenbergh *et al.*, 2006). In every iteration, these algorithms include the new datum in the dictionary and discard the oldest datum, thereby maintaining a dictionary of fixed size.

Error criterion. Instead of simply discarding the oldest datum, the error-based criterion determines the datum that will cause the least increase of the squared-error performance after it is pruned. This is a more sophisticated pruning strategy that was introduced by Csató and Opper (2002) and De Kruif and De Vries (2003) and requires quadratic complexity to evaluate, $\mathcal{O}(m^2)$. Interestingly, if the inverse kernel matrix is available, it is straightforward to evaluate this criterion. Given the ith element on the diagonal of the inverse kernel matrix, $[\boldsymbol{K}^{-1}]_{ii}$, and the ith expansion coefficient α_i, the squared error after pruning the ith center from a dictionary is $\alpha_i/[\boldsymbol{K}^{-1}]_{ii}$. The error-based pruning criterion therefore selects the index for which this quantity is minimized,

$$\arg\min_i \frac{\alpha_i}{[\boldsymbol{K}^{-1}]_{ii}}. \tag{9.67}$$

This criterion is used in the fixed-budget algorithms FB-KRLS (Van Vaerenbergh *et al.*, 2010) and KRLS-T (Van Vaerenbergh *et al.*, 2012b). An analysis performed by Lázaro-Gredilla *et al.* (2011) shows that the results obtained by this criterion are very close to the optimal approach, which is based on minimization of the KL divergence between the original and the approximate posterior distributions.

Currently, the most successful pruning criteria used in the kernel adaptive filtering literature have quadratic complexity, $\mathcal{O}(m^2)$ and therefore they can only be used in KRLS-type algorithms. Optimal pruning in KLMS is a particularly challenging problem, as it is hard to define a pruning criterion that can be evaluated with linear computational complexity. A simple criterion is found in (Rzepka, 2012), where the center with the least weight is pruned, and weight is determined by the associated expansion coefficient,

$$\arg\min_i |\alpha_i|. \tag{9.68}$$

The design of more sophisticated pruning strategies is currently an open topic in KLMS literature. Some recently proposed criteria can be found in Zhao *et al.* (2013, 2016).

9.8 Probabilistic Approaches to Kernel Adaptive Filtering

In many signal processing applications, the problem of signal estimation is addressed. Probabilistic models have proven to be very useful in this context (Arulampalam *et al.* 2002; Rabiner 1989). One of the advantages of probabilistic approaches is that they force the designer to specify all the prior assumptions of the model, and that they make a clear separation between the model and the applied algorithm. Another benefit is that they typically provide a measure of uncertainty about the estimation. Such an uncertainty estimate is not provided by classical kernel adaptive filtering algorithms, which produce a point estimate without any further guarantees.

In this section, we will review how the probabilistic framework of GPs allows us to extend kernel adaptive filters to probabilistic methods. The resulting GP-based algorithms not only produce an estimate of an unknown function, but also an entire probability distribution over functions; see Figure 9.8. Before we describe any probabilistic kernel adaptive filtering algorithms, it is instructive to take a step back to the nonadaptive setting, and consider the KRR problem in Equation 9.32. We will adopt the GP framework to analyze this problem from a probabilistic point of view.

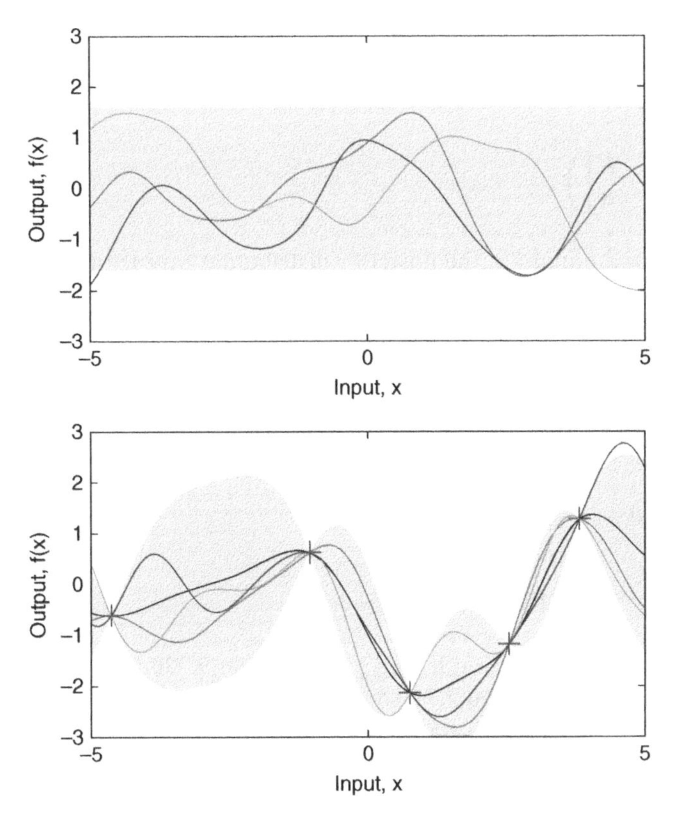

Figure 9.8 Functions drawn from a GP with a squared exponential covariance $K(\mathbf{x}, \mathbf{x}') = \exp(-\|\mathbf{x} - \mathbf{x}'\|^2/2\sigma^2)$. The 95% confidence interval is plotted as the shaded area. Up: draws from the prior function distribution. Down: draws from the posterior function distribution, which is obtained after five data points (blue crosses) are observed. The predictive mean is displayed in black.

9.8.1 Gaussian Processes and Kernel Ridge Regression

Let us assume that the observed data in a regression problem can be described by the model $d_n = f(\boldsymbol{x}_n) + \epsilon_n$, in which f represents an unobservable *latent function* and $\epsilon_n \sim \mathcal{N}(0, \sigma^2)$ is zero-mean Gaussian noise. Note that, unlike in previous chapters, here we use σ^2 for the noise variance and not for the Gaussian kernel length-scale parameter to avoid confusion with the time instant subscript n. We will furthermore assume a zero-mean GP prior on $f(\boldsymbol{x})$

$$f(\boldsymbol{x}) \sim \mathcal{GP}(m(\boldsymbol{x}), K(\boldsymbol{x}, \boldsymbol{x}')) \tag{9.69}$$

and a Gaussian prior on the noise ϵ, $\epsilon \sim \mathcal{N}(0, \sigma_n^2)$. In the GP literature, the kernel function $K(\boldsymbol{x}, \boldsymbol{x}')$ is referred to as the *covariance*, since it specifies the a priori relationship between values $f(\boldsymbol{x})$ and $f(\boldsymbol{x}')$ in terms of their respective locations, and its parameters are called *hyperparameters*.

By definition, the marginal distribution of a GP at a finite set of points is a joint Gaussian distribution, with its mean and covariance being specified by the functions

$m(\boldsymbol{x})$ and $K(\boldsymbol{x},\boldsymbol{x}')$ evaluated at those points (Rasmussen and Williams, 2006). Thus, the joint distribution of outputs $\boldsymbol{d} = [d_1, \dots, d_n]\mathrm{T}$ and the corresponding latent vector $\boldsymbol{f} = [f(\boldsymbol{x}_1), f(\boldsymbol{x}_2), \dots, f(\boldsymbol{x}_n)]\mathrm{T}$ is

$$\begin{bmatrix} \boldsymbol{d} \\ \boldsymbol{f} \end{bmatrix} \sim \mathcal{N}\left(\boldsymbol{0}, \begin{bmatrix} \boldsymbol{K} + \sigma^2 \boldsymbol{I} & \boldsymbol{K} \\ \boldsymbol{K} & \boldsymbol{K} \end{bmatrix}\right). \tag{9.70}$$

By conditioning on the observed outputs $\boldsymbol{y}$, the posterior distribution over the latent vector can be inferred

$$p(\boldsymbol{f}|\boldsymbol{d}) = \mathcal{N}(\boldsymbol{f}|\boldsymbol{K}(\boldsymbol{K}+\sigma^2\boldsymbol{I})^{-1}\boldsymbol{d}, \boldsymbol{K} - \boldsymbol{K}(\boldsymbol{K}+\sigma^2\boldsymbol{I})^{-1}\boldsymbol{K}) = \mathcal{N}(\boldsymbol{f}|\boldsymbol{\mu}, \boldsymbol{\Sigma}). \tag{9.71}$$

Assuming this posterior is obtained for the data up till time instant $n-1$, the predictive distribution of a new output d_n at location $\boldsymbol{x}_n$ is computed as

$$p(d_n|\boldsymbol{x}_n, \boldsymbol{d}_{n-1}) = \mathcal{N}(d_n|\mu_{\mathrm{GP},n}, \sigma^2_{\mathrm{GP},n}) \tag{9.72a}$$

$$\mu_{\mathrm{GP},n} = \boldsymbol{k}_n\mathrm{T}(\boldsymbol{K}_{n-1} + \sigma^2\boldsymbol{I})^{-1}\boldsymbol{d}_{n-1} \tag{9.72b}$$

$$\sigma^2_{\mathrm{GP},n} = \sigma^2 + k_{nn} - \boldsymbol{k}_n\mathrm{T}(\boldsymbol{K}_{n-1} + \sigma^2\boldsymbol{I})^{-1}\boldsymbol{k}_n. \tag{9.72c}$$

The mode of the predictive distribution, given by $\mu_{\mathrm{GP},n}$ in Equation 9.72b, coincides with the prediction of KRR, given by Equation 9.34, showing that the regularization in KRR can be interpreted as a noise power σ^2. Furthermore, the variance of the predictive distribution, given by $\sigma^2_{\mathrm{GP},n}$ in Equation 9.72c, coincides with Equation 9.40, which is used by the ALD dictionary criterion for KRLS.

9.8.2 Online Recursive Solution for Gaussian Processes Regression

A recursive update of the complete GP in Equation 9.72 was proposed by Csató and Opper (2002), as the sparse online GP (SOGP) algorithm. We will follow the notation of Van Vaerenbergh *et al.* (2012b), whose solution is equivalent to SOGP but whose choice of variables allows for an easier interpretation. Specifically, the predictive mean and covariance of the GP solution in Equation 9.72 can be updated as

$$p(\boldsymbol{f}_n|X_n, \boldsymbol{d}_n) = \mathcal{N}(\boldsymbol{f}_n|\boldsymbol{\mu}_n, \boldsymbol{\Sigma}_n) \tag{9.73a}$$

$$\boldsymbol{\mu}_n = \begin{bmatrix} \boldsymbol{\mu}_{n-1} \\ \hat{d}_n \end{bmatrix} + \frac{e_n}{\hat{\sigma}^2_{dn}} \begin{bmatrix} \boldsymbol{h}_n \\ \hat{\sigma}^2_{fn} \end{bmatrix} \tag{9.73b}$$

$$\boldsymbol{\Sigma}_n = \begin{bmatrix} \boldsymbol{\Sigma}_{n-1} & \boldsymbol{h}_n \\ \boldsymbol{h}_n\mathrm{T} & \hat{\sigma}^2_{fn} \end{bmatrix} - \frac{1}{\hat{\sigma}^2_{dn}} \begin{bmatrix} \boldsymbol{h}_n \\ \hat{\sigma}^2_{fn} \end{bmatrix} \begin{bmatrix} \boldsymbol{h}_n \\ \hat{\sigma}^2_{fn} \end{bmatrix} \mathrm{T}, \tag{9.73c}$$

where $\boldsymbol{X}_n$ contains the n input data, $\boldsymbol{h}_n = \boldsymbol{\Sigma}_{n-1}\boldsymbol{K}^{-1}_{n-1}\boldsymbol{k}_n$, and $\hat{\sigma}^2_{fn}$ and $\hat{\sigma}^2_{dn}$ are the predictive variances of the latent function and the new output respectively, calculated at the new input. Details can be found in Lázaro-Gredilla *et al.* (2011) and Van Vaerenbergh *et al.* (2012b). In particular, the update of the predictive mean can be shown to be equivalent to the KRLS update. The advantage of using a full GP model is that not only does it allow us to update the predictive mean, as does KRLS, but it keeps track of the entire predictive distribution of the solution. This allows us, for instance, to establish CIs when predicting new outputs.

Similar to KRLS, this online GP update assumes a stationary model. Interestingly, however, the Bayesian approach (and in particular its handling of the uncertainty) does allow for a principled extension that performs tracking, as we briefly discuss in what follows.

9.8.3 Kernel Recursive Least Squares Tracker

Van Vaerenbergh *et al.* (2012b) presented a KRLS-T algorithm that explicitly handles uncertainty about the data, based on the probabilistic GP framework. In stationary environments it operates identically to the earlier proposed SOGP from Csató and Opper (2002), though it includes a *forgetting mechanism* that enables it to handle nonstationary scenarios as well. During each iteration, KRLS-T performs a forgetting operation in which the mean and covariance are replaced through

$$\boldsymbol{\mu} \leftarrow \sqrt{\lambda}\boldsymbol{\mu} \tag{9.74a}$$

$$\boldsymbol{\Sigma} \leftarrow \lambda\boldsymbol{\Sigma} + (1-\lambda)\boldsymbol{K}. \tag{9.74b}$$

The effect of this operation on the predictive distribution is shown in Figure 9.9. For illustration purposes, the forgetting factor is chosen unusually low, $\lambda = 0.9$.

This particular form of forgetting corresponds to blending the informative posterior with a "noise" distribution that uses the same color as the prior. In other words, forgetting occurs by taking a step back toward the prior knowledge. Since the prior has zero mean, the mean is simply scaled by the square root of the forgetting factor λ. The covariance, which represents the posterior uncertainty on the data, is pulled toward the covariance of the prior. Interestingly, a regularized version of RLS (known as *extended RLS*) can be obtained by using a linear kernel with the B2P forgetting procedure.

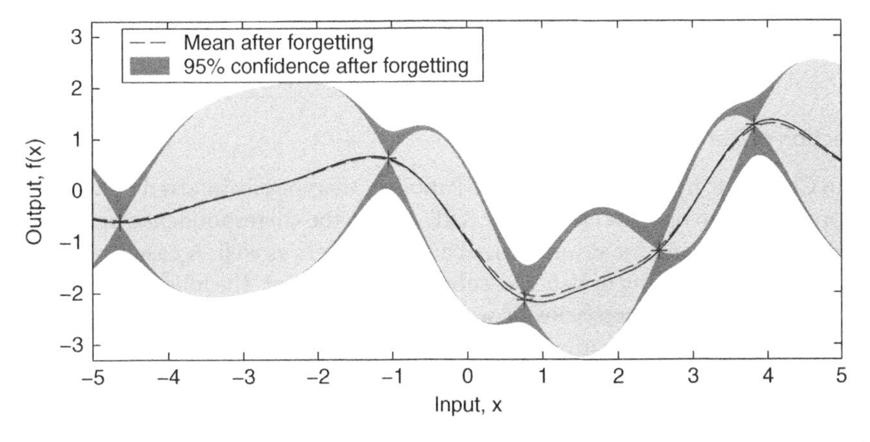

Figure 9.9 Forgetting operation of KRLS-T. The original predictive mean and variance are indicated as the black line and shaded gray area, as in Figure 9.8. After one forgetting step, the mean becomes the dashed red curve, and the new 95% CI is indicated in blue.

Standard RLS can be obtained by using a different forgetting rule; see Van Vaerenbergh *et al.* (2012b).

The KRLS-T algorithm can be seen as a probabilistic extension of KRLS that obtains CIs and is capable of adapting to time-varying environments. It obtains state-of-the-art performance in several nonlinear adaptive filtering problems – see Van Vaerenbergh and Santamaría (2013) and the results of Figure 9.6 – though it has a more complex formulation than most other kernel adaptive filters and it requires a higher computational complexity. We will explore these aspects through additional examples in Section 9.10.

9.8.4 Probabilistic Kernel Least Mean Squares

The success of the probabilistic approach for KRLS-like algorithms has led several researchers to investigate the design of probabilistic KLMS algorithms. The low complexity of KLMS-type algorithms makes them very popular in practical solutions. Nevertheless, this low computational complexity is also a limitation that makes the design of a probabilistic KLMS algorithm a particularly hard research problem.

Some advances have already been made in this direction. Specifically, Park *et al.* (2014) proposed a probabilistic KLMS algorithm, though it only considered the MAP estimate. Van Vaerenbergh *et al.* (2016a) showed that several KLMS algorithms can be obtained by imposing a simplifying restriction on the full SOGP model, thereby linking KLMS algorithms and online GP approaches directly.

9.9 Further Reading

A myriad of different kernel adaptive filtering algorithms have appeared in the literature. We described the most prominent algorithms, which represent the state of the art. While we only focused on their online learning operation, several other aspects are worth studying. In this section, we briefly introduce the most interesting topics that are the subject of current research.

9.9.1 Selection of Kernel Parameters

As we saw in Chapter 8, hyperparameters of GP models are typically inferred by type-II ML. This corresponds to optimal choices for KRR, due to the correspondence between GP regression and KRR, and for several kernel adaptive filters as well. A case study for KRLS and KRLS-T can be found in Van Vaerenbergh *et al.* (2012a). The following should be noted, however. In online scenarios, it would be interesting to perform an online estimation of the optimal hyperparameters. This, however, is a difficult open research problem for which only a handful of methods have been proposed; see for instance Soh and Demiris (2015). In practice, it is still more appropriate to perform type-II ML offline on a batch of training data, before running the online learning procedure using the hyperparameters found.

9.9.2 Multi-Kernel Adaptive Filtering

In the last decade, several methods have been proposed to consider multiple kernels instead of a single one (Bach *et al.*, 2004; Sonnenburg *et al.*, 2006a). The different kernels may correspond to different notions of similarity, or they may address information coming from multiple, heterogeneous data sources. In the field of linear adaptive filtering, it was shown that a convex combination of adaptive filters can improve the convergence rate and tracking performance (Arenas-García *et al.*, 2006) compared with running a single adaptive filter. Multi-kernel adaptive filtering combines ideas from the aforementioned two approaches (Yukawa, 2012). Its learning procedure activates those kernels whose hyperparameters correspond best to the currently observed data, which could be interpreted as a form of hyperparameter learning. Furthermore, the adaptive nature of these algorithms allows them to track the importance of each kernel in time-varying scenarios, possibly giving them an advantage over single-kernel adaptive filtering. Several multi-kernel adaptive filtering algorithms have been proposed in the recent literature (Gao *et al.* e.g., 2014; Ishida and Tanaka e.g., 2013; Pokharel *et al.* e.g., 2013; Yukawa e.g., 2012). While they show promising performance gains over single-kernel adaptive filtering algorithms, their computational complexity is much higher. This is an important aspect inherent to the combination of multiple kernel methods, and it is a topic of current research.

9.9.3 Recursive Filtering in Kernel Hilbert Spaces

Modeling and prediction of time series with kernel adaptive filters is usually addressed by time-embedding the data, thus considering each time lag as a different input dimension. As we have already seen in previous chapters, this approach presents some important drawbacks, which become more evident in adaptive, online settings: first, the optimal filter order may change over time, which would require an additional tracking mechanism; second, if the optimal filter order is high, as for instance in audio applications (Van Vaerenbergh *et al.*, 2016b), the method be affected by the curse of dimensionality. For some problems, the concept of an optimal filter order may not even make sense.

An alternative approach to modeling and predicting time series is to construct *recursive* kernel machines, which implement recursive models explicitly in the RKHS. Preliminary work in this direction considered the design of a *recursive kernel* in the context of infinite recurrent NNs (Hermans and Schrauwen, 2012). More recently, explicit recursive versions of the AR, MA, and gamma filters in RKHS were proposed (Tuia *et al.*, 2014), as we have revised before. By exploiting properties of functional analysis and recursive computation, this approach avoids the reduced-rank approximations that are required in standard kernel adaptive filters. Finally, a kernel version of the ARMA filter was presented by Li and Príncipe (2016).

9.10 Tutorials and Application Examples

This section presents experiments in which kernel adaptive filters are applied to time-series prediction and nonlinear system identification. These experiments are implemented using code based on the KAFBOX toolbox available on the book's web page (Van Vaerenbergh and Santamaría, 2013). Also, some examples of explicit recursivity in RKHS are provided.

9.10.1 Kernel Adaptive Filtering Toolbox

KAFBOX is a MATLAB benchmarking toolbox to evaluate and compare kernel adaptive filtering algorithms. It includes a large list of algorithms that have appeared in the literature, and additional tools for hyperparameter estimation and algorithm profiling, among others.

The kernel adaptive filtering algorithms in KAFBOX are implemented as objects using the `classdef` syntax. Since all KAF algorithms are online methods, each of them includes two basic operations: (1) obtaining the filter output, given a new input $\boldsymbol{x}_*$; and (2) training on a new data pair $(\boldsymbol{x}_n, d_n)$. These operations are implemented as the methods `evaluate` and `train` respectively.

As an example, we list the code for the KLMS algorithm in Listing 9.9. The object definition contains two sets of properties: one for the hyperparameters and one for the variables it will learn. The first method is the object's constructor method, which copies the specified hyperparameter settings. The second method is the `evaluate` function, which performs the operation $y_* = \sum_{i=1}^{n} \alpha_i K(\boldsymbol{x}_i, \boldsymbol{x}_*)$. It includes an `if` clause to check if the algorithm has at least performed one training step yet. If not, zeros are returned as predictions. Finally, the `train` method implements a single training step of the online learning algorithm. This method typically handles algorithm initialization as well, such that functions that operate on a KAF object do not have to worry about initializing. The training step itself is summarized in very few lines of MATLAB code, for many algorithms. For the following experiments, we will use v2.0 of KAFBOX.

```
% Kernel Least-Mean-Square algorithm
%    W. Liu, P.P. Pokharel, and J.C. Principe, "The Kernel ...
    Least-Mean-Square % Algorithm," IEEE Transactions on Signal ...
    Processing, vol. 56, no. 2, pp. % 543-554, Feb. 2008, ...
    http://dx.doi.org/10.1109/TSP.2007.907881
%
% Remark: implementation includes a maximum dictionary size M
% This file is part of the Kernel Adaptive Filtering Toolbox for Matlab.
% https://github.com/steven2358/kafbox/

classdef klms < handle
    properties (GetAccess = 'public', SetAccess = 'private')
        eta = .5; % learning rate
        M = 10000; % maximum dictionary size
        kerneltype = 'gauss'; % kernel type
        kernelpar = 1; % kernel parameter
    end
    properties (GetAccess = 'public', SetAccess = 'private')
        dict = []; % dictionary
        alpha = []; % expansion coefficients
    end
    methods
        function kaf = klms(parameters) % constructor
            if (nargin > 0) % copy valid parameters
```

```
                for fn = fieldnames(parameters)',
                    if ismember(fn,fieldnames(kaf)),
                        kaf.(fn{1}) = parameters.(fn{1});
                    end
                end
            end
        end
        function y_est = evaluate(kaf,x) % evaluate the algorithm
            if size(kaf.dict,1)>0
           k = kernel(kaf.dict,x,kaf.kerneltype,kaf.kernelpar);
                y_est = k'*kaf.alpha;
            else
                y_est = zeros(size(x,1),1);
            end
        end
        function train(kaf,x,y) % train the algorithm
            if (size(kaf.dict,1)<kaf.M), % avoid infinite growth
                y_est = kaf.evaluate(x);
                err = y - y_est;
                kaf.alpha = [kaf.alpha; kaf.eta*err]; % grow
                kaf.dict = [kaf.dict; x]; % grow
            end
        end
    end
end
```

Listing 9.9 MATLAB code for the KLMS algorithm object class, from KAFBOX.

9.10.2 Prediction of a Respiratory Motion Time Series

In the first experiment we apply KAF algorithms to predict a biomedical time series, more specifically a respiratory motion trace. These data come from robotic radiosurgery, in which a photon beam source is used to ablate tumors. The beam is operated by a robot arm that aims to move the beam source to compensate for the motion of internal organs. Traditionally, this is achieved by recording the motion of markers applied to the body surface and by using this motion to draw conclusions about the tumor position. Although this method significantly increases the targeting accuracy, the system delay arising from data processing and positioning of the beam results in a systematic error. This error can be decreased by predicting the motion of the body surface (Ernst, 2012).

The data were recorded at Georgetown University Hospital using CyberKnife® equipment, and it represents the recorded position of one of the markers attached to the body surface.[3] A snapshot of this motion trace is shown in Figure 9.10. The delay to compensate totals 115 ms, which, at a sampling frequency of 26 Hz, corresponds to three samples. The task thus consists of three-step ahead prediction. We use a time-embedding of eight samples. Since the breathing pattern may change over time, we employed only tracking algorithms. Their parameters are listed in Table 9.1. The MSE results of the four algorithms are displayed in the last column of last column of this

3 Data available at http://signals.rob.uni-luebeck.de/.

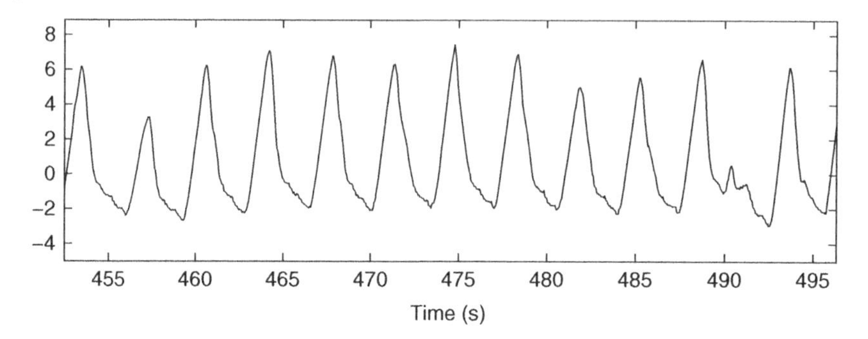

Figure 9.10 A snapshot of the respiratory motion trace.

Table 9.1 Parameters used for predicting the respiratory motion trace, MSE result for three-step ahead prediction, and measured training time.

Algorithm	Parameters	MSE performance (dB)	Training time (s)
NORMA	$\lambda = 10^{-4}$, $\tau = 30$	−5.78	0.16
QKLMS	$\eta = 0.99$, $\epsilon_{\mathbb{U}} = 1$	−8.14	0.18
SWKRLS	$c = 10^{-4}$, $m = 50$	−13.35	0.29
KRLS-T	$\sigma_n^2 = 10^{-4}$, $m = 50$, $\lambda = 0.999$	−18.16	0.54

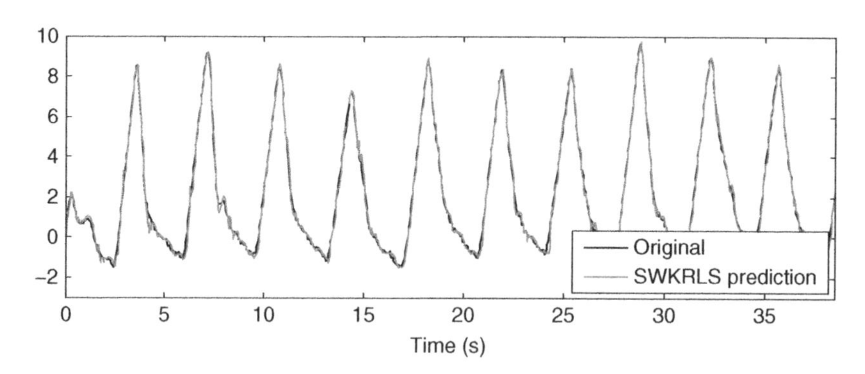

Figure 9.11 The respiratory motion trace and the three-step ahead prediction of a KAF algorithm.

table. A comparison of the original series and the predictions of one of the algorithms is shown in Figure 9.11. The code to reproduce these results can be found in Listing 9.10.

```
% 3-step ahead prediction on the respiratory motion time series.
clear, clc
currentdir = pwd; cd ../kafbox/; install; cd(currentdir)
%% Parameters
h = 3; % prediction horizon
L = 8; % embedding
n = 1000; % number of data
sigma = 7; % kernel parameter
% Uncomment one of the following algorithms. All use a Gaussian kernel.
```

```
% kaf = norma(struct('lambda',1E-4,'tau',30,'kernelpar',sigma,'eta'
                                                    ,0.99));
% kaf = qklms(struct('epsu',1,'kernelpar',sigma,'eta',0.99));
kaf = swkrls(struct('M',50,'kernelpar',sigma,'c',1E-4));
% kaf = krlst(struct('M',50,'lambda',0.999,'sn2',1E-4,'kernelpar',
                                                      sigma));
%% Prepare data
data = load('respiratorymotion3.dat');
X = zeros(n,L);
for i = 1:L,
    X(i:n,i) = data(1:n-i+1,1); % time embedding
end
y = data((1:n)+h);
%% Run algorithm
MSE = zeros(n,1);
y_est_all = zeros(n,1);
title_ = upper(class(kaf)); % store algorithm name
fprintf('Training %s',title_)
for i = 1:n,
    if ~mod(i,floor(n/10)), fprintf('.'); end % progress indicator
    xi = X(i,:);
    yi = y(i);
    y_est = kaf.evaluate(xi); % evaluate on test data
    MSE(i) = (yi-y_est)^2; % test error
    y_est_all(i) = y_est;
    kaf = kaf.train(xi,yi); % train with one input-output pair
end
fprintf('\n');
%% Output
fprintf('Mean MSE: %.2fdB\n\n',10*log10(mean(MSE)));
figure(1); clf; hold all;
t = (1:n)/26; % sample rate is 26 Hz
plot(t,y), plot(t,y_est_all,'r')
legend({'original',sprintf('%s prediction',title_)},'Location','SE');
```

Listing 9.10 MATLAB code for running KAF algorithms on the respiratory motion prediction problem.

9.10.3 Online Regression on the KIN40K Dataset

In the second experiment we train the online algorithms to perform regression of the KIN40K dataset (Ghahramani, 1996),[4] which is a standard regression problem in the machine-learning literature. The KIN40K data set is obtained from the forward kinematics of an eight-link all-revolute robot arm, similar to the one depicted in Figure 9.12. It contains 40 000 examples, each consisting of an eight-dimensional input vector and a scalar output. KIN40K was generated with maximum nonlinearity and little noise, representing a very difficult regression test.

In this experiment, we first determine the optimal hyperparameters for the kernel adaptive filters by running the tool `kafbox_parameter_estimation`, which is based on the GPML toolbox from Rasmussen and Williams (2006). We use 1000 randomly selected data points for the hyperparameter optimization. In the literature, an anisotropic kernel function, which has a different kernel width per dimension,

4 Data available at http://www.cs.toronto.edu/~delve/data/datasets.html.

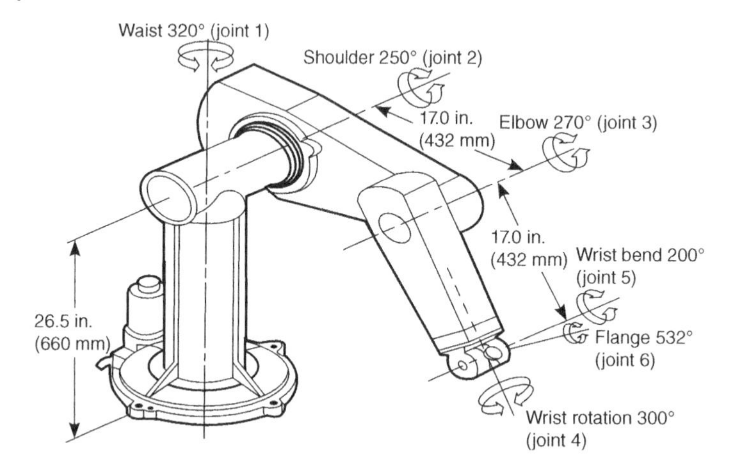

Figure 9.12 Sketch of a five-link all-revolute robot arm. The data used in the KIN40K experiment were generated by simulating an eight-link extension of this arm.

Table 9.2 Optimal hyperparameters found for the KIN40K regression problem.

Parameter	Optimal value
Kernel width σ	1.66
Regularization	2.47×10^{-6}
Forgetting factor λ	1

Table 9.3 Additional parameters used in the KIN40K regression experiment, final dictionary size, and measured training time.

Algorithm	Parameters	Final dictionary size	Training time (s)
QKLMS	$\eta = 0.99$, $\epsilon_{\mathbb{U}} = 1.2$	2750	18.51
KRLS	$\nu = 0.32$	510	12.09
FBKRLS	$m = 500$	500	33.49
KRLS-T	$m = 500$	500	86.41

is commonly used on these data. For simplicity, though, we employ an isotropic Gaussian kernel. The hyperparameters found by the optimization procedure are listed in Table 9.2. The forgetting factor is only used by KRLS-T, though its optimal value is determined to be 1, which indicates that no forgetting takes place in practice.

From the remaining data we randomly select 5000 data points for training and 5000 data for testing the regression. Apart from the hyperparameters that are determined automatically in this experiment, the kernel adaptive filters have some parameters relating to memory size and learning rate. The values chosen for these parameters are listed in Table 9.3. The values for QKLMS are chosen such that it obtains optimal

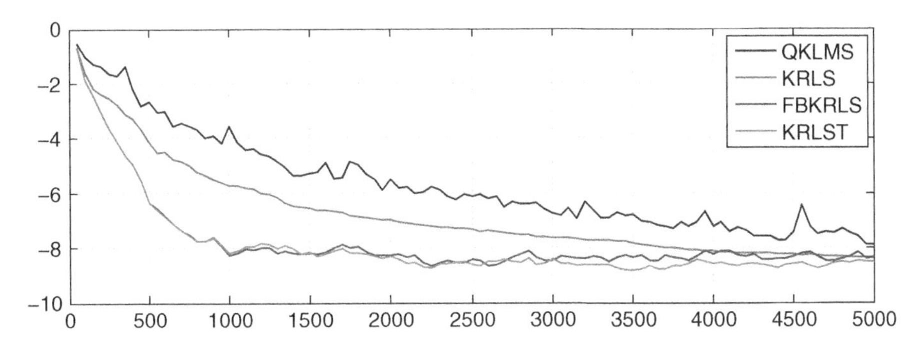

Figure 9.13 Learning curves of different algorithms on the KIN40K data.

performance after training with a dictionary size that is one order of magnitude larger than that of the KRLS algorithms. The precision parameter ν of KRLS is tuned to yield a dictionary size of around $m = 500$ at the end of the experiment, which is the budget of SWKRLS and KRLS-T. The learning curves for this experiment are shown in Figure 9.13. The code for reproducing this experiment is available on the book's web page and in the KAFBOX toolbox demos.

9.10.4 The Mackey–Glass Time Series

The Mackey–Glass time-series prediction is a benchmarking problem in nonlinear time-series modeling. We discussed this time series and the prediction results for KLMS and KRLS in Sections 9.4.3 and 9.5.3 respectively. The learning curves, shown in Figure 9.4, indicate that KLMS converges very slowly, and that KRLS can obtain a much lower MSE in less iterations. On the other hand, KRLS requires an order of magnitude more computation and memory. These results are in line with the intuitions from linear adaptive filtering, in which LMS and RLS represent two different choices in the compromise between complexity and convergence rate. Nevertheless, there is a fundamental difference between the complexity analysis of linear and kernel adaptive filtering algorithms. While in linear adaptive filters the complexity depends on the data dimension, in KAF algorithms it depends on the dictionary size. And, importantly, the latter is a parameter that can be controlled.

A KRLS-type algorithm with a large dictionary can converge faster than a KLMS-type algorithm with a similarly sized dictionary, at the expense of a higher computational complexity. But it would be instructive to ask how a KRLS-type algorithm with a small dictionary compares with a KLMS-type algorithm with a large dictionary. Can the KRLS algorithm obtain similar complexity as KLMS, while maintaining its better convergence rate? This question is answered in the diagrams of Figure 9.14. We have included two KAF algorithms, QKLMS and KRLS-T, that allow easy control over their dictionary size.

Figure 9.14 presents the results obtained by the KAF profiler tool included in KAFBOX. The MATLAB code to reproduce this figure is available on the book's web page. The profiler tool runs each algorithm several times with different configurations, whose parameters are shown in Table 9.4, producing one point in the plot per algorithm configuration. It calculates several variables, such as the number of FLOPS, the memory (in bytes) used, and execution time.

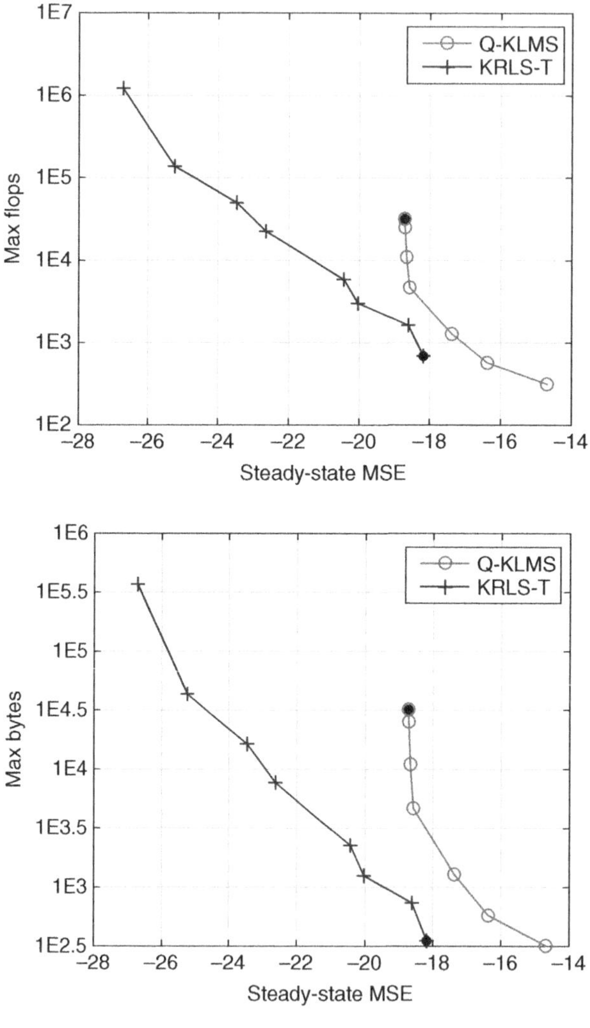

Figure 9.14 MSE versus complexity trade-off comparisons for prediction of the Mackey–Glass time series. Up: maximum number of floating point operations (FLOPS) per iteration as a function of the steady-state MSE. Down: maximum number of bytes per iteration as a function of the steady-state MSE. Each marker represents a single run of one of the algorithms with a single set of parameters. The start of each parameter sweep is indicated by a black dot.

Table 9.4 Parameters used in the Mackey–Glass time-series prediction. A Gaussian kernel with $\sigma = 1$ was used.

Algorithm	Fixed parameters	Varying parameter
QKLMS	$\eta = 0.5$	$\epsilon_{\mathbb{U}} \in \{10^{-4}, 10^{-3}, 10^{-2}, 0.1, 0.2, 0.3, 0.5, 0.7, 1\}$
KRLS-T	$\lambda = 1, \sigma_n^2 = 10^{-6}$	$m \in \{3, 5, 7, 10, 20, 30, 50, 150\}$

By plotting the MSE versus the FLOPS or memory, we get an idea of the resources required to obtain a desired MSE result. If, for instance, we are working in a scenario with a restriction on computational complexity, we should select the algorithm that performs best under this restriction by determining which performance curve is most to the left for the amount of FLOPS available. In the same manner, by fixing a maximum on MSE we obtain the FLOPS and memory required by each algorithm. In the left plot of Figure 9.14 we observe that if the available computational complexity is very limited, it may be more interesting to use QKLMS. In other cases, KRLS-T is preferred as a better MSE is obtained for the same amount of FLOPS. In terms of memory used, it appears that it is always advantageous to use KRLS-T, as can be seen in the right plot.

9.10.5 Explicit Recursivity on Reproducing Kernel in Hilbert Space and Electroencephalogram Prediction

This experiment deals with a real EEG signal prediction problem with four-samples ahead. This is a very challenging nonlinear problem with high levels of noise and uncertainty. We used file "SLP01A" from the MIT-BIH Polysomnographic Database.[5] All recordings in the database include an ECG signal, an invasive blood pressure signal, an EEG signal, and a respiration signal. In this experiment, we considered the prediction of the EEG signal. As for the MG experiments, we used $\{50, 100, 200\}$ training samples, while the next 2000 samples were used for prediction. Listing 9.11 gives a simple piece of code for the implementation of the recursive kernel γ-filter, which can be used to generate the appropriate code for obtaining the results.

```
% RECURSIVEKERNELMATRIX

% Kimn = recursivekernelmatrix(x1,x2,parameters);
%
% Inputs:
%   x1: data matrix with training samples in rows and features in columns
%   x2: data matrix with test samples in rows and features in columns % ...
    parameters: ker: {'lin' 'poly' 'rbf'}
%         p1
%             width of the RBF kernel
%             bias in the linear and polinomial kernel
%             degree in the polynomial kernel
%         P  filter order (tap)
%         mu gamma and Laguerre filters free parameter
%
% Output:
%   Kimn: recursive kernels (one per tap!)

function Kimn = recursivekernelmatrix(x1,x2,sigma,p,mu)

N1 = length(x1);      % Training samples
N2 = length(x2);      % Test samples
% Compute Kimn = K^i(m,n).
Kimn = zeros(p,N1,N2);
Kimn(1,:,:) = kernelmatrix('rbf',x1,x2,sigma*sigma);
```

5 http://www.physionet.org/physiobank/database/slpdb/slpdb.shtml.

```
for i = 2:p
    for m = 2:N1
        for n = 2:N2
            Kaux1 = 0;
            if m > 2
                for j = 2:(m-1)
                    Kaux1 = Kaux1 + (1-mu)^(j-1) * Kimn(i-1,m-j,n-1);
                end
            end
            Kaux2 = 0;
            if n > 2
                for j = 2:(n-1)
                    Kaux2 = Kaux2 + (1-mu)^(j-1) * Kimn(i-1,m-1,n-j);
                end
            end
            Kimn(i,m,n) = (1-mu)^2*Kimn(i,m-1,n-1) + ...
                mu^2 * Kimn(i-1,m-1,n-1) + mu^2 * (Kaux1+Kaux2);
        end
    end
end
```

Listing 9.11 MATLAB code for the recursive kernel γ-filter.

Figure 9.15 shows the KRR prediction of two different EEG segments in the test set for the different approaches; all the methods show the tendency to oversmooth data where peaks are observed. Nonetheless, the $\boldsymbol{K}_2$ kernel always outperforms the other approaches, confirming the numerical results.

9.10.6 Adaptive Antenna Array Processing

This last real example illustrates the method in a purely adaptive setting of high complexity. The problem deals with array processing using a planar antenna array model that consists of 3×3 circular printed patches tuned at a frequency of 1.87 GHz. The distance between patches has been set to 0.5λ. The chosen substrate was PVC. We computed the excitation voltage V_n of each element for five different angles of arrival. In order to compute these excitations, we assume that $V_n = V_n^+ + V_n^-$, where V_n^+ is the ideal excitation, and V_n^- is the one produced by the mutual coupling between antenna elements (Pozar, 2003). We run five finite-difference time-domain-based simulations using CST Microwave Studio 2008 (http://www.cst.com) to obtain the active reflection coefficients Γ_n, and then the excitation voltages V_n produced by each angle of arrival. The azimuth and elevation angles of arrival were 0°, −20°, 20°, 30° and −135° and 0°, −5°, 5°, −5° and −20° degrees respectively.

Transmitted signals consisted of frames containing 142 binary phase-shift keying symbols, of which the 26 central ones (midamble) are training samples, and the rest contain the transmitted information. Users are assumed to transmit frames synchronously. In the simulated scenario, 10 realizations contained 200 frames. At the beginning of each simulation, the five users transmitted independent signals with equal power P. At frame 66, users 3 and 5 switch off and users 1, 2, and 4 change their power to $0.64P$. At frame 131, user 3 switches on and all users transmit at power P. The SNR with respect to user 4 (taken as the desired one) was 10 dB all times.

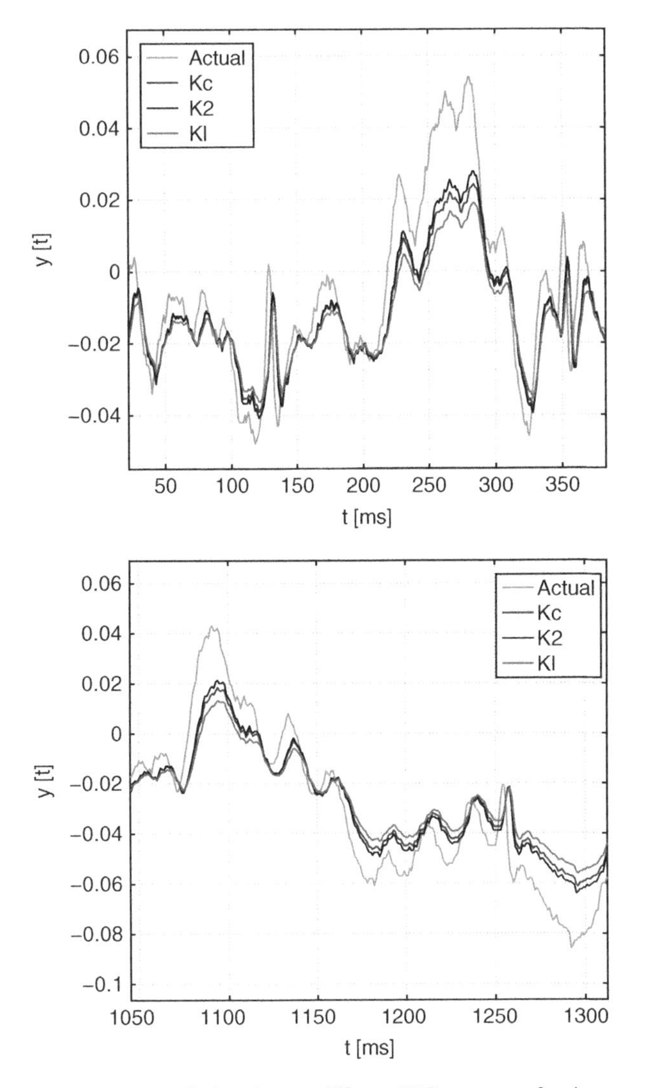

Figure 9.15 Predictions in two different EEG segments for the recursive $\boldsymbol{K}_c$, the standard KRR $\boldsymbol{K}_2$, and the lagged $\boldsymbol{K}_l$ kernels.

Figure 9.16 illustrates the BER as a function of time for the desired user. We compare our method (RKHS–KRLS) with the KLMS (Liu *et al.*, 2008), KRLS (Engel *et al.*, 2004), along with their sparse versions via "surprise criterion" (Liu *et al.*, 2009) using the standard RBF kernel. The regularization factor was set to $\eta = 0.001$, the forgetting factor to 0.99, and the threshold used in approximate linear dependency was set to 0.55. The recursive kernel used $\mu = 0.9$. The experiment was repeated 10 times and the average curves are shown. The recursive kernel has the same computational as embedding-based kernels (Table 9.5). The optimization computational time of the recursive kernel algorithm is $\mathcal{O}(t^2)$, which is the same as KRLS. The added computational burden comes from the computation of the kernels. Both algorithms need to compute a vector of

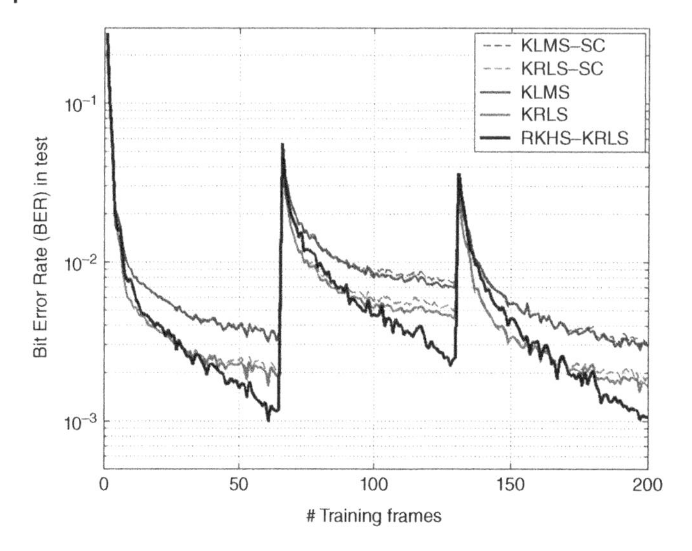

Figure 9.16 BER as a function of time (training frames) for the desired user.

Table 9.5 Computational complexity at iteration *t*.

Algorithm	LMS	KLMS	KRLS	Proposed RKHS recursivity
Computation	$\mathcal{O}(n)$	$\mathcal{O}(t)$	$\mathcal{O}(t^2)$	$\mathcal{O}((P+1)t^2)$
Memory	$\mathcal{O}(n)$	$\mathcal{O}(t)$	$\mathcal{O}(t^2)$	$\mathcal{O}((P+1)t^2)$

kernel products per iteration. The recursive kernel needs an additional computational time of $\mathcal{O}(Pt^2)$ to implement the recursion of Equation 9.65. A higher memory space is needed because the P kernel matrices of the recursion plus their combination need to be stored. Following Equation 9.65, kernel entries depend solely on previous kernels. In addition, the information introduced by each sample into the first kernel ($P = 1$) propagates through the following kernels in subsequent recursion steps with a decaying factor depending on $1 - \mu$. Also, the recursive kernel yields, in general, lower error bounds in steady state and shows competitive performance with KRLS when the channel changes. The method is implemented here using only scalars, and hence it shows a slightly longer convergence time than the rest of the methods. Extension of the method to considering deeper embeddings is straightforward and will be studied in the future. The recursive kernel used in this example is available on the book's web page.

9.11 Questions and Problems

Exercise 9.11.1 Discuss when it is more useful to use a KLMS-like algorithm and when a KRLS-like algorithm.

Exercise 9.11.2 Demonstrate that the computational complexity of KLMS is $\mathcal{O}(m)$, where m is the number of data in its dictionary.

Exercise 9.11.3 Demonstrate that the computational complexity of KRLS is $\mathcal{O}(m^2)$, where m is the number of data in its dictionary.

Exercise 9.11.4 List the advantages and disadvantages of using a sliding-window approach for determining a filter dictionary, as used for instance by SW-KRLS and NORMA.

Exercise 9.11.5 Demonstrate that the recursive kernel gamma filter is unique and bounded.

Exercise 9.11.6 In the tracking experiment of Section 9.5.5, the slope of the learning curve for some algorithms is less steep just after the switch than at the beginning of the experiment. Identify for which algorithms this happens, in Figure 9.6, and explain for each of these algorithms why this is the case.

Part III

Classification, Detection, and Feature Extraction

10

Support Vector Machine and Kernel Classification Algorithms

10.1 Introduction

This chapter introduces the basics of SVM and other kernel classifiers for pattern recognition and detection. We start by introducing the main elements and concept underlying the successful binary SVM. We analyze the issue of regularization and model sparsity. The latter serves to introduce the ν-SVM, a version of the SVM that allows us to control capacity of the classifier easily. A recurrent topic in classification settings is that of how to tackle problems involving several classes; for that we then introduce several available extensions to cope with multiclass and multilabel problems. Other kernel classifiers, such as the LSs SVM and the kernel Fisher's discriminant (KFD) analysis, are also summarized and compared experimentally in this chapter.

After this first section aimed at introducing the basic classification structures, we introduce more advanced topics in SVM for classification, including large margin filtering (LMF), SSL, active learning, and large-scale classification using SVMs. Section 10.4 has to do with large-scale implementations of SVMs, either involving new efficient algorithmic extensions or parallelization techniques.

10.2 Support Vector Machine and Kernel Classifiers

10.2.1 Support Vector Machines

This section describes the perhaps more useful and impactful machine-learning classifier presented in recent decades: the SVM (Boser *et al.* 1992; Cortes and Vapnik 1995; Gu and Sheng 2017; Gu *et al.* 2012; Schölkopf and Smola 2002; Vapnik 1998). The SVM for classification became very popular after its inception due to its outstanding performance, comparable to those of sophisticated NNs trained with high computational cost algorithms in complex tasks. SVMs were originally conceived as efficient methods for pattern recognition and classification (Vapnik, 1995). Right after their introduction, researchers used these algorithms in a plethora of DSP classification problems and applications, such as speech recognition (Picone *et al.*, 2007; Xing and Hansen, 2017), computer vision and image processing (Cremers *et al.*, 2003; Kim *et al.*, 2005; Xu *et al.*, 2017), channel equalization (Chen *et al.*, 2000), multiuser detection (Bai *et al.* 2003; Chen *et al.* 2001a,b), and array processing (Rohwer *et al.*, 2003). Adaptive SVM detectors and estimators for communication system applications were

Digital Signal Processing with Kernel Methods, First Edition. José Luis Rojo-Álvarez, Manel Martínez-Ramón, Jordi Muñoz-Marí, and Gustau Camps-Valls.

also introduced (Navia-Vázquez *et al.*, 2001). We will start the section by introducing the basic elements behind SVMs, then we introduce the notation and formulations, and end up the section with some useful variants: the ν-SVM introduced by (Schölkopf *et al.*, 2000), and the ℓ_1-norm SVM introduced by (Bradley and Mangasarian, 1998). In the next sections we will pay attention to other kernel methods for classification.

Support Vector Machine Elements

Let $(\boldsymbol{x}_1, y_1), \dots (\boldsymbol{x}_N, y_N) \in \mathcal{X} \times \mathcal{Y}$ be a labeled data training set where $\boldsymbol{x}_i$ represents an input data point and y_i its corresponding label. In a binary classification context, labels have values $\{-1, 1\}$, and in a *multiclass* context labels have arbitrary values corresponding to each one of the possible classes present in the problem. Another possible context is the *multilabel* one. In this context, examples $\boldsymbol{x}_i$ are associated simultaneously with multiple labels. For example, one can classify vehicles in an image. They simultaneously belong to different classes, as a function of their size (motorbike/car/truck), usage (police/first responder/private/cab), color, engine type, and so on.

Definition 10.2.1 (Nonlinear SVM binary classifier) Let $\boldsymbol{\phi}(\cdot)$ be a nonlinear operator that maps the samples $\boldsymbol{x}_i$ into a higher dimensional Hilbert space; that is, $\boldsymbol{\phi} : \mathcal{X} \to \mathcal{H}$. The SVM classifier is a linear operation of the transformed data into this Hilbert space as

$$\hat{y}_i = \langle \boldsymbol{w}, \boldsymbol{\phi}(\boldsymbol{x}_i) \rangle_{\mathcal{H}} + b, \tag{10.1}$$

where $\boldsymbol{w}$ is a weight vector inside the RKHS and b is the bias term.

In binary classification, the information regarding the predicted class for $\boldsymbol{x}_i$ is contained in the sign of the estimated label $\hat{y}_i$. Hence, the estimation function for classification, or classifier, is described as

$$\operatorname{sgn}(\hat{y}_i) = \operatorname{sgn}(\langle \boldsymbol{w}, \boldsymbol{\phi}(\boldsymbol{x}_i) \rangle_{\mathcal{H}} + b). \tag{10.2}$$

The SVM approach uses an *indicator function*: the classification is correct if its sign is positive and incorrect when its sign is negative. Then, a loss or *slack variable* ξ_i is introduced to work with these two options simultaneously:

$$\xi_i = 1 - y_i \left(\langle \boldsymbol{w}, \boldsymbol{\phi}(\boldsymbol{x}_i) \rangle_{\mathcal{H}} + b \right), \tag{10.3}$$

which is, obviously, always positive if the classification is correct. The magnitude of the error depends on the distance between the sample and the classification boundary described by the set $C := \{\boldsymbol{x} | \langle \boldsymbol{w}, \boldsymbol{\phi}(\boldsymbol{x}) \rangle_{\mathcal{H}} + b = 0\}$.

The idea underlying the SVM consists of the so-called SRM along with the empirical risk minimization described in Section 3.2.8 ((Boser *et al.*, 1992; Burges, 1998; Cortes and Vapnik, 1995)). While the empirical risk is defined as the mean of the cost function applied over the losses, the structural risk can be defined as the difference between the empirical risk over the training samples and the empirical risk over test samples. Thus, a small structural risk reveals a machine where the estimation function or classifier boundary is close to the optimal one over all possible choices of the class of estimation functions. If the structural risk is high, then the machine is suboptimal, which means that the classifier has been trained to minimize the empirical risk over the training set rather than to capture the statistical properties of the input samples, hence being limited to predict out of the sample. This effect is known as *overfitting*, and it is a consequence of two combined causes. The first one is that model's complexity may be too high, then having an expression capability able to classify any set of data regardless of its statistical properties. Then, the model will show a poor performance when used to classify data not present in the training set. The other cause is, obviously, a lack of statistical information in the training dataset. Indeed, the data used for training is finite, and the presence or not of enough information about its statistical structure will depend on the number of available samples and their representativity. The model must have a complexity according to the amount of statistical information of the data.

It has been proven that, in order to minimize the complexity of the classification machine (and thus the structural risk), one must minimize the norm of a model's parameters $\boldsymbol{w}$. Interestingly, this is in turn equivalent to the maximization of the classifier margin. The margin of a classifier is defined as the distance between the hyperplanes

$$\Pi_1 := \langle \boldsymbol{w}, \boldsymbol{\phi}(\boldsymbol{x})\rangle_{\mathcal{H}} + b = 1 \tag{10.4}$$

$$\Pi_{-1} := \langle \boldsymbol{w}, \boldsymbol{\phi}(\boldsymbol{x})\rangle_{\mathcal{H}} + b = -1. \tag{10.5}$$

The SVM approach is intended to minimize the empirical risk over the training samples and, at the same time, maximize the distance between these two hyperplanes. The distance vector between both hyperplanes is proportional to the parameter vector $\boldsymbol{w}$; that is, let $\boldsymbol{\phi}_0$ and $\boldsymbol{\phi}_1$ be two vectors belonging to each one of the hyperplanes:

$$\begin{aligned} \boldsymbol{w}^{\mathrm{T}}\boldsymbol{\phi}_0 + b &= 1 \\ \boldsymbol{w}^{\mathrm{T}}\boldsymbol{\phi}_1 + b &= -1 \end{aligned} \tag{10.6}$$

and force

$$\boldsymbol{\phi}_1 = \alpha \boldsymbol{w}^{\mathrm{T}} + \boldsymbol{\phi}_0. \tag{10.7}$$

If this is the case, then the distance between both planes is $d = \|\boldsymbol{\phi}_1 - \boldsymbol{\phi}_0\|_{\mathcal{H}}$. By changing vector $\boldsymbol{\phi}_1$ in Equation 10.6 by Equation 10.7, it is straightforward to see that $d = 2/\|\boldsymbol{w}\|_2^2$. This means that minimizing $\|\boldsymbol{w}\|_2^2$ is equivalent to maximizing the distance between the hyperplanes, or, in other words, the classifier margin (see Figure 10.1).

Support Vector Machine

Following the criterion described earlier, the binary SVM classification algorithm can be described as the minimization of the empirical error plus the minimization of the norm of the classifier parameters $\boldsymbol{w}$. The empirical error is measured as the sum of losses or slack variables defined in Equation 10.3 corresponding to the training samples. Then, a (primal) functional to be optimized can be constructed as

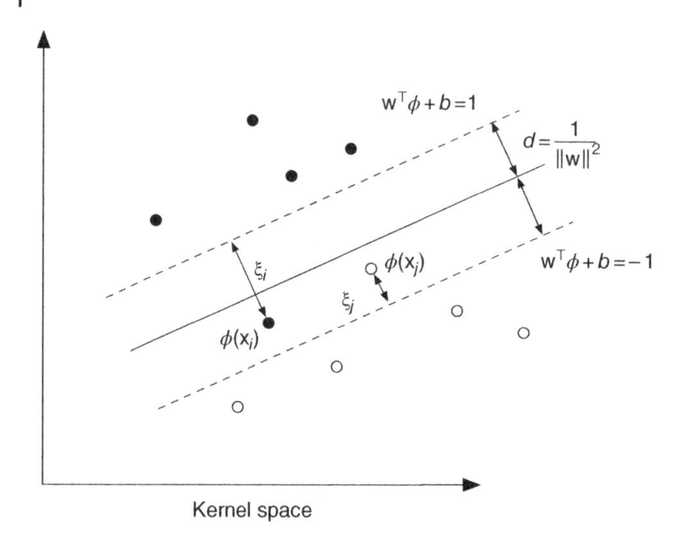

Figure 10.1 Example of margin in the kernel space. Sample $\phi(\boldsymbol{x}_i)$ is wrongly classified, so its associated slack variable ξ_i is higher than 1. Sample $\phi(\boldsymbol{x}_j)$ is correctly classified, but inside the margin, so its associated slack variable ξ_j is positive and less than 1.

$$
\begin{aligned}
&\min_{\boldsymbol{w},\xi_i} \left\{ \frac{1}{2}\|\boldsymbol{w}\|_2^2 + C\sum_{i=1}^{N}\xi_i \right\} \\
&\text{subject to:} \\
&\quad y_i\left(\langle \boldsymbol{w}, \boldsymbol{\phi}(\boldsymbol{x}_i)\rangle_{\mathcal{H}} + b\right) \geq 1 - \xi_i, \quad i = 1,\dots,N \\
&\quad \xi_i \geq 0, \quad i = 1,\dots,N,
\end{aligned} \tag{10.8}
$$

where the term 1/2 has been added to simplify the result. Note that this functional forces the losses or slack variables ξ_i to be positive or zero. That is, when the product between the label and the prediction is higher than unity this means that the corresponding sample has been correctly classified and it is out of the margin, so its distance to the classification hyperplane is higher than d. The implicit cost function is equal to the slack variable ξ_i (or absolute value of the difference between the prediction and the label) if the slack variable is positive, and zero if the slack variable is negative. This cost function is known as the *hinge loss* function, with the following definition.

Definition 10.2.2 (Hinge loss) Given a sample $\boldsymbol{x}_i, y_i$ and an estimator $\hat{f}(\boldsymbol{x}_i) = \hat{y}_i = \langle \boldsymbol{w}, \phi(\boldsymbol{x}_i)\rangle_{\mathcal{H}} + b)$, the hinge loss can be defined as

$$
H(y_i,\hat{f}(\boldsymbol{x}_i)) := \max(0, 1 - y_i\hat{f}(\boldsymbol{x}_i)) = \max(0, 1 - y_i\hat{y}_i), \tag{10.9}
$$

which is sometimes indicated as $H(y_i,\hat{f}(\boldsymbol{x}_i)) := [0, 1 - y_i\hat{f}(\boldsymbol{x}_i)]_+$.

Note that this definition matches that of slack variable ξ_i; therefore, it can be said that the cost function used in SVM is a hinge loss. Parameter C stands for the trade-off between the minimization of the structural and empirical risks, as in the functional for

SVR (see Section 4.6.2). This functional can be optimized using Lagrange optimization. Let positive-valued variables α_i and μ_i be the corresponding Lagrange coefficients for the two sets of constraints. Then, a Lagrangian functional can be constructed as

$$\min_{\boldsymbol{w},\xi_i} \left\{ \frac{1}{2}\|\boldsymbol{w}\|_2^2 + C\sum_{i=1}^{N} \xi_i - \sum_{i=1}^{N} \alpha_i(y_i(\langle \boldsymbol{w}, \boldsymbol{\phi}(\boldsymbol{x}_i)\rangle_{\mathcal{H}} + b) - 1 + \xi_i) - \sum_{i=1}^{N} \mu_i\xi_i \right\}. \tag{10.10}$$

Differentiating with respect to the primal variables $\boldsymbol{w}$ and b gives the results

$$\boldsymbol{w} = \sum_{i=1}^{N} \alpha_i y_i \boldsymbol{\phi}(\boldsymbol{x}_i), \tag{10.11}$$

$$\sum_{i=0}^{N} \alpha_i y_i = 0, \tag{10.12}$$

and

$$C - \alpha_i - \mu_i = 0. \tag{10.13}$$

Adding these three results to the Lagrangian (Equation 10.10) gives the dual problem to be maximized with respect the multipliers α_i only:

$$\max_{\boldsymbol{\alpha}} \left\{ -\frac{1}{2}\sum_{i=1}^{N}\sum_{j=1}^{N} \alpha_i y_i \langle \boldsymbol{\phi}(\boldsymbol{x}_i), \boldsymbol{\phi}(\boldsymbol{x}_j)\rangle_{\mathcal{H}} y_j \alpha_j + \sum_{i=1}^{N} \alpha_i \right\}, \tag{10.14}$$

where it can be observed the (kernel) dot product between samples $\langle \boldsymbol{\phi}(\boldsymbol{x}_i), \boldsymbol{\phi}(\boldsymbol{x}_j)\rangle_{\mathcal{H}} = K(\boldsymbol{x}_i, \boldsymbol{x}_j)$. The equation can be written in matrix notation as

$$L_d = -\boldsymbol{\alpha}^{\mathrm{T}}\boldsymbol{YKY\alpha} + \boldsymbol{\alpha}^{\mathrm{T}}\mathbf{1}, \tag{10.15}$$

where matrix $\boldsymbol{K}$ contains all dot products between the mapped samples, $\boldsymbol{\alpha}$ is a column vector containing all Lagrange multipliers, $\boldsymbol{Y}$ is a diagonal matrix containing all labels, and $\mathbf{1}$ is an all ones column vector. The result in Equation 10.13 gives a constraint over the Lagrange multipliers. Since μ_i is defined as positive, then

$$0 \leq \alpha_i \leq C. \tag{10.16}$$

Besides, the result $\mu_i\xi_i = 0$ implies that when $\xi_i = 0$, $\alpha_i = C$. In summary if the sample is correctly classified and it is out of the margin, then its corresponding slack variable is zero, which implies that its Lagrange multiplier α_i is zero too. If the sample is well classified and inside the margin, or wrongly classified inside or outside the margin, then its Lagrange multiplier will be strictly between zero and C. For those well-classified samples which are right in the margin, the Lagrange multiplier will be equal to C.

Appropriate choice of nonlinear mapping $\boldsymbol{\phi}$ guarantees that the transformed samples are more likely to be linearly separable in the (higher dimension) feature space. The regularization parameter C controls the generalization capability of the classifier, and it must be selected by the user. Note that the decision function for any test point $\boldsymbol{x}_*$ is given by

$$\hat{y}_* = f(\boldsymbol{x}_*) = \text{sgn}\left(\sum_{i=1}^{N} y_i \alpha_i K(\boldsymbol{x}_i, \boldsymbol{x}_*) + b \right), \tag{10.17}$$

where α_i are Lagrange multipliers obtained from solving the QP problem in Equation 10.14. The *support vectors* are those training samples $\boldsymbol{x}_i$ with nonzero Lagrange multipliers $\alpha_i \neq 0$. Typically, the bias term b is calculated by using the *unbounded* Lagrange multipliers as $b = 1/k \sum_{i=1}^{k} (y_i - \langle \boldsymbol{\phi}(\boldsymbol{x}_i), \boldsymbol{w} \rangle_{\mathcal{H}})$, where k is the number of *unbounded* Lagrange multipliers ($0 \leq \alpha_i < C$) and $\boldsymbol{w} = \sum_{i=1}^{N} y_i \alpha_i \boldsymbol{\phi}(\boldsymbol{x}_i)$ (Schölkopf and Smola, 2002).

The ν-Support Vector Machine for Classification

An interesting variation of the SVM is the ν-SVM introduced in (Schölkopf *et al.*, 2000). In the SVM formulation, the soft margin is controlled by parameter C, which may take any positive value. This makes it difficult to adjust when training the classifier. The idea of the ν-SVM is forcing the soft margin to lie in the range $[0, 1]$. This is carried out by redefining the problem to solve

$$\min_{\boldsymbol{w}, \xi_i} \left\{ \frac{1}{2} \|\boldsymbol{w}\|_2^2 + \nu \rho + \frac{1}{N} \sum_{i=1}^{N} \xi_i \right\} \tag{10.18}$$

subject to:

$$y_i(\langle \boldsymbol{\phi}(\boldsymbol{x}_i), \boldsymbol{w} \rangle_{\mathcal{H}} + b) \geq \rho - \xi_i \qquad \forall i = 1, \ldots, N \tag{10.19}$$

$$\rho \geq 0, \xi_i \geq 0 \qquad \forall i = 1, \ldots, N. \tag{10.20}$$

In this new formulation, parameter C has been removed and a new variable ρ with coefficient ν has been introduced. This new variable ρ adds another degree of freedom to the margin, the size of the margin increasing linearly with ρ. The old parameter C controlled the trade-off between the training error and the generalization error. In the ν-SVM formulation, this is done adjusting ν in the range $[0, 1]$, which acts as an upper bound on the fraction of margin errors, and it is also a lower bound on the fraction of support vectors.

The ℓ_1-Norm Support Vector Machine for Classification

The standard SVM works with the ℓ_2-norm on the model's weights, because it makes the problem more easily solvable in practice via QP routines. Other norms can be used though, such as the ℓ_1-norm. The ℓ_1-SVM was proposed by (Bradley and Mangasarian, 1998) and shows some advantages over the standard ℓ_2-norm SVM, especially when dealing with redundant noisy features. The formulation is as follows:

$$\min_{\boldsymbol{w}} \left\{ \|\boldsymbol{w}\|_1 + \sum_{i=1}^{N} H(y_i, \hat{f}(\boldsymbol{x}_i)) \right\}, \tag{10.21}$$

where $\|\boldsymbol{w}\|_1 := \sum_{i=1}^{d} |w_i|, \hat{f}(\boldsymbol{x}_i) = \boldsymbol{\phi}(\boldsymbol{x}_i)^{\mathrm{T}}\boldsymbol{w} + b$, and recall that $H(y, \hat{f}(\boldsymbol{x})) = \max(0, 1 - y\hat{f}(\boldsymbol{x}))$ is the hinge loss defined in Equation 10.9. Note that the ℓ_1-norm cost is not differentiable at zero, which explains why the machine can delete many noisy features by estimating their coefficients by zero. The model can be actually considered a better option when the underlying signal structure is sparse. However, optimization is not that easy, and several approaches have been proposed (Bradley and Mangasarian, 1998; Weston *et al.*, 2001; Zhu *et al.*, 2004).

10.2.2 Multiclass and Multilabel Support Vector Machines

Many methods have been proposed for multiclass and multilabel classification on SVM literature. The simplest approach consists of the so-called one-against-all (OAA) or one-vs-all, where each class is compared with the rest of the classes. This approach is very popular because of its simplicity and good results in general (Rifkin and Klautau, 2004). The OAA implies training $L-1$ classifiers for problems with L classes. This strategy gives rise to binary classifiers that will cope with biased problems: there will be $L - 1$ more examples in one class than in the other. In order to alleviate the potential inconveniences of this approach, the one-against-one (OAO) (Allwein *et al.* 2001; Friedman 1996; Hastie and Tibshirani 1998; Hsu and Lin 2002) strategy was proposed and popularized.[1] In this scheme, authors proposed to compare groups of two classes. During the training phase, the data are organized in all possible pairs of two classes, and a classifier is trained with each pair. During the test phase, each data point is presented to all classifiers, and a decision is taken based on the most consistent classification. In other words, the sample is classified as belonging to the most *voted* class. Friedman (1996) prove that the OAO classifier is equivalent to the Bayes rule when the class posterior probabilities are known. Hastie and Tibshirani (1998) improved the OAO classification performance by using probability estimates of the binary decision of each classifier and then combining these estimates in a joint probability estimator for all classes. This multiclass classifier needs the use of $L(L-1)/2$ binary classifiers; thus, its potential drawback is the computational cost, but this issue depends on the number of examples per class as well.

Single Optimization Approaches

The first single machine compact multiclass SVM was presented by (Weston and Watkins, 1999). The idea of the multiclass SVM consists of constructing L different classification machines. For a training input pattern $\boldsymbol{x}_i$ with an associated label $y_i \in 1, \ldots, L$, each machine has the form

$$f_l(\boldsymbol{x}_i) = \langle \boldsymbol{w}_l, \boldsymbol{\phi}(\boldsymbol{x}_i) \rangle + b_l, \tag{10.22}$$

and it is intended to produce a positive output if $y_i = l$ and a negative output otherwise. This can also be viewed as an OAA machine, but constructed using a

1 For instance, the popular LIBSVM toolbox implements the OAO strategy by default.

compact formulation. Since all machines are optimized at the same time, this alleviates the effect of the unbalanced sets. These classification machines can be used to construct a primal function, given by

$$\min_{\mathbf{w},\xi_i^l} \left\{ \frac{1}{2} \sum_{l=1}^{L} \|\mathbf{w}_l\|^2 + C \sum_{i=1}^{N} \sum_{l \neq y_i} \xi_i^l \right\} \tag{10.23}$$

subject to

$$\begin{aligned} &\langle \mathbf{w}_{y_i}, \boldsymbol{\phi}(\mathbf{x}_i) \rangle + b_{y_i} \geq \langle \mathbf{w}_l, \boldsymbol{\phi}(\mathbf{x}_i) \rangle + b_l - \xi_i^l \\ &\xi_i^l \geq 0 \quad l \in 1, \ldots, L \setminus y_i. \end{aligned} \tag{10.24}$$

The constraints force the output of the classifier to which the sample belongs to be higher than the rest of the machines. The differences between the machine corresponding to the class y_i and the rest of the classes are the slack variables ξ_i^l. If the difference is negative, this means that the machine y_k produces a higher response than the machine $l \neq y_k$, and then the slack variable is constrained to zero. When the difference is positive, the sample has been wrongly classified.

The primal functional can be optimized using Lagrange optimization. See Weston and Watkins (1999) for a complete derivation of the dual. The optimization must include variables

$$c_i^l = \begin{cases} 1 & \text{if } y_i = l \\ 0 & \text{if } y_i \neq l \end{cases} \quad \text{and} \quad A_i = \sum_{l=1}^{L} \alpha_i^l. \tag{10.25}$$

The dual formulation for the optimization problem is then

$$\min_{\alpha_i^l} \left\{ 2 \sum_{i,l} \alpha_i^l + \sum_{i,j,l} \left(-\frac{1}{2} c_j^{y_i} A_i A_j + \alpha_i^l \alpha_j^{y_i} - \frac{1}{2} \alpha_i^l \alpha_j^l \right) K(\mathbf{x}_i, \mathbf{x}_j) \right\} \tag{10.26}$$

with the linear constraints

$$\sum_{i=1}^{N} \alpha_i^l - \sum_{i=1}^{N} c_i^l A_i = 0, \quad l \in 1, \ldots, L \tag{10.27}$$

$$0 \leq \alpha_i^l \leq C, \quad \alpha_i^{y_i} = 0. \tag{10.28}$$

As pointed out by Rifkin and Klautau (2004), other approaches that solve the classification problem using a single optimization scheme have been introduced that present differences with the aforementioned multiclass SVM. The machine presented by Crammer and Singer (2002) simplifies the formulation by including in the constraints only the machine belonging to the class y_i and the machine with largest output among the remaining machines. Lee *et al.* (2004) proved that while the standard SVM tends to the Bayes solution when the number of training patterns tends to infinity, the OAA machines do not have this property. Thus, in this work, a multiclass SVM is formulated with this asymptotic property. Other approaches based on a single machine can be found in Bredensteiner and Bennett (1999), where the standard regularization of the SVM is changed by the regularization $\|\mathbf{w}_i - \mathbf{w}_j\|$. Finally, the Clifford SVM presented in Bayro-Corrochano and Arana-Daniel (2010) is worth mentioning; here, the authors reformulate the multiclass problem using a d-dimensional algebra and a geometric (or Clifford) product to obtain a compact formulation of the optimization problem.

Error-Correcting Approaches

An alternative approach presented by Dietterich and Bakiri (1995) is rooted in the idea of the error correction codes (ECCs) widely used in communications. The error-correcting approach takes the classification procedure as a channel whose input is the pattern, and whose output is a code associated with the pattern label. In the error-correcting approach, each label is then associated with a binary sequence. If there are L different classes and the binary string has length S, then a codebook can be constructed with Hamming distances equal to $d = S - L + 1$. Then, classification machines can be constructed to produce codes as a response of each input pattern. When a response is produced with an error with respect to the actual label which is less than half the Hamming distance (i.e., less than or equal to $\lfloor (d-1)/2 \rfloor$) the error can be corrected to produce a correct classification. In other cases, at least, the error can be detected and the classification identified as wrong. In particular, the approach considers a binary matrix $\mathbf{M}$ of size $L \times S$, where S is equal to the string length and this equals the number of classifiers to be trained. As in ECCs used in communications (e.g., Cover and Thomas, 2006), here one has to satisfy two main properties. First, codewords (or the rows of matrix $\mathbf{M}$) should be well separated in terms of the Hamming distance. Also, each bit of each classifier should be uncorrelated with the same bit of other classifiers. This can be achieved by forcing the Hamming distance between each column of matrix $\mathbf{M}$ and the other columns to be large and forcing the Hamming distance between each and the complement of the other columns to be large.

In a code with L classes, a set of L sequences with 2^L bits can be constructed. In order to satisfy both conditions, one must remove the columns with all zeros and all ones, and the columns that are complementary to other columns. For example, a codebook for $L = 3$ is depicted in Table 10.1. Each column of this table codes digits from 0 to 7 in a binary notation. In this table, columns l and $7 - l$ are complementary, so one put of each two can be removed, so only half of the columns are useful. However, one of the columns will be all zeros or ones, so only three columns are left (see Table 10.2). Obviously, in that case, the Hamming distance will be one and there will be no possibility

Table 10.1 Code for a three class problem (Dietterich and Bakiri, 1995).

Class	f_0	f_1	f_2	f_3	f_4	f_5	f_6	f_7
1	0	0	0	0	1	1	1	1
2	0	0	1	1	0	0	1	1
3	0	1	0	1	0	1	0	1

Table 10.2 Pruned code for a three-class problem.

Class	f_0	f_1	f_2	f_3
1	0	0	0	0
2	0	0	1	1
3	0	1	0	1

of error correction. In general case is at most $2^{L-1} - 1$. For $L = 4$, codes of length 7 are available, and for $L = 5$, the corresponding length will be 15. In that case, the Hamming distance will be $d = 11$.

The error-correcting code approach uses a maximum number of classifiers that exponentially grows with the number of classes. Nevertheless, in this approach, this number must be taken as an upper bound, since the number of outputs will determine the Hamming distance of the codes used that, in turn, will determine the error correction ability of the classifier. For example, for $L = 7$, the maximum number of outputs will be 63, with a Hamming distance $d = 57$, which allows one to correct up to 28 bits. In most applications, a much lower correcting code will be good enough. Once a set of machines is trained to output the desired code for each sample, the test is performed simply by computing the corresponding output of each machine for a given sample. Then a decision is made as a function of the distance between the output obtained and each one of the codes contained in the codebook.

The idea of margin maximization in error-correcting codes was introduced by Allwein *et al.* (2001). They, in particular, introduced a general framework for multiclass classifiers that minimizes loss functions expressed in terms of $y_i f(\boldsymbol{x}_i)$, including what is known as the AdaBoost algorithm, which minimizes a cost function that is a function of $\exp(-y_i f(\boldsymbol{x}_i))$. In these studies, the output matrix contains label 0 in addition to -1 and $+1$, in order to automatically exclude all examples with label equal to 0 during the training. This trick allows us to provide a unifying theory that includes OAO, OAA, and ECC classifiers. See Figure 10.2 for a numerical comparison between the different multiclass strategies.

The multiclass estimation function that uses the Hamming distance for the decoding can be constructed as

$$f(\boldsymbol{x}) = \arg \min_{l \in 1,\dots,L} \sum_{i=1}^{S} \left(\frac{1 - \operatorname{sgn}(M_{l,i} f_i(\boldsymbol{x}))}{2} \right), \tag{10.29}$$

where $M_{l,i}$ is the element (l, i) of code matrix $\boldsymbol{M}$. Here, if $M_{l,i} = 0$ then its contribution to the cost is 1/2. If the signs of $f_i \boldsymbol{x}$ and $M_{l,i}$ match, then the contribution is zero, and one otherwise. This output only accounts for the possible events of prediction agreement, disagreement with the label, plus offering the possibility of ignoring the output, but provided that only the sign of the prediction is used, this approach discards the magnitude of the predictions, which can be an approach to the confidence of the output. A better estimation function is then

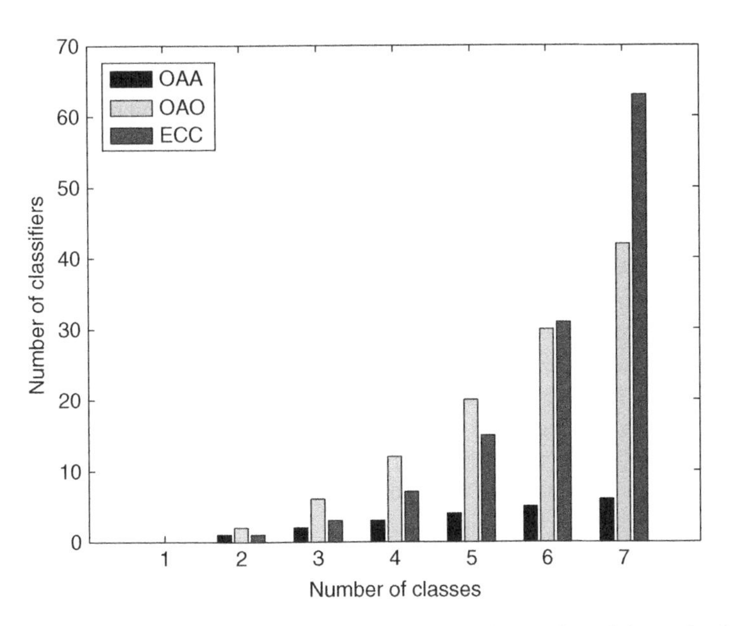

Figure 10.2 Number of classifiers as a function of the number of classes for the various multiclass classification strategies. In an L class classification problem, the OAA and the single machine approaches have $L-1$ outputs, while the OAO uses $(1/2)N(N-1)$ outputs. The ECC approach has a maximum number of outputs equal to of $2^{L-1}-1$.

$$f(\boldsymbol{x}) = \arg\min_{l\in 1,\dots,L} \sum_{i=1}^{S} L(M_{l,i} f_i(\boldsymbol{x})), \tag{10.30}$$

where the sgn function has been removed. In an OAA approach, where only $\boldsymbol{M}_{l,i} = 1$ and the rest of the labels are negative, then the function can be rewritten as $\arg\min_{l\in 1,\dots,L} L(f_r(\boldsymbol{x})) - \sum_{i\neq l} SL(-f_i(\boldsymbol{x}))$.

Multilabel Support Vector Machine Classifiers

A multilabel classifier considers the problem of classifying data that simultaneously belongs to more than one class. Multilabel classification methods are required in bioinformatics problems; for example, see the protein function classification example in Zhang and Zhou (2005) and text categorization in Li *et al.* (2006), among many others. In these problems, each instance is not associated with a single label as in binary or multiclass classification, but rather with a set labels. In general, the number of labels of each pattern is variable. For example, in a text mining task, a single text may belong to several topics (e.g., politics and religion), while others will belong to a single topic only (e.g., soccer). Schapire and Singer (2000) introduced a boosting multilabel system for text categorization. In this work, they state that overfitting is very important in this class of problems, so regularization is fundamental in multilabel problems. Elisseeff and Weston (2001) introduced a maximum margin method intended to provide a good control of the complexity. The study introduces a class of SVM multilabel classifiers that can easily be kernelized.

Assume a set of examples $\boldsymbol{x}_i$, $1 \leq i \leq N$, which can be multilabeled among L different labels. The maximum number of possible multilabels is then 2^L, and they can be represented with binary vectors $\boldsymbol{y}_i$ of L components, in a way similar to the multiclass OAA labeling, but in this case more than one position of the vector can hold a value +1 due to the multilabel feature of the patterns. One straightforward approach to solve the multilabel problem consists of setting up L different binary classifiers, one for each element of the label vectors (Joachims, 1998b). Nevertheless, as is pointed out by Elisseeff and Weston (2001), McCallum (1999), and Pavlidis *et al.* (2001), this approach does not take into account the possible dependencies between labels; that is, with such an approach the information that can be captured about the statistical structure of the data will be partial unless the labels are independent in practice. This model is clearly incomplete in a wide variety of real-world problems.

The approach considered by Elisseeff and Weston (2001) is based on a ranking procedure, where all classes are jointly taken into account. The ranking machine simply sorts the data depending on an estimation of the probability of belonging to each one of the classes. Assume a set of L linear binary classifiers

$$f_l(\boldsymbol{x}_i) = \langle \boldsymbol{w}_l, \boldsymbol{\phi}(\boldsymbol{x}_i) \rangle + b_l \tag{10.31}$$

constructed to produce a ranked output. In order to produce a maximum margin training strategy, the following criterion is applied to the parameters of the classifiers:

$$\min_{j \in \boldsymbol{y}, k \in \bar{\boldsymbol{y}}} \frac{f_j(\boldsymbol{x}_i) - f_k(\boldsymbol{x}_i)}{\|\boldsymbol{w}_j - \boldsymbol{w}_k\|^2}, \tag{10.32}$$

where $\bar{\boldsymbol{y}}$ represents the complementary of label $\boldsymbol{y}$, j represents the index of the correct label, and k the set of incorrect labels. Moreover, if one assumes that all training samples are correctly labeled and well ranked, then the difference between outputs is approximated by the lower bound

$$f_j(\boldsymbol{x}_i) - f_k(\boldsymbol{x}_i) \geq 1. \tag{10.33}$$

Then, the criterion in Equation 10.32 can be approximated as in Elisseeff and Weston (2001) by the functional

$$\min_{j \in \boldsymbol{y}, k \in \bar{\boldsymbol{y}}} \frac{1}{\|\boldsymbol{w}_j - \boldsymbol{w}_k\|^2} \approx \min \sum_{j=1}^{L} \|\boldsymbol{w}_j\|^2. \tag{10.34}$$

In general, the patterns will not be linearly separable; that is, some of the patterns will be inside the margin or wrongly classified. Then, a nonnegative slack variable must be included, which leads to the constraint

$$f_j(\boldsymbol{x}_i) - f_k(\boldsymbol{x}_i) \geq 1 - \xi_{ikj}. \tag{10.35}$$

Hence, the optimization criterion for the multilabel classification reduces to

$$\begin{aligned} &\min_{\boldsymbol{w},\xi_{ikj}} \left\{ \sum_{j=1}^{L} \|\boldsymbol{w}_j\|^2 + C \sum_{i=1}^{L} \frac{1}{\|\boldsymbol{y}_i\|\|\hat{\boldsymbol{y}}_i\|} \sum_{j,k \in \boldsymbol{y}_i \times \bar{\boldsymbol{y}}_i} \xi_{ikj} \right\} \\ &\text{subject to} \quad f_j(\boldsymbol{x}_i) - f_k(\boldsymbol{x}_i) \geq 1 - \xi_{ikj} \\ &\qquad\qquad\quad \xi_{ikj} \geq 0. \end{aligned} \tag{10.36}$$

This functional can be expressed in terms of dual variables, and then *kernelized*, though it has been suggested to use approximate methods due to the high computational burden of quadratic methods.

10.2.3 Least-Squares Support Vector Machine

The LS-SVM (Suykens and Vandewalle, 1999, 2000) uses an alternative quadratic form for the empirical risk while maintaining the structural risk and the margin-related constraints, although now including equality constraints. The primal functional for the LSSVM is

$$\begin{aligned} &\min_{\boldsymbol{w}} \left\{ \frac{1}{2}\|\boldsymbol{w}\|^2 + \frac{1}{2} C \sum_{i=1}^{N} e_i^2 \right\} \\ &\text{subject to} \quad y_i(\langle \boldsymbol{w}, \boldsymbol{\phi}(\boldsymbol{x}_i)\rangle + b) = 1 - e_i. \end{aligned} \tag{10.37}$$

A Lagrange analysis similar to that of the standard SVM can be applied here. The Lagrangian consists of simply solving

$$\min_{\boldsymbol{w},e_i,\alpha_i,b} \left\{ \frac{1}{2}\|\boldsymbol{w}\|^2 + \frac{1}{2} C \sum_{i=1}^{N} e_i^2 - \sum_{i=1}^{N} \alpha_i \left(y_i(\langle \boldsymbol{w}, \boldsymbol{\phi}(\boldsymbol{x}_i)\rangle + b) - 1 + e_i \right) \right\}. \tag{10.38}$$

This Lagrangian is optimized by computing its gradient with respect to all primal and dual variables, to obtain the solution

$$\begin{pmatrix} \boldsymbol{I} & \boldsymbol{0} & \boldsymbol{0} & -\boldsymbol{\Phi}^{\mathrm{T}}\boldsymbol{Y} \\ \boldsymbol{0} & 0 & \boldsymbol{0} & -\boldsymbol{y}^{\mathrm{T}} \\ \boldsymbol{0} & \boldsymbol{0} & C\boldsymbol{I} & -\boldsymbol{I} \\ \boldsymbol{Y}\boldsymbol{\Phi} & \boldsymbol{y} & \boldsymbol{I} & \boldsymbol{0} \end{pmatrix} \begin{pmatrix} \boldsymbol{w} \\ b \\ \boldsymbol{e} \\ \boldsymbol{\alpha} \end{pmatrix} = \begin{pmatrix} \boldsymbol{0} \\ 0 \\ \boldsymbol{0} \\ \boldsymbol{1} \end{pmatrix}, \tag{10.39}$$

where $\boldsymbol{y}$ is a column vector containing all training labels, $\boldsymbol{Y}$ is a diagonal matrix containing $\boldsymbol{y}$, $\boldsymbol{e}$ is a column vector containing all errors, $\boldsymbol{0}$ represents a vector of zeros, and $\boldsymbol{1}$ is a column vector of ones. A dual solution of the equations is found by eliminating $\boldsymbol{w}$ and $\boldsymbol{e}$ as

$$\begin{pmatrix} 0 & -\boldsymbol{y}^{\mathrm{T}} \\ \boldsymbol{y} & \boldsymbol{YKY} + C\boldsymbol{I} \end{pmatrix} \begin{pmatrix} b \\ \boldsymbol{\alpha} \end{pmatrix} = \begin{pmatrix} 0 \\ \boldsymbol{1} \end{pmatrix}. \tag{10.40}$$

This equation can be solved in block instead of using QP, with the advantage of a much less computational burden. Note that the matrix to be inverted is positive definite; hence, the solution exists and is unique. However, the values of the dual variables are proportional to the errors; hence, the solution is not sparse. Authors provide solutions to obtain sparse approximations to these solutions via pruning when the quantity of training samples is not affordable for the application at hand.

10.2.4 Kernel Fisher's Discriminant Analysis

A related formulation to the LS-SVM is the KFD proposed by Mika *et al.* (1999). Assume that N_1 out of N training samples belong to class -1 and N_2 to class $+1$. Let $\boldsymbol{\mu}$ be the mean of the whole set, and $\boldsymbol{\mu}_-$ and $\boldsymbol{\mu}_+$ the means for classes -1 and $+1$ respectively. Analogously, let $\boldsymbol{\Sigma}$ be the covariance matrix of the whole set, and $\boldsymbol{\Sigma}_-$ and $\boldsymbol{\Sigma}_+$ the covariance matrices for the two classes. The linear Fisher's discriminant (LFD) seeks for projections that maximize the interclass variance and minimize the intraclass variance (Fisher, 1936; Hastie *et al.*, 2001). By defining the *between-class scatter matrix* $\boldsymbol{S}_{\mathrm{B}} = (\boldsymbol{\mu}_- - \boldsymbol{\mu}_+)(\boldsymbol{\mu}_- - \boldsymbol{\mu}_+)^{\mathrm{T}}$ and the *within-class scatter matrix* $\boldsymbol{S}_{\mathrm{W}} = \boldsymbol{\Sigma}_- + \boldsymbol{\Sigma}_+$, the problem reduces to maximizing the ratio

$$J(\boldsymbol{w}) = \frac{\boldsymbol{w}^{\mathrm{T}}\boldsymbol{S}_{\mathrm{B}}\boldsymbol{w}}{\boldsymbol{w}^{\mathrm{T}}\boldsymbol{S}_{\mathrm{W}}\boldsymbol{w}}. \tag{10.41}$$

The problem is usually reformatted as follows. If $\boldsymbol{w}$ is a solution of Equation 10.41, any scalar multiple of it will be also. To avoid multiple solutions the arbitrary constraint $\boldsymbol{w}^{\mathrm{T}}\boldsymbol{S}_{\mathrm{B}}\boldsymbol{w} = 4$ is imposed. This is equivalent to imposing $\boldsymbol{w}^{\mathrm{T}}(\boldsymbol{\mu}_- - \boldsymbol{\mu}_+) = 2$. The optimization problem becomes minimizing

$$\boldsymbol{w}T\boldsymbol{S}_{\mathrm{W}}\boldsymbol{w} \quad \text{subject to} \quad \boldsymbol{w}^{\mathrm{T}}(\boldsymbol{\mu}_- - \boldsymbol{\mu}_+) = 2. \tag{10.42}$$

Solving the Lagrange function associated with the above problem, a closed-form solution for $\boldsymbol{w}$ is obtained, $\boldsymbol{w} = 2\lambda\boldsymbol{S}_{\mathrm{W}}^{-1}(\boldsymbol{\mu}_- - \boldsymbol{\mu}_+)$, where λ is the Lagrange multiplier obtained by

$$\lambda = \frac{1}{(\boldsymbol{\mu}_- - \boldsymbol{\mu}_+)^{\mathrm{T}}\boldsymbol{S}_{\mathrm{W}}^{-1}(\boldsymbol{\mu}_- - \boldsymbol{\mu}_+)}. \tag{10.43}$$

When classes are normally distributed with equal covariance, $\boldsymbol{w}$ is in the same direction as the discriminant in the corresponding Bayes optimal classifier. Hence, for this special case, LFD is equivalent to the Bayes optimal classifier (the one defined by linear discriminant analysis (LDA)).

The KFD is obtained by defining the LFD in a high-dimensional *feature* space $\mathcal{H}$. Now, the problem reduces to maximizing

$$J(\boldsymbol{w}) = \frac{\boldsymbol{w}^{\mathrm{T}}\boldsymbol{S}_{\mathrm{B}}^{\phi}\boldsymbol{w}}{\boldsymbol{w}^{\mathrm{T}}\boldsymbol{S}_{\mathrm{W}}^{\phi}\boldsymbol{w}}, \tag{10.44}$$

where $\boldsymbol{w}$, $\boldsymbol{S}_{\mathrm{B}}^{\phi}$ and $\boldsymbol{S}_{\mathrm{W}}^{\phi}$ are defined in $\mathcal{H}$, $\boldsymbol{S}_{\mathrm{B}}^{\phi} = (\boldsymbol{\mu}_-^{\phi} - \boldsymbol{\mu}_+^{\phi})(\boldsymbol{\mu}_-^{\phi} - \boldsymbol{\mu}_+^{\phi})^{\mathrm{T}}$, and $\boldsymbol{S}_{\mathrm{W}}^{\phi} = \boldsymbol{\Sigma}_-^{\phi} + \boldsymbol{\Sigma}_+^{\phi}$. We need to express Equation 10.44 in terms of dot products only. According to the representer theorem (Schölkopf and Smola, 2002), any solution $\boldsymbol{w} \in \mathcal{H}$ can be represented as a linear combination of training samples in $\mathcal{H}$. Therefore, $\boldsymbol{w} = \sum_{i=1}^{N} \alpha_i \boldsymbol{\phi}(\boldsymbol{x}_i)$, and then

$$\langle \boldsymbol{w}, \boldsymbol{\mu}_i^{\phi} \rangle = \frac{1}{N_i} \sum_{j=1}^{N} \sum_{k=1}^{N_i} \alpha_j K(\boldsymbol{x}_j, \boldsymbol{x}_k^i) = \boldsymbol{\alpha}^{\mathrm{T}} \boldsymbol{M}_i, \tag{10.45}$$

where $\boldsymbol{x}_k^i$ represents samples $\boldsymbol{x}_k$ of class i, and $(\boldsymbol{M}_i)_j = (1/N_i) \sum_{k=1}^{N_i} K(\boldsymbol{x}_j, \boldsymbol{x}_k^i)$. Taking the definition of $\boldsymbol{S}_{\mathrm{B}}^{\phi}$ and Equation 10.45, the numerator of Equation 10.44 can be rewritten as $\boldsymbol{w}^{\mathrm{T}} \boldsymbol{S}_{\mathrm{B}}^{\phi} \boldsymbol{w} = \boldsymbol{\alpha}^{\mathrm{T}} \boldsymbol{M} \boldsymbol{\alpha}$, and the denominator as $\boldsymbol{w}^{\mathrm{T}} \boldsymbol{S}_{\mathrm{W}}^{\phi} \boldsymbol{w} = \boldsymbol{\alpha}^{\mathrm{T}} \boldsymbol{N} \boldsymbol{\alpha}$, where

$$\boldsymbol{M} = (\boldsymbol{M}_- - \boldsymbol{M}_+)(\boldsymbol{M}_- - \boldsymbol{M}_+)^{\mathrm{T}}, \tag{10.46}$$

$$\boldsymbol{N} = \sum_{j=\{-1,+1\}} \boldsymbol{K}_j (\boldsymbol{I} - \mathbf{1}_{n_j}) \boldsymbol{K}_j^{\mathrm{T}}. \tag{10.47}$$

$\boldsymbol{K}_j$ is an $N \times N_j$ matrix with $(\boldsymbol{K}_j)_{nm} = K(\boldsymbol{x}_n, \boldsymbol{x}_m^j)$ (the kernel matrix for class j), $\boldsymbol{I}$ is the identity, and $\mathbf{1}_{N_j}$ is a matrix with all entries set to $1/N_j$. Finally, the FLD in $\mathcal{H}$ is solved by maximizing

$$J(\boldsymbol{\alpha}) = \frac{\boldsymbol{\alpha}^{\mathrm{T}} \boldsymbol{M} \boldsymbol{\alpha}}{\boldsymbol{\alpha}^{\mathrm{T}} \boldsymbol{N} \boldsymbol{\alpha}}, \tag{10.48}$$

which is solved as in the linear case by finding the leading eigenvector of $\boldsymbol{N}^{-1}\boldsymbol{M}$. The projection of a new sample $\boldsymbol{x}$ onto the discriminant $\boldsymbol{w}$ can be computed through the kernel function implicitly:

$$\langle \boldsymbol{w}, \phi(\boldsymbol{x}) \rangle = \sum_{i=1}^{N} \alpha_i K(\boldsymbol{x}_i, \boldsymbol{x}). \tag{10.49}$$

The difference between the SVM and the Fisher discriminant is that the latter forces the maximum separation between class distributions (see Figure 10.3 for an illustrative comparison between KFD and SVM). In order to use these projections for classification, one needs to find a suitable threshold which can be chosen as the mean of the average projections of the two classes.

The dimension of the feature space is equal to, or higher than, the number of training samples, which makes regularization necessary, otherwise overfitting is guaranteed. Since KFD analysis minimizes the variance of data along the projection and maximizes the distance between average outputs for each class, one can equivalently solve a QP problem that contains a regularizer $\|\boldsymbol{w}\|^2$ of the form (Mika *et al.*, 1999)

$$\min_{\boldsymbol{w},\xi_i,b} \left\{ \frac{1}{2} \|\boldsymbol{w}\|_2^2 + C \sum_{i=1}^{N} \xi_i^2 \right\} \tag{10.50}$$

constrained to

$$\phi(\boldsymbol{x}_i)\boldsymbol{w} + b = y_i + \xi_i, \quad i = 1, \ldots, N. \tag{10.51}$$

This minimization procedure can be intuitively interpreted by noting that KFD attempts to obtain a regularized solution (term $\|\boldsymbol{w}\|_2^2$ in Equation 10.50) in which the output

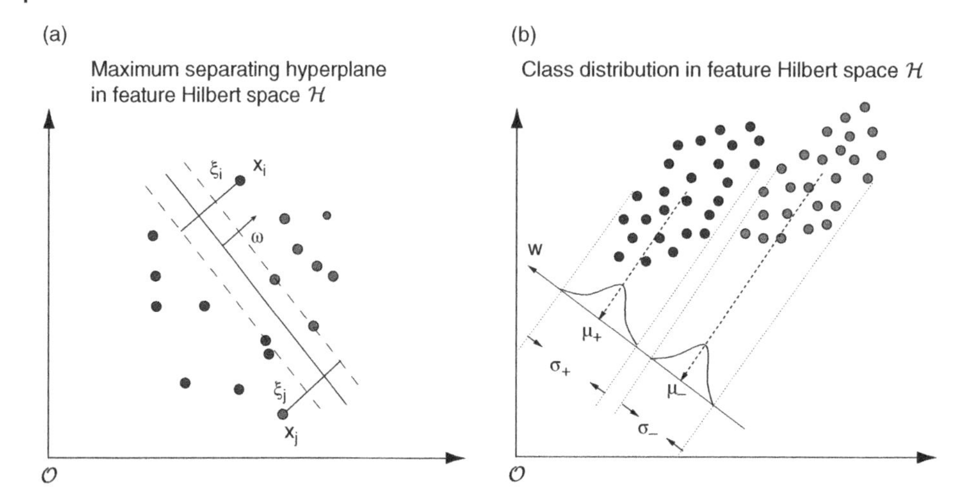

Figure 10.3 Comparison between two kernel classifiers. (a) SVM: linear decision hyperplanes in a nonlinearly transformed, feature space, where *slack* variables ξ_i are included to deal with errors. (b) KFD: separates the classes by projecting them onto a hyperplane where the difference of the projected means (μ_+, μ_-) is large, and the variance around means σ_+ and σ_- is small.

for each sample is forced to move forward its corresponding class label (restriction in Equation 10.51), the variance of the errors is minimized (term $\sum_i \xi_i^2$ in Equation 10.50).

The solution of this primal problem leads to a dual with the solution

$$w = \frac{1}{C}\sum_{i=1}^{N} \alpha_i \boldsymbol{\phi}(\boldsymbol{x}) \tag{10.52}$$

and a dual optimization problem

$$\max_{\boldsymbol{\alpha}} \left\{ -\frac{1}{2}\boldsymbol{\alpha}^{\mathrm{T}}\boldsymbol{\alpha} - \frac{1}{2C}\boldsymbol{\alpha}^{\mathrm{T}}\boldsymbol{K}\boldsymbol{\alpha} + \boldsymbol{y}^{\mathrm{T}}\boldsymbol{\alpha} \right\} \tag{10.53}$$

subject to

$$\boldsymbol{\alpha}^{\mathrm{T}}\mathbf{1} = 0. \tag{10.54}$$

Minimizing the functional in Equation 10.50 leads to a nonsparse solution; that is, all training samples are taken into account and weighted in the solution obtained. This may be a dramatic problem when working with a high number of labeled samples, inducing problems of high computational cost and memory requirements. This fact is illustrated in Figure 10.4, where the distribution of nonzero Lagrange multipliers illustrates the concept of sparsity for the SVM and KFD. Moreover, the issue of sparsity poses the question of the computational burden. For large datasets, the evaluation of $\boldsymbol{w}^{\mathrm{T}}\boldsymbol{\phi}(\boldsymbol{x})$ in KFD is very computationally demanding, and thus the optimization becomes more difficult. In fact, clever tricks like chunking (Osuna *et al.*, 1997) or sequential minimal optimization (SMO) (Platt, 1999a) cannot be applied, or only at a much higher

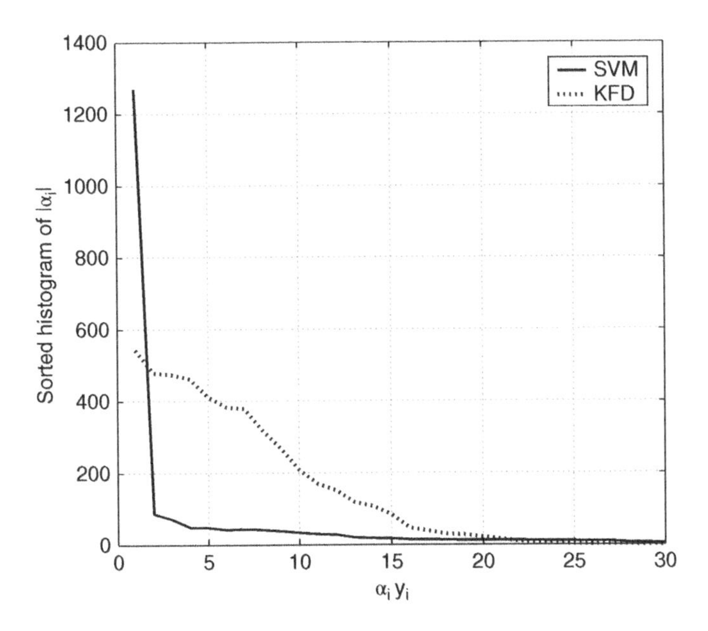

Figure 10.4 Illustration of the sorted density of Lagrange multipliers α_i for the best SVM and KFD classifiers. The concept of sparsity in SVMs and the nonsparse solution offered by KFD are evident.

computational cost. This problem has been previously pointed out, and sparse versions of the KFD have been proposed (Mika *et al.*, 1999).

Experimental Comparison

Here we compare the performance of ν-SVM, LFD, and KFD methods in a remote-sensing multisource image classification problem: the identification of classes "urban" and "nonurban." The images used are from ERS2 synthetic aperture radar (SAR) and Landsat TM sensors acquired in 1999 over the area of Naples, Italy (Gómez-Chova *et al.*, 2006). The dataset has seven Landsat bands, two SAR backscattering intensities (0–35 days), and the SAR interferometric coherence. Since these features come from different sensors, the first step was to perform a specific processing and conditioning of optical and SAR data, and to co-register all images. Then, all features were stacked at a pixel level. A small area of the image of 400×400 pixels was selected.

We used 10 randomly selected pixels (samples) of each class to train the classifiers (only 10 "urban" samples for the one-class experiment). Except for the LFD, the other classifiers have free parameters that must be tuned in the training process. To do this, the training set was split following a ν-fold strategy. For all methods, we used the RBF kernel where σ was tuned in the range $[10^{-3}, 10^{3}]$ in logarithmic increments of 10. The ν-SVM has an additional parameter to tune: ν was varied in the range $[0.1, 0.5]$ in increments of 0.1. Experiments were repeated 10 times with different random realizations of the training sets. Averaged results are shown using four different error measures obtained from the confusion matrices: the estimated κ statistic (Congalton and Green, 1999); the precision P, defined as the ratio between the number of true positives and the sum of true positives and false positives; and the recall R, defined as the ratio between the

Table 10.3 Mean and standard deviation of estimated κ statistic, precision, recall, F-measure, and rate of support vectors for the 10 realizations. Best results are in bold.

Method	κ	Precision	Recall	F-Measure	% SVs
ν-SVM lin	0.81 ± 0.06	0.83 ± 0.07	$\mathbf{0.90 \pm 0.07}$	0.86 ± 0.04	33 ± 0.13
ν-SVM RBF	0.80 ± 0.07	0.86 ± 0.08	0.85 ± 0.10	0.85 ± 0.05	36 ± 0.24
LFD	0.72 ± 0.06	0.76 ± 0.08	0.84 ± 0.05	0.79 ± 0.04	—
KFD	$\mathbf{0.82 \pm 0.03}$	0.87 ± 0.04	0.86 ± 0.05	$\mathbf{0.86 \pm 0.02}$	—

number of true positives and the sum of true positives and false negatives. The last one is the F-measure (or unbiased F-score), computed as $F = 2(PR)/P + R$, which combines both measures.

From Table 10.3, the linear kernel yields slightly better results, but the differences with the RBF kernel are not statistically significant. On the contrary, KFD is better than the linear kernel LFD. Algorithms based on SVM using the RBF kernel yield slightly lower performance than the KFD algorithm, probably due to the low number of training samples used.

10.3 Advances in Kernel-Based Classification

This section introduces advanced kernel machines for classification. We first introduce the LMF method that performs both signal filtering and classification simultaneously by learning the most appropriate filters, then we pay attention to the manifold learning framework and introduce SSL with SVMs that exploit the information contained in both labeled and unlabeled examples. We also review useful kernel developments such as the MKL, which allows to combine different pieces of information in SVM classification. When the outputs are related, structured learning (SL) SVMs can be very useful, and we will exemplify their use as well. We finish the section summarizing the field of active learning (AL), where the training set is optimized such that the SVM classification accuracy is optimal with a reduced number of informative points.

10.3.1 Large Margin Filtering

Many signal processing problems are tackled by first filtering the signal to obtain useful features, and subsequently performing a feature classification or regression. Both steps are critical and need to be designed carefully to deal with the particular statistical characteristics of both the signal and the noise. Signal *sequence labeling* is a paradigmatic example where one aims at tagging speech. Another example is encountered in biomedical engineering where one typically filters electrocardiographic signals before performing arrhythmia prediction. However, optimal design of the filter and the classifier are typically tackled in a separated way, thus leading to suboptimal classification schemes. The LMF method presented in Flamary *et al.* (2012) is an efficient methodology to learn an optimal signal filter and an SVM classifier jointly. After introducing the basic formulation, we illustrate the performance of the method in a challenging real-life dataset of brain–computer interface (BCI) time-series classification.

The setting in which LMF is placed is as follows. Imagine we want to predict a sequence of labels either from a multichannel signal or from multichannel features extracted from that signal by learning from examples. We consider that the training samples are gathered in a matrix $\boldsymbol{X} \in \mathbb{R}^{N\times d}$ containing d channels and N samples. $\boldsymbol{X}_{i,j}$ is the value of channel j for the ith sample ($\boldsymbol{X}_{i,\cdot}$). The vector $\mathbf{y} \in \{-1,1\}^N$ contains the class for each sample. In the following, multiclass problems are handled by means of pairwise binary classifiers. Let us define the filter applied to $\boldsymbol{X}$ by the matrix $\boldsymbol{F} \in \mathbb{R}^{\tau\times d}$. Each column of $\boldsymbol{F}$ is a filter for the corresponding channel in $\boldsymbol{X}$, and τ is the size of the FIR filters. We define the filtered data matrix $\tilde{\boldsymbol{X}}$ by

$$\tilde{\boldsymbol{X}}_{i,j} = \sum_{m=1}^{\tau} \boldsymbol{F}_{m,j}\boldsymbol{X}_{i+1-m+n_0,j} = \boldsymbol{X}_{i,j} \otimes \boldsymbol{F}_{\cdot,j}, \tag{10.55}$$

where the sum is a unidimensional convolution ($\otimes$) of each channel by the filter in the appropriate column of $\boldsymbol{F}$. In this setting, the problem that LMF solves is

$$\min_{g,\boldsymbol{F}} \left\{ \frac{1}{2}\|g\|_{\mathcal{H}}^2 + \frac{C}{N}\sum_{i=1}^{N} H(\mathbf{y}_i, g(\tilde{\boldsymbol{X}}_{i,\cdot}))^p + \lambda\Omega(\boldsymbol{F}) \right\}, \tag{10.56}$$

where λ is a regularization parameter and $\Omega(\cdot)$ represents a differentiable regularization function of $\boldsymbol{F}$. Note that the two leftmost parts of Equation 10.56 reduce to a standard SVM for filtered samples $\tilde{\boldsymbol{X}}$ as defined in Equation 10.55. However, here, $\boldsymbol{F}$ is a variable to be minimized instead of being a fixed filter structure. When jointly optimizing over the decision function g and the filter $\boldsymbol{F}$, the objective function is typically nonconvex. However, the problem defined by Equation 10.56 is convex with respect to $g(\cdot)$ for any fixed filter $\boldsymbol{F}$, and in such a case it boils down to solving the SVM problem. Therefore, in order to take into account this specific structure of the problem, authors proposed to solve the problem through the following min–max approach:[2]

$$\min_{\boldsymbol{F}}\{J(\boldsymbol{F})\} = \min_{\boldsymbol{F}}\{J'(\boldsymbol{F}) + \lambda\Omega(\boldsymbol{F})\}, \tag{10.57}$$

where $J'(\boldsymbol{F})$ is the objective value of the following primal problem:

$$J'(\boldsymbol{F}) = \min_{g} \left\{ \frac{1}{2}\|g\|_{\mathcal{H}}^2 + \frac{C}{N}\sum_{i=1}^{N} H(y_i, g(\tilde{\boldsymbol{X}}_{i,\cdot})) \right\}, \tag{10.58}$$

---

2 Instead of solving the problem in Equation 10.56 through a min–max approach, one could have considered a gradient-descent approach on joint parameters $\boldsymbol{F}$ and $g(\cdot)$. However, such an approach presents several disadvantages over the chosen one. First of all, it does not take into account the structure of the problem, which is the well-studied SVM optimization problem for a fixed $\boldsymbol{F}$. Hence, by separating the optimization over $\boldsymbol{F}$ and over $g(\cdot)$, one takes advantage of the SVM optimization framework and any improvements made to SVM solvers. Furthermore, as stated in Chapelle (2007), addressing the nonlinear SVM problem directly in the primal does not lead to improved computational efficiency; therefore, no speed gain should be expected by solving the problem in Equation 10.56 directly.

where H is the hinge loss function (see Equation 10.9), and $g(\cdot)$ implements the classifier over the filtered signal; that is, $g(\tilde{X}'_{i,\cdot}) = \sum_{j=1}^{N} \alpha_j y_j K(\tilde{X}'_{i,\cdot}, \tilde{X}_{j,\cdot}) + b$. Note that, owing to the strong duality of the SVM problem, $J'(\cdot)$ can be expressed in either its primal or dual form. For solving the optimization problem, authors proposed a CG descent algorithm along $\boldsymbol{F}$ with a line search method (Flamary *et al.*, 2012). We will refer to the method as the kernel filtering SVM (KF-SVM) in the following (see example in Section 10.5.3).

10.3.2 Semi-supervised Learning

Signal classification can be a difficult task because very often only a small number of labeled points are available (Hughes, 1968). In this setting, SSL naturally appears as a promising tool for combining labeled and unlabeled information (Chapelle *et al.* 2006; Zhu 2005). SSL techniques rely on the assumption of *consistency*, in terms of nearby points likely having the same label, and points on the same data structure (cluster or manifold) likely having the same label. This argument is often called the *cluster assumption* (Chapelle *et al.* 2003; Seeger 2001). Traditional SSL methods are based on generative models estimating the conditional density, and they have been widely used in signal and image applications (Jackson and Landgrebe, 2001).

Recently, more attention has been paid to *discriminative* approaches, such as: (1) the transductive SVM (TSVM) (Vapnik, 1998), which maximizes the margin for labeled and unlabeled samples simultaneously; (2) graph-based methods, in which each example spreads its label information to its neighbors until a global steady state is achieved on the whole dataset (Camps-Valls *et al.* 2007b; Zhou *et al.* 2004); and (3) the Laplacian SVM (LapSVM) (Belkin *et al.*, 2006), which deforms the kernel matrix of a standard SVM with the relations found by building the graph Laplacian (we already saw a regression instantiation of this approach in Chapter 8). Also, the design of cluster and bagged kernels (Chapelle *et al.*, 2003) has shown successful results in general signal processing problems, and in image processing in particular (Gómez-Chova *et al.* 2010; Tuia and Camps-Valls 2009); here, the essential idea is to modify the eigenspectrum of the kernel matrix via clustering the data. Figure 10.5 illustrates a typical SSL situation where the distribution of unlabeled samples helps improve the generalization of the classifier.

Manifold-Based Regularization Framework

Regularization helps in producing smooth decision functions that avoid overfitting to the training data. Since the work of Tikhonov (1963), many regularized algorithms have been proposed to control the capacity of the classifier (Evgeniou *et al.*, 2000; Schölkopf and Smola, 2002). As we have already seen, regularization becomes strictly necessary when few labeled samples are available compared with the high dimensionality of the problem. In the last decade, the most paradigmatic case of regularized nonlinear algorithm is the SVM; as we have seen, maximizing the margin is equivalent to applying a kind of regularization to model weights (Camps-Valls and Bruzzone, 2005; Schölkopf and Smola, 2002). These regularization methods are especially appropriate when a low number of samples are available, but are not concerned about the geometry of the marginal data distribution.

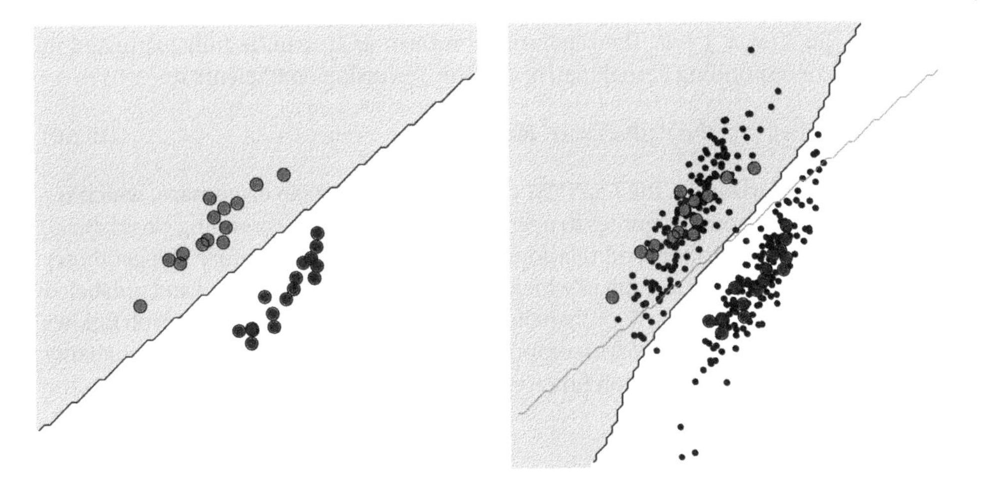

Figure 10.5 Left: classifier obtained using labeled data (grey and black circles denote different classes). Right: classifier obtained using labeled data plus unlabeled data distribution (black points denote unlabeled data).

Semi-supervised Regularization Framework

The classical regularization framework has been recently extended to the use of unlabeled samples (Belkin *et al.*, 2006) as follows. Notationally, we are given a set of l labeled sample pairs $\{(\boldsymbol{x}_i, y_i)\}_{i=1}^{l}$ and a set of u unlabeled samples $\{\boldsymbol{x}_i\}_{i=l+1}^{l+u}$. Let us now assume a general-purpose decision function f. The regularized functional to minimize is

$$\mathcal{L} = \frac{1}{l}\sum_{i=1}^{l} V(\boldsymbol{x}_i, y_i, f) + \gamma_L \|f\|_{\mathcal{H}}^2 + \gamma_M \|f\|_{\mathcal{M}}^2, \tag{10.59}$$

where V represents a generic cost function of the committed errors on the labeled samples (next, we will replace V with the hinge loss H), γ_L controls the complexity of f in the associated Hilbert space $\mathcal{H}$, and γ_M controls its complexity in the intrinsic geometry of the data distribution. For example, if the probability distribution is supported on a low-dimensional manifold, $\|f\|_{\mathcal{M}}^2$ penalizes f along that manifold $\mathcal{M}$. Note that this SSL framework allows us to develop many different algorithms just by playing around with the loss function V and the regularizers $\|f\|_{\mathcal{H}}^2$ and $\|f\|_{\mathcal{M}}^2$.

Laplacian Support Vector Machines

Here, we briefly review the LapSVM as an instantiation of the previous framework. More details can be found in Belkin *et al.* (2006), and its application to image classification in Gómez-Chova *et al.* (2008). The LapSVM uses the same hinge loss function as the traditional SVM; that is, $V(\boldsymbol{x}_i, y_i, f) = \max(0, 1 - y_i f(\boldsymbol{x}_i))$, where f represents the decision function implemented by the selected classifier and the predicted labels are $y_* = \text{sgn}(f(\boldsymbol{x}_*))$. Hereafter, unlabeled or test samples are highlighted with $*$.

The decision function used by the LapSVM is $f(\boldsymbol{x}_*) = \langle \boldsymbol{w}, \boldsymbol{\phi}(\boldsymbol{x}_*)\rangle + b$, where $\boldsymbol{\phi}(\cdot)$ is a nonlinear mapping to a higher dimensional Hilbert space $\mathcal{H}$, and $\boldsymbol{w}$ and b define a linear decision function in that space. The decision function is given by

$f(\boldsymbol{x}_*) = \sum_{i=1}^{l+u} \alpha_i K(\boldsymbol{x}_i, \boldsymbol{x}_*) + b$. The regularization term $\|f\|_{\mathcal{H}}^2$ can be fully expressed in terms of the corresponding kernel matrix and the expansion coefficients $\boldsymbol{\alpha}$:

$$\|f\|_{\mathcal{H}}^2 = \|\boldsymbol{w}\|^2 = (\boldsymbol{\Phi}\boldsymbol{\alpha})^{\mathrm{T}}(\boldsymbol{\Phi}\boldsymbol{\alpha}) = \boldsymbol{\alpha}^{\mathrm{T}}\boldsymbol{K}\boldsymbol{\alpha}. \tag{10.60}$$

For manifold regularization, the LapSVM relies on the Laplacian eigenmaps, which try to map nearby input points/samples to nearby outputs (e.g., corresponding class labels), thus preserving the neighborhood relations between samples. Therefore, the geometry of the data is modeled with a graph in which nodes represent both labeled and unlabeled samples connected by weights W_{ij} (Chapelle *et al.*, 2006). Regularizing the graph follows from the *smoothness* (or *manifold*) assumption and is intuitively equivalent to penalizing "rapid changes" of the classification function f evaluated between nearby samples in the graph:

$$\|f\|_{\mathcal{M}}^2 = \frac{1}{(l+u)^2} \sum_{i,j=1}^{l+u} W_{ij}(f(\boldsymbol{x}_i) - f(\boldsymbol{x}_j))^2 = \frac{\boldsymbol{f}^{\mathrm{T}}\boldsymbol{L}\boldsymbol{f}}{(l+u)^2}, \tag{10.61}$$

where $\boldsymbol{L} = \boldsymbol{D} - \boldsymbol{W}$ is the graph Laplacian, whose entries are sample- and graph-dependent; $\boldsymbol{D}$ is the diagonal degree matrix of $\boldsymbol{W}$ given by $D_{ii} = \sum_{j=1}^{l+u} W_{ij}$ and $D_{ij} = 0$ for $i \neq j$; the normalizing coefficient $1/(l+u)^2$ is the natural scale factor for the empirical estimate of the Laplace operator (Belkin *et al.*, 2006); and $\boldsymbol{f} = [f(\boldsymbol{x}_1), \dots, f(\boldsymbol{x}_{l+u})]^{\mathrm{T}} = \boldsymbol{K}\boldsymbol{\alpha}$, where we deliberately dropped the bias term b.

Now, by plugging Equations 10.60 and 10.61 into Equation 10.59, we obtain the regularized function to be minimized:

$$\min_{\substack{\xi_i \in \mathbb{R}^l \\ \boldsymbol{\alpha} \in \mathbb{R}^{l+u}}} \left\{ \frac{1}{l} \sum_{i=1}^{l} \xi_i + \gamma_L \boldsymbol{\alpha}^{\mathrm{T}}\boldsymbol{K}\boldsymbol{\alpha} + \frac{\gamma_M}{(l+u)^2} \boldsymbol{\alpha}^{\mathrm{T}}\boldsymbol{K}\boldsymbol{L}\boldsymbol{K}\boldsymbol{\alpha} \right\} \tag{10.62}$$

subject to:

$$y_i \left(\sum_{j=1}^{l+u} \alpha_j K(\boldsymbol{x}_i, \boldsymbol{x}_j) + b \right) \geq 1 - \xi_i, \; i = 1, \dots, l \tag{10.63}$$

$$\xi_i \geq 0, \; i = 1, \dots, l, \tag{10.64}$$

where ξ_i are slack variables to deal with committed errors in the labeled samples. Introducing the restrictions in Equations 10.63 and 10.64 into the primal functional in Equation 10.62 through Lagrange multipliers, β_i and η_i, and taking derivatives with respect to b and ξ_i, we obtain

$$\min_{\boldsymbol{\alpha},\boldsymbol{\beta}} \left\{ \frac{1}{2} \boldsymbol{\alpha}^{\mathrm{T}} \left(2\gamma_L \boldsymbol{K} + \frac{2\gamma_M}{(l+u)^2} \boldsymbol{K}\boldsymbol{L}\boldsymbol{K} \right) \boldsymbol{\alpha} - \boldsymbol{\alpha}^{\mathrm{T}}\boldsymbol{K}\boldsymbol{J}^{\mathrm{T}}\boldsymbol{Y}\boldsymbol{\beta} + \sum_{i=1}^{l} \beta_i \right\}, \tag{10.65}$$

where $\boldsymbol{J} = [\boldsymbol{I}\; \boldsymbol{0}]$ is an $l \times (l+u)$ matrix with $\boldsymbol{I}$ as the $l \times l$ identity matrix (the first l points are labeled) and $\boldsymbol{Y} = \mathrm{diag}(y_1, \dots, y_l)$. Taking derivatives again with respect to $\boldsymbol{\alpha}$, we obtain the solution (Belkin *et al.*, 2006)

$$\boldsymbol{\alpha} = \left(2\gamma_L \boldsymbol{I} + 2\frac{\gamma_M}{(l+u)^2}\boldsymbol{L}\boldsymbol{K}\right)^{-1} \boldsymbol{J}^{\mathrm{T}}\boldsymbol{Y}\boldsymbol{\beta}^*. \tag{10.66}$$

Now, substituting again Equation 10.66 into the dual functional Equation 10.65, we obtain the following QP problem to be solved:

$$\boldsymbol{\beta}^* = \max_{\beta} \left\{ \sum_{i=1}^{l} \beta_i - \frac{1}{2}\boldsymbol{\beta}^{\mathrm{T}}\boldsymbol{Q}\boldsymbol{\beta} \right\} \tag{10.67}$$

subject to $\sum_{i=1}^{l} \beta_i y_i = 0$ and $0 \leq \beta_i \leq 1/l$, $i = 1, \ldots, l$, where

$$\boldsymbol{Q} = \boldsymbol{Y}\boldsymbol{J}\boldsymbol{K}\left(2\gamma_L \boldsymbol{I} + 2\frac{\gamma_M}{(l+u)^2}\boldsymbol{L}\boldsymbol{K}\right)^{-1} \boldsymbol{J}^{\mathrm{T}}\boldsymbol{Y}. \tag{10.68}$$

Therefore, the basic steps for obtaining the weights α_i are: (1) build the weight matrix $\boldsymbol{W}$ and compute the graph Laplacian $\boldsymbol{L} = \boldsymbol{D} - \boldsymbol{W}$; (2) compute the kernel matrix $\boldsymbol{K}$; (3) fix regularization parameters γ_L and γ_M; and finally (4) compute $\boldsymbol{\alpha}$ using Equation 10.66 after solving the problem Equation 10.67.

Transductive Support Vector Machine

The TSVM was originally proposed by Vapnik (1998), and aims at choosing a decision boundary that maximizes the margin on both labeled and unlabeled data. The TSVM optimizes a loss function similar to Equation 10.59, but $\gamma_M \|f\|^2_{\mathcal{M}}$ is replaced by a term related to the distance of unlabeled samples to the margin. The TSVM functional to be minimized is

$$\mathcal{L} = \frac{1}{l}\sum_{i=1}^{l} V(\boldsymbol{x}_i, y_i, f) + \gamma_L \|f\|^2_{\mathcal{H}} + \lambda \sum_{j=l+1}^{l+u} L^*(f(\boldsymbol{x}_j^*)), \tag{10.69}$$

where λ is a free parameter that controls the relevance of unlabeled samples, and L^* is the symmetric hinge loss function:

$$L^*(f(\boldsymbol{x}^*)) = \max(0, 1 - |f(\boldsymbol{x}^*)|). \tag{10.70}$$

The optimization of L^* can be seen as "self-learning"; that is, we use the prediction for $\boldsymbol{x}^*$ for training the mapping for that same example. Minimizing Equation 10.70 pushes away unlabeled samples from the margin, either negative or positive, thus minimizing the absolute value.

Graph-Based Label Propagation

Let us review a semi-supervised classifier that solely relies on the graph Laplacian (Zhou and Schölkopf, 2004). Given a dataset $\mathcal{X} = \{\boldsymbol{x}_1, \ldots, \boldsymbol{x}_l, \boldsymbol{x}_{l+1}, \ldots, \boldsymbol{x}_N\} \subset \mathbb{R}^d$, and a label set $\mathcal{L} = \{1, \ldots, c\}$, the first l points $\boldsymbol{x}_i$ $(i \leq l)$ are labeled as $y_i \in \mathcal{L}$ and the remaining points $\boldsymbol{x}_u$ $(l+1 \leq u \leq N)$ are unlabeled. The goal in SSL is to predict the labels of the unlabeled points.

Let $\mathcal{F}$ denote the set of $N \times c$ matrices with nonnegative entries. A matrix $\boldsymbol{F} = [\boldsymbol{f}_1 | \cdots | \boldsymbol{f}_N]^{\mathrm{T}}$ corresponds to a classification on the dataset $\mathcal{X}$ by labeling each point $\boldsymbol{x}_i$ with a label $y_i = \arg\max_{j \leq c} F_{ij}$. We can understand $\boldsymbol{F}$ as a vectorial function $F : \mathcal{X} \to \mathbb{R}^c$ which assigns a vector $\boldsymbol{f}_i$ to each point $\boldsymbol{x}_i$. Define an $N \times c$ matrix $\boldsymbol{Y} \in \mathcal{F}$ with $Y_{ij} = 1$ if $\boldsymbol{x}_i$ is labeled as $y_i = j$ and $Y_{ij} = 0$ otherwise. Note that $\boldsymbol{Y}$ is consistent with the initial labels assigned according to the decision rule. At each iteration t, the algorithm can be summarized as follows:

1) Calculate the affinity matrix $\boldsymbol{W}$, for instance using the RBF kernel:

$$W_{ij} \equiv W(\boldsymbol{x}_i, \boldsymbol{x}_j) = \exp(-\|\boldsymbol{x}_i - \boldsymbol{x}_j\|^2 / 2\sigma^2), \qquad \forall i \neq j \tag{10.71}$$

and make $W_{ii} = 0$ to avoid self-similarity.

2) Construct the matrix

$$\boldsymbol{S} = \boldsymbol{D}^{-1/2} \boldsymbol{W} \boldsymbol{D}^{-1/2}, \tag{10.72}$$

where $\boldsymbol{D}$ is a diagonal matrix with its (i, i) element equal to the sum of the ith row of $\boldsymbol{W}$. Note that this step corresponds to the normalization in feature spaces. Certainly, if we consider a semi-definite kernel matrix formed by the dot products of mapped samples, $W_{ij} = \langle \boldsymbol{\phi}(\boldsymbol{x}_i), \boldsymbol{\phi}(\boldsymbol{x}_j) \rangle$, the normalized version is given by

$$\hat{W}(\boldsymbol{x}_i, \boldsymbol{x}_j) = \left\langle \frac{\boldsymbol{\phi}(\boldsymbol{x}_i)}{\|\boldsymbol{\phi}(\boldsymbol{x}_i)\|}, \frac{\boldsymbol{\phi}(\boldsymbol{x}_j)}{\|\boldsymbol{\phi}(\boldsymbol{x}_j)\|} \right\rangle = \frac{W(\boldsymbol{x}_i, \boldsymbol{x}_j)}{\sqrt{W(\boldsymbol{x}_i, \boldsymbol{x}_i) W(\boldsymbol{x}_j, \boldsymbol{x}_j)}}. \tag{10.73}$$

3) Iterate the following spreading function until convergence:

$$\boldsymbol{F}(t+1) = \alpha \boldsymbol{S} \boldsymbol{F}(t) + (1 - \alpha) \boldsymbol{Y}, \tag{10.74}$$

where α is a parameter in $(0, 1)$.

These three steps should be iteratively repeated until convergence. Now, if F^* denotes the limit of the sequence $\{\boldsymbol{F}(t)\}$, the predicted labels for each point $\boldsymbol{x}_i$ are done using

$$y_i = \arg\max_{j \leq c} F^*_{ij}. \tag{10.75}$$

However, it is worth noting here that one can demonstrate (Zhou *et al.*, 2004) that in the limit

$$\boldsymbol{F}^* = \lim_{t \to \infty} \boldsymbol{F}(t) = (1 - \alpha)(\boldsymbol{I} - \alpha \boldsymbol{S})^{-1} \boldsymbol{Y}, \tag{10.76}$$

and thus the final estimating function $\boldsymbol{F}^*$ can be computed directly without iterations.

This algorithm can be understood intuitively in terms of spreading activation networks from experimental psychology (Anderson, 1983; Shrager *et al.*, 1987), and explained as random walks on graphs (Zhou and Schölkopf, 2004). Basically, the method can be interpreted as a graph $G = (V, E)$ defined on $\mathcal{X}$, where the vertex set V

is just $\mathcal{X}$ and the edges E are weighted by W. In the second step, the weight matrix $\boldsymbol{W}$ of $\boldsymbol{G}$ is normalized symmetrically, which is necessary for the convergence of the following iteration. The first two steps are exactly the same as in spectral clustering (Ng *et al.*, 2001). During the third step, each sample receives the information from its neighbors (first term), and also retains its initial information (second term).

With regard to the free parameter α, one can see that it specifies the relative amount of the information from its neighbors and its initial label information. It is worth noting that *self-reinforcement* is avoided since the diagonal elements of the affinity matrix are set to zero in the first step. Moreover, the information is spread *symmetrically* since $\boldsymbol{S}$ is a symmetric matrix. Finally, the label of each unlabeled point is set to be the class of which it has received most information during the iterative process.

Relations between Semi-supervised Classifiers

The LapSVM is intimately related to other unsupervised and semi-supervised classifiers. This is because the method incorporates both the concepts of kernels and graphs in the same classifier, thus having connections with transduction, clustering, graph-based, and label propagation methods. The minimizing functional used in the standard TSVM considers a different regularization parameter for labeled and unlabeled samples, which is the case in this framework; see Equation 10.59. Also, LapSVM is directly connected with the soft-margin SVM ($\gamma_M = 0$), the hard margin SVM ($\gamma_L \to 0, \gamma_M = 0$), the graph-based regularization method ($\gamma_L \to 0, \gamma_M > 0$), the label-propagation regularization method ($\gamma_L \to 0, \gamma_M \to 0, \gamma_M \gg \gamma_L$), and spectral clustering ($\gamma_M = 1$). In conclusion, by optimizing parameters γ_L and γ_M over a wide enough range, the LapSVM theoretically outperforms the aforementioned classifiers. See Belkin *et al.* (2006) for deeper details and theoretical comparison.

Experimental Results for Semi-supervised Classification

This section presents the experimental results of semi-supervised methods in a real image classification problem. We are here concerned about a pixelwise binary classification problem to distinguish between the urban and nonurban class using satellite image pixels as inputs (Gómez-Chova *et al.*, 2006). Different sets of labeled and unlabeled training samples were used.Training and validation sets consisting of $l = 400$ labeled samples (200 samples per class) were generated, and $u = 400$ unlabeled (randomly selected) pixels (samples) from the analyzed images were added to the training set for the LapSVM and TSVM. We focus on the ill-posed scenario and vary the rate of both labeled and unlabeled samples independently; that is, $\{2, 5, 10, 20, 50, 100\}$% of the labeled/unlabeled samples of the training set were used to train the models in each experiment.

Both linear and RBF kernels were used in the SVM, LapSVM, and TSVM. The graph Laplacian, $\boldsymbol{L}$, consisted of $l + u$ nodes connected using k-NNs, and computed the edge weights W_{ij} using the Euclidean distance among samples. Free parameters γ_L and γ_M were varied in steps of one decade in the range $\{10^{-4}, 10^4\}$, the number of neighbors k used to compute the graph Laplacian was varied from three to nine, and the Gaussian width was tuned in the range $\sigma = \{10^{-3}, \ldots, 10\}$ for the RBF kernel. The selection of the best subset of free parameters was done by cross-validation.

Figure 10.6 shows the validation results for the analyzed SVM-based classifiers. Several conclusions can be obtained. First, LapSVM classifiers produce better classification

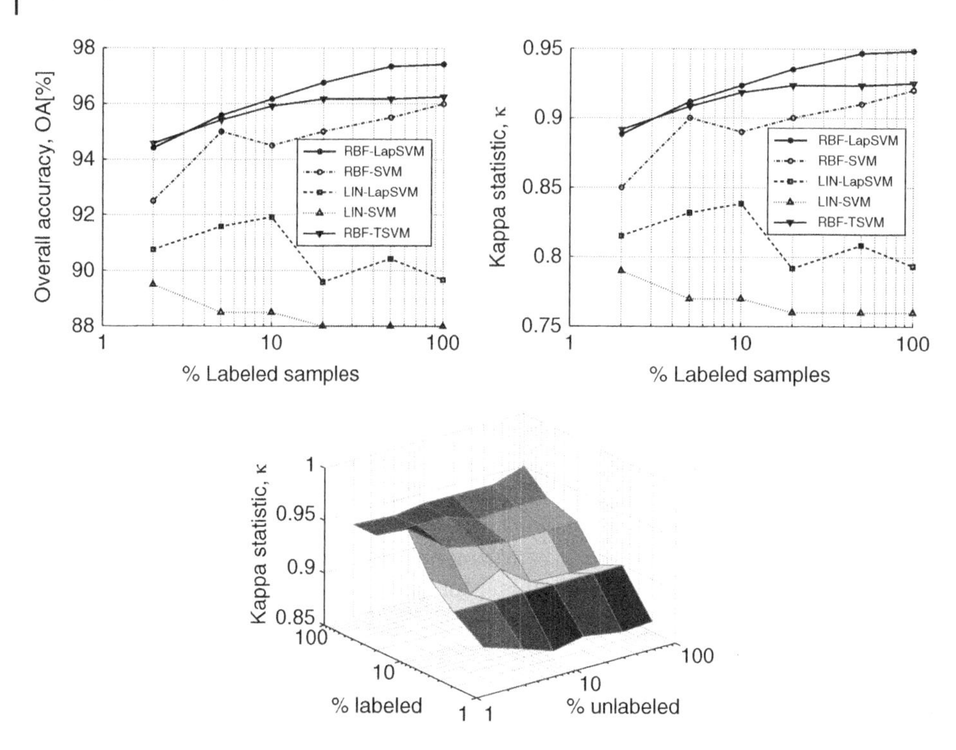

Figure 10.6 Results for the urban classification. First: Overall accuracy OA[%]; Second: κ statistic over the validation set as a function of the rate of labeled training samples used to build models. Third: κ statistic surface over the validation set for the best RBF-LapSVM classifier as a function of the rate of both labeled and unlabeled training samples.

results than SVM in all cases (note that SVM is a particular case of the LapSVM for $\gamma_M = 0$) for both the linear and the RBF kernels. LapSVM also produces better classification results than TSVM when the number of labeled samples is increased. Differences among methods are numerically very similar when a low number of labeled samples are available. The κ surface for the LapSVM highlights the importance of the labeled information in this problem.

10.3.3 Multiple Kernel Learning

The idea of combining kernels referring to different sources of data was previously explored in the *composite kernels* framework (Camps-Valls *et al.*, 2006a) (see Chapter 4), and extensively exploited in regression and system identification problems in previous chapters. The idea exploits the summation and scaling properties of kernels; for example, to define

$$K(\boldsymbol{x}_i, \boldsymbol{x}_j) = \mu_1 K_1(\boldsymbol{x}_i, \boldsymbol{x}_j) + \mu_2 K_2(\boldsymbol{x}_i, \boldsymbol{x}_j), \tag{10.77}$$

where one accounts for the relative importance of kernels by tuning the scalings $\mu_{1,2} \geq 0$. Composite kernels have been shown to work well in practice, as they provide an intuitive way of trading off the importance of different feature sets, used to compute

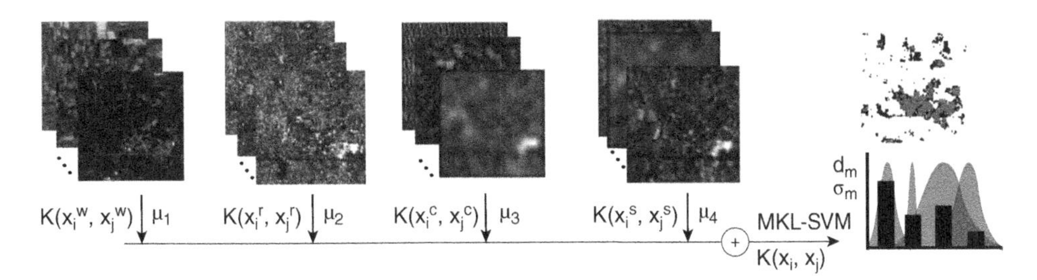

Figure 10.7 General idea behind MKLg. Given different sources of registered data, a linear combination of the different similarity matrices (the kernels) is found.

each kernel, or even to discover the best kernel function with a fixed feature set but different hyperparameters. However, they are not suitable for cases involving more than a few kernels, since the tuning of a set of μ weights by cross-validation would become computationally expensive. The MKL framework (Rakotomamonjy *et al.*, 2008) answers to this call, as it aims at *learning* (i.e., via optimization) the optimal linear combination of μ weights.

Simple Multiple Kernel Learning

The idea of MKL is summarized in Figure 10.7. We have $\mathcal{V} = 1, \ldots, m, \ldots, M$ views of the same data (M blocks of features), which can be spectral bands, groups of bands, image time sequences, or spatial filters of different scale or nature. For each view we build a separate kernel indexed by m, each one of the most appropriate type and with the most appropriate parameters. For example, an MKL system can easily combine a histogram kernel on bag of words features with a Gaussian kernel dedicated to particular subsets of features. We aim at finding the best combination of the form

$$\begin{aligned} K(\boldsymbol{x}_i, \boldsymbol{x}_j) &= \sum_{m=1}^{M} \mu_m K_m(\boldsymbol{x}_i, \boldsymbol{x}_j), \\ \text{s.t.} \quad & \mu_m \geq 0 \\ & \sum_{m=1}^{M} \mu_m = 1. \end{aligned} \tag{10.78}$$

MKL aims at optimizing a convex linear combination of kernels (i.e., the μ_m weights) at the same time as it trains the classifier. In the case of the SVM, the optimization of the μ_m weights involves gradient descent over the SVM objective value (Rakotomamonjy *et al.*, 2008). Globally, we adopt a minimization strategy alternating two steps: first, we solve an SVM with the composite kernel defined by current μ_m and then we update μ_m by gradient descent.

If we use the kernel in Equation 10.78, and then plug it into the SVM dual formulation, we obtain the following problem:

$$\max_{\boldsymbol{\alpha}} \left\{ \sum_{i=1}^{N} \alpha_i - \frac{1}{2} \sum_{i,j=1}^{N} \alpha_i \alpha_j y_i y_j \sum_{m=1}^{M} \mu_m K_m(\boldsymbol{x}_i, \boldsymbol{x}_j) \right\} \tag{10.79}$$

constrained to $0 \leq \alpha_i \leq C$, $\sum_i \alpha_i y_i = 0$, $\forall i = 1, \dots, n$, $\sum_m \mu_m = 1$, and $\mu_m \geq 0$. The dual corresponds to a standard SVM in which the kernel is composed of a linear combination of sub-kernels as in Equation 10.78. One can show (see Rakotomamonjy *et al.* (2008) for details) that maximizing the dual problem in Equation 10.79 is equivalent to solving the problem

$$\min_{\mu} J(\boldsymbol{\mu}) \quad \text{such that} \quad \sum_{m=1}^{M} \mu_m = 1, \quad \mu_m \geq 0 \tag{10.80}$$

where

$$J(\boldsymbol{\mu}) = \begin{cases} \min_{\mathbf{w},b,\xi} & \frac{1}{2} \sum_{m=1}^{M} \frac{1}{\mu_m} \|\mathbf{w}_m\|^2 + C \sum_{i=1}^{N} \xi_i \\ \text{s.t.} & y_i(\sum_{m=1}^{M} \langle \mathbf{w}_m, \boldsymbol{\phi}_m(\mathbf{x}_i) \rangle + b) \geq 1 - \xi_i \\ & \xi_i \geq 0 \end{cases} \tag{10.81}$$

and $\mathbf{w}_m$ represents the weights of the partial decision function of the subproblem m with associated kernel mapping $\boldsymbol{\phi}_m(\mathbf{x}_i)$. In other words, we have now an objective to optimize by gradient descent over the vector of possible μ_m values. We therefore alternate the solution of an SVM (providing the current error) and the optimization of μ_m. The ℓ_1 norm constraint on the kernel weights forces d_m coefficients to be zero, thus encouraging sparsity of the solution and, in turn, a natural feature selection.

Experimental Results for Multiple Kernel Learning

Let us illustrate the performance of the simpleMKL in a computer vision application: the classification of flower types from color images. In particular, we are interested in improving the classification by combining different features of inherently different nature such as features related to color, shape, and attention. For this example, we used the Oxford Flowers dataset.[3] Feature extraction involved computing descriptors (bag-of-words histograms) that account for shape (through SIFT), color, self-similarity (SS), and color attention (CA). A total of 17 classes with 80 each images yielded a total of 1360 images. Several images were used to train an SVM with multiple kernel combinations of features. The results are shown in Figure 10.8 for different settings: simpleMKL with all features, only with CA and SS, a linear SVM, a different combinations of kernels using CA and SIFT only. The simpleMKL solution with all kernels was further analyzed in Figurre 10.8b by plotting the weights for all features, which allows some problem insight. The results suggest that the most useful information is contained in CA and SIFT.

10.3.4 Structured-Output Learning

Traditional kernel classifiers assume independence among the classification outputs. As a consequence, each misclassification receives the same weight in the loss function. Moreover, the kernel function only takes into account the similarity between input

3 Available at http://www.robots.ox.ac.uk/~vgg/data/flowers/17/index.html.

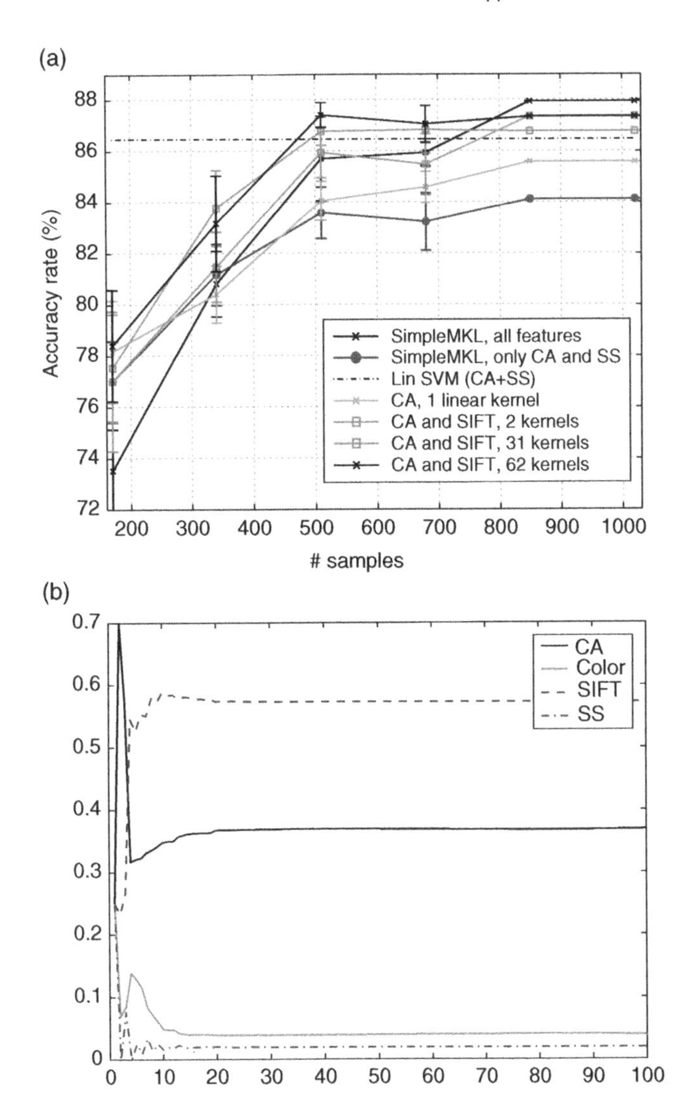

Figure 10.8 (a) Accuracy rates for different training set sizes. (b) Weights of the four kernels obtained for the four groups of features.

values and ignores possible relationships between the classes to be predicted. These assumptions are not consistent for most of real-life DSP problems; for example, in speech recognition, computer vision, or image processing this is not a good assumption either. For instance, segmentation of images often deals with a supervised setting in which a predefined set of classes of interest is defined. In this situation, classes share strong relationships, both *colorimetrically* (e.g., a tree is naturally more similar to grass than to water) and *semantically* (e.g., a chair is more likely to be next to a table).

Structured-output learning tries to encode those output relations in a classifier. In what follows, we review the structured SVM (StructSVM), where the output space structure is encoded using a hierarchical tree, and these relations are added to the model in both the kernel and the loss function. The methodology gives rise to a set of new tools for structured classification, and generalizes the traditional nonstructured classification methods. Several ways can be considered to introduce interdependencies in the output space. The first attempts to develop structured SVM classifiers are found in Joachims *et al.* (2009), Tsochantaridis *et al.* (2004, 2005), and Zien *et al.* (2007). Altun *et al.* (2007) presented an excellent review of the state of the art in structured learning. Despite the confinement of these methods in the machine-learning community, the first applications of structured-output learning appeared in other DSP-related disciplines, such as natural language learning (Joachims *et al.*, 2009), object localization (Blaschko and Lampert, 2008), and image segmentation (Tuia *et al.*, 2011b).

The classical SVM method assumes independence between the outputs. This choice is justified by the fact that, in principle, no assumptions about the distribution of the outputs may be done. However, a given sample can be associated with a *structured output* that considers a more complex relational information, for instance, through a hierarchical tree structure showing several classification levels. In the literature, this output information is very often exploited indirectly through the use of cost matrices and balancing constraints, typically encoded a priori. This approach is not useful when few (or not representative) labeled samples are available. Besides, these are second-order output class relations, and they are not learned from data but fixed before learning. Structured-output learning (SL) (Altun *et al.*, 2007) formalizes the problem of output space relations. This SL framework aims at predicting complex objects, such as trees, sequences, or web queries, where, contrarily to usual machine-learning algorithms, the relationship between the associated outputs plays a role in the prediction.

Notationally, we may define the aim of SL for classification as that of learning a function $h : \mathcal{X} \rightarrow \mathcal{Y}$, where $\mathcal{X}$ is the space of inputs and $\mathcal{Y}$ is the space of structured outputs. Using an i.i.d. training sample $\mathcal{X} = \{(\boldsymbol{x}_1, y_1), (\boldsymbol{x}_2, y_2), \dots, (\boldsymbol{x}_N, y_N)\}$, h is the function minimizing the empirical risk

$$R_E^{\Delta}(h) = \frac{1}{N} \sum_{i=1}^{N} \Delta(y_i, \hat{y}_i), \tag{10.82}$$

where $\hat{y}_i$ is the prediction and y is the correct class assignment. The quality of the model is evaluated by a loss function: $\Delta(y_i, \hat{y}_i) = 0$ if the label is correctly assigned and $\Delta(y_i, \hat{y}_i) \geq 0$ otherwise. Note that the classical 0/1 loss is a particular case of this function returning the loss 1 for each wrong output. Coming back to the h functions, they are of the form

$$h(\boldsymbol{x}) : \arg\max_{y \in \mathcal{Y}} f(\boldsymbol{x}, y), \tag{10.83}$$

where $f : \mathcal{X} \times \mathcal{Y} \rightarrow \mathbb{R}$ is a joint function between inputs and outputs evaluating how well a certain prediction matches the observed output. The joint function f can be represented as $f(\boldsymbol{x}, y) = \boldsymbol{w}^{\mathrm{T}} \boldsymbol{\Psi}(\mathbf{x}, y)$, where $\boldsymbol{w}$ is a weight vector and the joint input–output mapping $\boldsymbol{\Psi}$ relates inputs $\boldsymbol{x}$ and outputs y. In order to account for a structured output, two main ingredients must be modified: the *mapping* $\boldsymbol{\Psi}$ and the *loss function* Δ.

Figure 10.9 Toy example of hierarchical structure: the classes of interest are the three leaves in the tree, which can be semantically grouped in superclasses.

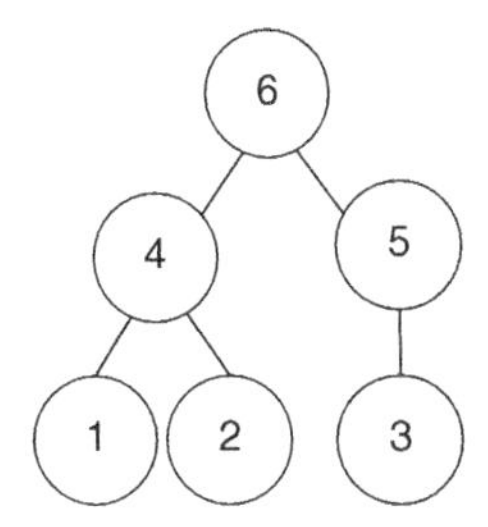

They will lead to structured-output SVM (SSVM) formulations that are application dependent.

The Joint Input–Output Mapping

The goal is to encode the structure of the tree into the mapping $\boldsymbol{\Phi}(\mathbf{x})$, resulting into a joint input–output mapping $\boldsymbol{\Psi}(\boldsymbol{x}, y)$. Altun *et al.* (2007) proposed a mapping considering tree-structures for taxonomies. Consider a taxonomy as a set of elements $\mathcal{Z} \supseteq \mathcal{Y}$ ordered by a partial order $\prec$, and let $\beta_{(y,z)}$ be a measure of similarity respecting the order $\prec$. The representation of the outputs $\boldsymbol{\Lambda}(y)$ can be generalized as

$$\lambda_z(y) = \begin{cases} \beta_{(y,z)} & \text{if } y \prec z \text{ or } y = z \\ 0 & \text{otherwise} \end{cases} \tag{10.84}$$

This way, the similarity between two outputs sharing common superclasses will be higher than between outputs that are distant in the tree structure. Then, we can define the joint input–output feature map via a tensor product:

$$\boldsymbol{\Psi}(\boldsymbol{x}, y) = \boldsymbol{\Phi}(\boldsymbol{x}) \otimes \boldsymbol{\Lambda}(y). \tag{10.85}$$

This formulation introduces a weight vector $\boldsymbol{w}_z$ for every node in the hierarchy. The inner product of the joint feature map decomposes into kernels over input and output space (using the properties proposed in Schölkopf and Smola (2002)):

$$\begin{aligned} \left\langle \boldsymbol{\Psi}(\boldsymbol{x}, y), \boldsymbol{\Psi}(\boldsymbol{x}', y') \right\rangle &= K_{\otimes}((\boldsymbol{\Phi}(\boldsymbol{x}), \boldsymbol{\Lambda}(y)), (\boldsymbol{\Phi}(\boldsymbol{x}'), \boldsymbol{\Lambda}(y')) \\ &= \left\langle \boldsymbol{\Lambda}(y), \boldsymbol{\Lambda}(y') \right\rangle K(\boldsymbol{x}, \boldsymbol{x}'). \end{aligned} \tag{10.86}$$

In order to illustrate this principle, consider a three-class problem with the structure shown in Figure 10.9. Classes 4, 5, and 6 are the superclasses giving the tree structure.

In the nonstructured version of the algorithm (equivalent to the usual multiclass classification), the $\boldsymbol{\Lambda}$ and $\boldsymbol{\Psi}$ matrices take the form

$$\boldsymbol{\Lambda}(\mathrm{y}) = \begin{pmatrix} 1 & 0 & 0 \\ 0 & 1 & 0 \\ 0 & 0 & 1 \end{pmatrix} \qquad \boldsymbol{\Psi}(\mathrm{y}) = \begin{pmatrix} \boldsymbol{x} & 0 & 0 \\ 0 & \boldsymbol{x} & 0 \\ 0 & 0 & \boldsymbol{x} \end{pmatrix} \tag{10.87}$$

Taking into account the structure shown in Figure 10.9, $\boldsymbol{\Lambda}$ and $\boldsymbol{\Psi}$ become

$$\boldsymbol{\Lambda}(\mathrm{y}) = \begin{pmatrix} 1 & 0 & 0 & 1 & 0 & 1 \\ 0 & 1 & 0 & 1 & 0 & 1 \\ 0 & 0 & 1 & 0 & 1 & 1 \end{pmatrix} \qquad \boldsymbol{\Psi}(\mathrm{y}) = \begin{pmatrix} \boldsymbol{x} & 0 & 0 & \boldsymbol{x} & 0 & \boldsymbol{x} \\ 0 & \boldsymbol{x} & 0 & \boldsymbol{x} & 0 & \boldsymbol{x} \\ 0 & 0 & \boldsymbol{x} & 0 & \boldsymbol{x} & \boldsymbol{x} \end{pmatrix}. \quad (10.88)$$

The linear dot product between the two first classes will result in $2\langle \boldsymbol{x}, \boldsymbol{x}' \rangle$, while between the classes 1 and 3 (and 2 and 3) it is of $\langle \boldsymbol{x}, \boldsymbol{x}' \rangle$ only. Thus, using a joint input–output mapping, output structure participates in the similarity between samples.

The Loss Function

To define the loss function, we can modify the classical 0/1 loss by exploiting the tree-based output structure. The proposed tree-based loss assumes a common superclass in the tree at level $l = \{1, ..., L\}$ as follows:

$$\Delta(y, \hat{y}) = \begin{cases} (l-1)/L & \text{if} \quad y^l = \hat{y}^l \\ 1 & \text{otherwise} \end{cases} \quad (10.89)$$

Using this loss, errors predicting "far away" classes are penalized more than "close" errors. A class predicted correctly will receive a loss of zero ($l - 1 = 0$), while the prediction of a class not sharing any superclass with the true class will receive a loss of 1. The loss function presented in Equation 10.89 assumes equal distance between the classes and their superclasses; this can be refined by constructing ad hoc class distances from the labeled data or by learning interclass distances through, for example, clustering.

The *N*-Slack and 1-Slack Structured-Output Support Vector Machine

The modification of the loss and the mapping allows us the integration of output-space similarities into the kernel function. However, to exploit this new source of information, the whole SVM must be reformulated: it is easy to see that the mapping $\boldsymbol{\Psi}(\mathbf{x}, y)$ cannot be computed for test points, for which the class membership is obviously unknown. In order to solve this general problem in structured learning, specific SVM formulations must be developed. Several strategies have been proposed for the SSVM (Altun *et al.*, 2003; Joachims *et al.*, 2005; Taskar *et al.*, 2003; Tsochantaridis *et al.*, 2004, 2005), but the formulation of Tsochantaridis *et al.* (2005) is the most general as it includes the rest as particular cases. This formulation is usually referred to as the N-slack SSVM (N-SSVM), since it assigns a different slack variable to each of the N training examples. Specifically, in the margin-rescaling version of Tsochantaridis *et al.* (2005), the position of the hinge is adapted while the slope is fixed. Each possible output is considered and the model is constrained iteratively by adding constraints on the $(\boldsymbol{x}, \boldsymbol{y})$ pairs that most violate the SVM solution (note that $\boldsymbol{y}$ has become a vector containing all possible outputs). In other words, a sort of regularization is done, restricting the set of possible functions h. This way, the formulation becomes

$$\min_{w,\xi} \left\{ \frac{1}{2}\|w\|^2 + \frac{C}{N}\sum_{i=1}^{N} \xi_i \right\} \tag{10.90}$$

$$\forall i : \xi_i \geq 0$$
$$\underbrace{\forall \hat{y} \in \mathcal{Y},\ \forall i}_{\text{(a)}} : \underbrace{\langle w, \Psi(\mathbf{x}_i, y_i)\rangle}_{\text{(b)}} - \underbrace{\langle w, \Psi(x_i, \hat{y}_i)\rangle}_{\text{(c)}} \geq \underbrace{\Delta(y_i, \hat{y})}_{\text{(d)}} - \xi_i. \tag{10.91}$$

The objective is the conventional regularized risk used in SVMs. The constraints state that for each incorrect label and for each training example (x_i, y_i) (a), the score $\langle w, \Psi(x_i, y_i)\rangle$ of the correct structure y_i (b) must be greater than the score $\langle w, \Psi(\mathbf{x}_i, \hat{y}_i)\rangle$ of all incorrect structures $\hat{y}$ (c) by a required margin (d). If the margin is violated, the slack variable ξ_i becomes nonzero.

This quadratic program involves a very large, possibly infinite, number of linear inequality constraints, which makes it impossible to be optimized explicitly. Alternatively, the problem is solved by using *delayed constraint generation*, where only a finite and small subset of constraints is taken into account. An efficient method to solve the problem is the N-slack algorithm originally reformulated by Joachims *et al.* (2009): the 1-slack SSVM (1-SSVM). The model has a unique slack variable ξ applied to all the constraints. The interested reader can find the proof for the equivalence with the N-slack formulation in Joachims *et al.* (2009). The n cutting planes of the previous model are replaced by a single cutting plane for the sum of the hinge losses. In this sense, Equations 10.90 and 10.91 can be replaced by

$$\min_{w,\xi} \left\{ \frac{1}{2}\|w\|^2 + \frac{C}{N}\xi \right\} \tag{10.92}$$

$$\forall \hat{y} \in \mathcal{Y} : \frac{1}{N}\sum_{i=1}^{N} \left(\langle w, \Psi(x_i, y_i)\rangle - \langle w, \Psi(x_i, \hat{y}_i)\rangle \right) \geq \frac{1}{N}\sum_{i=1}^{N} \Delta(y_i, \hat{y}) - \xi, \tag{10.93}$$

where $\xi = 1/N \sum_i \xi_i$. The 1-SSVM, as proposed in Joachims *et al.* (2009), starts with an empty working set of constraints $\mathcal{W} = \emptyset$, and then the solution is computed over the current $\mathcal{W}$, finding the most violated constraint (just one for all the training points) and adding it up to the working set. The algorithm terminates when no constraint is added in the previous iteration; that is, when all the constraints are fulfilled up to a precision ϵ. Unlike the N-SSVM, only one constraint is added at each iteration. This new formulation has $|\mathcal{Y}|^N$ constraints, one for each possible combination of labels $[y_1, \ldots, y_N]$, but only one slack variable ξ is shared across all constraints. See Joachims *et al.* (2009) for details.

An illustrative example using a linear SVM and the SVM^{struct} toolbox (see pointers on the book web page) is given in Listing 10.1. Complete operational source code can be found in the toolbox file `test_svm_struct_learn.m` available from the links before. The goal here is to learn a predictor function $x \mapsto y$ that smoothly fits the training data. The structured SVM formulation of such a prediction function is

$$\hat{y}(x) = \text{argmax}_{y \in \{-1,+1\}} F(x, y; w), \quad F(x, y; w) = \langle w, \Psi(x, y)\rangle,$$

where w is the parameter vector to be learned, $\Psi(x, y) \in \mathbb{R}^2$ is the *joint feature map*, and $F(x, y; w)$ is an auxiliary function usually interpreted as a *compatibility score* between input $\boldsymbol{x}$ and output y. In this example, a simple $\boldsymbol{\Psi}(\boldsymbol{x}, y) = \boldsymbol{x}y/2$ is defined, which makes the structured SVM formulation equivalent to a standard binary SVM (see `featureCB` function). Then we define a loss function $\Delta(y, \hat{y})$ measuring how well the prediction $\hat{y}$ matches the ground truth y. We use the 01-loss here, which, for our binary labels, writes $\Delta(y, \hat{y}) = (1-y\hat{y})/2$ (see `lossCB` function). Finally, one defines the constraint generation function which captures the structure of the problem. Generating a constraint for an input–output pair $(\boldsymbol{x}, y)$ means identifying what is the most incorrect output $\hat{y}$ that the current model still deems to be compatible with the input $\boldsymbol{x}$ (see `constraintCB` function). Function weights $\boldsymbol{w}$ can be accessed in `model.w`. The SVM-struct call sets the C constant of the SVM to 1 (`-c 1.0`), allows slack rescaling (`-o 1`), and verbosity is activated using (`-v 1`).

```
function ssvmcode

% Code snippet simplifying the function test_svm_struct_learn.m from
% Andrea Vedaldi available in the SVM-Struct toolbox in
%     http://www.robots.ox.ac.uk/~vedaldi/svmstruct.html#download
                                                    -and-install
%
% - Assume have a set of input data X with structured outputs Y

% Initialize parameters
parm.patterns = X;                        % cellarray of patterns (inputs).
parm.labels = Y;                          % cellarray of labels (outputs).
parm.lossFn = @lossCB;                    % loss function callback.
parm.constraintFn  = @constraintCB; % constraint generation callback.
parm.featureFn = @featureCB;              % feature map callback.
parm.dimension = 2;                       % feature dimension.
% Run SVM struct
model = svm_struct_learn(' -c 1.0 -o 1 -v 1 ', parm);
% Return model weights
w = model.w ;

% Some SVM struct callbacks
function psi = featureCB(param, x, y)
    psi = sparse(y*x/2) ;
end
function delta = lossCB(param, y, ybar)
    delta = double(y ~= ybar) ;
end
function yhat = constraintCB(param, model, x, y)
% slack resaling: argmax_y delta(yi, y) (1 + <psi(x,y), w> - <psi(x,yi), ...
    w>)
% margin rescaling: argmax_y delta(yi, y) + <psi(x,y), w>
    if dot(y*x, model.w) > 1, yhat = y ; else yhat = - y ; end
end
```

Listing 10.1 Illustration of the structured SVM (ssvmcode.m).

10.3.5 Active Learning

When designing a supervised classifier, the performance of the model depends strongly on the quality of the labeled information available. This constraint makes the generation of an appropriate training set a difficult and expensive task requiring extensive manual human interaction. Therefore, in order to make the models as efficient as possible, the training set should be kept as small as possible and focused on the samples that really help to improve the performance of the model. The basic idea is that a classifier trained on a small set of well-chosen examples can perform as well as a classifier trained on a larger number of randomly chosen examples (Cohn *et al.* 1994, 1996; MacKay 1992).

AL aims at constructing effective and compact training sets. AL uses the interaction between the user and the classifier to achieve this goal. The model returns to the user the samples whose classification outcomes are the most uncertain, the user provides their label, and the samples are reused to improve accuracy. This way, the model is optimized on well-chosen difficult examples, maximizing its generalization capabilities.

Active Learning Concepts and Definitions

Let $X = \{(\boldsymbol{x}_i, y_i)\}_{i=1}^{l}$ be a training set of labeled samples, with $\boldsymbol{x}_i \in \mathcal{X}$ and $y_i = \{1, \ldots, l\}$. $\mathcal{X}$ is the d-dimensional input space $\mathbb{R}^d$. Let also $U = \{\boldsymbol{x}_i\}_{i=l+1}^{l+u} \in \mathcal{X}$, with $u \gg l$, be the *pool of candidates*, a set of unlabeled patterns to be sampled. AL algorithms are iterative sampling schemes, where a classification model is adapted regularly by feeding it with new labeled samples corresponding to the most beneficial ones to improve the model's performance. These samples are usually the ones lying in the areas of *uncertainty* of the model, and their inclusion in the training set forces the model to solve such uncertainty regions of low confidence. For a given iteration t, the algorithm selects, from the pool U^t, the q candidates that simultaneously maximize the performance and reduce model uncertainty if they are added to the current training set X^t. Once the batch of samples $S^t = \{\boldsymbol{x}_m\}_{m=1}^{q} \subset U$ has been selected, it is labeled by an oracle (usually a human expert); that is, the labels $\{y_m\}_{m=1}^{q}$ are discovered. Finally, the set S^t is both added to the current training set ($X^{t+1} = X^t \cup S^t$) and removed from the pool ($U^{t+1} = U^t \backslash S^t$), and the process is iterated until a stopping criterion is met. Algorithm 1 summarizes the active selection process. From now on, iteration index t will be omitted for the sake of clarity.

Algorithm 1 General AL algorithm.

Inputs
- Initial training set $X^t = \{\boldsymbol{x}_i, y_i\}_{i=1}^{l}$ ($X \in \mathcal{X}, t = 1$).
- Pool of candidates $U^t = \{\boldsymbol{x}_i\}_{i=l+1}^{l+u}$ ($U \in \mathcal{X}, t = 1$).
- Number of samples q to add at each iteration (define set S).

1: **repeat**
2: Train a model with current training set X^t.
3: **for** each candidate in U^t **do**
4: Evaluate a user-defined *heuristic*
5: **end for**
6: Rank the candidates in U^t according to the score of the heuristic
7: Select the q most interesting samples. $S^t = \{\boldsymbol{x}_k\}_{k=1}^{q}$
8: The user assigns a label to selected samples. $S^t = \{\boldsymbol{x}_k, y_k\}_{k=1}^{q}$
9: Add the batch to the training set $X^{t+1} = X^t \cup S^t$.
10: Remove the batch from the pool of candidates $U^{t+1} = U^t \backslash S^t$
11: $t = t + 1$
12: **until** a stopping criterion is met.

An AL process requires interaction between the oracle and the model: the former provides the labeled information and the knowledge about the desired classes, while the latter provides both its interpretation of the classes distribution and the most relevant samples that would be needed in the future in order to solve the encountered discrepancies. This is the key point for the success of an AL algorithm: the machine needs a *heuristic* to rank the samples in the pool U. The *heuristics* are what differentiate the algorithms proposed in the next sections, and can be divided into three main families: (1) *committee*-based heuristics; (2) *large-margin*-based heuristics; and (3) *posterior probability* based heuristics. Let us now review each one of the families.

Committee-Based Active Learning

Committee-based AL methods quantify the uncertainty of a sample by considering a committee of learning models (Freund *et al.* 1997; Seung *et al.* 1992). Each model of the committee is based on different hypotheses about the classification problem, and thus provides different labels for the samples in the pool of candidates. The committee-based heuristic is based on selecting the samples showing maximal disagreement between the different classification models in the committee. Examples of these methods are the one presented in Freund *et al.* (1997), which uses Gibbs sampling to build the committees, or methods based on bagging and boosting (Abe and Mamitsuka, 1998). The main advantage of these methods is that they are applicable to any kind of classifiers.

Normalized Entropy Query-by-Bagging

Abe and Mamitsuka (1998) proposed bagging (*bootstrap aggregation* (Breiman, 1994)) to build the committee. First, k training sets built drawing with replacement of the original data are defined. These sets account for a part of the available labeled samples only. Then, each set is used to train a classifier and to predict the u labels of the candidates. At the end of the procedure, k predictions are provided for each candidate $\boldsymbol{x}_i \in U$. Once we have these k predictions, a *heuristic* has to be defined in order to *rank* them. In Tuia *et al.* (2009), the entropy H^{BAG} of the distribution of the predictions provided by the k classifiers for each sample $\boldsymbol{x}_i$ in U is used as the heuristic. In Copa *et al.* (2010), this measure is normalized in order to bound it with respect to the number of classes predicted by the committee and avoid hot spots of the value of uncertainty in regions where several classes overlap. The *normalized entropy query-by-bagging* (nEQB) heuristic is defined as

$$\hat{\boldsymbol{x}}^{\text{nEQB}} = \arg\max_{\boldsymbol{x}_i \in U} \left\{ \frac{H^{\text{BAG}}(\boldsymbol{x}_i)}{\log(N_i)} \right\}, \tag{10.94}$$

where

$$H^{\text{BAG}}(\boldsymbol{x}_i) = -\sum_{\omega=1}^{N_i} p^{\text{BAG}}(y_i^* = \omega|\boldsymbol{x}_i) \log[p^{\text{BAG}}(y_i^* = \omega|\boldsymbol{x}_i)] \tag{10.95}$$

$$\text{with} \quad p^{\text{BAG}}(y_i^* = \omega|\boldsymbol{x}_i) = \frac{\sum_{m=1}^{k} \delta(y_{i,m}^*, \omega)}{\sum_{m=1}^{k} \sum_{j=1}^{N_i} \delta(y_{i,m}^*, \omega_j)}$$

is an empirical measure of entropy, y_i^* is the prediction for the sample $\boldsymbol{x}_i$, and $p^{\text{BAG}}(y_i^* = \omega|\boldsymbol{x}_i)$ is the observed probability to have the class ω predicted using the training set X by the committee of k models for the sample $\boldsymbol{x}_i$. N_i is the number of classes predicted for sample $\boldsymbol{x}_i$ by the committee, with $1 \leq N_i \leq N$. The $\delta(y_{i,m}^*, \omega)$ operator returns 1 if the classifier using the mth bag classifies the sample $\boldsymbol{x}_i$ into class ω and 0 otherwise. Entropy maximization is a natural multiclass heuristic. A candidate for which all the classifiers in the committee agree is associated with null entropy, and its inclusion in the training set will not provide additional information. On the other hand, a candidate with maximum disagreement between the classifiers results in maximum entropy, and including it in the training set will provide much information.

Adaptive Maximum Disagreement

When working with high-dimensional datasets it is useful to construct the committee by splitting the feature space into a number of subsets, or *views* (Muslea, 2006). Di and Crawford (2010) exploit this principle to generate different views of a particular image on the basis of the block-diagonal structure of the covariance matrix. By generating views corresponding to the different blocks, independent classifications of the same sample can be generated and an entropy-based heuristic can be used similarly to nEQB.

Given a partition of the d-dimensional input space into V disjoint views accounting for data subsets $\boldsymbol{x}^v$ such that $\bigcup_{v=1}^{V} \boldsymbol{x}^v = \boldsymbol{x}$, the *adaptive maximum disagreement* (AMD) heuristic selects candidates according to

$$\hat{\boldsymbol{x}}^{\text{AMD}} = \arg\max_{\boldsymbol{x}_i \in U} H^{\text{MV}}(\boldsymbol{x}_i), \tag{10.96}$$

where the multiview entropy H^{MV} is defined for each view v as

$$H^{\text{MV}}(\boldsymbol{x}_i) = -\sum_{\omega=1}^{N_i} p^{\text{MV}}(y_{i,v}^* = \omega|\boldsymbol{x}_i^v) \log[p^{\text{MV}}(y_{i,v}^* = \omega|\boldsymbol{x}_i^v)] \tag{10.97}$$

$$\text{where} \quad p^{\text{MV}}(y_i^* = \omega|\boldsymbol{x}_i^v) = \frac{\sum_{v=1}^{V} W^{t-1}(v,\omega)\delta(y_{i,v}^*, \omega)}{\sum_{v=1}^{V}\sum_{j=1}^{N_i} W^{t-1}(v,\omega)},$$

where the $\delta(y_{i,v}^*, \omega)$ operator returns 1 if the classifier using the view v classifies the sample y_i into class ω and 0 otherwise. $\boldsymbol{W}^{t-1}$ is an $N \times V$ weighting matrix accounting for the abilities of discrimination of the views in the different classes. At each iteration, $\boldsymbol{W}^{t-1}$ is updated using the true labels of the samples obtained at iteration $t-1$:

$$W^t(v,\omega) = W^{t-1}(v,\omega) + \delta(y_{i,v}, \omega), \quad \forall i \in S, \tag{10.98}$$

and its columns are normalized to unitary sum. This matrix weights the confidence of each view to predict a given class.

Large-Margin-Based Active Learning

This family of methods is specific to margin-based classifiers such as SVM. SVM and similar methods are naturally good base methods for large-margin AL: the distance to the separating hyperplane is a straightforward way of estimating the classifier confidence on an unseen sample. Let us first consider a binary problem; the distance of a sample $\boldsymbol{x}_i$ from the SVM hyperplane is given by

$$f(\boldsymbol{x}_i) = \sum_{j=1}^{N} \alpha_j y_j K(\boldsymbol{x}_j, \boldsymbol{x}_i) + b, \tag{10.99}$$

where $K(\boldsymbol{x}_j, \boldsymbol{x}_i)$ is the kernel, y_j are the labels of the support vectors α_j, and b is the bias term.

This evaluation of the distance is the base ingredient of almost all large-margin heuristics. Roughly speaking, these heuristics use the intuition that a sample away from the decision boundary (with a high $f(\boldsymbol{x}_i)$) has a high confidence about its class assignment and is thus not interesting for future sampling.

In the particular case of SVM, since SVMs rely on a sparse representation of the data, large-margin-based heuristics aim at finding the samples in U that are most likely to receive a nonzero α_i weight if added to X. That is, the points more likely to become support vectors are the ones lying within the margin of the current model (Tong and Koller, 2001). The heuristic based on this idea is called margin sampling (MS) (Campbell *et al.*, 2000; Schohn and Cohn, 2000). A variation on this idea, aiming at minimizing the risk of selecting points that will not become support vectors, can be found in Zomer *et al.* (2004) and Cheng and Shih (2007).

Margin Sampling

MS is based on SVM geometrical properties; in particular, on the fact that unbounded support vectors are labeled samples that lie on the margin with a decision function value of exactly 1 (Boser *et al.*, 1992; Schölkopf and Smola, 2002). For example, look at the pool of candidates of Figure 10.10a referring to a three-class toy problem. Considering a multiclass OAA setting, the distance to each hyperplane is represented by Figure 10.10d–f.

The MS heuristic selects the samples from the pool of candidates that minimize

$$\hat{\boldsymbol{x}}^{\text{MS}} = \arg\min_{\boldsymbol{x}_i \in U} \{\min_{\omega} |f(\boldsymbol{x}_i, \omega)|\}, \tag{10.100}$$

where $f(\boldsymbol{x}_i, \omega)$ is the distance of the sample to the hyperplane defined for class ω in an OAA setting for multiclass problems. The MS heuristic for the toy problem is reported in Figure 10.10b. The MS heuristic can be found in the literature with different names, such as *most ambiguous* (Ferecatu and Boujemaa, 2007), *binary-level uncertainty* (Demir *et al.*, 2011), or $\text{SVM}_{\text{SIMPLE}}$ (Di and Crawford, 2010).

Multiclass Level Uncertainty

Demir *et al.* (2011) extended the idea of MS to multiclass uncertainty; see Vlachos (2008). The MCLU considers the difference between the distance to the margin for the two most probable classes

$$\hat{\boldsymbol{x}}^{\text{MCLU}} = \arg\min_{\boldsymbol{x}_i \in U} \{f(\boldsymbol{x}_i)^{\text{MC}}\} \tag{10.101}$$

$$\text{where} \quad f(\boldsymbol{x}_i)^{\text{MC}} = \max_{\omega \in N} \{f(\boldsymbol{x}_i, \omega)\} - \max_{\omega \in N \setminus \omega^+} \{f(\boldsymbol{x}_i, \omega)\}, \tag{10.102}$$

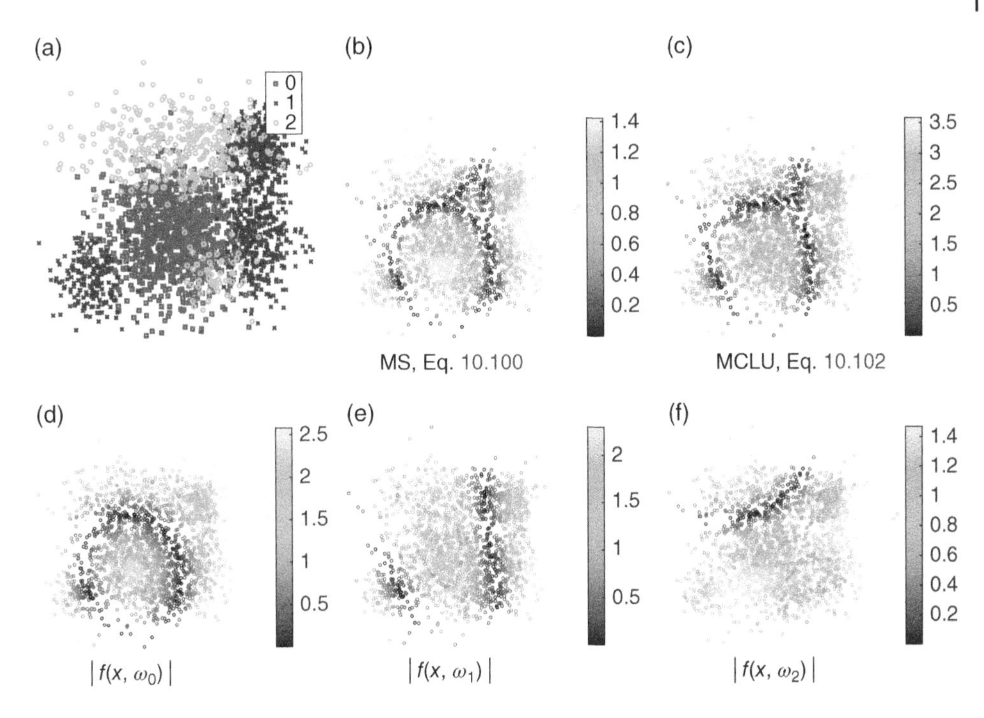

Figure 10.10 (a) Large-margin heuristics for a three-class toy example. The color intensity represents the distance from the hyperplane: (b) MS heuristic; (c) multiclass level uncertainty (MCLU) heuristic; darker areas are the ones with maximal uncertainty, minimizing Equation 10.100 or 10.102 respectively. (d)–(f) Absolute values of per-class distances.

and ω^+ is the class showing maximal confidence. In contrast with the MS heuristic, which only considers the most uncertain class of the SVM, the MCLU assesses the uncertainty between the two most likely classes. A high value of this heuristic corresponds to samples assigned with high certainty to the most confident class, whereas a low value represents the opposite. Figure 10.10c shows this heuristic compared with MS. Although very similar, MCLU shows differences in areas where the three classes mix.

Significance Space Construction

In the significance space construction (SSC) heuristic (Pasolli and Melgani, 2010), the training samples are used to define a second classification function $f(\boldsymbol{x})^{\text{SSC}}$ where the training samples with $\alpha_j > 0$ (support vectors) are classified against the training samples with $\alpha_j = 0$ (non-support vectors). When applied to the pool of candidates, this second classifier predicts which samples are likely to become support vectors:

$$\hat{\boldsymbol{x}}^{\text{SSC}} = \arg_{\boldsymbol{x}_i \in U}\{f(\boldsymbol{x}_i)^{\text{SSC}} > 0\}. \tag{10.103}$$

The candidates more likely to become support vectors are added to the training set X.

Algorithm 2 General diversity-based heuristic (for a single iteration).

Inputs
- Current training set $X^t = \{\boldsymbol{x}_i, y_i\}_{i=1}^l$ $(X \in \mathcal{X})$.
- Subset of candidates pool minimizing Eq. 10.100 or 10.101 $U^* = \{\boldsymbol{x}_i\}$ $(U^* \in \mathcal{X}$ and $U^* \subset U^t)$.
- Number of samples q to add at each iteration (define set S).

1: Add the sample minimizing Eq. 10.100 or 10.101 to S.
2: **repeat**
3: Compute the diversity criterion between samples in U^* and in S.
4: Select the sample $\boldsymbol{x}_{\mathrm{D}}$ maximizing diversity with current batch.
5: Add $\boldsymbol{x}_{\mathrm{D}}$ to current batch $S = S \cup \boldsymbol{x}_{\mathrm{D}}$.
6: Remove $\boldsymbol{x}_{\mathrm{D}}$ to current list of cadidates $U^* = U^* \setminus \boldsymbol{x}_{\mathrm{D}}$.
7: **until** batch S contains q elements.
8: The user labels the selected samples $S = \{\boldsymbol{x}_k, y_k\}_{k=1}^q$.
9: Add batch to the training set $X^{t+1} = X^t \cup S$.
10: Remove batch from the pool of candidates $U^{t+1} = U^t \backslash S$.

Diversity Criteria

In addition to have a heuristic to rank and select samples among the pool of candidates, before adding them to the training set it is also important to ensure they are diverse (Brinker, 2003). A diversity criterion allows one to reject candidates that rank high according to the heuristic but are redundant between each other. Algorithm 2 shows the general diversity heuristic.

In Ferecatu and Boujemaa (2007) the margin sampling heuristic is constrained with a measure of the angle between candidates in the feature space. This heuristic, named *most ambiguous and orthogonal* (MAO) is as follows: (1) samples are selected by MS; (2) from this subset, the algorithm iteratively selects the samples minimizing the highest values between the candidates and the samples already included in the training set. That is, each new sample in selected using

$$\hat{\boldsymbol{x}}^{\mathrm{MAO}} = \arg\min_{\boldsymbol{x}_i \in U^{\mathrm{MS}}} \{\max_{\boldsymbol{x}_j \in S} K(\boldsymbol{x}_i, \boldsymbol{x}_j)\}. \tag{10.104}$$

Demir *et al.* (2011) applied the MAO criterion among a subset of U maximizing the MCLU criterion (instead of MS), defining the MCLU angle-based diversity (MCLU-ABD) heuristic. But more importantly, the authors generalize the MAO heuristic to any type of kernels by including normalization in the feature space:

$$\hat{\boldsymbol{x}}^{\mathrm{MCLU\text{-}ABD}} = \arg\min_{\boldsymbol{x}_i \in U^{\mathrm{MCLU}}} \left\{ \lambda f(\boldsymbol{x}_i)^{\mathrm{MC}} + (1-\lambda) \max_{\boldsymbol{x}_j \in S} \frac{K(\boldsymbol{x}_i, \boldsymbol{x}_j)}{\sqrt{K(\boldsymbol{x}_i, \boldsymbol{x}_i) K(\boldsymbol{x}_j, \boldsymbol{x}_j)}} \right\} \tag{10.105}$$

where $f(\boldsymbol{x}_i)^{\mathrm{MC}}$ is the uncertainty function defined by Equation 10.102.

Another diversity criterion is the one proposed in Tuia *et al.* (2009), where first the MS heuristic is used to build a subset, U^{MS}, and then the diversity is obtained by choosing samples near to different support vectors. This approach improved the diversification of the MS heuristic selecting samples as a function of the geometrical distribution of the support vectors. However, this approach does not guarantee diversity among the selected samples, because a pair of samples associated with different support vectors can be in fact close to each other. The *closest support vector* diversity heuristic is defined as

$$\hat{\boldsymbol{x}}^{\mathrm{cSV}} = \arg\min_{\boldsymbol{x}_i \in U^{\mathrm{MS}}} \{ |f(\boldsymbol{x}_i, \omega)| \mid \mathrm{cSV}_i \notin \mathrm{cSV}_\theta \} \tag{10.106}$$

where $\theta = [1, \dots, q-1]$ are the indices of the already selected candidates and cSV is the set of selected closest support vectors.

Another way of ensuring diversity is using clustering in the feature space. Kernel k-means (Dhillon *et al.* 2005; Girolami 2002a; Shawe-Taylor and Cristianini 2004) was used by Demir *et al.* (2011) to cluster the samples selected by MCLU and select diverse batches. First, a set of samples selected using the MCLU criterion, U^{MCLU}, is obtained. Then this set is partitioned into q clusters using kernel k-means. Finally, the MCLU enhanced cluster-based diversity (MCLU-ECBD) selects a single sample per cluster minimizing the following function:

$$\hat{\boldsymbol{x}}^{\mathrm{MCLU\text{-}ECBD}} = \arg\min_{\boldsymbol{x}_i \in c_m} \{ f(\boldsymbol{x}_i)^{\mathrm{MC}} \}, \quad m = [1, \dots, q], \boldsymbol{x}_i \in U^{\mathrm{MCLU}} \tag{10.107}$$

where c_m is one among the q clusters.

A binary hierarchical variant of this method is proposed in Volpi *et al.* (2012) that aims to exclude the less informative samples from the selected batch, which are the ones more likely to become bounded support vectors. Cluster kernel k-means is used to reduce the redundancy among samples in the feature space, whereas building clusters excluding bounded support vectors maximizes the informativeness of each added sample. At each iteration, the *informative hierarchical margin cluster sampling* (hMCS-i) builds a dataset composed of a subset of samples selected using the MCLU criterion, U^{MCLU}, and the bounded support vectors obtained at the previous iteration. Then this subset is iteratively partitioned using kernel k-means in a binary way. In each partition the biggest cluster is always selected. The partition process continues until q clusters with no bounded support vectors are obtained. Then candidates are selected among these q clusters according to

$$\hat{\boldsymbol{x}}^{\mathrm{hMCS\text{-}i}} = \arg\min_{\boldsymbol{x}_i \in c_m} \{ f(\boldsymbol{x}_i)^{\mathrm{MC}} \}, \quad m = [1, \dots, q | n_{c_m}^{\mathrm{bSV}} = 0], \boldsymbol{x}_i \in U^{\mathrm{MCLU}} \tag{10.108}$$

Posterior-Probability-Based Active Learning

The third class of AL methods uses the estimation of posterior probabilities of class membership (i.e., $p(y|\boldsymbol{x})$) to rank the candidates. The main drawback of this method is that it can only be applied to classifiers providing probabilistic outputs.

KL-max

The idea of these methods is to select samples whose inclusion in the training set would maximize the changes in the posterior distribution. In (Roy and McCallum, 2001), the proposed heuristic maximizes the KL divergence between the distributions before and after adding the candidate. Each candidate is removed from U and included in the training set using the label maximizing its posterior probability. The KL divergence is then computed between the posterior distributions of the models with and without the candidate. After computing this measure for all candidates, those maximizing the criterion KL-max are selected:

$$\hat{\boldsymbol{x}}^{\text{KL-max}} = \arg\max_{\boldsymbol{x}_i \in U} \left\{ \sum_{\omega \in N} \frac{1}{(u-1)} \text{KL}(p^+(\omega|\boldsymbol{x})||p(\omega|\boldsymbol{x}))p(y_i^* = \omega|\boldsymbol{x}_i) \right\} \tag{10.109}$$

where

$$\text{KL}(p^+(\omega|\boldsymbol{x})||p(\omega|\boldsymbol{x})) = \sum_{\boldsymbol{x}_j \in U \backslash \boldsymbol{x}_i} p^+(y_j^* = \omega|\boldsymbol{x}_j) \log \frac{p^+(y_j^* = \omega|\boldsymbol{x}_j)}{p(y_j^* = \omega|\boldsymbol{x}_j)} \tag{10.110}$$

and $p^+(\omega|\boldsymbol{x})$ is the posterior distribution for class ω and sample $\boldsymbol{x}$, estimated using the increased training set $X^+ = [X, (\boldsymbol{x}_i, y_i^*)]$, with y_i^* being the class maximizing the posterior. Jun and Ghosh (2008) extended this approach proposing to use boosting to weight samples that were previously selected but were no longer relevant for the current classifier. These heuristics are only feasible when using classifiers with low computational cost, since in each iteration $u + 1$ models have to be trained.

Breaking Ties

This strategy, similar to nEQB presented in Section 10.3.5, consists of assessing the conditional probability of predicting a given label $p(y_i^* = \omega|\boldsymbol{x}_i)$ for each candidate $\boldsymbol{x}_i \in U$. This is estimated as the predictions for all the candidates $y_i^* = \arg\max_{\omega \in N} f(\boldsymbol{x}_i, \omega)$. Some classification algorithms already provide these estimations, such as probabilistic NNs or ML classifiers. For SVM, Platt's probability estimations can be used (Platt, 1999b). Once the posterior probabilities are obtained, it is straightforward to assess the uncertainty of the class membership for each candidate. The breaking ties (BT) heuristic chooses candidates showing a near-uniform probability of belonging to each class; that is, $p(y_i^* = \omega|\boldsymbol{x}_i) = 1/N, \ \forall \omega \in N$.

The BT heuristic for a binary problem is based on the smallest difference of the posterior probabilities for each sample (Luo *et al.*, 2005). In a multiclass setting, and independently of the number of classes, the difference between the two highest probabilities reveals how a sample is considered by the classifier. If the two highest probabilities for a given sample are similar, then the classifier's confidence on assigning that sample to one of the classes is low. The BT heuristic is formulated as

$$\hat{\boldsymbol{x}}^{\text{BT}} = \arg\min_{\boldsymbol{x}_i \in U} \left\{ \max_{\omega \in N} \{p(y_i^* = \omega|\boldsymbol{x})\} - \max_{\omega \in N \backslash \omega^+} \{p(y_i^* = \omega|\boldsymbol{x})\} \right\} \tag{10.111}$$

where ω^+ is the class showing maximal probability. Comparing Equation 10.101 with Equation 10.111, the relationship between BT and the MCLU heuristic presented in Section 10.3.5 is clear.

10.4 Large-Scale Support Vector Machines

This section pays attention to the important caveat of SVM and kernel classifiers: the huge computational cost involved when many input data are available. We will summarize then the main algorithmic approximations existing nowadays in the literature to alleviate such computationally demanding cost. We will pay special attention to approaches that either break a large QP problem into a series of manageable QP subproblems or approximate the kernel with random feature projections. For the former we will study the SMO (Platt, 1999a) and other related approaches, while for the latter we will review the method of random Fourier features (Rahimi and Recht, 2007). We will finish the +section by paying attention to standard techniques for parallelizing SVMs.

10.4.1 Large-Scale Support Vector Machine Implementations

Some alternatives to solve the QP problem implied in SVMs exist. Note that, in the case of the SVM, the QP problem is directly expressed as a function of the training kernel matrix $\boldsymbol{K}$, which contains the similarity (distance) among all training samples. Thus, solving this problem requires storing the matrix and making operations with it. One of the most effective interior point methods (IPMs) for solving the QP problem with linear constraints is the primal–dual IPM (Mehrotra, 1992), which essentially tries to remove inequality constraints using a bounding function, and then exploit the iterative Newton's method to solve the KKT conditions related to the Hessian matrix $\boldsymbol{Q}$. This is very computationally demanding, as it grows as $\mathcal{O}(N^3)$ in time and $\mathcal{O}(N^2)$ in space, where N is the number of training samples.

Some alternative algorithms have appeared to solve the problem in reasonable time and amount of memory for several thousands of training samples. A powerful approach to scale up SVM training is by using decomposition methods (Osuna *et al.*, 1997), which break a large QP problem into a series of manageable QP subproblems. The most popular decomposition method is the SMO (Platt, 1999a), in which smaller QP problems are solved analytically, which avoids using a time-consuming numerical QP optimization as an inner loop. The amount of memory required for SMO is linear in the training set size, which allows SMO to handle very large training sets. Without kernel caching, SMO scales somewhere between linear and quadratic in the training set size for various test problems, while a standard projected CG chunking algorithm scales somewhere between linear and cubic in the training set size. SMO's computation time is dominated by SVM evaluation; hence, SMO is fastest for linear SVMs and sparse data sets. Despite these advantages, the SMO algorithm still requires a large amount of computation time for solving large-scale problems, and also the kernel storage, which is of the size of the square of training points.

In recent years, other scale-up strategies have been proposed. The SimpleSVM algorithm (Vishwanathan *et al.*, 2003) makes use of a greedy working set selection strategy to identify the most relevant samples (support vectors) for incremental updating of the kernel submatrix. Bordes *et al.* (2005) proposed an online SVM learning together with active example selection. In Sonnenburg *et al.* (2006b), special data structures for fast computations of the kernel in their chunking algorithm are exploited. In addition, instead of maximizing the dual problem as is usually done, Chapelle (2007) proposed directly minimizing the primal problem. Some other powerful algorithms for solving

SVM have recently appeared, such as the Core SVM or the ball vector machine. However, large-scale problems involving $N>10^5$ samples require more efficiency that is still needed. However, the reader should consult Bottou *et al.* (2007) for a more up-to-date treatment of large-scale SVMs.

Since these initial approaches, many efforts have been made to deliver large-scale versions of kernel machines able to work with several thousands of examples (Bottou *et al.*, 2007). They typically resort to reducing the dimensionality of the problem by decomposing the kernel matrix using a subset of basis: for instance, using Nyström eigendecompositions (Sun *et al.*, 2015), sparse and low-rank approximations (Arenas-García *et al.*, 2013; Fine *et al.*, 2001), or smart sample selection (Bordes *et al.*, 2005). However, there is no clear evidence that these approaches work in general, given that they are mere approximations to the kernel computed with all (possibly millions of) samples.

10.4.2 Random Fourier Features

Let us now review a different approximation to deal with the kernel parameterization and optimization simultaneously. Selecting and optimizing a kernel function is very challenging even with moderate amounts of data. An alternative pathway to that of optimization has recently captured much attention in machine learning: rather than *optimization*, one trusts to *randomization*. While it may appear odd at first glance, the approach has surprisingly yielded competitive results in recent years, being able to exploit many samples at a fraction of the computational cost. Besides its practical convenience, the approximation of the kernel with random basis is theoretically consistent as well. The seminal work of Rahimi and Recht (2007) presented the randomization framework, which relies on an approximation to the empirical kernel mapping of the form

$$K(\boldsymbol{x}, \mathbf{y}) = \langle \boldsymbol{\phi}(\boldsymbol{x}), \boldsymbol{\phi}(\boldsymbol{y}) \rangle \approx \boldsymbol{z}(\boldsymbol{x})^{\mathrm{T}} \boldsymbol{z}(\boldsymbol{y}),$$

where the implicit mapping $\boldsymbol{\phi}(\cdot)$ is replaced with an *explicit* (low-dimensional) feature mapping $\boldsymbol{z}(\cdot)$ of dimension D.

Consequently, one can simply transform the input with $\boldsymbol{z}$, and then apply fast linear learning methods to approximate the corresponding nonlinear kernel machine. This approach not only provides extremely fast learning algorithms, but also excellent performance in the test phase: for a given test point $\boldsymbol{x}$, instead of $f(\boldsymbol{x}) = \sum_{i=1}^{N} \alpha_i K(\boldsymbol{x}_i, \boldsymbol{x})$, which requires $\mathcal{O}(Nd)$ operations, one simply does a linear projection $f(\boldsymbol{x}) = \boldsymbol{w}^{\mathrm{T}} \boldsymbol{z}$, which requires $\mathcal{O}(D+d)$ operations. The question now is how to construct efficient and sensible $\boldsymbol{z}$ mappings. Rahimi and Recht (2007) also introduced a particular technique to do so-called *random kitchen sinks* (RKSs).

Bochner's theorem (Reed and Simon, 1981) states that a continuous kernel $K(\boldsymbol{x}, \boldsymbol{x}') = K(\boldsymbol{x} - \boldsymbol{x}')$ on $\mathbb{R}^d$ is positive definite if and only if $\boldsymbol{K}$ is the FT of a nonnegative measure. If a shift-invariant kernel $\boldsymbol{K}$ is properly scaled, its FT $p(\boldsymbol{w})$ is a proper probability distribution. This property is used to approximate kernel functions and matrices with linear projections on a number of D random features as follows:

$$K(\boldsymbol{x}, \boldsymbol{x}') = \int_{\mathbb{R}^d} p(\boldsymbol{w}) \mathrm{e}^{-j\boldsymbol{w}^{\mathrm{T}}(\boldsymbol{x}-\boldsymbol{x}')} \mathrm{d}\boldsymbol{w} \approx \sum_{i=1}^{D} \frac{1}{D} \mathrm{e}^{-j\boldsymbol{w}_i^{\mathrm{T}}\boldsymbol{x}} \mathrm{e}^{j\boldsymbol{w}_i^{\mathrm{T}}\boldsymbol{x}'},$$

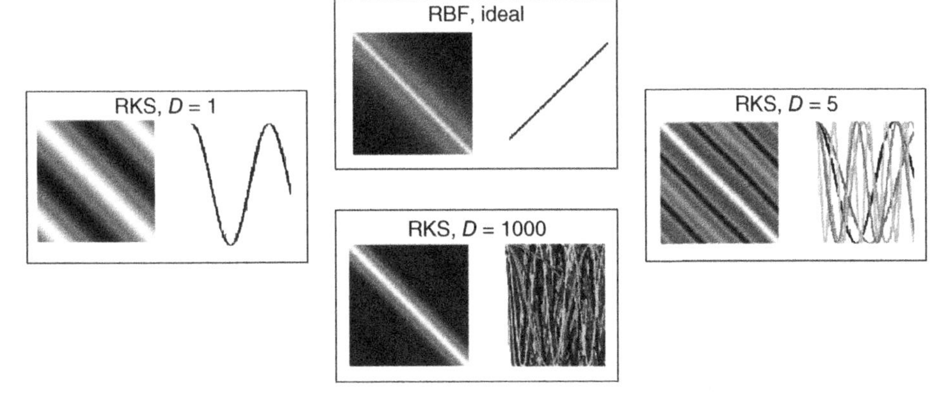

Figure 10.11 Illustration of the effect of randomly sampling D bases from the Fourier domain on the kernel matrix. With sufficiently large D, the kernel matrix generated by an RKS approximates that of the RBF kernel, at a fraction of the time.

Table 10.4 Computational and memory costs for different approximate kernel methods in problems with d dimensions, D features, N samples.

Method	Train time	Test time	Train memory	Test memory
Naive (Shawe-Taylor and Cristianini, 2004)	$\mathcal{O}(N^2d)$	$\mathcal{O}(Nd)$	$\mathcal{O}(Nd)$	$\mathcal{O}(Nd)$
Low rank (Fine *et al.*, 2001)	$\mathcal{O}(NDd)$	$\mathcal{O}(Dd)$	$\mathcal{O}(Dd)$	$\mathcal{O}(Dd)$
RKS (Rahimi and Recht, 2007)	$\mathcal{O}(NDd)$	$\mathcal{O}(Dd)$	$\mathcal{O}(Dd)$	$\mathcal{O}(Dd)$

where $p(\boldsymbol{w})$ is set to be the inverse FT of $\boldsymbol{K}$, and $\boldsymbol{w}_i \in \mathbb{R}^d$ is randomly sampled from a data-independent distribution $p(\boldsymbol{w})$ (Rahimi and Recht, 2009). Note that we can define a D-dimensional *randomized* feature map $\boldsymbol{z}(\boldsymbol{x}) : \mathbb{R}^d \rightarrow \mathbb{R}^D$, which can be explicitly constructed as $\boldsymbol{z}(\boldsymbol{x}) := [\exp(j\boldsymbol{w}_1^{\mathrm{T}}\boldsymbol{x}), \dots, \exp(j\boldsymbol{w}_D^{\mathrm{T}}\boldsymbol{x})]^{\mathrm{T}}$. Other definitions are possible; one could, for instance, expand the exponentials in pairs $[\cos(\boldsymbol{w}_i^{\mathrm{T}}\boldsymbol{x}), \sin(\boldsymbol{w}_i^{\mathrm{T}}\boldsymbol{x})]$, but this increases the mapped data dimensionality to $\mathbb{R}^{2D}$, while approximating exponentials by $[\cos(\boldsymbol{w}_i^{\mathrm{T}}\boldsymbol{x}+b_i)]$, where $b_i \sim \mathcal{U}(0, 2\pi)$, is more efficient (still mapping to $\mathbb{R}^D$) but has been revealed as less accurate, as noted in Sutherland and Schneider (2015).

In matrix notation, given N data points, the kernel matrix $\boldsymbol{K} \in \mathbb{R}^{N\times N}$ can be approximated with the explicitly mapped data, $\boldsymbol{Z} = [\boldsymbol{z}_1 \cdots \boldsymbol{z}_N]^{\mathrm{T}} \in \mathbb{R}^{N\times D}$, and will be denoted as $\hat{\boldsymbol{K}} \approx \boldsymbol{Z}\boldsymbol{Z}^{\mathrm{T}}$. This property can be used to approximate any shift-invariant kernel. For instance, the familiar SE Gaussian kernel $K(\boldsymbol{x}, \boldsymbol{x}') = \exp(-\|\boldsymbol{x} - \boldsymbol{x}'\|^2/(2\sigma^2))$ can be approximated using $\boldsymbol{w}_i \sim \mathcal{N}(\boldsymbol{0}, \sigma^{-2}\boldsymbol{I}), 1 \leq i \leq D$. An illustrative example of how an RKS approximates $\boldsymbol{K}$ with random bases is given in Figure 10.11. The method is very efficient in both speed and memory requirements, as shown in Table 10.4.

We can use such recipe to derive an approximated version of the LS kernel regression/classification, by simply replacing the mapped feature vectors (now explicitly) in the canonical *normal equations*. A simple code snippet is given in Listing 10.2 for illustration purposes.

```
% Random Kitchen Sinks (RKS) for Least squares kernel regression
%
% We are given the following training-test data matrices
% Xtrain: N x d, Xtest: M x d
% Ytrain: N x 1, Ytest: M x 1

% Training
D      = 100;                 % number of random features
lambda = 1e-3;                % regularization parameter
W      = randn(D,d);          % random projection matrix
Z      = exp(1i * Xtrain * W);     % explicitly projected features, N x D
alpha  = (Z'*Z + lambda*eye(D)) \ (Z'*Ytrain);   % primal weights, D x 1

% Testing
Ztest = exp(1i * Xtest * W);      % M x D
Ypred = real(Ztest * alpha);      % M x 1
```

Listing 10.2 Example of a kernel LS regression approximated with random features (rks.m).

The RKS algorithm can actually exploit other approximating functions besides Fourier expansions. Randomly sampling distribution functions impacts the definition of the corresponding RKHS: sampling the Fourier bases with $z_\omega(\boldsymbol{x}) = \sqrt{2}\cos(\omega_o^{\mathrm{T}}\boldsymbol{x}+b)$ actually leads to the Gaussian RBF kernel $K(\boldsymbol{x},\boldsymbol{y}) = \exp(-\|\boldsymbol{x}-\boldsymbol{y}\|^2/(2\sigma^2))$, while a random stump (i.e., sigmoid-shaped functions) sampling defined by $z_\omega(\boldsymbol{x}) = \mathrm{sgn}(\boldsymbol{x} - \omega)$ leads to the kernel $K(\boldsymbol{x},\boldsymbol{y}) = 1 - (1/a)\|\boldsymbol{x} - \boldsymbol{y}\|_1$.

10.4.3 Parallel Support Vector Machine

Thanks to the increase in computational capabilities of current CPUs and GPUs, as well as large facilities and commodity clusters, it is nowadays possible to train SVMs on large-scale problems in parallel. There are several SVM implementations. Let us just mention two of them. Cao *et al.* (2006) presented a parallel implementation of the SMO. The parallel SMO was developed using message passing interface by first partitioning the entire training set into smaller subsets and then simultaneously running multiple CPU processors to deal with each of the partitioned datasets. Experiments showed a great speedup. Nevertheless, the algorithm had to work with the samples, not with the kernel matrix, which may offer interesting opportunities of multimodal data assimilation.

An interesting alternative to the parallelization of the SMO has been presented by Chang *et al.* (2008). This parallel SVM algorithm (PSVM) is able to reduce the memory use and to parallelize both data loading and computation, and at the same time works directly with the precomputed kernel matrix. Given N training instances each with d dimensions, the PSVM first loads the training data in a round-robin fashion onto m machines. Next, PSVM performs a parallel row-based incomplete Cholesky factorization (ICF) on the loaded data (Golub and Van Loan, 1996). At the end of parallel ICF, each machine stores only a fraction of the factorized matrix. PSVM then performs parallel IPM to solve the QP optimization problem, while the computation and the memory demands are improved with respect to other decomposition-based algorithms, such as the SVMLight (Joachims, 1998a), libSVM (Chang and Lin, 2002), SMO (Platt, 1999a), or SimpleSVM (Vishwanathan *et al.*, 2003). Let us summarize the main aspects of this PSVM algorithm.

Parameters for Parallel Algorithm Design

There are multiple, sometimes conflicting, design goals to be considered when developing a parallel algorithm. One could maximize static runtime characteristics such as *speedup* (sequential execution runtime divided by parallel execution time) or *efficiency* (speedup divided by the number of processors employed). Another goal could be maximizing *scalability* – the ability to maintain constant efficiency when both the problem size and the number of processors are increased; that is, in an attempt to increase the longevity of a solution in the face of continuously improving computation power. Yet another dimension could be to maximize *productivity* (the usefulness of a solution as a function of time divided by its costs) or in particular *development time productivity*, which is defined as speedup divided by relative effort (the effort needed to develop a serial version divided by the effort needed to develop a parallel solution) (Funk *et al.*, 2005).

Parallel Support Vector Machine Via Cholesky Factorization

The PSVM method originally introduced by Chang *et al.* (2008) is aimed at reducing memory use through performing a row-based, approximate matrix factorization. The key step of PSVM is the parallel ICF. Traditional column-based ICF (Bach and Jordan, 2005a; Fine *et al.*, 2001) can reduce computational cost, but the initial memory requirement is not efficient for very large datasets. Alternatively, the PSVM performs parallel row-based ICF as its initial step, which loads training instances onto parallel machines and performs factorization simultaneously on these machines. Once parallel ICF has loaded N training data distributedly on m machines, and reduced the size of the kernel matrix through factorization, IPM can be solved on parallel machines simultaneously.

Efficient Implementation of the Algorithm

Notationally, ICF approximates the Hessian matrix $\boldsymbol{Q}$ (of size $N\times N$) by a smaller matrix $\boldsymbol{H}$ (of size $p\times N$); that is, $\boldsymbol{Q} = \boldsymbol{H}\boldsymbol{H}^{\mathrm{T}}$. ICF, together with the exploitation of the Sherman–Morrison–Woodbury formula,[4] can greatly reduce the computational complexity in solving an $N \times N$ linear system. The work of Fine *et al.* (2001) provides a theoretical analysis of how ICF influences the optimization problem in Equation 10.65.

The PSVM in Chang *et al.* (2008) iterates until the approximation of $\boldsymbol{Q}$ by $\boldsymbol{H}_k\boldsymbol{H}_k^{\mathrm{T}}$ (measured by $\mathrm{Tr}(\boldsymbol{Q} - \boldsymbol{H}_k\boldsymbol{H}_k^{\mathrm{T}})$) is satisfactory, or the predefined maximum iterations (or, say, the desired rank of the ICF matrix) p is reached. As suggested in Golub and Van Loan (1996), a parallelized ICF algorithm can be obtained by constraining the parallelized Cholesky factorization algorithm, iterating at most p times. However, in the proposed algorithm (Golub and Van Loan, 1996), matrix $\boldsymbol{H}$ is distributed by columns in a round-robin way on m machines (hence it is called "column-based parallelized ICF"). Such a column-based approach is optimal for the single-machine setting, but cannot gain full benefit from parallelization because of (i) the large memory requirement (as each machine must be able to store a local copy of the training data) and (ii) the limited parallelizable computation, since the summation of local inner product result and the vector update must be performed on one single machine. Therefore, rather

4 The Sherman–Morrison–Woodbury formula from linear algebra states that $(\boldsymbol{C} + \boldsymbol{AB})^{-1} = \boldsymbol{C}^{-1} - \boldsymbol{C}^{-1}\boldsymbol{A}$ $(\boldsymbol{I} + \boldsymbol{BC})^{-1}\boldsymbol{A})^{-1}\boldsymbol{BC}^{-1}$, where $\boldsymbol{C}$ is an invertible $N \times N$ matrix, $\boldsymbol{A} \in \mathbb{R}^{N\times m}$ and $\boldsymbol{B} \in \mathbb{R}^{m\times N}$.

than performing column-wise, a row-based approach starts by initializing variables and loading training data onto m machines in a round-robin fashion. The algorithm then performs the ICF main loop until the termination criteria are satisfied (e.g., the rank of matrix $\boldsymbol{H}$ reaches p).

At the end of the algorithm, $\boldsymbol{H}$ is stored distributedly on m machines, ready for parallel IPM (Chang *et al.*, 2008). Parallel ICF enjoys three advantages: (i) parallel memory use ($\mathcal{O}(Np/m)$), (ii) parallel computation ($\mathcal{O}(p^2N/m)$), and (iii) low communication overhead ($\mathcal{O}(p^2 log(m))$). Particularly on the communication overhead, its fraction of the entire computation time shrinks as the problem size grows. We will verify this in the experimental section. This pattern permits a larger problem to be solved on more machines to take advantage of parallel memory use and computation. More details can be found in Chang *et al.* (2008). The method loads only essential data to each machine to perform parallel computation. Let N denote the number of training instances, p the reduced matrix dimension after factorization (p is significantly smaller than N), and m the number of machines. PSVM reduces the memory requirement from $\mathcal{O}(N^2)$ to $\mathcal{O}(Np/m)$, and improves computation time to $\mathcal{O}(Np^2/m)$.

Example of Parallelized Support Vector Machine Classification

We illustrate the capabilities of the PSVM for classification of large-scale real data problems. In particular, here we are concerned with the classification of multispectral images in a large facility.[5] In particular, we build a PSVM to detect cloud presence in multispectral images.

Experiments were carried out using two MERIS Level 1b (L1b) images taken over Barrax (Spain), which are part of the data acquired in the framework of the SPARC 2003 and 2004 ESA campaigns (ESA-SPARC Project). These images were acquired on July 14 of two consecutive years (2003 and 2004). For our experiments, we used as input 13 spectral bands (MERIS bands 11 and 15 were removed since they are affected by atmospheric absorptions) and six physically inspired features extracted from MERIS bands in a previous study (Gómez-Chova *et al.*, 2007): cloud brightness and whiteness in the visible and near-infrared spectral ranges, along with atmospheric oxygen and water vapor absorption features. Cloud presence is considered as the target class. Cloud screening is specially well suited to semi-supervised approaches since cloud features change to a great extent depending on the cloud type, thickness, transparency, height, and background (being extremely difficult to obtain a representative training set); and cloud screening must be carried out before atmospheric correction (being the input data affected by the atmospheric conditions). In the selected image, the presence of snow can be easily confused with clouds, which increases the difficulty of the problem.

Scalability experiments were run with three large datasets obtained from randomly subsampling the MERIS image ($\{10^4, 10^5, 10^6\}$ samples). Note that, for the case $N = 10^6$,

---

5 MareNostrum comprises 2560 JS21 compute nodes (blades) and 42 p615 servers. Every blade has two processors at 2.3 GHz running Linux operating system with 8 GB of memory RAM and 36 GB local disk storage. All the servers provide a total of 280 TB of disk storage accessible from every blade through GPFS (Global Parallel File System). The networks that interconnect the MareNostrum are: (1) *Myrinet Network*: high-bandwidth network used by parallel applications communications; and (2) *Gigabit Network:* Ethernet network used by the blades to mount remotely their root file system from the servers and the network over which GPFS works. More information is available at http://www.bsc.es/.

a single machine cannot store the factorized matrix $\boldsymbol{H}$ in local memory, so we show results for the cases $m>10$. The running time consists of three parts: computation ("Comp"), communication ("Comm"), and synchronization ("Sync"). Figure 10.12 shows the scores for different numbers of machines and the three data sizes. The PSVM provides excellent performance and achieves a steady state in computation cost for $m>20$. For more than 100 machines, the performance deteriorates. In this case, especially important for $m>200$, smaller problems per machine have to be solved, which results in an overall increase of support vectors, and thus both training and test processing times increase.

Figure 10.13 shows the speedup curves for different data sizes, along with the linear speedup line. This parallel version of the SVM cannot achieve linear speedup beyond a limit, which depends on the number of machines and the size of the dataset. This result has already been reported in Chang *et al.* (2008). The fact that the problem is split in many machines also increases the time needed for communication and synchronization overheads. Communication time is incurred when message passing takes place between machines. Synchronization overhead is incurred when the master machine waits for task completion on the slowest machine. Note that, in homogeneous schemes, there are not obviously "slower" machines but harder tasks to be performed, as in our case. The computation speedup becomes sublinear when adding machines beyond a threshold (around 50 machines). This is due to the fact that the algorithm computes the inverse of a matrix whose size depends on the number of machines m; but fortunately, the larger the dataset is, the smaller is this (unparallelizable) time fraction. Therefore, more machines (larger m) can be employed for larger datasets (larger N) to gain speedup. Finally, we should note that the highest impact on speedup is the communication overhead, rather than the synchronization.

10.4.4 Outlook

Blink! And the Moore's Law breaks...

We are facing a new era in which data are flooding our processing capabilities. Kernel machines are hampered by their swallow architecture to cope with such big data problems. In this section we have reviewed the main approximations to computational efficiency using kernel machines. We have seen that there are algebraic solutions to decompose QP problems into smaller problems, we also have methods to approach the problem of large eigendecomposition and matrix inversion problems, many tricks to select the most relevant (informative) points entering the kernel machine, and to approximate the kernel expansion into a subset of basis functions. We have also seen that such basis can even be random features drawn from our favorite basis in signal processing, the Fourier basis. And we should not forget about the important issues of adaptive, online, and recursive algorithms reviewed in Chapter 9. In a nutshell, all these are very useful ways to solve these small- to medium- size problems. And then, we have reviewed recent approximations to resort not only to software efficiency but also to hardware efficiency; there are also parallelized versions of standard kernel machines like SVMs, and smart strategies to decompose a big problem into smaller ones following the divide-and-conquer strategy.

But even with all these advances, kernel machines are not still prepared to tackle really *big data* problems. However, there is something more fundamental than the

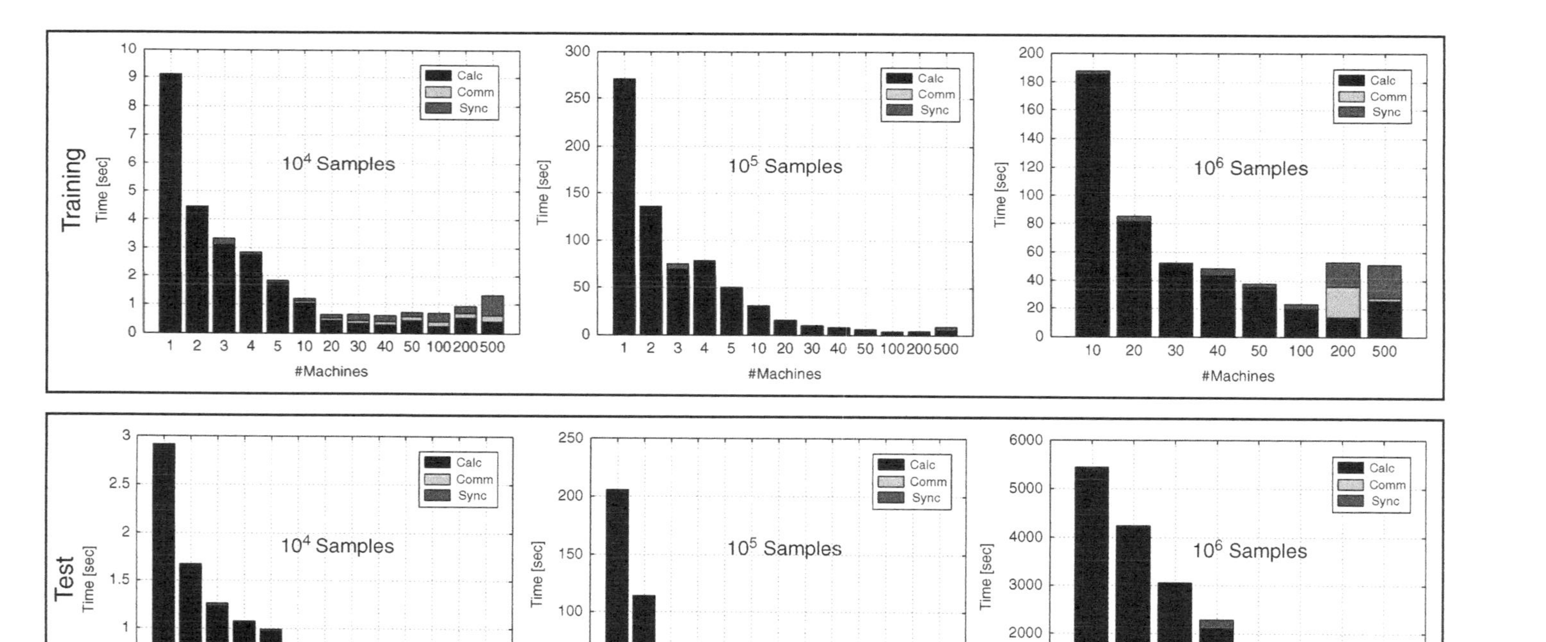

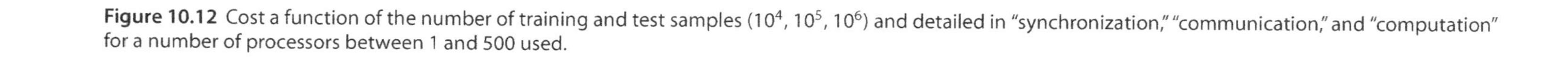

Figure 10.12 Cost a function of the number of training and test samples (10^4, 10^5, 10^6) and detailed in "synchronization," "communication," and "computation" for a number of processors between 1 and 500 used.

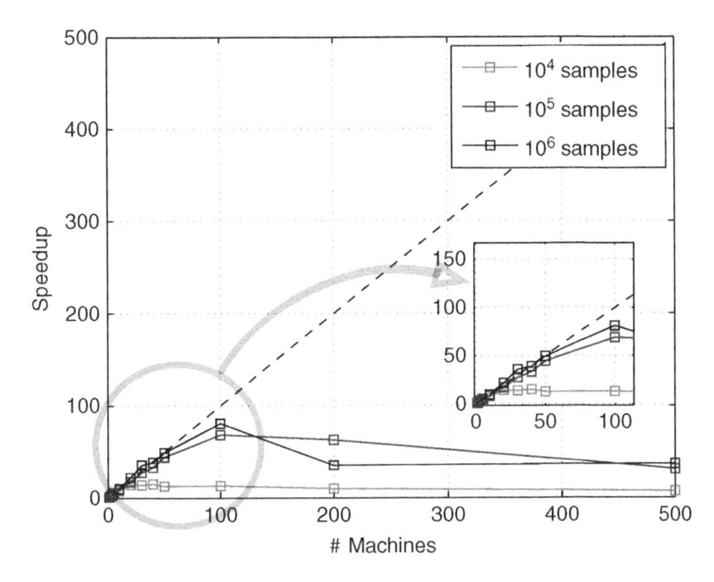

Figure 10.13 Speedup for different data sizes using the PSVM.

availability of such huge amounts of data ready to be exploited. One should question if all the acquired samples are useful or worth using. The answer is, of course, "No, it depends on the problem." We have seen how SVMs discard all samples not involved in the definition of the margin, while GPR assigns different relevance and predictive variance to different samples. Here, it seems we enter into contradiction with a current line of thought in machine learning by which bigger datasets are a blessing rather than a curse, and one should actually trust the unreasonable effectiveness of data rather than trying to develop a system that mimics the natural physical rules (Halevy *et al.*, 2009). But, even in big data processing, one often discards uninformative samples. Efficiency is not only a question of computationally capacity but also of smart strategy to learn the problem with as few data as possible: just like the human brain, we are fed with tons of sensory data that is massively and continuously screened and (a high portion of it) is discarded.

10.5 Tutorials and Application Examples

This section includes some examples and MATLAB code to reproduce applications in the previous section as well as to illustrate relevant concepts in kernel-based classification.

10.5.1 Examples of Support Vector Machine Classification

Empirical and Structural Risks

The following example illustrates the behavior of the empirical and structural risks as a function of the free parameter C. The synthetic data used in this example are embedded in an 10-dimension space. In this space, data are randomly generated from eight Gaussians whose centers form a 3D cube inside this space. The lengths of the cube edges are 2σ and the Gaussians are spherically symmetric with a standard deviation equal to σ. The data generated out of these Gaussians describing one face of the cube have been labeled as +1, and the rest as −1. The optimal separating surface is a hyperplane containing the origin, with an associate vector proportional to $\boldsymbol{u}^{\mathrm{T}} = [1, \dots, 1]$. The maximum complexity of a linear classifier is then the one that can be achieved in a space of 10 dimensions, where the complexity needed for an optimal classification is the one of a machine embedded in a space of three dimensions. Thus, a linear classifier inside this space is prone to overfit.

In order to illustrate this, and how an SVM works to reduce the overfitting, an SVM classifier is trained with 100 samples and then tested with another 1000 samples. The training and test errors have been averaged 1000 times for each of the 100 values of C ranging from 0.03 to 10, with a logarithmic spacing. Figure 10.14 shows the behavior of the errors. The training error, which is an estimation of the empirical risk, decreases as C increases. The effect of this parameter is to establish a trade-off between the minimization of the complexity and the minimization of the training error. Then, as C increases, the machine gives more importance to the training errors, thus tending to overfit. This can be seen by observing that, as the training error decreases, the test error (which is an estimation of the actual risk) increases. On the other side, when C decreases, the test error goes to lower values up to a minimum that is around $(6--7) \times 10^{-2}$. At this point, the value of C gives the optimal complexity to the machine. Figure 10.14 also shows the difference between both errors, which is an estimate of the structural risk. The example can be run using Listing 10.3. The implementation of the SVM used in this example is the well-known LIBSVM (Chang and Lin, 2002).

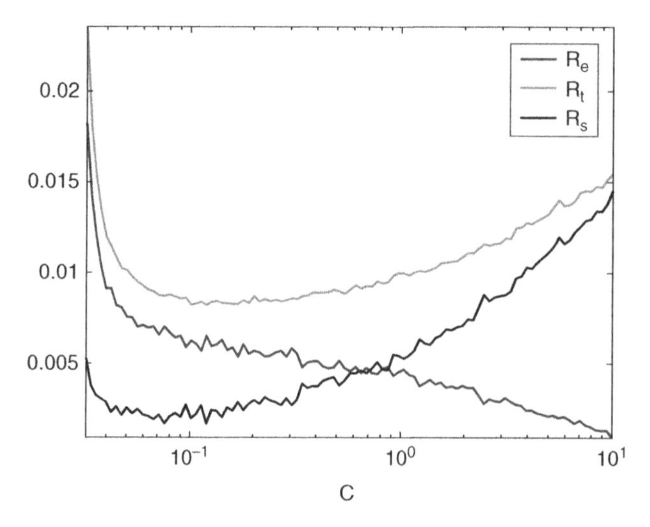

Figure 10.14 The decreasing line corresponds to the training error average (estimation of the empirical risk) over 1000 realizations, as a function of parameter C. The soft grey line corresponds to the test error average (estimation of the actual risk), and the black line is the difference between them (estimation of the structural risk).

```
function EmpStruct
addpath('../libsvm/')
% In this function we compute the Empirical risk and the test risk for
% different values of C. Changing C will change the complexity of the
% solution. The difference between the empirical and the test risk is a
% bound of the structural risk.
c = logspace(-1.5,1,100);          % This is the range of C that we sweep
Remp = zeros(1000,length(c));  % Matrix to store the Empirical Risk
R = Remp;                          % Matrix to store the Tesr risk
for i = 1:1000 % Average 1000 times
    for j = 1:length(c)
        [X,Y] = data(100,1);                          % Produce data for ...
            training
        options = ['-s 0 -t 0 -c ' num2str(c(j))]; % SVM options
        model = mexsvmtrain(Y,X,options);             % Train the SVM
        [~,p] = mexsvmpredict(Y,X,model);             % Test the SVM with ...
            the train data
        Remp(i,j) = 1-p(1)/100;                       % Compute the ...
            empirical risk
        [X,Y] = data(1000,1);                         % Produce test data
        [~,p] = mexsvmpredict(Y,X,model);             % Test with it
        R(i,j) = 1-p(1)/100;                          % Compute the test risk
    end
    % Plot figures
    figure(1)
    semilogx(c,mean(Remp(1:i,:)),'r'), hold on        % Empirical risk (red)
    semilogy(c,mean(R(1:i,:)))                        % Test (actual) risk ...
        (blue)
    % Difference between both risks, which is a bound on the Structural ...
        risk
    semilogy(c,mean(R(1:i,:))-mean(Remp(1:i,:)),'k'), hold off; drawnow
end

function [X,Y] = data(N,sigma)
w = ones(1,10)/sqrt(10); % A vector in a 10 dimension space
w1 = w .* [ 1  1  1  1  1 -1 -1 -1 -1 -1]; % One more orthogonal to the ...
    first
w2 = w .* [-1 -1  0  1  1 -1 -1  0  1  1]; % One more orthogonal to the ...
    previous ones
w2 = w2 / norm(w2);                         % Normalize
% The following four vectors are centers of four clusters forming a square
x(1,:) = zeros(1,10); x(2,:) = x(1,:) + sigma * w1;
x(3,:) = x(1,:) + sigma * w2; x(4,:) = x(3,:) + sigma * w1;
% The following data are the four clusters pf data labelled with +1 and -1
X1 = x + sigma * repmat(w,4,1) / 2; X2 = x - sigma * repmat(w,4,1) / 2;
% The previous eight clusters are the edges of a cube in a 3D space
X1 = repmat(X1,2*N,1); X2 = repmat(X2,2*N,1); X = [X1;X2];
Y = [ones(4*2*N,1) ; -ones(4*2*N,1)]; Z = randperm(8*2*N);
Z = Z(1:N); X = X(Z,:) + 0.2 * sigma * randn(size(X(Z,:))); Y = Y(Z);
```

Listing 10.3 Script for the example in Figure 10.14 (EmpStruct.m).

Consistency of Learning

A desirable property of a training algorithm is the *consistency* of learning. When the number of training data increases, the empirical error increases and the structural error should decrease as a consequence of the improvement of the generalization of the machine. If the learning machine is consistent, then the training error and the test error should uniformly converge to the same value. In the example of Figure 10.15, the training and test errors of the previous example are computed for a different number of training data. For small training sets, the training error is low, where the test error is high. This suggests that there is high overfitting. Actually, when the number of samples is too low, the machine tends to adapt to the sample, thus producing high-biased solutions that produce a high test error (high structural risk). When the number of samples increases, the SVM tends to better generalize, and both errors converge to the same value. The code for this example can be seen in Listing 10.4.

```
function consistency
addpath ../libsvm/
c = 1;                          % Set C = 1 in the functional
NN = 10:500;                    % Number of samples from 10 to 500
Remp = zeros(1000,190);         % Matrix to store empirical risc
R = Remp;                       % Matrix to store test risk
for i = 1:100
    for j = 1:length(NN)
        [X,Y] = data(NN(j),1);                          % Produce training data
        options = ['-s 0 -t 0 -c ' num2str(c)];         % SVM options
        model = mexsvmtrain(Y,X,options);               % Train the SVM
        [~,p] = mexsvmpredict(Y,X,model);               % Test the SVM with ...
            training data
        Remp(i,j) = 1-p(1)/100;                         % Compute the empirical ...
            risk
        [X,Y] = data(1000,1);                           % Produce test data
        [~,p] = mexsvmpredict(Y,X,model);               % Test with it
        R(i,j) = 1-p(1)/100;                            % Compute the test risk
    end
    % Plot the results
    figure(1), plot(NN,mean(Remp(1:i,:)),'r'), grid on, hold on
    plot(NN,mean(R(1:i,:))), hold off; drawnow
end
keyboard % this allows to inspect variables before exiting the function

function [X,Y] = data(N,sigma)
w = ones(1,10) / sqrt(10); % A vector in a 10 dimension space
w1 = w .* [ 1  1  1  1  1 -1 -1 -1 -1 -1]; % One more orthogonal to the ...
    first
w2 = w .* [-1 -1  0  1  1 -1 -1  0  1  1]; % One more orthogonal to the ...
    previous ones
w2 = w2 / norm(w2);                        % Normalize
% The following four vectors are centers of four clusters forming a square
x(1,:) = zeros(1,10); x(2,:) = x(1,:) + sigma * w1;
x(3,:) = x(1,:) + sigma * w2; x(4,:) = x(3,:) + sigma * w1;
% The following data are the four clusters pf data labelled with +1 and -1
X1 = x + sigma * repmat(w,4,1) / 2; X2 = x - sigma * repmat(w,4,1) / 2;
% The previous eight clusters are the edges of a cube in a 3D space
X1 = repmat(X1,2*N,1); X2 = repmat(X2,2*N,1); X = [X1;X2];
Y = [ones(4*2*N,1) ; -ones(4*2*N,1)]; Z = randperm(8*2*N); Z = Z(1:N);
X = X(Z,:) + 0.4 * sigma * randn(size(X(Z,:))); Y = Y(Z);
```

Listing 10.4 Script for the example in Figure 10.14 (consistency.m).

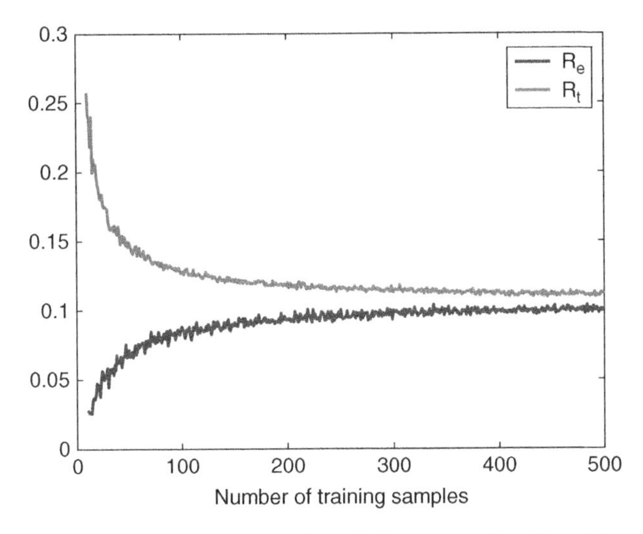

Figure 10.15 Training (R_e) and test (R_t) errors as a function of the number of training data for the example in Listing 10.4. It can be seen that, since the SVM is consistent, both errors converge to the same value.

XOR Problem and Overfitting

The example in Figure 10.16 shows the classical XOR problem. This problem consists of the binary classification data generated by four Gaussian distributions centered in four quadrants in a 2D space at positions $\pm 1, \pm 1$. The Gaussians centered at $[-1, -1]$ and $[1, 1]$ are labeled -1 and the other two are labeled $+1$.

An SVM using the Gaussian kernel with parameter $\sigma = 5$ was trained with two different values of C. High values of C will give high importance to the training error, then ignoring the minimization of the structural error. Such a solution may suffer from overfitting, while low values of C will produce solutions with a more balanced trade-off between structural and empirical risk. The left and central panels of Figure 10.16 show the classification boundary achieved with a value of C equal to 100 and 10. It is obvious that these boundaries are not close to the optimal value, but it is rather constructed to produce a small number of errors during the training phase; in other words, this is a graphical example of overfitting. With a value of C set to 0.1 (right panel of Figure 10.16), a more reasonable solution is produced that approaches the optimum separating surface. The script for data generation and figure representation is shown in the MATLAB Listing 10.5.

```
function xor_example
% Toy example of the solution of the XOR problem using kernels and SVMs
addpath('../libsvm/'), addpath('../simpleR/'), rng(1);
C = 0.1;          % Complexity parameter
sigma = 0.5;      % Kernel parameter
c = [ 1 -1 -1 1 ; 1 -1 1 -1 ];   % Centers of the 4 clusters
% Generation of 400 datapoints
x = [ repmat(c(:,1:2),1,100) repmat(c(:,3:4),1,100) ];
x = x + 0.4 * randn(size(x));
y = [ ones(1,200) -ones(1,200) ]';
K = kernelmatrix('rbf', x, x, sigma);      % Kernel matrix
```

```
options = ['-s 0 -t 4 -c ' num2str(C)];
model = mexsvmtrain(y,K,options);                    % Train the SVM
% alpha = pinv(K + gamma * eye(size(K))) * y; % Compute dual parameters
% Note the use of a diag matrix
[X1,X2] = meshgrid(-3:.05:3,-3:.05:3);             % Grid for representing data
% representation of the data
xt = [X1(:) X2(:)]';                                % Vectorize it to use it in test
Ktest = kernelmatrix('rbf', xt, x, sigma);
[~,~,z] = mexsvmpredict(zeros(size(xt,2),1), Ktest, model); % Predictions
Z = buffer(z, size(X1,1), 0);   % convert output vector into matrix
% Plot data
figure(1), clf
plot(x(1,y>0), x(2,y>0), 'bo', x(1,y<0),x(2,y<0), 'ro'); hold on
contour(X1, X2, Z, [0 0], 'LineWidth', 2, 'Color', 'Black'), hold off
figure(2), surfc(X1,X2,Z), shading interp, camlight right, view([-38 30])
xlabel('$x_1$','Interpreter','latex'); ...
    ylabel('$x_2$','Interpreter','latex')
```

Listing 10.5 Script for the example in Figure 10.16 (xor_example.m) using an SVM classifier.

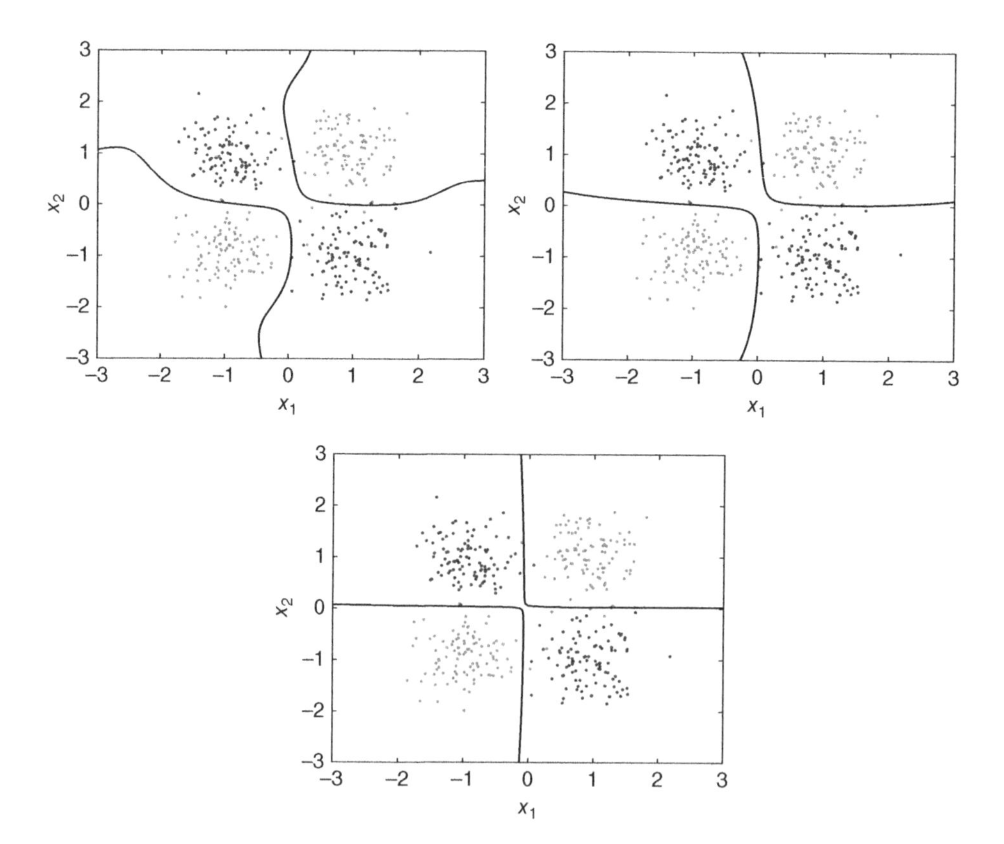

Figure 10.16 Results of the XOR example. The first and second panels show a solution where the SVM has been set to give too much importance to the training samples ($C = 100$ and 10), where the third panel shows a solution with $C = 0.1$.

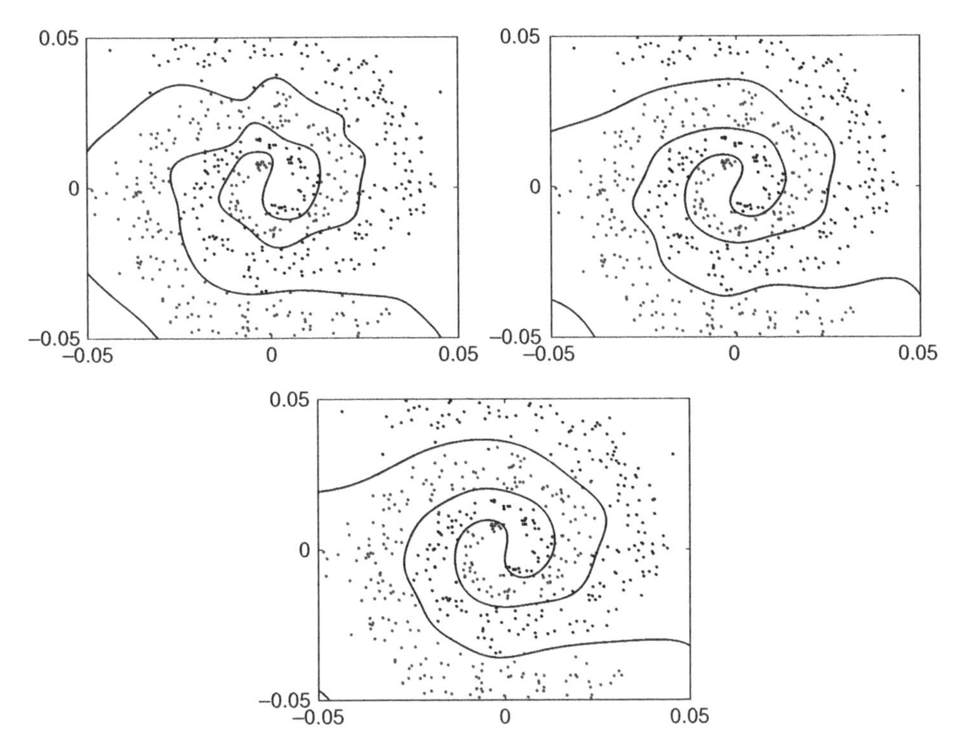

Figure 10.17 Result of the Fibonacci example for three different values of the SVM parameter C, from a high one (first) to a low one (second). Increasing the value of C increases the complexity of the solution, making it prone to overfitting, as is observed in the first panel.

Double Fibonacci Spiral

In some cases, the complexity needed by a machine to solve a classification problem depends on the position of the data in the space. This is the case of the toy problem called the double Fibonacci spiral. In this problem, two sets of data labeled with labels +1 and −1 are generated following two Fibonacci spirals. This spiral has a radius that increases exponentially, so the complexity needed to classify decreases when the radius increases (see Figure 10.17). The code to generate this example is in Listing 10.6.

```
function spirals
% Setup paths
addpath('../libsvm/'), addpath('../simpleR/')
% Generate and plot a set of data in two spirals
rng(1); sigma = 0.01;
[X1,X2,y] = fibonacci(300);
x = [X1 X2]; % Data stored in a matrix. Labels are stored in y
K = kernelmatrix('rbf', x, x, sigma); % Training kernel matrix

nfig = 1;
for C = [1000 100 10 1]
```

```
    options = ['-s 0 -t 4 -c ' num2str(C)];
     % Compute the dual parameters and store them in a column vector ...
         called alpha.
    model = mexsvmtrain(y,K,options);
    M = max(abs(x(:))) * 0.9;
    [X1_,X2_] = meshgrid(-M:M/100:M, -M:M/100:M); % Grid of points for
    xt = [X1_(:) X2_(:)]';  % Vectorize it to use it in test
    Ktest = kernelmatrix('rbf', xt, x, sigma);  % Test matrix
    % Classify the test data
    [~,~,z] = mexsvmpredict(zeros(size(xt,2),1), Ktest, model);
    % Convert the output vector into a matrix for representation
    Z = buffer(z, size(X1_,1), 0);
    % Plotting results
    figure(nfig), nfig = nfig + 1; clf
    plot(X1(1,:), X1(2,:), 'bo', X2(1,:), X2(2,:), 'ro'); hold on
    contour(X1_, X2_, Z, [0 0], 'LineWidth', 2, 'Color', 'Black'); hold ...
        off
    axis square; title(['C = ' num2str(C)])
end

function [X1,X2,y] = fibonacci(N)
theta = rand(1,N) * (log10(3.5*pi));
theta = 10.^theta + pi; theta = 4*pi - theta;
a = 0.01; b = 0.2; r = a * exp(b*theta);
x1 = r .* cos(theta); y1 = r .* sin(theta);
d = 0.14 * sqrt(x1.^2 + y1.^2);
x1 = x1 + randn(size(x1)) .* d; y1 = y1 + randn(size(y1)) .* d;
X1 = [x1;y1]; x2 = r .* cos(theta+pi); y2 = r .* sin(theta+pi);
d = 0.14 * sqrt(x2.^2 + y2.^2); x2 = x2 + randn(size(x2)) .* d;
y2 = y2 + randn(size(y2)) .* d; X2 = [x2;y2];
y = [ones(1,300) -ones(1,300)]';
```

Listing 10.6 Script for the example in Figure 10.17 (spirals.m).

10.5.2 Example of Least-Squares Support Vector Machine

We provide in Listing 10.7 a simple implementation of the LS-SVM. For the sake of brevity, we only list the two lines of code needed to train and test the LS-SVM, and the corresponding LS-SVM function. We let as an exercise for the reader the implementation of examples XOR Problem and Overfitting and Double Fibonacci Spiral (see Exercise 10.7.4). Note that this implementation simply reproduces Equation 10.40. The classifier has exactly the same formulation as the one of the standard SVM classifier; that is:

$$\hat{y}_* = f(\boldsymbol{x}_*) = \text{sgn}\left(\sum_{i=1}^{N} y_i \alpha_i K(\boldsymbol{x}_i, \boldsymbol{x}_*) + b\right),$$

In the implementation in Listing 10.7, the dual variables α_i are multiplied by the labels inside the function to add simplicity to the implementation.

```
function y = ls_svm(y,K,Ktest,lambda)
% Example of LS-SVM

% Train
```

```
w = pinv([ 0 -y'; y diag(y)*K*diag(y) + lambda*eye(size(K,1)) ]) * ...
        [ 0 ; ones(size(y)) ];

alpha = w(2:end) .* y;
b = w(1);

% Test
y = Ktest' * alpha + b;
```

Listing 10.7 Code snippets needed to implement the function LSSVM (ls_svm.m).

10.5.3 Kernel-Filtering Support Vector Machine for Brain–Computer Interface Signal Classification

The BCI problem considered here is one of the problems presented in *BCI Competition III* (Blankertz *et al.*, 2004). The objective is to obtain a sequence of labels out of brain activity signals for three different human subjects. The data consist of 96 channels containing PSD features for different band-pass (three training sessions and one test session, with $N \simeq 3000$ samples per session) and three possible labels (left arm, right arm, or a word). Several classes were tackled through a classical OAA strategy. For the nonlinear approaches, 30% of the available data was used for training. The regularization parameters were tuned using a grid search strategy on the third training session. In these experiments, N_0 was set to zero, as we are interested in predicting the current mental task with no delay.

The test error for different methods and filter lengths is summarized in Table 10.5. For the linear models, the best methods for all tested filter sizes are KF-SVM, SKF-SVM, and SWinSVM; see details in Flamary *et al.* (2012). This shows the advantage of taking into account the neighborhood of the samples for decisions and the importance of a proper regularization. Longer filtering provides the best results, especially in conjunction with regularization that helps to avoid overfitting. The best overall results are obtained by KF-SVM and SKF-SVM with the filter length $\tau = 50$. The results follow the same trends for the nonlinear models, showing that for this task a linear classifier is sufficient. However, one should keep in mind that, in these cases, the decision functions are learned from only 30% of the samples. In this case, the Avg-SVM performs well, since the noise is in the high frequencies and the nonlinearities that can be induced by overfiltering are handled by the Gaussian kernel.

Links to the MATLAB code for LMF are available on the book's web page. A simple demo illustrating how the functions are called is given in Listing 10.8. Essentially, one has to start with scaled data, initialize the options' MATLAB structure for the routines, and call each model with the corresponding training and testing functions. The code shows how to run a comparison between a canonical SVM working directly with raw data (no filtering), using SVM on filtered data, and the LMF-SVM method where the optimal filters are learned such that SVM classification accuracy is maximized.

```
% Sketch of the use of LMF in the BCI problem
%   - You need to download and install FilterSVM toolbox from
%       http://remi.flamary.com/soft/soft-filtersvm.html
%   - Assume that the data is stored in matrix X and Y (training),
```

```
%       Xte and Yte (test), and Xval and Yval (validation)
% Initialize parameters
f                               = 10; % filter length f (=tau)
options.f                       = f;
options.decy                    = 0;  % y time shift...
options.F                       = ones(f,d)/(f); % init filter weights
options.stopvarx                = 0;  % threshold for stopping on F changes
options.stopvarj                = 1e-3;
options.usegolden               = 1;
options.goldensearch_deltmax = .01;
options.goldensearch_stepmax = 1;
options.numericalprecision    = 1e-3;
options.multiclass              = 1;
linear = 1;  % let's try with a linear kernel
if linear == 1
    options.kernel = 'linear';
    options.kerneloption = 1;
    options.C = 1e0; % C parameter in the SVM
    options.regnorm2 = 1e0;
    options.regnorm12 = 0e-3;
    options.solver = 3;
    options.use_cg = 1;
else
    options.kernel = 'gaussian';
    options.kerneloption = 10;
    options.C = 1e3; % C parameter in the SVM
    options.regnorm2 = 1e1;
    options.regnorm12 = 0e2;
    options.solver = 3;
    options.use_cg = 1;
end
% No filtering at all
[svm,obj] = svmclass2(X,Y,options);
% Filtering and then classify
[svmf,obj] = svmclass2(mfilter(options,X),Y,options);
% LMF models: learn the optimal filters for classification
[sigsvm,obj] = filtersvmclass(X,Y,options);
% Evaluate in the test set
ysvm = svmval2(Xte,svm); prec_svm = get_precision(Yte,ysvm);
[ypred] = svmval2(mfilter(options,Xte),svmf); prec_filt = ...
    get_precision(Yte,ypred);
[ypred,yval] = filtersvmval(Xte,sigsvm); prec_filtsvm = ...
    get_precision(Yte,ypred);
```

Listing 10.8 Comparison between different approaches to classification under noisy data: SVM working without filtering the data, using SVM on filtered data, and the LMF-SVM where optimal filters are learned for optimal classification (demoLMF.m).

10.5.4 Example of Laplacian Support Vector Machine

Let us exemplify the use of the Laplacian SVM with a demo involving the familiar two moons example (see Listing 10.9). In this particular example we used $u = 300$ unlabeled data points and only $l = 5$ labeled points per class.

Table 10.5 Classification error rate for the BCI dataset for different methods (see details in Flamary *et al.* (2012)) and filter length τ. Results are given for the linear model (top) and for the nonlinear model (bottom). The three best methods are highlighted in bold.

	$\tau = 10$				$\tau = 20$				$\tau = 50$			
Method	**S1**	**S2**	**S3**	**Avg**	**S1**	**S2**	**S3**	**Avg**	**S1**	**S2**	**S3**	**Avg**
Linear model												
SVM	0.25	0.37	0.55	0.39	0.25	0.37	0.55	0.39	0.25	0.37	0.55	0.39
Avg-SVM	0.22	0.34	0.53	0.36	0.19	0.29	0.53	0.34	0.13	0.23	0.47	0.28
KF-SVM	0.20	0.30	0.51	**0.34**	0.18	0.26	0.42	**0.29**	0.12	0.23	0.42	**0.26**
SKF-SVM	0.20	0.29	0.47	**0.32**	0.18	0.26	0.48	**0.30**	0.12	0.22	0.43	**0.26**
WinSVM	0.21	0.31	0.54	0.35	0.19	0.28	0.53	0.33	0.14	0.22	0.48	0.28
SWinSVM	0.21	0.31	0.47	**0.33**	0.19	0.26	0.42	**0.29**	0.14	0.21	0.46	**0.27**
Nonlinear model (Gaussian kernel)												
SVM	0.23	0.35	0.48	0.35	0.23	0.35	0.48	0.35	0.23	0.35	0.48	0.35
Avg-SVM	0.21	0.33	0.47	0.34	0.19	0.29	0.44	0.31	0.12	0.23	0.45	**0.27**
KF-SVM	0.20	0.30	0.48	**0.33**	0.17	0.26	0.48	**0.30**	0.15	0.22	0.44	0.27
SKF-SVM	0.20	0.30	0.48	**0.33**	0.17	0.26	0.44	**0.29**	0.11	0.23	0.47	**0.27**
WinSVM	0.21	0.32	0.47	**0.33**	0.17	0.28	0.44	**0.30**	0.13	0.23	0.44	**0.26**

```
function [X,Y] = moons(n)

rng(2);
space = 1.2; noise = 0.1;

r = randn(n,1) * noise + 1; theta = randn(n,1) * pi;
r1 = 1.1 * r; r2 = r;
X1 = ([r1 .* cos(theta) abs(r2 .* sin(theta))]);
Y1 = ones(n,1); % labels

r = randn(n,1) * noise + 1; theta = randn(n,1) * pi + 2*pi;
r1 = 1.1 * r; r2 = r;
X2 = ([r1 .* cos(theta) + space*rand -abs(r2 .* sin(theta)) + 0.2 ]);
Y2 = -ones(n,1); % labels

X = [X1;X2]; Y = [Y1;Y2]; v = randperm(2*n); X = X(v,:); Y = Y(v);
```

Listing 10.9 Generating data drawn from the two moons manifold (moons.m).

The first operation is building a graph Laplacian, which uses all examples $l + u$. Links to the MATLAB code for this are available on the book's web page. After that we build the kernel matrix $\boldsymbol{K}$ using `kernelmatrix` for both training and testing, and solve the LapSVM equations (Equations 10.66–10.68). The code is given in the wrapper function in Listing 10.10. Note that the QP involved can be solved with the `LIBSVM` toolbox, and the penalization parameter C in the standard SVM algorithm is now embedded in the γ_A of the LapSVM.

```
function [alpha,b]=lapsvm(K,Y,L,lambda1,lambda2)

I = eye(size(K,1)); lab = find(Y); l = length(lab);

if isempty(L) || lambda2 == 0 % SVM
    G = I / (2*lambda1) ;        % Csvmmax = 1 / (2*lambda1*l) = 1/(2*gA*l)
else                             % Laplacian SVM
    G = (2*lambda1*I + 2*lambda2*L*K) \ I;
end
Gram = K*G; Gram = Gram(lab,lab); Ylab=Y(lab);
% Calling libSVM
C = 1; % C = 1 in Laplacian SVM
model = mexsvmtrain(Ylab, Gram, ['-t 4 -s 0 -c ' num2str(C)]);
betay = model.sv_coef; svs = model.SVs; nlab = model.Label;
% b = model.rho;
% nsv = model.nSV;
if nlab(1) == -1, betay = -betay; end;
Betay = zeros(l,1); Betay(svs+1) = betay';
alpha = G(:,lab) * Betay; b = 0; % Laplacian SVM dual weights
```

Listing 10.10 Learning the Laplacian SVM (lapsvm.m).

We put all these pieces together in a simple demo for running the LapSVM; see Listing 10.11. Several values of γ_L and γ_M were tuned to illustrate its effect while fixing the σ parameter of the Gaussian kernel function. Results and the decision surfaces are shown in Figure 10.18 for different combinations of regularization parameters. We can observe that when $\gamma_M = 0$, LapSVM disregards the information conveyed by the unlabeled data and returns the SVM decision boundary which is fixed by the location of

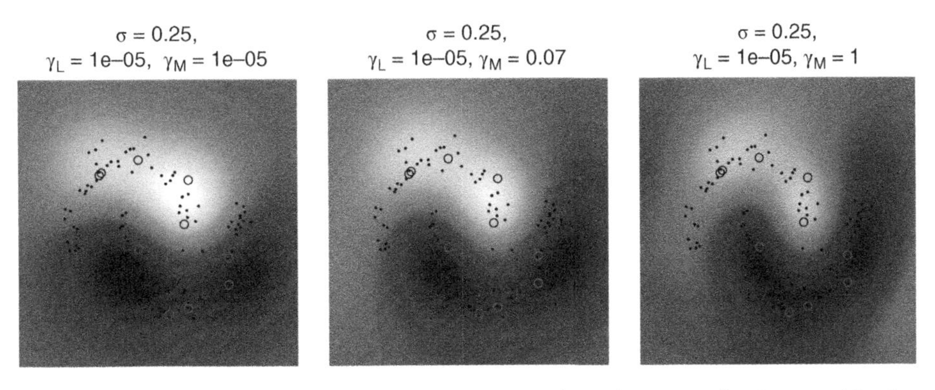

Figure 10.18 Laplacian SVM with RBF kernels for various values of $\gamma_M = \{10^{-5}, 0.07, 1\}$ and fixed $\gamma_L = 10^{-5}$. Labeled points are shown in circles, while unlabeled points are given in black dots.

the two labeled points. As γ_M is increased, the intrinsic regularizer $\|f\|^2_{\mathcal{M}}$ incorporates unlabeled data and causes the decision surface to appropriately adjust according to the geometry of the two classes.

```
%% Demo for Laplacian SVM
clear, clc; rng('default')
%% Setup paths
addpath('../simpleR/'), addpath('../libsvm/')
%% Training data
l = 2; u = 300; [Xl,Yl] = moons(l/2);
%% Unlabeled data
Xu = moons(u/2); Yu = zeros(u,1); X1 = [Xl;Xu]; Y1 = [Yl;Yu];
%% Test data
[X2,Y2] = moons(200);
%% Scale data
X1 = scale(X1); X2 = scale(X2);
%% Build the graph Laplacian L of the adjacency graph W with all data l+u
options.type = 'nn';      % nearest neighbors to obtain the adjacent points
options.NN = 12;          % number of neighbors
options.GraphDistanceFunction = 'euclidean'; % distance for the adjacency
options.GraphWeights = 'heat'; % heat function applied to the distances ...
    in W
options.GraphWeightParam = 1;  % width for heat kernel (t) if 'heat'
options.GraphNormalize = 1;    % normalizes the adjacencies of each point
L = laplacian(X1, 'nn', options);
%% Hyperparameters: sigma, gamma_A, gamma_I
sigma = 0.3;
gL = 0.05; lambda1 = gL;
gM = 5;    lambda2 = gM; % Use gM = 0 for standard SVM
%% Train LapSVM
K1 = kernelmatrix('rbf', X1', X1', sigma);
K2 = kernelmatrix('rbf', X2', X1', sigma);
[alpha,b] = lapsvm(K1, Y1, L, lambda1, lambda2);
Y2pred = sign(K2 * alpha - b);
OA = 100 * length(find(Y2pred == Y2)) / length(Y2)
%% Plot results
x = -0.25:0.05:1.25; y = -0.25:0.05:1.25; % x and y range
[xx,yy] = meshgrid(x,y); X2 = [xx(:) yy(:)];
K2 = kernelmatrix('rbf', X2', X1', sigma);
```

```
z  = (K2 * alpha - b); Z = reshape(z,length(x),length(y));
figure(1), clf
[cs,h] = contourf(xx, yy, Z, 1); shading flat; colormap(summer) ...
    %colormap(cmr)
hold on, plot(X1(Y1==0,1), X1(Y1==0,2), 'k.');
plot(X1(Y1==+1,1), X1(Y1==+1,2), 'bo', X1(Y1==-1,1), X1(Y1==-1,2), 'ro');
axis off square
```

Listing 10.11 Demo for the Laplacian SVM in the two moons example (demoLapSVM.m).

10.5.5 Example of Graph-Based Label Propagation

Listing 10.12 gives the code to run the graph-based label propagation algorithm in Zhou and Schölkopf (2004). The function arguments require the input matrix $\boldsymbol{X}$ where each row contains an example, a label matrix $\boldsymbol{Y}$ with 1-of-k encoding (i.e., if there are three classes, $k = 3$ and then the row vectors of $\boldsymbol{Y}$ are possibly $[1, 0, 0]$, $[0, 1, 0]$, $[0, 0, 1]$, and $[0, 0, 0]$ for the unlabeled points), a σ parameter for the width of the Gaussian affinity matrix, and the propagation parameter α between $[0, 1]$. The output of the function returns in $\boldsymbol{C}$ the final classification, where the elements of $\boldsymbol{C}$ are scalars from 1 to k, where k is the number of classes.

```
function  C = semi(X,Y,sigma,alpha)

N = size(X, 1);

% Step 1: Affinity matrix
M = zeros(N, N); % norm matrix
for i = 1:N % compute the pairwise norm
    for j = (i+1):N
        M(i, j) = norm(X(i, :) - X(j, :));
        M(j, i) = M(i, j);
    end
end
% Use a Gaussian to form an affinity matrix
K = exp(-M.^2/(2*sigma^2));
K = K - eye(N);

% Step 2: Symmetrical normalization
D = diag(1./sqrt(sum(K))); % inverse of the square root of the degree ...
    matrix
S = D*K*D; % normalize the affinity matrix

% Step 3(a): Compute the classification function
F = (eye(N) - alpha * S) \ Y;
% Step 3(b): Predictions
[~,C] = max(F, [], 2);
```

Listing 10.12 Semisupervised graph-based classification example (semi.m).

10.5.6 Examples of Multiple Kernel Learning

Besides the simpleMKL algorithm, another relevant study on kernel combination was presented in Cortes *et al.* (2012), where new and effective algorithms for learning kernels were introduced. The simplicity of the framework, and the good results obtained in general, are good reasons to be studied. The framework presents several techniques to learn the relevance of the features and how to exploit the discovered structure to combine dedicated kernels. Interestingly, these approaches do not involve an exhaustive training of SVMs or any particular classifier, but just simple matrix operations.

Let us consider just an approximation to the kernel combination, the so-called alignment maximization algorithm, in which the correlation between the base kernel matrices is taken into account. The goal is to determine a combination of kernels, one per dimension d; that is:

$$K_{\mu}(\boldsymbol{x}_i, \boldsymbol{x}_j) = \sum_{f=1}^{d} \mu_f K_f(x_i^f, x_j^f),$$

where x_i^f is the feature f of point $\boldsymbol{x}_i$. Let us assume a supervised problem in which we are given the input data matrix $\boldsymbol{X} \in \mathbb{R}^{n\times d}$ and the corresponding outputs (targets, labels) $\boldsymbol{y} \in \mathbb{R}^{n\times 1}$. Here, the aim is to learn a weight a_f per dimension. The approach considers aligning the kernel $\boldsymbol{K}_{\mu}$ with the labels kernel (sometimes referred as to the ideal kernel) expressed as $\boldsymbol{K}_y = \boldsymbol{y}\boldsymbol{y}^{\mathrm{T}}$. Let us recall the kernel matrix alignment operation between two kernel matrices $\boldsymbol{K}$ and $\boldsymbol{K}'$ is expressed as

$$A = \frac{\langle \boldsymbol{K}, \boldsymbol{K}' \rangle}{\|\boldsymbol{K}\|_F \|\boldsymbol{K}'\|_F},$$

which can be calculated easily in MATLAB using the code snippet in Listing 10.13.

```
function A = alignment(Ker,Y)

A = trace(Ker'*Y) / sqrt(trace(Ker'*Ker) * trace(Y'*Y));
```

Listing 10.13 Code snippet for calculating the kernel matrix alignment (alignment.m).

Cortes *et al.* (2012) proposed two particular strategies to learn the weights $\mu_f, f = 1, \dots, d$: one exploits an LS linear combination (according to Proposition 8 in Cortes *et al.* (2012)) and the other performs a convex combination by solving a QP problem (Proposition 9 in Cortes *et al.* (2012)).

Linear combination The linear combination approach considers optimizing

$$\boldsymbol{\mu}_{\mathrm{opt}} = \arg\max_{\boldsymbol{\mu}} \left\{ \frac{\langle \boldsymbol{K}_{\mu}, \boldsymbol{y}\boldsymbol{y}^{\mathrm{T}} \rangle}{\|\boldsymbol{K}_{\mu}\|_F} \right\} \quad \text{s.t. } \|\boldsymbol{\mu}\|_2 = 1,$$

where we should note that the denominator does not contain $\|\boldsymbol{y}\boldsymbol{y}^{\mathrm{T}}\|_F$ because this does not depend on $\boldsymbol{\mu}$. It can be shown that the solution to this problem is

$$\boldsymbol{\mu}_{\mathrm{opt}} = \frac{\boldsymbol{M}^{-1}\boldsymbol{a}}{\|\boldsymbol{M}^{-1}\boldsymbol{a}\|},$$

where $\boldsymbol{M}$ is an invertible matrix with entries $M_{kl} = \langle \tilde{\boldsymbol{K}}_l, \tilde{\boldsymbol{K}}_l \rangle$ for $k, l \in [1, d]$, and $\boldsymbol{a} := [\langle \tilde{\boldsymbol{K}}_1, \boldsymbol{y}\boldsymbol{y}^{\mathrm{T}} \rangle, \dots, \langle \tilde{\boldsymbol{K}}_d, \boldsymbol{y}\boldsymbol{y}^{\mathrm{T}} \rangle]^{\mathrm{T}}$, where the tilde indicates the matrix is centered (see Listing 10.14).

```
function K = centering(K)

[Ni,Nj] = size(K);
K = K - mean(K,2)*ones(1,Nj) - ones(Ni,1)*mean(K,1) + mean(K(:));
```

Listing 10.14 Centering a kernel matrix (centering.m).

Convex combination This approach constrains the values of $\boldsymbol{\mu}$ to be positive and sum to one, and solves the problem

$$\boldsymbol{\mu}_{\text{opt}} = \arg\max_{\boldsymbol{\mu}} \left\{ \frac{\boldsymbol{\mu}^{\text{T}}\boldsymbol{a}\boldsymbol{a}^{\text{T}}\boldsymbol{\mu}}{\boldsymbol{\mu}^{\text{T}}\boldsymbol{M}\boldsymbol{\mu}} \right\} \quad \text{s.t. } \|\boldsymbol{\mu}\|_2 = 1, \text{ and } \boldsymbol{\mu} \geq 0,$$

whose solution reduces to $\boldsymbol{\mu}_{\text{opt}} = \boldsymbol{v}_{\text{opt}}/\|\boldsymbol{v}_{\text{opt}}\|$, where $\boldsymbol{v}_{\text{opt}}$ is the solution of the QP problem

$$\boldsymbol{v}_{\text{opt}} = \arg\min_{\boldsymbol{v}\geq 0}\{\boldsymbol{v}^{\text{T}}\boldsymbol{M}\boldsymbol{v} - 2\boldsymbol{v}^{\text{T}}\boldsymbol{a}\}.$$

Listing 10.15 gives MATLAB code for these two multiple kernel combinations.

```
% Choose method
mklmethod = 'align'; % or alignf
% Given X and y
[n d] = size(X);
% Optimal kernel
Y = y * 2 - 1; Ky = Y * Y'; Kyc = centering(Ky);
% The best combination of kernels with different sigmas
ker = 'rbf';
Kc = zeros(size(K,1),size(K,2),d); a = zeros(1,d);
for dim = 1:d
    sigma = median(pdist(X(:,dim)));
    K = kernelmatrix(ker,X(:,dim)',X(:,dim)',sigma);
    Kc(:,:,dim) = centering(K);
    a(dim) = alignment(Kc(:,:,dim),Ky);
end
% Compute the M kernel matrix
M = zeros(d);
for i = 1:d
    for j = 1:d
        M(i,j) = sum(sum(Kc(:,:,i) .* Kc(:,:,j)));
    end
end
% Solve the problems
if strcmpi(mklmethod,'align') % Linear combination ("align", Prop. 8)
    v = M \ a';
elseif strcmpi(mklmethod, 'alignf') % Convex combination ("alignf", ...
    Prop. 9)
    options = optimset('Algorithm', 'interior-point-convex');
    v = quadprog(M,-a',-1*ones(1,length(a)),0,[],[],[],[],[],
                                              options);
end
mu = v/norm(v);
% The combined kernels
Kbest = zeros(n);
```

```
for dim = 1:d
    Kbest = Kbest + mu(dim) * squeeze(Kc(:,:,dim));
end
```

Listing 10.15 Combination of kernels via alignment and optimization (automaticSimplerMKL.m).

10.6 Concluding Remarks

This chapter has provided a summary of the SVM for classification. The SVM is a linear algorithm that is solved using a dual formulation. Hence, nonlinear versions of the SVM using the kernel trick can be readily obtained by changing the scalar dot product by any positive -definite kernel function. We have seen how the SVM also admits straightforward multiclass and multilabel, and several modifications on the regularizer, such as ℓ_p-norms or the ν-SVM to enforce all kinds of sparsity and thus control for the compactness of the solution. Also, we revised novel machine-learning paradigms that grew under the SVM framework, such as SSL, AL, MKL, and SL.

The reason for the success of the SVM is based on its ability to *control the complexity* and *regularization* of the solution through the inclusion of the SRM as part of the optimization criterion of the machine. Also an important property of SVM is the *convexity* of the functional to be solved, which is not shared by many other algorithms. In practice, the SRM translates into the minimization of the norm of the machine parameters $\boldsymbol{w}$, which, in turn, is equivalent to the maximization of the classification margin. This apparently simple criterion makes the nonlinear extension of the SVM to infinite-dimension Hilbert spaces, where the maximum possible complexity is equal to the number of training data, and where other approaches that do not control the complexity are guaranteed to overfit.

Besides the accuracy, sparsity, and flexibility of SVMs, the availability of many efficient toolboxes has also helped popularize the technique. This versatility, together with the simplicity of the formulation of the SVM and its moderate computational burden, mainly bounded by the number of training data (and not its dimension), allowed researchers to apply the SVM to problems that were previously solved using more computationally heavy solutions. Nevertheless, one of the main drawbacks of the original SVM for classification (yet also for regression problems and feature extraction) is that the computation time for the optimization increases dramatically with the number of data; as a consequence of that, the dual formulation involves the use of kernel matrices. This drawback has been tackled in numerous studies in the last decade, and we have revised the main current advances, either based on approximating the kernel matrices with randomized versions, Nyström approximations, or smart strategies for parallelizing the algorithms.

SVMs have played a key role for solving classification problems during this last decade, and it is expected that further developments will appear in the next years. We can only speculate about the directions for this evolution, as probably unexpected approaches will emerge naturally. Given the rise of deep learning techniques in the last years, it is expected that advanced and solid schemes will be given by integrating the best of both. Another naturally incoming scenario is the big data technology, where some advances have already started to be developed, as we have seen in this chapter. Another long-known yet unresolved topic is feature selection, and SVMs are well equipped to tackle it.

10.7 Questions and Problems

Exercise 10.7.1 Reproduce the example of Figure 10.17 using a classifier based on the ridge regression criterion and compare it with the SVM classifier. In order to reproduce the example, write a code that computes the dual variables α_i of the estimator and store them in a column vector. Insert the code in line 15 of the code provided in Listing 10.6. Then, write a code that classifies the training data stored in matrix xt and insert it in line 24. The function computes and represents the double spiral and the classification boundary.

For the ridge regression, use different values of the variable γ that regularizes the kernel matrix. For the SVM, use different values of C to force different levels of complexity.

Exercise 10.7.2 Represent the support vectors obtained with code in Listing 10.5 in the previous exercise.

Exercise 10.7.3 Derive Equation 10.39 corresponding to the solution of the LS-SVM Lagrangian. by optimizing it with respect to its parameters. Then derive Equation 10.40 by applying the assumption that the primal parameters are a linear combination of the data.

Exercise 10.7.4 Use the result of the exercise above to modify the script in Listing 10.6 and obtain a solution that uses the LS-SVM. In order to work out this exercise, the reader can use the code in Listing 10.7. In this code, the name `lambda` is used for the regularization parameter, instead of `C`. Represent the values of the dual parameters α_i as a function of the error. What relationship between both can be observed?

Exercise 10.7.5 Given a vector $\boldsymbol{x}$, a valid transformation into a higher dimension Hilbert space is

$$\phi(\boldsymbol{x}) = \begin{pmatrix} \boldsymbol{\phi}_1(\boldsymbol{x}) \\ \boldsymbol{\phi}_2(\boldsymbol{x}) \end{pmatrix}, \tag{10.112}$$

where $\boldsymbol{\phi}_1(\cdot)$ and $\boldsymbol{\phi}_2(\cdot)$ are two transformations into two different RKHSs endowed with kernels $K_1(\cdot,\cdot)$ and $K_2(\cdot,\cdot)$. Prove that the kernel of the transformation in Equation 10.112 is the one of Equation 10.77.

Exercise 10.7.6 Modify the code of Listing 10.2 to obtain a ridge regression version of the algorithm. Use it to reproduce the "XOR Problem and Overfitting" example. Compare their performances in accuracy and computational time for a large amount of data.

11

Clustering and Anomaly Detection with Kernels

Many problems in signal processing deal with the identification of the right subspace where signals can be better represented. Finding good and representative groups or clusters in the data is of course the main venue. We will find many kernel algorithms to approach the problem, by 'kernelizing' either the metric or the algorithm. When there is just one class of interest, alternative methods exist under the framework of one-class detection or anomaly detection. The field of anomaly detection has the challenge of finding the anomalous patterns in the data. Anomaly detection from data has many practical implications, such as the classification of rare events in time series, the identification of target signals buried in noise, or the online identification of changes and anomalies in the wild.

The problem of defining regularity, cluster, membership, and abnormality is a conceptually challenging one, for which there exist many lines of attack and assumptions involved. This chapter will review the recent advances of kernel methods in the fields of *clustering, domain description* (also known as *outlier identification* or *one-class classification*) and *subspace detectors*. It goes without saying that all these methods are *unsupervised* or *weakly supervised*, so the challenge is even bigger. The chapter will present examples dealing with synthetic data as well as real problems in all these domains.

11.1 Introduction

Detecting patterns and groups in data sets is perhaps the most challenging problem in machine learning. The problem is unsupervised and requires many assumptions to attain significant results. For example, the user should ask elusive questions, such as: How many (semantically meaningful) groups characterize the data distribution? What is the appropriate notion of distance? What is the intrinsic (subspace) dimensionality where the data live? How do you derive robust statistics for assessing the significance of the results and to measure the different error types? In this chapter, we review the main developments in the field of kernel methods to tackle these problems.

Nowadays, we have kernel methods for clustering data in different forms and sophistication. Kernel methods have been used to "kernelize" the metric, perform clustering explicitly in feature spaces by exploiting the familiar kernel trick on standard clustering algorithms, and to describe the data distribution via support vectors (Filippone *et al.*, 2008). Once the data distribution is described or distances between samples correctly

Digital Signal Processing with Kernel Methods, First Edition. José Luis Rojo-Álvarez, Manel Martínez-Ramón, Jordi Muñoz-Marí, and Gustau Camps-Valls.

captured, one can use the clustering solution for data visualization, classification, or anomaly detection. The latter implies that the incoming data points are identified as normal or anomalous by simply computing the distance (or similarity) to those data instances used in the optimization of the kernel decision function in the particular clustering algorithm.

Clustering and density estimation are tightly related machine-learning tasks, and sometimes are collectively referred as to *domain description* (Tax and Duin, 1999). When there is not an underlying statistical law for the anomalous/outlying patterns, or when we cannot assume one, nonparametric kernel methods can do the job efficiently. Actually, these two cases are quite common; very often the number of extreme/anomalous/rare events is very low, and most of the times they change over time, space, or both. This is the ideal situation to describe (the boundary of) the target class instead of tackling the more challenging problem of density estimation especially in high-dimensional feature spaces, since that requires large datasets and dealing with often intractable integrals. In any case, the use of kernel methods – which are in nature related to classical statistical techniques such as Parzen's windowing – for density estimation have witnessed an increasing interest in recent years, guided either by maximum variance and entropic kernel components, or by the introduction of an appealing family of density ratio estimation methods that measure the relevance of examples in feature spaces under a density estimation perspective.

The field of anomaly detection in signal processing is vast. Very often one is confronted with the task of deciding if an event is rare or unusual compared with the expected behavior of the system. Just to name a few examples, anomaly detection techniques have found wide application for fraud detection, intruder detection, fault detection in security systems, military surveillance, network traffic monitoring, and so on. Anomaly detection tries to find patterns in the data that do not follow the expected behavior of the system's mechanism generating the observations. These strange, anomalous nonconforming patterns are often referred to as anomalies, outliers, exceptions, aberrations, surprises, peculiarities, or contaminants in different application domains (Chandola *et al.*, 2009). Actually, one often uses the terms "anomalies" and "outliers" interchangeably. Either way, detecting anomalies should be put in the context of modeling what is *normal*, *expected*, *regular*, and *pervasive*, which boils down to properly modeling the *pdf* or to finding representative and compact groups/clusters in the data. The problem of anomaly detection can be tackled via unsupervised learning, supervised learning, and SSL. Unsupervised anomaly detection techniques detect anomalies in an unlabeled test dataset under the assumption that the majority of the instances in the data set are normal by looking for instances that seem to fit least to the remainder of the dataset. Supervised anomaly detection techniques require a data set that has been labeled as "normal" and "abnormal" and involves training a classifier. Semi-supervised anomaly detection techniques construct a model representing normal behavior from a given normal training dataset, and then testing the likelihood of a test instance to be generated by the learnt model. In this chapter, however, we will only focus on the unsupervised setting, as it is more common in DSP.

A different approach to anomaly detection is that of subspace learning. Since its invention, matched (subspace) filters have been revealed as efficient for a myriad of signal- and image-processing applications. Essentially, a matched filter (originally known as a North filter) is obtained by correlating a known signal, or template, with an unknown signal to detect the presence (or absence) of the template in the unknown signal.[1] Matched subspace detection exploits this and assumes a subspace mixture model for target detection where the target and background points are represented by their corresponding subspaces. A high number of kernel-subspace methods have been introduced, relying on different assumptions and criteria; for example, maximum variance subspace, orthogonality between anomalies and clutter, Reed–Xiaoli criterion (Kwon and Nasrabadi, 2005c), maximum separation of clusters, or maximum eigenspace separation (Nasrabadi, 2008). We will review the most representative kernel subspace matched detectors, including the kernel orthogonal subspace projection (KOSP) (Kwon and Nasrabadi, 2005b) and the kernel spectral angle mapper (KSAM) (Camps-Valls, 2016), as well as a family of kernel anomaly change detection (KACD) algorithms that include the popular RX and cronochrome anomaly change detectors as special cases, both under Gaussianity and elliptically contoured (EC) distributions (Longbotham and Camps-Valls, 2014).

The issue of testing whether or not a sample belongs to a given (contrast) distribution can be approached from *signal detection* concepts. This is the field of *hypothesis testing*. Kernel-based methods provide a rich and elegant framework for developing nonparametric detection procedures for signal processing. Several recently proposed methods can be simply described using basic concepts of RKHS embeddings of probability distributions, mainly mean elements and covariance operators (Harchaoui *et al.*, 2013). The field draws relationships with information divergences between distributions, and has connections with *information-theoretic learning* concepts (Principe, 2010) introduced in the field of signal processing contemporaneously. Intimately related problems to anomaly detection under the previous perspectives are those of *change detection* and *change-point identification*; here, given two subsequent instants of a data distribution, one may be interested in identifying the points that changed and if the change was significant or not.

All the previous families of approaches are interconnected and share common characteristics.[2] This chapter will review them organized in the following taxonomy:

1) *Clustering, domain description*, and *one-class classification* are concerned with the task of finding groups in the data, or to detect one class of interest (the normality class) and discard the rest (the anomalous class).
2) *Density estimation* focuses on modeling the distribution of the normal patterns.
3) *Matched subspace detection* tries to identify a subspace where target and background features can be separated.
4) *Change detection methods* tackle the problem of detecting changes in the data.
5) *Hypothesis testing* aims to detect changes in the data distributions under a certain hypothesis of data regularity.

---

1 This is equivalent to convolving the unknown signal with a conjugated time-reversed version of the template. The matched filter is the optimal linear filter for maximizing the SNR in the presence of additive stochastic noise.

2 Clustering and anomaly detection is related to the field of feature extraction and dimensionality reduction, for which a plethora of kernel methods has been proposed. The exploitation of kernel multivariate analysis methods (Arenas-García *et al.*, 2013) that seek for directions in feature spaces that highlight groups or anomalies, thus requiring projections on subspaces, will be the subject of Chapter 12.

We note some clear relations among the families and approaches, and for the sake of simplicity we grouped them in three main dedicated sections in the chapter; namely, clustering, domain description and matched subspace detection, and hypothesis testing. Illustrative methods from each family are theoretically studied and illustrated experimentally through both synthetic and real-life examples.

11.2 Kernel Clustering

As in the majority of kernel methods in this book, kernel clustering is based on reformulating existing clustering methods with kernels. Such reformulation, nevertheless, can take two different pathways: either "kernelize" a standard clustering algorithm that relies solely on dot products between samples or that relies on distances between samples. Either way, the key point is that given a mapping function $\boldsymbol{\phi} : \mathbb{R}^d \rightarrow \mathcal{H}$ that maps the input data to a high-dimensional feature space, we need to define a measure of distance between points therein in terms of dot products of mapped samples to readily replace them by kernel evaluations.

11.2.1 Kernelization of the Metric

A common place in clustering algorithms is to compute distances (e.g., Euclidean) between points. If we map data into $\mathcal{H}$, we can compute distances therein without explicitly knowing the mapping $\boldsymbol{\phi}$ or the explicit coordinates of the mapped data points as follows:

$$\begin{aligned}\|\boldsymbol{\phi}(\boldsymbol{x}_i) - \boldsymbol{\phi}(\boldsymbol{x}_j)\|_{\mathcal{H}}^2 &= (\boldsymbol{\phi}(\boldsymbol{x}_i) - \boldsymbol{\phi}(\boldsymbol{x}_j))^{\mathrm{T}}(\boldsymbol{\phi}(\boldsymbol{x}_i) - \boldsymbol{\phi}(\boldsymbol{x}_j)) \\ &= \boldsymbol{\phi}(\boldsymbol{x}_i)^{\mathrm{T}}\boldsymbol{\phi}(\boldsymbol{x}_i) + \boldsymbol{\phi}(\boldsymbol{x}_j)^{\mathrm{T}}\boldsymbol{\phi}(\boldsymbol{x}_j) - 2\boldsymbol{\phi}(\boldsymbol{x}_i)^{\mathrm{T}}\boldsymbol{\phi}(\boldsymbol{x}_j) \\ &= K(\boldsymbol{x}_i, \boldsymbol{x}_i) + K(\boldsymbol{x}_j, \boldsymbol{x}_j) - 2K(\boldsymbol{x}_i, \boldsymbol{x}_j). \end{aligned} \tag{11.1}$$

Actually, for any positive-definite kernel, we assume that the mapped data in $\mathcal{H}$ are distributed in a surface S smooth enough to be considered a Riemannian manifold (Burges, 1999). The line element of S can be expressed as

$$\mathrm{d}s^2 = g_{ab}\,\mathrm{d}\boldsymbol{\phi}^a(\boldsymbol{x})\,\mathrm{d}\boldsymbol{\phi}^b(\boldsymbol{x}) = g_{\mu\nu}\,\mathrm{d}x^\mu\,\mathrm{d}x^\nu, \tag{11.2}$$

where superscripts a and b correspond to the vector space $\mathcal{H}$, $g_{\mu\nu}$ is the induced metric, and the surface S is parameterized by x^μ. Note that Einstein's summation convention over repeated indices is used. Computing the components of the (symmetric) metric tensor only needs the kernel function:

$$g_{\mu\nu} = (1/2)\partial_{x_\mu}\partial_{x_\nu}K(\boldsymbol{x}, \boldsymbol{x}) - \{\partial_{x'_\mu}\partial_{x'_\nu}K(\boldsymbol{x}, \boldsymbol{x}')\}_{\boldsymbol{x}'=\boldsymbol{x}}. \tag{11.3}$$

For the RBF kernel with a given σ parameter, this metric tensor becomes flat, $g_{\mu\nu} = \delta_{\mu\nu}/\sigma^2$, and the squared geodesic distance between $\boldsymbol{\phi}(\boldsymbol{x})$ and $\boldsymbol{\phi}(\boldsymbol{x}')$ becomes

$$\|\boldsymbol{\phi}(\boldsymbol{x}_i) - \boldsymbol{\phi}(\boldsymbol{x}_j)\|_{\mathcal{H}}^2 = 2\left[1 - \exp\left(-\frac{\|\boldsymbol{x}_i - \boldsymbol{x}_j\|_{\mathcal{X}}^2}{2\sigma^2}\right)\right] = 2(1 - K(\boldsymbol{x}_i, \boldsymbol{x}_j)). \tag{11.4}$$

Note that the metric solely depends on the original data points yet computed implicitly in a higher dimensional feature space $\mathcal{H}$, whose notion of distance is controlled by the parameter σ: the larger σ is the smoother (linear) is the space. Actually, $\sigma \to \infty$ reduces the RBF kernel to approximately compute the Euclidean distance between vectors, which reduces the metric tensor to $g_{\mu\nu} = 1$.

Kernel Online Anomaly Detector

Let us imagine now that we receive incoming samples online. The problem now is to select the informative samples and discard the anomalous ones. This way the machine will not grow indefinitely and will retain only the information contained in the most useful incoming data points. Studying how rare a sample is becomes very complex with kernels because we do not have the inverse function from Hilbert space $\mathcal{H}$ to input space to measure normality in physically meaningful units. Remember that, even though we never actually map samples to Hilbert spaces explicitly, the model and the metric are both defined therein. However, we can actually estimate distances in Hilbert spaces implicitly via reproducing kernels. For this, one only has to check if, given a previous set of N examples $\boldsymbol{x}_i$, a test incoming point $\boldsymbol{x}_*$ is inside a sufficiently big ball of mean $\boldsymbol{\phi}_\mu \in \mathcal{H}$:

$$\|\boldsymbol{\phi}(\boldsymbol{x}_*) - \boldsymbol{\phi}_\mu\| \geq \max_{1\leq i\leq N} \{\|\boldsymbol{\phi}(\boldsymbol{x}_i) - \boldsymbol{\phi}_\mu\|\}. \tag{11.5}$$

Note that one has to estimate distances to empirical means in Hilbert spaces. On the one hand, it is straightforward to show that we can compute distances between points using kernels, as before:

$$d_{\mathcal{H}}(\boldsymbol{x}_i, \boldsymbol{x}_j) = \|\boldsymbol{\phi}(\boldsymbol{x}_i) - \boldsymbol{\phi}(\boldsymbol{x}_j)\|_{\mathcal{H}} = \sqrt{K(\boldsymbol{x}_i, \boldsymbol{x}_i) + K(\boldsymbol{x}_j, \boldsymbol{x}_j) - 2K(\boldsymbol{x}_i, \boldsymbol{x}_j)}. \tag{11.6}$$

On the other hand, one can readily define the distance to the empirical mean in $\mathcal{H}$ as $\boldsymbol{\phi}_\mu = \frac{1}{N}\sum_{j=1}^{N} \boldsymbol{\phi}(\boldsymbol{x}_j)$, and we can compute it via kernels:

$$d_{\mathcal{H}}(\boldsymbol{x}_i, \boldsymbol{\phi}_\mu)^2 := K(\boldsymbol{x}_i, \boldsymbol{x}_i) - \frac{2}{N}\sum_{j=1}^{N} K(\boldsymbol{x}_i, \boldsymbol{x}_j) + \frac{1}{N^2}\sum_{j=1}^{N}\sum_{k=1}^{N} K(\boldsymbol{x}_j, \boldsymbol{x}_k). \tag{11.7}$$

Therefore, we only have to test the following condition over $\boldsymbol{x}_*$ to check for largely anomalous points in Hilbert spaces:

$$K(\boldsymbol{x}_*, \boldsymbol{x}_*) - \frac{2}{N}\sum_{j=1}^{N} K(\boldsymbol{x}_*, \boldsymbol{x}_j) + \frac{1}{N^2}\sum_{j=1}^{N}\sum_{k=1}^{N} K(\boldsymbol{x}_j, \boldsymbol{x}_k) \geq \max_{1\leq i\leq N}\{\, d_{\mathcal{H}}(\boldsymbol{x}_i, \boldsymbol{\phi}_\mu)\}. \tag{11.8}$$

Therefore, from an operational point of view, one only has to store a scalar $\theta = \max_{1\leq i\leq N} d_{\mathcal{H}}(\boldsymbol{x}_i, \boldsymbol{\phi}_\mu)$ accounting for the maximum similarity (minimum distance) among all the training samples, and then to check if the new example $\boldsymbol{x}_*$ is beyond that threshold θ. This procedure is known as the KOAD, and has been successfully used in several applications, including the detection of anomalies in backbone routers communications (Ahmed *et al.*, 2007).

Difference Kernels for Change Detection

We can use the previous property of kernels to define particular kernels for change detection. Let us now define a temporal classification scenario, in which we aim to detect the presence or absence of changes in the acquisition. This is a standard problem in surveillance applications and change detection from satellite imagery (Camps-Valls *et al.*, 2008). A standard, extremely simple technique is called the *change vector analysis* in which a threshold on the difference vector between two consecutive acquisitions t and $t-1$ (i.e., $\boldsymbol{d} = \boldsymbol{x}^t - \boldsymbol{x}^{t-1}$) is applied to detect changes. This difference vector can be formulated in the *kernel feature space* by defining a proper kernel mapping function:

$$K(\boldsymbol{x}_i^t, \boldsymbol{x}_j^t) = K_t(\boldsymbol{x}_i^t, \boldsymbol{x}_j^t) + K_{t-1}(\boldsymbol{x}_i^{t-1}, \boldsymbol{x}_j^{t-1}) - K_{t,t-1}(\boldsymbol{x}_i^t, \boldsymbol{x}_j^{t-1}) - K_{t-1,t}(\boldsymbol{x}_i^{t-1}, \boldsymbol{x}_j^t).$$

We should note that the implementation is easy as it only involves application of kernel functions to define the *differential metric* among observations. Such a difference kernel can then be included in any supervised or unsupervised kernel classifier.[3]

11.2.2 Clustering in Feature Spaces

A complementary way to perform clustering in feature spaces is to follow the standard kernelization procedure. Let us introduce in this section the main concepts involved before defining the following kernel methods. Clustering methods try to obtain partitions of the data based on the optimization of a particular objective function, giving as result separation *hypersurfaces* among clusters. To handle nonlinearly separable clusters, the methods often define multiple *centroids* to provide a richer description of the dataset at hand.

Notationally, let us consider a set of data points $\mathcal{X} := \{\boldsymbol{x}_1, \dots, \boldsymbol{x}_N\}$ so $\boldsymbol{x}_i \in \mathbb{R}^d$. The set of centroids $\mathcal{V}$ for this dataset is called a *codebook*, and is defined as $\mathcal{V} := \{\boldsymbol{v}_1, \dots, \boldsymbol{v}_c\}$, with $c \ll N$. Each element of $\mathcal{V}$, $\boldsymbol{v}_i \in \mathbb{R}^d$, is called a *centroid* or *codevector*. Each centroid $\boldsymbol{v}_i$ defines the so-called *Voronoi region* $\mathcal{R}_i$ as the set of vectors fulfilling the following condition (Aurenhammer, 1991): $\mathcal{R}_i = \{\boldsymbol{z} \in \mathbb{R}^d | i = \arg\min_j \|\boldsymbol{z} - \boldsymbol{v}_i\|^2\}$. Each Voronoi region is convex and its boundaries are linear segments. A further definition we need is the *Voronoi set*, S_i, of the centroid $\boldsymbol{v}_i$, being the subset of $\mathcal{X}$ for which the centroid $\boldsymbol{v}_i$ is the nearest vector; that is, $S_i = \{\boldsymbol{x} \in \mathcal{X} | i = \arg\min_j \|\boldsymbol{x} - \boldsymbol{v}_j\|^2\}$. A partition of $\mathbb{R}^d$ induced by all Voronoi regions is called a *Voronoi tessellation* or *Dirichlet tessellation* (Aurenhammer, 1991).

Kernel *k*-Means

Let us start by summarizing the formulation and optimization of the standard k-means, and then we will show how to kernelize it. The goal of the canonical k-means is to construct a Voronoi tessellation by moving k centroids to the arithmetic mean of their Voronoi sets. To achieve this, the k-means algorithm searches the elements of $\mathcal{V}$ that jointly minimize the *empirical quantization error*

---

3 We should note that this kernel raises some issues about its validity because it may become non-positive definite. In such cases, extra regularization can be needed to make it positive definite.

$$E(\mathcal{X}) = \frac{1}{2N}\sum_{i=1}^{k}\sum_{\boldsymbol{x}\in S_i}\|\boldsymbol{x}-\boldsymbol{v}_i\|^2, \tag{11.9}$$

which can be achieved if each centroid $\boldsymbol{v}_i$ fulfills the *centroid condition*

$$\sum_{\boldsymbol{x}\in S_i}\frac{\partial\|\boldsymbol{x}-\boldsymbol{v}_i\|^2}{\partial\boldsymbol{v}_i} = 0. \tag{11.10}$$

In the case of using the Euclidean distance, such a condition reduces to

$$\boldsymbol{v}_i = \frac{1}{|S_i|}\sum_{\boldsymbol{x}\in S_i}\boldsymbol{x}, \tag{11.11}$$

where $|S_i|$ denotes the cardinality of S_i. The k-means algorithm is defined by the following steps: (1) choose the number k of clusters; (2) initialize $\mathcal{V}$ with a set of centroids $\boldsymbol{v}_i$ randomly selected; (3) compute the Voronoi set S_i for each centroid $\boldsymbol{v}_i$; (4) move each centroid $\boldsymbol{v}_i$ to the mean of S_i using Equation 11.11; and (5) stop the algorithm if no changes are observed, otherwise go to step 3. Note that the k-means algorithm is an example of an *EM* algorithm. Step 3 is the *expectation*, and step 4 is the *maximization*. The convergence of the k-means algorithm is guaranteed since each EM algorithm is always convergent to a local minimum (Bottou and Bengio, 1995).

In order to obtain the kernel k-means algorithm, we just translate the definitions and concepts before to a feature space. The first step is mapping our dataset $\mathcal{X}$ to a high-dimensional feature space using a nonlinear map ϕ. The *codebook* is defined in the feature space as $V_\phi = \{\boldsymbol{\phi}(\boldsymbol{v}_1), \dots, \boldsymbol{\phi}(\boldsymbol{v}_k)\}$. We have a *Voronoi region* in the feature space defined by $\mathcal{R}_i^\phi = \{\boldsymbol{\phi}(\boldsymbol{x}) \in F | i = \arg\min_j \|\boldsymbol{\phi}(\boldsymbol{x}) - \boldsymbol{\phi}(\boldsymbol{v}_j)\|^2\}$, and the Voronoi set S_i^ϕ of the centroid $\boldsymbol{\phi}(\boldsymbol{v}_i)$ defined as $S_i^\phi = \{\boldsymbol{x} \in \mathcal{X} \,| i = \arg\min_j \|\boldsymbol{\phi}(\boldsymbol{x}) - \boldsymbol{\phi}(\boldsymbol{v}_j)\|^2\}$. In this case, the Voronoi regions induce a Voronoi tessellation in the feature space.

The steps of the k-means algorithm in the feature space are the same as the ones described in Section 11.2.1. The only difference is that the *maximization* step is computed using the following equation:

$$\boldsymbol{\phi}(\boldsymbol{v}_i) = \frac{1}{|S_i^\phi|}\sum_{\boldsymbol{x}\in S_i^\phi}\boldsymbol{\phi}(\boldsymbol{x}). \tag{11.12}$$

The problem is that we do not explicitly know $\boldsymbol{\phi}$ and cannot move the centroid using Equation 11.12. This issue is solved using the representer theorem (Schölkopf and Smola, 2002), so that we can write each centroid in the feature space as a linear combination of the mapped data vectors, $\boldsymbol{\phi}(\boldsymbol{v}_i) = \sum_{j=1}^N \alpha_{ij}\boldsymbol{\phi}(\boldsymbol{x}_j)$. By replacing the expansion into Equation 11.12, and noting that α_{ij} should be zero if $\boldsymbol{x}_j \notin S_i^\phi$, we obtain

$$\begin{aligned}\|\boldsymbol{\phi}(\boldsymbol{x}_i) - \boldsymbol{\phi}(\boldsymbol{v}_j)\|_{\mathcal{H}}^2 &= \|\boldsymbol{\phi}(\boldsymbol{x}_i) - \sum_{j=1}^{N}\alpha_{ij}\boldsymbol{\phi}(\boldsymbol{x}_j)\|^2 \\ &= K(\boldsymbol{x}_i,\boldsymbol{x}_i) - 2\sum_{j=1}^{N}\alpha_{ij}K(\boldsymbol{x}_i,\boldsymbol{x}_j) + \sum_{l=1}^{N}\sum_{m=1}^{N}\alpha_{il}\alpha_{im}K(\boldsymbol{x}_l,\boldsymbol{x}_m),\end{aligned} \tag{11.13}$$

which allows us to compute the *closest feature space centroid* for each sample, and update the coefficients α_{ij} accordingly. As in the standard k-means algorithm, the procedure is repeated until all α_{ij} do not change significantly.

Kernel Fuzzy Clustering

Fuzzy clustering is essentially dealing with the elusive definition of *hard* and *fuzzy* partitions (Pal *et al.*, 2005). Clustering solutions, such those offered by k-means, do not offer disambiguation of cluster membership of samples that do not fit exactly to one single cluster but to several clusters in different degrees. Let us define a fuzzy-c partition (or clustering) as

$$\mathcal{U} := \left\{ \mathbf{U} \middle| 0 \leq U_{ij} \leq 1 \quad \forall i,j; \quad \text{s.t.} \sum_{i=1}^{c} U_{ij} = 1 \, \forall j; 0 < \sum_{j=1}^{N} U_{ij} < n \, , \forall i \right\}, \tag{11.14}$$

where the set $\mathcal{U}$ can be defined with the *membership matrix* $\mathbf{U}$ of size $c \times N$, with $2 \leq c < N$ and $U_{ij} \in \mathbb{R}$. Each element U_{ij} is the membership of the jth pattern to the *ith* cluster, and the constraints ensure that (1) the sum of the membership of a pattern to all clusters is one (*probabilistic constraint*), and (2) a cluster cannot by empty or contain all samples.

Now, given a codebook $\mathcal{V}$ and a membership set $\mathcal{U}$, the fuzzy c-means algorithm (Pal *et al.*, 2005) minimizes the functional

$$J(\mathcal{U}, \mathcal{V}) = \sum_{i=1}^{c} \sum_{j=1}^{N} U_{ij}^{m} \|\boldsymbol{x}_j - \boldsymbol{v}_i\|^2, \tag{11.15}$$

subject to $\sum_{i=1}^{c} U_{ij} = 1, \forall j$. The parameter m establishes the *fuzziness* of the membership. For $m = 1$ the classic k-means hard clustering algorithm is obtained, whereas large values of m tend to set all memberships the same, and hence no structure is found on the data.

The functional in Equation 11.15 is minimized defining a Lagrangian function for each sample $\boldsymbol{x}_j$:

$$L_j = \sum_{i=1}^{c} U_{ij}^{m} \|\boldsymbol{x}_j - \boldsymbol{v}_i\|^2 + \alpha_j \left(\sum_{i=1}^{c} U_{ij} - 1 \right), \tag{11.16}$$

and deriving with respect to $\boldsymbol{v}_i$ and U_{ij} and setting to zero we obtain the following two equations:

$$\boldsymbol{v}_i = \frac{\sum_{j=1}^{N} U_{ij}^{m} \boldsymbol{x}_j}{\sum_{j=1}^{N} U_{ij}^{m}}, \tag{11.17}$$

$$U_{ij} = \frac{(\|\boldsymbol{x}_j - \boldsymbol{v}_i\|)^{-2/(m-1)}}{\sum_{h=1}^{c} (\|\boldsymbol{x}_j - \boldsymbol{v}_h\|)^{-2/(m-1)}}. \tag{11.18}$$

This pair of equations is used with a *Picard iteration method* (Pal *et al.*, 2005) composed of two steps. In the first step, the membership variables U_{ij} are kept fixed and the codevectors $\mathbf{v}_i$ are optimized. Then, the second step optimizes the membership variables while the codevectors are fixed. These two steps are repeated until the variables change less than a predefined threshold. Equation 11.18 shows that the fuzzy membership of an input sample decreases as the distance increases. The sum in the denominator acts as a normalization factor.

Next, we show two approaches to obtain a kernel version of the fuzzy clustering algorithm. One is based in the *kernelization of the metric*, and the other directly solves the equations of *fuzzy c-means in feature space.*

Kernel Fuzzy Clustering Based on the Metric Kernelization

The functional to minimize is the same as in Equation 11.15 but in a high-dimensional feature space:

$$J_{\boldsymbol{\phi}}(\mathcal{U}, \mathcal{V}) = \sum_{j=1}^{N} \sum_{i=1}^{c} U_{ij}^{m} \|\boldsymbol{\phi}(\mathbf{x}_j) - \boldsymbol{\phi}(\mathbf{v}_i)\|^2, \tag{11.19}$$

subject to $\sum U_{ij} = 1, \forall i$. In order to minimize this cost function, we take derivatives with respect to $\mathbf{v}_j$ and U_{ij}. To be able to apply the *kernel trick* and obtain a practical solution, we explicitly compute the derivatives of the kernel function, which for the RBF kernel the result is as follows:

$$\frac{\partial K(\mathbf{x}_j, \mathbf{v}_i)}{\partial \mathbf{v}_i} = -\frac{(\mathbf{x}_j - \mathbf{v}_i)}{\sigma^2} K(\mathbf{x}_j, \mathbf{v}_i).$$

Computing all the derivatives and equating them to zero we obtain the following equations for the Picard iteration method:

$$\mathbf{v}_i = \frac{\sum_{j=1}^{N} U_{ij}^{m} K(\mathbf{x}_j, \mathbf{v}_i) \mathbf{x}_j}{\sum_{j=1}^{N} U_{ij}^{m} K(\mathbf{x}_j, \mathbf{v}_i)}, \tag{11.20}$$

$$U_{ij} = \frac{(1 - K(\mathbf{x}_j, \mathbf{v}_i))^{-1/(m-1)}}{\sum_{h=1}^{c} (1 - K(\mathbf{x}_j, \mathbf{v}_h))^{-1/(m-1)}}, \tag{11.21}$$

which are now solely based on evaluating kernel functions, and hence the method can be easily implemented.

Kernel Fuzzy Clustering in Feature Space

The functional to minimize is shown in Equation 11.19. Instead of obtaining a solution taking derivatives using $\mathbf{v}_i$ and U_{ij}, now we use the representer theorem to express $\boldsymbol{\phi}(\mathbf{v}_i)$ as a linear combination of samples in feature space; that is, $\boldsymbol{\phi}(\mathbf{v}_i) = \sum_{h=1}^{N} \beta_{ih} \boldsymbol{\phi}(\mathbf{x}_h)$. This will allow us to compute the distance appearing in Equation 11.19 just in terms of the kernel function as in Equation 11.1. Therefore, let us rewrite the functional as

$$J_{\phi}(\mathcal{U}, \mathcal{V}) = \sum_{j=1}^{N} \sum_{i=1}^{c} U_{ij}^{m} (K(\boldsymbol{x}_j, \boldsymbol{x}_j) - 2\boldsymbol{\beta}_i \boldsymbol{k}_j^{\mathrm{T}} + \boldsymbol{\beta}_i \boldsymbol{K} \boldsymbol{\beta}_i^{\mathrm{T}}), \tag{11.22}$$

where $\boldsymbol{k}_j$ is the jth row of the kernel; that is, $\boldsymbol{k}_j := [K(\boldsymbol{x}_j, \boldsymbol{x}_1), \dots, K(\boldsymbol{x}_j, \boldsymbol{x}_N)]^{\mathrm{T}}$, and $\boldsymbol{\beta} = [\beta_1, \dots, \beta_N]$ are the weights for the $\boldsymbol{\phi}(\boldsymbol{v})$ codevector, and $\boldsymbol{K}$ is the whole kernel matrix. This functional can be derived with respect to the variables $\boldsymbol{\beta}_i$ and U_{ij}, obtaining the equations we need for the Picard iteration method:

$$\boldsymbol{\beta}_i = \frac{\sum_{j=1}^{N} U_{ij}^{m} \boldsymbol{k}_j \boldsymbol{K}^{-1}}{\sum_{j=1}^{N} U_{ij}^{m}}, \tag{11.23}$$

$$U_{ij} = \frac{(K(\boldsymbol{x}_j, \boldsymbol{x}_j) - 2\boldsymbol{\beta}_i \boldsymbol{k}_j^{\mathrm{T}} + \boldsymbol{\beta}_i \boldsymbol{K} \boldsymbol{\beta}_i^{\mathrm{T}})^{-1/(m-1)}}{\sum_{h=1}^{c} (K(\boldsymbol{x}_j, \boldsymbol{x}_j) - 2\boldsymbol{\beta}_h \boldsymbol{k}_j^{\mathrm{T}} + \boldsymbol{\beta}_h \boldsymbol{K} \boldsymbol{\beta}_h^{\mathrm{T}})^{-1/(m-1)}}. \tag{11.24}$$

There is an alternative pair of equivalent equations that we can obtain using the following reasoning. Deriving the functional in Equation 11.19 directly with respect to $\boldsymbol{\phi}(\boldsymbol{v}_i)$ we will obtain, similarly to Equation 11.17, the expansion

$$\boldsymbol{\phi}(\boldsymbol{v}_i) = \frac{\sum_{j=1}^{N} U_{ij}^{m} \boldsymbol{\phi}(\boldsymbol{x}_j)}{\sum_{j=1}^{N} U_{ij}^{m}} = a_i \sum_{j=1}^{N} U_{ij}^{m} \boldsymbol{\phi}(\boldsymbol{x}_j), \tag{11.25}$$

where we have defined a_i as

$$a_i^{-1} = \sum_{h=1}^{N} (u_{ih})^{m}. \tag{11.26}$$

With Equations 11.26 and 11.25 we can rewrite Equation 11.24 as

$$U_{ij} = \frac{(K(\boldsymbol{x}_j, \boldsymbol{x}_j) - 2a_i \boldsymbol{u}_i \boldsymbol{k}_j^{\mathrm{T}} + \boldsymbol{u}_i K \boldsymbol{u}_i^{\mathrm{T}})^{-1/(m-1)}}{\sum_{h=1}^{c} (K(\boldsymbol{x}_j, \boldsymbol{x}_j) - 2a_h \boldsymbol{u}_h \boldsymbol{k}_j^{\mathrm{T}} + \boldsymbol{u}_h K \boldsymbol{u}_h^{\mathrm{T}})^{-1/(m-1)}}, \tag{11.27}$$

where $\boldsymbol{u}_i := [(u_{i1})^m, \dots, (u_{in})^m]^{\mathrm{T}}$. It is clear that Equations 11.24 and 11.27 are equivalent, noting that, by definition, $\beta_{ij} = a_i U_{ij}^{m}$. However, the Picard iteration method is easier to implement using Equations 11.26 and 11.27.

Kernel Self-Organizing Maps

A self-organized map (SOM) (Kohonen, 1990) is a type of artificial NN trained to obtain a low-dimensional (most of the times 2D) representation of the input data. The training phase is unsupervised where SOM defines a function that tries to preserve the topological properties of the input dataset. SOMs are typically used to obtain low-dimensional visualizations of high-dimensional data. As for other NNs, an SOM consists of a series of interconnected *nodes* or *neurons*. Each node has a weight vector of the same dimension as the input data samples, and a location in the map space, while each vector of the input space is placed on the map space by finding the closest/nearest node in the map space.

Let us define our *coodebook*, denoted $\mathcal{V}$, as the set of neurons (centroids) that will compose the map. The first step of the training phase is to randomly select the weights of the neurons in $\mathcal{V}$. Alternatively, Kohonen (1990) showed that it is possible to speed up the training phase by evenly sampling the subspace spanned by the two largest principal component eigenvectors, instead of randomly selecting the first neurons' weights. The second step is to take an input sample $\boldsymbol{x}$ from $\mathcal{X}$ and obtain the nearest neuron using a distance function d (Euclidean of Manhattan distances are the most often used). The shortest distance would be the one that solves $\min_{\boldsymbol{v}_j \in V} \|\boldsymbol{x} - \boldsymbol{v}_j\|$. Then the weights of the neurons are adapted using a standard gradient descent strategy:

$$\boldsymbol{v}_j(t+1) = \boldsymbol{v}_j(t) + \varepsilon(t)h(d)(\boldsymbol{x} - \boldsymbol{v}_j), \tag{11.28}$$

where h is a decreasing function of the distance d; for example, $h(d) = \exp(-d^2/(2\sigma^2(t)))$. Here, $\sigma(t)$ is a decreasing function of t, $\sigma(t) = \sigma_{\mathrm{i}}(\sigma_{\mathrm{f}}/\sigma_{\mathrm{i}})^{t/t_{\max}}$, $\varepsilon(t)$ is also a decreasing function of t, $\varepsilon(t) = \varepsilon_{\mathrm{i}}(\varepsilon_{\mathrm{f}}/\varepsilon_{\mathrm{i}})^{t/t_{\max}}$, and where the subscripts "i" and "f" indicate the initial and final values respectively. The adaptation of neurons $\boldsymbol{v}_j$ in Equation 11.28 is repeated until a predefined number of iterations is achieved; that is, $t = t_{\max}$.

In order to reformulate the SOM in the feature space, we need to first define our set of initial neurons $\boldsymbol{\phi}(\boldsymbol{v}_j)$. Since we do not know $\boldsymbol{\phi}$ explicitly, we use the representer theorem to write each neuron as a combination of mapped input samples:

$$\boldsymbol{\phi}(\boldsymbol{v}_j) = \sum_{l=1}^{N} \alpha_{jl}\boldsymbol{\phi}(\boldsymbol{x}_l), \tag{11.29}$$

and we randomly initialize the coefficients α_{jl}. Then, given an input sample $\boldsymbol{\phi}(\boldsymbol{x}_i)$, we need to compute the nearest neuron to it:

$$\underset{\boldsymbol{\phi}(\boldsymbol{v}_j)\in V_H}{\arg\min} \left\{ \|\boldsymbol{\phi}(\boldsymbol{x}_i) - \boldsymbol{\phi}(\boldsymbol{v}_j)\| \right\},$$

which, after applying the kernel trick, translates into

$$\underset{\boldsymbol{\phi}(\boldsymbol{v}_j)\in V_H}{\arg\min} \left(K(\boldsymbol{x}_i, \boldsymbol{x}_i) - 2\sum_{l} \alpha_{jl}K(\boldsymbol{x}_i, \boldsymbol{x}_l) + \sum_{r}\sum_{s} \alpha_{jr}\alpha_{js}K(\boldsymbol{x}_r, \boldsymbol{x}_s) \right). \tag{11.30}$$

Finally, the neurons in the feature space are updated as

$$\boldsymbol{\phi}(\boldsymbol{v}_j(t+1)) = \boldsymbol{\phi}(\boldsymbol{v}_j(t)) + \varepsilon(t)h(d)(\boldsymbol{\phi}(\boldsymbol{x}) - \boldsymbol{\phi}(\boldsymbol{v}_j)), \tag{11.31}$$

which according to Equation 11.29 can be expressed as

$$\sum_{l=1}^{N} \alpha_{jl}(t+1)\boldsymbol{\phi}(\boldsymbol{x}_l) = \sum_{l=1}^{N} \alpha_{jl}\boldsymbol{\phi}(\boldsymbol{x}_l) + \varepsilon(t)h(d)\left(\boldsymbol{\phi}(\boldsymbol{x}_i) - \sum_{l=1}^{N} \alpha_{jl}\boldsymbol{\phi}(\boldsymbol{x}_l) \right), \tag{11.32}$$

from which we can now obtain the update rule for α_{jl} as follows:

$$\alpha_{jl}(t+1) = \begin{cases} \alpha_{jl}t(1-\varepsilon(t)h(d)), & \text{if} \quad i \neq j, \\ \alpha_{jl}t(1-\varepsilon(t)h(d)) + \varepsilon(t)h(d), & \text{otherwise.} \end{cases} \tag{11.33}$$

Note that the updating rule can be obtained iteratively until convergence.

11.3 Domain Description Via Support Vectors

As an alternative to the previous approaches for clustering with kernels, the description of the domain can be done via support vectors. Instead of directly kernelizing the metric or the clustering algorithm, one attempts here to describe the data distribution in feature spaces with a minimum volume hypersphere enclosing all data points *but* the outliers, which translates into a hypersurface in the original input space. This idea leads to several algorithms, such as the one-class SVM (OC-SVM) and the related support vector domain description (SVDD). After computing the ball with these methods, the SVC method permits one to assign labels to patterns in input space enclosed by the same surface. In the next subsections we will review the formulation of these three main approaches.

11.3.1 Support Vector Domain Description

A different problem statement for classification is given by the SVDD (Tax and Duin, 1999). The SVDD is a method to solve one-class problems, where one tries to describe one class of objects, distinguishing them from all other possible objects.

The problem notation is defined as follows. Let $\{\boldsymbol{x}_i\}_{i=1}^N$ be a dataset belonging to a given *class of interest*. The purpose is to find a minimum volume *hypersphere* in a high-dimensional feature space $\mathcal{H}$, with radius $R > 0$ and center $\boldsymbol{a} \in \mathcal{H}$, which contains most of these data objects (Figure 11.1a). Since the training set may contain outliers, a set of *slack variables* $\xi_i \geq 0$ is introduced, and the problem becomes

$$\min_{R,\boldsymbol{a}} \left\{ R^2 + C \sum_{i=1}^{N} \xi_i \right\} \tag{11.34}$$

constrained to

$$\|\boldsymbol{\phi}(\boldsymbol{x}_i) - \mathbf{a}\|^2 \leq R^2 + \xi_i \qquad \forall i = 1, \dots, N \tag{11.35}$$

$$\xi_i \geq 0 \qquad \forall i = 1, \dots, N, \tag{11.36}$$

where parameter C controls the trade-off between the volume of the hypersphere and the permitted errors. Parameter ν, defined as $\nu = 1/(NC)$, can be used as a rejection fraction parameter to be tuned, as noted by Schölkopf *et al.* (1999).

The dual functional is a QP problem that yields a set of Lagrange multipliers (α_i) corresponding to constraints in Equation 11.35. When free parameter C is adjusted properly, most of the α_i are zero, giving a sparse solution. The Lagrange multipliers are also used to calculate the distance from a test point to the center $R(\mathbf{x}_*)$:

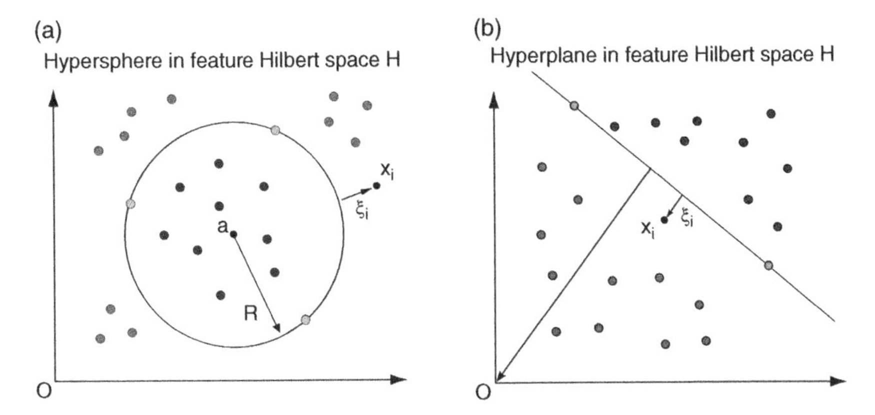

Figure 11.1 Illustration of the one-class kernel classifiers. (a) In the SVDD, the hypersphere containing the target data is described by center ***a*** and radius ***R***, and samples in the boundary and outside the ball are unbounded and bounded support vectors, respectively. (b) In the case of the OC-SVM, all samples from the target class are mapped with maximum distance to the origin.

$$R(\mathbf{x}_*) = K(\mathbf{x}_*, \mathbf{x}_*) - 2\sum_{i=1}^{N} K(\boldsymbol{x}_i, \mathbf{x}_*) + \sum_{i,j=1}^{N} K(\boldsymbol{x}_i, \boldsymbol{x}_j), \tag{11.37}$$

which is compared with ratio R. Unbounded support vectors are those samples $\boldsymbol{x}_i$ satisfying $0 \leqslant \alpha_i < C$, while bounded support vectors are samples whose associated $\alpha_i = C$, which are considered outliers.

11.3.2 One-Class Support Vector Machine

In the OC-SVM, instead of defining a hypersphere containing all examples, a hyperplane that separates the data objects from the origin with maximum margin is defined (Figure 11.1b). It can be shown that, when working with normalized data and the RBF Gaussian kernel, both methods yield the same solution (Schölkopf *et al.*, 1999).

Notationally, in the OC-SVM, we want to find a hyperplane $\boldsymbol{w}$ which separates samples $\boldsymbol{x}_i$ from the origin with margin ρ. The problem thus becomes

$$\min_{\boldsymbol{w},\rho,\xi} \left\{ \frac{1}{2}\|\boldsymbol{w}\|^2 - \rho + \frac{1}{\nu N}\sum_{i=1}^{N} \xi_i \right\} \tag{11.38}$$

constrained to

$$\langle \boldsymbol{w}, \boldsymbol{\phi}(\boldsymbol{x}_i)\rangle \geq \rho - \xi_i \qquad \forall i = 1, \ldots, N \tag{11.39}$$

$$\xi_i \geq 0 \qquad \forall i = 1, \ldots, N. \tag{11.40}$$

The problem is solved through its Lagrangian dual introducing a set of Lagrange multipliers α_i. The margin ρ can be computed as $\rho = \langle \boldsymbol{w}, \boldsymbol{\phi}(\boldsymbol{x}_i)\rangle = \sum_j \alpha_j K(\boldsymbol{x}_i, \boldsymbol{x}_j)$.

11.3.3 Relationship Between Support Vector Domain Description and Density Estimation

Let us now point out the connection between *pdf* estimation and the previous one-class kernel classifiers. The goal of kernel density estimation (KDE) is to estimate the *unknown pdf* $p(\boldsymbol{x})$ given N i.i.d. samples $\{\boldsymbol{x}_1, \dots, \boldsymbol{x}_N\}$ drawn from it. The kernel (Parzen) estimate of the *pdf* using an arbitrary kernel function $K_\sigma(\cdot)$ parameterized with a length-scale parameter σ is given by

$$\hat{p}(\boldsymbol{x}) = \frac{1}{N\sigma} \sum_{i=1}^{N} K\left(\frac{\boldsymbol{x} - \boldsymbol{x}_i}{\sigma}\right), \tag{11.41}$$

where K is typically a Gaussian kernel, and the hyperparameters of the model (e.g., the length scale or kernel width) can be obtained via ML estimation or fixed through ad hoc rule of thumb rules, such as Silverman's rule. Other kernel functions can be used: uniform, normal, triangular, bi-weighted, tri-weighted, just to name a few. The Epanechnikov kernel, for example, is optimal in an MSE sense. The interested reader can explore the MATLAB function `ksdensity.m` and the MATLAB KDE toolbox available oin the book's web page.

Density estimates can be used to characterize distributions and to design anomaly detection schemes. Note, for instance, that once $\hat{p}(\boldsymbol{x})$ is obtained, an anomaly detector can be readily defined simply as

$$\mathcal{A}(\boldsymbol{x}_*) = 1 - \hat{p}(\boldsymbol{x}_*). \tag{11.42}$$

Now, by plugging the previous Gaussian kernel estimate in here and defining a centroid in the feature space as the mean of the mapped samples, a KDE-based anomaly detector can be readily derived by simply kernelizing the distance between a test point $\boldsymbol{x}_*$ and all the training points (see Section 11.2.1):

$$\mathcal{A}(\boldsymbol{x}_*) = \text{constant} - \frac{2}{N} \sum_{i=1}^{N} K(\boldsymbol{x}_*, \boldsymbol{x}_i). \tag{11.43}$$

Note that definition of SVDD seen previously in Equations 11.34 and 11.35 leads to an SVDD anomaly detector in the form

$$\mathcal{A}(\boldsymbol{x}) = \|\boldsymbol{\phi}(\boldsymbol{x}) - \boldsymbol{a}\|^2 = \text{constant} - \sum_i \alpha_i K(\boldsymbol{x}, \boldsymbol{x}_i), \tag{11.44}$$

where $\alpha_i > 0$ are support vectors. Comparing Equation 11.43 with Equation 11.44 we see they are very similar, except that the SVDD anomaly detector places unequal weights on the training points $\boldsymbol{x}_i$. In other words, while the kernel KDE works as an unsupervised method where the centroid is placed at $\boldsymbol{\mu}_\phi = (1/N) \sum_{i=1}^{N} \boldsymbol{\phi}(\boldsymbol{x}_i)$, the SVDD is a supervised method that uses training (labeled) samples to place the centroid at $\boldsymbol{a} = \sum_i \alpha_i \boldsymbol{\phi}(\boldsymbol{x}_i)$. Figure 11.2 shows a toy example comparing the boundaries obtained by the SVDD (centroid at **a**) and KDE (centroid at $\boldsymbol{\mu}_\phi$).

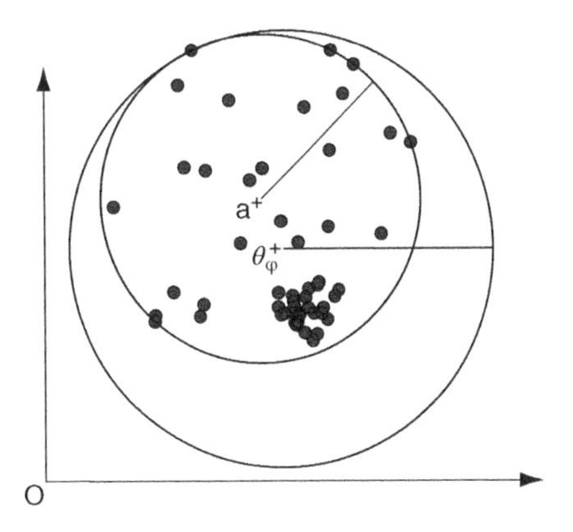

Figure 11.2 Boundary comparison between SVDD (sphere centered at **a**) and KDE (sphere centered at μ_ϕ).

11.3.4 Semi-supervised One-Class Classification

Exploiting unlabeled information in one-class classifiers can be done via semi-supervised versions of the OC-SVM. A particularly convenient instantiation is the biased-SVM (b-SVM) classifier, which was originally proposed by Liu *et al.* (2003) for text classification using only target and unlabeled data. The underlying idea of the method lies on the assumption that if the sample size is large enough, minimizing the number of unlabeled samples classified as *targets* while correctly classifying the labeled target samples will give an accurate classifier.

Let us consider again the dataset $\{\boldsymbol{x}_i\}_{i=1}^{l+u} \in \mathbb{R}^d$ made up of l labeled samples and u unlabeled samples ($N = l + u$). The l labeled samples all belong to the same class: the target class. Concerning unlabeled samples, they are treated by the algorithm as being all outliers, although their class is unknown. This characteristic makes b-SVM a sort of semi-supervised classifier. Under these assumptions, the b-SVM formulation is defined as

$$\min_{\boldsymbol{w},\xi} \left\{ \frac{1}{2}\|\boldsymbol{w}\|^2 + C_{\mathrm{t}} \sum_{i=1}^{l} \xi_i + C_{\mathrm{o}} \sum_{i=l+1}^{l+u} \xi_i \right\} \qquad \forall i = 1, \dots, N \tag{11.45}$$

subject to

$$\langle \boldsymbol{w}, \boldsymbol{\phi}(\boldsymbol{x}_i)\rangle \geq 1 - \xi_i, \tag{11.46}$$

where $\xi_i \geq 0$ are slack variables, and C_{t} and C_{o} are the costs assigned to errors on target and outlier (unlabeled) classes, respectively. The two cost values should be adjusted to fulfill the goal of classifying the target class correctly while at the same time trying to minimize the number of unlabeled samples classified as target class. To achieve this goal, C_{t} should have a large value, as initially we trust our labeled training set, and C_{o} a small value, because it is unknown whether the samples unlabeled are actually targets or outliers.

11.4 Kernel Matched Subspace Detectors

Target detection from data is of great interest in many applications. Detecting targets is typically described as a two-step methodology: first, a specific detector identifies anomalies; second, a classifier is aimed at identifying whether the anomaly is a target or natural clutter. In many application domains, this step is only possible if the target signature is known, which can be obtained from a library of "pure elements" or learned from data by using subspace matched filters. Several techniques have been proposed in the literature, such as the Reed–Xiaoli anomaly detector (Reed and Yu, 1990), the orthogonal subspace projection (OSP) (Harsanyi and Chang, 1994), the Gaussian mixture model (GMM) (Stein *et al.*, 2002), the cluster-based detector (Stein *et al.*, 2002), or the signal subspace processor (Ranney and Soumekh, 2006). In recent years, many detection algorithms based on spectral matched (subspace) filters have been reformulated under the kernel methods framework: matched subspace detector, orthogonal subspace detector, as well as adaptive subspace detectors (Kwon and Nasrabadi, 2005a). Certainly, the use of kernel methods alleviates several key problems and offers many advantages: they combat the high-dimensionality problem, make the method robust to noise, and allow for flexible nonlinear mappings with controlled (regularized) complexity (Shawe-Taylor and Cristianini, 2004).

This section deals with target/anomaly detection and its corresponding kernelized extensions under the viewpoint of matched subspace detectors. This topic has been very active in the last decade, and also in the particular field of signal and image anomaly detection. We pay attention to different kernel anomaly subspace detectors: KOSP, KSAM and the family of KACD that generalizes several anomaly detectors under Gaussian and EC distributions.

11.4.1 Kernel Orthogonal Subspace Projection

The OSP algorithm (Harsanyi and Chang, 1994) tries to maximize the SNR in the subspace orthogonal to the background subspace. OSP has been widely adopted in the community of automatic target recognition because of its simplicity (only depends on the noise second-order statistics) and its good behavior. In the standard formulation of the OSP algorithm (Harsanyi and Chang, 1994), a linear mixing model is assumed for each d-dimensional observed point $\boldsymbol{r}$ as follows:

$$\boldsymbol{r} = \boldsymbol{M}\boldsymbol{\alpha} + \boldsymbol{n},$$

where $\boldsymbol{M}$ is the matrix of size $(d \times p)$ containing the p *endmembers* (i.e., pure constituents of the observed mixture) contributing to the mixed point $\boldsymbol{r}$, $\boldsymbol{\alpha} \in \mathbb{R}^{p\times 1}$ contains the weights in the mixture (sometimes called *abundance* vector), and $\boldsymbol{n} \in \mathbb{R}^{d\times 1}$ stands for an additive zero-mean Gaussian noise vector. In order to identify a particular *desired* point $\boldsymbol{d}$ in a data set (e.g., a pixel in an image) with corresponding abundance α_p, the earlier expression can be organized by rewriting the $\boldsymbol{M}$ matrix in two submatrices $\boldsymbol{M} = [\boldsymbol{U}|\boldsymbol{d}]$ and $\boldsymbol{\alpha} = [\gamma, \alpha_p]^{\mathrm{T}}$, so that

$$\boldsymbol{r} = \boldsymbol{U}\gamma + \boldsymbol{d}\alpha_p + \boldsymbol{n}, \tag{11.47}$$

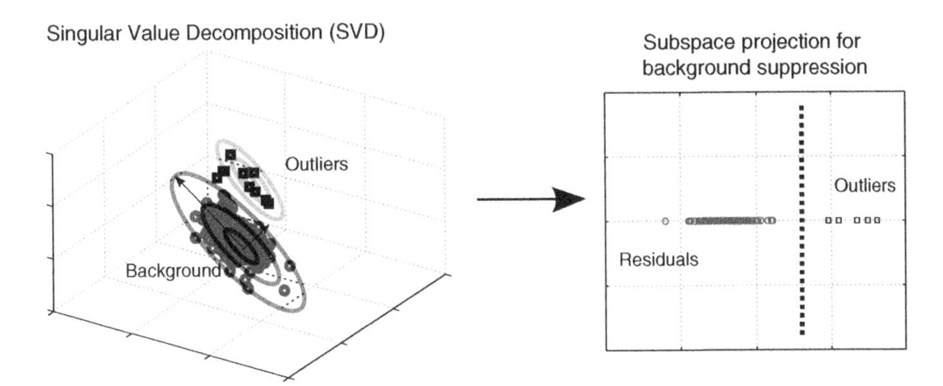

Figure 11.3 The standard (linear) OSP method first performs a linear transformation that looks for projections that identify the subspace spanned by the background, and then the background suppression is carried out by projecting the data onto the subspace orthogonal to the one spanned by the background components. The kernel version of this algorithm consists of the same procedure yet performed in the kernel feature space.

where the columns of $\boldsymbol{U}$ represent the *undesired* points and $\boldsymbol{\gamma}$ contains their abundances. The OSP operator that maximizes the SNR performs two steps. First, an annihilating operator rejects the background points so that only the desired point should remain in the data subspace. This operator is given by the $(d \times d)$ matrix $\boldsymbol{P}_{\boldsymbol{U}}^{\perp} = \boldsymbol{I} - \boldsymbol{U}\boldsymbol{U}^{\dagger}$, where $\boldsymbol{U}^{\dagger}$ is the right Moore–Penrose pseudoinverse of $\boldsymbol{U}$. The second step of the OSP algorithm is represented by the filter $\boldsymbol{w}$ that maximizes the SNR of the filter output; that is, the matched filter $\boldsymbol{w} = k\boldsymbol{d}$, where k is an arbitrary constant (Harsanyi and Chang, 1994; Kwon and Nasrabadi, 2005b). The OSP detector is given by $\boldsymbol{q}_{\mathrm{OSP}}^{\mathrm{T}} = \boldsymbol{d}^{\mathrm{T}}\boldsymbol{P}_{\boldsymbol{U}}^{\perp}\boldsymbol{r}$, and the output of the OSP classifier is

$$D_{\mathrm{OSP}} = \boldsymbol{q}_{\mathrm{OSP}}^{\mathrm{T}}\boldsymbol{r} = \boldsymbol{d}^{\mathrm{T}}\boldsymbol{P}_{\boldsymbol{U}}^{\perp}\boldsymbol{r}. \tag{11.48}$$

By using the singular-value decomposition of $\boldsymbol{U} = \boldsymbol{B}\boldsymbol{\Sigma}\boldsymbol{A}^{\mathrm{T}}$, where $\boldsymbol{B}$ is the eigenvector of $\boldsymbol{U}\boldsymbol{U}^{\mathrm{T}} = \boldsymbol{B}\boldsymbol{\Sigma}\boldsymbol{\Sigma}^{\mathrm{T}}\boldsymbol{B}^{\mathrm{T}}$ and $\boldsymbol{A}$ is the eigenvector of $\boldsymbol{U}^{\mathrm{T}}\boldsymbol{U} = \boldsymbol{A}\boldsymbol{\Sigma}^{\mathrm{T}}\boldsymbol{\Sigma}\boldsymbol{A}$. The annihilating projector operator becomes $\boldsymbol{P}_{\boldsymbol{U}}^{\perp} = \boldsymbol{I} - \boldsymbol{B}\boldsymbol{B}^{\mathrm{T}}$, where the columns of $\boldsymbol{B}$ are obtained from the eigenvectors of the covariance matrix of the background samples. This finally yields the expression

$$D_{\mathrm{OSP}} = \boldsymbol{d}^{\mathrm{T}}(\boldsymbol{I} - \boldsymbol{B}\boldsymbol{B}^{\mathrm{T}})\boldsymbol{r}. \tag{11.49}$$

Figure 11.3 shows the effect of the OSP algorithm on a toy data set.

A nonlinear kernel-based version of the OSP algorithm can be devised by defining a linear OSP in an appropriate Hilbert feature space where samples are mapped to through a kernel mapping function $\boldsymbol{\phi}(\cdot)$. Similar to the previous linear case, let us define the annihilating operator now in the feature space as $\boldsymbol{P}_{\boldsymbol{\Phi}}^{\perp} = \boldsymbol{I}_{\phi} - \boldsymbol{B}_{\phi}\boldsymbol{B}_{\phi}^{\mathrm{T}}$. Then, the output of the OSP classifier in the feature space is now trivially given by

$$D_{\mathrm{KOSP}} = \boldsymbol{\phi}(\boldsymbol{d})^{\mathrm{T}}(\boldsymbol{I}_{\phi} - \boldsymbol{B}_{\phi}\boldsymbol{B}_{\phi}^{\mathrm{T}})\boldsymbol{\Phi}(\boldsymbol{r}). \tag{11.50}$$

The columns of matrix $\boldsymbol{B}_{\phi}$ (e.g., denoted as $\boldsymbol{b}^{j}_{\phi}$) are the eigenvectors of the covariance matrix of the undesired background signatures. Clearly, each eigenvector in the feature space can be expressed as a linear combination of the input vectors in the feature space transformed through the function $\boldsymbol{\phi}(\cdot)$. Hence, one can write $\boldsymbol{b}^{j}_{\phi} = \boldsymbol{\Phi}_b \boldsymbol{\beta}^j$ – where $\boldsymbol{\Phi}_b$ contains the mapped background points $\boldsymbol{X}_b$, and $\boldsymbol{\beta}^j$ are the eigenvectors of the (centered) Gram matrix $\boldsymbol{K}_{bb'}$, whose entries are $K_{bb}(\boldsymbol{b}_i, \boldsymbol{b}_j) = \boldsymbol{\phi}(\boldsymbol{b}_i)^{\mathrm{T}}\boldsymbol{\phi}(\boldsymbol{b}_j)$, and $\boldsymbol{b}_i$ are background points.[4] The kernelized version in Equation 11.50 is given (Kwon and Nasrabadi, 2005b) by

$$D_{\mathrm{KOSP}} = \boldsymbol{k}^{\mathrm{T}}_{m,d} \boldsymbol{\Upsilon}\boldsymbol{\Upsilon}^{\mathrm{T}} \boldsymbol{k}_{m,r} - \boldsymbol{k}^{\mathrm{T}}_{b,d} \mathcal{B}\mathcal{B}^{\mathrm{T}} \boldsymbol{k}_{b,r}, \tag{11.51}$$

where $\boldsymbol{k}$ are column vectors referred to as the *empirical kernel maps* and the subscripts indicate the elements entering the corresponding kernel function: $\boldsymbol{d}$ for the desired target, $\boldsymbol{r}$ the observed point, and subscript m indicates the concatenation of both; that is, $\boldsymbol{X}_m = [\boldsymbol{X}_b|\boldsymbol{d}]$. Also $\mathcal{B}$ is the matrix containing the eigenvectors $\boldsymbol{\beta}^j$ described earlier; and $\boldsymbol{\Upsilon}$ is the matrix containing the eigenvectors $\boldsymbol{\upsilon}^j$, similar to $\boldsymbol{\beta}^j$ yet obtained from the centered kernel matrix $\boldsymbol{K}_{m,m}$ thus working with the expanded matrix $\boldsymbol{X}_m$. Listing 11.1 illustrates the simplicity of the code for the KOSP method.

```
function [D,sigmas] = KOSP(patterns,target,background)
ss = logspace(-2,2,21);j = 0;
for sigma = ss
    j = j+1;sigmas(j) = sigma; Xbd = [background;target];
    K_mm = kernelmatrix('rbf',Xbd',Xbd',sigma);
    K_bb = kernelmatrix('rbf',background',background',sigma);
    K_m = kernelmatrix('rbf',Xbd',target',sigma);
    K_mr = kernelmatrix('rbf',Xbd',patterns',sigma);
    K_bd = kernelmatrix('rbf',background',target',sigma);
    K_br = kernelmatrix('rbf',background',patterns',sigma);
    [Gamma values] = eig(K_mm);
    Gamma = Gamma./repmat(diag(values)',size(Gamma,1),1);
    [Beta values] = eig(K_bb);
    Beta = Beta./repmat(diag(values)',size(Beta,1),1);
    D(j,:) = (K_md'*(Gamma*Gamma')*K_mr)-(K_bd'*(Beta*Beta')*K_br);
end
```

Listing 11.1 MATLAB code for the KOSP method (KOSP.m).

11.4.2 Kernel Spectral Angle Mapper

The spectral angle mapper (SAM) is a method for directly comparing image spectra (Kruse *et al.*, 1993). Since its seminal introduction, SAM has been widely used in chemometrics, astrophysics, imaging, hyperspectral image analysis, industrial engineering, and computer vision applications (Ball and Bruce 2007; Cho *et al.* 2010; García-Allende *et al.* 2008; Hecker *et al.* 2008), just to name a few. The measure achieved widespread popularity thanks to its implementation in software packages, such as ENVI or ArcGIS. The reason is simple: SAM is invariant to (unknown) multiplicative

4 See Chapter 4 for the implementation of the centering and scaling operations in kernel methods.

scalings of spectra due to differences in illumination and angular orientation (Keshava, 2004). SAM is widely used for fast spectral classification, as well as to evaluate the quality of extracted endmembers, and the spectral quality of an image compression algorithm.

The popularity of SAM is mainly due to its simplicity and geometrical interpretability. However, the main limitation of the measure is that it only considers second-order angle dependencies between spectra. In the following, we will see how to generalize the SAM measure to the nonlinear case by means of kernels (Camps-Valls and Bruzzone, 2009). This short note aims to characterize KSAM theoretically, and to show its practical convenience over the widespread linear counterpart. KSAM is very easy to compute, and can be used in any kernel machine. KSAM also inherits all properties of the original distance, is a valid reproducing kernel, and is universal. The induced metric by the kernel will be easily derived by traditional tensor analysis, and tuning the kernel hyperparameter of the squared exponential kernel function used is shown to be stable. KSAM is tested here in a target detection scenario for illustration purposes. The spatially explicit detection and metric maps give us useful information for semiautomatic anomaly/target detection.

Given two d-dimensional samples, $\boldsymbol{x}, \boldsymbol{z} \in \mathbb{R}^d$, the SAM measures the angle θ between them:

$$\theta = \arccos\left(\frac{\boldsymbol{x}^{\mathrm{T}}\boldsymbol{z}}{\|\boldsymbol{x}\|\|\boldsymbol{z}\|}\right), \quad 0 \leq \theta \leq \frac{\pi}{2}. \tag{11.52}$$

KSAM requires the definition of a feature mapping $\boldsymbol{\phi}(\cdot)$ to a Hilbert space $\mathcal{H}$ endorsed with the kernel reproducing property. Now, if we simply map the original data to $\mathcal{H}$ with a mapping $\boldsymbol{\phi} : \mathbb{R}^d \rightarrow \mathcal{H}$, the SAM can be expressed in $\mathcal{H}$ as

$$\theta_K = \arccos\left(\frac{\boldsymbol{\phi}(\boldsymbol{x})^{\mathrm{T}}\boldsymbol{\phi}(\boldsymbol{z})}{\|\boldsymbol{\phi}(\boldsymbol{x})\|_{\mathcal{H}}\|\boldsymbol{\phi}(\boldsymbol{z})\|_{\mathcal{H}}}\right), \quad 0 \leq \theta_K \leq \frac{\pi}{2}. \tag{11.53}$$

Having access to the coordinates of the new mapped data is not possible unless one explicitly defines the mapping function $\boldsymbol{\phi}$. It is possible to compute the new measure implicitly via kernels:

$$\theta_K = \arccos\left(\frac{K(\boldsymbol{x},\boldsymbol{z})}{\sqrt{K(\boldsymbol{x},\boldsymbol{x})}\sqrt{K(\boldsymbol{z},\boldsymbol{z})}}\right), \quad 0 \leq \theta_K \leq \frac{\pi}{2}. \tag{11.54}$$

Popular examples of reproducing kernels are the linear kernel $K(\boldsymbol{x},\boldsymbol{z}) = \boldsymbol{x}^{\mathrm{T}}\boldsymbol{z}$, the polynomial $K(\boldsymbol{x},\boldsymbol{z}) = (\boldsymbol{x}^{\mathrm{T}}\boldsymbol{z}+1)^d$, and the RBF kernel $K(\boldsymbol{x},\boldsymbol{z}) = \exp(-(1/2\sigma^2)\|\boldsymbol{x}-\boldsymbol{z}\|^2)$. In the linear kernel, the associated RKHS is the space $\mathbb{R}^d$ itself and KSAM reduces to the standard linear SAM. In polynomial kernels of degree d, KSAM effectively only compares moments up to order d. For the Gaussian kernel, the RKHS is of infinite dimension and KSAM measures higher order feature dependences. In addition, note that for RBF kernels, self-similarity $K(\boldsymbol{x},\boldsymbol{x}) = 1$, and thus the measure simply reduces to $\theta_K = \arccos(K(\boldsymbol{x},\boldsymbol{z}))$, $0 \leq \theta_K \leq \pi/2$. Following the code in Listing 11.1, we show a simple (just one line) snippet of KSAM in Listing 11.2.

```
[errKRadians] = acos(kernelmatrix('rbf',target', patterns',sigma));
```

Listing 11.2 MATLAB code for the KSAM method (KSAM.m).

11.5 Kernel Anomaly Change Detection

Anomalous change detection (ACD) differs from standard change detection in that the goal is to find anomalous or rare changes that occurred between two datasets. The distinction is important both statistically and from the practitioner's point of view. Statistically speaking, standard change detection emphasizes *all* the differences between the two data distributions, while ACD is more interested in modeling accurately the tails of the difference distribution.

The interest in ACD is vast, and many methods have been proposed in the literature, ranging from regression-based approaches like in the *chronocrome* (Schaum and Stocker, 1997), where big residuals are associated with anomalies, to equalization-based approaches that rely on whitening principles (Mayer *et al.*, 2003), as well as multivariate methods (Arenas-García *et al.*, 2013) that reinforce directions in feature spaces associated with noisy or rare events (Green *et al.*, 1998; Nielsen *et al.*, 1998). Theiler and Perkins (2006) formalized the main differences between the two fields and introduced a framework for ACD. The framework assumes data Gaussianity, yet the derived detector delineates hyperbolic decision functions. Even though the Gaussian assumption reports some advantages (e.g., problem tractability and generally good performance) it is still an ad hoc assumption that it is not necessarily fulfilled in practice. This is the motivation in Theiler *et al.* (2010), where the authors introduce EC distributions that generalize the Gaussian distribution and prove more appropriate to modeling fatter tail distributions and thus detect anomalies more effectively. The EC decision functions are pointwise nonlinear and still rely on second-order feature relations. In this section we will see the extension of the methods in Theiler *et al.* (2010) to cope with higher order feature relations through the theory of reproducing kernels.

11.5.1 Linear Anomaly Change Detection Algorithms

A family of linear ACD can be framed using solely cross-covariance matrices, as illustrated in Theiler *et al.* (2010). Notationally, given two samples $\boldsymbol{x}, \boldsymbol{y} \in \mathbb{R}^d$, the decision of *anomalousness* is given by

$$\mathcal{A}(\boldsymbol{x}, \boldsymbol{y}) = \xi_z - \beta_x \xi_x - \beta_y \xi_y,$$

where $\xi_z = \boldsymbol{z}^\mathrm{T} \boldsymbol{C}_z^{-1} \boldsymbol{z}$, $\xi_x = \boldsymbol{x}^\mathrm{T} \boldsymbol{C}_x^{-1} \boldsymbol{x}$, and $\xi_y = \boldsymbol{y}^\mathrm{T} \boldsymbol{C}_y^{-1} \boldsymbol{y}$, $\boldsymbol{z} = [\boldsymbol{x}, \boldsymbol{y}] \in \mathbb{R}^{2d}$, and $\boldsymbol{C}_z = \boldsymbol{Z}^\mathrm{T} \boldsymbol{Z}$, $\boldsymbol{C}_x = \boldsymbol{X}^\mathrm{T} \boldsymbol{X}$, and $\boldsymbol{C}_y = \boldsymbol{Y}^\mathrm{T} \boldsymbol{Y}$ are the estimated covariance matrices with the available data. Parameters $\beta_x, \beta_y \in \{0, 1\}$ and the different combinations give rise to different anomaly detectors (see Table 11.1). These methods and some variants have been widely used in image analysis settings because of their simplicity and generally good performance (Chang and Lin 2002; Kwon *et al.* 2003; Reed and Yu 1990).

However, these methods are hampered by a fundamental problem: the assumption of Gaussianity that is implicit in the formulation. Accommodating other data distributions may not be easy in general. Theiler *et al.* (2010) introduced an alternative ACD to cope

Table 11.1 A family of ACD algorithms (HACD: hyperbolic ACD).

ACD algorithm	β_x	β_y
RX	0	0
Chronocrome $\boldsymbol{y}\vert\boldsymbol{x}$	0	1
Chronocrome $\boldsymbol{x}\vert\boldsymbol{y}$	1	0
HACD	1	1

with EC distributions (Cambanis *et al.*, 1981): roughly speaking, the idea is to model the data using the multivariate Student distribution. The formulation introduced in (Theiler *et al.*, 2010) gives rise to the following decision of *EC anomalousness*:

$$\begin{aligned}\mathcal{A}_{\mathrm{EC}}(\boldsymbol{x},\boldsymbol{y}) &= (2d+\nu)\log\left(1+\frac{\xi_z}{\nu-2}\right)\\ &-\beta_x(d+\nu)\log\left(1+\frac{\xi_x}{\nu-2}\right)\\ &-\beta_y(d+\nu)\log\left(1+\frac{\xi_y}{\nu-2}\right),\end{aligned} \tag{11.55}$$

where ν controls the shape of the generalized Gaussian: for $\nu \rightarrow \infty$ the solution approximates the Gaussian and for $\nu \rightarrow 2$ it diverges.

11.5.2 Kernel Anomaly Change Detection Algorithms

Note that the previous methods solely depend on estimating covariance matrices with the available data, and use them as a metric for testing. The methods are fast to apply, delineate pointwise nonlinear decision boundaries, but still rely on second-order statistics. We solve the issue as usual using reproducing kernel functions.

For the sake of simplicity, we just show how to estimate the anomaly term ξ_x in the Hilbert space, $\xi_x^{\mathcal{H}}$. The other terms are derived in the same way. Notationally, let us map all the observations to a higher dimensional Hilbert feature space $\mathcal{H}$ by means of the feature map $\boldsymbol{\phi} : \boldsymbol{x} \rightarrow \boldsymbol{\phi}(\boldsymbol{x})$. The mapped training data matrix $\boldsymbol{X} \in \mathbb{R}^{N\times d}$ is now denoted as $\boldsymbol{\Phi} \in \mathbb{R}^{N\times d_{\mathcal{H}}}$. Note that one could think of different mappings: $\boldsymbol{\phi} : \boldsymbol{x} \rightarrow \boldsymbol{\phi}(\boldsymbol{x})$ and $\boldsymbol{\psi} : \boldsymbol{y} \rightarrow \boldsymbol{\psi}(\boldsymbol{y})$. However, in our case, we are forced to consider mapping to the same Hilbert space because we have to stack the mapped vectors to estimate ξ_z; that is, $\mathcal{F} = \mathcal{H}$. The mapped training data to Hilbert spaces are denoted as $\boldsymbol{\Phi}, \boldsymbol{\Psi} \in \mathbb{R}^{N\times d_{\mathcal{H}}}$ respectively. In order to estimate $\xi_x^{\mathcal{H}}$ for a test example $\boldsymbol{x}_*$, we first map the test point $\boldsymbol{\phi}(\boldsymbol{x}_*)$ and then apply

$$\xi_x^{\mathcal{H}} = \boldsymbol{\phi}(\boldsymbol{x}_*)^{\mathrm{T}}(\boldsymbol{\Phi}^{\mathrm{T}}\boldsymbol{\Phi})^{-1}\boldsymbol{\phi}(\boldsymbol{x}_*).$$

Note that we do not have access to either the samples in the feature space or the covariance. We can, nevertheless, estimate the eigendecomposition of the covariance matrix in kernel feature space; that is, $\boldsymbol{C}_{\mathcal{H}} = \boldsymbol{\Phi}^{\mathrm{T}}\boldsymbol{\Phi} = \boldsymbol{V}_{\mathcal{H}}\boldsymbol{\Lambda}_{\mathcal{H}}\boldsymbol{V}_{\mathcal{H}}^{\mathrm{T}}$, where $\boldsymbol{\Lambda}_{\mathcal{H}}$ is a diagonal matrix containing the eigenvalues and $\boldsymbol{V}_{\mathcal{H}}$ contains the eigenvectors in columns. It is worth noting that the maximum number of eigenvectors equals the number of examples used N. Plugging the eigendecomposition of the covariance pseudoinverse $\boldsymbol{C}_{\mathcal{H}}^{\dagger} = \boldsymbol{V}_{\mathcal{H}}\boldsymbol{\Lambda}_{\mathcal{H}}^{-1}\boldsymbol{V}_{\mathcal{H}}^{\mathrm{T}}$, and after some linear algebra, one can express the term as

$$\xi_x^{\mathcal{H}} = \boldsymbol{\phi}(\boldsymbol{x}_*)^{\mathrm{T}}\boldsymbol{\Phi}^{\mathrm{T}} \ (\boldsymbol{\Phi}\boldsymbol{\Phi}^{\mathrm{T}}\boldsymbol{\Phi}\boldsymbol{\Phi}^{\mathrm{T}})^{-1}\boldsymbol{\Phi}\boldsymbol{\phi}(\boldsymbol{x}_*).$$

We can replace all dot products by reproducing kernel functions using the representer theorem (Kimeldorf and Wahba, 1971; Shawe-Taylor and Cristianini, 2004), and hence $\xi_x^{\mathcal{H}} = \boldsymbol{k}_*^{\mathrm{T}}(\boldsymbol{K}\boldsymbol{K})^{-1}\boldsymbol{k}_*$, where $\boldsymbol{k}_* \in \mathbb{R}^{N\times 1}$ contains the similarities between $\boldsymbol{x}_*$ and all training points $\boldsymbol{X}$; that is, $\boldsymbol{k}_* := [K(\boldsymbol{x}_*, \boldsymbol{x}_1), \ldots, K(\boldsymbol{x}_*, \boldsymbol{x}_N)]^{\mathrm{T}}$, and $\boldsymbol{K} \in \mathbb{R}^{N\times N}$ stands for the kernel matrix containing all training data similarities. The solution may need extra regularization $\xi_x^{\mathcal{H}} = \boldsymbol{k}_*^{\mathrm{T}}(\boldsymbol{K}\boldsymbol{K}+\lambda\boldsymbol{I})^{-1}\boldsymbol{k}_*$, $\lambda \in \mathbb{R}^+$. Therefore, in the kernel case, the decision of anomalousness is given by

$$\mathcal{A}^{\mathcal{H}}(\boldsymbol{x}, \boldsymbol{y}) = \xi_z^{\mathcal{H}} - \beta_x\xi_x^{\mathcal{H}} - \beta_y\xi_y^{\mathcal{H}},$$

where now $\xi_z^{\mathcal{H}} = \boldsymbol{k}_z^{\mathrm{T}}(\boldsymbol{K}_z\boldsymbol{K}_z + \lambda\boldsymbol{I})^{-1}\boldsymbol{k}_z$, $\xi_x^{\mathcal{H}} = \boldsymbol{k}_x^{\mathrm{T}}(\boldsymbol{K}_x\boldsymbol{K}_x + \lambda\boldsymbol{I})^{-1}\boldsymbol{k}_x$, and $\xi_y^{\mathcal{H}} = \boldsymbol{k}_y^{\mathrm{T}}(\boldsymbol{K}_y\boldsymbol{K}_y + \lambda\boldsymbol{I})^{-1}\boldsymbol{k}_y$.

Equivalently, including these expressions in Equation 11.55, one obtains the EC kernel versions $\mathcal{A}_{\mathrm{EC}}^{\mathcal{H}}(\boldsymbol{x}, \boldsymbol{y})$:

$$\begin{aligned} \mathcal{A}_{\mathrm{EC}}^{\mathcal{H}}(\boldsymbol{x}, \boldsymbol{y}) &= (2d+\nu)\log\left(1 + \frac{\xi_z^{\mathcal{H}}}{\nu - 2}\right) \\ &-\beta_x(d+\nu)\log\left(1 + \frac{\xi_x^{\mathcal{H}}}{\nu - 2}\right) \\ &-\beta_y(d+\nu)\log\left(1 + \frac{\xi_y^{\mathcal{H}}}{\nu - 2}\right). \end{aligned} \tag{11.56}$$

In the case of $\beta_x = \beta_y = 0$, the algorithm reduces to kernel RX which was introduced in Kwon and Nasrabadi (2005c). As in the linear case, one has to center the data before computing either covariance or Gram (kernel) matrices. This is done via the kernel matrix operation $\boldsymbol{K} \leftarrow \boldsymbol{H}\boldsymbol{K}\boldsymbol{H}$, where $\boldsymbol{H}_{ij} = \delta_{ij} - (1/n)$, and δ represents the Kronecker delta $\delta_{i,j} = 1$ if $i = j$ and zero otherwise.

Two parameters need to be tuned in the kernel versions: the regularization parameter λ and the kernel parameter σ. In the examples, we will use two different isotropic kernel functions, $K(\boldsymbol{x}_i, \boldsymbol{x}_j) = \exp(-d_{ij}^2/(2\sigma^2))$: the standard Gaussian kernel $d_{ij} = \|\boldsymbol{x}_i - \boldsymbol{x}_j\|$, and the SAM kernel $d_{ij} = \arccos(\boldsymbol{x}_i^{\mathrm{T}}\boldsymbol{x}_j/(\|\boldsymbol{x}_i\|\,\|\boldsymbol{x}_j\|))$. We should note that, when a linear kernel is used, $K(\boldsymbol{x}_i, \boldsymbol{x}_j) = \boldsymbol{x}_i^{\mathrm{T}}\boldsymbol{x}_j$, the proposed algorithms reduce to the linear counterparts proposed in Theiler *et al.* (2010). Working in the dual (or Q-mode) with the linear kernel instead of the original linear versions can be advantageous *only* in the case of higher dimensionality than available samples, $d > N$.

11.6 Hypothesis Testing with Kernels

Statistical hypothesis testing plays a crucial role in statistical inference. After all, the core of science and engineering is about testing hypotheses,[5] and in machine learning we do it through data analysis. Testing hypotheses is concurrently one of the key topics in statistical signal processing as well (Kay, 1993). A plethora of examples exists: from testing for equality of observations as in speaker verification (Bimbot *et al.*, 2004; Rabiner and Schafer, 2007), to assessing statistical differences between data acquisitions, such as in change-detection analysis (Basseville and Nikiforov, 1993) as seen before. Testing for a change-point in a time series is also an important problem in DSP (Fearnhead, 2006), such as in monitoring applications involving biomedical or geoscience time series.

Very often we aim to compare two real datasets, or a real dataset with a synthetically generated one by a model that encodes the system's governing equations. The result of such a test will tell us about the plausibility and the origin of the observations. In both cases, we need a solid measure to test the significance of their statistical difference, compared with an idealized null hypothesis in which no relationship is assumed (Kay, 1993). Hypothesis tests are used in determining what outcomes of a study would lead to a rejection of the null hypothesis for a prespecified level of significance. Imagine, presented with two datasets, we are interested in testing whether the underlying distributions are statistically different or not in the mean. Essentially, we have to test between two competing hypotheses, and decide upon them:

$$\mathrm{H}_0 : \mu_1 = \mu_2 \tag{11.57}$$

$$\mathrm{H}_\mathrm{A} : \mu_1 \neq \mu_2. \tag{11.58}$$

The hypothesis H_0 is referred to as the *null hypothesis*, and H_A is the *alternative hypothesis*. The process of distinguishing between the null hypothesis and the alternative hypothesis is aided by identifying two conceptually different types of errors (type I and type II), and by specifying parametric limits on, for example, how much type I error could be allowed. If the detector decides H_A but H_0 is true, we commit a type I error (also known as the false-alarm rate), whereas if we decide H_0 but H_A is true, we make a type II error (also known as the missed-detection rate).

Classical approaches to statistical hypothesis testing and detection (of changes and differences between samples) are parametric in nature (Basseville and Nikiforov, 1993). Procedures such as the CUSUM statistic[6] or Hotelling's T-squared test assume that the data is Gaussian, whereas the χ^2 and mutual information statistics can only be applied in finite-valued data. These test statistics are widely used due to their simplicity and good results in general when the assumptions about the underlying distributions are fulfilled. Alternatively, non-parametric hypothesis testing allows more robust and reliable results over larger classes of data distributions as this does not assume any

---

5 Roughly speaking, a statistical hypothesis will be testable on the basis of observing a process that is modeled via a set of random variables.

6 The CUSUM algorithm involves the calculation of a cumulative sum. This way, samples from a process x_i are assigned weights w_i, which are then summed up: $S_0 = 0$ and $S_{i+1} = \max(0, S_i + x_i - w_i)$. Detection occurs when the value of S is higher than a predefined threshold value.

particular form. Recently, nonparametric kernel-based methods have appeared in the literature (Harchaoui *et al.*, 2013).

In recent years, several kernel-based hypothesis testing methods have been introduced. For a recent sound review, the reader is addressed to Harchaoui *et al.* (2013). We next review some of these methods under the framework of distribution embeddings and the concepts of mean elements and covariance operators (Eric *et al.* 2008; Fukumizu *et al.* 2007). These test statistics rely on the eigenvector basis of particular covariance operators and, under the alternative hypothesis, the tests correspond to consistent estimators of well-known information divergences between probability distributions (Cover and Thomas, 2006). We give two instantiations of kernel hypothesis tests in dependence estimation and anomaly detection, while Chapter 12 will pay attention to a broader family of these kernel dependence measures for feature extraction.

11.6.1 Distribution Embeddings

Recall that our current motivation is to test statistical differences between distributions with kernels. Therefore, it is natural to ask ourselves about which are the appropriate descriptors of distributions embedded into possibly infinite-dimensional Hilbert spaces. The distributions' components therein can be described by the so-called *covariance operators* and *mean elements* (Eric *et al.* 2008; Fukumizu *et al.* 2007; Harchaoui *et al.* 2013; Vakhania *et al.* 1987).

Notationally, let us consider a random variable $\boldsymbol{x}$ taking values in $\mathcal{X}$ and a probability distribution $\mathbb{P}$. The *mean element* $\mu_{\mathbb{P}}$ associated with $\boldsymbol{x}$ is the unique element of the RKHS $\mathcal{H}$, such that, for all $f \in \mathcal{H}$, we obtain

$$\langle \mu_{\mathbb{P}}, f \rangle_{\mathcal{H}} = \mathbb{E}_{\mathbb{P}}[f(\boldsymbol{x})],$$

and the *covariance operator* $\boldsymbol{\Sigma}_{\mathbb{P}} : \mathcal{H} \otimes \mathcal{H} \to \mathbb{R}$ attached to $\boldsymbol{x}$ fulfills

$$\langle f, \boldsymbol{\Sigma}_{\mathbb{P}} g \rangle_{\mathcal{H}} := \mathrm{Cov}(f(\boldsymbol{x}), g(\boldsymbol{x})) = \mathbb{E}_{\mathbb{P}}[f(\boldsymbol{x}) g(\boldsymbol{x})] - \langle \mu_{\mathbb{P}}, f \rangle_{\mathcal{H}} \langle \mu_{\mathbb{P}}, g \rangle_{\mathcal{H}}.$$

In the empirical case, we are given a set of i.i.d. samples $\{\boldsymbol{x}_1, \ldots, \boldsymbol{x}_N\}$ drawn from $\mathbb{P}$, so we have to replace the previous population statistics by their empirical approximations to obtain the empirical mean $\hat{\mu}$

$$\langle \hat{\mu}, f \rangle_{\mathcal{H}} = \frac{1}{N} \sum_{i=1}^{N} f(\boldsymbol{x}_i)$$

and the empirical covariance operator $\hat{\boldsymbol{\Sigma}}$:

$$\langle f, \hat{\boldsymbol{\Sigma}} g \rangle_{\mathcal{H}} = \sum_{i=1}^{N} (f(\boldsymbol{x}_i) - \langle \hat{\mu}, f \rangle_{\mathcal{H}})(g(\boldsymbol{x}_i) - \langle \hat{\mu}, g \rangle_{\mathcal{H}}).$$

Interestingly, it can be shown that, for a large class of kernels, an embedding function $m(\cdot)$ that maps data into the kernel feature mean is injective (Harchaoui *et al.*, 2013). Now, if we consider two probability distributions $\mathbb{P}$ and $\mathbb{Q}$ on $\mathcal{X}$, and for all $f \in \mathcal{H}$

the equality $\langle \mu_{\mathbb{P}}, f \rangle_{\mathcal{H}} = \langle \mu_{\mathbb{Q}}, f \rangle_{\mathcal{H}}$ holds, then $\mathbb{P} = \mathbb{Q}$ for dense kernel functions K. This property has been exploited to define independence tests and hypothesis testing methods based on kernels, since it ensures that two distributions have the same RKHS mean *iff* they are the same, as we will see in Chapter 12.

11.6.2 Maximum Mean Discrepancy

Among several hypothesis testing algorithms, such as the kernel Fisher discriminant analysis and the kernel density-ratio test statistic revised in Harchaoui *et al.* (2013), a simple and effective one is the maximum mean discrepancy (MMD) statistic. Gretton *et al.* (2005b) introduced the MMD statistic for comparing the means of two samples in kernel feature spaces; see Figure 11.11. Notationally, we are given samples $\{\boldsymbol{x}_i\}_{i=1}^N$ from a so-called *source* distribution, $\mathbb{P}_x$, and $\{\boldsymbol{y}_i\}_{i=1}^M$ samples drawn from a second *target* distribution, $\mathbb{P}_y$. MMD reduces to estimate the distance between the two sample means in an RKHS $\mathcal{H}$ where data are embedded:

$$\mathrm{MMD}(\mathcal{X}, \mathcal{Y}) := \left\| \frac{1}{N} \sum_{i=1}^{N} \boldsymbol{\phi}(\boldsymbol{x}_i) - \frac{1}{M} \sum_{i=1}^{M} \boldsymbol{\phi}(\boldsymbol{y}_i) \right\|_{\mathcal{H}}^2 .$$

This estimate can be shown to be equivalent to

$$\mathrm{MMD}(\mathcal{X}, \mathcal{Y}) = \mathrm{tr}(\boldsymbol{K}\boldsymbol{L}),$$

where

$$\boldsymbol{K} = \begin{pmatrix} \boldsymbol{K}_{xx} & \boldsymbol{K}_{xy} \\ \boldsymbol{K}_{yx} & \boldsymbol{K}_{yy} \end{pmatrix},$$

and $L_{ij} = 1/N^2$ if $\boldsymbol{x}_i$ and $\boldsymbol{x}_j$ belong to the source domain, $\tilde{L}_{ij} = 1/M^2$ if $\boldsymbol{x}_i$ and $\boldsymbol{x}_j$ belong to the target domain, and $L_{ij} = -1/(NM)$ if $\boldsymbol{x}_i$ and $\boldsymbol{x}_j$ belong to the cross-domain. MMD tends asymptotically to zero when the two distributions $\mathbb{P}_x$ and $\mathbb{P}_y$ are the same.

It is worth mentioning that MMD was concurrently introduced in the field of signal processing under the formalism of information-theoretic learning (Principe, 2010). In such a framework, it can be shown that MMD corresponds to a *Euclidean divergence*. This can be shown easily. Let us assume two distributions $p(\boldsymbol{x})$ and $q(\boldsymbol{x})$ with N_1 and N_2 samples. The Euclidean divergence between them can be computed by exploiting the relation to the quadratic Rényi entropies (see Chapter 2 of Principe (2010)):

$$D_{\mathrm{ED}}(p, q) = \frac{1}{N_1^2} \sum_{i=1}^{N_1} \sum_{j=1}^{N_1} G(\boldsymbol{x}_i, \boldsymbol{x}_j) + \frac{1}{N_2^2} \sum_{i=1}^{N_2} \sum_{j=1}^{N_2} G(\boldsymbol{x}_i, \boldsymbol{x}_j) - \frac{2}{N_1 N_2} \sum_{i=1}^{N_1} \sum_{j=1}^{N_2} G(\boldsymbol{x}_i, \boldsymbol{x}_j),$$

where $G(\cdot)$ denotes Gaussian kernels with standard deviation σ. This divergence reduces to compute the norm of the difference between the means in feature spaces:

$$D_{\mathrm{ED}}(p, q) = \|\boldsymbol{\mu}_1\|^2 + \|\boldsymbol{\mu}_2\|^2 - 2\boldsymbol{\mu}_1^{\mathrm{T}} \boldsymbol{\mu}_2 = \|\boldsymbol{\mu}_1 - \boldsymbol{\mu}_2\|^2,$$

which is equivalent to the MMD estimate introduced from the theory of covariance operators in Gretton *et al.* (2005b).

11.6.3 One-Class Support Measure Machine

Let us give now an example of a kernel method for anomaly detection based on RKHS embeddings. A field of anomaly detection that has captured the attention recently is the so-called *grouped anomaly detection*. The interest here is that the anomalies often appear not only in the data themselves, but also as a result of their interactions. Therefore, instead of looking for pointwise anomalies, here one is interested in *groupwise anomalies*. A recent method, based on SVC, was been introduced by Muandet and Schölkopf (2013). An anomalous group could be defined as a group of anomalous samples, which is typically easy to detect. However, anomalous groups are often more difficult to detect because of their particular behavior as a core group and the higher order statistical relations among them. Importantly, detection can only be possible in the space of distributions, which can be characterized by recent kernel methods relying on concepts of RKHS embeddings of probability distributions, mainly mean elements and covariance operators. The method presented therein is called the one-class support measure machine (OC-SMM) and makes uses of the mean embedding representation, which is defined as follows.

Let $\mathcal{H}$ denote an RKHS of functions $f : \mathcal{X} \to \mathbb{R}$ endorsed with a reproducing kernel $K : \mathcal{X} \times \mathcal{X} \to \mathbb{R}$. The kernel mean map from the set of all probability distributions $\mathcal{B}_{\mathcal{X}}$ into $\mathcal{H}$ is defined as

$$\boldsymbol{\mu} : \mathcal{B}_{\mathcal{X}} \to \mathcal{H}, \quad \mathbb{P} \to \int_{\mathcal{X}} K(\boldsymbol{x}, \cdot)\mathrm{d}\mathbb{P}(\boldsymbol{x}).$$

Assuming that $K(\boldsymbol{x}, \cdot)$ is bounded for any $\boldsymbol{x} \in \mathcal{X}$, we can show that for any $\mathbb{P}$, letting $\boldsymbol{\mu}_{\mathbb{P}} = \boldsymbol{\mu}(\mathbb{P})$, the $\mathbb{E}_P[f] = \langle \boldsymbol{\mu}_{\mathbb{P}}, f\rangle_{\mathcal{H}}$, for all $f \in \mathcal{H}$.

The primal optimization problem for the OC-SMM can be formulated in an analogous way to the OC-SVM (or ν-SVM):

$$\min_{\boldsymbol{w},\rho,\xi} \left\{ \frac{1}{2}\|\boldsymbol{w}\|^2 - \rho + \frac{1}{\nu N}\sum_{i=1}^{N} \xi_i \right\} \tag{11.59}$$

constrained to

$$\langle \boldsymbol{\mu}_{\mathbb{P}_i}, \boldsymbol{w}\rangle \geq \rho - \xi_i \qquad \forall i = 1, \ldots, N \tag{11.60}$$

$$\xi_i \geq 0 \qquad \forall i = 1, \ldots, N, \tag{11.61}$$

where the meaning of the hyperparameters stands the same, but they are now instead associated with distributions (note the appearance of the mean distribution appears as a constraint in Equation 11.60). By operating in the same way, one obtains that the dual form is again a QP problem that depends on the inner product $\langle \boldsymbol{\mu}_{\mathbb{P}_i}, \boldsymbol{\mu}_{\mathbb{P}_j}\rangle_{\mathcal{H}}$ that can be computed using the *mean map kernel* function:

$$K(\hat{\mathbb{P}}_i, \hat{\mathbb{P}}_j) = \frac{1}{N_i N_j} \sum_{k=1}^{N_i} \sum_{l=1}^{N_j} K(\boldsymbol{x}_k^i, \boldsymbol{x}_l^j),$$

where $\boldsymbol{x}_k^i$ are the samples belonging to distribution i, and N_i represents its cardinality. Links to MATLAB source code for this method are available on the book's web page.

11.7 Tutorials and Application Examples

We next include a set of simple examples for providing insight on the topics of this chapter. Some of the functions are not included in the text, as indicated, but the interested reader can obtain the code from the book's repository.

11.7.1 Example on Kernelization of the Metric

Let us see the relation described in this chapter for kernelization of the metric in a simple code example, given in Listing 11.3. We generate the well-known *Swiss roll* dataset, then compute both the Euclidean distances in the original space and the Hilbert space (with a specific σ parameter for the kernel).

Figure 11.4 illustrates the behavior of the distance values in the input space against the distances computed in the Hilbert space. We can see how some of the distances have decreased values with respect to the input space in the range $[0, 1/2]$ and the remaining distances have increased in the Hilbert space. The shape of the curve shows the nonlinear relation between distances.

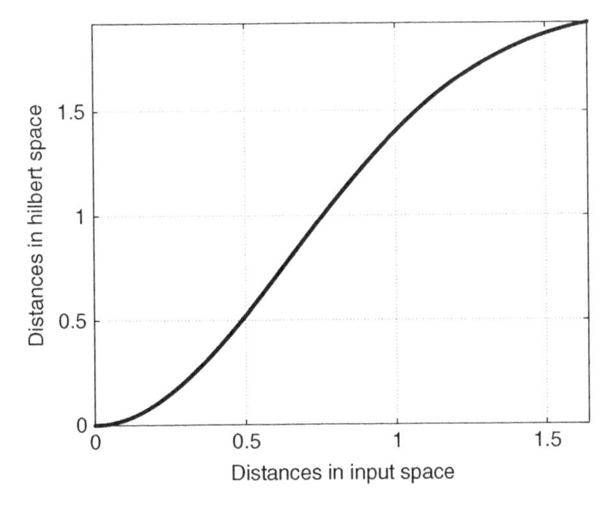

Figure 11.4 Distance values computed in input space against computed in the Hilbert space (through reproducing kernels). This figure is generated with Listing 11.3.

```
n = 100 % number of training points per class
[X,Y] = generate_toydata(n,'swissroll');

% Compute the RBF kernel matrix
```

```
sigma = median(pdist(X)); % heuristic
K = kernelmatrix('rbf',X',X',sigma);

% Compute distances between data points in Hilbert space from the kernel
for i=1:2*n
    for j=1:2*n
        Dist2_X(i,j) = norm(X(i,:)-X(j,:),'fro');
        Dist2_H(i,j) = K(i,i)+K(j,j)-2*K(i,j);
    end
end
figure, plot(Dist2_X,Dist2_H,'k.','markersize',10), grid on, axis tight
xlabel('Distances in input space'), ylabel('Distances in Hilbert space')
```

Listing 11.3 Comparing distances between the input and Hilbert feature spaces using an RBF kernel (DistancesXvsH.m).

11.7.2 Example on Kernel *k*-Means

The code in Listing 11.4 runs the k-means function of MATLAB and an implementation of kernel k-means for the problem of concentric rings. Figure 11.5 illustrates the final solution of both methods: k-means could not achieve a good solution (left) compared to kernel k-means which finally converges to the correct solution (right).

```
function kmeans_kernelkmeans
rand('seed',12345), randn('seed',12345)
close all, clear, clc
%% Generate data (2-class concentric rings problem)
N1=200;N2=500;d=2;
x1 = 2*pi*rand(N1,1);r1 = rand(N1,1);
x2 = 2*pi*rand(N2,1);r2 = 2+rand(N2,1);
X = [r1.*cos(x1) r1.*sin(x1);r2.*cos(x2) r2.*sin(x2)];
Y = [ones(n1,1);2*ones(n2,1)];
%% Some parameters
k = 2;[N] = size(X,1);
%% k-means (Matlab) function
[idx, V] = kmeans(X, k, 'Display', 'iter');c1 = idx == 1;c2 = idx == 2;
%% plot
figure(1),clf, scatter(X(c1,1),X(c1,2),'+'),hold on, grid on,
scatter(X(c2,1),X(c2,2),'ro')
%% Kernel k-means
sigma = 1;Ke  = kernelmatrix('rbf',X',X',sigma);
Si = [ones(N,1) zeros(N,1)];Si(1,:) = [0 1];Si(2,:) = [0 1];
Si_prev = Si + 1;
while norm(Si-Si_prev,'fro') > 0
    Si_prev = Si;Si = assign_cluster(Ke,Si,k);
end
%% Plot results
Si = logical(Si);
figure(2), clf, scatter(X(Si(:,1),1),X(Si(:,1),2),'+'),hold on, grid on,
scatter(X(Si(:,2),1),X(Si(:,2),2),'ro')
end

%% Callback function
function Si = assign_cluster(Ke,Si,kc)
[n] = size(Ke,1);Nk = sum(Si,1);dist = zeros(N,kc);
for k = 1:kc
```

```
    dist(:,k) = diag(Ke) - (2/(Nk(k)))*sum(repmat(Si(:,k)',N,1).*Ke,2) + ...
        Nk(k)^(-2)*sum(sum((Si(:,k)*Si(:,k)').*Ke));
end
Si = real(dist == repmat(min(dist,[],2),1,kc));
end
```

Listing 11.4 Simple code to perform kmeans and kernel kmeans (kmeans_kernelkmeans.m).

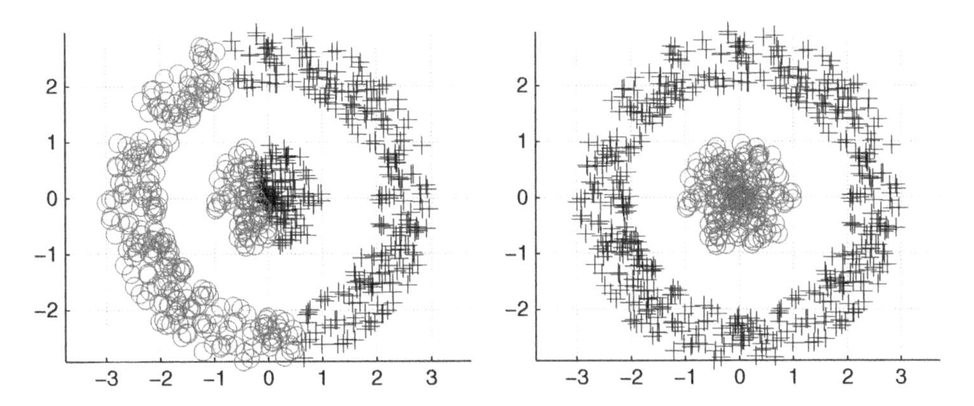

Figure 11.5 Results obtained by k-means (left) and kernel k-means (right) in the concentric ring problem.

11.7.3 Domain Description Examples

This section compares different standard one-class classifiers and kernel detectors: Gaussian (GDD), mixture of Gaussians (MoGDD), k-NN (KnnDD), and the SVDD (with the RBF kernel). All one-class domain description classifiers have a parameter to adjust, which is the fraction rejection, that controls the percentage of target samples the classifier can reject during training. Some other parameters need to be tuned depending on the classifier. For instance, the Gaussian classifier has a regularization parameter in the range $[0, 1]$ to estimate the covariance matrix, using all training samples or only diagonal elements.

With regard to the mixture of Gaussian classifier, four parameters must be adjusted: the shape of the clusters to estimate the covariance matrix (e.g., full, diagonal, or spherical), a regularization parameter in the range $[0, 1]$ used in the estimation of the covariance matrices, the number of clusters used to model the target class, and the number of clusters to model the outlier class. The most important advantage of this classifier with respect to the Gaussian classifier, apart from the obvious improvement of modeling a class with several Gaussian distributions, is the possibility of using outliers information when training the classifier. This allows the tracing of a more precise boundary around the target class, improving classification results notably. However, having a total of five parameters to adjust constitutes a serious drawback for this method, making it difficult to obtain a good working model.

With respect to the k-NN classifier, only the number of k neighbors used to compute the distance of each new sample to its class must be tuned. Finally, for the SVDD, the

Table 11.2 The coefficient of accuracy κ and the overall accuracy in parentheses (percent) obtained for the synthetic problems.

	GDD	MoGDD	KnnDD	SVDD
Two Gaussians	0.52 (76.0)	0.67 (83.5)	0.56 (78.1)	0.63 (81.7)
Mixed Gaussian–log	0.65 (82.7)	0.77 (88.8)	0.48 (74.1)	0.78 (89.4)

width of the RBF kernel (σ) has to be adjusted. In all the experiments, except the last one, σ is selected among the following set of discrete values between 5 and 500. It is worth stressing here that the SVDD is one of the methods among those considered, together with the mixture of Gaussians, that allows one to use outliers information to better define the target class boundary. In addition, it has the important advantage that only two free parameters have to be adjusted, thus making it relatively easier to define a good model.

In all the experiments, 30% of the samples available are used to train each classifier. In order to adjust free parameters, a cross-validation strategy with four folds is employed. Once the classifier is trained and adjusted, the final test is done using the remaining 70% of the samples. In all our experiments, we used the `dd_tools` and the `libSVM` software packages.

Two different synthetic problems are tested here. In each problem, binary classification is addressed, thus having two sets of targets and outliers. Each class contains 500 samples generated from different distributions. The two problems analyzed are mixture of two Gaussian distributions and a mixture of Gaussian and logarithmic distributions; both of them are strongly overlapped. Our goal here is testing the effectiveness of the different methods with standard and simple models but in very critical conditions on the overlapping of class distributions.

In order to measure the capability of each classifier to accept targets and reject outliers, the confusion matrix for each problem is obtained, and the kappa coefficient (κ) is estimated, which gives a good trade-off between the capability of the classifier to accept its samples (targets) and reject the others (outliers). Table 11.2 shows the results for the tests, and Figure 11.6 shows graphical representations of the classification boundary defined by each method. In each plot, the classifier decision boundary levels is represented with a blue curve. Listing 11.5 shows the code that has been used for generating these results. Functions `getsets.m`, `train_basic.m`, and `predict_basic.m` can be seen in the code (not included here for brevity).

```
function demo_svdd
% dependencies: assessment.m
% add paths to prtools and ddtols (suposed to be in the actual directory)
addpath(genpath('.'))
seed = 1;rand('state',seed);randn('state',seed);
N = 500; dif = 1;
XX =[randn(1,N)-dif randn(1,N)+dif; 2*randn(1,N)-dif 2*randn(1,N)+dif]';
YY =[2*ones(1,N) ones(1,N)]';
percent=0.3;vfold=4;par=[1 sqrt(10) 5 10 25 50 100];frac=[1e-2 .1:.1:.5];
% Sets of classes to train
trainclasses=1;w = cell(1,length(trainclasses));
```

```
% Construct validation and test sets
allclasses = unique(YY);
[train_set,test_set] = getsets(YY, allclasses, percent, 0);
% General training data
data.classifier = 'incsvdd';classifier = 'incsvdd';data.ktype = 'r';
% Reserve memory
kappa = zeros(vfold,1); oa = zeros(vfold,1);
% For every class to be trained
for cl = 1:length(trainclasses)
    % Reserve memory
    errors{cl} = zeros(length(par),length(frac),2);
    data.class = trainclasses(cl);
    % For every par value
    for pit = 1:length(par)
        fprintf('Par %f ...\n', par(pit))
        % For every fraction rejection value
        for fracit = 1:length(frac)
            % Train classifier
            fprintf('  frac %f ... ', frac(fracit))
            data.par  = par(pit);data.frac = frac(fracit);
            % Perform cross validation
            state = rand('state');rand('state',0);
            idx = randperm(length(train_set));
            rand('state',state);groups = ...
                ceil(vfold*idx/length(train_set));
            for vf = 1:vfold,
                in  = find(groups ~= vf);out = find(groups == vf);
                % Train
                data.XX = XX(train_set(in),:);data.YY = ...
                    YY(train_set(in),:);
                data.w  = train_basic(data);
                % Validation
                data.XX = XX(train_set(out),:);data.YY = ...
                    YY(train_set(out));
                [kappa(vf) oa(vf)] = predict_basic(data);
            end
            errors{cl}(pit,fracit,:) = [mean(kappa), mean(oa)];
            fprintf('  Mean K: %f \t OA: %f\n', errors{cl}(pit,fracit,:))
        end
    end
    % Show optimum values for Kappa
    [val,ifrac] = max(max([errors{:}(:,:,1)]));
    [val,ipar]  = max(max([errors{:}(:,:,1)]'));
    % Now train and test with best parameters
    data.XX   = XX(train_set,:); data.YY   = YY(train_set);
    data.par  = par(ipar);data.frac = frac(ifrac);
    data.w    = train_basic(data);data.XX  = XX(test_set,:);
    data.YY   = YY(test_set);
    [data.kappa(cl) data.oa(cl) data.yp(cl,:)] = predict_basic(data);
    fprintf('Results %s(%02d) Par: %f, f.r.: %f, K: %f, OA: %f\n', ...
        classifier, trainclasses(cl),par(ipar), frac(ifrac), data.kappa, ...
            data.oa)
end
% Plot the distributions
plot(XX(1:N,1),XX(1:N,2),'r.'),hold ...
    on,plot(XX(N+1:end,1),XX(N+1:end,2),'kx')
```

```
plotc(data.w,'b.',4),set(gca,'LineWidth',1)
end
function [ct,cv] = getsets(YY, classes, percent, vfold, randstate)
if nargin < 5;randstate = 0;end
s = rand('state');rand('state',randstate);ct = []; cv = [];
for ii=1:length(classes)
    idt = find(YY == classes(ii));
    if vfold
        ct = union(ct,idt);cv = ct;
    else
        lt  = fix(length(idt)*percent); idt = idt(randperm(length(idt)));
        idv = idt([lt+1:end]); % remove head
        idt = setdiff(idt,idv);ct  = union(ct,idt);cv  = union(cv,idv);
    end
end
rand('state',s);
end
%% Callback functions
function [w,res] = train_basic(d)
% Basic funcion to train a one-class classifier, used by xs
targets  = find(d.YY == d.class);outliers = find(d.YY ~= d.class);
x1 = gendatoc(d.XX(targets,:),d.XX(outliers,:));
w = incsvdd(x1,d.frac,d.ktype,d.par);
end
function [kappa,oa,yp] = predict_basic(d)
yt = d.YY;oc = gendatoc(d.XX);rr = oc * d.w;dd = +rr;
idx = find((dd(:,1) - dd(:,2)) > 0);yp = zeros(size(yt));yp(idx) = ...
    d.class;
yt(yt ~= d.class) = 0;
res = assessment(yt,yp,'class');kappa = res.Kappa;oa = res.OA;
end
```

Listing 11.5 Code to reproduce SVDD results over Gaussian synthetic set in Figure 11.6.

Several conclusions can be obtained from these tests. First, MoGDD and SVDD obtain the highest accuracies in our problems. When mixing two Gaussians, none of the classifiers shows a good behavior, as in our synthetic data the distributions are strongly overlapped. This is a common situation in the remote-sensing field when trying to classify classes very similar to each other. Mixture of overlapped Gaussian and logarithmic features is a difficult problem. In this synthetic example, we can see that the SVDD performs slightly better, since it does not assume any a priori data distribution.

11.7.4 Kernel Spectral Angle Mapper and Kernel Orthogonal Subspace Projection Examples

The following two experiments have the aim of comparing KSAM and KSOM (and also their linear versions) in terms of real data problems. In the former, a QuickBird image of a residential neighborhood of Zürich, Switzerland, is used for illustration purposes. The image size is (329 × 347) pixels. A total of 40 762 pixels were labeled by photointerpretation and assigned to nine land-use classes (see Figure 11.8). Four target detectors are compared in the task of detecting the class "Soil": OSP (Harsanyi and Chang, 1994), its kernel counterpart KOSP (Kwon and Nasrabadi, 2007), standard SAM (Kruse *et al.*, 1993), and the extension KSAM.

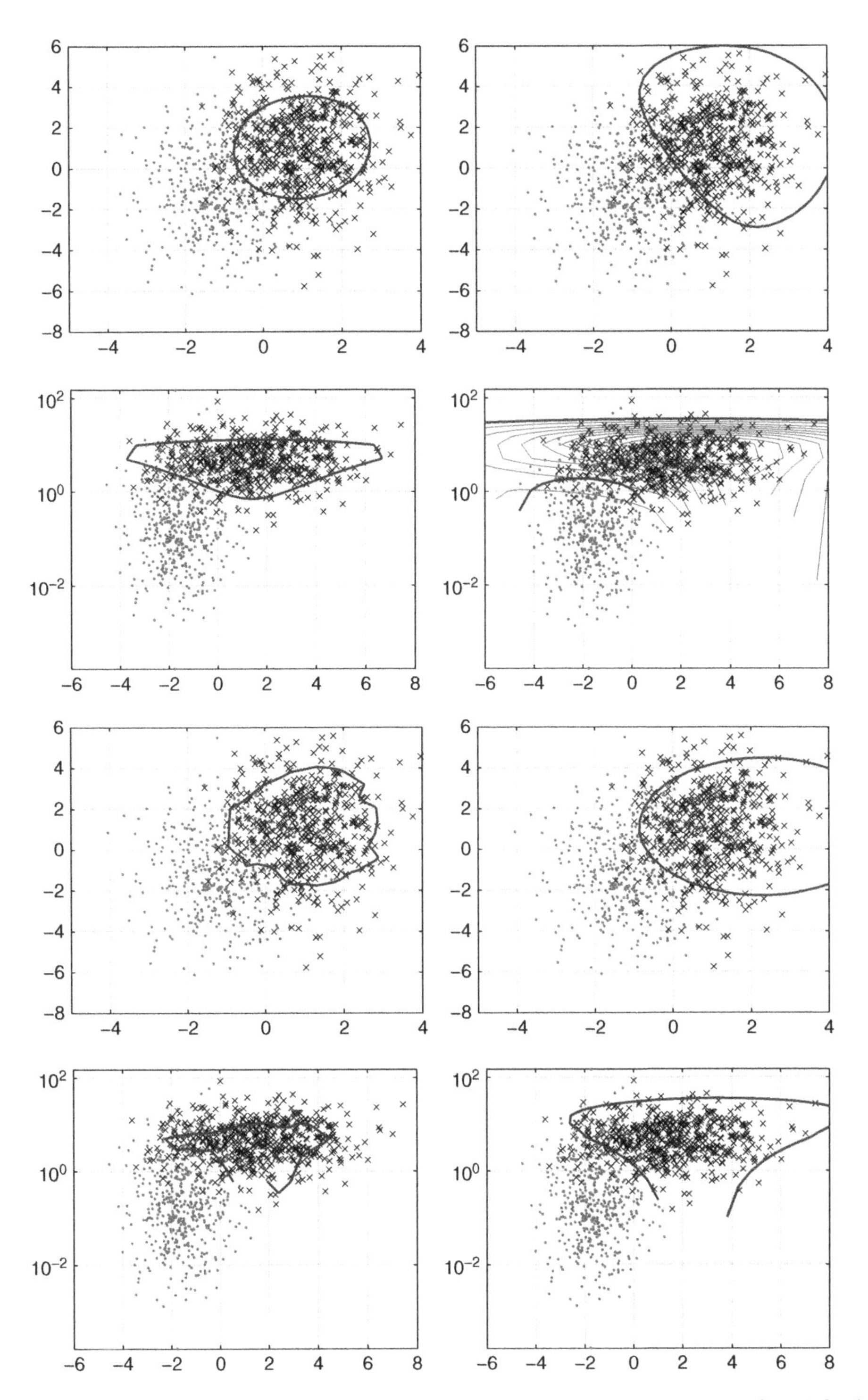

Figure 11.6 Plot of the decision boundaries obtained with GDD, MoGDD, KnnDD, and SVDD for the two Gaussians problem (top) and the mixed Gaussian–logarithm problem (bottom). Note the semilogarithmic scale in this latter case.

Figure 11.7 shows the receiver operating characteristic (ROC) curves and the area under the ROC curves (AUC) as a function of the kernel length-scale parameter σ. KSAM shows excellent detection rates, especially remarkable in the inset plot (note the logarithmic scale). Perhaps more importantly, the KSAM method is relatively insensitive to the selection of the kernel parameter compared with the KOSP detector, provided that a large enough value is specified.

The latter experiments allow us to use a traditional prescription to fix the RBF kernel parameter σ for both KOSP and KSAM as the mean distance among all spectra, d_M. Note that after data standardization and proper scaling, this is a reasonable heuristic $\sigma \approx d_M = 1$. The thresholds were optimized for all methods. OSP returns a decision function strongly contaminated by noise, while the KOSP detector results in a correct detection. Figure 11.8 shows the detection maps and the metric learned. The (linear) SAM gives rise to very good detection but with strong false alarms in the bottom left side of the image, where the roof of the commercial center saturates the sensor and thus returns a flat spectrum for its pixels in the morphological features. As a consequence, both the target and the roof vectors have flat spectra and their spectral angle is almost null. KSAM can efficiently cope with these (nonlinear) saturation problems. In addition, the metric space derived from the kernel suggests high discriminative (and spatially localized) capabilities.

11.7.5 Example of Kernel Anomaly Change Detection Algorithms

In this section we show the usage of the proposed methods in several simulated and real examples of pervasive and anomalous changes. Special attention is given to detection ROC curves, robustness to number of training samples, low-rank approximation of the solution, and the estimation of Gaussianity in kernel feature space. Comparisons between the hyperbolic (HACD), elliptical (EC), kernelized (K-HACD), and elliptical kernelized (K-EC-HACD) versions of the algorithms are included. Results can be reproduced using the script `demoKACD.m` available in the book's repository.

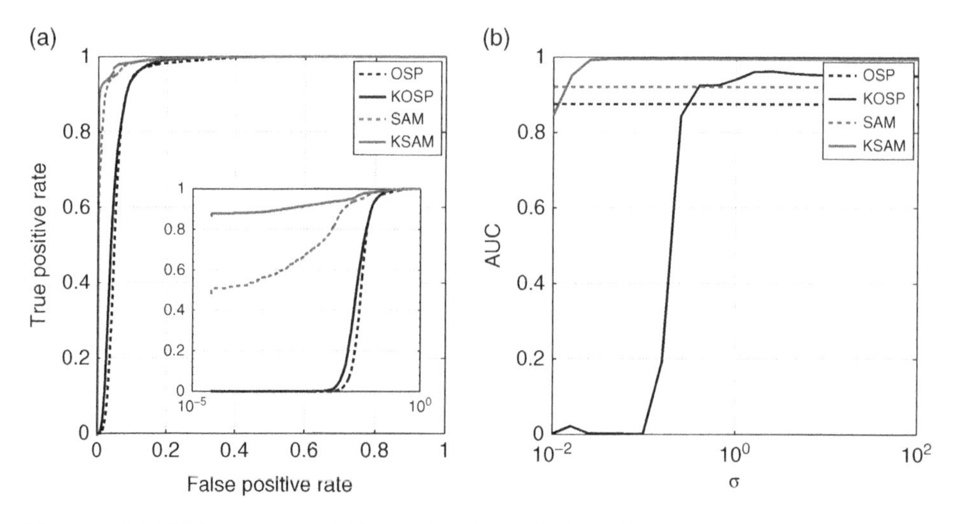

Figure 11.7 (a) ROC curves and (b) AUC as a function of the kernel length-scale parameter.

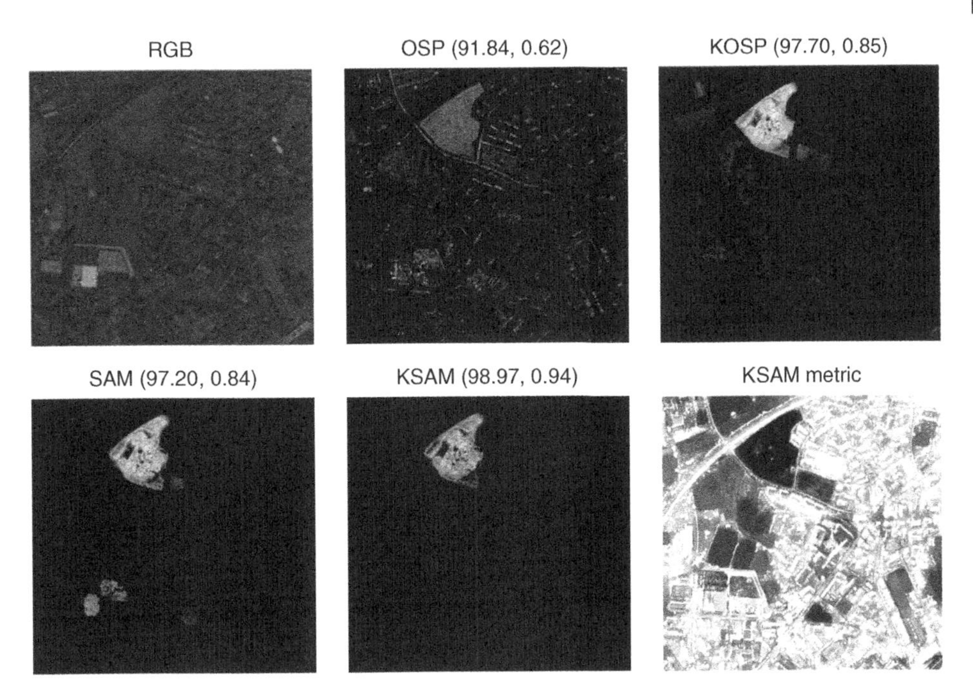

Figure 11.8 Detection results for different algorithms (accuracy, κ statistic).

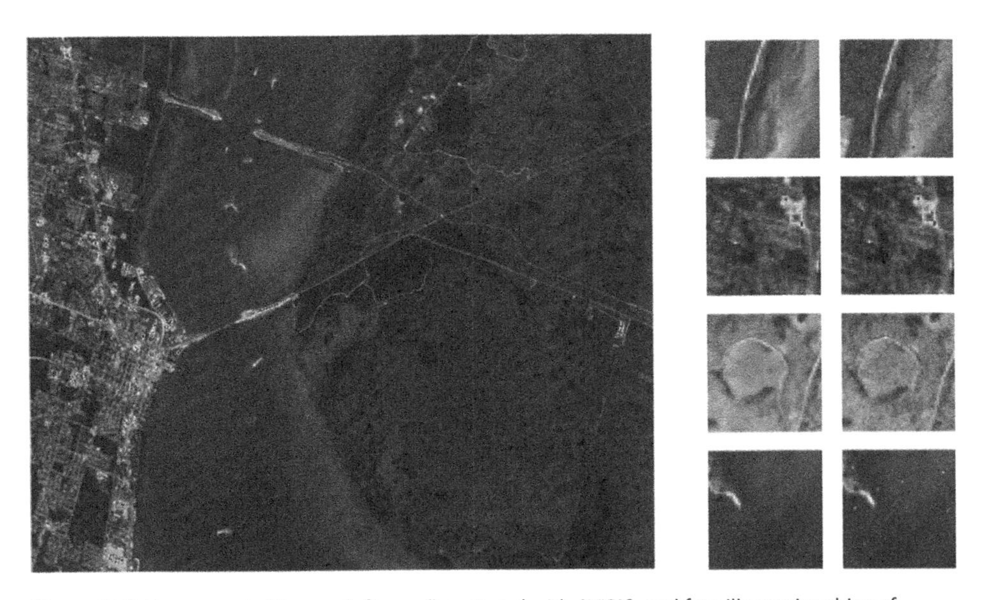

Figure 11.9 Hyperspectral image (left panel) captured with AVIRIS, and four illustrative chips of simulated changes (right panel). The original (leftmost) image is used to simulate an anomalous change image (rightmost) by adding Gaussian noise and randomly scrambling 1% of the pixels.

This example follows the simulation framework used in Theiler (2008). The dataset (see Figure 11.9) is an image acquired over the Kennedy Space Center, Florida, on March 23, 1996, by the Airbone Visible Infrared Imaging Spectrometer (AVIRIS). The

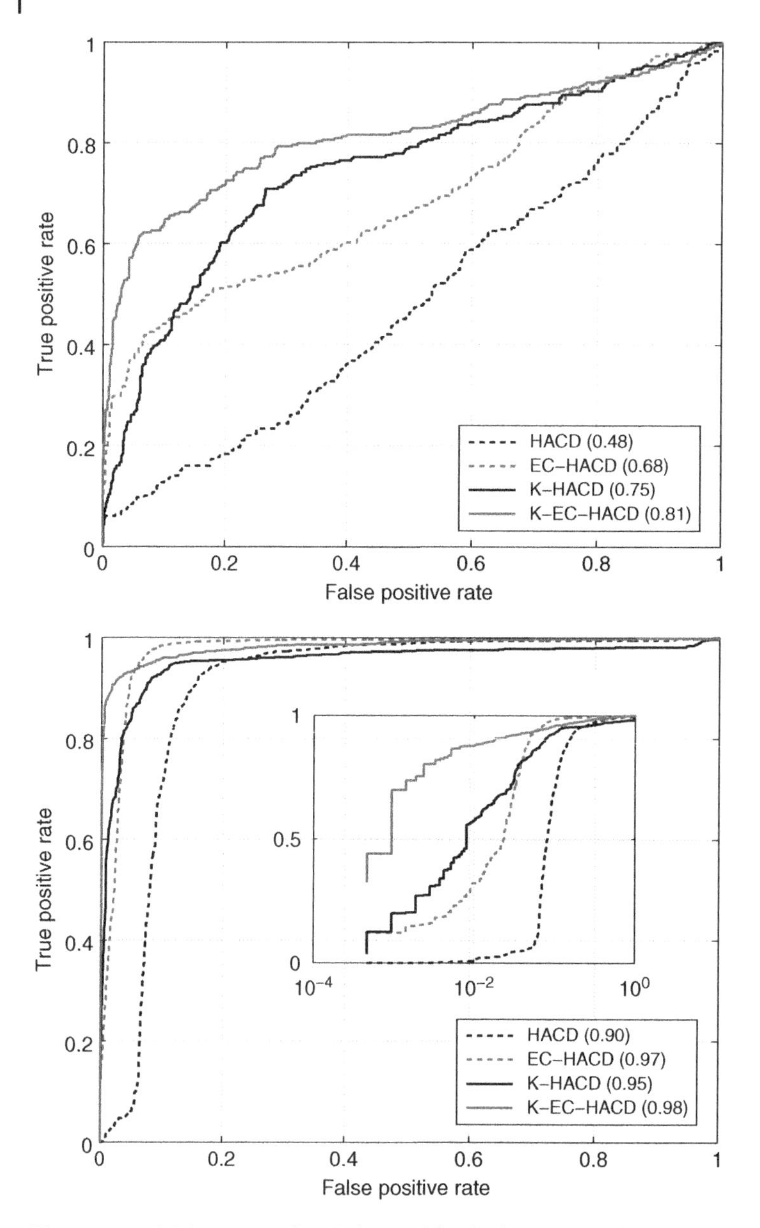

Figure 11.10 ROC curves and AUC obtained for the hyperspectral experiments by linear and kernel detectors for 100 (up) and 500 (down) training examples.

data were acquired from an altitude of 20 km and have a spatial resolution of 18 m. After removing low SNR and water absorption bands, a total of 176 bands remain for analysis. More information can be found at http://www.csr.utexas.edu/. To simulate a *pervasive change*, global Gaussian noise, $\mathcal{N}(0, 0.1)$, was added to all the pixels in the image to produce a second image. Then, *anomalous changes* are produced by scrambling the pixels in the target image. This yields anomalous change pixels whose components are not individually anomalous.

In all experiments, we used hyperbolic detectors (i.e., $\beta_x = \beta_y = 1$), as they generalize RX and chronocrome ACD algorithms, and have shown improved performance (Theiler *et al.*, 2010). In this example, we use the SAM kernel, which is more appropriate to capture hyperspectral pixel similarities than the standard RBF kernel. We tuned all the parameters involved (estimated covariance $\boldsymbol{C}_z$ and kernel $\boldsymbol{K}_z$, ν for the EC methods, length-scale σ parameter for the kernel versions) through standard cross-validation in the training set, and show results on the independent test set. We used $\lambda = 10^{-5}/N$, where N is the number of training samples, and the σ kernel length-scale parameter is tuned in the range of 0.05–0.95 percentiles of the distances between all training samples, d_{ij}.

Figure 11.10 shows the ROC curves obtained for the linear ACD and KACD methods. The dataset was split into small training sets of only 100 and 500 pixels, and results are given for 3000 test samples. The results show that the kernel versions improve upon

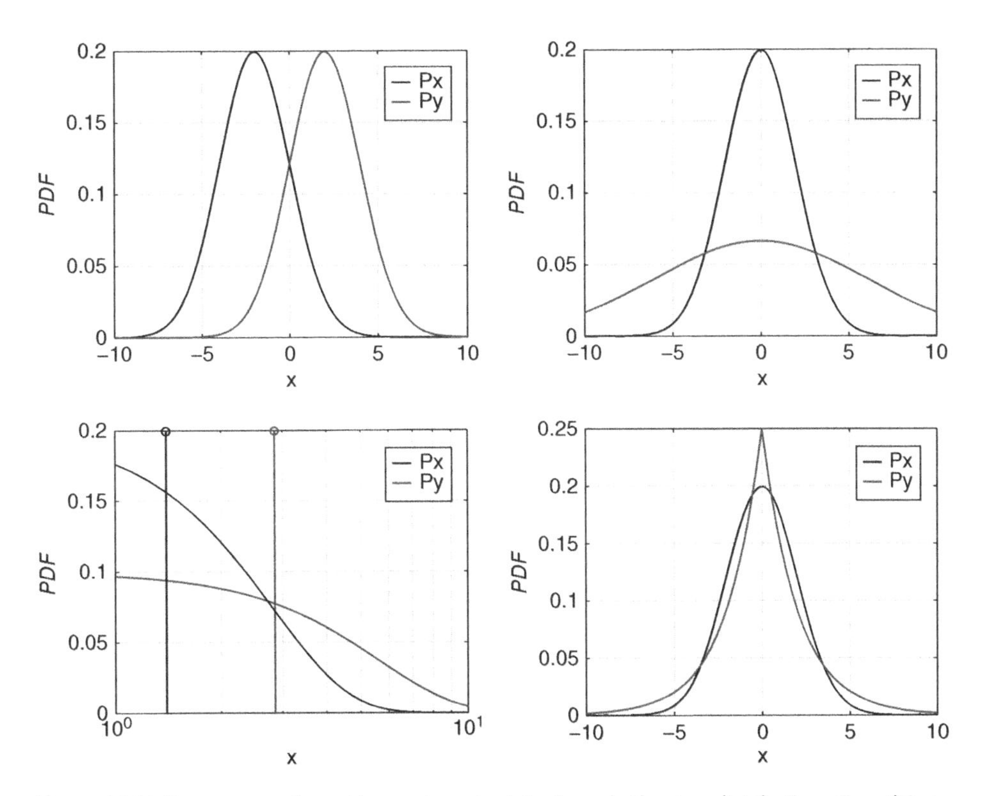

Figure 11.11 The two-sample problem reduces to detecting whether two distributions $\mathbb{P}_x$ and $\mathbb{P}_y$ are different or not. Up, left: whenever we have two Gaussians with different means one can assess statistical differences by means of a standard *t* test of different means (the distance between the empirical means is $d_{xy} := \|\mu_x - \mu_y\| = 3.91$). Up, right: when we have two Gaussians with the same means but different variance ($d_{xy} = 0.02$), the idea may be to look (down, left) at difference in means of transformed random variables (for the Gaussian case, second-order features of the form x^2 suffice for discrimination, $d_{xy} = 13.26$). Down, right: when distributions come from different *pdfs* (in this case, Gaussian and Laplace distributions) with the same mean and variance, one has to map the RVs to higher order features for discrimination ($d_{xy} = 0.06$ in the original domain and $d_{xy} = 75.80$ for a fifth-order mapping).

their linear counterparts (between 13–26% in Gaussian and 1–5% in EC detectors). The EC variants outperform their Gaussian counterparts, especially in the low-sized training sets (+30% over HACD and +18% over EC-HACD in AUC terms). Also noticeable is that results improve for all methods when using 500 training samples. The EC-HACD is very competitive compared with the kernel versions in terms of AUC, but still K-EC-HACD leads to longer tails of false-positive detection rates (right figure, inset plot in log-scale).

11.7.6 Example on Distribution Embeddings and Maximum Mean Discrepancy

Let us exemplify the use of the concept of distribution embeddings and MMD in a simple example comparing two distributions $\mathbb{P}_x$ and $\mathbb{P}_y$. We draw samples from (i) two Gaussians with different means, (ii) Gaussians with the same means but different variance, and (iii) from a Gaussian and a Laplace distributions. A code snippet is given in Listing 11.6 illustrating the ability of MMD to detect such differences. The results obtained are shown in Figure 11.11.

```
function mmd_demo

clc;close all; randn('seed',1234); rand('seed',1234)

% Two gaussians with different means
n=500;
x = linspace(-10,10,n);
mu1 = -2; si1 = +2; mu2 = +2; si2 = +2;
PX = normpdf(x,mu1,si1); PY = normpdf(x,mu2,si2);
Xe = mu1 + si1*randn(1,n); Ye = mu2 + si2*randn(1,n);
muXe = mean(Xe); muYe = mean(Ye); d = abs(muXe-muYe)
figure,plot(x,PX,'k-',x,PY,'r-'),
legend('Px','Py'), xlabel('x'), ylabel('PDF'), grid

% Two gaussians with the same mean but different variances
mu1 = 0; si1 = +2; mu2 = 0; si2 = +6;
PX = normpdf(x,mu1,si1); PY = normpdf(x,mu2,si2);
Xe = mu1 + si1*randn(1,n); Ye = mu2 + si2*randn(1,n);
muXe = mean(Xe); muYe = mean(Ye); d = abs(muXe-muYe)
figure,plot(x,PX,'k-',x,PY,'r-'),
legend('Px','Py'),xlabel('x'),ylabel('PDF'),grid

% Two gaussians with different variances, features transformed into x^2
mu1 = 0; si1 = +2; mu2 = 0; si2 = +4;
PX = normpdf(x.^2,mu1,si1); PY = normpdf(x.^2,mu2,si2);
Xe = mu1 + si1*randn(1,n); Ye = mu2 + si2*randn(1,n);
muXe = mean(Xe); muYe = mean(Ye); d = abs(muXe-muYe)
muXe2 = mean(Xe.^2); muYe2 = mean(Ye.^2); d = abs(muXe2-muYe2)
figure,semilogx(x.^2,PX,'k-',x.^2,PY,'r-'),
hold on, stem(log(muXe2),max(PX),'k'),stem(log(muYe2),max(PX),'r'),
legend('Px','Py'),xlabel('x'),ylabel('PDF'),grid, xlim([1,10])

% A gaussian and a Laplacian with same mean and variance
mu1 = 0; si1 = +2; mu2 = 0; si2 = +2;
PX = normpdf(x,mu1,si1); PY = lappdf(x,mu2,si2);
Xe = mu1 + si1*randn(1,n); Ye = mu2 + si2*randn(1,n);
```

```
muXe = mean(Xe); muYe = mean(Ye); d = abs(muXe-muYe)
figure,plot(x,PX,'k-',x,PY,'r-'),
legend('Px','Py'), xlabel('x'), ylabel('PDF'), grid

function x = randlap(mu,st,n)

u = rand(n) - 0.5;
x = mu - st*sign(u) .* log(1-2*abs(u));

function p = lappdf(x,mu,st)

p = 1/(2*st)*exp(-abs(x-mu)/st);
```

Listing 11.6 Comparing distributions of different nature with MMD (mmd_demo.m).

11.8 Concluding Remarks

This chapter treated the relevant topic of clustering and anomaly detection with kernels. The field is undoubtedly in the core of machine learning, and has many practical implications. The conceptual/philosophical problem of defining regularity, cluster, membership, and anomaly is an elusive one, so that there are the proper statistical definition of significant difference, hypothesis testing, and similarity measures between probabilities. Consequently, many methods and approaches have been proposed to tackle these problems. We reviewed several families available and related to a certain extent. We organized the kernel-based approaches systematically, and revised them organized in a simple taxonomy: (1) *clustering*, (2) *density estimation* (sometimes referred as to *domain description*), (3) *matched subspace detectors*, (4) *anomaly change detection*, and (5) *statistical hypothesis testing*.

The treatment in this chapter was far from exhaustive; rather, we tried to cover the main fields and to give an intuition on the relations among particular approaches. We observed clear links between clustering, subspace projection, and density estimation methods. The first essentially considered density characterization via *kernelization* of the metric or the algorithm, and the last relied on sound concepts of *distribution embeddings*.

We should also mention that the field of clustering and anomaly detection has arrived at a certain point of maturity with the introduction of relevant new paradigms from machine learning, such as SSL and AL, and from signal processing such as online and adaptive filtering. In many problems, introducing some labeled information properly helps in discerning the anomaly; also, including the user in the loop via active detection may help the detectors. The problem of anomaly detection in signal processing problems has to do very often with adaptation to changing environments/channels, so both adapting the classifier and/or the data representation has also been approached in the field. Other relevant issues to the DSP community, such as change-point detection, were intentionally omitted for the sake of conciseness.

The number of applications of the techniques presented will probably increase in the following years. The identification of events that are far from normal is an extremely appealing target for many real science and engineering applications, from health to traffic monitoring and climate science. In this setting, many big data environments or long-term monitoring from diverse kinds and nature will require DSP techniques running in parallel or even embedded with solid anomaly detection and domain description algorithms.

11.9 Questions and Problems

Exercise 11.9.1 Given an arbitrary dataset, obtain and compare the clustering solutions for $k = \{2, 3, 4, 5\}$ using k-means and kernel k-means with (a) the linear kernel and (b) the RBF Gaussian kernel for several values of the length-scale σ parameter.

Exercise 11.9.2 Obtain Equations 11.17 and 11.18.

Exercise 11.9.3 Demonstrate these latter two equations in Equation 11.20.

Exercise 11.9.4 Obtain the update rule for α_{jl} in Equation 11.33.

Exercise 11.9.5 Demonstrate that SVDD and OC-SVM reduce to the same equations when using the RBF kernel.

Exercise 11.9.6 Exploit the kernelization of the metric in the SAM similarity measure, which is commonly used in change detection, and that is given by

$$\theta = \arccos\left(\frac{\mathbf{x}^{\mathrm{T}}\mathbf{z}}{\|\mathbf{x}\|\|\mathbf{z}\|}\right), \tag{11.62}$$

where $0 \leq \theta \leq \pi/2$. Show that KSAM is a valid Mercer kernel and that KSAM is universal.

Exercise 11.9.7 A critical issue in kernel dependence estimation is about the selection of the kernel function and its hyperparameters. Very often one simply selects the SE kernel and sets the σ length-scale to the average distance between all samples. Test other ways based on the field of density estimation (e.g., Silverman's rule and related heuristics)). Figure out a way to tune Hilbert–Schmidt independence criterion (HSIC) parameters in a supervised manner.

Exercise 11.9.8 Derive the HSIC empirical estimate with regard to the σ parameter for the sake of optimization.

Exercise 11.9.9 Demonstrate the ECd kernel versions in Equations 11.55 and 11.56.

Exercise 11.9.10 The performance of the linear and kernel EC algorithms can be severely affected by the value of ν. Theiler *et al.* (2010) suggested a heuristic based on the ratio of moments of the distribution, $\nu = 2 + m\kappa_m/(\kappa_m - (d+m))$, where $\kappa_m = \langle r^{m+2}\rangle/\langle r^m\rangle$, $r = \xi_z^{1/2}$ and fixed $m = 1$ to control the impact of outliers. Discuss on the motivation, and its suitability in the kernel case. Can you say anything about the Gaussianity in feature spaces?

12

Kernel Feature Extraction in Signal Processing

Kernel-based feature extraction and dimensionality reduction are becoming increasingly important in advanced signal processing. This is particularly relevant in applications dealing with very high-dimensional data. Current methods tackle important problems in signal processing: from signal subspace identification to nonlinear blind source separation, as well as nonlinear transformations that maximize particular criteria, such as variance (KPCA), covariance (kernel PLS (KPLS)), MSE (kernel orthonormalized PLS (KOPLS)), correlation (kernel canonical correlation analysis (KCCA)), mutual information (kernel generalized variance) or SNR (kernel SNR), just to name a few. Kernel multivariate analysis (KMVA) has closed links to KFD analysis as well as interesting relations to information theoretic learning. The application of the methods is hampered in two extreme cases: when few examples are available the extracted features are either overfitted or meaningless, while in large-scale settings the computational cost is prohibitive. Semi-supervised and sparse learning have entered the field to alleviate these problems. Another field of intense activity is that of domain adaptation and manifold alignment, for which kernel method feature extraction is currently being used. All these topics are the subject of this chapter. We will review the main kernel feature extraction and dimensionality reduction methods, dealing with supervised, unsupervised and semi-supervised settings. Methods will be illustrated in toy examples, as well as real datasets.

12.1 Introduction

In the last decade, the amount and diversity of sensory data has increased vastly. Sensors and systems acquire signals at higher rate and resolutions, and very often data from different sensors need to be combined. In this scenario, appropriate data representation, adaptation, and dimensionality reduction become important concepts to be considered. Among the most important tasks of machine learning are feature extraction and dimensionality reduction, mainly due to the fact that, in many fields, practitioners need to manage data with a large dimensionality; for example, in image analysis, medical imaging, spectroscopy, or remote sensing, and where many heterogeneous feature components are computed from data and grouped together for machine-learning tasks.

Extraction of relevant features for the task at hand boils down to finding a proper representation domain of the problem. Different objectives will give different

Digital Signal Processing with Kernel Methods, First Edition. José Luis Rojo-Álvarez, Manel Martínez-Ramón, Jordi Muñoz-Marí, and Gustau Camps-Valls.

optimization problems. The field dates back to the early 20th century with the works of Hotelling, Wold, Pearson, and Fisher, just to name a few.[1] They originated the field of multivariate analysis (MVA) in its own right. The methods developed for MVA address the problems of dimensionality reduction in a very principled way, and typically resort to linear algebra operations. Owing to their simplicity and uniqueness of the solution, MVA methods have been successfully used in several scientific areas (Wold, 1966a).

Roughly speaking, the goal of MVA methods is to exploit dependencies among the variables to find a reduced set of them that are relevant for the learning task. PCA, PLS, or CCA are amongst the best known MVA methods. PCA uses only the correlation between the input dimensions in order to maximize the variance of the data over a chosen subspace, disregarding the target data. PLS and CCA choose subspaces where the projections maximize the covariance and the correlation between the features and the targets, respectively. Therefore, they should in principle be preferred to PCA for regression or classification problems. In this chapter, we consider also a fourth MVA method known as OPLS that is also well suited to supervised problems, with certain optimality in LS multiregression.

In any case, all these MVA methods are constrained to model linear relationships between the input and output; thus, they will be suboptimal when these relationships are actually nonlinear. This issue has been addressed by constructing nonlinear versions of MVA methods. They can be classified into two fundamentally different approaches (Rosipal, 2011): (1) the modified methods in which the linear relations among the latent variables are substituted by nonlinear relations (Laparra *et al.* 2014, 2015; Qin and McAvoy 1992; Wold *et al.* 1989); and (2) variants in which the algorithms are reformulated to fit a kernel-based approach (Boser *et al.* 1992; Schölkopf *et al.* 1998; Shawe-Taylor and Cristianini 2004). An appealing property of the resulting kernel algorithms is that they obtain the flexibility of nonlinear expressions using straightforward methods from linear algebra.

The field has captured the attention of many researchers from other fields mainly due to interesting relations to KFD analysis algorithms (a plethora of variants has been introduced in the last decade in the fields of pattern recognition and computer vision),

1 These are names every statistician should know quite well. Harold Hotelling was an American statistician highly influential in economics, and he is well known for his T-squared distribution in statistics and the perhaps most important method in data analysis and multivariate statistics: the PCA method. Herman Ole Andreas Wold was a Norwegian-born econometrician and statistician who developed his career in Sweden. Wold is known for his excellent works in mathematical economics, in time-series analysis, and in econometrics. He was ahead of his time: Wold contributed to the development of multivariate methods such as PLS, the Cramér–Wold theorem characterizing the normal distribution, and his advances in causal inference from empirical data. Karl Pearson was an English mathematician and biostatistician, whose contributions in statistics are outstanding: the correlation coefficient to measure association between random variables, the method of moments, the χ^2-distance, the p-value, and PCA are the most famous examples of contributions in the field. And what to say about Fisher! Sir Ronald Aylmer Fisher was an English statistician and biologist who combined mathematics and genetics, explaining most of the processes of natural selection. When it comes to signal processing and machine learning, however, he is well known because of the development of LDA, Fisher's information score, the F-distribution, the Student t-distribution or relevant contributions in sufficient statistics, just to name a few. Fisher was a prolific author, a controverted individual, a genius after all.

and the more appealing relations to particular instantiations of kernel methods for information theoretic learning (e.g., links between covariance operators in Hilbert spaces and information concepts such as mutual information or Rényi entropy). In Section 12.3 we will show some of the relations between KMVA and other feature extraction methods based on nonparametric kernel dependence estimates.

The fact that KMVA methods need the construction of a Gram matrix of kernel dot products between data precludes their direct use in applications involving large datasets, and the fact that in many applications only a small number of labeled samples is also strongly limiting. The first difficulty is usually minimized by the use of sparse or incremental versions of the algorithms, and the second has been overcome by the field of SSL. Approaches to tackle these problems include special types of *regularization*, guided either by selection of a reduced number of basis functions or by considering the information about the manifold conveyed by the unlabeled samples. This chapter reviews all these approaches (Section 12.4).

Domain adaptation, transfer learning, and manifold alignment are relevant topics nowadays. Note that classification or regression algorithms developed with data coming from one domain (system, source) cannot be directly used in another related domain, and hence adaptation of either the data representation or the classifier becomes strictly imperative (Quiñonero-Candela *et al.*, 2009; Sugiyama and Kawanabe, 2012). The fields of manifold alignment and domain adaptation have captured high interest nowadays in pattern analysis and machine learning. On many occasions, it is more convenient to adapt the data representations to facilitate the use of standard classifiers, and several kernel methods have been proposed trying to map different data sources to a common latent space where the different source manifolds are matched. Section 12.5 reviews some kernel methods to perform domain adaptation while still using linear algebra.

12.2 Multivariate Analysis in Reproducing Kernel Hilbert Spaces

This section fixes the notation used in the chapter, and reviews the framework of MVA both in the linear case and with kernel methods. An experimental evaluation on standard UCI datasets and real high-dimensional image segmentation problems illustrate the performance of all the methods.

12.2.1 Problem Statement and Notation

Assume any supervised regression or classification problem, were $\boldsymbol{X}$ and $\boldsymbol{Y}$ are column-wise centered input and target data matrices of sizes $N \times d$ and $N \times m$ respectively. Here, N is the number of training data points in the problem, and d and m are the dimensions of the input and output spaces respectively. The output response data (sometimes referred to as the target) $\boldsymbol{Y}$ are typically a set of variables in columns that need to be approximated in regression settings, or a matrix that encodes the class membership information in classification settings. The sample covariance matrices are given by $\boldsymbol{C}_x = (1/N)\boldsymbol{X}^{\mathrm{T}}\boldsymbol{X}$ and $\boldsymbol{C}_y = (1/N)\boldsymbol{Y}^{\mathrm{T}}\boldsymbol{Y}$, whereas the input–output cross-covariance is expressed as $\boldsymbol{C}_{xy} = (1/N)\boldsymbol{X}^{\mathrm{T}}\boldsymbol{Y}$.

The objective of standard linear multiregression is to adjust a linear model for predicting the output variables from the input ones; that is, $\hat{Y} = XW$, where W contains the regression model coefficients. The standard LS solution is $W = X^{\dagger}Y$, where $X^{\dagger} = (X^{\mathrm{T}}X)^{-1}X^{\mathrm{T}}$ is the Moore–Penrose pseudoinverse[2] of X. If some input variables are highly correlated, the corresponding covariance matrix will be rank deficient, and thus its inverse will not exist. The same situation will be encountered if the number N of data is less that the dimension d of the space. In order to turn these problems into well-conditioned ones it is usual to apply a Tikhonov regularization term; for instance, by also minimizing the Frobenius norm of the weights matrix $\|W\|_F^2$ one obtains the regularized LS solution $W = (X^{\mathrm{T}}X+\lambda I)^{-1}X^{\mathrm{T}}Y$, where parameter λ controls the amount of regularization.

MVA techniques solve the aforementioned problems by projecting the input data into a subspace that preserves the information useful for the current machine-learning problem. MVA methods transform the data into a set of features through a transformation $X' = XU$, where $U = [u_1 | \cdots | u_{n_f}]$ will be referred hereafter as the projection matrix, u_i being the ith projection vector and n_f the number of extracted features. Some MVA methods consider also a feature extraction in the output space, $Y' = YV$, with $V = [v_1 | \cdots | v_{n_f}]$.

Generally speaking, MVA methods look for projections of the input data that are "maximally aligned" with the targets. Different methods are characterized by the specific objectives they maximize. Table 12.1 summarizes the MVA methods that are being presented in this section. It is important to be aware of that MVA methods are based on the first- and second-order moments of the data, so these methods can be formulated as a generalized eigenvalue problem; hence, they can be solved using standard algebraic techniques.

12.2.2 Linear Multivariate Analysis

The best known MVA method, and probably the oldest one, is PCA (e.g., Pearson, 1901), also known as the *Hotelling transform* or the *Karhunen–Loève transform* (Jolliffe, 1986). PCA selects the maximum variance projections of the input data, imposing an orthonormality constraint for the projection vectors (see Table 12.1). This method assumes that the projections that contain the information are the ones that have the highest variance. PCA is an *unsupervised* feature extraction method. Even if supervised methods, which include the target information, may be preferred, PCA and its kernel version, KPCA, are used as preprocessing stages in many supervised problems, possibly because of their simplicity and ability to discard irrelevant directions (Abrahamsen and Hansen 2011; Braun *et al.* 2008).

In signal processing applications, maximizing the variance of the signal may not be a good idea per se, and one typically looks for transformations of the observed signal such that the SNR is maximized, or alternatively for transformations that minimize the fraction of noise. This is the case of the MNF transform (Green *et al.*, 1998), which

---

2 In MATLAB we have functions `inv` and `pinv` to compute the inverse and pseudoinverse of a matrix respectively.

Table 12.1 Summary of linear and KMVA methods. We state for all methods treated in this section the objective to maximize, constraints for the optimization, and maximum number of features n_p that can be extracted, and how the data are projected onto the extracted features.

Method	PCA	SNR	PLS	CCA	OPLS
$\max_{\boldsymbol{u}}$	$\boldsymbol{u}^{\mathrm{T}}\boldsymbol{C}_x\boldsymbol{u}$	$\boldsymbol{u}^{\mathrm{T}}\boldsymbol{C}_n^{-1}\boldsymbol{C}_x\boldsymbol{u}$	$\boldsymbol{u}^{\mathrm{T}}\boldsymbol{C}_{xy}\boldsymbol{v}$	$\boldsymbol{u}^{\mathrm{T}}\boldsymbol{C}_{xy}\boldsymbol{v}$	$\boldsymbol{u}^{\mathrm{T}}\boldsymbol{C}_{xy}\boldsymbol{C}_{xy}^{\mathrm{T}}\boldsymbol{u}$
s.t.	$\boldsymbol{U}^{\mathrm{T}}\boldsymbol{U}=\boldsymbol{I}$	$\boldsymbol{U}^{\mathrm{T}}\boldsymbol{C}_{\mathrm{n}}\boldsymbol{U}=\boldsymbol{I}$	$\boldsymbol{U}^{\mathrm{T}}\boldsymbol{U}=\boldsymbol{I}$ $\boldsymbol{V}^{\mathrm{T}}\boldsymbol{V}=\boldsymbol{I}$	$\boldsymbol{U}^{\mathrm{T}}\boldsymbol{C}_x\boldsymbol{U}=\boldsymbol{I}$ $\boldsymbol{V}^{\mathrm{T}}\boldsymbol{C}_y\boldsymbol{V}=\boldsymbol{I}$	$\boldsymbol{U}^{\mathrm{T}}\boldsymbol{C}_x\boldsymbol{U}=\boldsymbol{I}$
n_p	$r(\boldsymbol{X})$	$r(\boldsymbol{X})$	$r(\boldsymbol{X})$	$r(\boldsymbol{C}_{xy})$	$r(\boldsymbol{C}_{xy})$
$P(\boldsymbol{X}_*)$	$\boldsymbol{X}_*\boldsymbol{U}$	$\boldsymbol{X}_*\boldsymbol{U}$	$\boldsymbol{X}_*\boldsymbol{U}$	$\boldsymbol{X}_*\boldsymbol{U}$	$\boldsymbol{X}_*\boldsymbol{U}$
	KPCA	**KSNR**	**KPLS**	**KCCA**	**KOPLS**
$\max_{\boldsymbol{\alpha}}$	$\boldsymbol{\alpha}^{\mathrm{T}}\boldsymbol{K}_x^2\boldsymbol{\alpha}$	$\boldsymbol{\alpha}^{\mathrm{T}}(\boldsymbol{K}_{xn}\boldsymbol{K}_{nx})^{-1}\boldsymbol{K}_x^2\boldsymbol{\alpha}$	$\boldsymbol{\alpha}^{\mathrm{T}}\boldsymbol{K}_x\boldsymbol{Y}\boldsymbol{v}$	$\boldsymbol{\alpha}^{\mathrm{T}}\boldsymbol{K}_x\boldsymbol{Y}\boldsymbol{v}$	$\boldsymbol{\alpha}^{\mathrm{T}}\boldsymbol{K}_x\boldsymbol{Y}\boldsymbol{Y}^{\mathrm{T}}\boldsymbol{K}_x\boldsymbol{\alpha}$
s.t.	$\boldsymbol{A}^{\mathrm{T}}\boldsymbol{K}_x\boldsymbol{A}=\boldsymbol{I}$	$\boldsymbol{A}^{\mathrm{T}}\boldsymbol{K}_{xn}\boldsymbol{K}_{nx}\boldsymbol{A}=\boldsymbol{I}$	$\boldsymbol{A}^{\mathrm{T}}\boldsymbol{K}_x\boldsymbol{A}=\boldsymbol{I}$ $\boldsymbol{V}^{\mathrm{T}}\boldsymbol{V}\boldsymbol{V}=\boldsymbol{I}$	$\boldsymbol{A}^{\mathrm{T}}\boldsymbol{K}_x^2\boldsymbol{A}=\boldsymbol{I}$ $\boldsymbol{V}^{\mathrm{T}}\boldsymbol{C}_y\boldsymbol{V}=\boldsymbol{I}$	$\boldsymbol{A}^{\mathrm{T}}\boldsymbol{K}_x^2\boldsymbol{A}=\boldsymbol{I}$
n_p	$r(\boldsymbol{K}_x)$	$r(\boldsymbol{K}_x)$	$r(\boldsymbol{K}_x)$	$r(\boldsymbol{K}_x\boldsymbol{Y})$	$r(\boldsymbol{K}_x\boldsymbol{Y})$
$P_{\mathcal{H}}(\boldsymbol{X}_*)$	$\boldsymbol{K}_x(\boldsymbol{X}_*,\boldsymbol{X})\boldsymbol{A}$	$\boldsymbol{K}_x(\boldsymbol{X}_*,\boldsymbol{X})\boldsymbol{A}$	$\boldsymbol{K}_x(\boldsymbol{X}_*,\boldsymbol{X})\boldsymbol{A}$	$\boldsymbol{K}_x(\boldsymbol{X}_*,\boldsymbol{X})\boldsymbol{A}$	$\boldsymbol{K}_x(\boldsymbol{X}_*,\boldsymbol{X})\boldsymbol{A}$

Vectors $\mathbf{u}$ and $\boldsymbol{\alpha}$ are column vectors in matrices $\boldsymbol{U}$ and $\boldsymbol{A}$ respectively. $r(\cdot)$ denotes the rank of a matrix.

extends PCA by maximizing the signal variance while minimizing the estimated noise variance. Let us assume that observations $\boldsymbol{x}_i$ follow an *additive noise model*; that is, $\boldsymbol{x}_i = \boldsymbol{s}_i + \boldsymbol{n}_i$, where $\boldsymbol{n}_i$ may not necessarily be Gaussian. The observed data matrix is assumed to be the sum of a "signal" and a "noise" matrix, $\boldsymbol{X} = \boldsymbol{S} + \boldsymbol{N}$, being typically $N > d$. Matrix $\tilde{\boldsymbol{X}}$ indicates the centered version of $\boldsymbol{X}$, and $\boldsymbol{C}_x = (1/N)\tilde{\boldsymbol{X}}^{\mathrm{T}}\tilde{\boldsymbol{X}}$ represents the empirical covariance matrix of the input data. Very often, signal and noise are assumed to be orthogonal, $\boldsymbol{S}^{\mathrm{T}}\boldsymbol{N} = \boldsymbol{N}^{\mathrm{T}}\boldsymbol{S} = \boldsymbol{0}$, which is very convenient for SNR maximization and blind-source separation problems (Green *et al.*, 1998; Hundley *et al.*, 2002). Under this assumption, one can incorporate easily the information about the noise through the noise covariance matrix $\boldsymbol{C}_{\mathrm{n}} = (1/N)\tilde{\boldsymbol{N}}^{\mathrm{T}}\tilde{\boldsymbol{N}}$.

The PLS algorithm (Wold, 1966b) is a supervised method based on latent variables that account for the information in the cross-covariance matrix $\boldsymbol{C}_{xy}$. The strategy here consists of constructing the projections that maximize the covariance between the projections of the input and the output, while keeping certain orthonormality constraints. The procedure is solved iteratively or in block through an eigenvalue problem. In iterative schemes, datasets X and Y are recursively transformed in a process which subtracts the information contained in the already estimated latent variables. This process is usually called *deflation*, and it can be done in several ways that define the many variants of PLS existing in the literature.[3] The algorithm does not involve matrix inversion, and it is robust against highly correlated data. These properties made it usable in many fields, such as chemometrics and remote sensing, where signals typically are acquired in a range of highly correlated spectral wavelengths.

The main feature of CCA is that it maximizes the correlation between projected input and output data (Hotelling, 1936). This is why CCA can manage spatial dimensions in the input or output data that show high variance but low correlation between input and output, which would be emphasized by PLS. CCA is a standard method for aligning data sources, with interesting relations to information-theoretic approaches. We will come back to this issue in Sections 12.3 and 12.5.

An interesting method we will pay attention to is OPLS, also known as multilinear regression (Borga *et al.*, 1997) or semi-penalized CCA (Barker and Rayens, 2003). OPLS is optimal for performing multilinear LS regression on the features extracted from the training data; that is:

$$\boldsymbol{U}^* = \arg\min_{\boldsymbol{U}} \|\boldsymbol{Y} - \boldsymbol{X}'\boldsymbol{W}\|_{\mathrm{F}}^2, \tag{12.1}$$

with $\boldsymbol{W} = \boldsymbol{X}'^{\dagger}\boldsymbol{Y}$ being the matrix containing the optimal regression coefficients. It can be shown that this optimization problem is equivalent to the one stated in Table 12.1. This problem can be compared with maximizing a Rayleigh coefficient that takes into account the projections of input and output data, $\frac{(\boldsymbol{u}^{\mathrm{T}}\boldsymbol{C}_{xy}\boldsymbol{v})^2}{(\boldsymbol{u}^{\mathrm{T}}\boldsymbol{C}_x\boldsymbol{u})(\boldsymbol{v}^{\mathrm{T}}\boldsymbol{v})}$. This is why the method is called semi-penalized CCA. Indeed, it does not take into account the variance of the projected input data, but emphasizes those input dimensions that better predict large variance projections of the target data. This asymmetry makes sense in *supervised subspace learning* where matrix Y contains target values to be approximated from the extracted input features.

Actually, OPLS (for classification problems) is equivalent to LDA provided an appropriate labeling scheme is used for Y (Barker and Rayens, 2003). However, in "two-view learning" problems (also known as domain adaptation or manifold alignment), in which X and Y represent different views of the data (Shawe-Taylor and Cristianini, 2004, Section 6.5), one would like to extract features that can predict both data representations simultaneously, and CCA could be preferred to OPLS. We will further discuss this matter in Section 12.5.

Now we can establish a simple common framework for PCA, SNR, PLS, CCA, and OPLS, following the formalization introduced in Borga *et al.* (1997), where it was shown

---

3 Perhaps the most popular PLS method was presented by Wold *et al.* (1984). The algorithm, hereafter referred to as PLS2, assumes a linear relation between X and Y that implies a certain deflation scheme, where the latent variable of X is used to deflate also Y (Shawe-Taylor and Cristianini, 2004, : 182). Several other variants of PLS exist, such as "PLS Mode A" and PLS-SB; see Geladi (1988) for a discussion of the early history of PLS and Kramer and Rosipal (2006) for a well-written overview.

that these methods can be reformulated as (generalized) eigenvalue problems, so that linear algebra packages[4] can be used to solve them. Concretely:

$$
\begin{aligned}
\text{PCA}: &\quad \boldsymbol{C}_x \boldsymbol{u} = \lambda \boldsymbol{u} \\
\text{SNR}: &\quad \boldsymbol{C}_x \boldsymbol{u} = \lambda \boldsymbol{C}_n \boldsymbol{u} \\
\text{PLS}: &\quad \begin{pmatrix} 0 & \boldsymbol{C}_{xy} \\ \boldsymbol{C}_{xy}^{\mathrm{T}} & 0 \end{pmatrix} \begin{pmatrix} \boldsymbol{u} \\ \boldsymbol{v} \end{pmatrix} = \lambda \begin{pmatrix} \boldsymbol{u} \\ \boldsymbol{v} \end{pmatrix} \\
\text{OPLS}: &\quad \boldsymbol{C}_{xy} \boldsymbol{C}_{xy}^{\mathrm{T}} \boldsymbol{u} = \lambda \boldsymbol{C}_x \boldsymbol{u} \\
\text{CCA}: &\quad \begin{pmatrix} 0 & \boldsymbol{C}_{xy} \\ \boldsymbol{C}_{xy}^{\mathrm{T}} & 0 \end{pmatrix} \begin{pmatrix} \boldsymbol{u} \\ \boldsymbol{v} \end{pmatrix} = \lambda \begin{pmatrix} \boldsymbol{C}_x & 0 \\ 0 & \boldsymbol{C}_y \end{pmatrix} \begin{pmatrix} \boldsymbol{u} \\ \boldsymbol{v} \end{pmatrix}.
\end{aligned}
\tag{12.2}
$$

Note that CCA and OPLS require the inversion of matrices $\boldsymbol{C}_x$ and $\boldsymbol{C}_y$. Whenever these are rank deficient, it becomes necessary to first extract the dimensions with nonzero variance using PCA, and then solve the CCA or OPLS problems. Alternative approaches include adding extra regularization terms. A very common approach is to solve the aforementioned problems using a two-step iterative procedure: first, the projection vectors corresponding to the largest (generalized) eigenvalue are chosen, for which there exist efficient methods such as the power method; and then in the second step known as *deflation*, one removes from the data (or the covariance matrices) the covariance that can be already obtained from the features extracted in the first step. Equivalent solutions can be found by formulating them as regularized LS problems. As an example, sparse versions of PCA, CCA, and OPLS were introduced by adding sparsity promotion terms, such as LASSO or ℓ_1-norm on the projection vectors, to the LS functional (Hardoon and Shawe-Taylor 2011; Van Gerven *et al.* 2012; Zou *et al.* 2006).

12.2.3 Kernel Multivariate Analysis

The framework of KMVA algorithms is aimed at extracting nonlinear projections while actually working with linear algebra. Let us first consider a function $\boldsymbol{\Phi} : \mathbb{R}^d \rightarrow \mathcal{F}$ that maps input data into a Hilbert feature space $\mathcal{H}$. The new mapped dataset is defined as $\boldsymbol{\Phi} = [\boldsymbol{\Phi}(\boldsymbol{x}_1)| \cdots |\boldsymbol{\Phi}(\boldsymbol{x}_l)]^{\mathrm{T}}$, and the features extracted from the input data will now be given by $\boldsymbol{\Phi}' = \boldsymbol{\Phi}\boldsymbol{U}$, where matrix $\boldsymbol{U}$ is of size $\dim(\mathcal{F}) \times n_{\mathrm{f}}$. The direct application of this idea suffers from serious practical limitations when the $\mathcal{H}$ is very large. To implement practical KMVA algorithms we need to rewrite the equations in the first

4 Among the most well-known package to deal with linear algebra problems we find LAPACK http://www.netlib.org/lapack/. LAPACK is written in Fortran 90 and provides routines for solving systems of simultaneous linear equations, LSs solutions of linear systems of equations, eigenvalue problems, and singular value problems, as well as the associated matrix factorizations (LU, Cholesky, QR, SVD, Schur, generalized Schur). The routines in LAPACK can deal with dense and banded matrices. LAPACK has been sponsored by MathWorks and Intel for many years, which ensures lifelong updated routines.

half of Table 12.1 in terms of inner products in $\mathcal{H}$ only. For doing so, we rely on the availability of a kernel matrix $\boldsymbol{K}_x = \boldsymbol{\Phi}\boldsymbol{\Phi}^{\mathrm{T}}$ of dimension $N \times N$, and on the representer theorem (Kimeldorf and Wahba, 1971), which states that the projection vectors can be written as a linear combination of the training samples; that is, $\boldsymbol{U} = \boldsymbol{\Phi}^{\mathrm{T}}\boldsymbol{A}$, matrix $\boldsymbol{A} = [\boldsymbol{\alpha}_1| \cdots |\boldsymbol{\alpha}_{n_f}]$ being the new argument for the optimization.[5] Equipped with the *kernel trick*, one readily develops kernel versions of the previous linear MVA, as shown in Table 12.1.

For PCA, it was Schölkopf *et al.* (1998) who introduced a kernel version denoted KPCA. Lai and Fyfe in 2000 first introduced the kernel version of CCA denoted KCCA (Lai and Fyfe, 2000a; Shawe-Taylor and Cristianini, 2004). Later, Rosipal and Trejo (2001) presented a nonlinear kernel variant of PLS. In that paper, $\boldsymbol{K}_x$ and the $\boldsymbol{Y}$ matrix are deflated the same way, which is more in line with the PLS2 variant than with the traditional algorithm "PLS Mode A," and therefore we will denote it as KPLS2. A kernel variant of OPLS was presented in (Arenas-García *et al.*, 2007) and is here referred to as KOPLS. Recently, the kernel MNF was introduced in (Nielsen, 2011) but required the estimation of the noise signal in input space. This kernel SNR method was extended in (Gómez-Chova *et al.*, 2017) to deal with signal and noise relations in Hilbert space *explicitly*. That is, in the explicit KSNR (KSNRe) the noise is estimated in Hilbert spaces rather than in the input space and then mapped with an appropriate reproducing kernel function (which we refer to as implicit KSNR, KSNRi).

As for the linear case, KMVA methods can be implemented as (generalized) eigenvalue problems:

$$
\begin{aligned}
&\text{KPCA}: \quad \boldsymbol{K}_x\boldsymbol{\alpha} = \lambda\boldsymbol{\alpha} \\
&\text{KSNR}: \quad {\boldsymbol{K}_x}^2\boldsymbol{\alpha} = \lambda\boldsymbol{K}_{xn}\tilde{\boldsymbol{K}}_{xn}^{\mathrm{T}}\boldsymbol{\alpha} \\
&\text{KPLS}: \quad \begin{pmatrix} \boldsymbol{0} & \boldsymbol{K}_x\boldsymbol{Y} \\ \boldsymbol{Y}\boldsymbol{K}_x & \boldsymbol{0} \end{pmatrix}\begin{pmatrix} \boldsymbol{\alpha} \\ \boldsymbol{\nu} \end{pmatrix} = \lambda\begin{pmatrix} \boldsymbol{\alpha} \\ \boldsymbol{\nu} \end{pmatrix} \\
&\text{KOPLS}: \quad \boldsymbol{K}_x\boldsymbol{Y}\boldsymbol{Y}^{\mathrm{T}}\boldsymbol{K}_x\boldsymbol{\alpha} = \lambda\boldsymbol{K}_x\boldsymbol{K}_x\boldsymbol{\alpha} \\
&\text{KCCA}: \quad \begin{pmatrix} \boldsymbol{0} & \boldsymbol{K}_x\boldsymbol{Y} \\ \boldsymbol{Y}\boldsymbol{K}_x & \boldsymbol{0} \end{pmatrix}\begin{pmatrix} \boldsymbol{\alpha} \\ \boldsymbol{\nu} \end{pmatrix} = \lambda\begin{pmatrix} \boldsymbol{K}_x\boldsymbol{K}_x & \boldsymbol{0} \\ \boldsymbol{0} & \boldsymbol{C}_y \end{pmatrix}\begin{pmatrix} \boldsymbol{\alpha} \\ \boldsymbol{\nu} \end{pmatrix}.
\end{aligned}
\tag{12.3}
$$

Note that the output data could also be mapped to some feature space $\mathcal{H}$, as was considered for KCCA for a multiview learning case (Lai and Fyfe, 2000a). Here, we consider instead that it is the actual labels in $\boldsymbol{Y}$ which need to be well represented by the extracted input features, so we will deal with the original representation of the output data.

5 In this chapter, we assume that data are centered in feature space, which can easily be done through a simple modification of the kernel matrix (see Chapter 4).

12.2.4 Multivariate Analysis Experiments

In this section we illustrate through different application examples the use and capabilities of the kernel multivariate feature extraction framework. We start with a toy example where we can visualize the features, and therefore get intuition about what is doing each method. After that we compare the performance of linear and KMVA methods in a real classification problem of (high-dimensional) satellite images (Arenas-García and Camps-Valls 2008; Arenas-García *et al.* 2013).

Toy Example

In this first experiment we illustrate the methods presented using toy data. Figure 12.1 illustrates the features extracted by the methods for a toy classification problem with three classes. Figure 12.1 can be reproduced by downloading the SIMFEAT toolbox and executing the script `Demo_Fig_14_1.m` from the book's repository. A simplified version of the code is shown in Listing 12.1. Note how all the methods are based on eigendecompositions of covariance matrices or kernels.

```
%% Data
NN = 200; Nc = 2; YY = binarize([ones(NN,1); 2*ones(NN,1); 3*ones(NN,1)]);
XX(:,1) = [rand(3*NN,1)*2];
XX(:,2) = [(XX(1:NN,1)-1).^2; (XX(NN+1:NN*2,1)-1).^2+0.2 ; ...
    (XX(2*NN+1:NN*3,1)-1).^2+0.4]+0.04*randn(3*NN,1);
XX = XX-repmat(mean(XX),size(XX,1),1);
ii = randperm(3*NN); X = XX(ii(1:NN),:); Y = YY(ii(1:NN),:);
Xts = XX(ii(NN+1:3*NN),:); Yts = YY(ii(NN+1:3*NN),:);

%% Number of features (projections, components) to be extracted
nf = 2;
%% Covariances for Linear Methods
Cxx = cov(X); Cxy = X'*Y; Cyy = cov(Y);
% PCA
[U_pca D] = eig(Cxx);
XtsProj_PCA = Xts * U_pca(:,1:nf); XProj_PCA = X * U_pca(:,1:nf);
Ypred_PCA = classify(XtsProj_PCA,XProj_PCA,Y);
% PLS
U_pls = svds(Cxy);
XtsProj_PLS = Xts * U_pls(:,1:nf); XProj_PLS = X * U_pls(:,1:nf);
Ypred_PLS = classify(XtsProj_PLS,XProj_PLS,Y);
% OPLS
U_opls = gen_eig(Cxy*Cxy',Cxx);
XtsProj_OPLS = Xts * U_opls(:,1:nf); XProj_OPLS = X * U_opls(:,1:nf);
Ypred_OPLS = classify(XtsProj_OPLS,XProj_OPLS,Y);
% CCA
U_cca = gen_eig(Cxy*inv(Cyy)*Cxy',Cxx);
XtsProj_CCA = Xts * U_cca(:,1:nf); XProj_CCA = X * U_cca(:,1:nf);
Ypred_CCA = classify(XtsProj_CCA,XProj_CCA,Y);

%% Kernels for Nonlinear methods
sigmax = estimateSigma(X,X);
K = kernel('rbf',X,X,sigmax); Kc=kernelcentering(U_kpca.Ktrain);
Ktest=kernel('rbf',X,Xts,sigmax); Kctest=kernelcentering(Ktest,Kc);
% KPCA
npmax = min([nf,size(X,1)]);
U_kpca = eigs(Kc,npmax);
```

```
XProj_KPCA = Kc' * U_kpca; XtsProj_KPCA  = Kctest' * U_kpca;
Ypred_KPCA = classify(XtsProj_KPCA,XProj_KPCA,Y);
% KPLS
U_kpls=svds(Kc*Y,nf);
XProj_KPLS = Kc' * U_kpls; XtsProj_KPLS  = Kctest' * U_kpls;
Ypred_KPLS = classify(XtsProj_KPLS,XProj_KPLS,Y);
% KOPLS
npmax = min([nf,max(Y)]);
U_kopls = gen_eig(K*Y*Y'*K,K'*K,npmax);
XProj_OKPLS = Kc' * U_kopls; XtsProj_OKPLS  = Kctest' * U_kopls;
Ypred_OKPLS = classify(XtsProj_OKPLS,XProj_OKPLS,Y);
% KCCA
npmax = min([nf,size(X,1)]);
U_kcca = gen_eig(Kc*Y*inv(Cyy)*Y'*Kc,Cxx,npmax);
XProj_KCCA = Kc' * U_kcca; XtsProj_KCCA  = Kctest' * U_kcca;
Ypred_KCCA = classify(XtsProj_KCCA,XProj_KCCA,Y);
```

Listing 12.1 MATLAB code of a KMVA application.

The data were generated from three noisy parabolas, so that a certain overlap exists between classes. For the first extracted feature we show its sample variance, the largest correlation and covariance that can be achieved with a linear transformation of the output, and the optimum MSE of the feature when used to approximate the target. All these are shown above the scatter plot. The first projection for each method maximizes the variance (PCA), covariance (PLS), and correlation (CCA), while OPLS finds the MMSE projection. However, since these methods can just perform linear transformations of the data, they are not able to capture any nonlinear relations between the input variables. For illustrative purposes, we have included in Figure 12.1 the projections obtained using KMVA methods with an RBF kernel. Input data were normalized to zero mean and unit variance, and the kernel width σ was selected as the median of all pairwise distances between samples (Blaschko *et al.*, 2011). The same σ has been used for all methods, so that features are extracted from the same mapping of the input data. We can see that the nonlinear mapping improves class separability.

Figure 12.1 Toy example of MVA methods in a three-class problem. Top: features extracted using the training set. Figure shows the variance (var), MSE when the projection is used to approximate $\mathbf{y}$, and the largest covariance (cov) and correlation (corr) achievable using a linear projection of the target data. The first thing we see is that linear methods (top row) perform a rotation on the original data, while the features extracted by the nonlinear methods (second row) are based on more complicated transforms. There is a clear difference between the unsupervised methods (PCA and KPCA) and the supervised methods, which try to pull apart the data coming from different classes. Bottom: classification of test data using a linear classifier trained in the transformed domain, numerical performance is given by the overall accuracy (OA). Since in this case there is no dimensionality reduction, all the linear methods obtain the same accuracy. However, the nonlinear methods obtain very different results. In general, nonlinear methods obtain better results than the linear ones. A special case is KPCA, which obtains very poor results. KPCA is unsupervised and in this case looking for high-variance dimensions is not useful in order to discern between the classes. KPLS obtains lower accuracy than KOPLS and KCCA, which in turn obtain a similar result since their formulation is almost equivalent (only an extra normalization step is performed by KCCA).

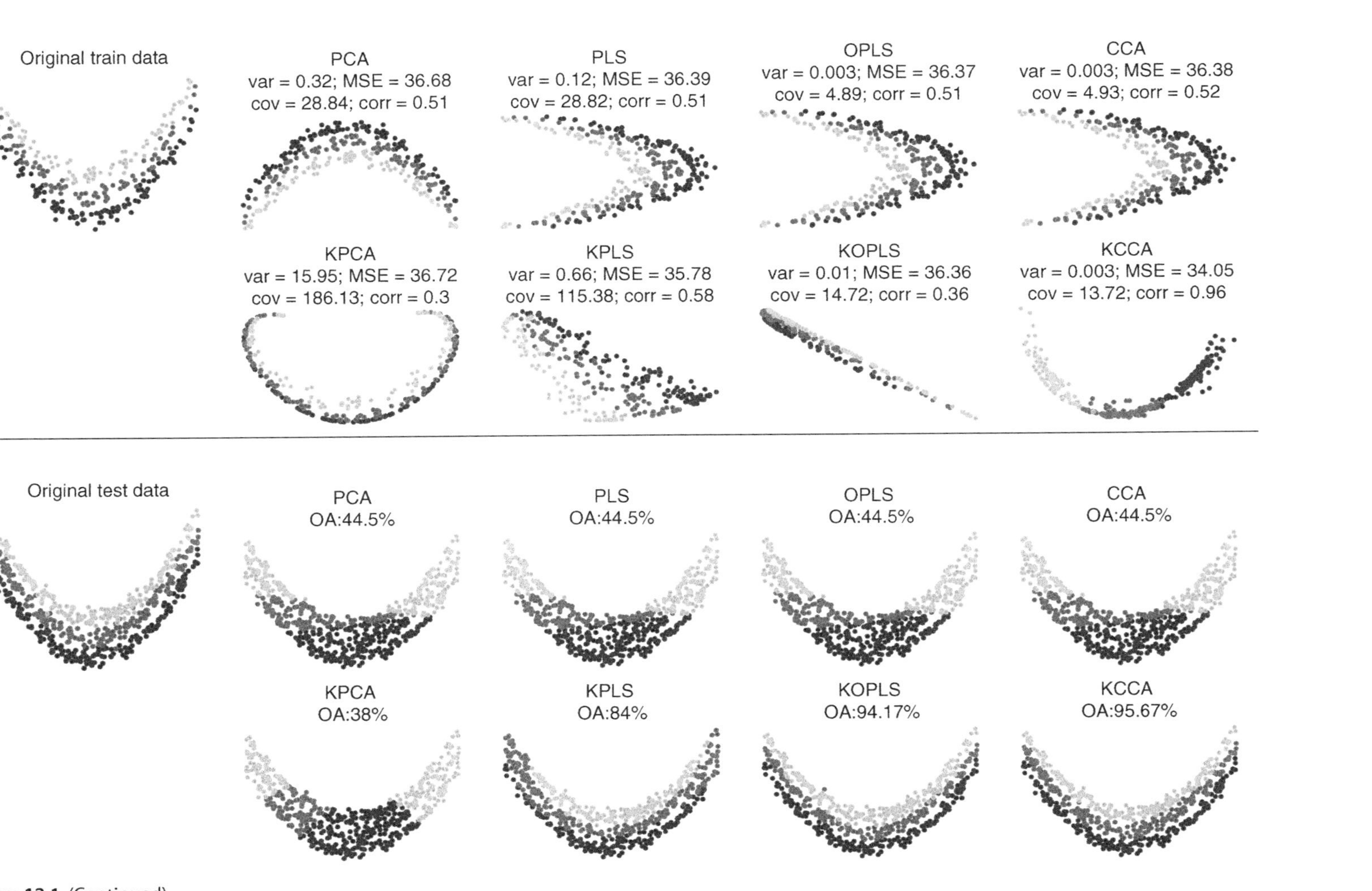

Figure 12.1 (Continued)

Satellite Image Classification

Nowadays, multi- and hyperspectral sensors mounted on satellite and airborne platforms may acquire the reflected energy by the Earth with high spatial detail and in several spectral wavelengths. Here, we pay attention to the performance of several KMVA methods for image segmentation of hyperspectral images (Arenas-García and Petersen, 2009). The input data $\boldsymbol{X}$ is the spectral radiance obtained for each pixel (sample) of dimension d equal to the number of spectral channels considered, and the output target variable $\boldsymbol{Y}$ corresponds to the class label of each particular pixel.

The data used is the standard AVIRIS image obtained from NW Indiana's Indian Pine test site in June 1992.[6] The 20 noisy bands corresponding to the region of water absorption have been removed, and we used pixels of dimension $d = 200$ spectral bands. The high number of narrow spectral bands induces a high collinearity among features. Discriminating among the major crop classes in the area can be very difficult. The image is 145×145 pixels and contains 16 quite unbalanced classes (ranging from 20 to 2468 pixels each). Among the available $N = 10\,366$ labeled pixels, 20% were used to train the feature extractors, where the remaining 80% were used to test the methods. The discriminative properties of the extracted features were assessed using a simple classifier constructed with a linear model adjusted with LS plus a "winner takes all" activation function.

The result of the classification accuracy during the test for different number n_{f} of extracted features is shown in Figure 12.2. For linear models, OPLS performs better than all other methods for any number of extracted features. Even though CCA provides similar results for $n_{\mathrm{f}} = 10$, it involves a slightly more complex generalized eigenproblem. When the maximum number of projections is used, all methods result in the same error. Nevertheless, while PCA and PLS2 require $n_{\mathrm{f}} = 200$ features (i.e., the dimensionality of the input space), CCA and OPLS only need $n_{\mathrm{f}} = 15$ features to achieve virtually the same performance.

We also considered nonlinear KPCA, KPLS2, KOPLS, and KCCA, using an RBF kernel. The width parameter of the kernel was adjusted with a fivefold cross-validation over the training set. The same conclusions obtained for the linear case apply also to MVA methods in kernel feature space. The features extracted by KOPLS allow us to achieve a slightly better overall accuracy than KCCA, and both methods perform significantly better than KPLS2 and KPCA. In the limit of n_{f}, all methods achieve similar accuracy. The classification maps obtained for $n_{\mathrm{f}} = 10$ confirm these conclusions: higher accuracies lead to smoother maps and smaller error in large spatially homogeneous vegetation covers.

Finally, we transformed all the 200 original bands into a lower dimensional space of 18 features and assessed the quality of the extracted features by standard PCA, SNR/MNF, KPCA, and KSNR approaches. Figure 12.3 shows the extracted features in descending order of relevance. In this unsupervised setting it is evident that incorporating the signal-to-noise relations in the extraction with KSNR provides some clear advantages in the form of more "noise-free" features than the other methods.

---

6 The calibrated data are available online (along with detailed ground-truth information) from http://dynamo.ecn.purdue.edu/~biehl/MultiSpec.

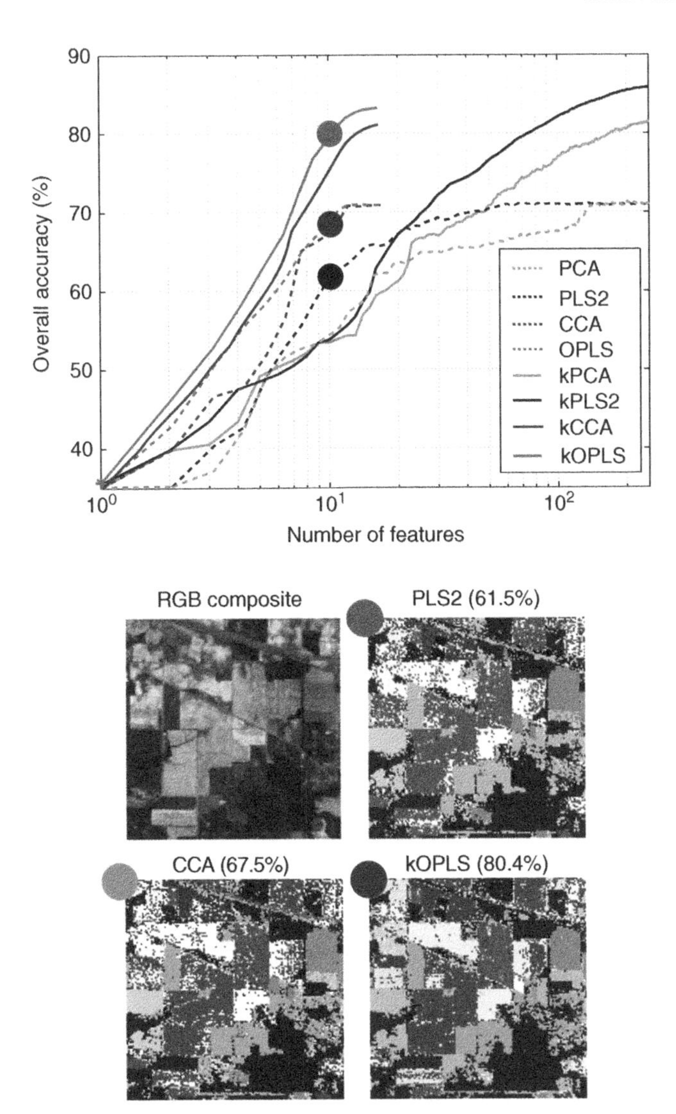

Figure 12.2 Average classification accuracy (percent) for linear and KMVA methods as a function of the number of extracted features, along some classification maps for the case of $n_f = 10$ extracted features.

12.3 Feature Extraction with Kernel Dependence Estimates

In this section we review the connections of feature extraction when using dependency estimation as optimization criterion. First, we present a popular method for dependency estimation based on kernels, the HSIC (Gretton *et al.*, 2005c). We will analyze the connections between HSIC and classical feature extraction methods. After that, we will review two methods that employ HSIC as optimization criterion to find interesting

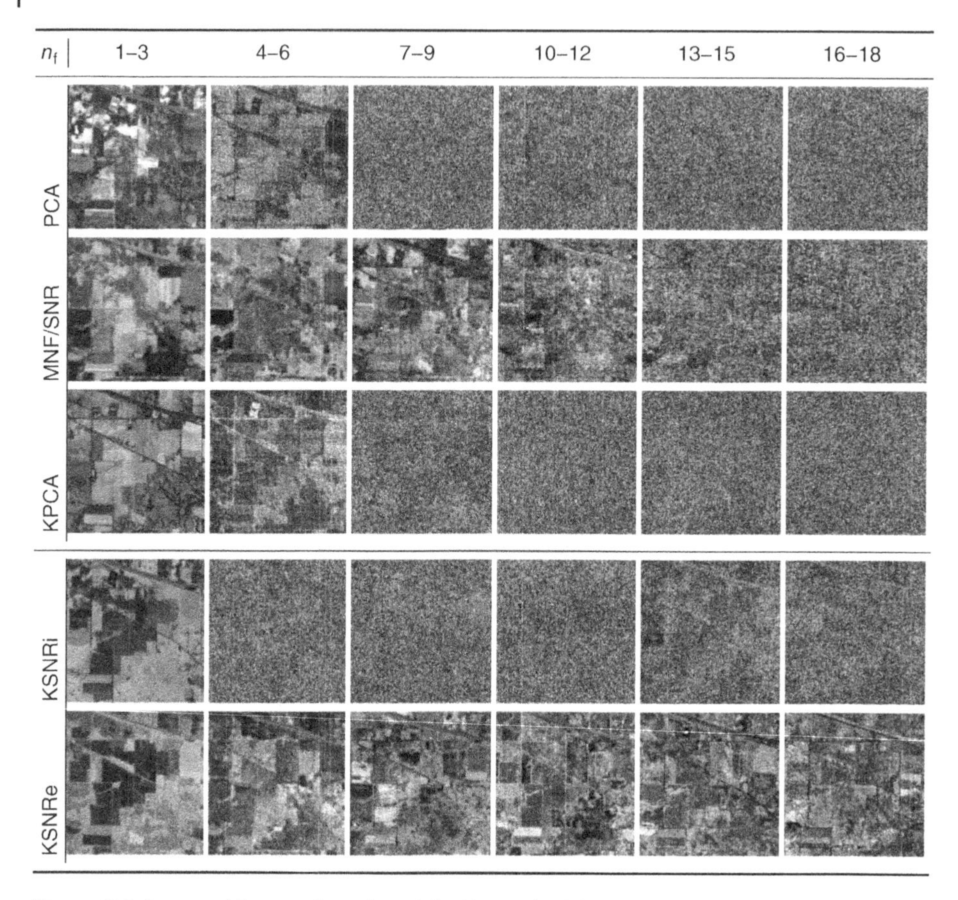

Figure 12.3 Extracted features from the original image by PCA, MNF/SNR, KPCA, standard (implicit) KSNR, and explicit KSNR in the kernel space for the first $n_f = 18$ components, plotted in RGB composite of three top components ordered in descending importance (images converted to grey scale for the book).

features: the Hilbert–Schmidt component analysis (HSCA) (Daniušis and Vaitkus, 2009) and the kernel dimensionality reduction (KDR) (Fukumizu *et al.* 2003, 2004). A related problem found in DSP is the BSS problem; here, one aims to find a transformation that separates the input observed signal into a set of independent signal components. We will review different BSS methods based on kernels.

12.3.1 Feature Extraction Using Hilbert–Schmidt Independence Criterion

MVA methods optimize different criteria mostly based on second-order relations, which is a limited approach in problems exhibiting higher order nonlinear feature relations. Here, we analyze different methods to deal with higher order statistics using kernels. The idea is to go beyond simple *association* statistics. We will build our discussion on the HSIC (Gretton *et al.*, 2005c), which is a measure of dependence between random variables. The goal of this measure is to discern if sensed variables are related and to measure the strength of such relations. By using this measure, we can look for features that maximize directly the amount of information extracted from the data instead of using a proxy like correlation.

Hilbert–Schmidt Independence Criterion

The HSIC method measures cross-covariances in an adequate RKHS by using the entire spectrum of the cross-covariance operator. As we will see, the HSIC empirical estimator has low computational burden and nice theoretical and practical properties. To fix notation, let us consider two spaces $\mathcal{X} \subseteq \mathbb{R}^{d_x}$ and $\mathcal{Y} \subseteq \mathbb{R}^{d_y}$, where input and output observations $(\boldsymbol{x}, \boldsymbol{y})$ are sampled from distribution $\mathbb{P}_{\boldsymbol{xy}}$. The covariance matrix is

$$C_{\boldsymbol{xy}} = \mathbb{E}_{\boldsymbol{xy}}(\boldsymbol{x}\boldsymbol{y}^{\mathrm{T}}) - \mathbb{E}_{\boldsymbol{x}}(\boldsymbol{x})\mathbb{E}_{\boldsymbol{y}}(\boldsymbol{y}^{\mathrm{T}}), \tag{12.4}$$

where $\mathbb{E}_{\boldsymbol{xy}}$ is the expectation with respect to $\mathbb{P}_{\boldsymbol{xy}}$, and $\mathbb{E}_{\boldsymbol{x}}$ is the marginal expectation with respect to $\mathbb{P}_{\boldsymbol{x}}$ (here and below, we assume that all these quantities exist). First-order dependencies between variables are included into the covariance matrix, and the Hilbert–Schmidt norm is a statistic that effectively summarizes the content of this matrix. The squared sum of its eigenvalues γ_i is equal to the square of this norm:

$$\|C_{\boldsymbol{xy}}\|_{\mathrm{HS}}^2 = \sum_i \gamma_i^2. \tag{12.5}$$

This quantity is zero only in the case where there is no first-order dependence between $\boldsymbol{x}$ and $\boldsymbol{y}$. Note that the Hilbert–Schmidt norm is limited to the detection of first-order relations, and thus more complex (higher order effects) cannot be captured.

The previous notion of covariance operator in the field of kernel functions and measures was proposed in (Gretton *et al.*, 2005c). Essentially, assume a (non-linear) mapping $\boldsymbol{\phi} : \mathcal{X} \to \mathcal{F}$ in such a way that the dot product between features is given by a positive definite kernel function $K_x(\boldsymbol{x}, \boldsymbol{x}') = \langle \boldsymbol{\phi}(\boldsymbol{x}), \boldsymbol{\phi}(\boldsymbol{x}') \rangle$. The feature space $\mathcal{F}$ has the structure of an RKHS. Assume another mapping $\boldsymbol{\psi} : \mathcal{Y} \to \mathcal{G}$ with associated kernel $K_y(\boldsymbol{y}, \boldsymbol{y}') = \langle \boldsymbol{\psi}(\boldsymbol{y}), \boldsymbol{\psi}(\boldsymbol{y}') \rangle$. Under these conditions, a conditions, a cross-covariance operator between these mappings can be defined, similar to the covariance matrix in Equation 12.4. The cross-covariance is a linear operator $C_{\boldsymbol{xy}} : \mathcal{G} \to \mathcal{F}$ of the form

$$C_{\boldsymbol{xy}} = \mathbb{E}_{\boldsymbol{xy}}[(\boldsymbol{\phi}(\boldsymbol{x}) - \mu_x) \otimes (\boldsymbol{\psi}(\boldsymbol{y}) - \mu_y)], \tag{12.6}$$

where $\otimes$ is the tensor product, $\mu_x = \mathbb{E}_{\boldsymbol{x}}[\boldsymbol{\phi}(\boldsymbol{x})]$, and $\mu_y = \mathbb{E}_{\boldsymbol{y}}[\boldsymbol{\psi}(\boldsymbol{y})]$. See more details in Baker (1973) and Fukumizu *et al.* (2004). The HSIC is defined as the squared norm of the cross-covariance operator, $\|C_{xy}\|^2_{\mathrm{HS}}$, and it can be expressed in terms of kernels (Gretton *et al.*, 2005c):

$$\begin{aligned}\mathrm{HSIC}(\mathcal{F}, \mathcal{G}, \mathbb{P}_{xy}) &= \|C_{xy}\|^2_{\mathrm{HS}} \\ &= \mathbb{E}_{xx'YY'}[K_x(\boldsymbol{x}, \boldsymbol{x}')K_y(\boldsymbol{y}, \boldsymbol{y}')] \\ &\quad + \mathbb{E}_{xx'}[K_x(\boldsymbol{x}, \boldsymbol{x}')]\mathbb{E}_{yy'}[K_y(\boldsymbol{y}, \boldsymbol{y}')] \\ &\quad - 2\mathbb{E}_{xy}[\mathbb{E}_{x'}[K_x(\boldsymbol{x}, \boldsymbol{x}')]\mathbb{E}_{y'}[K_y(\boldsymbol{y}, \boldsymbol{y}')]],\end{aligned}$$

where $\mathbb{E}_{xx'yy'}$ is the expectation over both $(\boldsymbol{x}, \boldsymbol{y}) \sim \mathbb{P}_{xy}$ and an additional pair of variables $(\boldsymbol{x}', \boldsymbol{y}') \sim \mathbb{P}_{xy}$ drawn independently according to the same law.

Now, given a sample dataset $Z = \{(\boldsymbol{x}_1, \boldsymbol{y}_1), \dots, (\boldsymbol{x}_N, \boldsymbol{y}_N)\}$ of size N drawn from $\mathbb{P}_{xy}$, an empirical estimator of HSIC is (Gretton *et al.*, 2005c)

$$\mathrm{HSIC}(\mathcal{F}, \mathcal{G}, \mathbb{P}_{xy}) = \frac{1}{N^2}\mathrm{Tr}(\boldsymbol{K}_x\boldsymbol{H}\boldsymbol{K}_y\boldsymbol{H}) = \frac{1}{N^2}\mathrm{Tr}(\boldsymbol{H}\boldsymbol{K}_x\boldsymbol{H}\,\boldsymbol{K}_y),$$

where Tr is the trace (the sum of the diagonal entries), $\boldsymbol{K}_x$ and $\boldsymbol{K}_y$ are the kernel matrices for the data $\boldsymbol{x}$ and the labels $\boldsymbol{y}$ respectively, and $H_{ij} = \delta_{ij} - (1/N)$ centers the data and the label features in $\mathcal{F}$ and $\mathcal{G}$. Here, δ represents the Kronecker symbol, where $\delta_{ij} = 1$ if $i = j$, and zero otherwise.

Summarizing, HSIC (Gretton *et al.*, 2005c) is a simple yet very effective method to estimate statistical dependence between random variables. HSIC corresponds to estimating the norm of the cross-covariance in $\mathcal{F}$, whose empirical (biased) estimator is HSIC $:= [1/(N-1)^2]\mathrm{Tr}(\boldsymbol{K}_x\boldsymbol{K}_y)$, where $\boldsymbol{K}_x$ and $\boldsymbol{K}_y$ are kernels working on data from sets $\mathcal{X}$ and $\mathcal{Y}$. It can be shown that, if the RKHS kernel is universal, such as the RBF or Laplacian kernels, HSIC asymptotically tends to zero when the input and output data are independent, and HSIC $= 0$ if $\mathbb{P}_x = \mathbb{P}_y$. Note that, in practice, the actual HSIC is the Hilbert–Schmidt norm of an operator mapping between potentially infinite-dimensional spaces, and thus would give rise to an infinitely large matrix. However, due to the kernelization, the empirical HSIC only depends on computable matrices of size $N \times N$. We will come back to the issue of computational cost later.

Relation between Hilbert–Schmidt Independence Criterion and Maximum Mean Discrepancy

In Chapter 11 we introduced the MMD for hypothesis testing. We can show that MMD is related to HSIC. Let us define the product space $\mathcal{F} \times \mathcal{G}$ with a kernel $\langle \boldsymbol{\phi}(\boldsymbol{x}, \boldsymbol{y}), \boldsymbol{\phi}(\boldsymbol{x}', \boldsymbol{y}') \rangle = K((\boldsymbol{x}, \boldsymbol{y}), (\boldsymbol{x}', \boldsymbol{y}')) = K(\boldsymbol{x}, \boldsymbol{y})L(\boldsymbol{x}', \boldsymbol{y}')$, and the *mean elements* are given by

$$\langle \mu_{x,y}, \boldsymbol{\phi}(\boldsymbol{x}, \boldsymbol{y}) \rangle = \mathbb{E}_{x',y'}\langle \boldsymbol{\phi}(\boldsymbol{x}', \boldsymbol{y}'), \boldsymbol{\phi}(\boldsymbol{x}, \boldsymbol{y}) \rangle = \mathbb{E}_{x,x'}K(\boldsymbol{x}, \boldsymbol{x}')L(\boldsymbol{y}, \boldsymbol{y}'),$$

and

$$\langle \mu_{x\perp y}, \boldsymbol{\phi}(\boldsymbol{x}, \boldsymbol{y}) \rangle = \mathbb{E}_{x',y'}\langle \boldsymbol{\phi}(\boldsymbol{x}', \boldsymbol{y}'), \boldsymbol{\phi}(\boldsymbol{x}, \boldsymbol{y}) \rangle = \mathbb{E}_{x'}K(\boldsymbol{x}, \boldsymbol{x}')\mathbb{E}_{y'}L(\boldsymbol{y}, \boldsymbol{y}').$$

Therefore, the MMD between these two mean elements is

$$\text{MMD}(\mathbb{P}, \mathbb{P}_x, \mathbb{P}_y, \mathcal{F} \times \mathcal{G}) = \|\mu_{x,y} - \mu_{x\perp y}\|^2_{\mathcal{F}\times\mathcal{G}} = \text{HSIC}(\mathbb{P}, \mathcal{F}, \mathcal{G}).$$

Relation between Hilbert–Schmidt Independence Criterion and Other Kernel Dependence Measures

HSIC was not the first kernel dependence estimate presented in the literature. Actually, Bach and Jordan (2002) derived a regularized correlation operator from the covariance and cross-covariance operators, and its largest singular value (the KCCA) was used as a statistic to test independence. Later, Gretton *et al.* (2005a) proposed the largest singular value of the cross-covariance operator as an efficient alternative that needs no regularization. This test is called the constrained covariance (COCO) (Gretton *et al.*, 2005a). A variety of empirical kernel quantities derived from bounds on the mutual information that hold near independence were also proposed, namely the kernel generalized variance (KGV) and the kernel mutual information (kMI) (Gretton *et al.*, 2005b). Later, HSIC (Gretton *et al.*, 2005c) was introduced, hence extending COCO by using the entire spectrum of the cross-covariance operator, not just the largest singular value. As we have seen, a related statistic is the MMD criterion, which tests for statistical differences between embeddings of probability distributions into RKHS. When MMD is applied to the problem of testing for independence, the test statistic reduces to HSIC. A simple comparison between all the methods treated in this section is given in Figure 12.4. Listing 12.2 gives a simple MATLAB implementation comparing Pearson's correlation and HSIC in this particular example. The book's repository provides MATLAB tools and links to relevant toolboxes on (kernel) dependence estimation.

```
% Number of examples
N = 1000;
% Distribution 1: correlation/linear association happens
X = rand(N,1); Y = X + 0.25*rand(N,1); X = X + 0.25*rand(N,1);
% Distribution 2: no correlation, but dependence happens
t = 2*pi*(rand(N,1)-0.5);
X = cos(t) + 0.25*rand(N,1); Y = sin(t) + 0.25*rand(N,1);
% Distribution 3: neither correlation nor dependence exist
X = 0.25*rand(N,1); Y = 0.25*rand(N,1);
% Get dependence estimates
C        = corr(X,Y); % Correlation
HSIClin  = hsic(X,Y,'lin',[]);
sigmax   = estimateSigma(X,X); HSICrbf = hsic(X,Y,'rbf',sigma);
MI       = mutualinfo(X,Y);

function h = hsic(X,Y,ker,sigma)

Kx = kernelmatrix(ker,X,X,sigma); Kxc = kernelcentering(Kx);
Ky = kernelmatrix(ker,Y,Y,sigma); Kyc = kernelcentering(Ky);
h  = trace(Kxc*Kyc)/(size(Kxc,1).^2);

function m = mutualinfo(X,Y)

binsx = round(sqrt(size(X,1))); [hh rr] = hist(X,binsx);
pp = hh/sum(hh); h1 = -sum(pp.*log2(pp)) + log2(rr(3)-rr(2));
binsy = round(sqrt(size(Y,1))); [hh rr] = hist(Y,binsy);
pp = hh/sum(hh); h2 = -sum(pp.*log2(pp)) + log2(rr(3)-rr(2));
```

```
[hh rr] = hist3([X Y],[binsx binsy]); pp = hh(:)/sum(hh(:));
h12 = -sum(pp.*log2(pp)) + log2((rr{1}(3)-rr{1}(2))*(rr{1}(2)-rr{2}(2)));
MI3 = h1 + h2 - h12;
```

Listing 12.2 MATLAB code comparing correlation and dependence estimation, using Pearson's correlation coefficient, mutual information, and the HSIC with both the linear and the RBF kernels.

Pearson's R	0.9571	0.0075	−0.0646
Mutual information	1.1564	0.7220	0.0190
HSIC (linear kernels)	0.8979	0.0000	0.0000
HSIC (RBF kernels)	0.0560	0.0048	0.0006
MMD	0.0018	0.0018	0.0058
COCO	0.2205	0.0715	0.0694
KGV	8.3427	6.0186	4.2353
kMI ($\theta = 1/2$)	7.8198	5.5939	3.8900

Figure 12.4 Dependence estimates for three examples revealing (left) high correlation (and hence high dependence), (middle) high dependence but null correlation, and (right) zero correlation and dependence. The Pearson correlation coefficient R and linear HSIC only capture second-order statistics (linear correlation), while the rest capture in general higher order dependences. Note that, for MMD, the higher the more divergent (independent), while KGV upper bounds kMI and mutual information.

Relations between Hilbert–Schmidt Independence Criterion and the Kernel Multivariate Analysis Framework

HSIC constitutes an interesting framework to study KMVA methods for dimensionality reduction. We summarize the relations between the methods under the HSIC perspective in Table 12.2. For instance, HSIC reduces to PCA for the case of linear kernels. Recall that PCA reduces to find an orthogonal projection matrix $\boldsymbol{V}$ such that the projected data, $\boldsymbol{XV}$, preserves the maximum variance; that is, $\max_{\boldsymbol{V}} \|\boldsymbol{XV}\|^2 = \max_{\boldsymbol{V}} \text{Tr}((\boldsymbol{XV})^{\text{T}}(\boldsymbol{XV}))$ subject to $\boldsymbol{V}^{\text{T}}\boldsymbol{V} = \boldsymbol{I}$, which may be written as follows:

$$\max_{\boldsymbol{V}} \text{Tr}(\boldsymbol{V}^{\text{T}}\boldsymbol{X}^{\text{T}}\boldsymbol{XV}) = \max_{\boldsymbol{V}} \text{Tr}(\boldsymbol{X}^{\text{T}}\boldsymbol{X}\ \ \boldsymbol{VV}^{\text{T}}) \quad \text{s.t.} \quad \boldsymbol{V}^{\text{T}}\boldsymbol{V} = \boldsymbol{I}.$$

Therefore, PCA can be interpreted as finding $\boldsymbol{V}$ such that its linear kernel has maximum dependence with the kernel obtained from the original data. Assuming that $\boldsymbol{X}$ is again centered, there is no need for including the centering matrix $\boldsymbol{H}$ in the HSIC estimate.

From HSIC, one can also reach KPCA (Schölkopf *et al.*, 1998) by using a kernel matrix $\boldsymbol{K}$ for the data $\boldsymbol{X}$ and requiring the second random variable $\boldsymbol{y}$ to be orthogonal and unit norm; that is:

$$\max_{\boldsymbol{Y}} \text{Tr}(\boldsymbol{KYY}^{\text{T}}) = \max_{\boldsymbol{Y}} \text{Tr}(\boldsymbol{Y}^{\text{T}}\boldsymbol{KY}) \quad \text{s.t.} \quad \boldsymbol{Y}^{\text{T}}\boldsymbol{Y} = \boldsymbol{I},$$

Table 12.2 Relation between dimensionality reduction methods and HSIC, HSIC $:= \mathrm{Tr}(\boldsymbol{H}\boldsymbol{K}_x\boldsymbol{H}\boldsymbol{K}_y)$, where all techniques use the orthogonality constraint $\boldsymbol{Y}^{\mathrm{T}}\boldsymbol{Y} = \boldsymbol{I}$.

Method	K_x	K_y	Comments
PCA (Jolliffe, 1986)	$\boldsymbol{X}\boldsymbol{X}^{\mathrm{T}}$	$\boldsymbol{Y}\boldsymbol{Y}^{\mathrm{T}}$	—
KPCA (Schölkopf *et al.*, 1998)	$\boldsymbol{K}_x$	$\boldsymbol{Y}\boldsymbol{Y}^{\mathrm{T}}$	Typically RBF kernels used
Maximum variance unfolding (MVU) (Song *et al.*, 2007)	$\boldsymbol{K}_x$	$\boldsymbol{L}\boldsymbol{L}^{\mathrm{T}}$	Subject to positive definiteness $\boldsymbol{K}_x \succeq 0$, preservation of local distance structure $\boldsymbol{K}_{ii} + \boldsymbol{K}_{jj} - 2\boldsymbol{K}_{ij} = d_{ij}^2$, and $\boldsymbol{L}$ is a coloring matrix accounting for the side information
Multidimensional scaling (MDS) (Cox and Cox, 1994)	$-\frac{1}{2}\boldsymbol{H}\boldsymbol{D}\boldsymbol{H}$	$\boldsymbol{Y}\boldsymbol{Y}^{\mathrm{T}}$	$\boldsymbol{D}$ is the matrix of all pairwise Euclidean distances
Isomap (Tenenbaum *et al.*, 2000)	$-\frac{1}{2}\boldsymbol{H}\boldsymbol{D}^g\boldsymbol{H}$	$\boldsymbol{Y}\boldsymbol{Y}^{\mathrm{T}}$	$\boldsymbol{D}^g$ is the matrix of all pairwise geodesic distances
Locally linear embedding (LLE) (Roweis and Saul, 2000)	$\boldsymbol{L}^{\dagger}$, or $\lambda_{\max}\boldsymbol{I} - \boldsymbol{L}$	$\boldsymbol{Y}\boldsymbol{Y}^{\mathrm{T}}$	$\boldsymbol{L} = (\boldsymbol{I}-\boldsymbol{V})(\boldsymbol{I}-\boldsymbol{V})^{\mathrm{T}}$, where $\boldsymbol{V}$ is the matrix of locally embedded weights
Laplacian eigenmaps (Belkin and Niyogi, 2003)	$\boldsymbol{L}^{\dagger}$, or $\lambda_{\max}\boldsymbol{I} - \boldsymbol{L}$	$\boldsymbol{Y}\boldsymbol{Y}^{\mathrm{T}}$	$\boldsymbol{L} = \boldsymbol{D} - \boldsymbol{W}$, where $\boldsymbol{W}$ is a positive symmetric affinity matrix and $\boldsymbol{D}$ is the corresponding degree matrix

where $\boldsymbol{Y}$ is any real-valued matrix of size $N \times d$. Note that KPCA does not require any other constraint imposed on $\boldsymbol{Y}$, but for other KMVA methods, as well as for clustering and metric learning techniques, one may require to design $\boldsymbol{Y}$ to include appropriate restrictions.

Maximizing HSIC has also been used in dimensionality reduction with (colored) MVU (Song *et al.*, 2007). This manifold learning method maximizes HSIC between source and target data, and accounts for the manifold structure by imposing local distance constraints on the kernel. In the original version of the MVU, a kernel $\boldsymbol{K}$ is learned via maximizing $\mathrm{Tr}(\boldsymbol{K})$ subject to some constraints. Assuming a similarity/dissimilarity matrix $\boldsymbol{B}$, one can modify the original MVU by maximizing $\mathrm{Tr}(\boldsymbol{K}\boldsymbol{B})$ instead of $\mathrm{Tr}(\boldsymbol{K})$ subject to the same constraints. This way, one simultaneously maximizes the dependence between the learned kernel $\boldsymbol{K}$ and the kernel $\boldsymbol{B}$ storing our side information. Note that, in this formulation, we center matrix $\boldsymbol{K}$ via one of the constraints in the problem, and thus can exclude matrix $\boldsymbol{H}$ in the empirical estimation of HSIC.

The HSIC framework can also help us to analyze relations to standard dimensionality reduction methods like MDS (Cox and Cox, 1994), Isomap (Tenenbaum *et al.*, 2000), LLE (Roweis and Saul, 2000) and Laplacian eigenmaps (Belkin and Niyogi, 2003) depending on the definition of the kernel $\boldsymbol{K}$ used in Equation 12.3.1. Ham *et al.* (2004) pointed out the relation between KPCA and these methods.

Hilbert–Schmidt Component Analysis

The HSIC empirical estimate can be specifically incorporated in feature extraction schemes. For example, the so-called HSCA method in Daniušis and Vaitkus (2009) iteratively seeks for projections that maximize dependence with the target variables and simultaneously minimize the dependence with previously extracted features, both in HSIC terms. This can be seen as a Rayleigh coefficient that leads to the iterative resolution of the following generalized eigendecomposition problem:

$$\boldsymbol{K}_x\boldsymbol{K}_y\boldsymbol{K}_x\boldsymbol{\Lambda} = \lambda\boldsymbol{K}_x\boldsymbol{K}_f\boldsymbol{K}_x\boldsymbol{\Lambda}, \tag{12.7}$$

where $\boldsymbol{K}_f$ is a kernel matrix of already extracted projections in the previous iteration. Note that if one is only interested in maximizing source–target dependence in HSIC terms, and uses a linear kernel for the targets $\boldsymbol{K}_y = \boldsymbol{Y}\boldsymbol{Y}^{\mathrm{T}}$, the problem reduces to

$$\boldsymbol{K}_x\boldsymbol{K}_y\boldsymbol{K}_x\boldsymbol{\Lambda} = \lambda\boldsymbol{K}_x\boldsymbol{K}_x\boldsymbol{\Lambda}, \tag{12.8}$$

which interestingly is the same problem solved by KOPLS (see Table 12.1). The equivalence means that features extracted with KOPLS are those that maximize statistical dependence (measured by HSIC) between the projected source data and the target data (Izquierdo-Verdiguier *et al.*, 2012).

Kernel Dimensionality Reduction

KDR is a supervised feature extraction method that seeks a linear transformation of the data such that it maximizes the conditional HSIC on the labels. Notationally, the input data matrix $\boldsymbol{X} \in \mathbb{R}^{N\times d}$, the output (label) matrix is $\boldsymbol{Y} \in \mathbb{R}^{N\times m}$, and $\boldsymbol{W} \in \mathbb{R}^{d\times r}$ is a projection matrix from the d-dimensional space to a r-dimensional space, $r \leq d$. Hence, the linear projection is $\boldsymbol{Z} = \boldsymbol{X}\boldsymbol{W}$, which is constrained to be orthogonal; that is, $\boldsymbol{W}^{\mathrm{T}}\boldsymbol{W} = \boldsymbol{I}$. The optimal $\boldsymbol{W}$ in HSIC terms is obtained by solving the constrained problem

$$\max_{\boldsymbol{W}} \mathrm{HSIC}(\boldsymbol{X}\boldsymbol{W}, \boldsymbol{Y}) = \max_{\boldsymbol{W}} \mathrm{Tr}(\tilde{\boldsymbol{K}}_{\boldsymbol{X}\boldsymbol{W}}\tilde{\boldsymbol{K}}_{\boldsymbol{Y}}) \quad \text{s.t.} \quad \boldsymbol{W}^{\mathrm{T}}\boldsymbol{W} = \boldsymbol{I}.$$

This approach has been followed for both supervised and unsupervised settings:

- Fukumizu *et al.* (2003, 2004) introduced KDR under the theory of covariance operators (Baker, 1973). KDR reduces to the KGV introduced by Gretton *et al.* (2005b) as a contrast function for ICA, in which the goal is to minimize mutual information. They showed that KGV is in fact an approximation of the mutual information among the recovered sources around the factorized distributions. The interest in KDR is different: the goal is to maximize the mutual information as a good proxy to the problem of dimensionality reduction. Interestingly, the computational problem involved in KDR is the same as in KICA (Bach and Jordan, 2002).
- The unsupervised KDR (Wang *et al.*, 2010) reduces to seeking $\boldsymbol{W}$ such that the signal $\boldsymbol{X}$ and its (regression-based) approximation $\hat{\boldsymbol{X}} = f(\boldsymbol{X}\boldsymbol{W})$ are mutually independent, given the projection; that is $\boldsymbol{X} \perp \hat{\boldsymbol{X}} \mid \boldsymbol{X}\boldsymbol{W}$. The unsupervised KDR method then reduces to

$$\max_{\boldsymbol{W}} \mathrm{HSIC}(\boldsymbol{X}\boldsymbol{W}, \boldsymbol{Y}) = \max_{\boldsymbol{W}} \mathrm{Tr}(\tilde{\boldsymbol{K}}_{\boldsymbol{X}}\tilde{\boldsymbol{K}}_{\boldsymbol{X}\boldsymbol{W}}) \quad \text{s.t.} \quad \boldsymbol{W}^{\mathrm{T}}\boldsymbol{W} = \boldsymbol{I}.$$

The problem is typically solved by gradient-descent techniques with line search, which constrains the projection matrix $\boldsymbol{W}$ to lie on the Grassmann–Stiefel manifold of $\boldsymbol{W}^{\mathrm{T}}\boldsymbol{W} = \boldsymbol{I}$.

Let us illustrate the performance of the supervised KDR compared with other linear dimensionality reduction methods for the Wine dataset, which is a 13-dimensional problem obtained from the UCI repository (http://archive.ics.uci.edu/ml/), where here it is used to illustrate projection onto a 2D subspace. Figure 12.5 shows the projection onto the 2D subspace estimated by each method. This example illustrates that optimizing HSIC to achieve linear separability is an alternative valid approach to maximize input–output covariance (PLS) or maximize correlation (CCA). Figure 12.5 can be reproduced by downloading the corresponding MATLAB toolboxes from the book's repository and executing the script `Demo_Fig_14_4.m` in the supplementary material provided. A simplified version of the code is shown in Listing 12.3.

```
%% Data
[XX Yb] = wine_dataset;
XX = XX';
XX = XX-repmat(min(XX),size(XX,1),1);
XX = XX./repmat(max(XX),size(XX,1),1);
[YY aa] = find(Yb==1);
ii = randperm(size(XX,1));
X = XX(ii(1:size(XX,1)),:);
Y = YY(ii(1:size(XX,1)),:);
Xts = XX(ii(size(XX,1)/2+1:size(XX,1)),:);
Yts = YY(ii(size(XX,1)/2+1:size(XX,1)),:);

%% Feature extraction settings and projections
nf = 2;
% PLS
[U_pls Ypred_PLS]=predictPLS(X,X,Y,nf);
XtsProj_PLS = Xts * U_pls.basis(:,1:nf);
XProj_PLS = X * U_pls.basis(:,1:nf);
Ypred_PLS = classify(XtsProj_PLS,XProj_PLS,Y);
% CCA
U_cca = cca(X,Y,nf);
XtsProj_CCA = Xts * U_cca.basis(:,1:nf);
XProj_CCA = X * U_cca.basis(:,1:nf);
Ypred_CCA = classify(XtsProj_CCA,XProj_CCA,Y);
% KDR
[U_kdr.basis, t] = KernelDeriv(X,Y,2,1,1,1);
XtsProj_KDR = Xts * U_kdr.basis(:,1:nf);
XProj_KDR = X * U_kdr.basis(:,1:nf);
Ypred_KDR = classify(XtsProj_KDR,XProj_KDR,Y);
```

Listing 12.3 MATLAB code comparing PLS, CCA, and KDR for feature extraction.

12.3.2 Blind Source Separation Using Kernels

Extracting sources from observed mixed signals without supervision is the problem of BSS; see Chapter 2 for an introductory review to the field. Several approaches exist in the literature to solve the problem, but most of them have relied on ICA (Comon 1994; Gutmann *et al.* 2014; Hyvärinen *et al.* 2001). ICA seeks for a basis system where the dimensions of the signal are as independent as possible. Kernel dependence

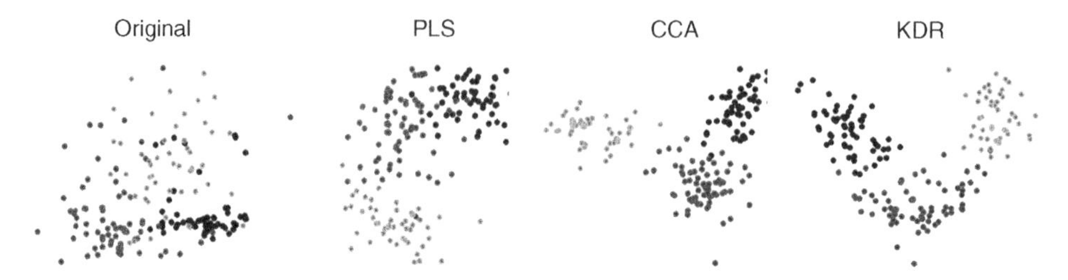

Figure 12.5 Projections obtained by different linear projection methods on the Wine data available at the UCI repository (http://archive.ics.uci.edu/ml/). Black and two different grey intensities represent three different classes.

estimates based either on KCCA or HSIC have given rise to kernelized versions of ICA, leveraging better performance at the cost of higher computational cost. An alternative approach to ICA-like techniques consists of exploiting second-order BSS methods in an RKH,S like finding orthonormal basis of the submanifold formed by kernel-mapped data (Harmeling *et al.*, 2003).

Independent Component Analysis

The instantaneous noise-free ICA model takes the form $\boldsymbol{X} = \boldsymbol{S}\boldsymbol{A}$, where $\boldsymbol{S} \in \mathbb{R}^{N\times s}$ is a matrix containing N observations of s sources, $\boldsymbol{A} \in \mathbb{R}^{s\times s}$ is the mixing matrix (assumed to have full rank), and $\boldsymbol{X} \in \mathbb{R}^{N\times s}$ contains the observed mixtures. We denote as $\boldsymbol{s}$ and $\boldsymbol{x}$ single rows of matrices $\boldsymbol{S}$ and $\boldsymbol{X}$ respectively, and s_i is the ith source in $\boldsymbol{s}$. ICA is based on the assumption that the components s_i of components s_i of $\boldsymbol{s}$, for all $i = 1, \dots, s$ are mutually statistically independent; that is, the observed vector $\boldsymbol{x}$ depends only on the source vector $\boldsymbol{s}$ at each instant and the source samples $\boldsymbol{s}$ are drawn independently and identically from the *pdf* $\mathbb{P}_s$. As a conclusion, the mixture samples $\boldsymbol{x}$ are likewise drawn independently and identically from the *pdf* $\mathbb{P}_x$. The task of ICA is to recover the independent sources via an estimate $\boldsymbol{B} \in \mathbb{R}^{s\times s}$ of the inverse of the mixing matrix $\boldsymbol{A}$ such that the recovered signals $\boldsymbol{Y} = \boldsymbol{S}\boldsymbol{A}\boldsymbol{B} \in \mathbb{R}^{N\times s}$ have mutually independent components. When the sources $\boldsymbol{s}$ are Gaussian, $\boldsymbol{A}$ can be identified up to an ordering and scaling of the recovered sources via a permutation matrix $\boldsymbol{P} \in \mathbb{R}^{s\times s}$ and a scaling matrix $\mathbf{D}$. Very often the mixtures $\boldsymbol{X}$ are pre-whitened via PCA, $\boldsymbol{W} = \boldsymbol{X}\boldsymbol{V} = \boldsymbol{S}\boldsymbol{A}\boldsymbol{V} \in \mathbb{R}^{N\times s}$, so $\mathbb{E}[\boldsymbol{w}\boldsymbol{w}^{\mathrm{T}}] = \boldsymbol{I}$, where $\boldsymbol{W}$ contains the whitened observations. Now, assuming that the sources s_i have zero mean and unit variance, $\boldsymbol{A}\boldsymbol{V}$ is orthogonal and the unmixing model becomes $\boldsymbol{Y} = \boldsymbol{W}\boldsymbol{Q}$, where $\boldsymbol{Q}$ is an orthogonal unmixing matrix ($\boldsymbol{Q}\boldsymbol{Q}^{\mathrm{T}} = \boldsymbol{I}$) and $\boldsymbol{Y} \in \mathbb{R}^{N\times s}$ contains the estimates of the sources. The idea of ICA can also be employed for optimizing CCA-like problems (for instance, see Gutmann *et al.* (2014)), where higher order relations are taken into account in order to find the basis to represent different datasets in a canonical space.

Kernel Independent Component Analysis Using Kernel Canonical Correlation Analysis

The original version of KICA was introduced by Bach and Jordan (2005b), and uses a KCCA-based contrast function to obtain the unmixing matrix $\boldsymbol{W}$. For "rich enough" kernels, such as the Gaussian kernel, it was shown that the components of the random vector $\boldsymbol{x}$ are independent if and only if their first canonical correlation in the

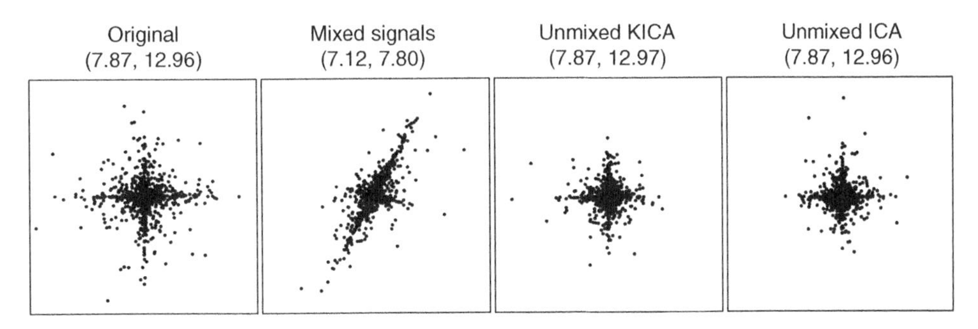

Figure 12.6 Demixing example using KICA. From left to right we show the original source signals, the original (linearly) mixed signals, and the results obtained by KICA and ICA. We indicate the kurtosis in every dimension.

corresponding RKHS is equal to zero. KICA utilizes a gradient-descent approach to minimize a KCCA-based contrast function, taking into account that whitening the features will have the effect of that the unmixing matrix $\boldsymbol{W}$ will have orthonormal vectors. KICA has better performance than the previous three algorithms and also it is more robust against outliers and near-Gaussianity of the sources. However, these performance improvements come at a higher computational cost. Figure 12.6 shows an example of using KICA to unmix signals that have been mixed using a linear combination. The example can be reproduced using the KICA toolbox linked in the book's repository and the code in Listing 12.4.

```
%% Example of ICA and Kernel ICA
% Number of samples
N = 1000;
s(1,:) = sign(randn(1,N)).*abs(randn(1,N)).^2;
s(2,:) = sign(randn(1,N)).*abs(randn(1,N)).^2;
randn('seed',2); W = randn(2);
% Mixing
x    = W*s;
% ICA Unmixing
[ss Aica Wica] = fastica(x); % function in fastICA toolbox
s_hat_ica   = Wica*x;
% kICA Unmixing
Wcca = kernel_ica(x); % function in Bach's toolbox
s_hat_kica   = Wcca*x;
```

Listing 12.4 MATLAB snippet for applying ICA and KICA.

Kernel Independent Component Analysis Using Hilbert–Schmidt Independence Criterion
KICA based on minimizing HSIC as a contrast function has been recently introduced (Shen *et al.* 2007, 2009). The motivation is clear: in the ICA model, the components s_i of the sources $\boldsymbol{s}$ are mutually statistically independent if and only if their *pdf* factorizes $\mathbb{P}_s = \prod_{i=1}^{s} \mathbb{P}_{s_i}$. The problem in this assumption is that pairwise independence does not imply mutual independence (while it holds vice versa). The solution here is that, in the ICA setting, unmixed components can be uniquely identified using only the

pairwise independence between components of the recovered sources $\boldsymbol{Y}$, since pairwise independence between components of $\boldsymbol{Y}$ in this case implies their mutual independence (and thus recovery of the sources $\boldsymbol{S}$). Hence, by summing all unique pairwise HSIC measures, an HSIC-based contrast function over the estimated signals $\boldsymbol{Y}$ is defined as

$$H(\boldsymbol{Q}) = \sum_{1\leq i<j\leq m}^{m} \mathbb{E}_{kl}[\boldsymbol{\phi}(\boldsymbol{q}_i^{\mathrm{T}}\boldsymbol{w}_{kl})\boldsymbol{\phi}(\boldsymbol{q}_j^{\mathrm{T}}\boldsymbol{w}_{kl})] \tag{12.9}$$

$$+ \mathbb{E}_{kl}[\boldsymbol{\phi}(\boldsymbol{q}_i^{\mathrm{T}}\boldsymbol{w}_{kl})]\mathbb{E}_{kl}[\boldsymbol{\phi}(\boldsymbol{q}_j^{\mathrm{T}}\boldsymbol{w}_{kl})] \tag{12.10}$$

$$- 2\mathbb{E}_{k}[\mathbb{E}_{l}\boldsymbol{\phi}(\boldsymbol{q}_i^{\mathrm{T}}\boldsymbol{w}_{kl})]\mathbb{E}_{k}[\boldsymbol{\phi}(\boldsymbol{q}_j^{\mathrm{T}}\boldsymbol{w}_{kl})], \tag{12.11}$$

where $\boldsymbol{w}_{kl} = \boldsymbol{w}_k - \boldsymbol{w}_l \in \mathbb{R}^m$ is the difference between the kth and the lth samples of the whitened observations, and $\mathbb{E}_{kl}[\cdot]$ represents the empirical expectation over all k and l. The expression of $H(\boldsymbol{Q})$ reduces to just summing over all possible HSIC estimates of $\boldsymbol{Y}$. Intuitively, one can see that the goal in KICA is to find an unmixing matrix $\boldsymbol{Q}$ such that the dependence between the estimated unmixed sources $\boldsymbol{Q}^{\mathrm{T}}\boldsymbol{W}$ is minimal in terms of the HSIC between all pairs of signals. The algorithm was first introduced by Bach and Jordan (2002), and later extended to a fast implementation known as FastKICA that uses an approximate Newton method to perform this optimization (Shen *et al.* 2007, 2009). The `fastKICA` package is a convenient implementation of the method.

Kernel Blind Source Separation

The previous approaches are based on introducing kernel dependence estimates as contrast functions in ICA. Alternatively, one can follow the standard "kernelization" procedure: first the data are (implicitly) mapped to a high (possibly infinite)-dimensional kernel feature space, and then a linear algorithm is defined and solved (implicitly) therein by means of the kernel trick. Let us review the two main approaches available in the literature, and their relations to KMVA.

Harmeling *et al.* (2003) followed the standard kernelization of BSS based on second-order statistics. In particular, the method exploits temporal decorrelation in Hilbert spaces via the time-delayed second-order projection (TDSEP) technique, which relies on simultaneous diagonalization techniques to perform linear blind source separation on the projected data. Therefore, one obtains a number of linear directions of separated nonlinear components in input space. The algorithm was coined kernel TDSEP. After embedding, data form a smaller submanifold in feature space, which is typically smaller than the number of training data points. The approach then tries to adapt to this sort of *effective dimension* as a preprocessing step and to construct an orthonormal basis of this submanifold. This first step performs standard dimensionality reduction.

In particular, Harmeling *et al.* (2003) proposed to construct an orthonormal basis in feature spaces from a subset of selected data points therein. Given T observed signals $\{\boldsymbol{x}_t \in \mathbb{R}^d\}_{t=1}^T$, and N vectors $\{\boldsymbol{v}_i \in \mathbb{R}^d\}_{i=1}^N$, we define the corresponding kernel matrices $\boldsymbol{K}_x = \boldsymbol{\Phi}_{\boldsymbol{x}}\boldsymbol{\Phi}_{\boldsymbol{x}}^{\mathrm{T}}$ and $\boldsymbol{K}_x = \boldsymbol{\Phi}_{\boldsymbol{v}}\boldsymbol{\Phi}_{\boldsymbol{v}}^{\mathrm{T}}$. Now let us assume that the span of $\boldsymbol{\Phi}_{\boldsymbol{x}}$ is the same as the span of $\boldsymbol{\Phi}_{\boldsymbol{v}}$ and the rank of $\boldsymbol{\Phi}_{\boldsymbol{v}}$ is N. Since $\boldsymbol{\Phi}_{\boldsymbol{v}}$ constitutes a basis, $\boldsymbol{K}_{\boldsymbol{v}}$ is full rank and has an inverse. The orthonormal basis can thus be defined as $\boldsymbol{V} = (\boldsymbol{\Phi}_{\boldsymbol{v}}\boldsymbol{\Phi}_{\boldsymbol{v}}^{\mathrm{T}})^{-1/2}\boldsymbol{\Phi}_{\boldsymbol{v}}$, and hence projecting new data onto the basis reduces simply to $\boldsymbol{\Psi}_{\boldsymbol{x}}[t] = (\boldsymbol{\Phi}_{\boldsymbol{v}}\boldsymbol{\Phi}_{\boldsymbol{v}}^{\mathrm{T}})^{-1/2}\boldsymbol{\Phi}_{\boldsymbol{v}}\boldsymbol{\Phi}_{\boldsymbol{x}}^{\mathrm{T}}[t] = \boldsymbol{K}_{\boldsymbol{v}}^{-1}\boldsymbol{K}_{\boldsymbol{vx}}$. The definition of the basis could be actually done extracting actually done extracting the most relevant directions in kernel feature spaces via the diagonalization of the full kernel matrix; that is, via KPCA. Nevertheless, the alternative approach to form the basis is very useful for making the subsequent application of BSS *linear methods* computationally and

Kernel Entropy Component Analysis

Kernel entropy component analysis (KECA) was proposed by Jenssen (2010, 2013) to implement a feature extractor according to the entropy components. KECA, like KPCA, is a spectral method based on the kernel similarity matrix. Nevertheless, KECA does not necessarily use the top eigenvalues and eigenvectors of the kernel matrix. Unlike KPCA, which preserves maximally the second-order statistics of the dataset, KECA is founded on information theory and tries to preserve the maximum Rényi entropy of the input space dataset. KECA has proven useful in, for example, remote sensing (Gómez-Chova *et al.*, 2012; Luo and Wu, 2012; Luo *et al.*, 2014), face recognition (Shekar *et al.*, 2011), and some other applications (Hu *et al.* 2013; Jiang *et al.* 2013; Xie and Guan 2012). Several extensions have been proposed for feature selection (Zhang and Hancock, 2012), class-dependent feature extraction (Cheng *et al.*, 2014), and SSL as well (Myhre and Jenssen, 2012).

The KECA algorithm relies on the eigendecomposition of the (uncentered) kernel matrix, and sorts the eigenvectors according to the entropy values of the projections. Given a dataset $\mathcal{D} = \{\boldsymbol{x}_1, \dots, \boldsymbol{x}_N\}$ of dimensionality d, the entropy may be estimated through kernel density estimation (Silverman, 1986) as $-\log V_2$, where V_2 is the so-called *information potential* (Principe, 2010):

$$V_2(\boldsymbol{x}) = \sum_{j=1}^{N_c} \left(\sum_{i=1}^{N} A_{ij} \right)^2. \tag{12.12}$$

In this expression, $N_c \leq N$ is the number or retained components and the matrix $\boldsymbol{A}$ is obtained from the $(N \times N)$ kernel or Gram matrix $\boldsymbol{K}$ whose entries are the kernel function values $K(\boldsymbol{x}_i, \boldsymbol{x}_j)$. Equation 12.12 is based on the kernel decomposition introduced in Jenssen (2010):

$$\boldsymbol{K} = \boldsymbol{A}\boldsymbol{A}^{\mathrm{T}} = (\boldsymbol{E}\boldsymbol{D}^{\frac{1}{2}})(\boldsymbol{D}^{\frac{1}{2}}\boldsymbol{E}^{\mathrm{T}}), \tag{12.13}$$

where $\boldsymbol{E}$ contains the eigenvectors in columns, $\boldsymbol{E} = [\boldsymbol{e}_1 | \cdots | \boldsymbol{e}_N]$, and $\boldsymbol{D}$ is a diagonal matrix containing the eigenvalues of $\boldsymbol{K}$; that is, $D_{ii} = \lambda_i$. An illustrative MATLAB code snippet for KECA is provided in Listing 12.5.

```
%% KECA implementation
sigma      = estimateSigma(X);
K          = kernelmatrix('rbf',X,X,sigma);
[E D]      = eig(K);
V2         = sum((E*(D.^0.5))).^2;
[V2 ind]   = sort(V2,'descend');
Vkeca      = E(:,ind);
```

Listing 12.5 MATLAB code example for the implementation of KECA.

KECA uses the Rényi entropy to sort the basis extracted by PCA. A novel proposal (Izquierdo-Verdiguier *et al.*, 2016), the optimized KECA (OKECA) method, looks directly for the basis that maximizes the Rényi entropy, therefore maximizing the information using as few components as possible. OKECA is based on a similar idea to ICA, and optimizes an extra rotation matrix on top of PCA directions. The new kernel matrix decomposition is expressed as

$$K = FF^{\mathrm{T}} = (ED^{\frac{1}{2}}Q)(Q^{\mathrm{T}}D^{\frac{1}{2}}E^{\mathrm{T}}), \tag{12.14}$$

where W is an orthonormal linear transformation; that is, $QQ^{\mathrm{T}} = I$. In order to solve the OKECA decomposition, OKECA resorts to a gradient-ascent approach. Note that OKECA is more computationally demanding than KECA because not only requires an eigendecomposition of a kernel matrix but also the gradient ascent procedure to refine the final features. An illustrative MATLAB code snippet for OKECA is provided in Listing 12.6.

```
%% OKECA implementation
% (Step 1) KECA-like part
% X    : input data
% Q    : extra rotation matrix
% gdit: #iterations in gradient descent
% dim : number of dimensions
Q = zeros(dim);
sigma = estimateSigma(X,X); K=kernelmatrix('rbf',X,X,sigma);
[E D] = eigs(K); [V2 ind] = sort(diag(D)); E = E(:,ind);
dD = diag(D); D = diag(dD(ind)); A = E*(D.^0.5);
% (Step 2) Optimization of the projections via gradient descent
tau = 1; % learning rate of the gradient descent procedure
for di = 1:size(K,1)
    m = rand(dim,1);
    m = m/sqrt(sum(m.^2));
    if di<dim
        for it = 1:gdit
            dJ = 2*(sum(A*m))*sum(A,1)';
            mn = m  + tau*dJ;
            mn = mn - sum((repmat((Q'*mn)',dim,1).*Q),2);
            mn = mn/sqrt(sum(mn.^2));
        end
    end
    Q(:,di) = mn;
end
F = A*Q;
```

Listing 12.6 MATLAB code example for OKECA.

In general, both KECA and OKECA are different from (but still intimately related to) KMVA methods. On the one hand, they maintain a probabilistic input space interpretation, seek to capture the entropy of the data in a reduced number of components, and constitute a convergence point between kernel methods and information theoretic learning (Jenssen, 2009). On the other hand, KPCA, KCCA, and KPLS maximize the variance, correlation, or alignment (covariance) with the output variables in kernel feature space respectively. The similarities between these methods arise from the use of a *kernel function* and from the exploitation of the spectral (eigenvalues and eigenvectors)

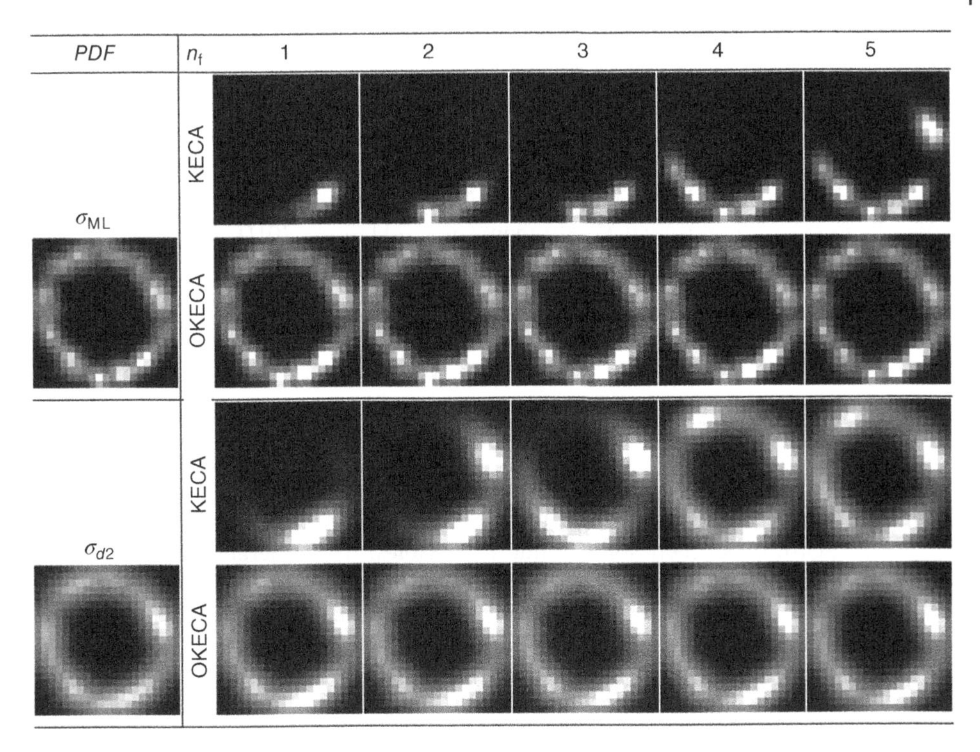

Figure 12.7 Density estimation for the ring dataset by KECA and OKECA using different number of extracted features n_f and estimates of the kernel length-scale parameter σ (ML or squared distance). Black color represents low *pdf* values and yellow color high *pdf* values.

properties of the corresponding kernel matrices. KECA and OKECA can be used for density estimation. As explained by Girolami (2002b), if the decomposition of the uncentered kernel matrix follows the form $\boldsymbol{K} = \boldsymbol{EDE}^{\mathrm{T}}$, where $\boldsymbol{E}$ is orthonormal and $\boldsymbol{D}$ is a diagonal matrix, then the kernel-based density estimation may be expressed as

$$\hat{p}(\mathbf{x}_*) = \mathbf{1}_N^{\mathrm{T}} \boldsymbol{E}_{n_f} \boldsymbol{E}_{n_f}^{\mathrm{T}} \boldsymbol{k}_*, \tag{12.15}$$

where $\boldsymbol{E}_{n_f}$ is the reduced version of $\boldsymbol{E}$ by keeping columns for $n_f < N$, and as usual $\boldsymbol{k}_* := [K(\boldsymbol{x}_*, \boldsymbol{x}_1), \ldots, K(\boldsymbol{x}_*, \boldsymbol{x}_N)]^{\mathrm{T}}$. Therefore, density estimation $\hat{p}$ can be readily done by fixing a number of extracted features n_f. Figure 12.7 illustrates the ability of KECA and OKECA for density estimation in a ring distribution. Note that OKECA concentrates most of the entropy information in the first projection. This agrees with the fact that few components are needed to obtain a good *pdf* estimation. On the contrary, KECA cannot estimate correctly the *pdf* using only the first component and actually needs at least five components. We show results for two standard ways to estimate the σ length-scale parameter: an ML strategy, σ_{ML} following Duin (1976), and setting σ to 15% of the median distance between points following prescriptions in Jenssen (2010). In both cases OKECA outperforms KECA.

12.4 Extensions for Large-Scale and Semi-supervised Problems

An important problem in kernel-based feature extraction is related to the computational cost. Since $\boldsymbol{K}_x$ is of size $N \times N$, method complexity scales quadratically with N in terms of memory, and cubically with respect to the computation time. The opposite situation often occurs as well: when N is small, the extracted features may be useless, especially for high-dimensional $\mathcal{F}$ (Abrahamsen and Hansen, 2011). These issues limit the applicability of kernel-based feature extraction methods in real-life scenarios with either very large or very small labeled datasets. We next summarize some extensions to deal with large-scale problems and semi-supervised situations in which few labeled data are available.

12.4.1 Efficiency with the Incomplete Cholesky Decomposition

Very often, feature extraction methods on kernels rely on the calculation of traces on products of kernel matrices; for instance, see HSCA or KICA that try to maximize the HSIC measure between some random variables. This operation is simple yet computationally demanding when a high number of samples are available. In order to speed this operation up, one can rely on sparse approximations of the basis spanning the solution (as we will see in the next section), by intelligent dataset subsampling (e.g., via AL), or Nyström approximations of the kernel matrices. A convenient alternative is to exploit low-rank approximations of the kernel matrices via Cholesky decomposition.

Notationally, given two kernel matrices $\boldsymbol{K}_1$ and $\boldsymbol{K}_2$, the cost associated with the operation $\text{Tr}(\boldsymbol{K}_1\boldsymbol{K}_2)$ can be greatly reduced via incomplete Cholesky estimation. An incomplete Cholesky decomposition of a Gram matrix $\boldsymbol{K}_i$ yields a low-rank approximation, $\boldsymbol{K}_i \approx \boldsymbol{G}_i\boldsymbol{G}_i^{\text{T}}$ that greedily minimizes $\text{Tr}(\boldsymbol{K}_i - \boldsymbol{G}_i\boldsymbol{G}_i^{\text{T}})$. The cost of computing the $N \times r$ matrix $\boldsymbol{G}_i$ is $\mathcal{O}(Nr^2)$, with $r \ll N$. Greater values of r result in a more accurate reconstruction of $\boldsymbol{K}_i$. Interestingly, it is well known that the spectrum of a Gram matrix based on the (RBF) Gaussian kernel generally decays rapidly, and a small r yields a very good approximation. Plugging the Cholesky eigendecomposition of the two kernels leads to

$$\text{Tr}(\boldsymbol{K}_1\boldsymbol{K}_2) = \text{Tr}(\boldsymbol{G}_1\boldsymbol{G}_1^{\text{T}}\boldsymbol{G}_2\boldsymbol{G}_2^{\text{T}}) = \text{Tr}(\boldsymbol{G}_2^{\text{T}}\boldsymbol{G}_1 \;\; \boldsymbol{G}_1^{\text{T}}\boldsymbol{G}_2),$$

which involves an equivalent trace of a much smaller matrix, so we can avoid computing the $N \times N$ matrices and just do $r_1 \times r_2$ matrix calculations. This technique has been widely used in KMVA and also in kernel dependence estimation methods such as the HSIC (Gretton *et al.*, 2005b).

12.4.2 Efficiency with Random Fourier Features

We have already seen the effectiveness of approximating shift-invariant kernels, using projections on random Fourier features (Rahimi and Recht, 2007, 2009). This technique was illustrated in classification and regression settings; see Section 10.4.2. Let us show the application of the approximation in HSIC presented in Pérez-Suay and Camps-Valls

(2017). Following previous notation, recall that $\boldsymbol{K} \in \mathbb{R}^{N\times N}$ can be approximated with the explicitly mapped data via $\boldsymbol{z}(\boldsymbol{x}) := [\exp(\mathbb{i}\boldsymbol{w}_1^{\mathrm{T}}\boldsymbol{x}), \dots, \exp(\mathbb{i}\boldsymbol{w}_D^{\mathrm{T}}\boldsymbol{x})]^{\mathrm{T}}$, and collectively grouped in the matrix $\boldsymbol{Z} = [\boldsymbol{z}_1|\cdots|\boldsymbol{z}_N]^{\mathrm{T}} \in \mathbb{R}^{N\times D}$, and will be denoted as $\hat{\boldsymbol{K}} \approx \boldsymbol{Z}\boldsymbol{Z}^{\mathrm{T}}$. The familiar SE Gaussian kernel $K(\boldsymbol{x}, \boldsymbol{x}') = \exp(-\|\boldsymbol{x}-\boldsymbol{x}'\|^2/(2\sigma^2))$ can be approximated using D random features, $\boldsymbol{w}_i \sim \mathcal{N}(\boldsymbol{0}, \sigma^{-2}\boldsymbol{I})$, $1 \le i \le D$. The randomized HSIC (RHSIC) for fast dependence estimation in large-scale problems is developed as follows: kernel matrices $\boldsymbol{K}_x$ and $\boldsymbol{K}_y$ are approximated using complex exponentials of projected data matrices, $\boldsymbol{Z}_x = \exp(\mathbb{i}\boldsymbol{X}\boldsymbol{W}_x)/\sqrt{D_x} \in \mathbb{C}^{N\times D_x}$ and $\boldsymbol{Z}_x = \exp(\mathbb{i}\boldsymbol{X}\boldsymbol{W}_x)/\sqrt{D_x} \in \mathbb{C}^{N\times D_x}$ and $\boldsymbol{Z}_y = \exp(\mathbb{i}\boldsymbol{Y}\boldsymbol{W}_y)/\sqrt{D_y} \in \mathbb{C}^{N\times D_y}$, where $\boldsymbol{W}_y = [\boldsymbol{w}_1^y|\cdots|\boldsymbol{w}_{D_y}^y] \in \mathbb{R}^{d_y\times D_y}$, both drawn from a Gaussian. Now, plugging the corresponding kernel approximations, $\hat{\boldsymbol{K}}_x = \tilde{\boldsymbol{Z}}_x\tilde{\boldsymbol{Z}}_x^{\mathrm{T}}$ and $\hat{\boldsymbol{K}}_y = \tilde{\boldsymbol{Z}}_y\tilde{\boldsymbol{Z}}_y^{\mathrm{T}}$, into Equation 12.3.1, and after manipulating terms, we obtain

$$\mathrm{RHSIC}(\mathcal{F}, \mathcal{G}, \mathbb{P}_{\mathbf{xy}}) = \frac{1}{N^2}\mathrm{Re}\{\mathrm{Tr}(\tilde{\boldsymbol{Z}}_x^{\mathrm{T}}\tilde{\boldsymbol{Z}}_y\tilde{\boldsymbol{Z}}_y^{\mathrm{T}}\tilde{\boldsymbol{Z}}_x)\}, \tag{12.16}$$

which corresponds to the Hilbert–Schmidt norm of the randomized cross-covariance operator, $\hat{C}_{\mathbf{xy}} = \mathrm{Re}\{\tilde{\boldsymbol{Z}}_x^{\mathrm{T}}\tilde{\boldsymbol{Z}}_y\} \in \mathbb{R}^{D_x\times D_y}$, which can be computed explicitly and just once. The computational complexity of RHSIC reduces considerably over the original HSIC. A naive implementation of HSIC runs $\mathcal{O}(N^2)$, while the RHSIC cost is $\mathcal{O}(ND^2)$, where $D = D_x = D_y$, since computing matrices $\boldsymbol{Z}$ only involves matrix multiplications and exponentials. We want to note that other KMVA methods have enjoyed use of this randomization technique to expedite performance, such as KCCA and KPCA (López-Paz *et al.*, 2014).

12.4.3 Sparse Kernel Feature Extraction

To address the problems of large-scale computation and dense solutions in KMVA, several solutions have been proposed. The family of KMVA sparse methods tries to obtain solutions that can be expressed as a combination of a reduced subset of the training data, and therefore require only r kernel evaluations per sample (being $r \ll N$) for feature extraction. In contrast to the many linear MVA algorithms that induce sparsity with respect to the original variables, we will only review methods attaining sparse solutions in terms of the samples (i.e., sparsity in the $\boldsymbol{\alpha}_i$ vectors).

Roughly speaking, sparsification methods can be classified into *low-rank* approximation methods that aim at working with reduced $r\times r$ matrices ($r \ll N$), and *reduced set* methods that work with $N\times r$ matrices. Following the first approach, the Nyström low-rank approximation of an $N\times N$ kernel matrix $\boldsymbol{K}_{NN}$ is expressed as $\tilde{\boldsymbol{K}}_{NN} = \boldsymbol{K}_{Nr}\boldsymbol{K}_{rr}^{-1}\boldsymbol{K}_{rN}$, where subscripts indicate row and column dimensions. This method was first used in Gaussian processes, and later by Hoegaerts *et al.* (2005) in order to approximate the feature mapping instead of the kernel, which leads to sparse versions of KPLS and KCCA.

Among the reduced set methods, a sparse KPCA (sKPCA) was proposed by Tipping (2001), where the sparsity in the representation is obtained by assuming a generative model for the data in $\mathcal{F}$ that follows a normal distribution and includes a noise term with variance v_{n}. The ML estimation of the covariance matrix is shown to depend on just a subset of the training data, and so it does the resulting solution. A sparse KPLS (sKPLS) was introduced by Momma and Bennett (2003). The method computes the projections with a reduced set of the training samples. Each one of the projections is found using a loss that is similar to the ε insensitive loss used in SVR. The sparsification is induced via a multistep adaptation with high computational burden.

The algorithms in Tipping (2001) and Momma and Bennett (2003), however, still require the computation of the full kernel matrix during the training.

A reduced complexity KOPLS (rKOPLS) was proposed by Arenas-García *et al.* (2007) by imposing sparsity in the projection vectors representation a priori, $\boldsymbol{U} = \boldsymbol{\Phi}_r^{\mathrm{T}}\boldsymbol{\beta}$, where $\boldsymbol{\Phi}_r$ is a subset of the training data containing r samples ($r \ll N$) and $\boldsymbol{\beta}$ is the new argument for the maximization problem, which now becomes

$$\begin{aligned} \max \quad & \boldsymbol{\beta}^{\mathrm{T}}\boldsymbol{K}_{rN}\boldsymbol{Y}\boldsymbol{Y}^{\mathrm{T}}\boldsymbol{K}_{rN}^{\mathrm{T}}\boldsymbol{\beta} \\ \text{subject to}: \quad & \boldsymbol{\beta}^{\mathrm{T}}\boldsymbol{K}_{rN}\boldsymbol{K}_{rN}^{\mathrm{T}}\boldsymbol{\beta} = 1, \end{aligned} \tag{12.17}$$

Since kernel matrix $\boldsymbol{K}_{rN} = \boldsymbol{\Phi}_r\boldsymbol{\Phi}^{\mathrm{T}}$ involves the inner products in $\mathcal{F}$ of all training points with the patterns in the reduced set, this method still takes into account all data during the training phase, and is therefore different from simple subsampling. This "sparsification" procedure avoids the computation of the full kernel matrix at any step of the algorithm. An additional advantage of this method is that matrices $\boldsymbol{K}_{rN}\boldsymbol{Y}\boldsymbol{Y}^{\mathrm{T}}\boldsymbol{K}_{rN}^{\mathrm{T}}$ and $\boldsymbol{K}_{rN}\boldsymbol{K}_{rN}^{\mathrm{T}}$ are both of size $r \times r$, and they can be computed as sums over the training data. This fact makes the storage space grow quadratically with r. Also, there is an implicit regularization imposed by the sparsification, which decreases the overfitting risk of the method. Reduced complexity versions of KPCA and KKCA are shown in the experimental section that use the same sparsification method, and will be referred as rKPCA and rKCCA.

Interestingly, the extension to KPLS2 is not straightforward, since the deflation step would still require the full kernel matrix $\boldsymbol{K}_{NN}$. Alternatively, two sparse KPLS schemes were presented by Dhanjal *et al.* (2009) under the name of sparse maximal alignment (SMA) and sparse maximal covariance (SMC). Here, KPLS iteratively estimates projections that either maximize the *kernel alignment* or the *covariance* of the projected data and the true labels.

Table 12.3 summarizes some computational and implementation issues of the aforementioned sparse KMVA methods, and of standard nonsparse KMVA and linear methods. Some aspects that can be helpful to choose an algorithm for a particular application can be obtained from an analysis of the properties of each algorithm. An important step is to choose an adequate kernel and parameters. KMVA methods can be adjusted using cross-validation to reduce overfitting at the expense of an increased computational burden, but regularization through sparsification can help to further reduce overfitting. Also, most methods can be implemented as either eigenvalue or generalized eigenvalue problems, whose complexity typically scales cubically with the size of the matrices analyzed. Therefore, both for memory and computational reasons, only linear MVA and the sparse approaches from Arenas-García *et al.* (2007) and Dhanjal *et al.* (2009) are affordable when dealing with large datasets. A final advantage of sparse KMVA is the reduced number of kernel evaluations to extract features for new out-of-the-sample data.

Table 12.3 Main properties of (K)MVA methods. Computational complexity and implementation issues are categorized for the considered dense and sparse methods in terms of the free parameters, number of kernel evaluations (KEs) during training, and storage requirements. Notation: N refers to the size of the labeled dataset, while d and m are the dimensions of the input and output spaces respectively.

Method	Parameters	KE (training)	Storage requirements
PCA	None	None	$\mathcal{O}(d^2)$
PLS	None	None	$\mathcal{O}((d+m)^2)$
CCA	None	None	$\mathcal{O}((d+m)^2)$
OPLS	None	None	$\mathcal{O}(d^2)$
KPCA	Kernel	N^2	$\mathcal{O}(N^2)$
KPLS	Kernel	N^2	$\mathcal{O}((N+m)^2)$
KCCA	Kernel	N^2	$\mathcal{O}((N+m)^2)$
KOPLS	Kernel	N^2	$\mathcal{O}(N^2)$
sKPCA (Tipping, 2001)	Kernel, ν_n	N^2	$\mathcal{O}(N^2)$
sKPLS (Momma and Bennett, 2003)	Kernel, ν, ε	N^2	$\mathcal{O}(N^2)$
rKPCA	Kernel, r	rN	$\mathcal{O}(r^2)$
rKCCA	Kernel, r	rN	$\mathcal{O}((r+m)^2)$
rKOPLS (Arenas-García *et al.*, 2007)	kernel, r	rN	$\mathcal{O}(r^2)$
SMA/SMC (Dhanjal *et al.*, 2009)	kernel, r	N^2	$\mathcal{O}(N^2)$

12.4.4 Semi-supervised Kernel Feature Extraction

When few labeled samples are available, the extracted features do not capture the structure of the data manifold well, which may lead to very poor results. SSL approaches have been introduced to alleviate these problems. Two approaches are encountered: the information conveyed by the unlabeled samples is either modeled with graphs or via kernel functions derived from generative clustering models.

Notationally, we are given l labeled and u unlabeled samples, a total of $N = l + u$. The semi-supervised KCCA (ss-KCCA) was introduced in Blaschko *et al.* (2011) by using the graph Laplacian. The method essentially solves the standard KCCA equations using kernel matrices computed with both labeled and unlabeled data, which are further *regularized* with the graph Laplacian:

$$\begin{pmatrix} \mathbf{0} & \boldsymbol{K}^x_{Nl}\boldsymbol{K}^y_{lN} \\ \boldsymbol{K}^x_{Nl}\boldsymbol{K}^y_{lN} & \mathbf{0} \end{pmatrix} \begin{pmatrix} \boldsymbol{\alpha} \\ \boldsymbol{\nu} \end{pmatrix} = \lambda \begin{pmatrix} \boldsymbol{K}^x_{Nl}\boldsymbol{K}^x_{lN} + \boldsymbol{R}^x_{NN} & \mathbf{0} \\ \mathbf{0} & \boldsymbol{K}^y_{Nl}\boldsymbol{K}^y_{lN} + \boldsymbol{R}^y_{NN} \end{pmatrix} \begin{pmatrix} \boldsymbol{\alpha} \\ \boldsymbol{\nu} \end{pmatrix},$$

where $\boldsymbol{R}^x_{NN} = \alpha_x \boldsymbol{K}^x_{NN} + \gamma_x \boldsymbol{K}^x_{NN}\boldsymbol{\mathcal{L}}^x_{NN}\boldsymbol{K}^x_{NN}$ and $\boldsymbol{R}^y_{NN} = \alpha_y \boldsymbol{K}^y_{NN} + \gamma_y \boldsymbol{K}^y_{NN}\boldsymbol{\mathcal{L}}^y_{NN}\boldsymbol{K}^y_{NN}$. For the sake of notation compactness, subindices indicate the size of the corresponding matrices while superscripts denote whether they involve input or output data. Parameters α_x, α_y, γ_x, and γ_y trade off the contribution of labeled and unlabeled samples, and

$\boldsymbol{L} = \boldsymbol{D}^{-1/2}(\boldsymbol{D} - \boldsymbol{M})\boldsymbol{D}^{1/2}$ represents the (normalized) graph Laplacian for the input and target domains, where $\boldsymbol{D}$ is the degree matrix whose entries are the sums of the rows of the corresponding similarity matrix $\boldsymbol{M}$; that is, $D_{ii} = \sum_j M_{ij}$. It should be noted that for $N = l$ and null regularization one obtains the standard KCCA (see Equation 12.3). Note also that this form of regularization through the graph Laplacian can be applied to any KMVA method. A drawback of this approach is that it involves tuning several parameters and working with larger matrices of size $2N \times 2N$, which can make its application difficult in practice.

Alternatively, *cluster kernels* – a form of generative kernel functions learned from data – have been used to develop semi-supervised versions of kernel methods in general, and of KMVA methods in particular. The approach was used for KPLS and KOPLS by Izquierdo-Verdiguier *et al.* (2012). Essentially, the method relies on combining a kernel function based on labeled information only, $K_s(\boldsymbol{x}_i, \boldsymbol{x}_j)$, and a *generative* kernel directly learned by clustering *all* (labeled and unlabeled) data, $K_c(\boldsymbol{x}_i, \boldsymbol{x}_j)$. Building K_c requires first running a clustering algorithm, such as EM assuming a GMM with different initializations, $q = 1, \dots, Q$, and with different number of clusters, $g = 2, \dots, G+1$. This results in $Q \cdot G$ cluster assignments where each sample $\boldsymbol{x}_i$ has its corresponding posterior probability vector $\boldsymbol{\pi}_i(q,g) \in \mathbb{R}^g$. The *probabilistic cluster* kernel K_c is computed by averaging all the dot products between posterior probability vectors:

$$K_c(\boldsymbol{x}_i, \boldsymbol{x}_j) = \frac{1}{Z} \sum_{q=1}^{Q} \sum_{g=2}^{G+1} \boldsymbol{\pi}_i(q,g)^{\mathrm{T}} \boldsymbol{\pi}_j(q,g),$$

where Z is a normalization factor. The final kernel function is defined as the weighted sum of both kernels, $K(\boldsymbol{x}_i, \boldsymbol{x}_j) = \beta K_s(\boldsymbol{x}_i, \boldsymbol{x}_j) + (1 - \beta) K_c(\boldsymbol{x}_i, \boldsymbol{x}_j)$, where $\beta \in [0, 1]$ is a scalar parameter to be adjusted. Intuitively, the cluster kernel accounts for *probabilistic* similarities at small and large scales (number of clusters) between all samples along the data manifold, and hence its name: the probabilistic cluster kernel (PCK) (Izquierdo-Verdiguier *et al.*, 2012). The method does not require computationally demanding procedures (e.g., current GMM clustering algorithms scale linearly with N), and the KMVA still relies just on the labeled data, and thus requires an $l \times N$ kernel matrix. All these properties are quite appealing from the practitioner's point of view. Listing 12.7 gives a MATLAB code snippet showing how to compute the PCK. Note that this particular this particular kernel function is semi-supervised and can be plugged into any kernel method.

```
%    Xtrain:   Train samples [samples x features]
%    Xtest:    Test samples  [samples x features]
%    C_tot:    Gaussian mixture parameter estimates
%    IDX_tot: Clusters from Gaussian mixture distribution
KbagUL = zeros(size(Xtrain,1),size(Xtest,1));
nk=0;
for k=Kvector % Number of clustering realizations
 nk=nk+1;
 for n=1:N    % Number of realizations
  if ~isempty(C_tot{nk,n})
   Pgmm = posterior(C_tot{nk,n},Xtrain);
   [ID_UL PP]= closerClusterGMM(Xtest,C_tot{nk,n},'resoft');
   idx_tot=IDX_tot{nk,n};
```

```
    idx_tot=idx_tot(1:size(Xtrain,1),:);
    for p=1:length(ID_UL)
        for q=1:length(idx_tot)
            KbagUL(q,p) = KbagUL(q,p) + PP(p,:)*Pgmm(q,:)';
        end
    end
  end
 end
end
KbagUL = KbagUL/max(max(KbagUL));
```

Listing 12.7 MATLAB code example for PCK.

Figure 12.8 shows an illustrative toy example evaluating the RBF, PCK, Fisher's (Jaakkola *et al.*, 1999), and Jensen–Shannon's (Bicego *et al.*, 2013) kernels. The data were generated by the composition of two normal distributions, $\mathcal{N}(3, 0.5)$ and $\mathcal{N}(5, 0.5)$. We look at the structure of the kernel matrix through the first eigenvectors, and compute the Frobenius norm of the residuals with the ideal kernel. The length-scale of the RBF was fixed to the average distance of all examples, while we used $Q = 10$ and $G = 25$ for K_c. The Fisher kernel used feature vectors extracted from a GMM using the same number of clusters as for the PCK, and the Jensen–Shannon kernel was built from divergence Jensen–Shannon obtained from Shannon entropy. Since the Fisher and Jensen–Shannon kernels are not intrinsically multiscale, here we implement a multiscale version of the Fisher and Jensen–Shannon kernels for the sake of a fair comparison. The figure shows that the PCK and the Jensen–Shannon return more discriminative eigenvectors and substantially different from the Fourier-like basis obtained by the RBF and the Fisher kernel. The PCK better captures the local structure than the Jensen–Shannon kernel does. This clearly draws from the visual comparison of the kernel matrices, which is also supported by the Frobenius norm of the differences to the ideal kernel (Figure 12.8, given in parentheses).

12.5 Domain Adaptation with Kernels

Domain adaptation constitutes a field of high interest nowadays in pattern analysis and machine learning. Classification or regression algorithms developed with data from one domain cannot be directly used in another related domain, and hence adaptation of either the data representation or the classifier becomes strictly imperative (Quiñonero-Candela *et al.*, 2009; Sugiyama and Kawanabe, 2012). There is actually strong evidence that a significant degradation in the performance of state-of-the-art image classifiers is due to test domain shifts such as changing image sensors and noise conditions (Saenko *et al.*, 2010), pose changes (Farhadi and Tabrizi, 2008), consumer versus commercial video (Duan *et al.*, 2012), and, more generally, datasets biased due to changing acquisition procedures (Torralba and Efros, 2011). Unlike in Hoffman *et al.* (2013), we focus on adapting data representations to facilitate the use of standard classifiers. Roughly speaking, domain adaptation solves a learning problem in a target domain by utilizing the training data in a different but related source domain. Intuitively, discovering a good feature representation across domains is crucial. The problem has captured the attention of researchers in many fields, also in the kernel methods community.

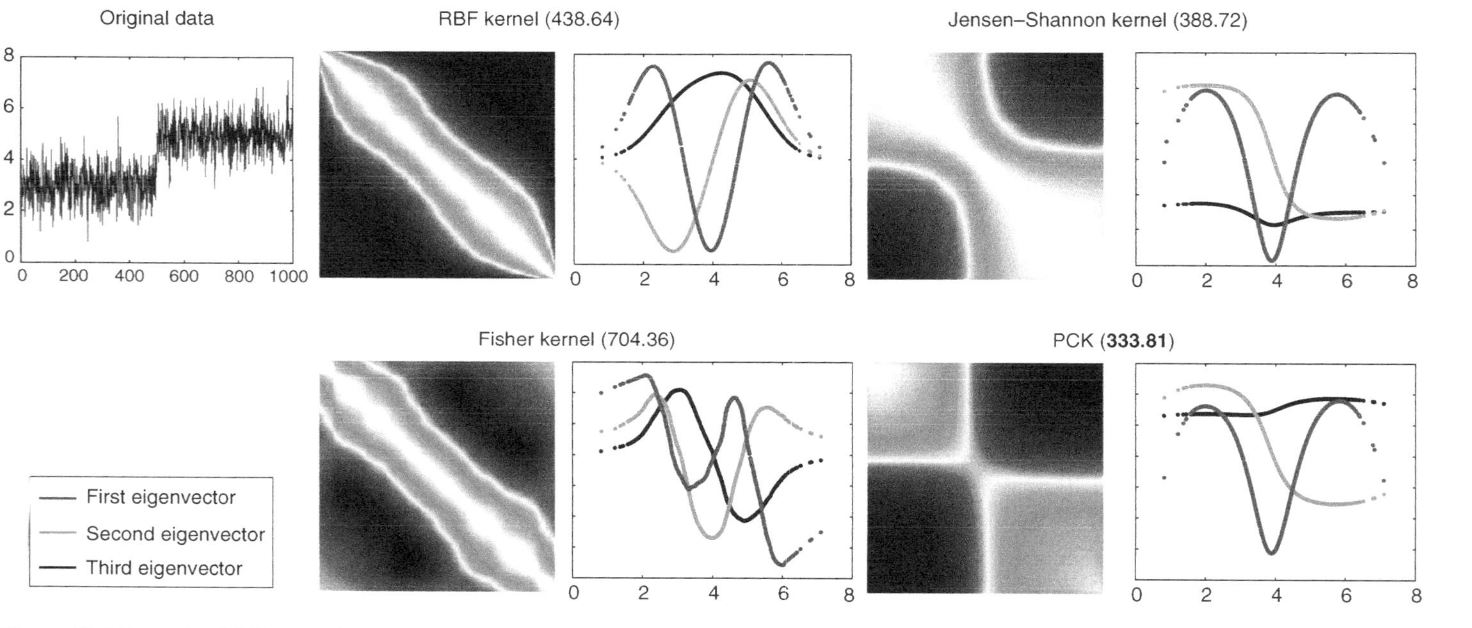

Figure 12.8 Example of differences between the local properties of the RBF, Jensen–Shannon kernel, Fisher's kernel, and the PCK. We indicate in parentheses the Frobenius norm of the residuals with the ideal kernel, $\|\boldsymbol{K} - \mathbf{y}\mathbf{y}^{\mathrm{T}}\|_{\mathrm{F}}^{2}$.

The signal- and image-processing communities have also greatly contributed to this community, yet under different names. Domain adaptation is sometimes referred to as simply *adaptation, transportability, inter-calibration,* or *feature invariance learning.* A collection of images must be compensated for illumination changes when a classifier must be used across domains, or speech signals from different users must be equalized for annotation and recognition steps. In general, data collected for the same problem in different instants, from different angles, by different systems and procedures show similar features but also reveal changes in their data manifold that need to be compensated before the direct exploitation. The problem is very challenging in high-dimensional and low-sized datasets, which is very often the case in signal-processing problems.

Recently, several approaches have been proposed to learn a common feature representation for domain adaptation. Daumé III and Marcu (2011) presented a simple heuristic nonlinear mapping function to map the data from both source and target domains to a high-dimensional feature space, where standard machine-learning methods are used to train classifiers. On the other hand, Blitzer *et al.* (2006) proposed the so-called structural correspondence learning algorithm to induce correspondences among features from the different domains. This method depends on the heuristic selections of *pivot features* that appear frequently in both domains. Although it is experimentally shown that structural correspondence learning can reduce the difference between domains based on the distance measure (Ben-David *et al.*, 2007), the heuristic criterion of pivot feature selection may be sensitive to different applications. Pan *et al.* (2008) proposed a new dimensionality reduction method, the so-called maximum mean discrepancy embedding (MMDE), for domain adaptation. The method aims at learning a shared *latent space* underlying the domains where distance between distributions can be reduced. However, MMDE suffers from two major limitations: (1) MMDE is transductive, and does not generalize to out-of-sample patterns; (2) MMDE learns the latent space by solving a semi-definite program (SDP), which is a very expensive optimization problem. Alternatives to these problems have been recently introduced in the literature.

Besides changing the data representation space via kernel feature extraction, another possibility is to correct for biases in the data distributions operating on the samples directly. The problem of *covariate shift* is intimately related to the problem of *adaptation* (Sugiyama and Kawanabe, 2012). Correcting for biases in the (changing) distributions can be addressed by upweighting and downweighting the relevance of the samples in the incoming target distribution. This technique has been widely exploited in many studies (Bickel *et al.* 2009; Huang *et al.* 2007; Kanamori *et al.* 2009): Huang *et al.* (2007) proposed the kernel mean matching (KMM) to reweight instances in an RKHS, Bickel *et al.* (2009) proposed to integrate the distribution correcting process into a kernelized logistic regression, and Kanamori *et al.* (2009) introduced a method called unconstrained LS importance fitting (uLSIF) to estimate sample relevance.

In this section, we review three examples of successful kernel methods for domain adaptation, each following different (yet obviously interconnected) philosophies: (1) sample reweighting strategies (KMM), (2) transfer component analysis (TCA), and (3) kernel manifold alignment (KEMA) for domain adaptation. Some illustrative examples of performance in signal processing tasks are given for illustration purposes.

12.5.1 Kernel Mean Matching

KMM is a kernel-based method for sample reweighting. Note that, in general, learning methods typically try to minimize the expected risk, $\mathcal{R}[\mathbb{P}, \theta, \ell]$, where $\mathbb{P}$ is the *pdf* where samples are drawn from, and $\ell(\boldsymbol{x}, y, \theta)$ is the empirical loss function that depends that depends on a parameter θ. Since one typically has access only to pairs of examples $(\boldsymbol{x}, y)$ drawn from a density $\mathbb{P}(\boldsymbol{x}, y)$, the problem is solved by computing the empirical average over all available data, $\mathcal{L}_{\text{emp}} = \sum_{i=1}^{N} \ell(\boldsymbol{x}_i, y_i, \theta)$, which is normally regularized to avoid overfitting to the training set. However, the problem turns out to be more complex when the training and test probability functions differ, even slightly; that is, $\mathbb{P}_{\text{train}}(\boldsymbol{x}, y) \neq \mathbb{P}_{\text{test}}(\boldsymbol{x}, y)$. Actually, what one would desire is to minimize the $\mathbb{P}_{\text{test}}(\boldsymbol{x}, y)$, but one has only access to data drawn from $\mathbb{P}_{\text{train}}(\boldsymbol{x}, y)$. The field of statistics known as *importance sampling* is concerned precisely with this problem: estimating properties of a particular distribution while only having samples generated from a estimating properties of a particular distribution while only having samples generated from a different distribution rather than the distribution of interest. The methods in this field are frequently used to estimate posterior densities or expectations in probabilistic models that are hard to treat analytically, as in

$$\mathcal{R}[\mathbb{P}_{\text{test}}, \theta, \ell(\boldsymbol{x}, y, \theta)] = \mathbb{E}_{(\boldsymbol{x},y)\sim\mathbb{P}_{\text{test}}}[\ell(\boldsymbol{x}, y, \theta)] = \mathbb{E}_{(\boldsymbol{x},y)\sim\mathbb{P}_{\text{train}}}\left[\frac{\mathbb{P}_{\text{test}}(\boldsymbol{x}, y)}{\mathbb{P}_{\text{train}}(\boldsymbol{x}, y)}\ell(\boldsymbol{x}, y\,\theta)\right]$$

$$= \mathcal{R}[\mathbb{P}_{\text{train}}, \theta, \beta(\boldsymbol{x}, y)\ell(\boldsymbol{x}, y, \theta)],$$

provided that the support of $\mathbb{P}_{\text{test}}$ is contained in the support of $\mathbb{P}_{\text{train}}$. Therefore, we can compute the test risk from a modified version of the train risk, and this modification essentially translates into estimating the relevance parameter β per training example. This approach, however, involves density estimation of $\mathbb{P}_{\text{train}}(\boldsymbol{x}, y)$ and $\mathbb{P}_{\text{test}}(\boldsymbol{x}, y)$, which in general is not an easy problem (recall Vapnik's *dictatum* in Chapter 2). In addition, a deficient estimation of such densities may give rise to inaccurate estimates for the weights β, and consequently it may happen that the algorithm with adaptation mechanism works substantially worse than one without it. An alternative to both problems is given by the empirical KMM optimization.

The KMM essentially tries to find weights $\beta \in \mathbb{R}^d$ in order to minimize the discrepancy between means subject to constraints $\beta_i \in [0, B]$ and $|(1/d)\sum_{i=1}^{d} \beta_i - 1| \leq \varepsilon$, which limits the discrepancy between $\mathbb{P}_{\text{train}}$ and $\mathbb{P}_{\text{test}}$, as well as ensure that $\beta(\boldsymbol{x})\mathbb{P}_{\text{train}}(\boldsymbol{x})$ approaches a probability distribution. The objective function is given by the discrepancy term between the two empirical means. Using kernels $[\boldsymbol{K}]_{ij} = K(\boldsymbol{x}_i, \boldsymbol{x}_j)$, and $\kappa_i = (d/d')\sum_{i=1}^{d'} K(\boldsymbol{x}_i, \boldsymbol{x}'_j)$, one can check that

$$\left\|\frac{1}{d}\sum_{i=1}^{d} \beta_i \boldsymbol{\phi}(\boldsymbol{x}_i) - \frac{1}{d'}\sum_{i=1}^{d'} \beta_i \boldsymbol{\phi}(\boldsymbol{x}'_i)\right\|^2 = \frac{1}{d^2}\boldsymbol{\beta}^{\text{T}}\boldsymbol{K}\boldsymbol{\beta} - \frac{2}{d^2}\boldsymbol{\kappa}^{\text{T}}\boldsymbol{\beta} + \text{constant terms},$$

which reduces to solving a QP problem to optimize the weight vector β:

$$\min_{\beta}\left\{\frac{1}{2}\beta^{\mathrm{T}}K\beta-\kappa^{\mathrm{T}}\beta\right\}\quad \text{s.t.}\quad \beta_i\in[0,B]\quad \text{and}\quad |\sum_{i=1}^{d}\beta_i-d|\leq m\varepsilon.$$

Note that KMM resembles the OC-SVM using the ν-trick, yet modified by the linear correction term involving κ, which measures sample relevance as well.

12.5.2 Transfer Component Analysis

TCA (Pan and Yang, 2011) tries to learn some transfer components across domains in an RKHS using the MMD measure (Borgwardt *et al.*, 2006) introduced in Chapter 11 and related to the HSIC introduced before. In the subspace spanned by these *transfer components*, data distributions in different domains are close to each other. As a result, with the new representations in this subspace, one can apply standard machine-learning methods to train classifiers or regression models in the source domain for use in the target domain. TCA does so in a very effective way. TCA essentially minimizes the distance between probability distributions of a source and a target domain. There are many distances in the literature to evaluate the difference between probability distributions: Kullback–Leibler divergence, Jensen–Shannon divergence, Bhattacharyya distance, and so on. However, these methods are affected by the data dimensionality, with the necessary probability density estimations becoming infeasible in high-dimensional spaces.

TCA considers a setting where the target domain has plenty of unlabeled data. TCA also assumes that some labeled data are available in a source domain s, while only unlabeled data are available in the target domain t. Data will be denoted as belonging to either source or target domains with the corresponding superscript; that is, x_i^s for the source and x_i^t for the target. The MMD estimate presented in Borgwardt *et al.* (2006) proposes a new indicator for comparing distributions based on the difference of the mean of the distributions computed in a common RKHS. This nonparametric measure is easily calculated no matter the number of variables describing the examples. Essentially, the empirical estimate of the MMD between the distribution of a given source dataset X_s and that of a related target dataset X_t is given by

$$\mathrm{MMD}(X_s,X_t)=\left\|\frac{1}{N_s}\sum_{i=1}^{N_s}\phi(x_i^s)-\frac{1}{N_t}\sum_{i=1}^{N_t}\phi(x_i^t)\right\|_{\mathcal{H}}^2.$$

The problem thus reduces to finding the nonlinear transformation $\phi(\cdot)$ explicitly, which in principle is not possible. Instead, by using the kernel trick, one defines a reproducing kernel as $K(x_i,x_j)=\phi(x_i)^{\mathrm{T}}\phi(x_j)$. It can be demonstrated that the problem of estimating the distance between the empirical means of the two domains reduces to

$$\mathrm{MMD}(X_s,X_t)=\mathrm{Tr}(KL),$$

where

$$K = \begin{pmatrix} K_{ss} & K_{st} \\ K_{ts} & K_{tt} \end{pmatrix},$$

is a kernel matrix with $N_s + N_t$ entries formed by three kernel matrices built on the data in the source domain $\boldsymbol{K}_{ss}$, target domain $\boldsymbol{K}_{tt}$, and cross domains, $\boldsymbol{K}_{st}$; and the matrix $\boldsymbol{L}$ is a positive semi-definite matrix with entries $L_{ij} = 1/N_s^2$ if $\boldsymbol{x}_i$ and $\boldsymbol{x}_j$ belong to the source domain, $L_{ij} = 1/N_t^2$ if $\boldsymbol{x}_i$ and $\boldsymbol{x}_j$ belong to the target domain, and $L_{ij} = 1/(N_s N_t)$ otherwise. Listing 12.8 gives an example for computing the MMD.

```
% Xs = data in source domain, Xt = data in the target domain
sigmax = estimateSigma(X,X);
Kss = kernel('rbf',Xs,Xs,sigmax);
Ktt = kernel('rbf',Xt,Xt,sigmax);
Kst = kernel('rbf',Xs,Xt,sigmax);
Kts = kernel('rbf',Xt,Xs,sigmax);
K = [Kss Kst;
     Kts Ktt];
L = [1/(length(Xs).^2)*ones(length(Xs)) ...
    1/(length(Xs)*length(Xt))*ones(length(Xs),length(Xt));
    1/(length(Xs)*length(Xt))*ones(length(Xt),length(Xs)) ...
        1/(length(Xt).^2)*ones(length(Xt))]
MMD = trace(K*L);
```

Listing 12.8 MATLAB code example for MMD.

In the transductive setting, learning the kernel can be solved by learning the kernel matrix $\boldsymbol{K}$ instead. In Pan *et al.* (2008), the resultant kernel matrix learning problem is formulated as an SDP, and PCA is then applied on the learned kernel matrix to find a low-dimensional latent space across domains. This is referred to as MMDE. Three main problems are devised with this approach. First, it is transductive and cannot generalize on unseen patterns. Second, the criterion to be optimized requires the kernel to be positive semi-definite and the resultant kernel learning problem has to be solved by expensive SDP solvers.

TCA instead relies on kernel feature extraction to solve the problem and simply introduces the considerations (1) we want to optimize a projection matrix such that the MMD is minimized after transformation, and (2) the extractor should respect the main properties of the source and target data. For the first condition, and starting from the kernel matrix $\boldsymbol{K}$ given before, it is possible to use a projection matrix $\boldsymbol{W}$ to compute the kernel matrix between mapped samples as $\boldsymbol{K} = \boldsymbol{K}\boldsymbol{W}\boldsymbol{W}^{\mathrm{T}}\boldsymbol{K}$. Afterwards, to obtain the MMD measure for the mapped samples, we simply have

$$\mathrm{MMD}(\boldsymbol{X}_s, \boldsymbol{X}_t) = \mathrm{Tr}(\boldsymbol{K}\boldsymbol{W}\boldsymbol{W}^{\mathrm{T}}\boldsymbol{K}\boldsymbol{L}) = \mathrm{Tr}(\boldsymbol{W}^{\mathrm{T}}\boldsymbol{K}\boldsymbol{L}\boldsymbol{K}\boldsymbol{W}),$$

which can be minimized with respect to $\boldsymbol{W}$. For the second condition, the mapping $\boldsymbol{\phi}$ should not harm the target supervised learning task by deforming too much the input space, so a regularization term is introduced able to preserve (and maximize) the initial data variance in the newly created subspace, $\|\boldsymbol{W}\|^2$. Combining both terms, TCA reduces to solve

$$\min_{\boldsymbol{W}}\{\mathrm{Tr}(\boldsymbol{W}^{\mathrm{T}}\boldsymbol{KLKW}) + \lambda\mathrm{Tr}(\boldsymbol{W}^{\mathrm{T}}\boldsymbol{W})\} \quad \text{s.t.} \quad \boldsymbol{\Sigma} = \boldsymbol{I},$$

where $\boldsymbol{\Sigma} = \boldsymbol{W}^{\mathrm{T}}\boldsymbol{KHKW}$ is the covariance matrix of the projected data, $\boldsymbol{H}$ is a centering matrix in Hilbert spaces, $\boldsymbol{H} = \boldsymbol{I} - 1/(N_s N_t)\boldsymbol{1}\boldsymbol{1}^{\mathrm{T}}$, and λ is a trade-off parameter for tuning the influence of the regularization term and thus controlling the complexity of $\boldsymbol{W}$. Such an optimization problem can be reformulated as a trace maximization problem yielding the following solution: the mapping matrix $\boldsymbol{W}$ is obtained by performing the eigendecomposition of

$$\boldsymbol{M} = (\boldsymbol{KLK} + \lambda\boldsymbol{I})^{-1}\boldsymbol{KHK},$$

and keeping the r eigenvectors associated with the r largest eigenvalues. Once $\boldsymbol{W}$ is available, one can readily compute the r coordinates (the r uncorrelated transfer components) of the mapped samples as $\mathcal{P}(\boldsymbol{X}) = \boldsymbol{KW}$. In this latent subspace where distribution differences are reduced it is now possible to train a supervised classifier on the mapped source labeled samples and subsequently use it to classify the target image embedded in the same subspace. The reader may find MATLAB code for TCA in the book's repository.

12.5.3 Kernel Manifold Alignment

The problem of aligning data manifolds reduces to finding projections between different sets of data, assuming that all lie on a common manifold. Roughly speaking, manifold alignment is a new form of MVA that dates back to the work of Hotelling in 1936 on CCA (Hotelling, 1936), where projections try to correlate the data sources onto a common target domain. The renewed concept of manifold alignment was first introduced by Ham *et al.* (2003), where a manifold constraint was added to the general problem of correlating sets of high-dimensional vectors. The main problem of CCA, its kernel counterpart KCCA (Lai and Fyfe, 2000b), and many other *domain adaptation* methods (Blitzer *et al.* 2006; Daumé III and Marcu 2011; Duan *et al.* 2009, 2012; Ham *et al.* 2003) when confronted with manifold alignment problems is that, unfortunately, points in different sources must be corresponding pairs, which is often hard to meet in real applications. Actually, methods typically assume that the source and target domains are represented by the same features and instances, and try to match the only changing entity, namely the data distributions. Methods in the related field of *transfer learning* also commonly assume the availability of enough labeled data shared across domains (Dai *et al.*, 2008; Pan and Yang, 2010); but again, this is a strong assumption in most of real-life problems.

The problem of manifold alignment without correspondences was tackled by Ham *et al.* (2005) and Wang and Mahadevan (2009) by means of mapping different domains to a new *latent space*, concurrently matching the corresponding instances and preserving the topology of each input domain. The exploitation of unlabeled samples in a semi-supervised setting improves the performance. Nevertheless, while appealing, these methods still require specifying a small amount of cross-domain correspondence relationships. This problem was addressed by Wang and Mahadevan (2011) essentially relaxing the constraint of paired correspondence between feature vectors with the constraint of having the same class labels in all domains. Hence, the algorithm, hereafter called semi-supervised manifold alignment (SSMA), tries to project data from D domains to a latent space $\mathcal{F}$ where samples belonging to the same class become closer, those of different classes are pushed far apart, and the geometry of each domain is preserved. The linear projection method performs well in general but cannot cope with strong nonlinear deformations of the manifolds and high-dimensional problems. Notationally, we are given D domains of data representation, and the corresponding data matrices defined therein, $\boldsymbol{X}_i \in \mathbb{R}^{N_i \times d_i}$, $i = 1, \dots, D$, containing N_i examples (labeled and unlabeled) of dimension d_i, and $N = \sum_i N_i$. The method maps all the data to a latent space $\mathcal{F}$ such that samples belonging to the same class become closer, those of different classes are pushed far apart, and the geometry of the data manifolds is preserved. Therefore, three entities have to be considered per domain: (1) a similarity matrix $\boldsymbol{W}_\text{s}$ will have components $W_\text{s}^{mn} = 1$ if $\boldsymbol{x}_m$ and $\boldsymbol{x}_n$ belong to the same class, and zero otherwise; (2) a dissimilarity matrix $\boldsymbol{W}_\text{d}$ will have entries $W_\text{d}^{mn} = 0$ if $\boldsymbol{x}_m$ and $\boldsymbol{x}_n$ belong to the same class, and one otherwise; and (3) a similarity matrix that represents the topology of each given domain, $\boldsymbol{W}$ (e.g., an RBF kernel.) The three different entities lead to three different graph Laplacians: $\boldsymbol{L}_\text{s}$, $\boldsymbol{L}_\text{d}$, and $\boldsymbol{L}$ respectively. Then, the embedding must minimize a joint cost function essentially given by the eigenvectors corresponding to the smallest nonzero eigenvalues of the generalized eigenvalue problem:

$$\boldsymbol{Z}^\text{T}(\boldsymbol{L} + \mu \boldsymbol{L}_\text{s})\boldsymbol{Z}\boldsymbol{V} = \lambda \boldsymbol{Z}^\text{T}\boldsymbol{L}_\text{d}\boldsymbol{Z}\boldsymbol{V},$$

where $\boldsymbol{Z}$ is a block diagonal matrix containing the data matrices $\boldsymbol{X}_i$ and $\boldsymbol{V}$ contains in the columns the eigenvectors organized in rows for the particular domain, $\boldsymbol{V} = [\boldsymbol{v}_1 | \cdots | \boldsymbol{v}_D]^\text{T}$. The method allows one to extract a maximum of $n_\text{f} = \sum_{i=1}^{D} d_i$ features, which serve for projecting the data to the common latent domain (hereafter called f) as $P_f(\boldsymbol{X}_i) = \boldsymbol{X}_i \boldsymbol{v}_i$. An interesting possibility of the SSMA method is that one can easily project data between domains j and i by simply mapping the source data in $\mathcal{X}_j$ to the latent domain and inverting back to the target domain in $\mathcal{X}_i$ as $P_i(\boldsymbol{X}_j) = \boldsymbol{X}_j \boldsymbol{v}_j \boldsymbol{v}_i^\dagger$, where † represents the pseudo-inverse of the eigenvectors of the target domain. Listing 12.9 gives an example for computing the SSMA.

```
% X = labeled data, U = unlabeled data
Z = blkdiag([X1,U1],[X2,U2]);    % (n1+n2) x (d1+d2)
[L,Ls,Ld] = Laplacians(X1,U1,Y1,X2,U2,Y2);
n = n1+n2; d = d1+d2;
% Combine graph Laplacians
mu = 0.1;
A  = L + mu*Ls; % (n1+n2) x (n1+n2)
B  = Ld;     % (n1+n2) x (n1+n2)
% Solve the generalized eigenproblem
[V D] = eigs(Z'*A*Z,Z'*B*Z,d,'SM');
nf = 4; % features
E1 = V(1:d1,1:nf); E2 = V(d1+1:end,1:nf);
% Project data
PX12 = X1*E1;
```

Listing 12.9 MATLAB code example for SSMA.

The kernelization of the previous method was presented by Tuia and Camps-Valls (2016) as KEMA. The KEMA method consists of the following steps: first, map the data to a Hilbert space, then apply the representer theorem and replace the dot products therein with reproducing kernel functions. Let us first map the D different datasets with D mapping functions to D in principle different Hilbert spaces $\mathcal{H}_i$ of dimension H_i, $\boldsymbol{\phi}_i(\cdot) : \boldsymbol{x} \rightarrow \boldsymbol{\phi}_i(\boldsymbol{x}) \in \mathcal{H}_i$, $i = 1, \ldots, D$. Now, by replacing all the samples with their mapped feature vectors, the problem becomes

$$\boldsymbol{\Phi}^{\mathrm{T}}(\boldsymbol{L} + \mu \boldsymbol{L}_{\mathrm{s}})\boldsymbol{\Phi}\boldsymbol{W} = \lambda \boldsymbol{\Phi}^{\mathrm{T}}\boldsymbol{L}_{\mathrm{d}}\boldsymbol{\Phi}\boldsymbol{W},$$

where $\boldsymbol{\Phi}$ is a block diagonal matrix containing the data matrices $\boldsymbol{\Phi}_i = [\boldsymbol{\phi}_i(\boldsymbol{x}_1)| \cdots |\boldsymbol{\phi}_i(\boldsymbol{x}_{N_i})]^{\mathrm{T}}$ and $\boldsymbol{W}$ contains in in the columns the eigenvectors organized in rows for the particular domain defined in Hilbert space $\mathcal{H}_i$, $\boldsymbol{W} = [\boldsymbol{w}_1| \cdots |\boldsymbol{w}_H]^{\mathrm{T}}$, where $H = \sum_i^D H_i$. This operation is possible thanks to the use of the direct sum of Hilbert spaces, a well-known theory (Reed and Simon, 1981). Note that the eigenvectors $\boldsymbol{w}_i$ are of possibly infinite dimension and cannot be explicitly computed. Instead, we resort to the definition of D corresponding Riesz representation theorems (see Chapter 4) so the eigenvectors can be expressed as a linear combination of mapped samples, $\boldsymbol{w}_i = \boldsymbol{\Phi}_i^{\mathrm{T}}\boldsymbol{\alpha}_i$, and in matrix notation $\boldsymbol{W} = \boldsymbol{\Phi}^{\mathrm{T}}\boldsymbol{\Lambda}$. After replacing in the linear counterpart equation, by premultiplying both sides by $\boldsymbol{\Phi}$, and replacing the dot products with the corresponding kernels, $\boldsymbol{K}_i = \boldsymbol{\Phi}_i\boldsymbol{\Phi}_i^{\mathrm{T}}$ we obtain the final solution:

$$\boldsymbol{K}(\boldsymbol{L} + \mu \boldsymbol{L}_{\mathrm{s}})\boldsymbol{K}\boldsymbol{\Lambda} = \lambda \boldsymbol{K}\boldsymbol{L}_{\mathrm{d}}\boldsymbol{K}\boldsymbol{\Lambda},$$

where $\boldsymbol{K}$ is a block diagonal matrix containing the kernel matrices $\boldsymbol{K}_i$. Now the eigenproblem becomes of size $N \times N$ instead of $d \times d$, and the number of extracted features becomes $n_{\mathrm{f}} = \sum_{i=1}^{D} N_i$. Listing 12.10 gives an example for computing the KEMA, while the full MATLAB KEMA toolbox is available on the book web page.

```
% X = labeled data, U = unlabeled data
K1 = kernelmatrix(X1,U1);    % n1 x n1
K2 = kernelmatrix(X2,U2);    % n2 x n
K  = blkdiag(K1,K2);    % (n1+n2) x (n1+n2)
[L,Ls,Ld] = Laplacians(X1,U1,Y1,X2,U2,Y2);
n  = n1+n2; d = d1+d2;
% Combine graph Laplacians
mu = 0.1;
A = L + mu*Ls;    % (n1+n2) x (n1+n2)
B = Ld; % (n1+n2) x (n1+n2)
% Solve the generalized eigenproblem
[A D] = eigs(K'*A*K,K'*B*K,d,'SM');
nf = 4; % n_f features
A1 = A(1:n1,1:nf); A2 = A(n1+1:end,1:nf);
% Project data
PX12 = K1*A1;
```

Listing 12.10 MATLAB code example for KEMA.

Let us illustrate the performance of KEMA in a toy example involving the alignment of data matrices $\boldsymbol{X}_1$ and $\boldsymbol{X}_2$, which are spirals with three classes. Then, a series of deformations are applied to the second domain: scaling, rotation, inversion of the order of the classes, the shape of the domain (spiral or line) or the data dimensionality. For

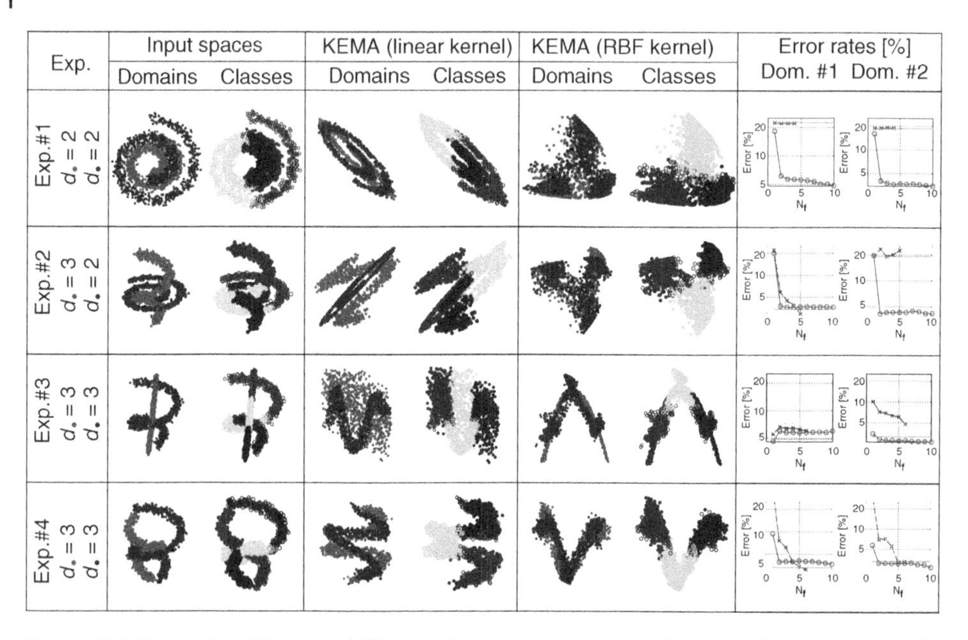

Figure 12.9 Illustration of linear and KEMA on the toy experiments. Left to right: data in the original domains (X1: •; X2: •) and per class (•, • and •), data projected with the linear and the RBF kernels, and error rates as a function of the extracted features when predicting data for the first (left inset) or the second (right inset) domain ($KEMA_{Lin}$, $KEMA_{RBF}$, SSMA, Baseline).

each experiment, 20 labeled pixels per class were sampled in each domain, as well as 1000 unlabeled samples that were randomly selected. Classification performance was assessed on 1000 held-out samples from each domain.

The projections found by KEMA together with a linear and an RBF kernel and the classification errors for the source and target domains can be seen in Figure 12.9. The linear version of KEMA (SSMA) produced good results in experiments #1 and #4, because they are linear transformations of the data. Nevertheless, its performance in experiments #2 and #3 is poor, where the manifolds have undergone stronger deformations. A nonlinear, more flexible version produced better performance, producing an unfolding plus alignment in all cases. The linear classifier trained on the projections of $KEMA_{lin}$ and SSMA does not do a good job in the classification of both domains, in spite of a correct alignment, while the $KEMA_{RBF}$ solution provides a latent space where both domains can be classified correctly. In experiment #2, the solution is quite different: the baseline error depicted by the green line is significantly lower in the source domain. This is because the dataset in three dimensions is linearly separable. Even if the classification of this first domain (•) is correct for all methods, classification after SSMA/$KEMA_{lin}$ projection of the second domain (•) is poor, since their projection in the latent space does not unfold the blue spiral. $KEMA_{RBF}$ provides the best result. Similar results as in experiment #2 can be seen in experiment #3. In experiment #4 one can see a very accurate baseline (both domains are linearly separable in the input spaces) and all methods provide accurate classification accuracies. Again, $KEMA_{RBF}$ provides the best match between the domains in the latent space.

Table 12.4 Properties of manifold alignment and domain adaptation methods.

Method	Supervised	Multisource	Unpaired	Nonlinear	Invertible	Cost
PCA (Jolliffe, 1986)	×	×	√	×	√	$\mathcal{O}(d^3)$
KPCA (Schölkopf *et al.*, 1998)	×	×	√	×	×	$\mathcal{O}(N^3)$
CCA (Hotelling, 1936)	×	√	×	√	√	$\mathcal{O}(d^3)$
KCCA (Lai and Fyfe, 2000b)	×	√	×	√	×	$\mathcal{O}(d^3)$
KTA (Pan and Yang, 2011)	√	×	×	√	√	$\mathcal{O}(3N^2)$
TCA (Pan and Yang, 2011)	×	×	√	×	×	$\mathcal{O}(N^3)$
GM (Tuia *et al.*, 2013)	×	×	√	×	√	$\mathcal{O}(dN^2)$
SSMA (Wang and Mahadevan, 2011)	√	√	√	×	√	$\mathcal{O}(d^3)$
KEMA (Tuia and Camps-Valls, 2016)	√	√	√	√	√	$\mathcal{O}(N^3)$

12.5.4 Relations between Domain Adaptation Methods

The properties of the methods analyzed are summarized and compared in Table 12.4. We include in the table some of the methods presented in Section 12.2 for comparison. The KMM method is not looking for a transformation in strict terms and depends on the optimization of a QP problem; thus, we do not include it in the table since comparison could be misleading. We have included KTA in the table, although it is usually used to fit parameters instead of to find a transformation. SSMA and KEMA, like CCA and KCCA, can in principle align multisource data, but importantly they do not need unpaired data points. This may be useful to align datasets of different features and number of samples, provided that a reasonable amount of labeled information is available. Contrarily to PCA, KPCA, TCA, and graph matching, KEMA does not necessarily require a leading source domain to which all the others are aligned to. TCA and KEMA, however, are more computationally demanding than the linear counterpart as the eigenproblem becomes $N \times N$ instead of $d \times d$. Nevertheless, two clarifications must be done here: first, both SSMA and KEMA involve the same construction of the graph Laplacian, which takes most of the time but can be computed just once and offline; and second, KEMA with linear kernels can efficiently solve the SSMA problem for high-dimensional data.

12.5.5 Experimental Comparison between Domain Adaptation Methods

We here evaluate numerically the standard manifold aligment algorithms. In particular, we compare KEMA (Tuia and Camps-Valls, 2016), SGF (Gopalan *et al.*, 2011), GFK (Gong *et al.*, 2012), OT-lab (Courty *et al.*, 2014), and MMDT (Hoffman *et al.*, 2013). The task is to find discriminative aligning projections for visual object recognition tasks. We use the dataset introduced in Saenko *et al.* (2010) in which we consider four domains Webcam (`W`), Caltech (`C`), Amazon (`A`), and DSLR (`D`), and selected the 10 common classes in the four datasets following Gong *et al.* (2012). By doing so, the domains contain 295 (Webcam), 1123 (Caltech), 958 (Amazon) and 157 (DSLR) images. The

Table 12.5 Accuracy in the visual object recognition study (C: Caltech, A: Amazon, D: DSLR, W: Webcam). 1-NN classification testing on all samples from the target domain.

	Train on source	Domain adaptation method								Train on target
		Unsupervised		Labeled from s only		Labeled from $s+t$				
	No adapt.	SGF*	GFK*	GFK*	OT-lab*	GFK†	MMDT†	KEMA K_l	KEMA K_{χ_2}	No. adapt.
N_S	0	0	0	20	20	20	20	20	20	0
N_T	0	0	0	0	0	3	3	3	3	8
C → A	21.4	36.8	36.9	40.4	43.5	44.7	49.4	47.1	47.9	35.4
C → D	12.3	32.6	35.2	41.1	41.8	57.7	56.5	61.5	63.4	65.1
A → C	19.9	35.3	35.6	37.9	35.2	36.0	36.4	29.5	30.4	28.4
A → W	17.5	31.0	34.4	35.7	38.4	58.6	64.6	65.4	66.5	63.5
W → C	24.2	21.7	27.2	29.3	35.5	31.1	32.2	32.9	32.4	28.4
W → A	27.0	27.5	31.1	35.5	40.0	44.1	47.7	44.9	45.9	35.4
D → A	19.0	32.0	32.5	36.1	34.9	45.7	46.9	44.2	45.2	35.4
D → W	37.4	66.0	74.9	79.1	84.2	76.5	74.1	64.1	66.7	63.5
Mean	22.34	35.36	38.48	41.89	44.19	49.30	50.98	48.70	49.80	44.39

N_{domain}: number of labels per class; $*$: results from Courty *et al.* (2014); †: results from Hoffman *et al.* (2013).

features were extracted as described in Saenko *et al.* (2010): we use an 800-dimensional normalized histogram of visual words obtained from a codebook constructed from a subset of the Amazon dataset on points of interest detected by the "speeded up robust features" method. We used the same experimental setting as Gopalan *et al.* (2011) and Gong *et al.* (2012), in order to compare with these unsupervised domain adaptation methods.

For all methods, we used 20 labeled pixels per class in the source domain for the C, A, and D domains and eight samples per class for the W domain. After alignment, an ordinary 1-NN classifier was trained with the labeled samples. The same labeled samples in the source domain were used to define the PLS eigenvectors for GFK and OT-lab. For all the methods using labeled samples in the target domain (including KEMA), we used three labeled samples in the target domain to define the projections. In all cases, we used sensible kernels for this problem in KEMA: the (fast) histogram intersection kernel, $K_i(\mathbf{x}_j, \mathbf{x}_k) = \sum_d \min\{x_j^d, x_k^d\}$, and the χ_2 kernel, $K_{\chi_2}(\mathbf{x}_j, \mathbf{x}_k) = \exp[-\chi^2/(2\sigma^2)]$, with $\chi^2 = (1/2)\sum_d (x_j^d - x_k^d)^2/(x_j^d + x_k^d)$ (Sreekanth *et al.*, 2010). We used $u = 300$ unlabeled samples to compute the graph Laplacians, for which a k-NN graph with $k = 21$ was used.

The performances in all problems are shown in Table 12.5: KEMA has the best performance of all methods and, almost in all cases, it improves the results obtained by the semi-supervised methods using labels in the source domain only. In three of the eight settings, the best results are obtained by KEMA, when compared with state-of-the-art (semi-)supervised algorithms, and similar performance to state-of-the-art GFK in six out of the eight settings. The advantage of KEMA is that it handles domains of different dimensionality without requiring a discriminative classifier to align the domains, such as for MMDT.

12.6 Concluding Remarks

We reviewed the field of kernel feature extraction and dimensionality reduction. This field of signal processing and machine learning deals with the challenging problem of finding a proper feature representation for the data. We have seen that selecting different criteria to optimize gives rise to particularly different optimization problems, solutions, and methods. We started our trip under the framework of multivariate analysis. The techniques described in this chapter are used in real-world application with an increased level of popularity. Following the popular PCA technique, there is a large amount of linear and kernel-based methods that in general perform better in supervised applications, for they find projections that maximize the alignment with the target variables. The common features and differences between methods have been analyzed, as well as the relationships to existing kernel discriminative feature extraction and statistical dependence estimation approaches. We also studied recent methods to make kernel MVA more suitable to real-life applications, both for large-scale data sets and for problems with few labeled data. Solutions include sparse and SSL extensions. Actually, seeking for the appropriate features that facilitate classification or regression cuts to the heart of manifold learning. We provided examples that show the properties of the methods using challenging real-world data which exhibit complex manifolds. We completed the chapter by reviewing the field of domain adaptation and the different approaches to the problem from the kernel formulation point of view.

12.7 Questions and Problems

Exercise 12.7.1 When it is useful to use KPCA instead of PCA?

Exercise 12.7.2 In what situation is it useful to use PLS instead of PCA?

Exercise 12.7.3 Using the iris dataset from MATLAB, predict the fourth dimension using the remainder three, but reducing previously the dimensionality to two dimensions using the methods proposed in Section 12.2.

Exercise 12.7.4 In the glass dataset from MATLAB, compute the HSIC between each pair of variables.

Exercise 12.7.5 Develop mathematically how Equation 12.7 can be reduced to Equation 12.8.

Exercise 12.7.6 Reproduce Figure 12.5 using the crab dataset from MATLAB.

Exercise 12.7.7 In what situation is it useful to use KICA instead of linear ICA?

Exercise 12.7.8 Obtain the actualization rule for OKECA.

Exercise 12.7.9 Implement HSIC using Cholesky decomposition.

Exercise 12.7.10 Implement KPCA imposing sparsity in the projection vectors.

References

Abbate, A., Koay, J., Frankel, J., Schroeder, S., and Das, P. (1997). Signal detection and noise suppression using a wavelet transform signal processor: application to ultrasonic flaw detection. *IEEE Transactions on Ultrasonics, Ferroelectrics, and Frequency Control*, **44**(1), 14–26.

Abe, N. and Mamitsuka, H. (1998). Query learning strategies using boosting and bagging. In J. Shavlik, editor, *ICML '98 Proceedings of the Fifteenth International Conference on Machine Learning*, pages 1–9. Morgan Kaufmann, San Francisco, CA.

Abrahamsen, T. J. and Hansen, L. K. (2011). A cure for variance inflation in high dimensional kernel principal component analysis. *Journal of Machine Learning Research*, **12**, 2027–2044.

Adali, T. and Liu, X. (1997). Canonical piecewise linear network for nonlinear filtering and its application to blind equalization. *Signal Processing*, **61**(2), 145–155.

Aharon, M., Elad, M., and Bruckstein, A. (2006). K-SVD: an algorithm for designing overcomplete dictionaries for sparse representation. *IEEE Transactions on Signal Processing*, **54**(11), 4311–4322.

Ahmed, T., Oreshkin, B., and Coates, M. (2007). Machine learning approaches to network anomaly detection. In J. Chase and I. Cohen, editors, *SYSML'07 Proceedings of the 2nd USENIX Workshop on Tackling Computer Systems Problems with Machine Learning Techniques*, SYSML'07, pages 7:1–7:6. USENIX Association, Berkeley, CA.

Aizerman, M. A., Braverman, E. M., and Rozoner, L. (1964). Theoretical foundations of the potential function method in pattern recognition learning. *Automation and Remote Control*, **25**, 821–837.

Akaike, H. (1974). A new look at the statistical model identification. *IEEE Transactions on Automatic Control*, **19**(6), 716–723.

Allwein, E. L., Schapire, R. E., and Singer, Y. (2001). Reducing multiclass to binary: a unifying approach for margin classifiers. *Journal of Machine Learning Research*, **1**, 113–141.

Alper, P. (1965). A consideration of the discrete volterra series. *IEEE Transactions on Automatic Control*, **10**(3), 322–327.

Altun, Y., Tsochantaridis, I., and Hofmann, T. (2003). Hidden Markov support vector machines. In T. Fawcett and N. Mishra, editors, *Proceedings of the Twentieth International Conference on Machine Learning (ICML-2003), Washington, DC*, pages 3–10. AAAI Press, Menlo Park, CA.

Digital Signal Processing with Kernel Methods, First Edition. José Luis Rojo-Álvarez, Manel Martínez-Ramón, Jordi Muñoz-Marí, and Gustau Camps-Valls.

Altun, Y., Hofmann, T., and Tsochantaridis, I. (2007). Support vector learning for interdependent and structured output spaces. In G. Bakır, T. Hofmann, B. Schölkopf, A. J. Smola, and S. Vishwanathan, editors, *Predicting Structured Data*, pages 85–105. MIT Press.

Anderson, J. R. (1983). *The Architecture of Cognition*. Harvard University Press, Cambridge, MA, USA.

Andrieu, C., de Freitas, N., Doucet, A., and Jordan, M. (2003). An introduction to MCMC for machine learning. *Machine Learning*, **50**, 5–43.

Anguita, D. and Gagliolo, M. (2002). MDL based model selection for relevance vector regression. In J. Dorronsoro, editor, *Artificial Neural Networks – ICANN 2002*, pages 468–473. Springer, Berlin.

Arenas-García, J. and Camps-Valls, G. (2008). Efficient kernel orthonormalized PLS for remote sensing applications. *IEEE Transactions on Geoscience and Remote Sensing*, **46**, 2872–2881.

Arenas-García, J. and Figueiras-Vidal, A. R. (2009). Adaptive combination of proportionate filters for sparse echo cancellation. *IEEE Transactions on Audio, Speech & Language Processing*, **17**(6), 1087–1098.

Arenas-García, J. and Petersen, K. B. (2009). Kernel multivariate analsis in remote sensing feature extraction. In G. Camps-Valls and L. Bruzzone, editors, *Kernel Methods for Remote Sensing Data Analysis*. John Wiley & Sons.

Arenas-García, J., Figueiras-Vidal, A. R., and Sayed, A. H. (2006). Mean-square performance of a convex combination of two adaptive filters. *IEEE Transactions on Signal Processing*, **54**(3), 1078–1090.

Arenas-García, J., Petersen, K. B., and Hansen, L. K. (2007). Sparse kernel orthonormalized PLS for feature extraction in large data sets. In B. Schölkopf, J. C. Platt, and T. Hoffman, editors, *NIPS'06 Proceedings of the 19th International Conference on Neural Information Processing Systems*. MIT Press, Cambridge, MA.

Arenas-García, J., Petersen, K. B., Camps-Valls, G., and Hansen, L. K. (2013). Kernel multivariate analysis framework for supervised subspace learning. *Signal Processing Magazine*, **30**(4), 16–29.

Aronszajn, N. (1950). Theory of reproducing kernels. *Transactions of the American Mathematical Society*, **68**(3), 337–404.

Arulampalam, M. S., Maskell, S., Gordon, N., and Clapp, T. (2002). A tutorial on particle filters for online nonlinear/non-Gaussian Bayesian tracking. *IEEE Transactions on Signal Processing*, **50**(2), 174–188.

Aschbacher, E. and Rupp, M. (2005). Robustness analysis of a gradient identification method for a nonlinear Wiener system. In *Proceedings of the 2005 IEEE/SP 13th Workshop on Statistical Signal Processing (SSP 2005)*, Bordeaux, France.

Atkinson, K. (1978). *An Introduction to Numerical Analysis*. John Wiley & Sons.

Aurenhammer, F. (1991). Voronoi diagrams – a survey of a fundamental geometric data structure. *ACM Computing Surveys*, **23**(3), 345–405.

Babaie-Zadeh, M., Jutten, C., and Nayebi, K. (2002). A geometric approach for separating post nonlinear mixtures. In *Proceedings of the 11th European Signal Processing Conference (EUSIPCO)*, volume II, pages 11–14. IEEE.

Bach, F. and Jordan, M. (2004). Blind one-microphone speech separation: a spectral learning approach. In L. Saul, Y. Weiss, and L. Bottou, editors, *Proceedings of the 17th*

Annual Conference on Neural Information Processing Systems (NIPS 2004), pages 65–72. MIT Press, Cambridge, MA.

Bach, F. and Jordan, M. (2005a). Predictive low-rank decomposition for kernel methods. In *ICML '05 Proceedings of the 22nd International Conference on Machine Learning*. ACM, New York.

Bach, F. and Jordan, M. I. (2002). Kernel independent component analysis. *Journal of Machine Learning Research*, **3**, 1–48.

Bach, F. R. and Jordan, M. I. (2005b). Predictive low-rank decomposition for kernel methods. In *ICML '05 Proceedings of the 22nd International Conference on Machine Learning*, pages 33–40. ACM, New York.

Bach, F. R., Lanckriet, G. R., and Jordan, M. I. (2004). Multiple kernel learning, conic duality, and the SMO algorithm. In *ICML '04 Proceedings of the 21st International Conference on Machine Learning*, page 6. ACM, New York.

Bahl, P. and Padmanabhan, V. (2000). RADAR: a in-building RF based user location and tracking system. In *IEEE INFOCOM '2000 Proceedings of the 19th Annual Joint Conference of the IEEE Computer and Communications Societies (Cat. No.00CH37064)*, pages 775–784. IEEE Press.

Bai, E.-W. (1998). An optimal two stage identification algorithm for Hammerstein–Wiener nonlinear systems. In *Proceedings of the 1998 American Control Conference*, volume 5, pages 2756–2760.

Bai, W., He, C., Jiang, L. G., and Li, X. X. (2003). Robust channel estimation in MIMO–OFDM systems. *Electronics Letters*, **39**(2), 242–244.

Bajwa, W. U., Haupt, J., Sayeed, A. M., and Nowak, R. (2010). Compressed channel sensing: a new approach to estimating sparse multipath channels. *Proceedings of the IEEE*, **98**(6), 1058–1076.

Baker, C. (1973). Joint measures and cross-covariance operators. *Transactions of the American Mathematical Society*, **186**, 273–289.

Bakır, G., Hofmann, T., Schölkopf, B., Smola, A., Taskar, B., and Vishwanathan, S., editors (2007). *Predicting Structured Data*. MIT Press, Cambridge, MA.

Balestrino, A., Landi, A., Ould-Zmirli, M., and Sani, L. (2001). Automatic nonlinear auto-tuning method for Hammerstein modeling of electrical drives. *IEEE Transactions on Industrial Electronics*, **48**(3), 645–655.

Ball, J. E. and Bruce, L. M. (2007). Level set hyperspectral image classification using best band analysis. *IEEE Transactions on Geoscience and Remote Sensing*, **45**(10), 3022–3027.

Banham, M. and Katsaggelos, A. (1997). Digital image restoration. *IEEE Signal Processing Magazine*, **14**, 24–41.

Barker, M. and Rayens, W. (2003). Partial least squares for discrimination. *Journal of Chemometrics*, **17**, 166–173.

Basseville, M. and Nikiforov, I. V. (1993). *Detection of Abrupt Changes: Theory and Application*. Prentice-Hall, Upper Saddle River, NJ.

Bayro-Corrochano, E. J. and Arana-Daniel, N. (2010). Clifford support vector machines for classification, regression, and recurrence. *IEEE Transactions on Neural Networks*, **21**(11), 1731–1746.

Belkin, M. and Niyogi, P. (2003). Laplacian eigenmaps for dimensionality reduction and data representation. *Neural Computation*, **15**(6), 1373–1396.

Belkin, M. and Niyogi, P. (2004). Semi-supervised learning on Riemannian manifolds. *Machine Learning, Special Issue on Clustering*, **56**, 209–239.

Belkin, M., Niyogi, P., and Sindhwani, V. (2006). Manifold regularization: a geometric framework for learning from labeled and unlabeled examples. *Journal of Machine Learning Research*, **7**, 2399–2434.

Bell, A. J. and Sejnowski, T. J. (1995). An information-maximization approach to blind separation and blind deconvolution. *Neural Computation*, **7**(6), 1129–1159.

Belouchrani, A. and Amin, M. G. (1998). Blind source separation based on time–frequency signal representations. *IEEE Transactions on Signal Processing*, **46**(11), 2888–2897.

Ben-David, S., Blitzer, J., Crammer, K., and Pereira, O. (2007). Analysis of representations for domain adaptation. In B. Schölkopf, J. Platt, and T. Hofmann, editors, *Advances in Neural Information Processing Systems 19, Proceedings of the 2006 Conference*. MIT Press, Cambridge, MA.

Bermejo, J., Antoranz, J., Burwash, I., Rojo-Álvarez, J. L., Moreno, M., García-Fernández, M., and Otto, C. (2002). In-vivo analysis of the instantaneous transvalvular pressure difference in aortic valve stenosis. implications of unsteady fluid-dynamics for the clinical assessment of disease severity. *Journal of Heart Valve Disease*, **11**(4), 557–566.

Bicego, M., Ulaş, A., Castellani, U., Perina, A., Murino, V., Martins, A. F. T., Aguiar, P. M. Q., and Figueiredo, M. A. T. (2013). Combining information theoretic kernels with generative embeddings for classification. *Neurocomputing*, **101**, 161–169.

Bickel, S., Brückner, M., and Scheffer, T. (2009). Discriminative learning under covariate shift. *Journal of Machine Learning Research*, **10**, 2137–2155.

Billings, S. (1980). Identification of nonlinear systems: a survey. *Proceedings of IEE, Part D*, **127**, 272–285.

Billings, S. A. and Fakhouri, S. Y. (1977). Identification of nonlinear systems using the Wiener model. *Electronics Letters*, **13**(17), 502–504.

Billings, S. A. and Fakhouri, S. Y. (1979). Nonlinear system identification using the Hammerstein model. *International Journal of System Sciences*, **10**(5), 567–578.

Billings, S. A. and Fakhouri, S. Y. (1982). Identification of systems containing linear dynamic and static nonlinear elements. *Automatica*, **18**, 15–26.

Bimbot, F., Bonastre, J.-F., Fredouille, C., Gravier, G., Magrin-Chagnolleau, I., Meignier, S., Merlin, T., Ortega-García, J., Petrovska-Delacrétaz, D., and Reynolds, D. A. (2004). A tutorial on text-independent speaker verification. *EURASIP Jounal of Applied Signal Processing*, **2004**, 430–451.

Bishop, C. (2006). *Pattern Recognition and Machine Learning*. Springer.

Bishop, C. and Tipping, M. (2000). Variational relevance vector machines. In C. Boutilier and M. Goldszmidt, editors, *Proceedings of 16th Conference on Uncertainty in Artificial Intelligence*, pages 46–53. Morgan Kaufmann, San Francisco, CA.

Blankertz, B., Muller, K.-R., Curio, G., Vaughan, T., Schalk, G., Wolpaw, J., Schlogl, A., Neuper, C., Pfurtscheller, G., Hinterberger, T., Schroder, M., and Birbaumer, N. (2004). The BCI competition 2003: progress and perspectives in detection and discrimination of EEG single trials. *IEEE Transactions on Biomedical Engineering*, **51**(6), 1044–1051.

Blaschko, M. and Lampert, C. (2008). Learning to localize objects with structured output regression. In D. Forsyth, P. Torr, and A. Zisserman, editors, *Computer Vision: ECCV 2008*, pages 2–15. Springer.

Blaschko, M., Shelton, J., Bartels, A., Lampert, C., and Gretton, A. (2011). Semi-supervised kernel canonical correlation analysis with application to human fMRI. *Pattern Recognition Letters*, **32**, 1572–1583.

Blattberg, R. and Neslin, S. (1993). Sales promotion models. In J. Eliashberg and G. Lilien, editors, *Handbooks in Operations Research and Management Science: Marketing Models*, pages 553–609. North-Holland, Amsterdam.

Blitzer, J., McDonald, R., and Pereira, F. (2006). Domain adaptation with structural correspondence learning. In *EMNLP '06 Proceedings of the 2006 Conference on Empirical Methods in Natural Language Processing*, pages 120–128. Association for Computational Linguistics, Stroudsburg, PA.

Bloomfield, P. and Steiger, W. (1980). Least absolute deviations curve-fitting. *SIAM Journal on Scientific Computing*, **1**(2), 290–301.

Bofill, P. and Zibulevsky, M. (2001). Underdetermined blind source separation using sparse representations. *Signal Processing*, **81**(11), 2353–2362.

Bookstein, F. (1989). Principal warps: thin-plate splines and the decomposition of deformations. *IEEE Transactions on Pattern Analysis and Machine Intelligence*, **11**(6), 567–585.

Bordes, A., Ertekin, S., Weston, J., and Bottou, L. (2005). Fast kernel classifiers with online and active learning. *Journal of Machine Learning Research*, **6**, 1579–1619.

Borga, M., Landelius, T., and Knutsson, H. (1997). A unified approach to PCA, PLS, MLR and CCA. Technical Report LiTH-ISY-R, 1400-3902, Linköping University.

Borgwardt, K., Gretton, A., Rasch, M., Kriegel, H.-P., Schoelkopf, B., and Smola, A. (2006). Integrating structured biological data by kernel maximum mean discrepancy. *Bioinformatics (ISMB)*, **22**(14), e49–e57.

Bose, N. K. and Basu, S. (1978). Multidimensional systems theory: matrix Padé approximants. In *1978 IEEE Conference on Decision and Control including the 17th Symposium on Adaptive Processes*, pages 653–657. IEEE Press.

Boser, B. E., Guyon, I., and Vapnik, V. N. (1992). A training algorithm for optimal margin classifiers. In *COLT '92 Proceedings of the Fifth Annual Workshop on Computational Learning Theory*, pages 144–152. ACM.

Bottou, L. and Bengio, Y. (1995). Convergence properties of the k-means algorithms. In G. Tesauro, D. S. Touretzky, and T. K. Leen, editors, *Advances in Neural Information Processing Systems 7*, pages 585–592. MIT Press, Cambridge, MA.

Bottou, L., Chapelle, O., DeCoste, D., and Weston, J., editors (2007). *Large Scale Kernel Machines*. MIT Press, Cambridge, MA.

Bouboulis, P. and Theodoridis, S. (2010). The complex Gaussian kernel LMS algorithm. In K. Diamantaras, W. Duch, and L. Iliadis, editors, *Artificial Neural Networks – ICANN 2010*, volume 6353 of *Lecture Notes in Computer Science*, pages 11–20. Springer, Berlin.

Bouboulis, P. and Theodoridis, S. (2011). Extension of Wirtinger's calculus to reproducing kernel Hilbert spaces and the complex kernel LMS. *IEEE Transactions on Signal Processing*, **59**(3), 964–978.

Bouboulis, P., Theodoridis, S., and Mavroforakis, M. (2012). The augmented complex kernel LMS. *IEEE Transactions on Signal Processing*, **60**(9), 4962–4967.

Boyd, S. and Chua, L. (1985). Fading memory and the problem of approximating nonlinear operators with Volterra series. *IEEE Transactions on Circuits and Systems*, **32**(11), 1150–1161.

Bradley, D. and Bagnell, J. (2008). Differentiable sparse coding. In D. Koller, D. Schuurmans, Y. Bengio, and L. Bottou, editors, *NIPS'08 Proceedings of the 21st International Conference on Neural Information Processing Systems*, pages 113–120. Curran Associates.

Bradley, P. S. and Mangasarian, O. L. (1998). Feature selection via concave minimization and support vector machines. In J. Shavlik, editor, *ICML '98 Proceedings of the Fifteenth International Conference on Machine Learning*, pages 82–90. Morgan Kaufmann, San Francisco, CA.

Braun, M. L., Buhmann, J. M., and Müller, K.-R. (2008). On relevant dimensions in kernel feature spaces. *Journal of Machine Learning Research*, **9**, 1875–1908.

Bredensteiner, E. J. and Bennett, K. P. (1999). Multicategory classification by support vector machines. *Computational Optimization and Applications*, **12**(1), 53–79.

Breiman, L. (1994). Bagging predictors. Technical Report 421, University of California at Berkley.

Brilliant, M. B. (1958). Theory of the analysis of nonlinear systems. RLE Technical Report 345, MIT.

Brinker, K. (2003). Incorporating diversity in active learning with support vector machines. In *ICML'03 Proceedings of the Twentieth International Conference on International Conference on Machine Learning*. AAAI Press.

Bruls, J., Chou, C., Haverkamp, B., and Verhaegen, M. (1999). Linear and nonlinear system identification using separable least-squares. *European Journal of Control Engineering Practice*, **5**(1), 116–128.

Bugallo, M. F., Martino, L., and Corander, J. (2015). Adaptive importance sampling in signal processing. *Digital Signal Processing*, **47**, 36–49.

Burg, J. P. (1967). Maximum entropy spectral analysis. In *Proceedings of 37th Meeting of Society of Exploration Geophysicists*, Oklahoma City, OK.

Burges, C. (1998). A tutorial on support vector machines for pattern recognition. *Data Mining and Knowledge Discovery*, **2**(2), 1–32.

Burges, C. J. C. (1999). Geometry and invariance in kernel based methods. In B. Schölkopf, C. J. C. Burges, and A. J. Smola, editors, *Advances in Kernel Methods – Support Vector Learning*, pages 89–116. MIT Press, Cambridge, MA.

Burrus, C. S. and Parks, T. W. (1970). Time domain design of recursive digital filters. *IEEE Transactions on Audio and Electroacoustics*, **18**(2), 137–141.

Butzer, P. L. and Stens, R. L. (1992). Sampling for not necessarily band-limited functions: a historical overview. *SIAM Review*, **34**(1), 40–53.

Byrne, C. L. and Fitzgerald, R. (1983). An approximation theoretic approach to maximum entropy spectral analysis. *IEEE Transactions on Acoustics, Speech and Signal Processing*, **31**(3), 734–736.

Cambanis, S., Huang, S., and Simons, G. (1981). On the theory of elliptically contoured distributions. *Journal of Multivariate Analysis*, **11**(3), 368 – 385.

Campbell, C., Cristianini, N., and Smola, A. (2000). Query learning with large margin classifiers. In P. Langley, editor, *ICML '00 Proceedings of the Seventeenth International Conference on Machine Learning*, pages 111 – 118. Morgan Kaufmann, San Francisco, CA.

Camps-Valls, G. (2016). Kernel spectral angle mapper. *Electronics Letters*, **52**(14), 1218–1220.

Camps-Valls, G. and Bruzzone, L. (2005). Kernel-based methods for hyperspectral image classification. *IEEE Transactions on Geoscience and Remote Sensing*, **43**, 1351–1362.

Camps-Valls, G. and Bruzzone, L., editors (2009). *Kernel methods for Remote Sensing Data Analysis*. Wiley & Sons, UK.

Camps-Valls, G., Soria-Olivas, E., Pérez-Ruixo, J., Artés-Rodríguez, A., Pérez-Cruz, F., and Figueiras-Vidal, A. (2001). A profile-dependent kernel-based regression for cyclosporine concentration prediction. In *Neural Information Processing Systems, NIPS'01. Workshop on New Directions in Kernel-based Learning Methods.*

Camps-Valls, G., Soria-Olivas, E., Pérez-Ruixo, J., Artés-Rodríguez, A., Pérez-Cruz, F., and Figueiras-Vidal, A. (2002). Cyclosporine concentration prediction using clustering and support vector regression methods. *Electronics Letters*, **38**(12), 568–570.

Camps-Valls, G., Martínez-Ramón, M., Rojo-Álvarez, J. L., and Soria-Olivas, E. (2004). Robust γ-filter using support vector machines. *Neurocomputing*, **62**, 493–499.

Camps-Valls, G., Gómez-Chova, L., Muñoz-Marí, J., Vila-Francés, J., and Calpe-Maravilla, J. (2006a). Composite kernels for hyperspectral image classification. *IEEE Geoscience and Remote Sensing Letters*, **3**(1), 93–97.

Camps-Valls, G., Gómez-Chova, L., Vila-Francés, J., Amorós-López, J., Muñoz-Marí, J., and Calpe-Maravilla, J. (2006b). Retrieval of oceanic chlorophyll concentration with relevance vector machines. *Remote Sensing of Environment*, **105**(1), 23–33.

Camps-Valls, G., Bruzzone, L., Rojo-Álvarez, J. L., and Melgani, F. (2006c). Robust support vector regression for biophysical parameter estimation from remotely sensed images. *IEEE Geoscience and Remote Sensing Letters*, **3**(3), 339–343.

Camps-Valls., G., Rojo-Álvarez, J. L., and Martínez-Ramón, M., editors (2007). *Kernel Methods in Bioengineering, Signal, and Image Processing*. Idea Group Inc., Hershey, PA.

Camps-Valls, G., Martínez-Ramón, M., Rojo-Álvarez, J. L., and Muñoz-Marí, J. (2007a). Non-linear system identification with composite relevance vector machines. *IEEE Signal Processing Letters*, **14**(4), 279–282.

Camps-Valls, G., Bandos, T., and Zhou, D. (2007b). Semi-supervised graph-based hyperspectral image classification. *IEEE Transactions on Geoscience and Remote Sensing*, **45**(10), 2044–3054.

Camps-Valls, G., Gómez-Chova, L., Muñoz-Marí, J., Rojo-Álvarez, J. L., and Martínez-Ramón, M. (2008). Kernel-based framework for multitemporal and multisource remote sensing data classification and change detection. *IEEE Transactions on Geoscience and Remote Sensing*, **46**(6), 1822–1835.

Camps-Valls, G., Muñoz-Marí, J., Gómez-Chova, L., Richter, K., and Calpe-Maravilla, J. (2009a). Biophysical parameter estimation with a semisupervised support vector machine. *IEEE Geoscience and Remote Sensing Letters*, **6**(2), 248–252.

Camps-Valls, G., Muñoz-Marí, J., Martínez-Ramón, M., Requena-Carrion, J., and Rojo-Álvarez, J. L. (2009b). Learning non-linear time-scales with kernel gamma-filters. *Neurocomputing*, **72**(4–6, SI), 1324–1328.

Candes, E. J. and Wakin, M. B. (2008). An introduction to compressive sampling. *IEEE Signal Processing Magazine*, **25**(2), 21–30.

Candes, E. J., Romberg, J., and Tao, T. (2006). Robust uncertainty principles: exact signal reconstruction from highly incomplete frequency information. *IEEE Transactions on Information Theory*, **52**(2), 489–509.

Cao, L. J., Keerthi, S. S., Ong, C.-J., Zhang, J. Q., Periyathamby, U., Fu, X. J., and Lee, H. P. (2006). Parallel sequential minimal optimization for the training of support vector machines. *IEEE Transactions on Neural Networks*, **17**, 1039–1049.

Capon, J. (1969). High-resolution frequency-wavenumber spectrum analysis. *Proceedings of the IEEE*, **57**(8), 1408–1418.

Cardoso, J. (1998). Blind signal separation: statistical principles. *Proceedings of the IEEE*, **9**(10), 2009–2025.

Cardoso, J.-F. and Souloumiac, A. (1993). Blind beamforming for non-Gaussian signals. *IEE Proceedings F – Radar and Signal Processing*, **140**(6), 362–370.

Casdagli, M. and Eubank, S., editors (1992). *Nonlinear Modeling and Forecasting*, volume XII. Addison-Wesley, Reading, MA.

Cawley, G. C. and Talbot, N. L. C. (2002). Reduced rank kernel ridge regression. *Neural Processing Letters*, **16**(3), 293–302.

Chandola, V., Banerjee, A., and Kumar, V. (2009). Anomaly detection: a survey. *ACM Compututing Surveys*, **41**(3), 15:1–15:58.

Chang, C.-C. and Lin, C.-J. (2002). Training nu-support vector regression: theory and algorithms. *Neural Computation*, **14**(8), 1959–1978.

Chang, E., Zhu, K., Wang, H., Bai, H., Li, J., Qiu, Z., and Cui, H. (2008). Parallelizing support vector machines on distributed computers. In J. C. Platt, D. Koller, Y. Singer, and S. Roweis, editors, *Advances in Neural Information Processing Systems 20*, pages 257–264. MIT Press, Cambridge, MA.

Chapelle, O. (2007). Training a support vector machine in the primal. *Neural Computation*, **19**(5), 1155–1178.

Chapelle, O., Weston, J., and Schölkopf, B. (2003). Cluster kernels for semi-supervised learning. In S. Becker, S. Thrun, and K. Obermayer, editors, *Advances in Neural Information Processing Systems 15*, pages 601–608. MIT Press, Cambridge, MA.

Chapelle, O., Schölkopf, B., and Zien, A. (2006). *Semi-Supervised Learning*. MIT Press, Cambridge, MA.

Chen, B., Zhao, S., Zhu, P., and Príncipe, J. C. (2012). Quantized kernel least mean square algorithm. *IEEE Transactions on Neural Networks and Learning Systems*, **23**(1), 22–32.

Chen, H.-W. (1995). Modeling and identification of parallel nonlinear systems: structural classification and parameter estimation methods. *Proceedings of the IEEE*, **83**(1), 39–66.

Chen, S., Billings, S. A., and Luo, W. (1989). Orthogonal least squares methods and their application to non-linear system identification. *International Journal of Control*, **50**, 1873–1896.

Chen, S., Gunn, S., and Harris, C. (2000). Decision feedback equalizer design using support vector machines. *IEE Proceedings – Vision, Image and Signal Processing*, **147**(3), 213–219.

Chen, S., Sanmigan, A. K., and Hanzo, L. (2001a). Adaptive mutiuser receiver using a support vector machine technique. In *IEEE VTS 53rd Vehicular Technology Conference, Spring 2001. Proceedings (Cat. No.01CH37202)*, pages 604–608. IEEE Press, Piscataway, NJ.

Chen, S., Sanmigan, A. K., and Hanzo, L. (2001b). Support vector machine multiuser receiver for DS-CDMA signals in multipath channels. *Neural Networks*, **12**(3), 604–611.

Chen, S. S., Donoho, D. L., Michael, and Saunders, A. (1998). Atomic decomposition by basis pursuit. *SIAM Journal on Scientific Computing*, **20**, 33–61.

Cheng, M., Pun, C.-M., and Tang, Y. Y. (2014). Nonnegative class-specific entropy component analysis with adaptive step search criterion. *Pattern Analysis and Applications*, **17**(1), 113–127.

Cheng, S. and Shih, F. Y. (2007). An improved incremental training algorithm for support vector machines using active query. *Pattern Recognition*, **40**, 964 – 971.

Cherkassky, V. and Ma, Y. (2004). Practical selection of SVM parameters and noise estimation for SVM regression. *Neural Networks*, **17**(1), 113–126.

Cherkassky, V. S. and Mulier, F. (1998). *Learning from Data: Concepts, Theory, and Methods*. John Wiley & Sons, Inc., New York, NY.

Cho, M. A., Debba, P., Mathieu, R., Naidoo, L., van Aardt, J., and Asner, G. P. (2010). Improving discrimination of savanna tree species through a multiple-endmember spectral angle mapper approach: canopy-level analysis. *IEEE Transactions on Geoscience and Remote Sensing*, **48**(11), 4133–4142.

Choi, H. and Munson, D. C. (1998a). Analysis and design of minimax-optimal interpolators. *IEEE Transactions on Signal Processing*, **46**(6), 1571–1579.

Choi, H. and Munson, D. C. (1998b). Direct-Fourier reconstruction in tomography and synthetic aperture radar. *International Journal of Imaging Systems and Technology*, **9**(1), 1–13.

Choi, H. and Willians, W. (1989). Improved time–frequency representation of multicomponent signals using exponential kernels. *IEEE Transactions on Acoustics, Speech, and Signal Processing*, **37**(6), 862–871.

Christodoulou, C. and Georgiopoulos, M. (2000). *Applications of Neural Networks in Electromagnetics*. Artech House.

Cipollini, P., Corsini, G., Diani, M., and Grass, R. (2001). Retrieval of sea water optically active parameters from hyperspectral data by means of generalized radial basis function neural networks. *IEEE Transactions on Geoscience and Remote Sensing*, **39**, 1508–1524.

Cohen, L. (1966). Generalized phase-space distribution functions. *Journal of Mathematical Physics*, **7**, 781–786.

Cohn, D., Atlas, L., and Ladner, R. (1994). Improving generalization with active learning. *Machine Learning*, **15**(2), 201–221.

Cohn, D., Ghaharamani, Z., and Jordan, M. I. (1996). Active learning with statistical models. *Journal of Artificial Intelligence Research*, **4**, 129–145.

Comon, P. (1994). Independent component analysis – a new concept? *Signal Processing*, **36**, 287–314.

Comon, P., Jutten, C., and Herault, J. (1991). Blind separation of sources, part II: problem statement. *Signal Processing*, **24**, 11–21.

Conde-Pardo, P., Guerrero-Curieses, A., Rojo-Álvarez, J. L, Yotti, R., Requena-Carrion, J., Antoranz, J., and Bermejo, J. (2006). A new method for single-step robust post-processing of flow color Doppler M-mode images using support vector machines. In *2006 Computers in Cardiology*, pages 525–528. IEEE Press, Piscataway, NJ.

Congalton, R. and Green, K. (1999). *Assessing the Accuracy of Remotely Sensed Data: Principles and Practices*. Lewis Publishers, Boca Raton, FL.

Constantinides, A. G. (1970). Spectral transformations for digital filters. *Proceedings of the IEE*, **117**, 1585–1590.

Copa, L., Tuia, D., Volpi, M., and Kanevski, M. (2010). Unbiased query-by-bagging active learning for VHR image classification. In *Proceedings of the SPIE Remote Sensing Conference*, Toulouse, France.

Cortes, C. and Vapnik, V. (1995). Support Vector Networks. *Machine Learning*, **20**, 273–97.

Cortes, C., Haffner, P., and Mohri, M. (2003). Positive definite rational kernels. In B. Schölkopf and M. Warmuth, editors, *Proceedings of the 16th Annual Conference on*

Computational Learning Theory (COLT 2003), volume 2777 of *Lecture Notes in Computer Science*, pages 41–56. Springer, Berlin.

Cortes, C., Mohri, M., and Rostamizadeh, A. (2012). Algorithms for learning kernels based on centered alignment. *Journal of Machine Learning Research*, **13**(1), 795–828.

Courty, N., Flamary, R., and Tuia, D. (2014). Domain adaptation with regularized optimal transport. In T. Calders, F. Esposito, E. Hüllermeier, and R. Meo, editors, *Domain Adaptation with Regularized Optimal Transport*, volume 8724 of *Lecture Notes in Computer Science*, pages 274–289. Springer, Berlin.

Cousseau, J. E., Figueroa, J. L., Werner, S., and Laakso, T. I. (2007). Efficient nonlinear Wiener model identification using a complex-valued simplicial canonical piecewise linear filter. *IEEE Transactions on Signal Processing*, **55**(5), 1780–1792.

Cover, T. M. and Thomas, J. A. (2006). *Elements of Information Theory*. Wiley-Interscience.

Cox, T. and Cox, M. (1994). *Multidimensional Scaling*. Chapman and Hall.

Crammer, K. and Singer, Y. (2002). On the algorithmic implementation of multiclass kernel-based vector machines. *Journal of Machine Learning Research*, **2**, 265–292.

Cremers, D., Kohlberger, T., and Schnörr, C. (2003). Shape statistics in kernel space for variational image segmentation. *Pattern Recognition*, **36**(9), 1929–1943.

Cross, B., Garlitski, A., and Estes, III, N. (2015). Advances in electroanatomic mapping systems. In S. Saksena, R. Damiano, Jr, N. Estes III, and F. Marchlinski, editors, *Interventional Cardiac Electrophysiology: A Multidisciplinary Approach*, pages 35–44. Cardiotext Publishing, Minneapolis, MN.

Csató, L. and Opper, M. (2002). Sparse online Gaussian processes. *Neural Computation*, **14**(3), 641–668.

Dai, W., Chen, Y., Xue, G.-R., Yang, Q., and Yu, Y. (2008). Translated learning: transfer learning across different feature spaces. In D. Koller, D. Schuurmans, Y. Bengio, and L. Bottou, editors, *NIPS'08 Proceedings of the 21st International Conference on Neural Information Processing Systems*, pages 353–360. Curran Associates.

Daniušis, P. and Vaitkus, P. (2009). Supervised feature extraction using Hilbert–Schmidt norms. In E. Corchado and H. Yin, editors, *Intelligent Data Engineering and Automated Learning – IDEAL 2009*, volume 5788 of *Lecture Notes in Computer Science*, pages 25–33, Berlin. Springer-Verlag.

Daumé III, H. and Marcu, D. (2011). Domain adaptation for statistical classifiers. *Journal of Artificial Intelligence Research*, **26**(1), 101–126.

de Berg, M., Cheong, O., van Kreveld, M., and Overmars, M. (2008). *Computational Geometry: Algorithms and Applications*. Springer-Verlag, Berlin.

De Freitas, N., Milo, M., Clarkson, P., Niranjan, M., and Gee, A. (1999). Sequential support vector machines. In *Neural Networks for Signal Processing IX: Proceedings of the 1999 IEEE Signal Processing Society Workshop (Cat. No.98TH8468)*, pages 31–40. IEEE Press.

De Kruif, B. J. and De Vries, T. J. A. (2003). Pruning error minimization in least squares support vector machines. *IEEE Transactions on Neural Networks*, **14**(3), 696–702.

De Vries, B. and Principe, J. C. (1992). Short term memory structures for dynamic neural networks. In *Conference Record of the Twenty-Sixth Asilomar Conference on Signals, Systems & Computers*. IEEE Press, Piscataway, NJ.

Del Moral, P. (1996). Non-linear filtering: interacting particle resolution. *Markov Processes and Related Fields*, **2**(4), 555–581.

Demir, B., Persello, C., and Bruzzone, L. (2011). Batch mode active learning methods for the interactive classification of remote sensing images. *IEEE Transactions on Geoscience and Remote Sensing*, **49**(3), 1014–1031.

Dempsey, E. J. and Westwick, D. T. (2004). Identification of Hammerstein models with cubic spline nonlinearities. *IEEE Transactions on Biomedical Engineering*, **51**(2), 237–245.

Demuth, H. and Beale, M. (1993). *Neural Network Toolbox: For Use with MATLAB: User's Guide*. The Mathworks, Natick, MA.

Deng, L. and Li, X. (2013). Machine learning paradigms for speech recognition: an overview. *IEEE Transactions on Audio, Speech, and Language Processing*, **21**(5), 1060–1089.

Desbrun, M., Hirani, A., and Leok, M.and Marsden, J. (2005). Discrete exterior calculus. arXiv preprint math/0508341.

Desobry, F. and Févotte, C. (2006). Kernel PCA based estimation of the mixing matrix in linear instantaneous mixtures of sparse sources. In *2006 IEEE International Conference on Acoustics Speech and Signal Processing Proceedings*, volume 5. IEEE Press, Piscataway, NJ.

Dhanjal, C., Gunn, S., and Shawe-Taylor, J. (2009). Efficient sparse kernel feature extraction based on partial least squares. *IEEE Transactions on Pattern Analysis and Machine Intelligence*, **31**(8), 1347–1361.

Dhillon, I., Guan, Y., and Kulis, B. (2005). A unified view of kernel k-means, spectral clustering and graph cuts. Technical Report UTCS Technical Report No. TR-04-25, University of Texas, Austin, Departement of Computer Science.

Di, W. and Crawford, M. (2010). Multi-view adaptive disagreement based active learning for hyperspectral image classification. In *2010 IEEE International Geoscience and Remote Sensing Symposium*. IEEE Press, Piscataway, NJ.

Diethron, E. and Munson, Jr, D. (1991). A linear time-varying system framework for noniterative discrete-time band-limited signal extrapolation. *IEEE Transactions on Signal Processing*, **39**(1), 55–68.

Dietterich, T. and Bakiri, G. (1995). Solving multiclass learning problems via error-correcting output codes. *Journal of Artificial Intelligence Research*, **2**, 263–286.

Donoho, D. and Johnstone, I. (1995). Adapting to unknown smoothness via wavelet shrinkage. *Journal of the American Statistical Association*, **90**(432), 1200–1224.

Donoho, D. L. (2006). Compressed sensing. *IEEE Transactions on Information Theory*, **52**(4), 1289–1306.

Dorffner, G. (1996). Neural networks for time series processing. *Neural Network World*, **6**, 447–468.

Doucet, A. and Wang, X. (2005). Monte Carlo methods for signal processing. *IEEE Signal Processing Magazine*, **22**(6), 152–170.

Drezet, P. and Harrison, R. (1998). Support vector machines for system identification. In *Proceedings of UKACC International Conference on Control'98*, pages 688–692. Institution of Electrical Engineers, London.

D'Souza, A., Vijayakumar, S., and Schaal, S. (2004). The Bayesian backfitting relevance vector machine. In *ICML '04 Proceedings of the Twenty-First International Conference on Machine Learning*. ACM, New York, NY.

Duan, L., Tsang, I. W., Xu, D., and Chua, T.-S. (2009). Domain adaptation from multiple sources via auxiliary classifiers. In *ICML '09 The 26th Annual International Conference on Machine Learning held in conjunction with the 2007 International Conference on Inductive Logic Programming*, pages 289–296. ACM, New York, NY.

Duan, L., Xu, D., Tsang, I. W.-H., and Luo, J. (2012). Visual event recognition in videos by learning from web data. *IEEE Transactions on Pattern Analysis and Machine Intelligence*, **34**(9), 1667–1680.

Duda, R. O., Hart, P. E., and Stork, D. G. (1998). *Pattern Classification and Scene Analysis: Part I Pattern Classification*. John Wiley & Sons, 2nd edition.

Dudgeon, D. E. and Merserau, R. M. (1984). *Multidimensional Signal Processing*. Signal Processing Series. Prentice Hall, Englewood Cliffs, NJ.

Duin, R. P. W. (1976). On the choice of smoothing parameters for Parzen estimators of probability density functions. *Computers, IEEE Transactions on*, **C-25**(11), 1175–1179.

Dyer, R., Zhang, R. H., Möller, T., and Clements, A. (2007). An investigation of the spectral robustness of mesh Laplacians. Technical report, SFU CS School, Vancouver, Canada.

Edfors, O., Sandell, M., van de Beek, J., Wilson, S., and Börjesson, P. O. (1996). Analysis of DFT-based channel estimators for OFDM. Research report TULEA 1996:17, Div Sig Proc, Luleå University of Technology, Sweden.

Efron, B. and Tibshirani, R. J. (1998). *An Introduction to the Bootstrap*. Chapman & Hall, New York.

Eiwen, D., Taubock, G., Hlawatsch, F., and Feichtinger, H. G. (2010a). Group sparsity methods for compressive channel estimation in doubly dispersive multicarrier systems. In *2010 IEEE 11th International Workshop on Signal Processing Advances in Wireless Communications (SPAWC)*, pages 1–5. IEEE Press, Piscataway, NJ.

Eiwen, D., Taubock, G., Hlawatsch, F., Rauhut, H., and Czink, N. (2010b). Multichannel-compressive estimation of doubly selective channels in MIMO–OFDM systems: exploiting and enhancing joint sparsity. In *2010 IEEE International Conference on Acoustics Speech and Signal Processing*, pages 3082–3085. IEEE Press, Piscataway, NJ.

Eldar, Y. C. and Oppenheim, A. V. (1999). Filterbank reconstruction of bandlimited signals from nonuniform and generalized samples. *IEEE Transactions on Signal Processing*, **47**(10), 2768–2782.

Elisseeff, A. and Weston, J. (2001). Kernel methods for multi-labelled classification and categorical regression problems. In *In Advances in Neural Information Processing Systems 14*, pages 681–687. MIT Press.

Engel, Y., Mannor, S., and Meir, R. (2004). The kernel recursive least-squares algorithm. *IEEE Transactions on Signal Processing*, **52**(8), 2275–2285.

Erceg, V., Hari, K., Smith, M., Baum, D., Soma, P., Greenstein, L., Michelson, D., Ghassemzadeh, S., Rustako, A., Roman, R., Sheikh, K., Tappenden, C., Costa, J., Bushue, C., Sarajedini, A., Scwartz, R., Branlund, D., Kaitz, T., and Trinkwon, D. (2003). Channel models for fixed wireless applications, IEEE 802.16a-03/01. http://wirelessman.org/tga/docs/80216a-03_01.pdf.

Erdogmus, D., Rende, D., Príncipe, J. C., and Wong, T. F. (2001a). Nonlinear channel equalization using multilayer perceptrons with information-theoric criterion. In *Neural Networks for Signal Processing XI: Proceedings of the 2001 IEEE Signal Processing Society Workshop (IEEE Cat. No.01TH8584)*, pages 401–451. IEEE Press, Piscataway, NJ.

Erdogmus, D., Vielva, L., and Príncipe, J. C. (2001b). Nonparametric estimation and tracking of the mixing matrix for underdetermined blind source separation. In *Proceedings of the 3rd International Conference on Independent Component Analysis and Blind Signal Separation (ICA 2001), San Diego, CA, USA*, pages 189–194, San Diego, California, USA.

Eric, M., Bach, F. R., and Harchaoui, Z. (2008). Testing for homogeneity with kernel Fisher discriminant analysis. In J. C. Platt, D. Koller, Y. Singer, and S. T. Roweis, editors, *Advances in Neural Information Processing Systems 20*, pages 609–616. Curran Associates.

Ernst, F. (2012). *Compensating for Quasi-periodic Motion in Robotic Radiosurgery*. Springer.

Espinoza, M., Suykens, J. A., and De Moor, B. (2005). Imposing symmetry in least squares support vector machines regression. In *Proceedings of the 44th IEEE Conference on Decision and Control*, pages 5716–5721, Piscataway, NJ. IEEE Press.

Evgeniou, T., Pontil, M., and Poggio, T. (2000). Regularization networks and support vector machines. *Advances in Computational Mathematics*, **13**(1), 1–50.

Farhadi, A. and Tabrizi, M. K. (2008). Learning to recognize activities from the wrong view point. In D. Forsyth, P. Torr, and A. Zisserman, editors, *Computer Vision – ECCV 2008*, volume 5302 of *Lecture Notes in Computer Science*, pages 154–166. Springer, Berlin.

Farid, H. and Simoncelli, E. (2004). Differentiation of discrete multi-dimensional signals. *IEEE Transactions on Image Processing*, **13**(4), 496–508.

Fearnhead, P. (2006). Exact and efficient bayesian inference for multiple changepoint problems. *Statistics and Computing*, **16**(2), 203–213.

Feher, K. (1983). *Digital Communications: Satellite/Earth Station Engineering*. Prentice-Hall, Englewood Cliffs, NJ.

Feijoo, J., Rojo-Álvarez, J. L., Cid-Sueiro, J., Conde-Pardo, P., and Mata-Vigil-Escalera, J. (2010). Modeling link events in high reliability networks with support vector machines. *IEEE Transactions on Reliability*, **59**, 191–202.

Ferecatu, M. and Boujemaa, N. (2007). Interactive remote sensing image retrieval using active relevance feedback. *IEEE Transactions on Geoscience and Remote Sensing*, **45**(4), 818–826.

Fernández-Getino, M., Rojo-Álvarez, J. L., Alonso-Atienza, F., and Martínez-Ramón, M. (2006). Support vector machines for robust channel estimation in OFDM. *IEEE Signal Processing Letters*, **13**, 397–400.

Fernández-Getino García, M. J., Páez-Borrallo, J. M., and Zazo, S. (2001). DFT-based channel estimation in 2D-pilot-symbol-aided OFDM wireless systems. In *IEEE VTS 53rd Vehicular Technology Conference, Spring 2001. Proceedings (Cat. No.01CH37202)*, volume 2, pages 815–819. IEEE Press, Piscataway, NJ.

Figueiras-Vidal, A., Docampo-Amoedo, D., Casar-Corredera, J., and Artés-Rodríguez, A. (1990). Adaptive iterative algorithms for spiky deconvolution. *IEEE Transactions on Acoustics, Speech, and Signal Processing*, **38**, 1462–1466.

Figueiredo, M. A. T. and Nowak, R. D. (2001). Wavelet-based image estimation: an empirical Bayes approach using Jeffrey's noninformative prior. *IEEE Transactions on Image Processing*, **10**(9), 1322–1331.

Figuera, C., Mora-Jiménez, I., Guerrero-Curieses, A., Rojo-Álvarez, J. L., Everss, E., Wilby, M., and Ramos-López, J. (2009). Nonparametric model comparison and uncertainty evaluation for signal strength indoor location. *IEEE Trans. Mobile Computing*, **8**(9), 1250–1264.

Figuera, C., Rojo-Álvarez, J. L., Mora-Jiménez, I., Guerrero-Curieses, A., Wilby, M., and Ramos-López, J. (2011). Time–space sampling and mobile device calibration for WiFi indoor location systems. *IEEE Transactions on Mobile Computing*, **10**, 913–926.

Figuera, C., Rojo-Álvarez, J. L., Wilby, M., Mora-Jiménez, I., and Caamaño, A. J. (2012). Advanced support vector machines for 802.11 indoor location. *Signal Processing*, **92**(9), 2126–2136.

Figuera, C., Barquero-Pérez, Ó., Rojo-Álvarez, J. L., Martínez-Ramón, M., Guerrero-Curieses, A., and Caamaño, A. J. (2014). Spectrally adapted Mercer kernels for support vector nonuniform interpolation. *Signal Processing*, **94**, 421–433.

Filippone, M., Camastra, F., Masulli, F., and Rovetta, S. (2008). A survey of kernel and spectral methods for clustering. *Pattern Recognition*, **41**(1), 176–190.

Fine, S., Scheinberg, K., Cristianini, N., Shawe-Taylor, J., and Williamson, B. (2001). Efficient SVM training using low-rank kernel representations. *Journal of Machine Learning Research*, **2**, 243–264.

Fisher, R. A. (1936). The use of multiple measurements in taxonomic problems. *Annals of Eugenics*, **7**, 179–188.

Flamary, R., Tuia, D., Labbé, B., Camps-Valls, G., and Rakotomamonjy, A. (2012). Large margin filtering. *IEEE Transactions on Signal Processing*, **60**(2), 648–659.

Fletcher, R. (1987). *Practical Methods of Optimization*. John Wiley & Sons, Inc. 2nd Edition.

Floater, M. S. (2003). Mean value coordinates. *Computer Aided Geometric Design*, **20**(1), 19–27.

Franz, M. O. and Schölkopf, B. (2006). A unifying view of Wiener and Volterra theory and polynomial kernel regression. *Neural Computation*, **18**(12), 3097–3118.

Fréchet, M. (1910). Sur les fonctionelles continues. *Annales Scientifiques de L'École Normale Supérieure*, **27**, 193–216.

Freund, Y., Seung, H. S., Shamir, E., and Tishby, N. (1997). Selective sampling using the query by committee algorithm. *Machine Learning*, **28**, 133–168.

Friedman, J., Hastie, T., and Tibshirani, R. (2001). *The Elements of Statistical Learning: Data Mining, Inference, and Prediction*. Springer Series in Statistics. Springer, Berlin, 1 edition.

Friedman, J. H. (1996). Another approach to polychotomous classification. Technical report, Department of Statistics, Stanford University, Stanford, CA.

Frieß, T.-T. and Harrison, R. F. (1999). A kernel-based adaline. In *Proceedings of the 7th European Symposium on Artificial Neural Networks (ESANN 1999), Bruges, Belgium*, pages 245–250.

Friston, K. J., Holmes, A. P., Worsley, K. J., Poline, J.-P., Frith, C. D., and Frackowiak, R. S. J. (1995). Statistical parametric maps in functional imaging: a general linear approach. *Human Brain Mapping*, **2**, 189–210.

Fukumizu, K., Bach, F. R., and Jordan, M. I. (2003). Kernel dimensionality reduction for supervised learning. In S. Thrun, L. Saul, and P. Schölkopf, editors, *Advances in Neural Information Processing Systems 16*. MIT Press, Cambridge, MA.

Fukumizu, K., Bach, F. R., and Jordan, M. I. (2004). Dimensionality reduction for supervised learning with reproducing kernel Hilbert spaces. *Journal of Machine Learning Research*, **5**, 73–99.

Fukumizu, K., Bach, F., and Gretton, A. (2007). Statistical consistency of kernel canonical correlation analysis. *Journal of Machine Learning Research*, **8**, 361–383.

Funk, A., Basili, V., Hochstein, L., and Kepner, J. (2005). Application of a development time productivity metric to parallel software development. In *SE-HPCS '05: Proceedings of the Second International Workshop on Software Engineering for High Performance Computing System Applications*, pages 8–12. ACM, New York, NY.

Gabor, D. (1946). Theory of communications. *Journal of the Institution of Electrical Engineers (London)*, **93**, 429–457.

Gangeh, M., Ghodsi, A., and Kamel, M. S. (2013). Kernelized supervised dictionary learning. *IEEE Transactions on Signal Processing*, **61**(19), 4753–4767.

Gao, S., Tsang, I. W.-H., and Chia, L.-T. (2010). Kernel sparse representation for image classification and face recognition. In K. Daniilidis, P. Maragos, and N. Paragios, editors, *Computer Vision – ECCV 2010*, volume 6314 of *Lecture Notes in Computer Science*, pages 1–14. Springer, Berlin.

Gao, W., Richard, C., Bermudez, J.-C. M., and Huang, J. (2014). Convex combinations of kernel adaptive filters. In *2014 IEEE International Workshop on Machine Learning for Signal Processing (MLSP)*, pages 1–5. IEEE Press, Piscataway, NJ.

García, A. (2000). Orthogonal sampling formulas: a unified approach. *SIAM Reviews*, **42**(3), 499–512.

García-Allende, P. B., Conde, O. M., Mirapeix, J., Cubillas, A. M., and Lopez-Higuera, J. M. (2008). Data processing method applying principal component analysis and spectral angle mapper for imaging spectroscopic sensors. *IEEE Sensors Journal*, **8**(7), 1310–1316.

Gardiner, A. B. (1973). Identification of processes containing single-valued nonlinearities. *International Journal of Control*, **18**(5), 1029–1039.

Geladi, P. (1988). Notes on the history and nature of partial least squares (PLS) modelling. *Journal of Chemometrics*, **2**, 231–246.

Gersho, A. and Gray, R. (1991). *Vector Quantization and Signal Compression*. Kluwer, Norwell, MA.

Ghahramani, Z. (1996). The kin datasets.

Ghosh, M. (1996). Analysis of the effect of impulse noise on multicarrier and single carrier QAM systems. *IEEE Transactions on Communications*, **44**(2), 145–147.

Giannakis, G. and Serpedin, E. (2001). A bibliography on nonlinear system identification. *Signal Processing*, **81**(3), 533–580.

Girolami, M. (2002a). Mercer kernel-based clustering in feature space. *IEEE Transactions on Neural Nets*, **13**(3), 780–784.

Girolami, M. (2002b). Orthogonal series density estimation and the kernel eigenvalue problem. *Neural Computation*, **14**(3), 669–688.

Girosi, F., Jones, M., and Poggio, T. (1995). Regularization theory and neural networks architectures. *Neural Computation*, **7**(2), 291–269.

Goethals, I., Pelckmans, K., Suykens, J. A. K., and Moor, B. D. (2005a). Identification of MIMO Hammerstein models using least squares support vector machines. *Automatica*, **41**(7), 1263–1272.

Goethals, I., Pelckmans, K., Suykens, J. A. K., and Moor, B. D. (2005b). Subspace identification of Hammerstein systems using least squares support vector machines. *IEEE Transactions on Automatic Control*, **50**(10), 1509–1519.

Goldberg, P., Williams, C., and Bishop, C. (1998). Regression with input-dependent noise: a Gaussian process treatment. In *NIPS '97 Proceedings of the 1997 Conference on Advances in Neural Information Processing Systems 10*, pages 493–499. MIT Press, Cambridge, MA.

Golden, R. M. and Kaiser, J. F. (1964). Design of wideband sampled-data filters. *The Bell System Technical Journal*, **43**(4), 1533–1546.

Golub, G. H. and Van Loan, C. F. (1996). *Matrix Computations*. The Johns Hopkins University Press.

Gómez, J. C. and Baeyens, E. (2007). Subspace-based blind identification of IIR Wiener systems. In *Proceedings of the 15th European Signal Processing Conference (EUSIPCO)*, Poznań, Poland.

Gómez-Chova, L., Fernández-Prieto, D., Calpe, J., Soria, E., Vila-Francés, J., and Camps-Valls, G. (2006). Urban monitoring using multitemporal SAR and multispectral data. *Pattern Recognition Letters*, **27**(4), 234–243.

Gómez-Chova, L., Camps-Valls, G., Calpe, J., Guanter, L., and Moreno, J. (2007). Cloud-screening algorithm for ENVISAT/MERIS multispectral images. *IEEE Transactions on Geoscience and Remote Sensing*, **45**(12), 4105–4118.

Gómez-Chova, L., Camps-Valls, G., Muñoz-Marí, J., and Calpe-Maravilla, J. (2008). Semi-supervised image classification with Laplacian support vector machines. *IEEE Geoscience and Remote Sensing Letters*, **5**(3), 336–340.

Gómez-Chova, L., Camps-Valls, G., Bruzzone, L., and Calpe-Maravilla, J. (2010). Mean map kernel methods for semisupervised cloud classification. *IEEE Transactions on Geoscience and Remote Sensing*, **48**(1), 207–220.

Gómez-Chova, L., Jenssen, R., and Camps-Valls, G. (2012). Kernel entropy component analysis for remote sensing image clustering. *IEEE Geoscience and Remote Sensing Letters*, **9**(2), 312–316.

Gómez-Chova, L., Nielsen, A. A., and Camps-Valls, G. (2017). Explicit signal to noise ratio in reproducing kernel Hilbert spaces. *IEEE Transactions on Neural Networks and Learning Systems.*

Gong, B., Shi, Y., Sha, F., and Grauman, K. (2012). Geodesic flow kernel for unsupervised domain adaptation. In *Proceedings of the IEEE Conference on Computer Vision and Pattern Recognition (CVPR)*, pages 2066–2073. IEEE Press, Piscataway, NJ.

Gonnouni, A. E., Martínez-Ramón, M., Rojo-Álvarez, J. L., Camps-Valls, G., Figueiras-Vidal, A. R., and Christodoulou, C. G. (2012). A support vector machine music algorithm. *IEEE Transactions on Antennas and Propagation*, **60**(10), 4901–4910.

Gopalan, R., Li, R., and Chellappa, R. (2011). Domain adaptation for object recognition: an unsupervised approach. In *ICCV '11 Proceedings of the 2011 International Conference on Computer Vision*, pages 999–1006. IEEE Computer Society, Barcelona, Spain.

Gorinevsky, D. and Boyd, S. (2006). Optimization-based design and implementation of multidimensional zero-phase iir filters. *IEEE Transactions on Circuits and Systems I: Regular Papers*, **53**(2), 372–383.

Goshtasby, A. (2005). *2-D and 3-D Image Registration for Medical, Remote Sensing, and Industrial Applications*, volume 1. John Wiley & Sons, Inc., Hoboken, NJ.

Grauman, K. and Darrell, T. (2005). The pyramid match kernel: discriminative classification with sets of image features. In *ICCV '05: Proceedings of the Tenth IEEE International Conference on Computer Vision*, pages 1458–1465. IEEE Computer Society, Washington, DC.

Greblicki, W. (1997). Nonparametric approach to Wiener system identification. *IEEE Transactions on Circuits and Systems I: Fundamental Theory and Applications*, **44**(6), 538–545.

Greblicki, W. (2004). Nonlinearity recovering in Wiener system driven with correlated signal. *IEEE Transactions on Automatic Control*, **49**(10), 1805–1812.

Green, A. A., Berman, M., Switzer, P., and Craig, M. D. (1998). A transformation for ordering multispectral data in terms of image quality with implications for noise removal. *IEEE Transactions on Geoscience and Remote Sensing*, **26**(1), 65–74.

Gretton, A., Davy, M., Doucet, A., and Rayner, P. J. W. (2001a). Nonstationary signal classification using support vector machines. In *11th IEEE Workshop on Statistical Signal Processing*, pages 305–308. IEEE Signal Processing Society, Piscataway, NY.

Gretton, A., Doucet, A., Herbrich, R., Rayner, P., and Schölkopf, B. (2001b). Support vector regression for black-box system identification. In *11th IEEE Workshop on Statistical Signal Processing*, pages 341–344. IEEE Signal Processing Society, Piscataway, NY.

Gretton, A., Smola, A., Bousquet, O., Herbrich, R., Belitski, A., Augath, M., Murayama, Y., Pauls, J., Schölkopf, B., and Logothetis, N. (2005a). Kernel constrained covariance for dependence measurement. In R. G. Cowell and Z. Ghahramani, editors, *Proceedings of the Tenth International Workshop on Artificial Intelligence and Statistics*, pages 112–119. Society for Artificial Intelligence and Statistics.

Gretton, A., Herbrich, R., and Hyvärinen, A. (2005b). Kernel methods for measuring independence. *Journal of Machine Learning Research*, **6**, 2075–2129.

Gretton, A., Bousquet, O., Smola, A., and Schölkopf, B. (2005c). Measuring statistical dependence with Hilbert-Schmidt norms. In S. Jain, H. Simon, and E. Tomita, editors, *Algorithmic Learning Theory. ALT 2005*, volume 3734 of *Lecture Notes in Computer Science*, pages 63–77. Springer, Berlin.

Grinspun, E., Desbrun, M., Polthier, K., Schröder, P., and Stern, A. (2006). Siggraph 2006 course notes. discrete differential geometry: An applied introduction, 7. http://geometry.caltech.edu/pubs/GSD06.pdf.

Gu, B. and Sheng, V. (2017). A robust regularization path algorithm for ν-support vector classification. *IEEE Transactions on Neural Networks and Learning Systems*, **28**(5), 1241–1248.

Gu, B., Wang, J. D., Zheng, G., and Yu, Y. (2012). Regularization Path for ν-Support Vector classification. *IEEE Transactions on Neural Networks and Learning Systems*, **23**(5), 800–811.

Gutiérrez, J., Ferri, F., and Malo, J. (2006). Regularization operators for natural images based on nonlinear perception models. *IEEE Transactions on Image Processing*, **15**(1), 189–200.

Gutmann, M. U., Laparra, V., Hyvärinen, A., and Malo, J. (2014). Spatio-chromatic adaptation via higher-order canonical correlation analysis of natural images. *PLoS ONE*, **9**(2), 1–21.

Halevy, A., Norvig, P., and Pereira, F. (2009). The unreasonable effectiveness of data. *IEEE Intelligent Systems*, **24**(2), 8–12.

Halmos, P. R. (1974). *Measure Theory*. Springer.

Ham, J., Lee, D. D., and Saul, L. K. (2003). Learning high-dimensional correspondences from low-dimensional manifolds. In *Workshop on The Continuum from Labeled to Unlabled Data in Machine Learning and Data Mining at Twentieth International Conference on Machine Learning*, pages 34–39.

Ham, J., Lee, D. D., Mika, S., and Scholkopf, B. (2004). A kernel view of the dimensionality reduction of manifolds. In *ICML '04 Proceedings of the Twenty-First International Conference on Machine Learning*, page 47. ACM, New York, NY.

Ham, J., Lee, D., and Saul, L. (2005). Semisupervised alignment of manifolds. In *10th International Workshop on Artificial Intelligence and Statistics*, pages 120–127.

Haralick, R. M., Sternberg, S. R., and Zhuang, X. (1987). Image analysis using mathematical morphology. *IEEE Transactions on Pattern Analysis and Machine Intelligence*, **9**(4), 532–550.

Harchaoui, Z., Bach, F., Cappe, O., and Moulines, E. (2013). Kernel-based methods for hypothesis testing: a unified view. *IEEE Signal Processing Magazine*, **30**(4), 87–97.

Hardoon, D. R. and Shawe-Taylor, J. (2011). Sparse canonical correlation analysis. *Machine Learning*, **83**, 331–353.

Harmeling, S., Ziehe, A., Kawanabe, M., and Müller, K.-R. (2003). Kernel-based nonlinear blind source separation. *Neural Computation*, **15**(5), 1089–1124.

Harsanyi, J. C. and Chang, C. I. (1994). Hyperspectral image classification and dimensionality reduction: an orthogonal subspace projection approach. *IEEE Transactions on Geoscience and Remote Sensing*, **32**(4), 779–785.

Hassibi, B., Stork, D. G., and Wolff, G. J. (1993). Optimal brain surgeon and general network pruning. In *IEEE International Conference on Neural Networks*, pages 293–299. IEEE.

Hastie, T. and Tibshirani, R. (1998). Classification by pairwise coupling. In M. Jordan, M. Kearns, and S. Solla, editors, *Advances in Neural Information Processing Systems 10*, pages 507–513. MIT Press, Cambridge, MA.

Hastie, T., Tibishirani, R., and Friedman, J. (2001). *The Elements of Statistical Learning*. Springer-Verlag, New York, NY.

Hastie, T., Tibshirani, R., and Friedman, J. H. (2009). *The Elements of Statistical Llearning: Data Mining, Inference, and Prediction*. Springer-Verlag, New York, NY, 2nd edition.

Hayes, M. (1996). *Statistical Digital Signal Processing and Modeling*. John Wiley & Sons, Inc., New York, NY.

Haykin, S. (1999). *Neural Networks – A Comprehensive Foundation*. Prentice Hall, 2nd edition.

Haykin, S. (2001). *Adaptive Filter Theory*. Prentice Hall, 4th edition.

Hecker, C., van der Meijde, M., van der Werff, H., and van der Meer, F. D. (2008). Assessing the influence of reference spectra on synthetic sam classification results. *IEEE Transactions on Geoscience and Remote Sensing*, **46**(12), 4162–4172.

Heerde, H. V., Leeflang, P., and Wittink, D. (2001). Semiparametric analysis to estimate the deal effect curve. *Journal of Marketing Research*, **38**(2), 197–215.

Helms, H. D. (1967). Fast fourier transform method of computing difference equations and simulating filters. *IEEE Transactions on Audio and Electroacoustics*, **15**(2), 85–90.

Hermans, M. and Schrauwen, B. (2012). Recurrent kernel machines: computing with infinite echo state networks. *Neural Computation*, **24**(1), 104–133.

Hertz, D. and Zeheb, E. (1984). Sufficient conditions for stability of multidimensional discrete systems. *Proceedings of the IEEE*, **72**(2), 226–226.

Hoegaerts, L., Suykens, J., Vandewalle, J., and De Moor, B. (2004). A comparison of pruning algorithms for sparse least squares support vector machines. In N. Pal, N. Kasabov, R. Mudi, S. Pal, and S. Parui, editors, *Neural Information Processing. ICONIP 2004*, volume 3316 of *Lecture Notes in Computer Science*, pages 1247–1253. Springer.

Hoegaerts, L., Suykens, J. A. K., Vanderwalle, J., and Moor, B. D. (2005). Subset based least squares subspace regression in RKHS. *Neurocomputing*, **63**, 293–323.

Hoerl, A. E. and Kennard, R. W. (1970). Ridge regression: biased estimation for nonorthogonal problems. *Technometrics*, **12**, 55–67.

Hoffman, J., Rodner, E., Donahue, J., Saenko, K., and Darrell, T. (2013). Efficient learning of domain invariant image representations. In *Proceedings of International Conference on Learning Representations (ICLR 2013), Scottsdale, Arizona*.

Hofmann, T., Schölkopf, B., and Smola, A. J. (2008). Kernel methods in machine learning. *Annals of Statistics*, **36**, 1171–1220.

Hotelling, H. (1936). Relations between two sets of variates. *Biometrika*, **28**(3/4), 321–377.

Hsu, C.-W. and Lin, C.-J. (2002). A comparison of methods for multiclass support vector machines. *IEEE Transactions on Neural Networks*, **13**(2), 415–425.

Hu, Y. D., Pan, J. C., and Tan, X. (2013). High-dimensional data dimension reduction based on KECA. *Applied Mechanics and Materials*, **303–306**, 1101–1104.

Huang, B. and Wang, Y. (2009). QRS complexes detection by using the principal component analysis and the combined wavelet entropy for 12-lead electrocardiogram signals. In *CIT09. Ninth IEEE International Conference on Computer and Information Technology, 2009.*, volume 1, pages 246–251. IEEE Press, Piscataway, NJ.

Huang, J., Smola, A. J., Gretton, A., Borgwardt, K. M., and Scholkopf, B. (2007). Correcting sample selection bias by unlabeled data. In B. Schölkopf, J. Platt, and T. Hoffman, editors, *Advances in Neural Information Processing Systems 19*, pages 601–608. MIT Press, Cambridge, MA.

Hughes, G. F. (1968). On the mean accuracy of statistical pattern recognizers. *IEEE Transactions on Information Theory*, **14**(1), 55–63.

Hundley, D. R., Kirby, M. J., and Anderle, M. (2002). Blind source separation using the maximum signal fraction approach. *Signal Processing*, **82**(10), 1505–1508.

Hunter, I. W. and Korenberg, M. J. (1986). The identification of nonlinear biological systems: Wiener and Hammerstein cascade models. *Biological Cybernetics*, **55**(2), 135–144.

Hyvärinen, A. and Oja, E. (1997). A fast fixed-point algorithm for independent component analysis. *Neural Computation*, **9**(7), 1483–1492.

Hyvärinen, A., Karhunen, J., and Oja, E. (2001). *Independent Component Analysis*. Wiley Interscience.

Ishida, T. and Tanaka, T. (2013). Multikernel adaptive filters with multiple dictionaries and regularization. In *2013 Asia-Pacific Signal and Information Processing Association Annual Summit and Conference (APSIPA)*, pages 1–6. IEEE Press, Piscataway, NJ.

Izquierdo-Verdiguier, E., Arenas-García, J., Muñoz-Romero, S., Gómez-Chova, L., and Camps-Valls, G. (2012). Semisupervised kernel orthonormalized partial least squares. In *2012 IEEE International Workshop on Machine Learning for Signal Processing*. IEEE Press, Piscataway, NJ.

Izquierdo-Verdiguier, E., Laparra, V., Jenssen, R., Gómez-Chova, L., and Camps-Valls, G. (2016). Optimized kernel entropy components. *IEEE Transactions on Neural Networks and Learning Systems*, **28**(6), 1466–1472.

Jaakkola, T. and Haussler, D. (1999). Exploiting generative models in discriminative classifiers. In M. Kearns, S. Solla, and D. Cohn, editors, *Advances in Neural Information Processing Systems 11*, pages 487–493. MIT Press, Cambridge, MA.

Jaakkola, T., Diekhans, M., and Haussler, D. (1999). Using the Fisher kernel method to detect remote protein homologies. In *Proceedings of the Seventh International Conference on Intelligent Systems for Molecular Biology*, pages 149–158. AAAI Press.

Jackson, J. I., Meyer, C. H., Nishimura, D. G., and Macovski, A. (1991). Selection of a convolution function for Fourier inversion using gridding. *IEEE Transactions on Medical Imaging*, **10**(3), 473–478.

Jackson, Q. and Landgrebe, D. (2001). An adaptive classifier design for high-dimensional data analysis with a limited training data set. *IEEE Transactions on Geoscience and Remote Sensing*, **39**(12), 2664–2679.

Jenssen, R. (2009). Information theoretic learning and kernel methods. In F. Emmert-Streib and M. Dehmer, editors, *Information Theory and Statistical Learning*, pages 209–230. Springer US.

Jenssen, R. (2010). Kernel entropy component analysis. *IEEE Transactions on Pattern Analysis and Machine Intelligence*, **32**(5), 847–860.

Jenssen, R. (2013). Entropy-relevant dimensions in the kernel feature space: cluster-capturing dimensionality reduction. *IEEE Signal Processing Magazine*, **30**(4), 30–39.

Jerri, A. J. (1977). The Shannon sampling theorem – its various extensions and applications: a tutorial review. *Proceedings of the IEEE*, **65**(11), 1565–1595.

Jiang, Q., Yan, X., Lv, Z., and Guo, M. (2013). Fault detection in nonlinear chemical processes based on kernel entropy component analysis and an angular structure. *Korean Journal of Chemical Engineering*, **30**(6), 1181.

Joachims, T. (1998a). Making large-scale support vector machine learning practical. In B. Schölkopf, C. Burges, and A. Smola, editors, *Advances in Kernel Methods: Support Vector Machines*, pages 169–184. MIT Press, Cambridge, MA.

Joachims, T. (1998b). Text categorization with suport vector machines: learning with many relevant features. In *ECML '98 Proceedings of the 10th European Conference on Machine Learning*, pages 137–142, London, UK, UK. Springer-Verlag.

Joachims, T., Galor, T., and Elber, R. (2005). Learning to align sequences: a maximum-margin approach. In B. Leimkuhler, C. Chipot, R. Elber, A. Laaksonen, A. Mark, T. Schlick, C. Schütte, and R. Skeel, editors, *New Algorithms for Macromolecular Simulation*, volume 49 of *Lecture Notes in Computational Science and Engineering*, pages 55–79. Springer.

Joachims, T., Finley, T., and Yu, C.-N. J. (2009). Cutting-plane training of structural SVMs. *Machine Learning*, **77**, 27.

Joho, M., Mathis, H., and Lambert, R. (2000). Overdetermined blind source separation: using more sensors than source signals in a noisy mixture. In *Proceedings of the 2nd International Conference on Independent Component Analysis and Blind Signal Separation (ICA 2000)*, pages 81–86, Helsinki, Finland.

Jolliffe, I. T. (1986). *Principal Component Analysis*. Springer-Verlag, Berlin.

Julier, S. J. and Uhlmann, J. K. (1996). A general method for approximating nonlinear transformations of probability distributions. Technical report, Robotics Research Group, Department of Engineering Science, University of Oxford.

Jun, G. and Ghosh, J. (2008). An efficient active learning algorithm with knowledge transfer for hyperspectral remote sensing data. In *IGARSS 2008 – 2008 IEEE International Geoscience and Remote Sensing Symposium*. IEEE Press, Piscataway, NJ.

Juneja, R. (2009). Radiofrequency ablation for cardiac tachyarrhythmias: principles and utility of 3D mapping systems. *Current Science*, **97**(3), 416–424.

Kaiser, J. (1972). Design methods for sampled data filters. In L. Rabiner and C. Rader, editors, *Digital Signal Processing*, pages 20–34. IEEE Press, New York, NY. Reprinted from *Proceedings of the First Annual Allerton Conference on Circuit Systems Theory*, 1963.

Kakazu, G. and Munson, Jr, D. (1989). A frequency-domain characterization of interpolation from nonuniformly spaced data. In *IEEE International Symposium on Circuits and Systems*, pages 288–291. IEEE Press, Piscataway, NJ.

Kalman, R. E. (1960). A new approach to linear filtering and prediction problems. *Journal of Basic Engineering*, **82**(1), 35–45.
Kalouptsidis, N. and Theodoridis, S. (1993). *Adaptive System Identification and Signal Processing Algorithms*. Prentice-Hall, Englewood Cliffs, NJ.
Kanamori, T., Hido, S., and Sugiyama, M. (2009). A least-squares approach to direct importance estimation. *Journal of Machine Learning Research*, **10**, 1391–1445.
Kassam, S. A. and Poor, H. V. (1985). Robust techniques for signal processing: a survey. *Proceedings of the IEEE*, **73**(3), 433–482.
Kawakami Harrop Galvão, R., Hadjiloucas, S., Izhac, A., Becerra, V. M., and Bowen, J. W. (2007). Wiener-system subspace identification for mobile wireless mm-wave networks. *IEEE Transactions on Vehicular Technology*, **56**(4), 1935–1948.
Kay, S. and Marple, Jr., S. (1981). Spectrum analysis – a modern perspective. *Proceedings of the IEEE*, **69**(11), 1380–1419.
Kay, S. M. (1993). *Fundamentals of Statistical Signal Processing: Detection Theory*, volume 2. Prentice Hall, Upper Saddle River, NJ.
Kechriotis, G., Zarvas, E., and Manolakos, E. S. (1994). Using recurrent neural networks for adaptive communication channel equalization. *IEEE Transactions on Neural Networks*, **5**, 267–278.
Kersting, K., Plagemann, C., Pfaff, P., and Burgard, W. (2007). Most likely heteroscedastic Gaussian processes regression. In *ICML '07 Proceedings of the 24th International Conference on Machine Learning*, pages 393–400.
Keshava, N. (2004). Distance metrics and band selection in hyperspectral processing with applications to material identification and spectral libraries. *IEEE Transactions on Geoscience and Remote Sensing*, **42**(7), 1552–1565.
Kim, K. I., Franz, M. O., and Scholkopf, B. (2005). Iterative kernel principal component analysis for image modeling. *IEEE Transactions on Pattern Analysis and Machine Intelligence*, **27**(9), 1351–1366.
Kimeldorf, G. and Wahba, G. (1971). Some results on Tchebycheffian spline functions. *Journal of Mathematical Analysis and Applications*, **33**(1), 82–95.
Kirkwood, J. (1933). Quantum statistics of almost classical ensembles. *Physical Review*, **44**, 31–37.
Kivinen, J., Smola, A. J., and Williamson, R. C. (2004). Online learning with kernels. *IEEE Transactions on Signal Processing*, **52**(8), 2165–2176.
Kohonen, T. (1990). The self-organizing map. *Proceedings of the IEEE*, **78**(9), 1464–1480.
Koller, D. and Friedman, N. (2009). *Probabilistic Graphical Models: Principles and Techniques*. MIT Press.
Korenberg, M. J. (1989). A robust orthogonal algorithm for system identification and time-series analysis. *Biological Cybernetics*, **60**(4), 267–276.
Kormylo, J. and Mendel, J. M. (1982). Maximum likelihood detection and estimation of Bernoulli–Gaussian processes. *IEEE Transactions on Information Theory*, **28**, 482–488.
Kormylo, J. J. and Mendel, J. M. (1983). Maximum-likelihood seismic deconvolution. *IEEE Transactions on Geoscience and Remote Sensing*, **GE-21**(1), 72–82.
Kramer, N. and Rosipal, R. (2006). Overview and recent advances in partial least squares. In C. Saunders, M. Grobelnik, S. Gunn, and J. Shawe-Taylor, editors, *Subspace, Latent Structure and Feature Selection Techniques*, volume 3940 of *Lecture Notes in Computer Science*, pages 34–51. Springer, Berlin.

Kreutz-Delgado, K., Murray, J. F., Rao, B. D., Engan, K., Lee, T.-W., and Sejnowski, T. J. (2003). Dictionary learning algorithms for sparse representation. *Neural Computation*, **15**(2), 349–396.

Kruse, F. A., Lefkoff, A. B., Boardman, J. W., Heidebrecht, K. B. Shapiro, A. T., Barloon, P. J., and Goetz, A. F. H. (1993). The spectral image processing system (SIPS) – interactive visualization and analysis of imaging spectrometer data. *Remote Sensing of Environment*, **44**(2–3), 145–163.

Kuo, J. M. and Principe, J. (1994). Noise reduction in state space using the focused gamma model. In *Proceedings of IEEE International Conference on Acoustics, Speech, and Signal Processing*, volume 2, pages 533–536. IEEE Press, Piscataway, NJ.

Kuo, J. M., Celebi, S., and Principe, J. (1994). Adaptation of memory depth in the gamma filter. In *Proceedings of IEEE International Conference on Acoustics, Speech, and Signal Processing*, volume 5, pages 373–376. IEEE Press, Piscataway, NJ.

Kuss, M. and Rasmussen, C. (2005). Assessing approximate inference for binary Gaussian process classification. *Machine LearningResearch*, **6**, 1679–1704.

Kwok, J. T. (2000). The evidence framework applied to support vector machines. *IEEE Transactions in Neural Networks*, **11**(5), 1162–1173.

Kwok, J. T. (2001). Linear dependency between ε and the input noise in ε-support vector regression. In G. Dorffner, H. Bischof, and K. Hornik, editors, *Artificial Neural Networks – ICANN 2001*, volume 2130 of *Lecture Notes in Computer Science*, pages 405–410. Springer, Berlin.

Kwon, H. and Nasrabadi, N. (2007). A comparative analysis of kernel subspace target detectors for hyperspectral imagery. *EURASIP Journal of Advances in Signal Processing*, **2007**, 029250.

Kwon, H. and Nasrabadi, N. M. (2005a). Kernel matched signal detectors for hyperspectral target detection. In *2005 IEEE Computer Society Conference on Computer Vision and Pattern Recognition (CVPR'05) – Workshops*, page 6. IEEE Press, Piscataway, NJ.

Kwon, H. and Nasrabadi, N. M. (2005b). Kernel orthogonal subspace projection for hyperspectral signal classification. *IEEE Transactions on Geoscience and Remote Sensing*, **43**(12), 2952–2962.

Kwon, H. and Nasrabadi, N. M. (2005c). Kernel RX-algorithm: a nonlinear anomaly detector for hyperspectral imagery. *IEEE Transactions on Geoscience and Remote Sensing*, **43**(2), 388–397.

Kwon, H., Der, S. Z., and Nasrabadi, N. M. (2003). Adaptive anomaly detection using subspace separation for hyperspectral imagery. *Optical Engineering*, **42**(11), 3342–3351.

Laguna, P., Moody, G., and Mark, R. (1998). Power spectral density of unevenly sampled data by least-square analysis: performance and application to heart rate signals. *IEEE Transactions on Biomedical Engineering*, **45**, 698–715.

Lai, P. L. and Fyfe, C. (2000a). Kernel and non-linear canonical correlation analysis. *International Journal of Neural Systems*, **10**, 365–377.

Lai, P. L. and Fyfe, C. (2000b). Kernel and nonlinear canonical correlation analysis. In *Proceedings of the IEEE–INNS–ENNS International Joint Conference on Neural Networks. IJCNN 2000. Neural Computing: New Challenges and Perspectives for the New Millennium*, pages 614–619. IEEE Press, Piscataway, NJ.

Lal, T. N., Schroder, M., Hinterberger, T., Weston, J., Bogdan, M., Birbaumer, N., and Scholkopf, B. (2004). Support vector channel selection in BCI. *IEEE Transactions on Biomedical Engineering*, **51**(6), 1003–1010.

Laparra, V., Gutiérrez, J., Camps-Valls, G., and Malo, J. (2010). Image denoising with kernels based on natural image relations. *Journal of Machine Learning Research*, **11**, 873–903.

Laparra, V., Jiménez, S., Tuia, D., Camps-Valls, G., and Malo, J. (2014). Principal polynomial analysis. *International Journal of Neural Systems*, **24**(7), 1440007.

Laparra, V., Malo, J., and Camps-Valls, G. (2015). Dimensionality reduction via regression in hyperspectral imagery. *IEEE Journal on Selected Topics in Signal Processing*, **9**(6), 1026–1036.

Larocque, J. R. and Reilly, P. (2002). Reversible jump MCMC for joint detection and estimation of sources in colored noise. *IEEE Transactions on Signal Processing*, **50**(2), 231–240.

Lawrence, N. (2005). Probabilistic non-linear principal component analysis with Gaussian process latent variable models. *Machine Learning Research*, **6**, 1783–1816.

Lázaro-Gredilla, M. (2012). Bayesian warped Gaussian processes. In F. Pereira, C. Burges, L. Bottou, and K. Weinberger, editors, *Advances in Neural Information Processing Systems 25*, pages 1628–1636. Curran Associates.

Lázaro-Gredilla, M. and Titsias, M. K. (2011). Variational heteroscedastic Gaussian process regression. In *ICML'11 Proceedings of the 28th International Conference on Machine Learning*, pages 841–848, Madison, WI. Omnipress.

Lázaro-Gredilla, M., Candela, J. Q., Rasmussen, C. E., and Figueiras-Vidal, A. R. (2010). Sparse spectrum Gaussian process regression. *Journal of Machine Learning Research*, **11**, 1865–1881.

Lázaro-Gredilla, M., Van Vaerenbergh, S., and Santamaría, I. (2011). A Bayesian approach to tracking with kernel recursive least-squares. In *2011 IEEE International Workshop on Machine Learning for Signal Processing*, pages 1–6. IEEE Press, Piscataway, NJ.

Lázaro-Gredilla, M., Titsias, M. K., Verrelst, J., and Camps-Valls, G. (2014). Retrieval of biophysical parameters with heteroscedastic gaussian processes. *IEEE Geoscience and Remote Sensing Letters*, **11**(4), 838–842.

Lazebnik, S., Schmid, C., and Ponce, J. (2006). Beyond bags of features: spatial pyramid matching for recognizing natural scene categories. In *CVPR '06: Proceedings of the 2006 IEEE Computer Society Conference on Computer Vision and Pattern Recognition*, pages 2169–2178. IEEE Computer Society, Washington, DC.

LeCun, Y., Denker, J. S., Solla, S. A., Howard, R. E., and Jackel, L. D. (1989). Optimal brain damage. In D. Touretzky, editor, *Advances in Neural Information Processing Systems 2*, pages 598–605. Morgan Kaufmann, San Francisco, CA.

Lee, J., McManus, D. D., Merchant, S., and Chon, K. H. (2012). Automatic motion and noise artifact detection in Holter ECG data using empirical mode decomposition and statistical approaches. *IEEE Transactions on Biomedical Engineering*, **59**(6), 1499–1506.

Lee, T. W., Lewicki, M. S., Girolami, M., and Sejnowski, T. J. (1999). Blind source separation of more sources than mixtures using overcomplete representations. *IEEE Signal Processing Letters*, **6**, 87–90.

Lee, Y. and Nelder, J. A. (1996). Hierarchical generalized linear models. *Journal of the Royal Statistical Society. Series B (Methodological)*, **58**(4), 619–678.

Lee, Y., Lin, Y., and Wahba, G. (2004). Multicategory support vector machines: theory and application to the classification of microarray data and satellite radiance data. *Journal of the American Statistical Association*, **99**(465), 67–82.

Lévy, B. (2006). Laplace–Beltrami eigenfunctions towards an algorithm that "understands" geometry. In *SMI '06 Proceedings of the IEEE International Conference on Shape Modeling and Applications 2006*, page 13. IEEE Computer Society, Washington, DC.

Lewis, F. L., Xie, L., and Popa, D., editors (2007). *Optimal and Robust Estimation: With an Introduction to Stochastic Control Theory*, volume 25 of *Automation and Control Engineering*. CRC Press.

Li, K. and Príncipe, J. C. (2016). The kernel adaptive autoregressive–moving-average algorithm. *IEEE Transactions on Neural Networks and Learning Systems*, **27**(2), 334–346.

Li, T., Zhu, S., and Ogihara, M. (2006). Using discriminant analysis for multi-class classification: an experimental investigation. *Knowledge and Information Systems*, **10**(4), 453–472.

Lid Hjort, N., Holmes, C., Muller, P., and Walker, S. (2010). *Bayesian Nonparametrics*. Cambridge University Press, New York, NY.

Lin, Y., Lee, Y., and Wahba, G. (2000). Support vector machines for classification in nonstandard situations. Department of Statistics TR 1016, University of Wisconsin-Madison.

Liu, B., Dai, Y., Li, X., Lee, W. S., and Yu, P. S. (2003). Building text classifiers using positive and unlabeled examples. In *ICDM '03 Proceedings of the Third IEEE International Conference on Data Mining*, page 179, Washington, DC. IEEE Computer Society.

Liu, J. S. (2004). *Monte Carlo Strategies in Scientific Computing*. Springer.

Liu, W., Pokharel, P. P., and Príncipe, J. C. (2008). The kernel least-mean-square algorithm. *IEEE Transactions on Signal Processing*, **56**(2), 543–554.

Liu, W., Park, I., Wang, Y., and Príncipe, J. C. (2009). Extended kernel recursive least squares algorithm. *IEEE Transactions on Signal Processing*, **57**(10), 3801–3814.

Liu, W., Príncipe, J. C., and Haykin, S. (2010). *Kernel Adaptive Filtering: A Comprehensive Introduction*. John Wiley & Sons.

Ljung, L. (1999). *System Identification: Theory for the User*. Prentice-Hall, Upper Saddle River, NJ.

Ljung, L. (2008). Perspectives on system identification. In *Plenary talk at the Proceedings of the 17th IFAC World Congress, Seoul, South Korea*.

Lomb, N. (1976). Least-squares frequency analysis of unequally spaced data. *Astrophysics and Space Science*, **39**, 447–462.

Longbotham, N. and Camps-Valls, G. (2014). A family of kernel anomaly change detectors. In *IEEE Workshop on Hyperspectral Image and Signal Processing, Whispers*, Lausanne, Switzerland.

López-Paz, D., Sra, S., Smola, A. J., Ghahramani, Z., and Schölkopf, B. (2014). Randomized nonlinear component analysis. In *ICML'14 Proceedings of the 31st International Conference on International Conference on Machine Learning*, volume 32, pages II-1359–II-1367. JMLR.org.

Lowe, D. G. (1999). Object recognition from local scale-invariant features. In *The Proceedings of the Seventh IEEE International Conference on Computer Vision, 1999*, volume 2, pages 1150–1157. IEEE Computer Society, Washington, DC.

Luengo, D., Santamaría, I., and Vielva, L. (2005). A general solution to blind inverse problems for sparse input signals. *Neurocomputing*, **69**(1–3), 198–215.

Luo, T., Kramer, K., Golgof, D. B., Hall, L. O., Samson, S., Remsen, A., and Hopkins, T. (2005). Active learning to recognize multiple types of plankton. *J. Mach. Learn. Res.*, **6**, 589–613.

Luo, X. Q. and Wu, X. J. (2012). Fusing remote sensing images using a statistical model. *Applied Mechanics and Materials*, **263–266**, 416–420.

Luo, X. Q., Wu, X. J., and Zhang, Z. (2014). Regional and entropy component analysis based remote sensing images fusion. *Journal of Intelligent and Fuzzy Systems*, **26**(3), 1279–1287.

MacKay, D. J. C. (1992). Information based objective functions for active data selection. *Neural Computation*, **4**(4), 590–604.

Madanayake, A., Wijenayake, C., Dansereau, D. G., Gunaratne, T. K., Bruton, L. T., and Williams, S. B. (2013). Multidimensional (MD) circuits and systems for emerging applications including cognitive radio, radio astronomy, robot vision and imaging. *IEEE Circuits and Systems Magazine*, **13**(1), 10–43.

Madanayake, A., Wijenayake, C., Lin, Z., and Dornback, N. (2015). Recent advances in multidimensional systems and signal processing: an overview. In *2015 IEEE International Symposium on Circuits and Systems (ISCAS)*, pages 2365–2368.

Mallat, S. and Zhang, Z. (1993). Matching pursuit with time-frequency dictionaries. *IEEE Transactions on Signal Processing*, **41**, 3397–3415.

Margenau, H. and Hill, R. N. (1961). Correlation between measurements in quantum theory. *Progress in Theoretical Physics*, **26**, 722–738.

Maritorena, S. and O'Reilly, J. (2000). OC2v2: update on the unitial operational SeaWiFS chlorophyll *a* algorithm. In S. Hooker and E. Firestone, editors, *SeaWiFS Postlaunch Calibration and Validation Analyses*, volume 11 of *NASA Technical Memorandum 2000–206892*, pages 3–8. John Wiley & Sons.

Marple, S. (1987). *Digital Spectral Analysis with Applications*. Prentice-Hall, Englewood Cliffs, NJ.

Martínez-Ramón, M., Xu, N., and Christodoulou, C. (2005). Beamforming using support vector machines. *IEEE Antennas and Wireless Propagation Letters*, **4**, 439–442.

Martínez-Ramón, M., Rojo-Álvarez, J. L., Camps-Valls, G., Navia-Vázquez, A., Muñoz-Marí, J., Soria-Olivas, E., and Figueiras-Vidal, A. (2006). Support vector machines for nonlinear kernel ARMA system identification. *IEEE Transactions on Neural Networks*, **17**, 1617–1622.

Martínez-Ramón, M., Rojo-Álvarez, J. L., Camps-Valls, G., and Christodoulou, C. (2007). Kernel antenna array processing. *IEEE Transactions on Antennas and Propagation*, **55**(3), 642–650.

Martino, L., Elvira, V., Luengo, D., and Corander, J. (2015a). An Adaptive Population Importance Sampler: Learning from the uncertanity. *IEEE Transactions on Signal Processing*, **63**(16), 4422–4437.

Martino, L., Yang, H., Luengo, D., Kanniainen, J., and Corander, J. (2015b). The FUSS algorithm: a fast universal self-tuned sampler within Gibbs. *Digital Signal Processing*, **47**, 68–83.

Martino, L., Elvira, V., Luengo, D., Corander, J., and Louzada, F. (2016). Orthogonal parallel MCMC methods for sampling and optimization. *Digital Signal Processing*, **58**, 64–84.

Martino, L., Elvira, V., Luengo, D., and Corander, J. (2017). Layered adaptive importance sampling. *Statistics and Computing*, **27**(3), 599–623.

Mattera, D. (2005). Support vector machines for signal processing. In L. Wang, editor, *Support Vector Machines: Theory and Applications*, volume 177 of *Studies in Fuzziness and Soft Computing*, pages 321–342. Springer, Berlin.

Mayer, R., Bucholtz, F., and Scribner, D. (2003). Object detection by using "whitening/dewhitening" to transform target signatures in multitemporal hyperspectral and multispectral imagery. *IEEE Transactions on Geoscience and Remote Sensing*, **41**(5), 1136–1142.

McCallum, A. K. (1999). Multi-label text classification with a mixture model trained by EM. In *AAAI-99 Workshop on Text Learning*.

Mehrotra, S. (1992). On the implementation of a primal-dual interior point method. *SIAM Journal on Optimization*, **2**(4), 575–601.

Meigering, E. (2002). A chronology of interpolation: from ancient astronomy to modern signal and image processing. *Proceedings of the IEEE*, **90**(3), 319–342.

Mendel, J. and Burrus, C. S. (1990). *Maximum-Likelihood Deconvolution: A Journey into Model-Based Signal Processing*. Springer-Verlag, New York, NY.

Mendel, J. M. (1986). Some modeling problems in reflection seismology. *IEEE ASSP Magazine*, **3**(2), 4–17.

Meyer, M., Desbrun, M., Schröder, P., and Barr, A. H. (2003). Discrete differential-geometry operators for triangulated 2-manifolds. In H.-C. Hege and K. Polthier, editors, *Visualization and Mathematics III*, pages 35–57. Springer.

Mika, S., Raetsch, G., Weston, J., Schölkopf, B., and Müller, K.-R. (1999). Fisher discriminant analysis with kernels. In *Neural Networks for Signal Processing IX: Proceedings of the 1999 IEEE Signal Processing Society Workshop (Cat. No.98TH8468)*, pages 41–48. IEEE Press, Piscataway, NJ.

Mika, S., Rätsch, G., Weston, J., Schölkopf, B., Smola, A., and Müller, K. (2003). Constructing descriptive and discriminative nonlinear features: Rayleigh coefficients in kernel feature spaces. *IEEE Transactions on Pattern Analysis and Machine Intelligence*, **25**, 623–628.

Molgedey, L. and Schuster, H. G. (1994). Separation of a mixture of independent signals using time delayed correlations. *Physical Review Letters*, **72**, 3634–3637.

Momma, M. and Bennett, K. (2003). Sparse kernel partial least squares regression. In B. Schölkopf and M. Warmuth, editors, *Learning Theory and Kernel Machines*, volume 2777 of *Lecture Notes in Computer Science*, pages 216–230. Springer, Berlin.

Morady, F. (1999). Radio-frequency ablation as treatment for cardiac arrhythmias. *New England Journal of Medicine*, **340**(7), 534–544.

Mouattamid, M. and Schaback, R. (2009). Recursive kernels. *Analysis in Theory and Applications*, **25**, 301–316.

Muandet, K. and Schölkopf, B. (2013). One-class support measure machines for group anomaly detection. In A. Nicholson and P. Smyth, editors, *UAI'13 Proceedings of the Twenty-Ninth Conference on Uncertainty in Artificial Intelligence*, pages 449–458. AUAI Press, Arlington, VA.

Mukherjee, S., Osuna, E., and Girosi, F. (1997). Nonlinear prediction of chaotic time series using a support vector machine. In J. Principe, L. Gile, N. Morgan, and E. Wilson, editors, *Neural Networks for Signal Processing VII. Proceedings of the 1997 IEEE Signal Processing Society Workshop*, pages 511–520, New York. IEEE Press.

Murphy, K. (2012). *Machine Learning. A Probabilistic Perspective*. MIT Press, Cambridge, MA.

Muslea, I. (2006). Active learning with multiple views. *Journal of Artificial Intelligence Research*, **27**, 203–233.

Myhre, J. and Jenssen, R. (2012). Mixture weight influence on kernel entropy component analysis and semi-supervised learning using the LASSO. In *IEEE International Workshop in Machine Learning for Signal Processing*, pages 1–6, Piscataway, NJ. IEEE Press.

Narendra, K. and Gallman, P. (1966). An iterative method for the identification of nonlinear systems using a Hammerstein model. *IEEE Transactions on Automatic Control*, **11**(3), 546–550.

Narendra, K. S. and Parthasarathy, K. (1990). Identification and control of dynamical systems using neural networks. *IEEE Transactions of Neural Networks*, **1**(1), 4–27.

Nasrabadi, N. M. (2008). Regularization for spectral matched filter and RX anomaly detector. In S. Shen and P. Lewis, editors, *Algorithms and Technologies for Multispectral, Hyperspectral, and Ultraspectral Imagery XIV*, volume 6966 of *Proceedings of SPIE*, page 696604. SPIE Press, Bellingham, WA.

Navia-Vázquez, A., Pérez-Cruz, F., Artés-Rodríguez, A., and Figueiras-Vidal, A. R. (2001). Weighted least squares training of support vector classifiers leading to compact and adaptive schemes. *IEEE Transactions on Neural Networks*, **12**(5), 1047–1059.

Naylor, P. A., Cui, J., and Brookes, M. (2006). Adaptive algorithms for sparse echo cancellation. *Signal Processing*, **86**(6), 1182–1192.

Nelles, O. (2000). *Nonlinear System Identification*. Springer-Verlag, Berlin.

Ng, A., Jordan, M., and Weiss, Y. (2001). On spectral clustering: analysis and an algorithm. In T. G. Dietterich, S. Becker, and Z. Ghahramani, editors, *Advances in Neural Information Processing Systems 14*, pages 849–856. MIT Press, Cambridge, MA.

Ngia, K. S. H. and Sjobert, J. (1998). Nonlinear acoustic echo cancellation using a Hammerstein model. In *Proceedings of the 1998 IEEE International Conference on Acoustics, Speech, and Signal Processing, 1998*, volume 2, pages 1229–1232. IEEE Press, Piscataway, NJ.

Nguyen, H. V., Patel, V. M., Nasrabadi, N. M., and Chellappa, R. (2012). Kernel dictionary learning. In *2012 IEEE International Conference on Acoustics, Speech and Signal Processing*, pages 2021–2024. IEEE Press, Piscataway, NJ.

Nielsen, A. A. (2011). Kernel maximum autocorrelation factor and minimum noise fraction transformations. *IEEE Transactions on Image Processing*, **20**(3), 612–624.

Nielsen, A. A., Conradsen, K., and Simpson, J. J. (1998). Multivariate alteration detection (MAD) and MAF post-processing in multispectral bi-temporal image data: new approaches to change detection studies. *Remote Sensing of Environment*, **64**(1), 1–19.

Nikolaev, N. and Tino, P. (2005). Sequential relevance vector machine learning from time series. In *Proceedings of International Joint Conference on Neural Networks*, volume 2, pages 1308–1313. IEE Press, Montreal, Canada.

O'Brien, S., Sinclair, A., and Kramer, S. (1994). Recovery of a sparse spike time series by L_1 norm deconvolution. *IEEE Transactions on Signal Processing*, **42**, 3353–3365.

Ogunfunmi, T. and Paul, T. (2011). On the complex kernel-based adaptive filter. In *2011 IEEE International Symposium on Circuits and Systems (ISCAS)*, pages 1263–1266. IEEE Press, Piscataway, NJ.

O'Hagan, A. and Kingman, J. F. C. (1978). Curve fitting and optimal design for prediction. *Journal of the Royal Statistical Society. Series B (Methodological)*, **40**(1), 1–42.

Ohm, J.-R. (2005). Advances in scalable video coding. *Proceedings of the IEEE*, **93**(1), 42–56.

Olofsson, T. (2004). Semi-sparse deconvolution robust to uncertainties in the impulse responses. *Ultrasonics*, **42**, 969–975.

Olshausen, B. A. and Fieldt, D. J. (1997). Sparse coding with an overcomplete basis set: a strategy employed by V1? *Vision Research*, **37**, 3311–3325.

Olshausen, B. A., Sallee, P., and Lewicki, M. S. (2003). Learning sparse multiscale image representations. *Advances in Neural Information Processing Systems*, **15**, 1327–1334.

Oppenheim, A. V. and Schafer, R. W. (1989). *Discrete-Time Signal Processing*. Prentice Hall, Englewood Cliffs, NJ.

O'Reilly, J. E. and Maritorena, S. (1997). SeaBAM evaluation data set. In *The SeaWiFS Bio-Optical Algorithm Mini-Workshop (SeaBAM)*. https://seabass.gsfc.nasa.gov/seabam/pub/maritorena_oreilly_schieber/SBAMset.ps (accessed July 7, 2017).

O'Reilly, J. E., Maritorena, S., Mitchell, B. G., Siegel, D. A., Carder, K., Garver, S. A., Kahru, M., and McClain, C. (1998). Ocean color chlorophyll algorithms for SeaWiFS. *Journal of Geophysical Research*, **103**(C11), 24937–24953.

Orr, G. B. and Müller, K.-R. (1998). *Neural Networks: Tricks of the Trade.* Springer-Verlag, Berlin.

Osuna, E., Freund, R., and Girosi, F. (1997). Improved training algorithm for support vector machines. In *Neural Networks for Signal Processing VII. Proceedings of the 1997 IEEE Signal Processing Society Workshop*, pages 276–285. IEEE Press, Piscataway, NJ.

Page, C. (1952). Instantaneous power spectra. *Journal of Applied Physics*, **23**, 103–106.

Pajunen, G. A. (1992). Adaptive control of Wiener type nonlinear systems. *Automatica*, **28**(4), 781–785.

Pal, N. R., Pal, K., Keller, J. M., and Bezdek, J. C. (2005). A possibilistic fuzzy c-means clustering algorithm. *IEEE Transactions on Fuzzy Systems*, **13**(4), 517–530.

Palkar, M. and Principe, J. (1994). Echo cancellation with the gamma filter. In *1994 IEEE International Conference on Acoustics, Speech, and Signal Processing*, volume 3, pages III/369–III/372. IEEE Press, Piscataway, NJ.

Pan, S. J. and Yang, Q. (2010). A survey on transfer learning. *IEEE Transactions on Knowledge and Data Engineering*, **22**(10), 1345–1359.

Pan, S. J. and Yang, Q. (2011). Domain adaptation via transfer component analysis. *IEEE Transactions on Neural Networks*, **22**, 199–210.

Pan, S. J., Kwok, J. T., and Yang, Q. (2008). Transfer learning via dimensionality reduction. In *AAAI'08 Proceedings of the 23rd National Conference on Artificial Intelligence*, volume 2, pages 677–682. AAAI Press.

Papoulis, A. (1991). *Probability Random Variables and Stochastic Processes*. McGraw-Hill, New York, NY, 3 edition.

Park, I. M., Seth, S., and Van Vaerenbergh, S. (2014). Probabilistic kernel least mean squares algorithms. In *2014 IEEE International Conference on Acoustics, Speech and Signal Processing*, pages 8272–8276. IEEE Press, Piscataway, NJ.

Parker, S. and Girard, P. (1976). Correlated noise due to roundoff in fixed point digital filters. *IEEE Transactions on Circuits and Systems*, **23**(4), 204–211.

Pasolli, E. and Melgani, F. (2010). Model-based active learning for SVM classification of remote sensing images. In *IEEE International Geoscience and Remote Sensing Symposium, IGARSS*, Hawaii, USA.

Pavlidis, P., Weston, J., Cai, J., and Grundy, W. N. (2001). Gene functional classification from heterogeneous data. In *RECOMB '01 Proceedings of the Fifth Annual International Conference on Computational Biology*, pages 249–255. ACM, New York, NY.

Pavlov, A., van de Wouw, N., and Nijmeijer, H. (2007). Frequency response functions for nonlinear convergent systems. *IEEE Transactions on Automatic Control*, **52**(6), 1159–1165.

Pawlak, M., Hasiewicz, Z., and Wachel, P. (2007). On nonparametric identification of Wiener systems. *IEEE Transactions on Signal Processing*, **55**(2), 482–492.

Pearson, K. (1901). On lines and planes of closest fit to systems of points in space. *Philosophical Magazine*, **2**, 559–572.
Pérez-Cruz, F. and Artés-Rodríguez, A. (2001). A new optimizing procedure for ν-support vector regressor. In *2001 IEEE International Conference on Acoustics, Speech, and Signal Processing. Proceedings (Cat. No.01CH37221)*, volume 2, pages 1265–1268. IEEE Press, Piscataway, NJ.
Pérez-Cruz, F. and Bousquet, O. (2004). Kernel methods and their potential use in signal processing. *IEEE Signal Processing Magazine*, **21**(3), 57–65.
Pérez-Cruz, F., Camps-Valls, G., Soria-Olivas, E., Pérez-Ruixo, J. J., Figueiras-Vidal, A. R., and Artés-Rodríguez, A. (2002). Multi-dimensional function approximation and regression estimation. In J. Dorronsoro, editor, *Artificial Neural Networks – ICANN 2002*, volume 2415 of *Lecture Notes in Computer Science*, pages 757–782. Springer-Verlag, Berlin.
Pérez-Cruz, F., Van Vaerenbergh, S., Murillo-Fuentes, J. J., Lazaro-Gredilla, M., and Santamaria, I. (2013). Gaussian processes for nonlinear signal processing: an overview of recent advances. *IEEE Signal Processing Magazine*, **30**(4), 40–50.
Pérez-Suay, A. and Camps-Valls, G. (2017). Sensitivity maps of the Hilbert–Schmidt independence criterion. *Applied Soft Computing*. in press. https://doi.org/10.1016/j.asoc.2017.04.024.
Pi, H. and Peterson, C. (1994). Finding the embedding dimension and variable dependencies in time series. *Neural Computation*, **6**(3), 509–520.
Picone, J., Ganapathiraju, A., and Hamaker, J. (2007). Applications of kernel theory to speech recognition. In G. Camps-Valls, J. L. Rojo-Álvarez, and M. Martínez-Ramón, editors, *Kernel Methods in Bioengineering, Signal and Image Processing*, pages 224–245. Idea Group Publishing, Hershey, PA.
Pinkall, U. and Polthier, K. (1993). Computing discrete minimal surfaces and their conjugates. *Experimental Mathematics*, **2**(1), 15–36.
Plackett, R. L. (1950). Some theorems in least squares. *Biometrika*, **37**, 149–157.
Platt, J. (1991). A resource-allocating network for function interpolation. *Neural computation*, **3**(2), 213–225.
Platt, J. (1999a). Fast training of support vector machines using sequential minimal optimization. In B. Schölkopf, C. J. C. Burges, and A. J. Smola, editors, *Advances in Kernel Methods – Support Vector Learning*, pages 185–208, Cambridge, MA. MIT Press.
Platt, J. (1999b). Probabilistic outputs for support vector machines and comparisons to regularized likelihood methods. In A. J. Smola and P. J. Bartlett, editors, *Advances in Large Margin Classifiers*, pages 61–74. MIT Press, Cambridge, MA.
Poggio, T. and Smale, S. (2003). The mathematics of learning: dealing with data. *Notices of the American Mathematical Society*, **50**(5), 537–544.
Pokharel, P. P., Liu, W., and Príncipe, J. C. (2009). Kernel least mean square algorithm with constrained growth. *Signal Processing*, **89**(3), 257–265.
Pokharel, R., Seth, S., and Principe, J. C. (2013). Mixture kernel least mean square. In *The 2013 International Joint Conference on Neural Networks (IJCNN)*, pages 1–7. IEEE Press, Piscataway, NJ.
Politis, D. N., Romano, J. P., and Lai, T.-L. (1992). Bootstrap confidence bands for spectra and cross-spectra. *IEEE Transactions on Signal Processing*, **40**(5), 1206–1215.
Powell, M. (1994). Some algorithms for thin plate spline interpolation to functions of two variables. In H. Dikshit and C. Micchelli, editors, *Proceedings of the Conference on*

Advances in Computational Mathematics: New Delhi, India, pages 303–319. World Scientific, Singapore.

Pozar, D. M. (2003). A relation between the active input impedance and the active element pattern of a phased array. *IEEE Transactions on Antennas and Propagation*, **51**(9), 2486–2489.

Press, W. H., Teukolsky, S. A., Vetterling, W. T., and Flannery, B. P. (1992). *Numerical Recipes in C*. Cambridge University Press, 2nd edition.

Principe, J. C. (2010). *Information Theoretic Learning: Renyi's Entropy and Kernel Perspectives*. Springer.

Principe, J. C., deVries, B., and deOliveira, P. G. (1993). The gamma filter – a new class of adaptive IIR filters with restricted feedback. *IEEE Transactions on Signal Processing*, **41**(2), 649–656.

Proakis, J. G. and Manolakis, D. K. (2006). *Digital Signal Processing*. Prentice Hall, Upper Saddle River, NJ, 4th edition.

Proakis, J. G. and Salehi, M. (2004). *Fundamentals of Communication Systems*. Prentice Hall, Upper Saddle River, NJ, 1st edition.

Qi, H. and Hughes, S. M. (2011). Using the kernel trick in compressive sensing: accurate signal recovery from fewer measurements. In *2011 IEEE International Conference on Acoustics, Speech and Signal Processing*, pages 3940–3943. IEEE Press, Piscataway, NJ.

Qin, S. and McAvoy, T. (1992). Non-linear PLS modelling using neural networks. *Computers & Chemical Engineering*, **16**, 379–391.

Quiñonero-Candela, J. (2004). *Learning with Uncertainty – Gaussian Processes and Relevance Vector Machines*. Ph.D. thesis, Technical University of Denmark, Informatics and Mathematical Modelling, Kongens Lyngby, Denmark.

Quiñonero-Candela, J. and Rasmussen, C. E. (2005). A unifying view of sparse approximate gaussian process regression. *Journal of Machine Learning Research*, **6**, 1939–1959.

Quiñonero-Candela, J., Sugiyama, M., Schwaighofer, A., and Lawrence, N. D. (2009). *Dataset Shift in Machine Learning*. Neural Information Processing Series. MIT Press, Cambridge, MA.

Rabiner, L. (1989). A tutorial on hidden Markov models and selected applications in speech recognition. *Proceedings of the IEEE*, **77**(2), 257–286.

Rabiner, L. R. and Schafer, R. W. (2007). Introduction to digital speech processing. *Foundations and Trends® in Signal Processing*, **1**(1–2), 1–194.

Rabiner, L. R., Kaiser, J. F., Herrmann, O., and Dolan, M. T. (1974). Some comparisons between FIR and IIR digital filters. *The Bell System Technical Journal*, **53**(2), 305–331.

Rahimi, A. and Recht, B. (2007). Random features for large-scale kernel machines. In J. C. Platt, D. Koller, Y. Singer, and S. T. Roweis, editors, *NIPS'07 Proceedings of the 20th International Conference on Neural Information Processing Systems*, pages 1177–1184. Curran Associates.

Rahimi, A. and Recht, B. (2009). Weighted sums of random kitchen sinks: Replacing minimization with randomization in learning. In D. Koller, D. Schuurmans, Y. Bengio, and L. Bottou, editors, *Advances in Neural Information Processing Systems 21*, pages 1313–1320. Curran Associates.

Rakotomamonjy, A., Bach, F., Grandvalet, Y., and Canu, S. (2008). SimpleMKL. *Journal of Machine Learning Research*, **9**, 2491–2521.

Ralaivola, L. and d'Alche Buc, F. (2004). Dynamical modeling with kernels for nonlinear time series prediction. In S. Thrun, L. K. Saul, and B. Schölkopf, editors, *Advances in Neural Processing Systems 16*, pages 129–136. MIT Press, Cambridge, MA.

Ralaivola, L. and d'Alche Buc, F. (2005). Time series filtering, smoothing and learning using the kernel kalman filter. In *IJCNN '05. Proceedings. 2005 IEEE International Joint Conference on Neural Networks*, volume 3, pages 1449–1454.

Randen, T. and Husoy, J. (1999). Filtering for texture classification: a comparative study. *IEEE Transactions on Pattern Analysis and Machine Intelligence*, **21**(4), 291–310.

Ranney, K. I. and Soumekh, M. (2006). Hyperspectral anomaly detection within the signal subspace. *IEEE Geoscience and Remote Sensing Letters*, **3**(3), 312–316.

Rasmussen, C. E. and Quiñonero-Candela, J. (2005). Healing the relevance vector machine by augmentation. In *Proceedings of the 22nd International Conference on Machine Learning*, pages 689–696. ACM, New York, NY.

Rasmussen, C. E. and Williams, C. K. I. (2006). *Gaussian Processes for Machine Learning*. The MIT Press, Cambridge, MA.

Reed, I. S. and Yu, X. (1990). Adaptive multiple-band CFAR detection of an optical pattern with unknown spectral distribution. *IEEE Transactions on Acoustics, Speech and Signal Processing*, **38**(10), 1760–1770.

Reed, M. and Simon, B. (1981). *Functional Analysis*, volume 1 of *Methods of Modern Mathematical Physics*. Academic Press, 1 edition.

Remondo, D., Srinivasan, R., Nicola, V. F., van Etten, W. C., and Tattje, H. E. P. (2000). Adaptive importance sampling for performance evaluation and parameter optimization of communication systems. *IEEE Transactions on Communications*, **48**(4), 557–565.

Richard, C., Bermudez, J. C. M., and Honeine, P. (2009). Online prediction of time series data with kernels. *IEEE Transactions on Signal Processing*, **57**(3), 1058–1067.

Riesz, F. and Nagy, B. S. (1955). *Functional Analysis*. Frederick Ungar Publishing Co.

Rifkin, R. and Klautau, A. (2004). In defense of one-versus-all classification. *Journal of Machine Learning Research*, **5**, 101–141.

Rihaczek, W. (1968). Signal energy distribution in time and frequency. *IEEE Transactions on Information Theory*, **14**, 369–374.

Rissanen, J. (1978). Modeling by shortest data description. *Automatica*, **14**(5), 465–471.

Robert, C. P. and Casella, G. (2004). *Monte Carlo Statistical Methods*. Springer.

Rohwer, J. A., Abdallah, C. T., and Christodoulou, C. G. (2003). Least squares support vector machines for direction of arrival estimation with error control and validation. In *Global Telecommunications Conference, 2003. GLOBECOM '03. IEEE*, volume 4, pages 2172–2176. IEEE Press, Piscataway, NJ.

Rojo-Álvarez, J. L., Martínez-Ramón, M., Figueiras-Vidal, A., García-Armada, A., and Artés-Rodríguez, A. (2003). A robust support vector algorithm for nonparametric spectral analysis. *IEEE Signal Processing Letters*, **10**(11), 320–323.

Rojo-Álvarez, J. L., Arenal-Maiz, A., and Artes-Rodriguez, A. (2002). Discriminating between supraventricular and ventricular tachycardias from EGM onset analysis. *IEEE Engineering in Medicine and Biology Magazine*, **21**(1), 16–26.

Rojo-Álvarez, J. L., Martínez-Ramón, M., de Prado-Cumplido, M., Artés-Rodríguez, A., and Figueiras-Vidal, A. R. (2004). Support vector method for robust ARMA system identification. *IEEE Transactions on Signal Processing*, **52**(1), 155–164.

Rojo-Álvarez, J. L., Camps-Valls, G., Martínez-Ramón, M., Soria-Olivas, E., A. Navia Vázquez, and Figueiras-Vidal, A. R. (2005). Support vector machines framework for linear signal processing. *Signal Processing*, **85**(12), 2316–2326.

Rojo-Álvarez, J. L., Figuera-Pozuelo, C., Martínez-Cruz, C. E., Camps-Valls, G., Alonso-Atienza, F., and Martínez-Ramón, M. (2007). Nonuniform interpolation of noisy signals using support vector machines. *IEEE Transactions on Signal Processing*, **55**(8), 4116–4126.

Rojo-Álvarez, J. L., , Martínez-Ramón, M., Muñoz-Marí, J., Camps-Valls, G., Cruz, C. E. M., and Figueiras-Vidal, A. R. (2008). Sparse deconvolution using support vector machines. *EURASIP Journal on Advances in Signal Processing*, **2008**, 816507.

Rojo-Álvarez, J. L., Martínez-Ramón, M., Jordi, Muñoz-Marí, and Camps-Valls, G. (2014). A unified SVM framework for signal estimation. *Digital Signal Processing*, **6**, 1–20.

Rosenberg, S. (1997). *The Laplacian on a Riemannian Manifold: An Introduction to Analysis on Manifolds*. Number 31 in London Mathematical Society Student Texts. Cambridge University Press.

Rosipal, R. (2011). Nonlinear partial least squares: an overview. In H. Lodhi and Y. Yamanishi, editors, *Chemoinformatics and Advanced Machine Learning Perspectives: Complex Computational Methods and Collaborative Techniques*, pages 169–189. Medical Information Science Reference, Hershey, PA.

Rosipal, R. and Trejo, L. J. (2001). Kernel partial least squares regression in reproducing hilbert spaces. *Journal of Machine Learning Research*, **2**, 97–123.

Roweis, S. T. and Saul, L. K. (2000). Nonlinear dimensionality reduction by locally linear embedding. *Science*, **290**(5500), 2323–2326.

Roy, N. and McCallum, A. (2001). Toward optimal active learning through sampling estimation of error reduction. In C. Brodley and A. Danyluk, editors, *ICML '01 Proceedings of the Eighteenth International Conference on Machine Learning*, pages 441–448. Morgan Kaufmann, San Francisco, CA.

Ruppert, D., Wand, M. P., and Carroll, R. J. (2003). *Semiparametric Regression*. Number 12 in Cambridge Series in Statistical and Probabilistic Mathematics. Cambridge University Press.

Rzepka, D. (2012). Fixed-budget kernel least mean squares. In *2012 IEEE 17th International Conference on Emerging Technologies & Factory Automation (ETFA)*, pages 1–4. IEEE Press, Piscataway, NJ.

Saenko, K., Kulis, B., Fritz, M., and Darrell, T. (2010). Adapting visual category models to new domains. In K. Daniilidis, P. Maragos, and N. Paragios, editors, *Computer Vision – ECCV'10: 11th European Conference on Computer Vision, Heraklion, Crete, Greece, September 5–11, 2010, Proceedings, Part IV*, pages 213–226. Springer-Verlag, Berlin.

Sampson, P. and Guttorp, P. (1992). Nonparametric estimation of nonstationary spatial covariance structure. *Journal of the American Statistical Association*, **87**(417), 108–119.

Sánchez-Fernández, M., de Prado-Cumplido, M., Arenas-García, J., and Pérez-Cruz, F. (2004). Support vector machine techniques for nonlinear equalization. *IEEE Transactions on Signal Processing*, **52**(8), 2298–2307.

Sands, N. P. and Cioffi, J. M. (1993). Nonlinear channel models for digital magnetic recording. *IEEE Transactions on Magnetism*, **29**, 3996–3998.

Santamaría-Caballero, I., Pantaleón-Prieto, C. J., and Artés-Rodríguez, A. (1996). Sparse deconvolution using adaptive mixed-Gaussian models. *Signal Processing*, **54**, 161–172.

Saunders, C., Gammerman, A., and Vovk, V. (1998). Ridge regression learning algorithm in dual variables. In J. W. Shavlik, editor, *ICML '98 Proceedings of the Fifteenth International Conference on Machine Learning*, pages 515–521. Morgan Kaufmann, San Francisco, CA.

Sayed, A. H. (2003). *Fundamentals of Adaptive Filtering*. Wiley–IEEE Press.

Schapire, R. E. and Singer, Y. (2000). BoosTexter: a boosting-based system for text categorization. *Machine Learning*, **39**(2–3), 135–168.

Schaum, A. and Stocker, A. (1997). Long-interval chronochrome target detection. In *Proceedings of the International Symposium on Spectral Sensing Research*, pages 1760–1770.

Schetzen, M. (1980). *The Volterra and Wiener Theories of Nonlinear Systems*. Krieger Publishing, New York, NY.

Schohn, G. and Cohn, D. (2000). Less is more: active learning with support vector machines. In P. Langley, editor, *ICML '00 Proceedings of the Seventeenth International Conference on Machine Learning*, pages 839–846. Morgan Kaufmann, San Francisco, CA.

Schölkopf, B. (1997). *Support Vector Learning*. R. Oldenbourg Verlag, Munich.

Schölkopf, B. and Smola, A. (2002). *Learning with Kernels – Support Vector Machines, Regularization, Optimization and Beyond*. MIT Press, Cambridge, MA.

Schölkopf, B., Smola, A., and Müller, K. (1998). Nonlinear component analysis as a kernel eigenvalue problem. *Neural Computation*, **10**(5), 1299–1319.

Schölkopf, B., Burges, C. J. C., and Smola, A. J., editors (1999). *Advances in Kernel Methods: Support Vector Learning*. MIT Press, Cambridge, MA.

Schölkopf, B., Bartlett, P. L., Smola, A., and Williamson, R. (1999). Shrinking the tube: a new support vector regression algorithm. In M. S. Kearns, S. Solla, and D. A. Cohn, editors, *Advances in Neural Information Processing Systems 11*, pages 330–336, Cambridge, MA. MIT Press.

Schölkopf, B., Williamson, R. C., Smola, A., and Shawe-Taylor, J. (1999). Support vector method for novelty detection. In T. K. L. Sara A. Solla and K.-R. Müller, editors, *Advances in Neural Information Processing Systems 12*, pages 582–588. MIT Press, Cambridge, MA.

Schölkopf, B., Smola, A. J., Williamson, R. C., and Bartlett, P. L. (2000). New support vector algorithms. *Neural Computation*, **12**(5), 1207–1245.

Schölkopf, B., Herbrich, R., and Smola, A. J. (2001). A generalized representer theorem. In D. Helmbold and B. Williamson, editors, *Computational Learning Theory. COLT 2001*, volume 2111 of *Lecture Notes in Computer Science*, pages 416–426. Springer, Berlin.

Seeger, M. (2001). Learning with labeled and unlabeled data. Technical Report TR.2001, Institute for Adaptive and Neural Computation, University of Edinburgh.

Selva, J. (2009). Functionally weighted Lagrange interpolation of band-limited signals from nonuniform samples. *IEEE Transactions on Signal Processing*, **57**(1), 168–181.

Seung, H. S., Opper, M., and Sompolinsky, H. (1992). Query by committee. In *COLT '92 Proceedings of the Fifth Annual Workshop on Computational Learning Theory*, pages 287–294. ACM, New York, NY.

Shannon, C. E. (1949). *The Mathematical Theory of Communication*. University of Illinois Press, 1st edition.

Shannon, C. E. (1998). Classic paper: Communication in the presence of noise. *Proceedings of the IEEE*, **86**(2), 447–457.

Shawe-Taylor, J. and Cristianini, N. (2004). *Kernel Methods for Pattern Analysis*. Cambridge University Press.

Shekar, B. H., Kumari, M. S., Mestetskiy, L. M., and Dyshkant, N. F. (2011). Face recognition using kernel entropy component analysis. *Neurocomputing*, **74**(6), 1053–1057.

Shen, H., Jegelka, S., and Gretton, A. (2007). Fast kernel ICA using an approximate Newton method. *Proceedings of Machine Learning Research*, **PLMR-2**, 476–483.

Shen, H., Jegelka, S., and Gretton, A. (2009). Fast kernel-based independent component analysis. *IEEE Transactions on Signal Processing*, **57**(9), 3498–3511.

Sheng, Y. (1996). Wavelet transform. In A. D. Poularika, editor, *The Transforms and Applications Handbook*, pages 747–827. CRC Press, Boca Raton, FL.

Shpun, S., Gepstein, L., Hayam, G., and Ben-Haim, S. A. (1997). Guidance of radiofrequency endocardial ablation with real-time three-dimensional magnetic navigation system. *Circulation*, **96**(6), 2016–2021.

Shrager, J., Hogg, T., and Huberman, B. A. (1987). Observation of phase transitions in spreading activation networks. *Science*, **236**, 1092–1094.

Silverman, B. (1985). Some aspects of the spline smoothing approach to non-parametric curve fitting. *Journal of the Royal Statistical Society*, **B-47**, 1–52.

Silverman, B. (1986). *Density Estimation for Statistics and Data Analysis*. Chapman and Hall, London.

Silvia, M. T. and Robinson, E. A. (1979). *Deconvolution of Geophysical Time Series in the Exploration for Oil and Natural Gas*, volume 10 of *Developments in Petroleum Science*. Elsevier Scientific, Amsterdam.

Sindhwani, V., Niyogi, P., and Belkin, M. (2005). Beyond the point cloud: from transductive to semi-supervised learning. In *ICML '05: Proceedings of the 22nd International Conference on Machine Learning*, pages 824–831, New York, NY. ACM.

Sinha, D. and Tewfik, A. (1993). Low bit rate transparent audio compression using adapted wavelets. *IEEE Transactions on Signal Processing*, **41**(12), 3463–3479.

Sjöberg, J., Zhang, Q., Ljung, L., Benveniste, A., Deylon, B., Glorennec, P.-Y., Hjalmarsson, H., and Juditsky, A. (1995). Nonlinear black-box modeling in system identification: a unified overview. *Automatica*, **31**, 1691–1724.

Smola, A. J. and Schölkopf, B. (2004). A tutorial on support vector regression. *Statistics and Computing*, **14**, 199–222.

Smola, A. J., Murata, N., Schölkopf, B., and Müller, K.-R. (1998). Asymptotically optimal choice of ε-loss for support vector machines. In L. Niklasson, M. Bodén, and T. Ziemke, editors, *ICANN 98*, Perspectives in Neural Computing, pages 105–110. Springer, London.

Snelson, E. and Ghahramani, Z. (2007). Local and global sparse Gaussian process approximations. *Proceedings of Machine Learning Research*, **PMLR-2**, 524–531.

Snelson, E., Rasmussen, C., and Ghahramani, Z. (2004). Warped Gaussian processes. In S. Thrun, L. Saul, and P. Schölkopf, editors, *Advances in Neural Information Processing Systems 16 (NIPS 2003)*, pages 337–344. MIT Press, Cambridge, MA.

Söderström, T. and Stoica, P. (1987). *System Identification*. Prentice-Hall, Englewood Cliffs, NJ.

Soguero-Ruiz, C., Gimeno-Blanes, F.-J., Mora-Jiménez, I., Martínez-Ruiz, M. P., and Rojo-Álvarez, J.-L. (2012). On the differential benchmarking of promotional efficiency with machine learning modeling (I): principles and statistical comparison. *Expert Systems with Applications*, **39**(17), 12772–12783.

Soguero-Ruiz, C., Hindberg, K., Rojo-Álvarez, J. L., Skrøvseth, S., Godtliebsen, F., Mortensen, K., Revhaug, A., Lindsetmo, R.-O., Mora-Jiménez, I., Augestad, K., and Jenssen, R. (2014a). Bootstrap resampling feature selection and support vector machine for early detection of anastomosis leakage. In *2014 IEEE–EMBS International Conference on Biomedical and Health Informatics (BHI)*, pages 577–580. IEEE Press, Piscataway, NJ.

Soguero-Ruiz, C., Hindberg, K., Rojo-Álvarez, J. L., Skrøvseth, S. O., Godtliebsen, F., Mortensen, K., Revhaug, A., Lindsetmo, R.-O., Augestad, K. M., and Jenssen, R. (2014b). Support vector feature selection for early detection of anastomosis leakage from bag-of-words in electronic health records. *IEEE Journal of Biomedical and Health Informatics*, **20**(5), 1404–1415.

Soguero-Ruiz, C., Palancar, F., Bermejo, J., Antoranz, J. L., and Rojo-Álvarez, J. (2016a). Autocorrelation kernel support vector machines for Doppler ultrasound M-mode images denoising. In *2016 Computing in Cardiology Conference (CinC)*, volume 43, pages 469–472. IEEE Press, Piscataway, NJ.

Soguero-Ruiz, C., Hindberg, K., Mora-Jiménez, I., Rojo-Álvarez, J. L., Skrøvseth, S. O., Godtliebsen, F., Mortensen, K., Revhaug, A., Lindsetmo, R.-O., Augestad, K. M., *et al.* (2016b). Predicting colorectal surgical complications using heterogeneous clinical data and kernel methods. *Journal of Biomedical Informatics*, **61**, 87–96.

Soh, H. and Demiris, Y. (2015). Spatio-temporal learning with the online finite and infinite echo-state Gaussian processes. *IEEE Transactions on Neural Networks and Learning Systems*, **26**(3), 522–536.

Solazzi, M., Parisi, R., and Uncini, A. (2001). Blind source separation in nonlinear mixtures by adaptive spline neural networks. In *Proceedings of the 3rd International Conference on Independent Component Analysis and Blind Signal Separation (ICA 2001), San Diego, California, USA*, pages 254–259.

Song, L., Smola, A. J., Borgwardt, K. M., and Gretton, A. (2007). Colored maximum variance unfolding. In J. C. Platt, D. Koller, Y. Singer, and S. T. Roweis, editors, *Advances in Neural Information Processing Systems 20 (NIPS 2007)*, pages 1385–1392. Curran Associates.

Sonnenburg, S., G. Rätsch, C. S., and Schölkopf, B. (2006a). Large scale multiple kernel learning. *Journal of Machine Learning Research*, **7**, 1531–1565.

Sonnenburg, S., Rätsch, G., Schäfer, C., and Schölkopf, B. (2006b). Large scale multiple kernel learning. *Journal of Machine Learning Research*, **7**, 1531–1565.

Sorkine, O. (2005). Laplacian mesh processing. In Y. Chrysanthou and M. Magnor, editors, *EUROGRAPHICS '05 STAR – State of the Art Report*, pages 53–70. The Eurographics Association.

Sreekanth, V., Vedaldi, A., Jawahar, C. V., and Zisserman, A. (2010). Generalized RBF feature maps for efficient detection. In F. Labrosse, R. Zwiggelaar, Y. Liu, and B. Tiddeman, editors, *British Machine Vision Conference*, pages 2.1–2.11. BMVA Press.

Stanford University (2016a). PLY files an ASCII polygon format. http://people.sc.fsu.edu/ jburkardt/data/ply/ply.html (accessed July 3, 2017).

Stanford University (2016b). The Stanford 3D scanning repository. http://graphics.stanford.edu/data/3Dscanrep/ (accessed July 3, 2017).

Stein, D. W. J., Beaven, S. G., Hoff, L. E., Winter, E. M., Schaum, A. P., and Stocker, A. D. (2002). Anomaly detection from hyperspectral imagery. *IEEE Signal Processing Magazine*, **19**(1), 58–69.

Stoica, P. and Sharman, K. C. (1990). Novel eigenanalysis method for direction estimation. *IEE Proceedings F, Radar and Signal Processing*, **137**(1), 19–26.

Strohmer, T. (2000). Numerical analysis of the non-uniform sampling problem. *Journal of Computational and Applied Mathematics*, **122**(1–2), 297–316.

Sugiyama, M. and Kawanabe, M. (2012). *Machine Learning in Non-Stationary Environments: Introduction to Covariate Shift Adaptation*. MIT Press, Cambridge, MA.

Sun, S., Zhao, J., and Zhu, J. (2015). A review of Nyström methods for large-scale machine learning. *Information Fusion*, **26**(C), 36–48.

Sutherland, D. J. and Schneider, J. G. (2015). On the error of random Fourier features. http://arxiv.org/abs/1506.02785.

Suykens, J., Vandewalle, J., and Moor, B. D. (2001a). Intelligence and cooperative search by coupled local minimizers. *International Journal of Bifurcation and Chaos*, **11**(8), 2133–2144.

Suykens, J. A. K. (2001). Support vector machines: a nonlinear modelling and control perspective. *European Journal of Control*, **7**(2–3), 311–327.

Suykens, J. A. K. and Vandewalle, J. (1999). Least squares support vector machine classifiers. *Neural Processing Letters*, **9**(3), 293–300.

Suykens, J. A. K. and Vandewalle, J. (2000). Recurrent least squares support vector machines. *IEEE Transactions on Circuits and Systems-I*, **47**(7), 1109–1114.

Suykens, J. A. K., Vandewalle, J., and Moor, B. D. (2001b). Optimal control by least squares support vector machines. *Neural Networks*, **14**(1), 23–35.

Suykens, J. A. K., Gestel, T. V., Brabanter, J. D., Moor, B. D., and Vandewalle, J., editors (2002). *Least Squares Support Vector Machines*. World Scientific, Singapore.

Takens, F. (1981). Detecting strange attractors in turbulence. *Dynamical Systems and Turbulence*, **898**, 366–381.

Taleb, A. and Jutten, C. (1999). Source separation in post-nonlinear mixtures. *IEEE Transactions on Signal Processing*, **47**, 2807–2820.

Taleb, A., Solé, J., and Jutten, C. (2001). Quasi-nonparametric blind inversion of Wiener systems. *IEEE Transactions on Signal Processing*, **49**(5), 917–924.

Tan, Y. and Wang, J. (2001). Nonlinear blind source separation using higher order statistics and a genetic algorithm. *IEEE Transactions on Evolutionary Computation*, **5**(6), 600–612.

Task Force of the European Society of Cardiology and the North American Society of Pacing and Electrophysiology (1996). Heart rate variability – standards of measurement, physiological interpretation, and clinical use. *Circulation*, **93**(5), 1043–1065.

Taskar, B., Guestrin, C., and Koller, D. (2003). Max-margin Markov networks. In S. Thrun, L. Saul, and P. Schölkopf, editors, *Advances in Neural Information Processing Systems 16*, pages 25–32. MIT Press, Cambridge, MA.

Taubin, G. (1995). A signal processing approach to fair surface design. In S. Mair and R. Cook, editors, *Proceedings of the 22nd Annual Conference on Computer Graphics and Interactive Techniques*, pages 351–358. ACM, New York, NY.

Tax, D. and Duin, R. P. W. (1999). Support vector domain description. *Pattern Recognition Letters*, **20**, 1191–1199.

Tenenbaum, J. B., de Silva, V., and Langford, J. C. (2000). A global geometric framework for nonlinear dimensionality reduction. *Science*, **290**(5500), 2319.

Tertinek, S. and Vogel, C. (2008). Reconstruction of nonuniformly sampled bandlimited signals using a differentiator–multiplier cascade. *IEEE Transactions on Circuits and Systems I*, **55**(8), 2273–2286.

Theiler, J. (2008). Quantitative comparison of quadratic covariance-based anomalous change detectors. *Applied Optics*, **47**(28), 12–26.

Theiler, J. and Perkins, S. (2006). Proposed framework for anomalous change detection. In *ICML Workshop on Machine Learning Algorithms for Surveillance and Event Detection*, pages 7–14.

Theiler, J., Scovel, C., Wohlberg, B., and Foy, B. R. (2010). Elliptically contoured distributions for anomalous change detection in hyperspectral imagery. *IEEE Geoscience and Remote Sensing Letters*, **7**(2), 271–275.

Theis, F. J. and Amari, S. (2004). Postnonlinear overcomplete blind source separation using sparse sources. In C. Puntonet and A. Prieto, editors, *Independent Component Analysis and Blind Signal Separation. ICA 2004*, volume 3195 of *Lecture Notes in Computer Science*, pages 718–725. Springer, Berlin.

Tibshirani, R. (1994). Regression shrinkage and selection via the lasso. *Journal of the Royal Statistical Society, Series B*, **58**, 267–288.

Tikhonov, A. N. (1963). Regularization of incorrectly posed problems. *Soviet Mathematics – Doklady*, **4**, 1624–1627.

Tikhonov, A. N. and Arsenin, V. I. A. (1977). *Solutions of Ill-Posed Problems*. Scripta Series in Mathematics. John Wiley & Sons, Ltd, Chichester. Translated by F. John.

Tipping, M. E. (2000). The relevance vector machine. In S. A. Solla, T. K. Leen, and K.-R. Müller, editors, *Advances in Neural Information Processing Systems 12*, pages 652–658. MIT Press, Cambridge, MA.

Tipping, M. E. (2001). Sparse Bayesian learning and the relevance vector machine. *Journal of Machine Learning Research*, **1**, 211–244.

Tong, S. and Koller, D. (2001). Support vector machine active learning with applications to text classification. *Journal of Machine Learning Research*, **2**, 45–66.

Torralba, A. and Efros, A. A. (2011). Unbiased look at dataset bias. In *CVPR'11 Proceedings of the 2011 IEEE Conference on Computer Vision and Pattern Recognition*, pages 1521–1528. IEEE Computer Society, Washington, DC.

Tosic, I. and Frossard, P. (2011). Dictionary learning. *IEEE Signal Processing Magazine*, **28**(2), 27–38.

Tropp, J. A. (2004). Greed is good: algorithmic results for sparse approximation. *IEEE Transactions on Information Theory*, **50**, 2231–2242.

Tropp, J. A. and Wright, S. J. (2010). Computational methods for sparse solution of linear inverse problems. *Proceedings of the IEEE*, **98**(6), 948–958.

Tsochantaridis, I., Hofmann, T., Joachims, T., and Altun, Y. (2004). Support vector machine learning for interdependent and structured output spaces. In *ICML '04 Proceedings of the Twenty-First International Conference on Machine Learning*, page 104. ACM, New York, NY.

Tsochantaridis, I., Finley, T., Joachims, T., Hofmann, T., and Altun, Y. (2005). Large margin methods for structured and interdependent output variables. *Journal of Machine Learning Research*, **6**, 1453–1484.

Tuia, D. and Camps-Valls, G. (2009). Semisupervised remote sensing image classification with cluster kernels. *IEEE Geoscience and Remote Sensing Letters*, **6**(2), 224–228.

Tuia, D. and Camps-Valls, G. (2016). Kernel manifold alignment for domain adaptation. *PLoS One*, **11**(2), e0148655.

Tuia, D., Ratle, F., Pacifici, F., Kanevski, M., and Emery, W. J. (2009). Active learning methods for remote sensing image classification. *IEEE Transactions on Geoscience and Remote Sensing*, **47**(7), 2218–2232.

Tuia, D., Verrelst, J., Alonso, L., Pérez-Cruz, F., and Camps-Valls, G. (2011a). Multioutput support vector regression for remote sensing biophysical parameter estimation. *IEEE Geoscience and Remote Sensing Letters*, **8**(4), 804–808.

Tuia, D., Muñoz-Marí, J., Kanevski, M., and Camps-Valls, G. (2011b). Structured output SVM for remote sensing image classification. *Journal of Signal Processing Systems*, **65**(3), 301–310.

Tuia, D., Muñoz-Marí, J., Gómez-Chova, L., and Malo, J. (2013). Graph matching for adaptation in remote sensing. *IEEE Transactions on Geoscience and Remote Sensing*, **51**(1), 329–341.

Tuia, D., Muñoz-Marí, J., Rojo-Álvarez, J. L., Martínez-Ramón, M., and Camps-Valls, G. (2014). Explicit recursive and adaptive filtering in reproducing kernel Hilbert spaces. *IEEE Transactions on Neural Networks and Learning Systems*, **25**(7), 1413–1419.

Unser, M. (2000). Sampling – 50 years after Shannon. *Proceedings of the IEEE*, **88**(4), 569–587.

Vaidyanathan, P. P. (2001). Generalizations of the sampling theorem: Seven decades after Nyquist. *IEEE Transactions on Circuits and Systems I*, **48**(9), 1094–1108.

Vakhania, N., Tarieladze, V., and Chobanyan, S. (1987). *Probability Distributions on Banach Spaces*. Springer Science & Business Media.

Vallet, B. and Lévy, B. (2008). Spectral geometry processing with manifold harmonics. *Computer Graphics Forum*, **27**(2), 251–260.

Vallet, B. and Lévy, B. (2007). Manifold harmonics. Technical report, INRIA – ALICE Project Team.

Van Gerven, M., Chao, Z., and Heskes, T. (2012). On the decoding of intracranial data using sparse orthonormalized partial least squares. *Journal of Neural Engineering*, **9**(2), 026017.

Van Huffel, S. and Vandewalle, J. (1991). *The Total Least Squares Problem*. Society for Industrial and Applied Mathematics.

Van Trees, H. (2002). *Detection, Estimation, and Modulation Theory*. John Wiley & Sons, Inc., New York, NY.

Van Vaerenbergh, S. and Santamaría, I. (2006). A spectral clustering approach to underdetermined postnonlinear blind source separation of sparse sources. *IEEE Transactions on Neural Networks*, **17**(3), 811–814.

Van Vaerenbergh, S. and Santamaría, I. (2013). A comparative study of kernel adaptive filtering algorithms. In *2013 IEEE Digital Signal Processing Workshop and Signal Processing Education Meeting. DSP/SPE 2013, Proceedings*, pages 181–186. IEEE Press, Piscataway, NJ.

Van Vaerenbergh, S., Vía, J., and Santamaría, I. (2006). A sliding-window kernel RLS algorithm and its application to nonlinear channel identification. In *2006 IEEE International Conference on Acoustics Speech and Signal Processing Proceedings*, volume 5, pages 789–792. IEEE Press, Piscataway, NJ.

Van Vaerenbergh, S., Vía, J., and Santamaría, I. (2008). A kernel canonical correlation analysis algorithm for blind equalization of oversampled Wiener systems. In *MLSP 2008: Proceedings of the 2008 IEEE Workshop on Machine Learning for Signal Processing*, pages 20–25. IEEE Press, Piscataway, NJ.

Van Vaerenbergh, S., Santamaría, I., Liu, W., and Príncipe, J. C. (2010). Fixed-budget kernel recursive least-squares. In *2010 IEEE International Conference on Acoustics, Speech and Signal Processing*, pages 1882–1885. IEEE Press, Piscataway, NJ.

Van Vaerenbergh, S., Santamaría, I., and Lázaro-Gredilla, M. (2012a). Estimation of the forgetting factor in kernel recursive least squares. In *2012 IEEE International Workshop on Machine Learning for Signal Processing*, pages 1–6. IEEE Press, Piscataway, NJ.

Van Vaerenbergh, S., Lázaro-Gredilla, M., and Santamaría, I. (2012b). Kernel recursive least-squares tracker for time-varying regression. *IEEE Transactions on Neural Networks and Learning Systems*, **23**(8), 1313–1326.

Van Vaerenbergh, S., Fernández-Bes, J., and Elvira, V. (2016a). On the relationship between online Gaussian process regression and kernel least mean squares algorithms. In *2016 IEEE 26th International Workshop on Machine Learning for Signal Processing (MLSP)*, pages 1–6. IEEE Press, Piscataway, NJ.

Van Vaerenbergh, S., Comminiello, D., and Azpicueta-Ruiz, L. A. (2016b). A split kernel adaptive filtering architecture for nonlinear acoustic echo cancellation. In *2016 24th European Signal Processing Conference (EUSIPCO)*, pages 1768–1772. IEEE Press, Piscataway, NJ.

Vanbeylen, L., Pintelon, R., and Schoukens, J. (2008). Application of blind identification to nonlinear calibration. *IEEE Transactions on Instrumentation and Measurement*, **57**(8), 1771–1778.

Vapnik, V. (1998). *Statistical Learning Theory, Adaptive and Learning Systems for Signal Processing, Communications, and Control*. John Wiley & Sons.

Vapnik, V. N. (1995). *The Nature of Statistical Learning Theory*. Springer-Verlag, New York, NY.

Vedaldi, A. and Zisserman, A. (2010). Efficient additive kernels via explicit feature maps. In *2010 IEEE Computer Society Conference on Computer Vision and Pattern Recognition*, pages 3539–3546. IEEE Press, Piscataway, NJ.

Vepa, R. (1993). A review of techniques for machine learning of real-time control strategies. *Intelligent Systems Engineering*, **2**(2), 77–90.

Verdú, S. (2003). *Multiuser Detection*. Cambridge University Press, Cambridge, UK.

Vielva, L., Erdogmus, D., and Príncipe, J. C. (2001). Underdetermined blind source separation using a probabilistic source sparsity model. In *Proceedings of the 3rd International Conference on Independent Component Analysis and Blind Signal Separation (ICA 2001), San Diego, California, USA*, pages 675–679.

Vielva, L., Erdogmus, D., Pantaleon, C., Santamaría, I., Pereda, J., and Príncipe, J. C. (2002). Underdetermined blind source separation in a time-varying environment. In *Proceedings of the 2002 IEEE International Conference on Acoustics, Speech, and Signal Processing*, volume 3, pages 3049–3052. IEEE Press, Piscataway, NJ.

Ville, J. (1948). Thèorie et applications de la notion de signal analytique. *Cables et Transmission*, **2A**, 61–74.

Vishwanathan, S., Smola, A., and Murty, M. (2003). SimpleSVM. In T. Fawcett and N. Mishra, editors, *ICML'03 Proceedings of the Twentieth International Conference on International Conference on Machine Learning*, pages 760–767. AAAI Press.

Vlachos, A. (2008). A stopping criterion for active learning. *Computer Speech & Language*, **22**(3), 295–312.

Volpi, M., Tuia, D., and Kanevski, M. (2012). Memory-based cluster sampling for remote sensing image classification. *IEEE Transactions on Geoscience and Remote Sensing*, **50**(8), 3096–3106.

Volterra, V. (1887a). Sopra le funzioni che dipendono da altre funzioni. nota 1. *Atti della Reale Accademia dei Lincei, Serie IV*, **3**(Semestre 2), 97–105.

Volterra, V. (1887b). Sopra le funzioni che dipendono da altre funzioni. nota 2. *Atti della Reale Accademia dei Lincei, Serie IV*, **3**(Semestre 2), 141–146.

Volterra, V. (1887c). Sopra le funzioni che dipendono da altre funzioni. nota 3. *Atti della Reale Accademia dei Lincei, Serie IV*, **3**(Semestre 2), 153–158.

Vuolo, F., D'Urso, G., and Dini, L. (2006). Cost-effectiveness of vegetation biophysical parameters retrieval from remote sensing data. In *2006 IEEE International Symposium on Geoscience and Remote Sensing*, pages 1949–1952. IEEE Press, Piscataway, NJ.

Wang, C. and Mahadevan, S. (2009). A general framework for manifold alignment. In *AAAI Fall Symposium: Manifold Learning and Its Applications*, pages 79–86. AAAI Press.

Wang, C. and Mahadevan, S. (2011). Heterogeneous domain adaptation using manifold alignment. In T. Walsh, editor, *IJCAI'11 Proceedings of the Twenty-Second International Joint Conference on Artificial Intelligence – Volume Two*, pages 1541–1546. AAAI Press.

Wang, J., Sano, A., Chen, T., and Huang, B. (2007). Blind Hammerstein identification for MR damper modeling. In *2007 American Control Conference*, pages 2277–2282. IEEE Press, Piscataway, NJ.

Wang, M., Sha, F., and Jordan, M. I. (2010). Unsupervised kernel dimension reduction. In J. D. Lafferty, C. K. I. Williams, J. Shawe-Taylor, R. S. Zemel, and A. Culotta, editors, *NIPS'10 Proceedings of the 23rd International Conference on Neural Information Processing Systems*, pages 2379–2387. Curran Associates.

Wardetzky, M., Mathur, S., Kälberer, F., and Grinspun, E. (2007). Discrete Laplace operators: no free lunch. In *SGP '07 Proceedings of the Fifth Eurographics Symposium on Geometry Processing*, pages 33–37. Eurographics Association.

Welch, P. D. (1967). The use of fast Fourier transform for the estimation of power spectra: a method based on time averaging over short, modified periodograms. *IEEE Transactions on Audio and Electroacoustics*, **15**(2), 70–73.

Weston, J. and Watkins, C. (1999). Support vector machines for multiclass pattern recognition. In *Proceedings of the Seventh European Symposium on Artificial Neural Networks*.

Weston, J., Elisseeff, A., and Schölkopf, B. (2001). Use of the ℓ_0-norm with linear models and kernel methods. Technical report, Biowulf Technologies.

Weston, J., Chapelle, O., Elisseeff, A., Schölkopf, B., and Vapnik, V. (2003). Kernel dependency estimation. In S. T. S. Becker and K. Obermayer, editors, *Advances in Neural Information Processing Systems 15*, pages 873–880. MIT Press, Cambridge, MA.

Westwick, D. T. and Kearney, R. E. (1998). Nonparametric identification of nonlinear biomedical systems, part 1: theory. *Critical Reviews in Biomedical Engineering*, **26**, 153–226.

Westwick, D. T. and Kearney, R. E. (2000). Identification of a Hammerstein model of the stretch reflex EMG using separable least squares. In *Proceedings of the 22nd Annual International Conference of the IEEE Engineering in Medicine and Biology Society (Cat. No.00CH37143)*, volume 3, pages 1901–1904. IEEE Press, Piscataway, NJ.

Weyman, A. (1994). *Principles and Practice of Echocardiography.* Lea & Febiger, 2nd edition.

Widrow, B., McCool, J., and Ball, M. (1975). The complex LMS algorithm. *Proceedings of the IEEE*, **63**(4), 719–720.

Wiener, N. (1958). *Nonlinear Problems in Random Theory.* MIT Press/John Wiley & Sons, Inc., Cambridge, MA/New York, NY, first edition.

Wiener, N. (1975). *Extrapolation, Inerpolation, and Smoothing of Stationary Time Series.* MIT Press, Cambridge, MA.

Wigner, E. (1932). On the quantum correction for thermodynamic equilibrium. *Physical Review*, **40**, 749–759.

Wigren, T. (1994). Convergence analysis of recursive identification algorithms based on the nonlinear Wiener model. *IEEE Transactions on Automatic Control*, **39**(11), 2191–2206.

Wilkinson, W. A. and Cox, M. D. (1996). Discrete wavelet analysis of power system transients. *IEEE Transactions on Power Systems*, **11**(4), 2038–2044.

Williams, C. and Rasmussen, C. (1996). Gaussian processes for regression. In D. S. Touretzky, M. C. Mozer, and M. E. Hasselmo, editors, *Advances in Neural Information Processing Systems 8*, pages 598–604. MIT Press, Cambridge, MA.

Wilson, A. G. and Nickisch, H. (2015). Kernel interpolation for scalable structured gaussian processes (KISS-GP). In *Proceedings of the 32nd International Conference on Machine Learning, ICML 2015, Lille, France, 6-11 July 2015*, pages 1775–1784.

Wipf, D., Palmer, J., and Rao, B. (2004). Perspectives on sparse Bayesian learning. In L. K. S. Sebastian Thrun and and B. Schölkopf, editors, *Advances in Neural Information Processing Systems 16*, pages 249–256. MIT Press, Cambridge, MA.

Wold, H. (1966a). Estimation of principal components and related models by iterative least squares. In P. R. Krishnaiah, editor, *Multivariate Analysis*, pages 391–420. Academic Press.

Wold, H. (1966b). Nonlinear estimation by iterative least squares procedures. In F. David and E. Fix, editors, *Research Papers in Statistics: Festschrift for J. Neyman*, page 411–444. John Wiley & Sons, London.

Wold, S., Albano, C., Dunn, III, W., Edlund, U., Esbensen, K., Geladi, P., Hellberg, S., Johansson, E., Lindberg, W., and Sjöström, M. (1984). Multivariate data analysis in chemistry. In B. Kowalski, editor, *Chemometrics: Mathematics and Statistics in Chemistry*, volume 138 of *Nato Science Series C*, pages 17–95. Reidel, Dordrecht.

Wold, S., Kettaneh-Wold, N., and Skagerberg, B. (1989). Nonlinear PLS modeling. *Chemometrics and Intelligent Laboratory Systems*, **7**, 53–65.

Xie, Z. and Guan, L. (2012). Multimodal information fusion of audio emotion recognition based on kernel entropy component analysis. In *2012 IEEE International Symposium on Multimedia*, pages 1–8. IEEE Press, Piscataway, NJ.

Xing, H. and Hansen, J. (2017). Single sideband frequency offset estimation and correction for quality enhancement and speaker recognition. *IEEE/ACM Transactions on Audio, Speech, and Language Processing*, **25**(1), 124–136.

Xu, M., Jiang, L., Sun, X., Ye, Z., and Wang, Z. (2017). Learning to detect video saliency with HEVC features. *IEEE Transactions on Image Processing*, **26**(1), 369–385.

Yaghoobi, M., Daudet, L., and Davies, M. E. (2009). Parametric dictionary design for sparse coding. *IEEE Transactions on Signal Processing*, **57**(12), 4800–4810.

Yeh, S. and Stark, H. (1990). Iterative and one-step reconstruction from nonuniform samples by convex projections. *Journal of the Optical Society of America*, **7**(3), 491–499.

Yen, J. L. (1956). On nonuniform sampling of bandwidth-limited signals. *IRE Transactions on Circuit Theory*, **CT-3**, 251–257.

Ying, L. and Munson, D. C. (2000). Approximation of the minmax interpolator. In *2000 IEEE International Conference on Acoustics, Speech, and Signal Processing. Proceedings (Cat. No.00CH37100)*, volume 1, pages 328–331. IEEE Press, Piscataway, NJ.

Yuan, X., Lu, Z., and Yue, C. Z. (2013). A novel adaptive importance sampling algorithm based on Markov chain and low-discrepancy sequence. *Aerospace Science and Technology*, **29**, 253–261.

Yukawa, M. (2012). Multikernel adaptive filtering. *IEEE Transactions on Signal Processing*, **60**(9), 4672–4682.

Zhan, H., Shi, P., and Chen, C. (2003). Retrieval of oceanic chlorophyll concentration using support vector machines. *IEEE Transactions on Geoscience and Remote Sensing*, **41**(12), 2947–2951.

Zhang, H., Van Kaick, O., and Dyer, R. (2010). Spectral mesh processing. *Computer Graphics Forum*, **29**(6), 1865–1894.

Zhang, L., Weida, Z., and Jiao, L. (2004). Wavelet support vector machine. *IEEE Transactions on Systems, Man, and Cybernetics B*, **34**(1), 34–39.

Zhang, M.-L. and Zhou, Z.-H. (2005). A k-nearest neighbor based algorithm for multi-label classification. In *2005 IEEE International Conference on Granular Computing*, volume 2, pages 718–721. IEEE Press, Piscataway, NJ.

Zhang, Z. and Hancock, E. (2012). Kernel entropy-based unsupervised spectral feature selection. *International Journal of Pattern Recognition and Artificial Intelligence*, **26**(5), 1260002.

Zhao, Q. and Principe, J. C. (2001). Support vector machines for SAR automatic target recognition. *IEEE Transactions on Aerospace and Electronic Systems*, **37**(2), 643–654.

Zhao, S., Chen, B., Zhu, P., and Príncipe, J. C. (2013). Fixed budget quantized kernel least-mean-square algorithm. *Signal Processing*, **93**(9), 2759–2770.

Zhao, S., Chen, B., Cao, Z., Zhu, P., and Principe, J. C. (2016). Self-organizing kernel adaptive filtering. *EURASIP Journal on Advances in Signal Processing*, **2016**(1), 106.

Zhou, D. and Schölkopf, B. (2004). A regularization framework for learning from graph data. In *ICML Workshop on Statistical Relational Learning and its Connections to other Fields, Banff, Alberta, Canada*, pages 132–137.

Zhou, D., Bousquet, O., Lal, T. N., Weston, J., and Schölkopf, B. (2004). Learning with local and global consistency. In S. Thrun, L. K. Saul, and B. Schölkopf, editors, *Advances in Neural Information Processing Systems 16*, pages 321–328. MIT Press, Cambridge, MA.

Zhou, D., Huang, J., and Schölkopf, B. (2007). Learning with hypergraphs: clustering, classification, and embedding. In B. Schölkopf, J. Platt, and T. Hoffman, editors, *Advances in Neural Information Processing Systems 19*, pages 1601–1608. MIT Press, Cambridge, MA.

Zhu, J., Rosset, S., Hastie, T., and Tibshirani, R. (2004). 1-norm support vector machines. In S. Thrun, L. K. Saul, and B. Schölkopf, editors, *Advances in Neural Information Processing Systems 16*, pages 49–56. MIT Press, Cambridge, MA.

Zhu, X. (2005). Semi-supervised learning literature survey. Technical Report 1530, Computer Sciences, University of Wisconsin-Madison, USA.

Zien, A., Brefeld, U., and Scheffer, T. (2007). Transductive support vector machines for structured variables. In Z. Ghahramani, editor, *ICML '07 Proceedings of the 24th International Conference on Machine Learning*, pages 1183–1190. ACM, New York, NY.

Zomer, S., Sànchez, M. N., Brereton, R. G., and Pérez-Pavón, J. (2004). Active learning support vector machines for optimal sample selection in classification. *Journal of Chemometrics*, **18**(6), 294–305.

Zou, H., Hastie, T., and Tibshirani, R. (2006). Sparse principal component analysis. *Journal of Computational and Graphical Statistics*, **15**, 265–286.

Index

Digital Signal Processing with Kernel Methods, First Edition. José Luis Rojo-Álvarez, Manel Martínez-Ramón, Jordi Muñoz-Marí, and Gustau Camps-Valls.

d

Printed and bound by CPI Group (UK) Ltd, Croydon, CR0 4YY
16/04/2025
14658554-0008